Topics in . . .

Chemical Instrumentation

A volume of reprints from the Journal of Chemical Education

GALEN W. EWING
Editor

Chemical Education Publishing Co.
Easton, Pa. 18042

QD
53
E9
v. 1

Printed in the United States of America by the
Mack Printing Company, Easton, Pa. 18042

FOREWORD

The "Topics in Chemical Instrumentation" column in the JOURNAL OF CHEMICAL EDUCATION was originated in 1959 by Dr. Seymour Z. Lewin. The objective was "to survey the variety of commercial instruments currently available and to compare and contrast their characteristics in order to provide the reader with a basis for intelligently choosing his equipment." Dr. Lewin personally wrote some sixteen articles before opening the column to guest authors in 1962. Since then there have been many contributions, from both academic and industrial authors, including several more from Dr. Lewin's prolific pen. I was honored by an invitation to take over the column in 1967, when Dr. Lewin asked to be relieved.

There has long been a desire to reprint these articles, or some of them, in book form. There were many reasons why this has previously seemed impracticable. Now, at long last, it can be done. The delay, however, raises problems of selection. We cannot possibly republish *all* the articles because of sheer bulk, and also because many of the older contributions are too long out-of-date.

The final selection of which items to include has been on the basis of wide interest and current validity. In each case the author was given an opportunity to update his article. Revision was accomplished by means of supplements rather than by making alterations in the text. (There are a few exceptions to this, which are individually marked.)

The articles have been reproduced by photo-offset from the reprints made at the time of original publication, with the supplementary material newly set in type. The lists of references and addresses of manufacturers have in most cases been combined following the supplements.

This method of reproduction produces a need for filler material, since most articles must start on new pages. For this purpose selected wood-cuts have been taken from *The Book of Signs* by Rudolf Koch, with the kind permission of Dover Publications, Inc.

I wish to express my appreciation to all of the authors for their cooperation; to the editor and former editor of the JOURNAL OF CHEMICAL EDUCATION, Dr. W. T. Lippincott, and Dr. William F. Kieffer, for their wholehearted support of this project; to the business manager, Mr. Frank J. Altschul, Jr., and to the production manager, Mrs. Frances F. Kuenzler, for their extensive assistance. Mrs. Kuenzler, particularly, has spent many hours on this work, over and above her normal duties.

Galen W. Ewing

TABLE OF CONTENTS

Topics In...
Chemical Instrumentation

Edited by **S. Z. LEWIN,** New York University, New York 3, N. Y.

...a  feature

The Journal of Chemical Education
Vol. 43, pages A7, A103
January and February, 1966

XXV. Instrumentation for Atomic Absorption

Herbert L. Kahn, *Perkin-Elmer Corporation, Norwalk, Connecticut*

Herbert L. Kahn is the Product Specialist in Atomic Absorption at the Perkin-Elmer Corporation, Norwalk, Connecticut. He was project engineer of the Model 303 Atomic Absorption Spectrophotometer, and has published many papers on aspects of this new technique. In previous years, he designed ultraviolet and infrared spectroscopic equipment.

Mr. Kahn was graduated from the Cooper Union, New York City, in 1954 with a Bachelors Degree in electrical engineering. Subsequently, he received his Masters Degree from Columbia University. Before coming to Perkin-Elmer, his experience included work on the Sidewinder air to air guided missile.

During the past year, atomic absorption spectrophotometry has made a very important stride as a technique for determining the concentration of metallic elements. It has ceased to be the wave of the future and become the wave of the present. At this time, well over 1,000 atomic absorption spectrophotometers are in use in North America, more than double the number employed 12 months ago.

Shortly after the first publication on atomic absorption in 1955 by the Australian physicist Alan Walsh (*1*), isolated researchers built the simple instrumentation required to test the technique. However, there was little interest outside Australia until commercial equipment became available, and problems involving burners and emission sources were solved. Even today, the progress of the atomic absorption method is due mainly to government and commercial interest; university scientists have been making only relatively minor contributions.

The reason for this quiescence on the part of educators may be that atomic absorption is, at least on the surface, too simple to be interesting. Instrumentally, even the most advanced atomic absorption unit is trivial when compared to a mass spectrometer or neutron activation equipment. Analytically, the major advantage of atomic absorption is that it has no spectral and few chemical interferences with its determinations. Chemical separations are thus rarely necessary, and are simple when they are required; there is no such thing as "interpretation" of the results.

Instrumental Principle

The basic principle of atomic absorption can well be described as the inverse of that of emission methods for determining metallic elements. In all emission techniques (flame, arc and spark, X-ray fluorescence, and neutron activation)

the sample is somehow excited in order to make it give off radiation of interest. At the same time, the sample cannot be prevented from giving off radiation which is not of interest. The appropriate type of filtering system is employed to select the radiation which the analyst wants; the radiation intensity is measured; and by comparision with standards, the concentration of the desired element in the sample is found.

In atomic absorption, the opposite process takes place. The element of interest in the sample is not excited, but is merely dissociated from its chemical bonds, and placed into an unexcited, un-ionized "ground" state. It is then capable of absorbing radiation at discrete lines of narrow bandwidth—the same lines as would be emitted if the element were excited.

The literature describes a number of proposed ways of dissociating the elements of interest from their chemical bonds, but at present, with very few exceptions, the dissociation is always achieved by burning the sample in a flame. The design of the burner is one of the critical factors of atomic absorption instrumentation, and a great deal of effort has been devoted to it.

The narrow emission lines which are to be absorbed by the sample are generally provided by a hollow cathode lamp—a source filled with neon or argon at a low pressure, which has a cathode made of the element being sought. Such a lamp emits only the spectrum of the desired element, together with that of the filler gas. The considerations affecting the design and choice of lamps will be treated later in some detail.

Figure 1 is a sketch of the atomic absorption process. In 1A, the emission spectrum of a hollow-cathode lamp is shown, with emission lines whose half-width is typically less than 0.01 Å. For most practical purposes, the desired element in the sample can be considered as being able to absorb only the "resonance"

lines, whose wavelengths correspond to transitions from the minimum energy state to some higher level. In 1B, the sample is shown to absorb an amount "x" which corresponds to the concentration of the element of interest. As seen in Figure 1C, after the flame, the resonance line is reduced, while the others are unaffected. In order to screen out the undesired emission, the radiation is now passed through a filter or monochromator (1D) which is tuned to pass the line of interest but screen out the others. The photodetector (1E) then sees only the diminished resonance line.

Advantages

From the sketch of Figure 1, it is possible to deduce some of the advantages of the atomic absorption technique. First, properly designed atomic absorption equipment is free from spectral interferences.

Spectral interferences occur in emission, when the radiation from the sample is closely surrounded by radiation from other elements or molecules. The desired radiation cannot be separated from the unwanted output, and the resultant reading depends on the concentration of more than one element. In properly designed atomic absorption instrumentation, spectral interference cannot occur. In the unlikely event that an unsought element in the sample has an absorption line within the passband of the monochromator (dotted line in Fig. 1B), it still has nothing to absorb, since the hollow-cathode lamp does not emit radiation at that wavelength. (Some care must be taken to make this true; it was once thought that copper interfered with the determination of selenium, until it was remembered that the selenium hollow cathode was made by melting it into a copper cup.)

Atomic absorption is also relatively free from chemical interferences, largely because the requirements of sampling are much easier than in emission techniques. For atomic absorption, the element of interest need only be dissociated from its chemical bonds, but need not be excited. Many elements can be completely dissociated at readily achievable temperatures. The compounds of zinc, for example, are completely dissociated far below the temperature of an air-acetylene flame (3000°K) while only one zinc atom in 10^9 is excited. Small differences in

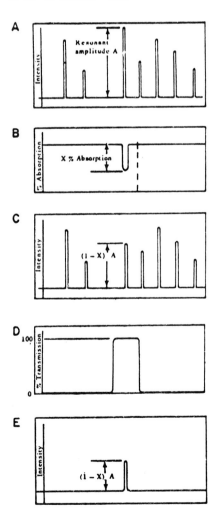

Figure 1. (A) Hollow cathode lamp emits line spectrum of element to be determined. (B) Sample absorbs energy at the resonance line. (C) Resultant spectrum after absorption by sample. (D) Monochromator isolates resonance wavelength, and rejects all others. (E) Photo-detector sees only the resonance line, diminished by sample absorption.

flame temperature or chemistry will therefore have negligible influence on zinc determinations.

In emission and other types of spectroscopy, on the other hand, the atom must be excited, and for most elements, at presently achievable energies, there is no such thing as complete excitation. One is always working on a rather steep slope in the energy-excitation curve, so that instrumental and chemical conditions have a strong bearing on the analytical result.

Furthermore, atomic absorption generally works with a larger number of atoms than do other techniques. Sodium, for instance, is one of the elements most favorably determined by flame emission; yet only 1.5% of the atoms are excited at obtainable flame temperatures. Most of the remaining atoms are dissociated from their compounds and unexcited, and thus available for atomic absorption. For calcium under similar circumstances, the ratio in favor of atomic absorption is much greater. For every atom that is excited and available for flame emission, at least 1000 are dissociated and accessible to atomic absorption.

It cannot be deduced from this ratio that atomic absorption is 1000 times more sensitive to calcium than is flame emission.

Instrumentally, it is easier to work with an excited atom than with a ground-state atom. The difference is similar to that between seeing a target and hitting it. More important, however, is what happens if the flame should change so dramatically that twice as many atoms are excited as before. Now, two atoms are available for flame emission, and 999 for atomic absorption. An event which has caused a change of 100% in the emission value has only reduced the atomic absorption result by 0.1%. It can therefore be deduced that atomic absorption is a technique from which high precision can be expected. In fact, with optimum equipment, a standard deviation of 0.2% is achievable.

Analytical Possibilities

The energy required can theoretically be supplied in a rather large number of ways; plasma jet, electrical discharge, laser, furnace, etc. However, with very few exceptions, present-day instrumentation always employs a flame to effect the dissociation. Mercury, which has an appreciable vapor pressure at room temperature, is often determined by merely heating the sampling cell; in fact the mercury vapor detector is the oldest form of analytical atomic absorption. Articles have been published on the determination of uranium through the use of a second, demountable hollow-cathode lamp as the sampling device. Nevertheless, nonflame sampling devices, though very promising and interesting, have not yet achieved general analytical importance and will not be discussed further in this article.

The use of a flame implies the presence of flame materials, and these, in turn, limit the usable range for atomic absorption to the region where they do not absorb. The resonance line for arsenic at 1937 Å is presently the lower limit at

which atomic absorption can be carried out. The element with its resonance line at the longest wavelength is cesium, at 8651 Å. The wavelength range for atomic absorption is thus similar to that of ultraviolet-visible spectrophotometry. It includes all metals and semi-metals, but excludes such elements as sulfur, phosphorus, the halogens, and other gases whose resonance wavelengths lie in the atmospheric-absorption region. Methods exist for the indirect determination of phosphorus, sulfur, and chlorine, by compounding them with a metal and then analyzing for the metal, but discussion of these is beyond the scope of this work.

Until very recently, it was not analytically useful to use flames developing temperatures higher than that of the air-acetylene combination. (This is discussed in more detail in the section on burners.) The temperature of the air-acetylene burner is not sufficient to dissociate the compounds of refractory elements like titanium, aluminum, and vanadium, and for several years these elements were considered to be out of reach of analytical atomic absorption, though some data could be obtained by special and rather cumbersome methods.

In early 1965, the Australian chemist John Willis (2) showed that a burner using nitrous oxide and acetylene would develop enough energy to dissociate the compounds of many of the refractory elements. In the intervening months, analytically useful work has been done on silicon, boron, rhenium, tungsten, and many of the rare earths, and it is now thought that all the metals can be determined when nitrous oxide is used.

Interferences

There is a considerable literature on atomic absorption interferences, which fall generally into three types, and can be controlled to some extent by choosing the proper sampling system. The three forms of interference are classifiable as "chemical," "ionization," and "bulk," or "matrix."

Chemical interferences are usually the result of incomplete dissociation of the compounds of the element being determined. When the dissociation is incomplete, some compounds are dissociated less completely than others, and the analytical result for a metal may depend on the other elements and radicals present in the solution. A flame of higher temperature often removes such interferences completely. For example, early workers in England, using air-coal gas flames, reported a multiplicity of chemical interferences with iron. When acetylene came into use, however, most of these interferences disappeared.

Figure 2 shows the effect of the interference of silicon and aluminum with strontium. A small concentration of either causes a profound depression in the absorbance of the strontium, when an air-acetylene flame is used. The upper line in the figure represents the result of a common measure taken to remove this particular interference—the addition of an excess of lanthanum, which pref-

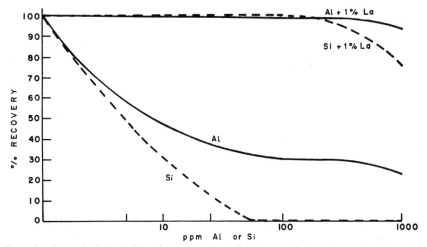

Figure 2. Removal of chemical interferences. Al and Si both severely depress strontium absorption. The addition of lanthanum to the solution removes the interference by binding the interfering elements.

erentially binds the interferents, and thus leaves the strontium unbound.

Preliminary recent work indicates that the chemical interferences with calcium disappear when a nitrous oxide-acetylene flame is used, and the calcium compounds are completely dissociated. However, an interference of the second kind, ionization, is introduced instead.

Ionization interference takes place when a substantial proportion of the atoms in the sample become ionized, causing them to absorb at a different radiation line. Since the different line may be outside the passband of the monochromator, and may not even be present in the lamp; such atoms are in practice lost to the determination. Since the degree of ionization depends on the flame temperature, burning conditions have a real effect on the determination of easily ionized elements.

Ionization interference is usually controlled by adding a large excess of a more easily ionized element—for the calcium problem described above, a substantial amount of sodium might be added. More generally, it is not always useful to operate with the hottest possible flame; a temperature a little above the point of complete dissociation would probably be ideal from the viewpoint of interferences.

Bulk or matrix interferences are changes in the analytical result caused by the viscosity or nature of the sample solution. One common matrix effect is the enhancement caused by organic solutions—a given concentration of an element in an organic solvent absorbs between two and four times as much as it does in water. There is not complete agreement on the reason for the improvement, but it is clear that the effect of the organic solvent improves the efficiency of the burner.

Another matrix effect is caused by different concentrations of dissolved solids in the sample solution. As the solution becomes more concentrated, it flows more slowly through the burner, and absorption is therefore decreased. Figure 3 shows such an interference, which can be distinguished from chemical interferences in that its slope is much more gradual, and that it does not reach an asymptote.

Both the number and the intensity of interferences seem to depend greatly on

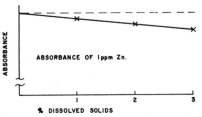

Figure 3. Bulk or viscosity interference. The presence of an increasing concentration of dissolved solids depresses sample absorption. Degree of depression depends upon burner design.

the design of the burner being employed. There is therefore not always general agreement on the presence or absence of interferences in specific analyses. More detail on this problem is given in the section on burners.

Sensitivity and Detection Limits

Before instrumentation is discussed in detail, a clear distinction should be made between the meanings of *sensitivity* and *detection limit*.

Sensitivity, in atomic absorption, is defined as the concentration of an element in water solution which will produce an absorption of one per cent, and is generally expressed as ppm/1%. The chief usefulness of this quantity lies in the estimation of optimum concentration for the sample, and as a test of proper performance of an instrument. When it is possible to choose the approximate concentration of the element to be determined, it should lie between 10 times and 100 times the sensitivity. An instrument or analysis yielding a given sensitivity should continue to yield it as time passes, and another instrument of the same design should yield about the same sensitivity.

Sensitivity is also one factor, among others, in determining the *detection limit* for an element (Fig. 4). Other factors in the detection limit are stability, and the signal-to-noise ratio. While there is no official agreement, an increasingly accepted definition for the detection limit is the concentration in water solution which gives a signal twice the size of the variability of the background. In statisti-

cal terms, the concentration of an element at the detection limit can be determined with a coefficient of variation of 50%.

In practice, it is often possible to increase the sensitivity and simultaneously to degrade the detection limits, by sacrificing stability. In the mixing chamber of the Perkin-Elmer burner, for example, is a baffle or flow spoiler which can readily be removed, giving a 50% improvement in sensitivity. However, very often the stability lost exceeds the sensitivity gained, so that the bargain is a poor one.

When the hollow-cathode lamps are sufficiently bright and the instrumental system sufficiently stable, it is possible to detect changes in concentration far smaller than one per cent, particularly with the aid of scale expansion. The detection limit for an element is therefore often a far smaller concentration than the

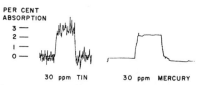

Figure 4. Comparison of sensitivity and detection limit. Thirty ppm each of tin (left) and mercury are determined in an air-acetylene flame. The sensitivity of both elements is approximately the same; the detection limit is clearly different.

sensitivity, when the instrumentation and burner are well designed.

Available Instrumentation

Instrumentation for atomic absorption is being offered by a variety of manufacturers. In the U.S.A., Perkin-Elmer and Jarrell-Ash are producing instruments, while Beckman has announced equipment to be available late in 1965. Overseas, at least Techtron, Zeiss, Unicam, Evans Electroselenium, and Hilger & Watts, have announced atomic absorption instrumentation. The U.S. firms of Atabel and Block have also indicated their intention of producing equipment. Details on these spectrophotometers, where available, are presented in a later section.

Atomic Absorption Systems

In Figure 1, the principle of atomic absorption is shown. The simplest possible system to carry out this principle appears in Figure 5a. Light from the hollow-cathode lamp passes through the flame, after which the resonance wavelength is isolated by a monochromator or filter, and then falls onto the photodetector and electronics. Some early commercial atomic absorption instrumentation was indeed built in this way, and functioned well enough to indicate that the method had great inherent usefulness. The system of Figure 5a is called the single-beam dc system, because the light from both the flame and the lamp is unchopped and therefore produces a direct current, or dc, in the detector.

Early workers soon found problems in the dc system for certain real analyses,

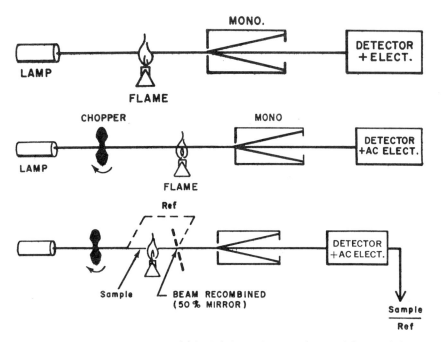

Figure 5. Atomic absorption systems. (a) In single-beam dc system, lamp and flame emission are both unchopped. (b) In single-beam ac system, lamp light is chopped, flame light is unchopped. (c) In double-beam ac system, lamp light alternately passes through flame (sample beam), and past flame (reference beam). Output is ratio of beams.

due to emission from the flame. In atomic absorption, the flame is ideally no more than a heated sampling cell, which should not emit radiation; however, it does. Calcium, for example, has a radiation continuum around the resonance line of sodium. When the dc system is used in the analysis for sodium in the presence of much calcium, the detector cannot tell whether a given photon that strikes it is a legitimate sodium photon from the lamp, or an illegitimate calcium photon from the flame. The results from the analysis therefore depend on the sodium concentration as well as on the calcium; therefore calcium interferes spectrally with sodium in the dc system.

Research therefore soon turned to the system of Figure 5b, in which the source light is chopped and the flame light left unchopped. Since the chopped light produces an alternating current in the detector, this is known as the single-beam ac system. The electronics are designed to amplify alternating current only, thereby causing the light from the flame to be ignored. The method of chopping is shown schematically in Figure 5b as a rotating chopper, but in actual instrumentation the same effect can be achieved by supplying the lamp with a chopped electrical current.

After problems were encountered with early commercial instrumentation using the dc system, the dc approach was abandoned and for several years only ac equipment was offered. At present, Beckman has announced the revival of the dc system in its atomic absorption accessory to the DU and DU-2 ultraviolet-visible spectrophotometers. Data have yet to be made public whether skillful design has succeeded in minimizing the spectral interferences in this approach, or whether the accessory is designed for those analyses where radiation from the flame is no great problem.

Most other equipment utilizes the single-beam ac system, including the accessory to the Beckman DB spectrophotometer, the Techtron AA-3 and AA-100, the three Jarrell-Ash instruments, and the Perkin-Elmer Model 290. The only important exception is the Perkin-Elmer Model 303, which employs a double-beam ac system, as shown in Figure 5c. Here, the chopper becomes a rotating sector mirror, which alternatively passes the source light through the flame and past the flame. The "sample" and "reference" beams are recombined, and their ratio is taken electronically.

The advantages and limitations of this type of double-beam system are worth discussing. Clearly, the double-beam system cannot overcome instability and noise from the burner, since the burner is only in one of the two beams. There had therefore been no point in using a double-beam system until 1962, when a burner was developed which was instrumentally and audibly quiet, with a constant aspiration rate and little tendency to clog. An examination of the single-beam atomic absorption system then showed that the burner was no longer the limiting factor but that the stability of the hollow-cathode source had become the weakest link. The lamp may take a half-hour to come to stable emission (less than 1% drift in five minutes), and even then the lamp emission changes due to a variety of factors, including current, ambient temperature, and the length of time it has been on.

The double-beam system can overcome the effects of lamp drift and change in detector sensitivity, as shown in Figure 6. In Figure 6a, a Westinghouse Ca-Mg-Al hollow-cathode lamp is warmed up at the wavelength for calcium in the single-beam mode, with the burner off. Time must be allowed for the emission to come to rest, and the output must be adjusted from time to time as samples are being run. In Figure 6b, the same lamp is used in

the double-beam mode, again from a cold start. At point A, the burner is turned on, and distilled water is aspirated. At point B, a sample of 1 ppm calcium is inserted, and removed again at point C.

The double-beam system thus gives a stable baseline almost immediately, and has the following advantages: lamps can usually be inserted and operated at once, with little or no wait for warm-up; the stability of the baseline makes it possible to see small departures from it, thereby improving detection limits over the single-beam approach; and a stable baseline implies good analytical precision. Among the liabilities of the double-beam approach are the lengthened optical path which is required, and the increased complexity of the instrumentation. A substantial percentage of the analyses for which atomic absorption is being used and contemplated can be adequately performed by the simpler single-beam equipment.

Burners

The burner system is probably the most important and controversial portion of atomic absorption instrumentation, and active work is taking place on various improvements. Some of the criteria of good burner design are:

(1) *Stability.* The absorption for a given concentration should remain constant, preferably even from day to day.

(2) *Sensitivity.* Absorption for a given concentration should be high.

(3) *Quietness.* The burner should be audibly and instrumentally quiet, and not induce flicker in the output.

(4) *Ability to burn concentrated solutions.* In real samples, the detection limit is often set by the concentration of solution that can be tolerated.

(5) *Freedom from memory.* The content of one sample should not affect the result from the next.

(6) *Freedom from background.* There should be little or no absorption from the flame itself, or from blank solutions free from the element of interest.

(7) *Linearity.* Working curves of concentration versus absorbance should be straight over as large a range as possible.

(8) *Versatility.* A large number of elements and types of sample should be handled with the same burner.

(9) *Speed of response.* In sample-limited situations, full absorption should be established rapidly after sample introduction.

(10) *Minimal emission.* In an ac system, emission from the burner will not produce a photometric error. However, high emission will contribute to the flicker of the output, because the noise current from the photomultipler detector increases as a function of the total light falling on it. A very bright flame will therefore tend to produce a fluctuating output.

Other obvious requirements are ease of cleaning, freedom from corrosion, and ease of adjustment.

Total Consumption versus Premix

Two basic types of burners have been

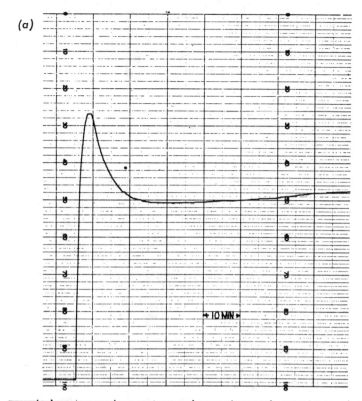

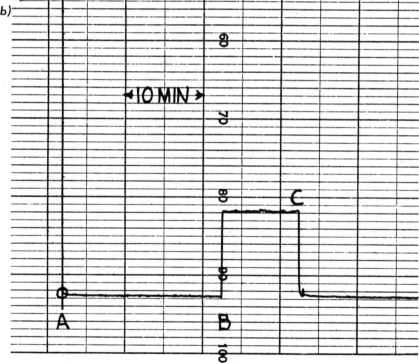

Figure 6. Single and double beam operation. Ca-Mg-Al lamp is operated from a cold start a resonance line for Ca, 4227 Å. Recorder speed is 0.1 in./min. (a) Single-beam run, burner off. (b)t Double-beam run. Burner is lit at point A, distilled water aspirated. One ppm Ca aspirated at point B, removed at point C.

used for atomic absorption. The one is the "total consumption" or "diffusion" burner, of the type long used by Beckman and others for flame emission work. The other is the "premix" type, in a variety of designs, in which sample, fuel, and oxidant are mixed in a chamber before entering the flame.

The total consumption burner, modifications of which are available from Beckman, Zeiss, Jarrell-Ash, Hitachi, Jobin-Yvon, and Chaffee-Keyes, is too well known to need much description. In essence, the fuel, oxidant, and sample are all passed through separate channels to a single opening from which the flame emerges. The flame is turbulent and relatively small in cross-section, with the hottest portion probably no more than 1 cm across. This form has been considered advantageous for flame photometry, in which the source of emission should be as small as practicable.

Opinion on its merits for atomic absorption, on the other hand, is sharply divided. The Australian workers and Perkin-Elmer did not find it very useful, and designed their own premix burners, while Jarrell-Ash continues to rely on only the total consumption type. According to Beckman literature, Beckman will offer both a premix burner and a bank of three total-consumption units in a row.

Despite the high efficiency implied by the name of the total consumption burner, both it and most of the premix burners are only about 5% efficient. In the total consumption burner, the entire sample passes into the flame, but in the rapid transit of the droplets through the hot region only the smallest droplets have time to be dried and burned. In the Perkin-Elmer burner, shown in Figure 7, the large droplets in the mixing chamber are collected on the walls and pass down the drain, while only the small droplets travel to the flame.

For atomic absorption, the turbulence of the total consumption burner, and the presence of unburned droplets in the flame, must count as disadvantages. The burner should be regarded as a sampling cell, and a turbulent flame corresponds to a sampling cell of unpredictably varying length. Unburned droplets will scatter incident light, so that a blank solution will produce apparent absorption which must be subtracted from the result.

The design of the Perkin-Elmer premix burner also presents a number of other advantages and disadvantages as compared to the total-consumption type. The Perkin-Elmer flame is not very luminous, and flicker and turbulence are quite low, so that for many elements the flame contributes no apparent noise to the output (Fig. 6b). Furthermore, there is rather little dependence of absorption upon sample flow rate; apparently there is an inverse ratio between burner efficiency and sample input. This is of benefit in two ways. First, the length of sample capillary, and its depth of immersion in the solution, are not very critical, so that samples can be aspirated from any vessel. For total consumption burners, by contrast, Petri dishes or very small sample containers are often recommended. Second, viscosity interferences due to variations in sample concentration are minimized, though not eliminated. In the Perkin-Elmer burner, when the sample flow rate is cut by a factor of 2, absorption is reduced by approximately 4%.

The total consumption burner has the advantages of being somewhat easier to clean, and having, in principle, a faster response time, a lower memory effect, and a greater versatility regarding the choice of oxidant. The Perkin-Elmer burner, though probably slower to come to equilibrium, nevertheless does so in under one second, as shown in Figure 8.

The chamber of the Perkin-Elmer burner is made of remarkably inert Penton plastic, so that, in very extensive analytical work involving hundreds of instruments, cross-contamination of samples has not been a problem. For the sake of certainty, a blank is usually run for a few seconds after every sample. When very high concentrations of an element are run,

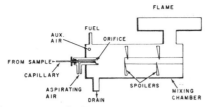

Figure 7. Atomizer and burner system. In this premix burner the sample is aspirated through a thin capillary tube by the air flowing into the atomizer section. The air-sample mixture emerges from the atomizer as a fine spray of droplets, which is then mixed with the fuel, usually acetylene. The mixture is rendered turbulent by the flow spoilers, and is then forced up into the burner head.

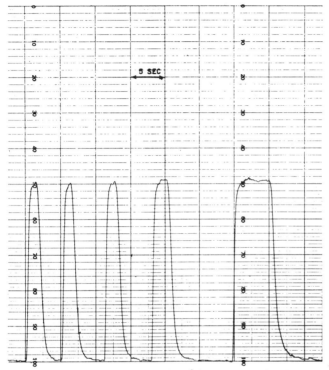

Figure 8. Recorder traces of the absorbance of 1.0 µg/ml magnesium with aspiration times of 1 sec each. 5-sec trace at right represents steady-state value for comparison purposes. Chart speed is 12 in./min. Burner typically consumes 3 ml/min; therefore sample consumption is approximately 0.05 ml here.

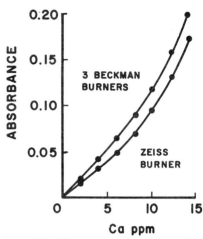

Figure 10. Working curves for calcium. Total consumption burners in 5-pass arrangement were used with Jarrell-Ash Model 82-362. Reprinted with permission from Sunderman and Carroll (4).

different purposes, with all heads fitting the same mixing-chamber-atomizer system. Two of these heads, which are described in Table 1, are capable of burning rather concentrated solutions without clogging. The "high-solids" head, which has a wider slot than the standard one, can burn 1:1 dilutions of blood serum, or 10% sugar solutions, indefinitely. Its only disadvantage is that it whistles on warm-up, before becoming as inaudible as the standard head.

Reaction Zones and Multiple-Passing

In principle, it is possible to improve the sensitivity of an atomic absorption determination by passing the sample beam several times through the flame. Whether multi-passing has real merit is again the subject of a considerable divergence of opinion between manufacturers. Jarrell-Ash includes a system for five passes through the flame; Beckman has announced a three-pass system, while Perkin-Elmer and Techtron do not provide multiple passing. The Jarrell-Ash system is shown in Figure 11.

For most elements, the reaction zone in the flame where useful absorption takes

there will be some persistence. For example, after several per cent sodium are aspirated for some time, traces of sodium will persist in the flame up to a few hours, unless the burner is taken apart and cleaned.

When the Perkin-Elmer burner becomes clogged, it is cleaned by passing a razor blade long the slot where the flame emerges. However, some analysts report a need to disassemble and wash the burner every week or two, a procedure that requires about 20 minutes. Aspiration rates for the two burners are comparable: about 3 ml/min for the Perkin-Elmer, about 1.5 ml/min for the total-consumption.

The Perkin-Elmer burner has the advantage in linearity, as shown by the contrast between Figures 9 and 10, both of which represent the determination of calcium in serum. In Figure 9, reproduced from Zettner and Seligson (3), a Perkin-Elmer Model 214 instrument and a premix burner were used, while the data of Figure 10, from Sunderman and Carroll (4), were obtained with a Jarrell-Ash 82-362 using two types of total-consumption burners. Since linearity also depends on the instrument system and the hollow-

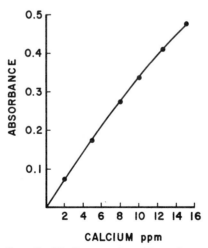

Figure 9. Working curve for calcium. A premix burner was used with Perkin-Elmer Model 214. Reprinted with permission from Zettner and Seligson (3).

cathode lamp, the above comparison cannot be considered absolute, but the shapes of the two curves are definitely characteristic of two types of burners.

In the past year, a number of different burner heads have become available for

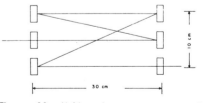

Figure 11. Multi-passing arrangement of Jarrell-Ash Model 82-362.

place is by no means uniform, is often quite small, and frequently depends on the fuel-air ratio quite critically. A careful study of reaction zones was undertaken by Rann and Hambly (5), using a Techtron premix burner. Their results for two elements are shown in Figure 12. Evidently, the reaction zone for copper is high enough that multiple-passing could improve the sensitivity, if the beam can be kept very narrow. For molybdenum, by

Table 1. Burner Heads

Name, Description	Features
303-0023 Standard burner head, 4 in. × 0.015 in. slot	For air-acetylene or air-hydrogen. Standard on Model 303
Propane burner head, 4 in. × 0.06 in. slot	For propane, coal gas, or natural gas with air
High solids burner head, 4 in. × 0.025 in. slot	Similar to 303-0023, except that it can burn concentrated solutions without clogging. Whistles while warming up
290 Burner head, 2 in. × 0.020 in. slot	For air-acetylene or air-hydrogen. Despite shorter length, provides approximately equal sensitivity to 303-0023. Can burn moderately concentrated solutions. Can be turned 90° in its mount, reducing sensitivity by 10–20 times
Boling burner head, three parallel 4 in. × 0.025 in. slots	For air-acetylene or air-hydrogen. Can burn concentrated solutions without clogging. Provides 1.5–2 times better sensitivity for many metals, is less critical to adjust than standard burner head for molybdenum, chromium, and calcium
Nitrous oxide head, 4 in. × 0.015 in. slot. (Slot width can be increased to 0.019 in. with shim provided)	For nitrous oxide with acetylene or hydrogen. Required for refractory elements. Can also be used with air, for non-refractory elements. Somewhat more difficult to clean than other heads

contrast, multipassing is clearly ineffective.

However, sensitivity is only one consideration, which happens to be the most easily measured. With a multiple-passing system, the sample beam passes through a variety of heights and flame temperatures, including areas far from the burner head where the flame is therefore not under as good control as in the optimum region. This sometimes leads to problems of interferences and instability, reducing whatever advantages are gained from the multipassing scheme. Therefore, it is not obvious whether the sensitivity improvement sometimes available from multipassing is worth the additional difficulties.

The Long Tube and the Heated Chamber

Dramatic improvements in sensitivity have been reported through the use of two devices which are not yet commercially available. The design principles of the two are in marked contrast.

In one scheme, first reported by Fuwa and Vallee (6), a Beckman total-consumption burner is tilted so that its flame is directed down a ceramic tube up to 90 cm long and 1 cm across. Sample atoms therefore stay in the beam for an extended time, and their opportunity to absorb is improved. Elements whose compounds dissociate at low temperatures, such as zinc and lead, produce absorptions perhaps 10 times higher than with conventional burners. Especially for situations where low concentrations must be determined in a limited sample, the long-tube system should be very useful. However, it is more difficult to use and to clean than conventional systems and, due to its considerable temperature gradient, more prone to interferences. Fuwa and Vallee, for example, note a phosphate interference with zinc.

The "heated chamber" burner, which has been described by Beckman and more recently by Perkin-Elmer, has promise of more general application, if various problems associated with it can be overcome. It represents an improvement on the premix burner, in that the mixing chamber is heated to a temperature of 300°C or more, causing the solvent to be vaporized.

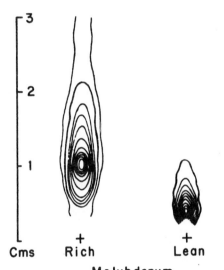

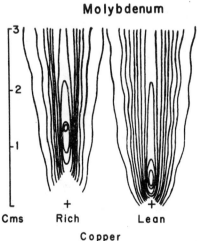

Figure 12. Flame absorption profiles for copper and molybdenum with Techtron premix burner. Central peak represents absorbance of 1.0. Contour lines represent reductions of 0.1 absorbance units. Reprinted with permission from Rann and Hambly (5).

The sample and solvent then pass through the cold trap, in which a large proportion of the solvent is drained away. The portion of the sample which reaches the flame is therefore highly concentrated. For many elements, an improvement of five or more times in sensitivity has been achieved.

Although the principle has been described and sensitivity measurements given, no real analytical data have yet been reported. The reaction of the laminar flow burner to real samples (organic solvents, concentrated solutions), its stability in the face of temperature and sample flow changes, and its ability to be cleaned, are as yet unknown quantities.

Variations in Solvents, Fuel, and Oxidant

For premix burners, the most common supply gases are air and acetylene, while the total-consumption burners generally use air and hydrogen. All detection limits are usually given in water solution.

When organic solvents are used, such as alcohol or methyl isobutyl ketone, the absorption for a given element concentration is enhanced by factors between 2 and 5. As nearly everything else having to do with burners, the reasons for the improvement are under debate. One plausible explanation is that the organic solvents, having a lower surface tension, produce smaller droplets and thus promote higher efficiency.

The effects of different fuels for the premix burner have been studied in some detail. At air-acetylene temperatures, the alkalies and alkaline earths ionize to an appreciable extent, and ionized atoms are lost to the determination. Also, for elements whose compounds are dissociated at low temperatures, such as lead and zinc, the absorption is higher when lower-temperature fuels are used, such as propane or natural gas. As flame temperature is reduced, pressures in the flame become lower, and more atoms remain in the optical path. Ionization interferences between alkali elements are also reduced.

However, the experience at Perkin-Elmer indicates that low-temperature fuels should be used with great circumspection if at all. The propane flame is not as stable as the acetylene flame, so that the improvement in sensitivity cannot generally be translated into an advantage in detection limit. Moreover, a large number of elements show interferences when low-temperature flames are used, which disappear when acetylene is employed.

For the premix burner, the best fuel for all elements seems to be acetylene, with the mysterious exception of tin. For tin, hydrogen improves the absorption severalfold over acetylene. The reason is not understood.

Fuel economy is also on the side of acetylene, since a 300-ft³ tank lasts approximately three working weeks. A similar tank of hydrogen, required by the total-consumption burner, is consumed in about three hours. Ordinary welding-grade acetylene is adequate for atomic absorption. The air supply must be free from dust and oil.

Refractory Elements: the Nitrous Oxide Burner

The air-acetylene combination is not hot enough to dissociate most compounds of a considerable number of elements, such as aluminum, boron, and silicon, and it incompletely dissociates those of other metals like chromium, molybdenum, and barium. An additional problem is that the "refractory" elements are quick to form stable oxides.

Some years ago it was discovered that the refractory elements could be made to absorb in a fuel-rich oxy-acetylene flame. The oxygen provided a higher temperature, and the excess of acetylene kept the flame sufficiently reducing to delay recombination.

The high burning speed and explosive potential of oxy-acetylene flames defeated efforts to produce a usable premix oxy-acetylene burner. The total-consumption burner, on the other hand, while capable of burning such a fuel combination, produced such a brilliant and turbulent flame that it was difficult to obtain analytical results with it. Various efforts to modify the total-consumption burner met with only moderate success.

So matters stood until John Willis (2) in Australia found that a combination of nitrous oxide and acetylene would not only dissociate refractory compounds, but could also be used in premix burners with minor modifications. The burner head was changed in only two ways: the gas flow patterns were made less abrupt, and the material was thickened to withstand the higher temperature.

With the aid of nitrous oxide and acetylene flames, many previously impossible elements have been determined, as summarized in Table 2. Here again an organic solvent provides the same improvement in absorption. In Table 2, sensitivities and detection limits are given for refractory elements in water solution. The Perkin-Elmer Model 303 with the premix nitrous oxide burner is used.

It is now believed that all metallic elements can be determined by atomic absorption with presently available commercial equipment.

Sources for Atomic Absorption

Many of the virtues of atomic absorption stem from the fact that the desired element absorbs only at very narrow lines, with a half-width of the order of 0.01 Å. Instrumental design is therefore greatly simplified if the emission sources emit lines no wider than this, and narrower if possible. Discharge lamps and hollow cathodes have been found to fill the requirements, and particularly the latter have been developed to a considerable level of sophistication.

Other sources have been proposed and in some cases developed. It has been suggested that a flame containing a high concentration of the element of interest would produce a simple, versatile, and inexpensive source, but difficulty was encountered in achieving the requisite stability and freedom from background radiation. Fassel and Kniseley of Iowa State have used the bright continuum of a xenon arc as a source (7), depending on the

Table 2 Conditions for Refractory Elements

Element	Spectral SIH(Å)	Wavelength (A)	Sensitivity (ppm/1% in H_2O)	Detection Limit in H_2O
Al	2	3093	1 ppm	0.1 ppm
Be	2	2349	0.1	0.02
B	7	2497	50	10
Dy	2	4212	1.5	0.7
Er	2	4008	0.7	0.2
Ho	2	4163	2.0	0.3
La	2	3574	106	20
Nd	2	4634	30	7
Pr	2	4951	72	15
Re	2	3460	25	1.5
Sm	2	4298	15	6
Si	2	2516	3	1
Ta	2	2715	60	6
Tb	2	4326	15	3
Ti	2	3643	1.5	0.5
W	2	2944	28	9
U	2	3514	250	30
V	7	3184	2	0.7
Yt	2	3988	0.2	0.05
Y	2	4077	2	0.5
Zr	2	3601	20	5

monochromator to achieve a narrow wavelength range. Working with an instrument of about 0.2 Å resolution, they achieved useful detection limits. Despite the promise offered by the continuous source to make possible "scanning" or multi-element atomic absorption, the severe instrumental difficulties associated with it have thus far discouraged commercial development or analytical use. A system promised by Block Associates, based on a high-resolution interferometer, may increase the usefulness of the continuous source.

Single-Element Hollow Cathodes

At the present time, all the elements except the alkalis and possibly mercury are determined most effectively and commonly with hollow-cathode sources. Discharge lamps are best for the alkalis, though hollow-cathode lamps for them are available. For mercury, the OZ4 "germicidal" lamp is a popular source, probably about equal to newly available hollow cathodes.

A single-element hollow-cathode lamp is shown in Figure 13. The active components are a cathode made of or lined with the element of interest, generally in the shape of a cylinder closed at one end, about 1 cm deep and 1 cm across, and an anode which is merely a straight metal wire. Neon or argon is used at a few millimeters pressure as a filler gas. The front face of the lamp is made of quartz or any of several types of glass, depending on the wavelength range it must transmit and also on the preference of the manufacturer.

When the current flows, metal atoms are sputtered from the cathode into the area within and in front of the cup. Collisions with the neon or argon ions cause a proportion of the metal atoms to become excited and emit their characteristic radiation. The choice of filler gas depends on the element: lead, iron, and nickel perform far better with neon than with argon; however, neon is not suitable with some elements, such as lithium and

arsenic, because a strong neon emission line is close to the best resonance line. For many elements, there is little to choose between the two gases.

Optimum lamp currents vary between

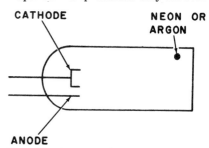

Figure 13. Schematic of Hollow Cathode Lamp. The cup-like cathode is made of or contains the element of interest. The envelope is filled with argon or neon at a low pressure.

elements and designs, from 5 to 100 ma at 100–200 volts. Manufacturers' rated maxima represent values above which the cathodes are in danger of destruction. Below the maximum, the lamp emission increases as the current is raised, improving the signal-to-noise ratio and reducing

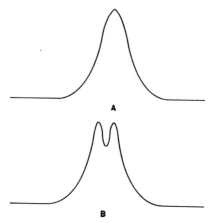

Figure 14. Resonance Line Emission of a Hollow Cathode Lamp. (a) Normal Emission. (b) Self-Reversed Emission, caused by excessive current, or flaws in design.

the flicker in the instrument output reading. However, for many elements the absorption for a given concentration of sample is reduced as the current rises over a certain value, since the emission line becomes somewhat self-absorbed. In self-absorption, sputtered ground state atoms in the lamp itself absorb the lamp radiation, causing the emission to assume the form of Figure 14b. Recommended lamp currents represent a compromise between emission intensity and sensitivity (absorption for a given concentration).

Manufacturers and Lamp Life

Currently, three lamp manufacturers are of importance in North America: Westinghouse, Perkin-Elmer, and the Australian firm Atomic Spectral Lamps Pty. (ASL). A number of other companies, including Unicam and Hilger and Watts, produce hollow-cathode lamps, but these are not yet in wide use in the USA and little is known here about their performance.

Perkin-Elmer and ASL produce lamps whose diameter is about two inches, and whose length is approximately six inches. Both manufacturers argue that lamp life is limited by the clean-up rate of the filler gas, and that a high lamp volume therefore contributes to long life. Westinghouse, on the other hand, makes considerably slimmer lamps with the claim that a large gas volume is not required if the cathode size is reduced. Perkin-Elmer, which sells lamps of all three producers, guarantees them for six months or five ampere-hours (typically 500 hours).

In general, there is little to choose between the price and performance of the three makes, although at the fast pace of lamp technology, one maker or another may produce a better lamp for a given element. Perkin-Elmer, for example, recently introduced a lamp for arsenic, the first to produce reasonable emission for an acceptable length of time, while Westinghouse makes a particularly good mercury hollow cathode. Prices for individual lamps vary between $75 and $200. Table 3 shows the lamp types suitable for the products of the different instrument manufacturers.

Lamps produced during the past two years, for all but a few particularly recalcitrant elements, have shown such long life that it has been all but impossible to gather statistical data on failure rates. For example, if one determination per minute is made on a double-beam instrument which does not expend any lamp life in warm-up, a 500-hour performance corresponds to 30,000 determinations.

Until recently, the greatest problem in producing, using, or maintaining lamps was the possible presence of hydrogen within the metal cathodes. Hydrogen produces a strong emission continuum in the ultraviolet region, and also has strong emission in the visible. The effect of its presence is to reduce the intensity of the resonance line, and to produce a pronounced flattening and bending of the working curve, as shown in Figure 15.

When a lamp is failing, the most common effects are a reduction in analytical sensitivity, and an increase in output fluctuation. If the instrument behaves properly for other elements, it can be deduced that the lamp is at fault. Particularly if the trouble occurs after the lamp has not been used for some time, it

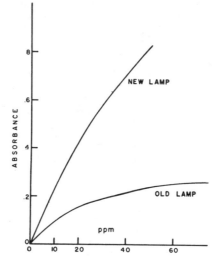

Figure 15. Effect of Background Hydrogen Emission on the Working Curve. Instead of assuming shape of the upper line, the working curve shows a lower sensitivity, has early curvature, and reaches an asymptote. Similar effects could be caused by the presence of unabsorbable lines within the passband of the monochromator.

can sometimes be cured by operating the lamp at maximum current overnight.

It should be noted that the permissible current varies with the lamp manufacturer and the type of instrument. Westinghouse and Perkin-Elmer lamps are rated for full-wave operation, while those of ASL are rated for an unfiltered half-wave supply. In practice, this means the following: Westinghouse and Perkin-Elmer lamps can be run at rated current in the Jarrell-Ash and Beckman equipment and in the Perkin-Elmer Model 303, and at one-half rated current in the Techtron instruments and the P-E Model 290. ASL lamps can be run at rated current in the Techtron and the 290, and at twice their rated current in the other equipment.

Vapor Discharge Lamps

Vapor discharge lamps produce emission by passing an electric current through a vapor composed at least partly of the element of interest. Such lamps are produced by Osram in Germany and Philips in Holland. The Osram lamps, which have found far more use in atomic absorption than have the Philips, are available in North America from a number of suppliers, including Perkin-Elmer and Edmund Scientific. Osram lamps exist for the elements mercury, thallium, zinc, cadmium, and the alkalis.

Osram lamps for mercury cannot be used in atomic absorption, because they contain vapor at so high a pressure that the emission line is almost completely self-absorbed, and the sensitivity is very small. It is also generally agreed that hollow-cathode lamps for thallium, zinc and cadmium are superior to the discharge lamps. However, for sodium, potassium, cesium and rubidium, the Osram lamps are superior, despite the fact that they are more troublesome to use.

Osram lamps require special mounts, and a special power supply capable of delivering a 1-ampere current at a starting voltage of about 300 volts and a running voltage of about 50. Furthermore, even in double-beam equipment, they require a wait of about ten minutes for warm-up, while the time in single-beam instruments is in excess of a half hour. Perkin-Elmer and Techtron provide Osram lamp mounts in their equipment, while Jarrell-Ash does not. There appears to be no reference to vapor discharge lamps in the literature from Beckman.

For the alkalis, Osram lamps are about 100 times brighter than hollow-cathode lamps. This is particularly important for potassium, rubidium, and cesium, which have their most sensitive resonance lines in the near infrared spectral region, where monochromator efficiency and detector sensitivity are at their lowest. The brightness of Osram lamps is also important when ultimate detection limits and high precision are required. For sodium and potassium, it is often desirable to work at the secondary resonance lines, which are far less sensitive than the primary lines, and thereby minimize the need for dilution of a concentrated sample. The secondary lines (3302 Å for sodium, 4044 Å for potassium) are readily usable with the Osram lamps, but barely detectable with the hollow cathodes.

This section should not close without mentioning that some workers take the view that the alkalis should be determined with flame emission, rather than by atomic absorption. However, the author believes that atomic absorption is capable of better precision and greater freedom from interferences than emission, if the equipment is well designed. This is taken up in more detail later on.

Multi-Element Lamps

The thought of building several elements into one cathode occurred to workers in atomic absorption almost immediately after equipment became available. Initial efforts were defeated mainly by the metallurgical naivete of the experiments. During the past year, however, multi-element hollow-cathode lamps became available from Westinghouse and from Perkin-Elmer, and have been promised soon by ASL. Depending on the metals involved, the cathodes are made from alloys, intermetallic compounds, or mixtures of powders sintered together.

The importance and value of multi-element lamps should be appraised realistically and not over-estimated. Many presently available combinations can be used without disadvantage as compared to single-element lamps, and can usually be obtained at a financial saving. Other combinations present problems due to line selection and potential spectral interferences, but may nevertheless be worth having if the limitations are understood.

All commercial atomic absorption equipment offers easy interchange of lamps, usually in some sort of pre-focused mount. In single-beam instruments, where warm-up time delays are appreciable, a multi-element lamp is attractive because all its elements are ready when one is. In double-beam equipment, lamp warm-up time is not a factor. Here, the charm of multi-element lamps is confined to their lower cost, and the smaller requirement for storage.

Certain multi-element lamps are excellent bargains. The standard Perkin-Elmer chromium and manganese lamps, for example, are both made with strong admixtures of copper, so each of them is as good a source of copper radiation as is the standard copper lamp. Recently, Perkin-Elmer developed a zinc lamp with much brighter and purer radiation than had

been previously available—with a cathode of an intermetallic compound between zinc and calcium. The new standard zinc lamp is thus perfectly usable for calcium. One of the earliest Westinghouse multi-element sources works well for calcium, magnesium, and aluminum.

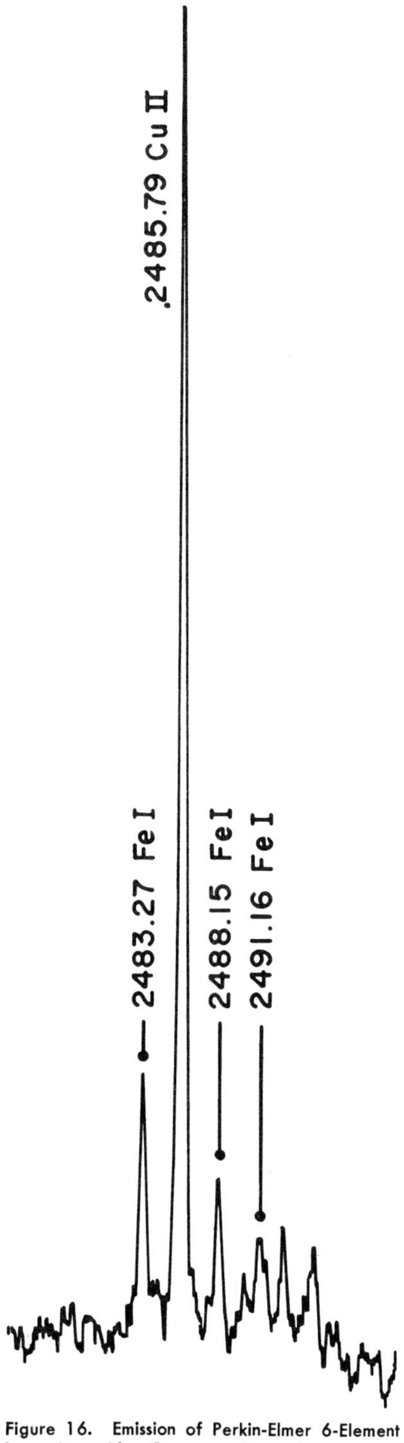

On the other hand, the Perkin-Elmer copper - chromium - cobalt - nickel - manganese-iron lamp falls into the limited-usefulness category. Its complicated emission spectrum (a portion of which is shown in Figure 16) makes it somewhat difficult to find the resonance lines. Furthermore, the energy is reduced and the detection limits are poorer than with single-element lamps, especially for cobalt and iron. A similar problem exists for the Westinghouse zinc-copper-lead-silver lamp, in which the most sensitive resonance line for lead (2170 Å) cannot be used due to strong neighboring copper lines, and which gives relatively weak emission at the other lead line (2833 Å).

High-Brightness Lamps

The emission intensity of ordinary hollow-cathode lamps is limited by the fact that only a fraction of the metallic atoms sputtered off the cathode are excited. In most analyses, a higher emission intensity would be very welcome, since it leads to a better signal-to-noise ratio and thus to better precision and detection limits. The tin determination in Figure 4 is a particularly poignant example of the effect of a weak lamp.

Within the limits of the conventional design, manufacturers have made remarkable advances. Lead, iron, nickel, and zinc hollow-cathode lamps are all far brighter than they were two years ago. For tin, Perkin-Elmer has recently made available a lamp containing tin in a swaged cup made of titanium, in which it is possible to obtain high emission by running the tin in its molten state. However, it was again at the Australian CSIRO laboratories that a definitive advance in lamp design was made (8).

In the "high-intensity" lamp design, shown schematically in Figure 17, a pair of auxiliary electrodes is mounted across and slightly in front of the cathode. A

cites them. A 10 to 100-fold increase in radiation intensity results. At the same time, metallic ion radiation is actually suppressed, so that for many elements the working curve becomes straighter and more sensitive.

In Figure 18 are shown the emission spectra of an ordinary nickel lamp and a high-intensity lamp. In the ordinary lamp, a strong ion line at 2319.76 Å is almost superimposed on the resonance line at 2320.03 Å, and cannot be resolved even by a good monochromator. The ion line does not absorb, so that the working curve (Fig. 19) is flattened and bent. There is another ion line at 2316 Å. The ion lines are suppressed in the high-intensity lamp while the resonance line is enhanced. A substantial improvement can be noted in the working curve.

High-intensity lamps are very useful for many metals, particularly those with complex emission spectra, and those which must be determined with bright flames, like vanadium and titanium. The lamps require an auxiliary power supply for the secondary electrodes.

The high-intensity lamps are being made in pilot quantities by ASL and by Perkin-Elmer. Perkin-Elmer has made successful lamps for nickel, vanadium, tin, and tungsten, while ASL produces at least nickel, iron, and cobalt in the high-intensity version. The lamps of both makers are in their standard size.

Interchange of Lamps

The different manufacturers offer varying systems for changing one lamp for another, all employing kinematic mounts. All are about equally convenient for the rapid installation of a lamp in proper focus. Since hollow-cathode lamps do not grow uncomfortably warm to the touch, there is no problem in exchanging them by hand.

In the Beckman system, three lamps in kinematic, "universal" mounts can be warmed up simultaneously in the source space. Without being removed from their mounts, they can be manually interchanged.

The Jarrell-Ash 82-362 includes a rotating turret which houses six lamps. As the turret is rotated, one lamp after another is placed at the source position. No warm-up supply is included, however. As a lamp reaches the source position, time must be allowed for it to reach steady emission.

Perkin-Elmer mounts its lamps in prealigned holders, which are inserted in the source position as desired. Since lamps for different elements require different currents, the Perkin-Elmer instruments provide several current ranges, protecting lamps from accidental burn-out (Fig. 20). In the double-beam Model 303, a warm-up supply is not required; with the Model 290, an external 3-lamp supply is offered as an accessory.

The Techtron AA-3 also provides kinematic mounts. The lamp supply is capable of warming up six lamps simultaneously.

Monochromators and Detectors

The wavelength range of atomic ab-

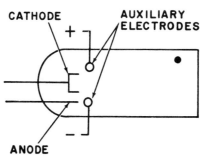

CATHODE **AUXILIARY ELECTRODES**

ANODE

Figure 17. Schematic of High-Intensity Lamp. Auxiliary electrodes, supplied with high current at a relatively low voltage, cause the filler gas to excite a large proportion of the sputtered metal atoms.

direct current of several hundred milliamps at a low voltage flows between the auxiliary electrodes. A stream of ionized gas atoms is produced, which collides with the sputtered metal atoms and ex-

.2485.79 Cu II

2483.27 Fe I

2488.15 Fe I

2491.16 Fe I

Figure 16. Emission of Perkin-Elmer 6-Element Lamp Around Iron Resonance Line. A very strong Cu II line is shown only 2.5Å away. This causes problems with iron determination when the six-element lamp is used. The scan was made with the Perkin-Elmer Model 303, equipped with emission accessory.

sorption is the same as that of flame emission and of ultraviolet spectrophotometry; as a result, it is possible to work successfully with atomic absorption accessories to equipment originally designed for other purposes. However, in many respects the requirements of atomic absorption are different.

The specifications for bandpass, for example, are easy to write. The monochromator must be able to pass the resonance line, and yet screen out the nearest non-resonant lines, in order to permit a straight working curve. Among the elements whose resonance lines are most closely surrounded by other lines are iron, nickel, and cobalt, and these require a monochromator of about 2 Å bandpass. A larger bandpass will cause the absorbance curve to flatten and bend; a smaller bandpass will produce a reduction

in the available light and therefore a degradation of the signal-to-noise ratio.

The large majority of elements can be determined at slit widths corresponding to bandpasses of 7 to 40 Å. As the slits are widened, more light is passed through the monochromator, and this can be translated into an improvement in precision and detection limit. A good monochromator for atomic absorption should therefore be capable of being operated at wide slit widths, and have as high a light transmission as possible.

The useful wavelength range of atomic absorption currently runs between the 1937 Å arsenic line and the 8521 Å cesium line. The monochromator-detector combination should have good energy over this range.

To achieve the above goals with reasonable economy, certain features required of

monochromators for different purposes can be sacrificed. A first-rate UV spectrophotometer, for example, has a wavelength accuracy of about 1 Å, and a stray light level of 0.0001%. In atomic absorption the wavelength can be "tuned in," so that a 10 Å wavelength accuracy is sufficient. Furthermore, stray light levels of perhaps 1% can be tolerated. In flame emission, spectral interferences are minimized by improved resolution, and commercial flame emission monochromators, such as the Jarrell-Ash 0.5 meter unit, have resolution specifications as low as 0.2 Å. In atomic absorption, a bandpass as low as this is not needed. (However when the same instrument is required to do atomic absorption and flame emission, the low bandpass capability is of course desirable.)

Design Differences Between Instruments

The two most popular monochromator designs are the Littrow and the Ebert-Fastie, shown schematically in Figure 21. The merits of the two systems have been compared for many years; for atomic absorption purposes, there seems to be little to choose between them.

The available energy passing through a system is the product of several variables; including grating efficiency, the dispersion of the monochromator, and the cathode efficiency of the photomultiplier detector. The dispersion is generally expressed in Angstroms/mm; the lower the number, the better the dispersion. As the number becomes lower, it is possible to achieve a given bandpass with wider slits, thus admitting more light. The light transmission of a monochromator also varies directly with the surface area or a prism or grating, and with focal length.

In photomultipliers, the choice is generally between the Bi-O-Ag and Cs-Sb types of cathode. The popular and widely used RCA 1P28 has a Cs-Sb cathode; the more expensive Bi-O-Ag material has advantages in the far UV (arsenic and selenium), and even greater advantages in the red (potassium, rubidium and cesium). At the potassium wavelength (7665 A) the Bi-O-Ag surface has perhaps ten times the efficiency of the Cs-Sb. Between 2200 and 5800 Å, the two surfaces are about equal.

In Table 4, a numerical comparison of some of the published characteristics of various instruments is given. Some of the more important points are listed below.

Beckman DB: 260-mm Littrow prism monochromator. Resolution is 5 Å in the UV and 15 Å in the visible range. 1P28 photomultiplier. Wavelength range is given as 2050-7000 Å.

Beckman DU-2: 500-mm Littrow prism monochromator. Resolution is 3 Å in the UV and 10 Å in the visible range. AC-line operated version includes 1P28, plus red-sensitive phototube. Wavelength range is given as 1900–10000 Å. (Note: Older DU's may have a much more restricted wavelength range, depending on age, optics, and detector.)

Jarrell-Ash 82–362 or 363: 500-mm grating Ebert monochromator. Resolution is 0.2 Å. Standard detector is 1P28.

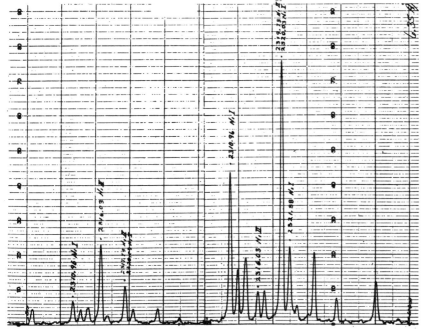

Figure 18. Emission Spectra of Perkin-Elmer Nickel Lamps. (a) Ordinary lamp shows presence of very strong ion line at 2316.03Å. (b) High intensity lamp shows strong enhancement for ground state line at 2320.03Å., and suppression for the ion line at 2316.03Å.

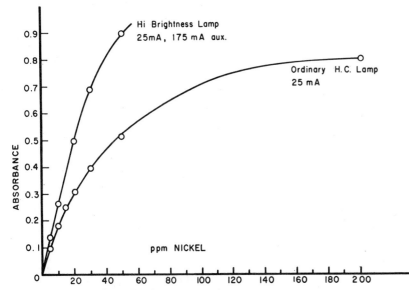

Figure 19. Working Curves for Nickel. The lower curve was achieved with an ordinary lamp. The upper curve represents the results from a high-intensity lamp.

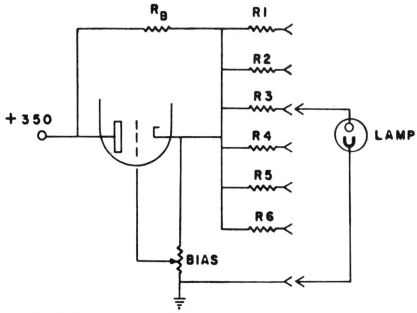

Figure 20. Simplified Schematic of Perkin-Elmer Lamp Supply. The different resistance ranges are intended to prevent accidental burnout of lamps.

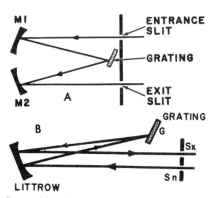

Figure 21. Simplified Schematics of Monochromator Designs. (a) Ebert-Fastie. Minor modification of this is known as Czerny-Turner. (b) Littrow.

To extend the wavelength range toward the red and the UV, it is possible to remove the standard grating and detector in order to substitute others. Slit assemblies up to a bandpass of 1.6 Å can be inserted in the interests of higher energy. Grating has a ruled surface 52 × 52 mm.

Reciprocal linear dispersion is 16 Å/mm.

Perkin-Elmer Model 290: 267-mm grating Littrow monochromator. Lowest slit width gives 2 Å resolution. 7 and 20 Å slits are also supplied. Standard detector is 1P28 with cathode modified by RCA to provide improved response in the red region. Grating has ruled surface 64 × 64 mm. Reciprocal linear dispersion is 16 Å/mm. Wavelength range is given as 2000–8700 A.

Perkin-Elmer Model 303: 400-mm grating Ebert monochromator. Nominal resolution at lowest slit width is 0.2 Å in UV. Resolution is variable by switch selection from nominal 0.2 to 20 Å in the UV, 0.4 to 40 Å in the VIS region. Standard detector is EMI 9592B, endwindow photomultiplier with Bi-O-Ag cathode. Wavelength range is given as 1890–8700 Å. Two separate gratings, mounted back-to-back, blazed for the VIS and UV ranges, are switch-selectable. Reciprocal linear dispersion is 13 Å/mm in VIS, 6.5 Å/mm in the UV range. Gratings have a ruled surface 64 × 64 mm.

Techtron AA-3: 500-mm grating Ebert monochromator. Widest bandpass 10 Å, narrowest 3.3Å. Continuously variable slits. Standard detector is 1P28. Wavelength range given as 1860–10,000 Å. Grating has a ruled surface 50 × 50 mm. Reciprocal linear dispersion is 33 Å/mm.

Table 3. Usability of Various Hollow-Cathodes with Instruments from Different Manufacturers

Lamp Type	Usable in Beckman	Usable in Jarrell-Ash	Usable in Perkin-Elmer	Usable in Techtron
Westinghouse	Yes	Standard	Yes, with adapter supplied by W. or P-E.	No
ASL	Yes	No	Yes	Standard
Perkin-Elmer	Yes, with adapter from Beckman	No	Yes	No

Table 4. Statistics for Different Instruments

Characteristic	Beckman DU2	Beckman DB	J-A 82-362	Perkin-Elmer 303	Perkin-Elmer 290	Techtron AA3
Type of system	Single beam DC	Single beam AC	Single beam AC	Double beam AC	Single beam AC	Single beam AC
Burner (preferred)	3 Total Consump	3 Total Consump	1 Total Consump	Premix	Premix	Premix
Passes through flame	3	3	5	1	1	1
Monochromator	Littrow	Littrow	Ebert	Ebert	Littrow	Ebert
Disperser	Prism	Prism	Grating	Grating	Grating	Grating
Resolution (Å)	3 UV 10 VIS	5 UV 15 VIS	0.2	0.2	2	3.3
Mono focal length (mm)	500	260	500	400	267	500
Grating lines/mm	...	...	1200	UV 2880 VIS 1440	1800	600
Dispersion (Å/mm)	Not given	Not given	16	UV 6.5 VIS 13	16	33
Maximum bandpass (Å)	Not given	Not given	1.6	UV 20 VIS 40	20	10
Nominal Wavelength range (mμ)	190–1000	205–700	Not given	190–870	200–870	186–1000
Grating surface mm	...	...	52 × 52	64 × 64	64 × 64	50 × 50
Detector	1P28 and Red photo tube	1P28	1P28	EMI 9592B Bi-O-Ag	1P28 B	1P28
Standard Readout	Meter	Meter	Meter	Counter	Meter	Meter
Linearity	% Trans.	% Trans.	% Trans.	% Absorption	Concentration	% absorption
Scale Expansion	No	No	No	1X, 2X, 5X 10X	1X-4X	1X-5X
Readout Accessory	Recorder	Recorder	Recorder	Digital Conc. Readout, also recorder	Recorder	Recorder
Bench Space (in.)	72 × 16	70 × 16	60 × 18	33 × 22	25 × 15	56 × 16

The Place of Flame Emission

Most commercial atomic absorption instruments are either designed to do flame emission also, or to employ flame emission accessories. At the present time, the availability of such units is a competitive necessity; however, estimates of the place of flame emission vary very widely indeed.

Professor Velmer Fassel of Iowa State University has pointed out that, according to published literature, flame emission achieves better detection limits than atomic absorption for more than half the elements for which data are available. Walter Slavin of Perkin-Elmer retorts that some of the quoted emission figures are optimistic in the extreme, and that the elements where flame emission limits are superior are those where no serious effort has yet been made by atomic absorption.

Jarrell-Ash personnel emphasize the flame emission capability of their equipment very strongly. They take the view that it is possible to make a qualitative analysis by scanning flame photometry; that sulfur and phosphorus are determinable by the emission of molecular bands; and that the alkalis, lithium, thallium, and some other metals are preferably done by flame.

Personnel at Perkin-Elmer disagree. They feel that the strong spectral, chemical, and inter-element interferences of flame emission make qualitative scanning, while theoretically attractive, rarely useful. Further, with a double-beam atomic absorption system, it is possible to interchange elements in less than one minute, so that one can make a quantitative analysis for fifteen elements in less than half an hour. Molecular band emission is also so fraught with interferences as to be analytically impractical, except in special cases.

The determination of the alkalis, lithium, and thallium is readily carried out by atomic absorption, if a premix laminar flow burner is used. With a turbulent-flow total-consumption burner, the flicker problems due to emission from the flame are much more severe, and may well form the basis of the Jarrell-Ash view. In the use of atomic absorption, spectral interferences disappear, chemical interferences are rare, and ionization effects are readily controlled. Therefore, the opinion at Perkin-Elmer is that all metallic elements are determined more precisely and accurately by atomic absorption than by flame emission. The usefulness of the scanning emission accessory is considered to be confined to physical research work.

Flame Emission Operation

Both the Beckman and the Jarrell-Ash equipment, which were designed for flame emission before conversion to atomic absorption, can easily be operated in the former mode. The DC system of the DU and DU-2 is particularly suitable for flame emission: all that is required is to switch off the hollow-cathode lamp. In the DB, the chopper is moved from in front of the flame to behind it, so that the flame light is chopped and can be amplified by the electronics. The Jarrell-Ash 82–362 is equally easily converted.

An eight-speed wavelength scanning accessory is available for the Jarrell-Ash unit.

For the Perkin-Elmer Model 303, an accessory package is available, which includes a chopper to be placed after the flame, a four-speed wavelength drive, and the electronics to convert the double-beam atomic absorption system to single-beam emission. The premix nitrous-oxide burner has shown itself to be quite a good emission source, when turned at 90° from its normal orientation, as shown for aluminum in the scan of Figure 22. A flame emission accessory is also available for the Model 290, as well as for the Techtron equipment.

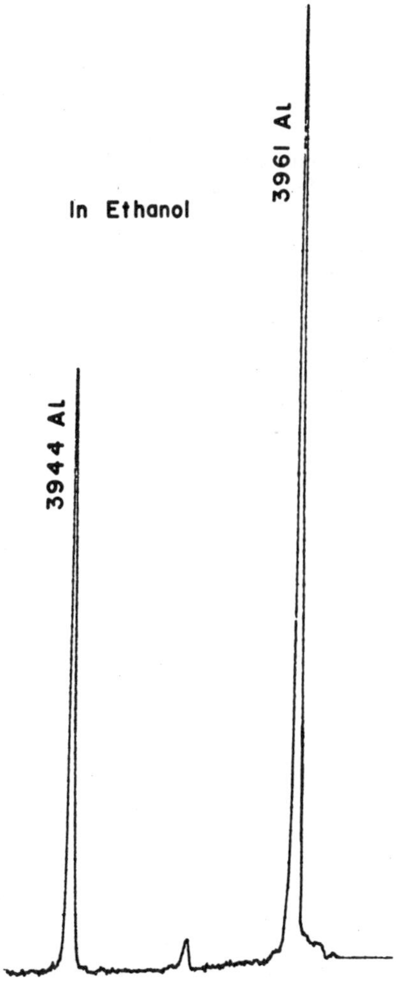

In Ethanol

3961 Al

3944 Al

Figure 22. Flame Emission Spectrum of Aluminum. The Model 303 flame emission accessory was used, with a premix nitrous-oxide burner head turned 90°. Concentration was 50 ppm aluminum in ethanol, bandpass was 0.2Å.

Readout Devices

Atomic absorption, when properly designed, obeys Beer's Law: that is, concentration is proportional to the negative logarithm of transmittance, over an appreciable concentration range. The readouts of many atomic absorption instruments are linear in transmittance or absorption, and the appropriate Beer's Law absorbance values are found by referring to a logarithmic chart or table. It is also possible to obtain readouts which are linear in absorbance, or even calibrated directly in concentration.

Furthermore, in some analyses, the precision and detection limit are limited by the readout, unless scale expansion is provided. Different readout devices therefore give scale expansion up to 100×. All commercial atomic absorption instruments can also be connected to recorders, which often produce improvements in performance.

The Jarrell-Ash instrument has the simplest readout, a 4-inch meter calibrated from 0 to 100% absorption, with no scale expansion. In order to improve readability, terminals are provided for a recorder.

The Beckman DB has a meter readout without scale expansion, linear in percent transmittance. For use with a recorder, a scale expansion accessory is offered, which can produce expansions of 2, 5, and 10×. The DU and DU-2 have a null-balance meter readout, linear in transmittance, from 0 to 100% transmittance.

The Techtron AA-3 has a meter readout which is linear in absorption (100%-transmittance). An accessory is offered, which together with suitable meter damping provides scale expansion variable between 1 and 5×.

The readout systems of the Perkin-Elmer Model 290 and 303 are described in the following sections.

The Model 290 Readout

The Model 290 readout differs from the foregoing in that it is linear in absorbance ($-\log T$), and that the scale can be expanded and the zero suppressed at will, up to about 4×. This design combination makes it possible to read the results of most analyses directly in concentration in any desired units.

The linear-to-logarithm conversion is achieved by taking advantage of the fact that the voltage across a semiconductor diode in the forward direction varies as the logarithm of the direct current passed through it. The temperature drift, with which such diodes are plagued, is overcome by chopping between the sample current and a fixed reference, and reading the difference. The output is presented on a $4\frac{1}{2}$ inch meter calibrated in arbitrary units, with a nominal range of 0–0.5 absorbance units. Continuously variable scale expansion can be employed, up to a limit where the meter scale has a length of about 0.12 absorbance units. Suitable meter-and-recorder damping capacitors are supplied, and any linear recorder also reads out the logarithm.

As shown in Figure 23, this system can read directly in concentration. If the range is to be 0–10 ppm calcium, one control is used to set zero when a blank is being aspirated, and another control sets the meter needle to 100 when the maximum standard is run. The intervening range is then usually linear.

Furthermore, the zero can be suppressed and the readings changed to any desired units. For example, clinical values for calcium in serum generally range between 3 and 7 milliequivalents per liter. A scale so calibrated can be purchased or drawn, and inserted under the meter glass over the original face. Two separate controls set the ends, and the meter scale is linear between them.

The Digital Concentration Readout for the Model 303

The built-in readout device for the Perkin-Elmer Model 303 is a conventional manual null. A control is turned until a null meter indicates zero, and the ratio between sample and reference beam intensities is indicated on a three-digit counter, calibrated in percent absorption from 000 to 99.9%. Scale expansion ratios of 2, 5, and 10× are built in, as is a null meter damping control.

In a double-beam system, some sort of automatic nulling must occur before the output can be fed to the recorder. Therefore, unlike single-beam units, the Model 303 requires a recorder readout accessory if a recorder is to be used. The accessory automatically takes the ratio between the sample and reference beams, and also makes available a variety of damping times between 0.5 second and 2 minutes. Scale expansions of 1, 3, 10, 30, and 100× are provided, with the ability to expand any portion of the scale. The use and performance of the accessory is shown in Figure 24, for a sample of 0.1 ppm manganese in water.

The most elaborate commercial readout device yet used for atomic absorption is the DCR-1 accessory to the Model 303, which presents concentration directly on four illuminated digits, and makes possible a speed of 15 element determinations per minute. It operates on the principle that the discharge of a capacitor is logarithmic with time. The reference channel is used to charge up a large capacitor, after which the large capacitor is discharged by repeatedly switching across it a very much smaller capacitor. The process continues until the voltage across the large capacitor is equal to the voltage of the sample channel. The reading is taken by counting the number of times the small capacitor has had to be switched across the large one. The readout is made proportional to concentration, in any desired units, by changing the size of the small capacitor.

By connecting a suitable printer, an entirely automatic concentration readout can be achieved. Further, improvements in precision and detection limit are attained by using various averaging modes: 4×, 8×, or 16×. In the 16× mode, for example, 16 readings of the same sample are taken automatically at 2-second intervals, and their average is computed and presented digitally.

When the working curve is bent in a typical fashion (absorbance increasing at a progressively slower rate than concentration), it is possible to regain linearity by employing the curve straightener of the DCR-1, which can be set to correct the output by an appropriate degree.

Detection Limits

In Table 5, operating conditions, best emission lines, and detection limits are shown for many of the elements that have been studied by atomic absorption.

(Those requiring nitrous oxide are shown in Table 2.) The most sensitive resonance line and the required bandpass for an element is usually the result of basic physical conditions, and generally not dependent on instrument design. However, the sensitivity (ppm/1% absorption) and the detection limit depend strongly on the lamp, the burner, and the instrument. All values are given with water as the solvent; organic solutions would improve the detection limits by a factor of about 3.

The detection limits given are for the Perkin-Elmer Model 303, equipped with a recorder readout accessory. When not otherwise stated, a premix burner is used with air and acetylene. Except where marked, commercially available hollow-cathode lamps are used. In Figure 25, the records from which the detection limit was deduced for aluminum are shown. In most cases, the elements requiring nitrous oxide have not been studied as exhaustively as the others, since the nitrous oxide flame has only recently come into use.

Commercial Atomic Absorption

Atabel Model 100. Single-beam AC system, total-consumption burner, Westinghouse lamps. Apparently not yet in production. Price $3995 less lamps.

Beckman Model 1300 for use with DB. Fully described in text. Accessory price about $3200. DB price from $2170.

Beckman Model 1301 for use with DU and DU-2. Accessory price about $2800. DU-2 price $3400.

Block Flame Line Discriminator. Intended as interferometer system with continuous source. Single beam. Not yet in production. Price not set.

Evans Electroselenium Ltd., London, England. Single-beam AC system, premix burner, grating monochromator.

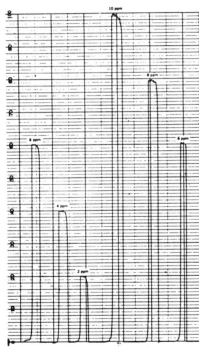

Figure 23. Various Concentrations of Calcium in Water, recorded directly in concentration with the Perkin-Elmer Model 290.

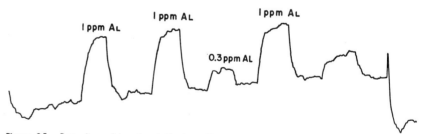

Figure 25. Recording of Low-Level Aluminum Absorption. From these traces for 0.3 and 1.0 ppm aluminum in water, the detection limit was deduced to be 0.1 ppm. Recorder Readout was used at 10 × scale expansion and Noise Suppression of 4.

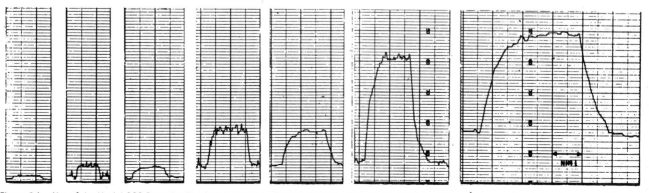

Figure 24. Use of the Model 303 Recorder Readout Accessory. Recorder traces of the absorption from 0.1 μg/ml manganese at various expansions and noise suppression response settings. Chart speed is constant.

Table 5. Relative Detection Limits Obtained with the Perkin-Elmer Model 303 Atomic Absorption Spectrophotometer

(Elements requiring nitrous oxide flames are listed in Table 2)

Metal	Rel. Det. Limit ug/ml	(A)	Slit (A)	ma	Metal	Rel. Det. Limit (ug/ml)	(A)	Slit (A)	ma
Ag	0.02	3281	7	12	Mg	0.003	2852	20	6
As	0.5	1937	7	12	Mn	0.01	2795	20	10
Au	0.1	2428	20	14	Mo	0.1	3133	2	30
Ba	1	5536	40	20	Na	0.005	5890	40	500
Bi	0.2	2231	2	10	Ni	0.03	2320	2	15
Ca	0.01	4227	20	10	Pb	0.05	2170	2	10
Cd	0.01	2288	7	4	Pd	1	2476	2	20
Co	0.05	2407	2	15	Pt	0.5	2659	2	25
Cr	0.01	3579	2	10	Rb[a]	0.005[d]	7800	40	400
Cs[a]	0.05[d]	8521	40	400	Rh	0.3	3435	2	20
Cu	0.01	3247	7	10	Ru	0.3	3499		
Fe	0.05	2483	2	40	Sb	0.2	2175	7	20
Ga	1.	2874		4	Se	1.	1961	2	10
Hg[c]	0.2	2537	20	20	Sn[e]	0.2	2246	7	30
In	0.5	3040	7	6	Sr	0.02	4607	13	10
K[a]	0.005	7665	40	400	Te	0.5	2143	7	16
Li	0.005	6708	40	15	Tl	0.2	2768	20	12
					Zn	0.005	2138	20	10

[a] Osram spectral lamp is used.
[c] Boling burner head used.
[d] In presence of high concentrations of another ionizable metal.
[e] Using an air-hydrogen flame.

Very compact. Price in U.S.: $2850.

Hilger and Watts Ltd., London, England, Model AA2/3. Single-beam AC system, premix burner, prism monochromator, Hilger lamps. Readout linear in absorbance. Has not yet been shown in U.S.

Jarrell-Ash Model 82–362. Fully described in text. Price from $5400.

Model 82–600, multichannel instrument with 750-mm grating monochromator, space for four hollow-cathode lamps. Single-beam AC operation, total consumption burner, Westinghouse lamps. Apparently not yet delivered, price not set.

Model 82–700. Single-beam AC system, 250-mm grating monochromator, total-consumption burner. Westinghouse lamps. Price $2900.

Perkin-Elmer Model 303. Fully described in text. Price $5920.

Model 290. Fully described in text. Price $2900.

Techtron, Melbourne, Australia. Model AA-3. Fully described in text. Price in U.S. $4995.

Model AA-100. Single-beam AC system, 250-mm grating monochromator, premix burner. Warm-up supply for one extra lamp. Price $2995.

Unicam Instruments, Ltd., Cambridge, England. Model SP900A. Accessory to SP900 flame photometer. Single-beam AC system, premix burner, prism monochromator. Unicam lamps. Apparently not yet exhibited in U.S.

Zeiss, Inc. Single-beam AC system, accessory to PMQ II UV-VIS spectrophotometer. Not yet exhibited in U.S. German lamps.

Atomic Absorption

Supplement

Commercial instrumentation has changed greatly since this article was written in late 1965. There are now at least 19 companies producing equipment, many of whom offer more than one instrument. Instruments and model numbers are continuing to change. Therefore, it would be doing a disservice to the reader to give a partial listing on so ephemeral a subject. We will therefore reluctantly forego the opportunity. The following points are worth noting:

1. All references in the body of the article to specific commercially available instruments should be ignored or at best treated very cautiously. Many of the instruments have been replaced or improved. Four manufacturers, rather than just one, now make double-beam instruments.

2. In the discussion of interferences, no mention was made of the occasional problems caused by background absorption. When low concentrations of particular elements are sought in a concentrated sample, the sample materials may produce nonspecific broad-band absorption, leading to erroneously high readings. When background absorption is encountered, it can be eliminated by one of three means:

• Use of a blank solution which is similar to the samples but does not contain the element of interest.
• Use of a nonabsorbable line, either of the same or of a different element, close to the same wavelength. Example: Background absorption for nickel at 2320 Å can be subtracted by using the Ni II line at 2316 Å.

• Use of a deuterium lamp as background corrector; this device continuously compares sample absorption of light from a hollow cathode lamp with that of the broadband emission of the deuterium arc (9).

3. Some systems and components described in the 1966 article have fallen into disuse. Only one manufacturer still offers a DC system, for example. Total-consumption burners are still being offered, but are very little used, and not much attempt is made to persuade others of their virtues. Much the same can be said of the practice of multi-passing. Neither the long tube nor the heated-chamber burner have found much application at this time, apparently because of problems which arise when the sample contains appreciable dissolved solids.

4. Considerations regarding light sources for atomic absorption have become greatly simplified. Excellent hollow-cathode lamps have been built for sodium, potassium, and mercury, and these are preferred over discharge lamps. Cesium and rubidium still require discharge lamps, but can also be determined by flame emission. Hollow-cathode lamps from several manufacturers have been greatly improved and are so bright for most elements (though exceptions still exist) that there would be no point in making them brighter. As a result, high-brightness or high-intensity designs, with their extra power requirements, are no longer recommended in general, though they are still used to a limited degree for special purposes. A large range of multielement lamps has become available, and many of these can be used without penalty in place of single-element designs.

The Place of Flame Emission

The most common use of flame emission has been, and continues to be, the determination of sodium and potassium. Both elements are easily excited, and produce high emission even in low-temperature flames; they can also be determined with the aid of filters, and do not require a monochromator.

Before the advent of atomic absorption, a large literature existed on the determination of other elements by flame emission, generally with a total-consumption burner and a prism monochromator. This emission system was subject to many chemical and spectral interferences. Atomic absorption, which was performed almost from the beginning with grating monochromators and premix burners, went far toward eclipsing flame emission for elements other than sodium and potassium, for several years.

More recently, it has been realized that many of the improvements in instrumentation, which were made for atomic absorption, also produce improvements in flame emission. It is now generally accepted that premixed nitrous oxide-acetylene flames are the best sources of flame emission, at least among commercially available burner systems(10). Furthermore, the availability of grating monochromators with better than 0.5 Å resolution, at manageable prices, has considerably reduced the spectral interferences with which emission had formerly been plagued.

For many years, proponents of flame emission have carried on a battle of detection limits against atomic absorption, pointing out that flame emission has better detection limits than atomic absorption for about 50% of the elements determinable by both techniques. Flame emission has an edge for the alkalis, some of the alkaline earths, and for most rare earth elements. Comparative detection limits for both techniques are shown in Table 6. Leaving detailed arguments about individual values to one side, Table 6 clearly shows that very many elements can be determined at useful levels by both techniques.

A flame emission capability is useful to perform the following functions:

(1) Determination of low levels of those elements where emission has better detection limits.
(2) Determination of elements for which no hollow cathode lamp happens to be available in the laboratory at the time.
(3) Determination of sulfur and phosphorus through molecular band emission in low-temperature flames. Chemical interferences, however, are severe.
(4) Qualitative as well as quantitative analysis of a sample, by scanning the wavelength. Figure 22, for example, represents a wavelength scan for a mixture of potassium, indium, chromium, and calcium.

Instrumentation for Flame Emission

In general, it is fair to state that it requires more skill and better instrumenta-

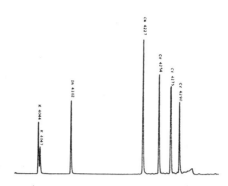

Figure 22. Flame emission survey spectrum of a solution containing potassium, indium, calcium, and chromium. A Perkin-Elmer Model 403 was used; scan speed was 200 Å/min, and the resolution setting was 0.7 Å.

tion to do a flame emission analysis than to determine the same element by atomic absorption.

Since flame emission depends on excitation of the element of interest, small variations in flame position, composition, and temperature, affect emission far more than they do absorption. Therefore, a good flame emission instrument must have very precise flame controls, and these must be watched carefully by the operator.

Also, spectral interferences are far more common in emission than in absorption. Atomic absorption is not, as had been thought earlier, completely free from spectral interferences, but there are nevertheless very few of analytical significance. In emission, on the other hand, there are far more lines to contend with, and the degree of freedom from spectral interferences depends critically on the resolution of the monochromator.

Instrument manufacturers have recognized the revived importance of flame

emission, and almost all atomic absorption spectrophotometers can be used for flame emission to some degree. However, the flame emission capabilities of different instruments vary widely. There is not space here to describe the peculiarities of individual instruments, but general rules can be given.

For first-class work, the monochromator should have a resolution of better than 0.5 Å. Wavelength scanning must be provided, and a range of speeds must be available. This range should encompass a speed of 10 Å per minute or less, for careful analytical work, to 100 Å per minute or more, to enable the operator to run survey spectra. For all but the simplest flame emission analyses, lock-in amplification is highly desirable in order to distinguish the fluctuations of the flame from the signal of interest.

Readout Devices

Atomic absorption, when properly designed, obeys Beer's law: that is, concentration is proportional to the negative logarithm of transmittance, over an appreciable concentration range. As concentration rises beyond the range, read-out becomes nonlinear; partly because of background emission in the light source, partly from stray light in the monochromator, and partly because of line-broadening effects in the flame. Most modern atomic absorption instruments make the logarithmic conversion electronically and present the result linearly in absorbance. Often, continuously variable scale expansion is also available, so that results may be set to read directly in concentration, over the range where the absorbance-concentration relationship is linear. Typically, the operator aspirates a "blank" solution, and sets his baseline or zero control to make the meter read zero. He then aspirates a standard solution containing, say 10 μg/ml, and sets his scale expansion or calibration control to make the meter read the desired value, such as 100 units. If his working curve is straight, samples containing less than 10 μg/ml can then be read directly in concentration.

Most instruments offer a basic working range of 0-1 absorbance unit, corresponding to a range from 0 to 90% absorption. For a limited number of elements, a larger range is useful. However, at the present state of the art, a range higher than 0-2 absorbance units is not usable.

When the working curve deviates from linearity, it does so, to a first approximation, in an exponential manner. Electronically, it is fairly simple to put in a "counterexponential" and linearize the working curve over an appreciable additional range. Several instruments use curvature correctors of various kinds.

Scale Expansion

The fact that most atomic absorption instruments now read out in absorbance makes necessary a new definition of scale expansion. When every instrument had a basic range of 0-100% transmittance, 10X scale expansion meant unequivocally that a 10% difference in transmittance covered full scale on the meter or recorder.

Table 6. Detection Limits for Atomic Absorption and Flame Emission, μg/ml

Element	Flame Emission	Atomic Absorption	Element	Flame Emission	Atomic Absorption
Aluminum (1)	0.005	0.03	Nickel	0.03	0.01
Antimony	...	0.1	Niobium (1)	1.	1.0
Arsenic (2) (4)	...	0.1	Osmium (1)	...	0.5
Barium (1) (4)	0.001	0.02	Palladium	0.05	0.02
Beryllium (1)	0.2	0.001	Phosphorus (1)	...	100.0
Bismuth	...	0.04	Platinum	2.0	0.1
Boron (1)	...	2.5	Potassium	...	0.005
Cadmium	...	0.001	Praseodymium (1)	1.	10.0
Calcium	0.0001	0.001	Rhenium (1)	0.2	1.0
Cesium	...	0.05	Rhodium	0.02	0.03
Chromium	0.005	0.003	Rubidium	...	0.005
Cobalt	0.05	0.01	Ruthenium	0.02	0.3
Copper	0.01	0.002	Samarium (2)	0.2	2.0
Dysprosium (1)	0.07	0.2	Scandium (1)	0.03	0.1
Erbium (1)	0.04	0.1	Selenium (2) (4)	...	0.1
Europium (1)	0.0006	0.04	Silicon (1)	...	0.08
Gadolinium (1)	2.	4.0	Silver	0.02	0.002
Gallium	0.01	0.1	Sodium	...	0.002
Germanium (1)	0.5	1.0	Strontium (1) (4)	0.0001	0.01
Gold	0.5	0.02	Tantalum (1)	5.	2.0
Hafnium (1)	...	8.0	Tellurium	...	0.09
Holmium (1)	0.02	0.1	Terbium (1)	0.4	3.0
Indium	0.002	0.05	Thallium	0.02	0.03
Iridium	30.	2.0	Thulium (1)	0.02	0.2
Iron	0.05	0.01	Tin (2) (4)	0.5	0.02
Lanthanum (1)	8.	2.0	Titanium (1)	0.2	0.09
Lead	0.2	0.02	Tungsten (1)	0.5	3.0
Lithium	0.00003	0.0006	Uranium (1)	...	30.0
Lutetium (1)	1.	3.0	Vanadium (1)	0.01	0.06
Magnesium	0.005	0.0001	Ytterbium (1)	0.002	0.04
Manganese	0.005	0.002	Yttrium (1)	0.4	0.3
Mercury	...	0.5	Zinc	...	0.002
Molybdenum (1) (4)	0.1	0.03	Zirconium (1)	3.	5.0
Neodymium (1)	0.2	2.0			

(1)-(4) refer to the atomic absorption determinations.
  (1) Nitrous oxide flame
  (2) Argon-hydrogen-air diffusion flame
  (3) Air-hydrogen flame
  (4) Can also be determined in the air-acetylene flame
The atomic absorption limits were determined with the Perkin-Elmer Model 403. The flame emission data are from Ref. 11.

Now let us suppose that an instrument readout has a basic range of 0–1 absorbance units. 10X scale expansion now means that 0.1 absorbance unit will read full scale. This value is not equivalent to the transmittance value, since 0.1 absorbance unit is equal to an absorption of 20.6%, not 10%. Furthermore, suppose now that the basic range of the instrument is 0–4 absorbance units; 10X scale expansion would now give a full-scale reading on the meter of 0.4 absorbance units. Clearly the "X" system of describing scale expansion is misleading and obsolete.

In this writer's opinion, it would be suitable to define scale expansion as the minimum absorbance range that can be made to cover full scale on a meter or recorder, or absorbance per digit when the readout is digital. For the Perkin-Elmer Model 403, for example, 0.01 absorbance unit can be made to cover full scale on the recorder, while one digit represents 0.00004 absorbance unit, or 0.01% absorption.

Noise Suppression

Any spectroscopic instrument carries within it certain noise sources; which result in unwanted fluctuations of the output reading. The use of scale expansion increases the size of the apparent signal, but increases the fluctuations by the same factor.

To reduce the fluctuations (noise), there are three approaches, all of which reduce noise at the expense of longer sampling time, and greater sample consumption.

- Damping, in which the response of the readout meter (or recorder) to rapid changes of signal is reduced, generally by fairly simple filter circuits.
- Integration, where the signal over some period (typically 1–10 sec) is stored, and then applied to the readout device.
- Digital averaging, where a number of separate readings are taken, and their average is computed automatically.

Either of the first two approaches can be combined with the third; that is, damped or integrated signals can be averaged. Instruments offering each of these techniques are available. There is little agreement among authorities as to which noise suppression system is superior; any meaningful discussion is soon deeply involved in mathematical statistics, and will not be attempted here. However, one general rule can be given. All three noise suppression techniques are fairly closely equivalent; given a certain sampling time, it is unlikely that substitution of one method for another will change a detection limit by as much as a factor of two.

Detection Limits

Detection limits were defined earlier in this article as the concentration, in water solution, giving a signal equal to twice the fluctuation of the background. When the sample is in an organic solution, the detection limit is generally improved by a factor of between two and five. When the sample is a solid, the detection limit is degraded by the dilution factor. That is, if one gram of sample must be dissolved in 10 ml of solvent, the detection limit value must be multiplied by ten.

It has now become not only possible but convenient to determine detection limits rigorously, through the use of a large number of measurements at concentrations near the expected detection limit, and their subsequent reduction by computer(*10*). Table 6 shows the detection limits achieved with a double-beam instrument under these circumstances, for atomic absorption. The flame emission limits are taken from Reference *11*.

Conclusion

Atomic absorption and the other areas of atomic spectroscopy still form a lively and developing field. These notes, appended in June 1970, by no means cover all the developments since the original article was written. They are only designed to keep the reader of the earlier work from being misled as to the present state of the art.

Literature Cited

(1) WALSH, A., *Spectrochim. Acta*, **7**, 108 (1955).
(2) WILLIS, J. B., *Nature*, **207**, 715 (1965).
(3) ZETTNER, A., AND SELIGSON, D., *Clin. Chem.*, **10**, 869 (1964).
(4) SUNDERMAN, F. W., JR., AND CARROLL, J. E., *Am. J. Clin. Path.*, **43**, 302 (1965).
(5) RANN, C. S., AND HAMBLY, A. N., *Anal. Chem.*, **37**, 879 (1965).
(6) FUWA, K., AND VALLEE, B. L., *Anal. Chem.*, **35**, 942 (1963).
(7) FASSEL, V. A., MOSSOTTI, V. G., GROSSMAN, W. E. L., AND KNISELEY, R. N., *Spectrochim. Acta*, **22**, 347 (1966).
(8) SULLIVAN J. V., and WALSH, A., *Spectrochim. Acta*, **21**, 721 (1965).
(9) KAHN, H. L. *At. Abs. Newsl.*, **7**, 40 (1968).
(10) PICKETT, E. E., and KOIRTYOHANN, S. R., *Anal. Chem.*, **41**, (14), 28A (Dec., 1969).
(11) KERBER, J. D., and BARNETT, W. B., *At. Abs. Newsl.*, **8**, 113 (1969).

Sulphur

Topics in...

Chemical Instrumentation

Edited by GALEN W. EWING, Seton Hall University, So. Orange, N. J. 07079

The Journal of Chemical Education
Vol. 45, pages A89, A169, A273
February to April, 1968

XXXVI. Recent Instrumentation for UV-Visible Spectrophotometry—Part I: Dual-Beam Spectrometers

PETER F. LOTT, *Chemistry Department, University of Missouri at Kansas City, Kansas City, Missouri 64110*

The commercially available spectrophotometers for measurements in the ultraviolet and visible region of the spectrum, as well as the principles of spectrophotometric instrumentation, were covered in this series by Lewin in 1960 (1). The purpose of this article is to report on recent and novel spectrophotometers which have been commercially marketed since that time. Many of the recently introduced instruments still retain the same basic design of the earlier models. Changes have been made, however, to incorporate advances in electronics possibly more than in optical design, and solid state circuitry is beginning to replace the vacuum tube. Deuterium light sources are now regularly offered in place of the hydrogen lamp as the source for the UV region, and the tungsten-iodide, quartz envelope, lamp which is of much higher light intensity is steadily replacing the familiar tungsten "headlight bulb" as the source for the visible near-IR region in recording spectrophotometers. New photodetectors have been developed which cover wider spectral spans. Advances have been made to meet existing needs such as spectrophotometers which will cover a wider region of the spectrum, instruments capable of measuring absorption changes or even scanning the whole spectrum in milliseconds, spectrophotometers capable of measuring the spectrum of a portion of a single biological cell, such as a blood cell, and instruments capable of measuring absorptivities with much higher precision.

The general criteria of instrument selection and the effect of such factors as signal-to-noise ratio, stray radiation, resolving power, and resolution were discussed by Lewin and may be found more fully described in the book by Bauman (2). Indeed, the much wider selection of instrumentation currently available in some respects complicates the choice, not only as to which type of instrument to acquire, whether it be single or double beam, but also as to what characteristics of resolution, wavelength range, accuracy, and additional functions need to be met. No single spectrophotometer will meet all needs. Rather, the choice of the instrument should be projected not only as to the type of measurements which are to be made, but should consider also how this instrument will supplement the instrumentation already existing in the laboratory and what future demands will have to be met. The recording spectrophotometers employed for spectral scanning generally automatically measure the difference in absorption between the reference material and the standard, and are equipped with a means of simultaneously changing the wavelength and recording the absorbance (or transmittance). The price of these instruments range from about $2500 to over $30,000. The vast array currently available should meet nearly any need. Only instruments not previously described in this series will be discussed.

Figure 1. Shimadzu Model MPS-50L recording spectrophotometer.

American Instrument Co.

The American Instrument Co. is the sole distributor in the United States of the Shimadzu Model MPS-50L recording spectrophotometer, a double beam, double detector, ratio recording instrument which has certain novel design features to permit recording of spectra of transparent, translucent, and opaque substances in the region from 190 mμ to 2.5μ. This instrument is shown in Figure 1 and an optical schematic is shown in Figure 2. A Littrow prism monochromator is the dispersing element; the mercury light source is provided for wavelength calibration. A unique feature of this instrument is that the sample is placed in close proximity to the large detecting surface of an end-on photomultiplier tube in order to minimize reflected light loss. This is of particular significance in recording the reflectance spectra of opaque materials or translucent

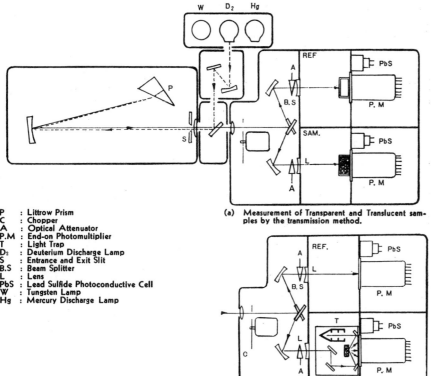

P : Littrow Prism
C : Chopper
A : Optical Attenuator
P.M : End-on Photomultiplier
T : Light Trap
D₂ : Deuterium Discharge Lamp
S : Entrance and Exit Slit
B.S : Beam Splitter
L : Lens
PbS : Lead Sulfide Photoconductive Cell
W : Tungsten Lamp
Hg : Mercury Discharge Lamp

(a) Measurement of Transparent and Translucent samples by the transmission method.

(b) Measurement of opaque and translucent samples by the reflection method.

Figure 2. Optical schematic for the Shimadzu Model MPS-50L spectrophotometer.

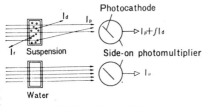

A) Non-uniform scattering

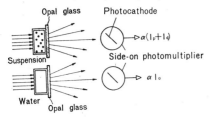

B) Uniform scattering

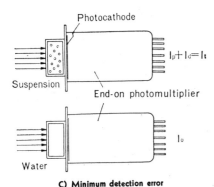

C) Minimum detection error

Figure 3. Light dispersion by various samples.

samples which contain suspended particulate matter.

In a conventional spectrophotometer employing a side-window photomultiplier (Fig. 3A) suspended matter reflects a significant amount of the light; thus primarily parallel-transmitted light passes to the photodetector. In order to overcome this defect, Shibata (3) introduced the idea of employing an opal glass surface next to the sample cell in order to attain uniform light diffusion (Fig. 3B). However, because of the distance of the sample from the photomultiplier, a significant light loss still occurs. To minimize this loss, the sample is placed in direct prox-

imity to the phototube (Fig. 3C), in the Shimadzu instrument. Spectra of opaque materials are recorded with a reflectance unit as shown in Figure 2B. The basic instrument costs about $12,600. A series of potentiometers are employed to adjust the base line; absorbance measurements can be made from 0.1 to 3 absorbance units directly, and from 3 to 5 absorbance units by means of an optical attenuator. A large number of accessories are offered such as a chromatogram scanner, fluorometry, flame photometry, and microspectrophotometry attachments. The last is of particular interest as with it one can measure the absorption spectrum of small crystals, biological cells, or subcellular particulates. The spectrum of a constituent as small as 2μ in diameter can be recorded. Figure 4 shows a microphotograph of *Porphyra* cells and Figure 5 the spectrum of that portion of cells positioned over the 2μ hole. In this spectrum, curve *A* is the spectrum of a living yellow cell, curve *B* the spectrum of a dead green cell, and curve *C* the base line.

Cary Instruments (a Varian subsidiary)

Previously, (1) the Cary Model 14 was described as a wide range recording instrument that covers the region from 186 mμ to 2.65μ. Since that time the Cary Models 15 and 16 spectrophotometers have been introduced. Both are double beam, high resolution, high wavelength-accuracy instruments. They both employ basically the same type of optical system, a double monochromator, double mirror-collimated optical system which employs two Littrow-type fused silica prisms as the dispersing element. The optical diagram is shown in Figure 6.

The model 15 instrument is designed for recording spectra and includes a strip chart recorder integral with the instrument (Fig. 7). This instrument covers the wavelength region from 185 to 800 mμ: a special version Model 15 is available which scans from 170 to 800 mμ. In the standard Model 15 the normal photometric range is from 0 to 1 or 1 to 2 absorbance units with a provision for ex-

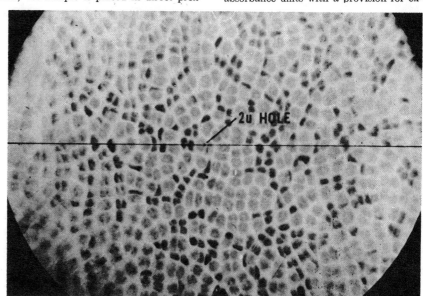

Figure 4. Microphotograph of *Porphyra* cells.

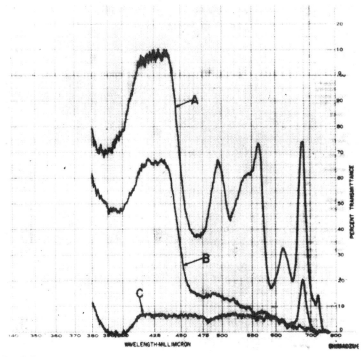

Figure 5. Cell spectra.

panding to full scale (10 in. of chart width) any 0.1 absorbance portion of the 0–0.3 part of the range. The photometric accuracy ranges from 0.002 absorbance units near zero absorbance to 0.008 absorbance units at 2.0. Wavelength dial accuracy ranges from 0.1 mμ at 300 mμ to 2 mμ at 800 mμ; the resolution ranges from 0.03 mμ at 200 mμ to 5 mμ at 700 mμ. The cost of the basic instrument is about $11,500.

The Model 16 instrument is designed primarily for accurate absorbance measurements (the basic instrument is not a recording spectrophotometer). It has a photometric accuracy of 0.00024 absorbance units near zero and 0.001 absorbance units near 1. The instrument covers the wavelength region from 186 to 800 mμ in the conventional model; a special model is available which employs a different detector and covers the region

from 170 to 600 mμ. Wavelength accuracy ranges from 0.3 mμ at 300 mμ to 2 mμ at 800 mμ. The resolving power varies from 0.03 mμ at 200 mμ to 2 mμ at 800 mμ. A large number of accessories are offered for both the Cary Models 15 and 16 instruments. Of particular interest is a recorder attachment which will adapt the Cary Model 16 for spectral recording of either absorbance or transmis-

Figure 7. Photograph of Cary Model 15 spectrophotometer.

sion data. This recording would normally be done for a series of samples at a fixed wavelength or to measure changes in absorbance in a sample undergoing chemical change. It is anticipated that a wavelength drive accessory will be made available in the near future which will then permit conventional recording of spectra with this instrument. The cost of the basic Cary Model 16 is about $8400.

Bausch and Lomb, Inc.

The Spectronic 600 instrument was designed by Bausch and Lomb specifically for those laboratories which use a few selected wavelengths for the greater share of their photometric work, and where the requirements for stability and reliability are high because of the great number of determinations which are run in a short time. Optically this instrument employs essentially the same components as were previously described by Lewin for the Spectronic 505. The 600 instrument features a meter readout of absorbance or transmittance; or if desired, through accessories, a digital readout. Spectral recording (wavelength scanning) is obtained through the use of an external recorder which may be plugged directly into a receptacle provided at the rear of the meter housing. A special signal generator is coupled to the wavelength drive to provide marking pips on the chart corresponding to 10 mμ intervals. The instrument scans the spectrum at fixed speeds in order to synchronize the wavelength drive with the paper travel of the external recorder. In contrast, the Spectronic 505 features an internal drum recorder and, in that instrument, the chart drive slows down automatically to allow the pen additional time to record the spectrum in rapidly changing areas of transmittance or absorbance.

The Spectronic 600 is shown in Figure 8. It will accept the same major accessories as the 505 such as a reflectance attachment, flame photometer accessory, or fluorescence accessory. The nominal wavelength range is from 200 to 600 mμ with the 1P28 photomultiplier supplied, but can be extended to 800 mμ with an end-on photomultiplier.[1] An automatic zero as well as 100% T adjustment is provided. The photometric range varies from −0.3 to 2 absorbance units and the

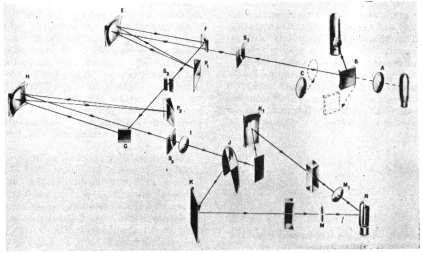

Mirror B selects either of two lamps; A and C are focusing lenses; S_1 to S_2, first monochromator (with prism P_1); S_2 to S_3, second monochromator (prism P_2); J, sectored mirror chopper; K and K_1, concave mirrors which direct light through sample and reference to photomultiplier N.

Figure 6. Optical diagram for the Cary Model 16 spectrophotometer.

[1]Rather than employing an end-on photomultiplier, which requires a special adapter, and changing phototubes according to the spectral region being scanned, some users have extended the normal wavelength response range of the B & L 600 or 505 instruments to about 750 mμ by substituting a wider response photomultiplier tube. Generally, Hamamatsu TV Co. R136 photomultipliers (with spectral response from 160 to 800 mμ), or R213 photomultipliers (with response from 185 to 800 mμ) have been employed. These tubes are distributed by the Kinsho-Mataichi Corp., 80 Pine St., New York, N. Y. 10005, and can be substituted directly for the 1P28 in any instrument. For comparison, the spectral response curve for the 1P28 tube is from 185 to 650 mμ.

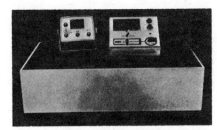

Figure 8. Photograph of Bausch and Lomb Spectronic 600 with digital readout accessory.

transmittance range from 20 to 200% T with a stated accuracy of 0.5% T. The cost of the Spectronic 600 is about $3000.

Beckman Instruments, Inc.

Two classes of double beam spectrophotometers are offered by this manufacturer, the DK series and the DB series. The DK series was first introduced in 1954 and has also been previously described by Lewin. Modifications have been made to improve the performance of the instruments and additional accessories have been introduced. As the same design has been retained, current wavelength specifications only will be listed. The distinguishing feature between the DK-1A and DK-2A instruments is the method of recording the spectra. The DK-1 employs a strip chart recorder external to the instrument, while the DK-2 instruments use a built-in flat-bed recorder. The DK-1A covers the wavelength region from 185 mμ to 3.5μ with linear wavelength presentation. Cost of the instrument is about $12,000; the UV DK-1A covers the wavelength region from 170 mμ to 3μ also with linear wavelength presentation and costs about $15,000. The DK-2A spectrophotometers are offered either with linear or nonlinear wavelength presentation. The nonlinear DK-2A instrument which expands the UV-visible region at the expense of the near-IR region costs about $9000. The linear wavelength DK-2A instrument is listed at approximately $10,000. A far-UV DK-2A which covers the region from 160 mμ to 3.5μ costs about $13,000. Both DK-1A and DK-2A instruments can be used for recording reflectance spectra through the use of an integrating sphere; flame spectra and fluorescence spectra can also be recorded with appropriate accessories. The resolution of the DK-1A and DK-2A instruments is stated to be better than 0.2 mμ at 220 mμ; for the far-UV instruments, the resolution is 0.02 mμ at 165 mμ. Wavelength accuracy ranges from 0.4 mμ in the UV to 8 mμ in the near-IR.

The DB series features low cost, double beam recording spectrophotometers. The spectrum is recorded with an external recorder which is offered as an instrument accessory. Currently two models are offered, the DB which employs a prism monochromator and the DB-G, a recently introduced instrument which employs a grating as the dispersing element rather than a prism. The electronics are essentially the same in both DB instruments, and since the DB has also been previously

described by Lewin, only specifications again are listed. The DB model covers the wavelength region from 205 to 770 mμ and the DB-G from 190 to 700 mμ. Both instruments have linear wavelength presentation and the cost of the basic in-

Figure 9. Photograph of Coleman Model EPS-3T.

strument ranges from $2200 to $2600. The DB-G instruments currently are employed in the Beckman atomic absorption instrument; consequently a DB-G instrument can be adapted for atomic absorption work by the purchase of a module which is offered as an instrument accessory.

Coleman Instruments Division of Perkin-Elmer Corporation

Two double beam spectrophotometers are offered. The Model EPS-3T, pictured in Figure 9, employs a quartz prism monochromator, a drum recorder, and is capable of scanning the wavelength region from 170 mμ to 2.6μ. The instrument employs solid state circuitry, a lead sulfide photocell for near-IR detection and a photomultiplier for UV-visible detection. Featured in the instrument is a bilateral slit and a 100% T compensator which permits flattening the base line electronically through a series of 15 adjustable potentiometers. The instrument may be employed for either single-beam or double-beam operation and records either transmittance or absorbance in five ranges varying from 0 to 200% T or −0.3 to 2 absorbance units with scale expansion for low concentration samples. The photometric accuracy of this instrument is stated to be 0.3% T; wavelength accuracy varies from 0.4 mμ in the UV to 8 mμ in the near IR; resolution is 0.1 mμ at 220 mμ. The cost of the basic instrument is about $10,000 and a number of accessory units are available, such as a thermostated cell holder and a spectrofluorometer accessory.

The Model 124 Coleman-Hitachi instrument (Fig. 10) is a grating spectrophotometer capable of scanning the spectrum from 190 to 800 mμ. An external recorder is employed (optional accessory) for wavelength scanning. Direct measurements of absorbance or transmittance are made on a meter with a mirrored scale.

Figure 10. Photograph of Coleman Model 124 Hitachi spectrophotometer.

The instrument employs solid state circuitry, and variable scanning speeds. Absorbance readings can be made from 0–1 or 0–2 absorbance units with greater

than usual accuracy since the readout is linear both in absorbance and transmittance units. The basic cost of the instrument is about $3500.

Durrum Instrument Corporation

The model PGS instrument manufactured by this newcomer to the spectrophotometer field offers several unique and interesting features. The instrument (Fig. 11) records the spectrum from 180 to 1000 mμ with a wavelength accuracy of 0.2 mμ and resolution of 0.1 mμ or better over the whole wavelength range on a 10 in. wide (absorbance axis) strip chart recorder which is integrally built into the spectrophotometer. Solid state electronics are employed and a single photomultiplier detector is used for photometric measurements throughout the whole wavelength region. According to the manufacturer the high resolution in this instrument is obtained by a novel double monochromator which features a prism dispersing element as well as two gratings mounted back-to-back on a reversible mount.[2] The

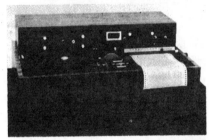

Figure 11. Photograph of Durrum Model PGS spectrophotometer.

reason for this combination is that a double prism monochromator tends to give poor dispersion in the visible and near-IR region while a double grating monochromator would require additional optical elements to eliminate second order spectra from both gratings. In the Durrum Instrument one grating is blazed at 200 mμ for use in the UV-visible spectral region and the other is blazed at 800 mμ for use in the near-IR. This construction permits the use of wider slits in order to attain maximum energy transfer per unit of bandwidth so that one photomultiplier can be employed for recording the whole spectral range. A deuterium light source is employed for the 180–350 mμ region and a tungsten-iodide lamp with quartz envelope is employed as the visible and near-IR source. An optically projected wavelength scale is also affixed to the monochromator camshaft to eliminate errors which are inherent in gear-coupled wavelength readouts and consequently permits the stated wavelength dial precision.

The optical diagram and principle of operation is shown in Figure 12. The Durrum PGS instrument features absorbance scale settings ranging from 0–0.1, 0–1, and 0–2 absorbance units; absorbance conformity is 0.002 near 0 absorbance and 0.003 at 0.4 absorbance. The transmittance scale is from 0–100% T; linearity is

[2] Note that the Cary Model 14 (described by Lewin) also utilizes a combined prism and grating double monochromator of this nature.

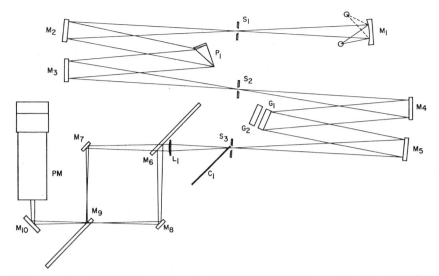

M1 source mirror; S1, S2, S3 monochromator slits; M2, M3, M4, M5 monochromator collimating mirrors; P1 prism (fused silica); G1, G2 gratings, blazed at 2000 and 8000 Å, on reversible mount; C1 chopper; L1 lens; M6, M9 rotating sector mirrors; M7, M8, M10 plane mirrors; PM end-on photomultiplier.

Figure 12. Optical diagram for Durrum PGS spectrophotometer.

0.3% and the 100% baseline flatness is 2% T throughout the spectral region. Bilateral adjustable slits are employed and the photomultiplier gain can be varied by a 10 turn potentiometer. The cost of the basic instrument is about $8500.

Perkin-Elmer Corporation

For recording in the UV-visible region, the Perkin-Elmer Corp. currently offers three instruments. The Model 450, a double-monochromator, fused silica prism instrument covers the wavelength region from 165 mμ to 2.7μ. Both electronically and optically the instrument appears quite similar to the Model 4000A previously described (1); the major change appears to be that the Model 450 will scan below 200 mμ. Nitrogen flushing is required for spectral measurements from 165 mμ to approximately 190 mμ. The instrument is a high-resolution spectrophotometer; photometric accuracy and specifications are very similar to the Model 4000A. The cost of the Model 450 is about $15,200. The same instrument is available with a less expensive photomultiplier covering the range from 193 mμ to 2.7μ at a cost of about $13,600.

A lower cost recording spectrophotometer for the visible-UV region is the Model 202 which employs a Littrow prism and covers the region from 190 to 750 mμ. The instrument records absorbance only; the presentation is linear on the chart scale. The wavelength scale is also linear and two rotations of the recorder drum are required to record the spectrum. Recording of the UV region is made from 190 to 390 mμ and visible recording from 350 to 750 mμ. The instrument employs an optical null balancing principle; resolution ranges from 0.2 mμ in the UV to 0.4 mμ in the visible region. The cost is about $5300. The Model 202 is pictured in Figure 13.

To satisfy the needs of the custom design enthusiasts, Perkin-Elmer offers a series of Model E-1 monochromators featuring an Ebert-mounted grating. This monochromator serves as the nucleus for building spectrophotometers to cover the region from 200 mμ to 40μ. Four series of monochromators are offered. The Model E-11 covers the region from 200 mμ to 1μ; the E-12 covers the near-IR region from 1 to 2.5μ; E-13 covers the mid IR from 2 to 40μ; and the Model E-14, a universal monochromator which covers the whole region from 200 mμ to 40μ. An optical bench is attached to the monochromator for mounting standard components such as light sources. Standard electronic and electrical components are added to the monochromator to serve as a means of driving the wavelength, beam chopping, spectral recording, etc. Obviously, this instrument could be exceedingly useful for specific spectral measurements which could not be obtained by conventional instruments. The basic cost of the monochromator with sufficient components for spectral recording would be about $19,200.

Unicam Instruments, Ltd.

The Unicam models 500 and 700 spectrophotometers (via Philips Electronic Instruments) have been described previously. The SP 800 instrument covers the wavelength region from 190 to 850 mμ. The normal absorbance range is from 0 to 2; a scale expansion accessory is offered (from 1/4 to 20×) to permit full scale recording for any portion of the absorbance region, for example, for the region from 1.0 to 1.1 absorbance units. A fused silica prism is employed as the dispersing element in the monochromator and a single photomultiplier as the detector for the whole wavelength region. Either linear wavelength or linear wavenumber versions of the instrument are available. Deuterium and tungsten light sources are employed. A unique feature in the instrument is an automatic change in light sources at 370 mμ. This change is carried out by the actuation of a solenoid-operated mirror. Accordingly, it is possible to leave the instrument entirely unattended for recording of the whole UV-visible range. A chart shift control which enables spectra to be presented side by

side also operates in conjunction with the automatic lamp change and enables any portion of the spectrum (i.e., 300–500 mμ, 190–400 mμ, etc.) to be continuously displayed on one chart. The monochromator slits are automatically programmed to provide a constant energy background throughout the spectral region and can be varied as needed to meet the desired operating conditions. A beam-balancing cam permits smoothing of the baseline throughout the whole range in about 2 min. The instrument is shown in Figure 14; it features a flat bed or strip chart recorder and numerous accessories are offered. Cost is about $6500.

Other Instruments

The coverage in this article is restricted to recording spectrophotometers introduced since the earlier article by Lewin. Many of these older instruments are still offered and have been continuously improved to maintain their positions of excellence. This would include recording spectrophotometers such as the General Electric, Optica Model CF4 recording spectrophotometer, Cary 14, Unicam Model SP-700, etc., as well as specialized

Figure 13. Perkin-Elmer Model 202.

components or instruments which may be employed for spectral recording such as certain types of spectrofluorophotometers (4). The fact that these instruments are not discussed or that modifications of earlier instrumentation are only briefly mentioned is not to be taken as any implication of poor design or obsolescence, but rather the fact that the basic principles of their design have been previously discussed. Furthermore, because of the vast variety of optional accessories offered for each of the instruments and the fact that these accessories are really quite similar (e.g., flame photometer attachments, reflectance attachments, fluorescence attachments), the discussion is limited to the basic spectrophotometer. Concurrently, the lengthier discussion of certain instruments in this article should not be considered as a bias in their favor, but only an attempt to describe what may be considered more novel instrumentation.

Warner and Swasey Company

Of different design and considerable interest is the Model 501 rapid scanning spectrophotometer made by Warner and Swasey Co. This instrument is designed primarily to study transient spectroscopic phenomena, such as the recording of the emission spectrum of a burning solid propellant, recording absorption spectra in kinetic studies, including reactions which occur through flash photolysis, etc. The spectrum can be scanned in as little time as 1 millisecond and the scan can be

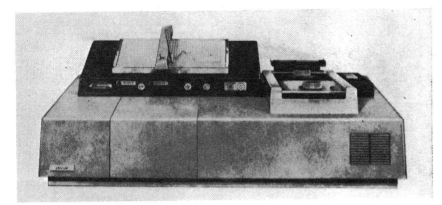

Figure 14. Unicam SP 800.

Figure 16. Warner and Swasey Model 501 rapid scanning spectrophotometer.

repeated every $1^1/_4$ millisecond with no degrading of the spectrum due to the rapid scanning speed. The instrument measures the spectrum from 250 mμ to 15 μ; either emission or absorption measurements can be made. The spectrum is presented on an oscilloscope or, if desired, the information can be played into a magnetic tape recorder. The magnetic tape recording is then played back into a strip chart recorder to give a plotting of the spectral phenomenon measured.

If desired, the tape recording can also be played back into the oscilloscope for re-observation of the phenomenon initially observed on the oscilloscope. Conventional photographing of the initial oscilloscope trace is also possible. The instrument has a very high resolution. For example, in measuring the neon emission spectrum the Model 501 resolved the 534.1 and 534.3 mμ lines showing a resolution of 0.22 mμ at a scanning speed of 100 milliseconds.

In contrast to conventional instruments no portion of the monochromator section moves during scanning. Instead, a series of corner mirrors are translated through an intermediate focal plane external to the monochromator. This permits rapid scanning without loss of resolution; the resolution thus approaches that theoretically obtainable with the 0.4 meter Czerny-Turner monochromator which is employed. The optical diagram of the instrument is shown in Figure 15. The instrument is designed to employ a variety of detectors such as the 1P28, 4473, 7102 photomultipliers or, if desired, InAs, InSb, GeAu, liquid nitrogen-cooled semiconductor detectors. Two detectors are used concurrently to get the best wavelength coverage. The Model 501 instrument is pictured in Figure 16; the basic cost is about $23,000.

The Model 702 instrument is unique in that it permits the recording of emission or absorption phenomena *in situ*. This instrument employs a unique optical collection system which can be focused or "zoomed in" on the sample which may be located anywhere from 28 cm to infinity from the monochromator. The instrument may be employed as a scanning spectrometer for measuring absolute or relative emission, spectra of samples such as combustion gases, solar phenomena, reflection spectra from solid surfaces, etc. Or, if desired, the instrument can also be operated at a fixed wavelength to record spectral changes as a function of time. The instrument covers the spectral region

from 250 mμ to 7.5 μ, or up to 15 μ on special order. The cost of the basic instrument is about $19,000.

Dual Wavelength Spectrophotometers

Dual wavelength spectrophotometers permit two different wavelengths of light to pass through the sample. Both the American Instrument Co. and Phoenix

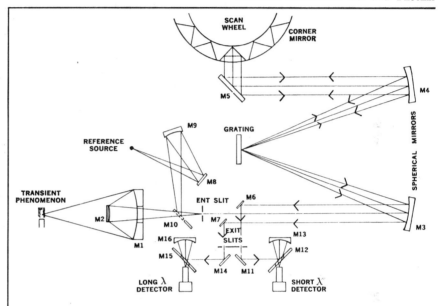

Figure 15. Optical diagram for Warner and Swasey Co. Model 501 rapid scanning spectrophotometer.

Radiant energy from the phenomenon (for an emission measurement) or from the source, modified by absorption from the sample (for an absorption measurement), is collected by the Cassegrainian optical system (M1 & M2), and imaged on the entrance slit of the Czerny-Turner monochromator. Energy from the entrance slit is collimated by mirror M3 and directed toward the grating, where it is dispersed as a function of wavelength, and, after reflection from mirrors M4 and M5, is brought to a focus on the scan wheel. Thus, a spectrum is focused in an intermediate focal plane. The wavelength scanning occurs in this focal surface. Mirror M5 is large enough to intercept the entire useful spectral range of the grating.

A corner mirror (two mirror surfaces exactly at 90° to each other) is used to direct the energy back a second time through the monochromator. The basic property of a corner mirror is to laterally displace any ray incident upon it and send it back parallel to itself. This causes a left-to-right reversal of the spectrum, and the second pass dispersion by the grating adds to the first pass instead of cancelling, as would be the case for a flat mirror. The result is a double-pass spectrum formed in the area of the entrance slit. The corner mirror, due to its symmetry, laterally displaces half of the spectrum to the left of the entrance slit and the other half to the right. The two spectral regions are intercepted by mirrors M6 and M7, directed toward the two exit slits, and ultimately to the two detectors used. Twenty-four corner mirrors are mounted on the periphery of the scan wheel, which is driven at a constant rate of speed by a hysteresis-synchronous motor. As each

corner mirror passes through the intermediate focal plane, the wavelengths intercepted by the corner mirror change, causing the spectrum to be scanned. The shorter wavelength half of the spectrum is scanned across one exit slit at the same time that the longer wavelength portion of the spectrum is scanned across the other exit slit.

The availability of two exit slits is very useful when scanning over a broad spectral region since it allows the detectors to be matched to the region scanned, and provides a wavelength overlap useful in checking system accuracy. The spectral region scanned is determined by the gratings, filters, and detectors used, and measurements can be made in selected regions from the near ultraviolet at 300 mu to the intermediate infrared at 9 microns. Overlapping orders are eliminated by the use of sharp cut-on filters which are selected to match the grating in use.

A reference radiation source is included in the Model 501 for intensity calibration. This source is a feedback controlled silicon carbide element for use in the infrared region or a tungsten strip lamp for use in the visible and near ultraviolet regions. Mirrors M8, M9, and M10 are used to focus radiation from the reference source onto the entrance slit of the monochromator. Mirror M10 is a flipping mirror which moves into the proper position to focus radiation from the reference source on the entrance slit when the instrument is set on CALIBRATE. When the instrument is set on OPERATE this mirror moves out of position, and energy from the phenomenon under study is focused on the entrance slit.

24

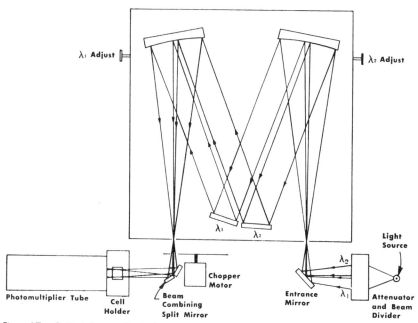

Figure 17. Optical diagram for dual wavelength spectrophotometers.

SPECTROPHOTOMETERS FOR CHROMATOGRAPHY

Canal Industrial Corporation

The Canalco Model M spectrophotometer is designed primarily for automated flow measurements; for example, to continuously monitor the composition of the eluent from a chromatographic column. The pure eluting solvent is employed as the reference standard and passes through a flow cell in the reference beam. The eluent passes through another flow cell in the sample beam. The instrument can be employed to monitor the chromatographic process either at a fixed wavelength or, if desired, to scan a previously selected spectral region. In this automatic scan mode of operation, the instrument is initially set to monitor the absorption change at a selected wavelength. However, as soon as the absorption exceeds a preset threshold value, the instrument scans the spectrum only for absorption values exceeding this threshold. Consequently, when fractions of increasing concentration emerge from the column, the instrument increases the wavelength range over which it scans and when the concentration decreases, the wavelength region through which the instrument records is correspondingly decreased. An automatic scale expansion feature is also provided by which all materials showing an absorbance up to 0.173 absorbance units are recorded at three times their actual strength. When the absorbance exceeds this level, the instrument automatically reduces the span to its normal setting. In addition, a manual over-ride switch is provided which will disconnect the automatic scan feature without otherwise disturbing the program. Direct readout up to 2 absorbance units is shown on a panel-mounted meter. It is also possible to make connections to an auxiliary pen on the recorder which is actuated each time a fraction collector advances. A typical chart is shown in Figure 20. The basic instrument is the manually set spectrophotometer which covers the wavelength region from 210 to 380 mμ. This spectrophotometer, which costs about

Precision Instruments offer such instruments. In using a dual wavelength spectrophotometer, a conventional spectrum is first recorded with the instrument to ascertain points of maximum light absorption. Once this information has been obtained, it is then possible to select two appropriate fixed wavelengths. These two different wavelengths are alternated in sequence through a single cuvet. The two light beams are initially balanced and any disturbance of this balance is recorded or may be monitored with an oscilloscope. Generally the one wavelength is preselected to an isosbestic point or other point of unchanging absorbance. The other wavelength is one which is capable of

absorbance units have been measured. The instruments cover the wavelength region from about 200 to 800 mμ; cost ranges from \$13,000 to \$16,000. The Aminco-Chance instrument is pictured in Figure 18, the instrument made by Phoenix Precision Instruments in Figure 19.

Figure 19. Photograph of Phoenix Precision Instruments Co. dual wavelength scanning spectrophotometer.

Figure 18. Photograph of Aminco-Chance dual wavelength spectrophotometer.

detecting changes in the compound undergoing transformation in the system.

Basically these two wavelengths are obtained through the use of an additional grating in the monochromator of the instrument, Figure 17. The use of two wavelengths tends to overcome the problem of light being scattered in a zigzag manner when it is passed through highly turbid samples. This changes the effective path length in the cuvet. Since both wavelengths of light pass through the same sample, they are scattered to the same extent and subject to the same path length. The dual wavelength instruments thus overcome the problem of trying to match the sample and reference cell as is done in normal spectrophotometry. Using dual wavelength spectrophotometers differences of 0.0002 absorbance units superimposed on a total background of 2

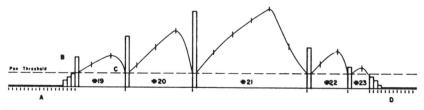

Figure 20. Chart presentation, Canalco Model M Spectrophotometer.

Monitor wavelength: 280 mμ
Scan limits: 240–300 mμ
Pen gate threshold: 0.03 absorbance
(A) shows passage of 15 nonabsorbing fractions, as indicated by the event marker pen actuated each time the fraction collector advances. Three fractions follow which do not trigger the chart drive, other than to record the absorption at the monitor wavelength. (Spectral scans take place in all 18 of these fractions, but are not recorded.) At (B), absorption at 280 mμ rises above the threshold value. When fraction #19 scan starts, that portion below the threshold is

not recorded; as absorbance rises above 0.03 the chart begins to move, with a pen blip each ten millimicrons to show interval scanned. At (C), absorbance drops below the threshold and recording stops. When the scan is complete, the monochromator resets itself again at 280 mμ. Subsequent curves, run on denser fractions, show higher peaks and wider scans (#20–21). As concentration decreases (#22), scanned curve shrinks both in height and width, finally dropping below the threshold entirely at (D) where it stays until another absorbing compound emerges from the column.

Figure 21. The Canalco Model M Spectrophotometer.

$3200, is shown in Figure 21. The automated functions are performed with optional accessories and the cost of the instrument with all accessories would be about $6200.

Schoeffel Instrument Corporation

The Schoeffel SD 3000 spectrodensitometer is designed to make spectral measurements through various chromatographic surfaces such as thin-layer chromatographic plates, paper or gel chromatograms. The instrument does not record a spectrum as such, but moves the chromatographic surface through a light path of preselected wavelength. The optical diagram is shown in Figure 22. A 150 watt xenon lamp is used to illuminate two closely adjacent areas of the sample (e.g., a TLC plate), one of which serves as a reference. The illuminating assembly contains a quartz prism monochromator and beam dividing optics. The two light paths then pass through the sample to the photomultipliers. The signals are amplified separately and fed into a ratio device, the output from which is in transmittance units and may be recorded. An accessory is available which will convert the transmission values to absorbance units. This same device will

may also be employed. Accessories for fluorescence and quenching applications on TLC, paper, and other materials, are available. By using flow cells in the sample compartment, the instrument can monitor the effluent in column chromatography. The basic instrument is shown in Figure 23; it covers the region from 200 to 700 mμ and costs about $4000.

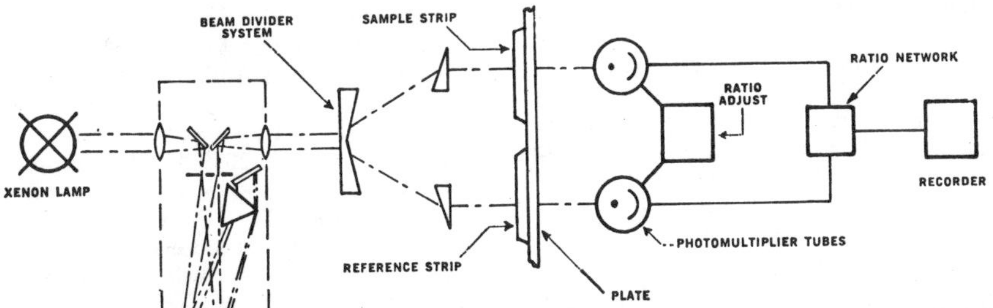

Part 2: Single-Beam Spectrophotometers

In contrast to the double-beam recording spectrophotometers which feature a continuous change in wavelength and an automatic comparison of the light intensities of the sample and the reference material, in single-beam spectrophotometers there is only one light path from the source to the detector. Accordingly, the single-beam spectrophotometer is operated at a fixed wavelength. The instruments are thus primarily employed for the quantitative determination of the concentration of a single component when a large number of samples are to be analyzed. In many respects they are the "work horses" of quantitative spectrophotometry. In using a single-beam instrument, the absorption maximum of the material must be known in advance. The wavelength is set to this absorption peak; the reference material (that is, the solvent blank) is then positioned into the light path and the instrument is adjusted to read 0% transmittance when no light passes to the detector (shutter closed) and 100% with the shutter open. After this adjustment has been made, the sample is placed in the light path and the absorbance or transmittance is read and related

strument stability is usually not a limiting factor.

The absorption spectrum of the compound can be measured with a single-beam spectrophotometer, but it requires point-by-point measurements throughout the whole spectral range. Similarly, double-beam instruments are not limited to recording spectra but may also be employed for measuring concentrations, although single-beam spectrophotometers are specifically designed to perform this task more rapidly.

In some respects, the recent advances in single-beam spectrophotometer design have centered more on performing sequential measurements more rapidly and automatically through the use of rapid sampling accessories rather than on major changes in either optical design or electronics. These accessory units rapidly fill and evacuate the spectrophotometer cell and make it possible to analyze several hundred samples per hour on a routine basis with an instrument costing perhaps only several hundred dollars. Other single-beam spectrophotometers have been developed primarily for educational purposes, and one spectrophotometer has been specifically designed to mix rapidly the sample and a reagent, and measure the absorbance change over a short period of time for kinetic studies.

A. R. F. Products, Inc.

A. R. F. Products features an integrated group of instruments which are primarily designed for instruction in analytical chemistry. The complete group of instruments permits coverage of electroanalytical measurements such as coulometry, polarography, pH, etc., as well as optical measurements. In the case of the spectrophotometric unit, one must purchase the spectrophotometer, an operational amplifier, and a meter. The monochromator of the spectrophotometric unit is specifically designed to be taken apart so that the student can see how the monochromator functions; the electronic components which are required for spectrophotometric operation are connected externally to make it easier to understand their operation. The spectrophotometric unit is shown in Figure 24. The mono-

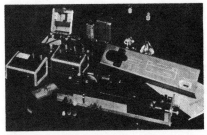

Figure 24. A.R.F. Products Model ATS-8 Absorptiometer, with cover removed.

chromator covers the wavelength region from 450 to 750 mμ and employs a modified Littrow mounting to permit interchange of a prism or a grating as the light dispersing element. A tungsten lamp which is powered by a constant voltage transformer is employed as the light source. The dispersed light beam, after emerging from the 1 mm fixed exit slit, is split by a pair of mirrors into two beams

Figure 22. Optical diagram for the Schoeffel Model SD 3000 Spectrodensitometer.

also integrate the area beneath the curve between any two points selected by the operator. In addition, a print-out device

Figure 23. The Schoeffel Model SD 3000 Spectrodensitometer.

to the concentration either through the use of calibration curves or by the use of appropriate algebraic methods. Depending upon the instrument used, the readout may be viewed directly by observing the deflection of a meter, a "null balance" scale setting, a recorder, or possibly printed out directly through the use of a digital printer.

An obvious requirement of single-beam spectrophotometers is a high degree of stability both of the light source and the detector system, as fluctuations would cause erroneous readings. However, in routine work where errors of ±1% of the measured quantity are insignificant, in which traverse two cuvets for double-beam operation. Two photovoltaic cells are fixed behind the cuvets and serve as the light detectors. Single-beam operation is made possible by repositioning the mirror so that only one photocell is illuminated. To further broaden the instructional approach, the electronic components which are required for the spectrophotometric measurements are encased in transparent plastic boxes. The operational amplifier module is employed as a dc signal amplifier and the output is read on a multiple-range voltmeter. The cost of the spectrophotometer unit (monochromator, operational amplifier, and meter) is about $1160.

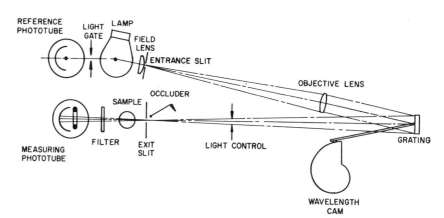

Figure 26. Optical path in Bausch and Lomb Spectronic 20 Spectrophotometer.

Figure 25. Bausch and Lomb Precision Spectrophotometer.

Bausch and Lomb, Inc.

The Bausch and Lomb Precision spectrophotometer, illustrated in Figure 25, employs a dual grating monochromator as the dispersing element and covers the wavelength region from 190 to 650 mμ. The wavelength region can be extended to 800 mμ by a special phototube. Since the instrument employs a grating monochromator, the band pass remains fixed throughout the whole spectral region and depends on the slit width. By simply moving a lever, slit widths which provide a band pass of 0.2, 0.5, or 2 mμ may be selected and filters are employed to reduce second-order radiations. A warning light shows when the wrong filter is positioned for a particular wavelength region. A deuterium lamp is employed as the light source for the UV region and a tungsten iodine, quartz envelope, lamp is employed for the visible near-IR region. In addition, a mercury calibration lamp is included in the lamp housing to permit checking of the wavelength accuracy, which is better than ±0.5 mμ for the whole range. Absorbance or transmittance readings are made on a sloping meter dial; five photometric ranges are provided (10 and 100%T; 0–2, 0.5–2.5, and 1–3 absorbance ranges). The instrument features a large cell compartment (6 × 6 × 8 in.) and the standard cell holder accepts four 1-cm *square* cells, microcells, or a flow cell. A special cell holder is available which will accommodate cylindrical cells from 1 to 10 cm long. A panel on the side of the instrument will accept accessories such as a recorder read-out unit, cell temperature controller, cell temperature monitor, and an automatic 100% control. The last unit sets the instrument to 100%T or 0 absorbance whenever the sample selector lever is put in the first position or when a button on the instrument panel is pressed. In addition to these accessories, adapter units are available which will allow additional functions such as reflectance, fluorescence, or flame photometric mea-

surements, to be made. The basic cost of the Precision Spectrophotometer is $3060.

Also made by Bausch and Lomb is the Spectronic 20 which has received wide acceptance since its initial introduction more than 10 years ago. On current instruments, the wavelength range is from 340 to 625 mμ and this can be extended to 950 mμ by substituting an infrared-sensitive phototube and filter. The bandwidth is 20 mμ for the whole wavelength range; a diffraction grating is employed as the dispersing element. Currently three versions of the instrument are available: a battery operated model, the standard model (recommended when variations in the line voltage and frequency are less than 1%), and a regulated model which is the most popular. With this regulated instrument, a 10-v change in the supply voltage causes less than a 1% variation in readings. Furthermore, because an untuned power transformer is used, it is not only insensitive to frequency changes but will operate equally well at 50 or 60 Hz.

Figure 26 shows the optical path in the regulated instrument. Constant voltage is obtained at the light source (a 6-v tungsten lamp) by monitoring the lamp output with a reference phototube. A portion of the lamp output is allowed to pass to the phototube through the light gate, the opening of which is set and cemented at the factory. A bridge circuit is used to balance and compare the monitoring phototube output with a constant voltage from the power supply. Both solid-state and vacuum-tube circuitry are employed. Constant voltage for the instrument is obtained through the use of voltage regulator tubes in the power supply. The differential output from the monitoring bridge circuit is transistor-amplified and transformed into conductance variations of two transistors operating on the center-tapped secondary winding of the power transformer. The transistors thus act to change the load of the power transformer. This load change is reflected back to the primary winding and consequently causes a readjustment of the primary load voltage; since the transformer also applies power to the lamp, the lamp output becomes independent of line voltage variations.

The basic Spectronic-20 regulated model (Model 4) costs $395. A number of accessories are offered for all three models by Bausch and Lomb. These include a reflectance attachment, digital read-out accessory, and a data acquisition

system. The last is specifically designed for routine analysis work. It incorporates a recorder for absorbance read-out and a flow cell cuvet with a pumping system which permits about 4 determinations to be made per minute on a continuing basis.

Figure 27. Burrell Corp. No. 950-446 Titration Pump and Bausch and Lomb Spectronic 20.

The Burrell Corp. has recently introduced their model 950-446 titration pump selling for $11.50 which adapts the Spectronic 20 or similar instruments for spectrophotometric titrations. The pump is operated by a magnetic stirrer and circulates the solution through the spectrophotometer cell during the addition of titrant to the titration vessel. Figure 27 shows a Spectronic 20 and a Burrell pump set up for point-to-point titration curve measurements. The system is ideally adaptable also for automatic recording of spectrophotometric titration curves since the Spectronic 20 has a recorder outlet. Thus, in addition to a recorder, all that is needed is a method of continuously adding the titrant at a fixed rate which can be done, for example, by a motor driven syringe. e.g., Sage Instruments, Inc.

Beckman Instruments, Inc.

The model DU-2 has been introduced by Beckman Instruments since the initial article on spectrophotometry by Lewin. This instrument employs essentially the same optical system as the earlier DU but has been improved to cover the wavelength region from 190 to 1000 mμ. Both battery-operated and ac-line operated versions of the DU-2 are offered. Changes made include the use of sloping panels for greater ease in reading dials and manipulating controls. In addition, an

Figure 31. Coleman Junior II Spectrophotometer.

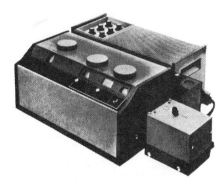

Figure 28. Beckman AC-Line-Operated DU-2 Spectrophotometer.

photometer attachment, fluorescence attachment, spectral energy recording attachment for the presentation of emission spectra in flame photometry, or fixed wavelength recording of absorbance changes which would be of interest in kinetic studies. The instrument is pictured in Figure 28; the ac line-operated instrument costs approximately $3600.

A simple battery-operated educational spectrophotometer is also made by Beckman Instruments in their West Germany plant. The instrument is usable from 400 to 800 mµ; it is pictured in Figures 29 and 30. The dispersing prism remains in a fixed position; the sample, detector, and collimating assembly are rotated as a unit for wavelength selection. Readout is in percent transmission on a galvanometer. Because a prism has a greater dispersion at lower wavelengths, the spectral bandwidth varies from 2 mµ at 400 mµ to 18 mµ at 800 mµ and the wavelength accuracy ranges from less than 1 to 5 mµ for the same wavelength range. The instrument costs approximately $300 and an absorption standard is supplied with the instrument for wavelength calibration. The Macalaster Scientific Corp. is the exclusive distributor for this instrument in the United States and Canada.

added expansion scale from 90 to 100%T has been added. The cell compartment will accept either standard 1 cm or microcells without auxiliary adapters. As with the earlier DU, a large number of accessories are offered; among these are a flame

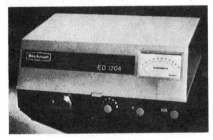

Figure 29. The Beckman Educational Spectrophotometer.

Coleman Instruments Corporation Division of the Perkin-Elmer Corporation

Four new single beam spectrophotometers are offered. The least expensive, the Coleman Junior II instrument (Fig. 31)

is an improved version of the earlier Junior model. The instrument covers the wavelength region from 325 to 825 mµ with a choice of either 20 or 35 mµ bandpass. The instrument is designed for routine analytical laboratory work. Featured is a jack which permits the instrument to be directly connected to a recorder, digital read-out unit, or printer. An accessory which has been recently introduced by Coleman is their "Vacuette" assembly. This attachment includes a spectrophotometer cell which is retained in the instrument for all samples. The sample is poured into the cuvet which has a funnel top. After the sample has been read, a touch on the vacuum valve removes the sample in 2–4 seconds and, according to the manufacturer, flushing and rinsing between samples is rarely required. Two cuvet sizes are offered: the 1.75 ml cell has a 1 cm light path and the 5.5 ml cell has a 1.7 cm light path. Using either cell and the Vacuette assembly, samples can be analyzed at the rate of hundreds per hour. The basic cost of the Coleman Junior II is $595; the Vacuette attachment may also be used with certain other Coleman instruments.

The Model 101 spectrophotometer is a more sophisticated instrument which covers the region from 220 to 900 mµ. A single grating blazed at 300 mµ is employed in the monochromator and fixed slits provide a constant bandpass of 10 mµ. A tungsten lamp covers the region from 340-900 mµ, and a deuterium lamp the region from 220 to 375 mµ. A novel feature (see below) is the use of a single photodetector to cover the whole wavelength range and to eliminate the inconvenience of shifting phototubes. A four-position holder will accept cells from 5 to 20 mm path length. The cost of the UV-visible model 101 is about $1675; the instrument is pictured in Figure 32.

Coleman Instruments also offers the Hitachi Perkin-Elmer model 139 spectrophotometer. This is a direct reading instrument covering the region from 195 to 950 mµ with a wavelength accuracy of ±0.5 mµ. The monochromator employs a grating as the dispersing element and has adjustable slits. Modular construction is employed which permits great flexibility and speed in mounting accessory units. Accessories offered include a flame photometer, atomic absorption unit, fluorophotometer, reflectance unit, spectral energy recorder, time drive recorder, and various types of sample cells. The photometer employs the same Hitachi dual-range phototube as the model 101. The phototube has two light detectors built into the same envelope and covers the region from 195 to 950 mµ. As the wavelength crosses the region from the

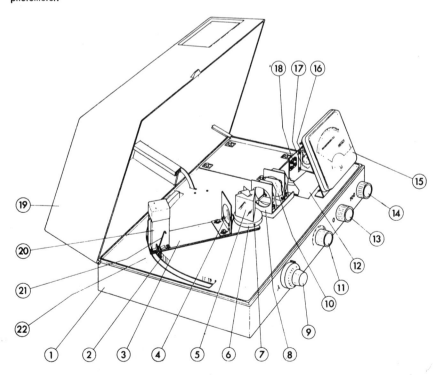

Figure 30. Inside view of Beckman Educational Spectrophotometer.

1. Base	8. Filter holder	15. Meter
2. Circular groove	9. Wavelength control	16. Tungsten lamp
3. Rotating arm	10. Collimating lens	17. Rectangular diaphragm
4. Collimating lens	11. Sensitivity control	18. Entrance slit
5. Prism	12. Light shutter	19. Hinged cover
6. Prism holder	13. Zero adjustment	20. Detector housing
7. Securing screw	14. 100% adjustment	21. Locking lever
		22. Alignment mark

28

MODEL 101 OPTICAL DIAGRAM

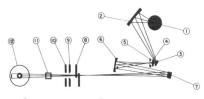

1. LIGHT SOURCES
2. SOURCE MIRROR
3. MIRROR
4. SLIT LENS
5. ENTRANCE SLIT
6. CONCAVE MIRROR
7. GRATING
8. DIAPHRAGM
9. EXIT SLIT
10. FILTERS
11. ABSORPTION CELL
12. PHOTOTUBE

Figure 32. Optical diagram and photograph of Coleman Model 101 UV-Visible Spectrophotometer.

visible to the near infrared, one detector gradually takes over from the other. For greater detection sensitivity, as is needed for atomic absorption and flame photometric work for example, a photomultiplier accessory can be added to the instrument to replace the conventional phototube. A photograph and optical diagram of the model 139 is shown in Figure 33; the cost of the basic instrument is $3100.

Durrum Instrument Corporation

Unique among single beam spectrophotometers is the Durrum Stopped-Flow spectrophotometer. This instru-

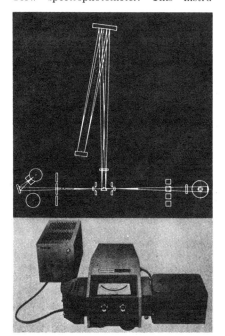

Figure 33. Optical diagram and photograph of Hitachi Perkin-Elmer Model 139 Spectrophotometer.

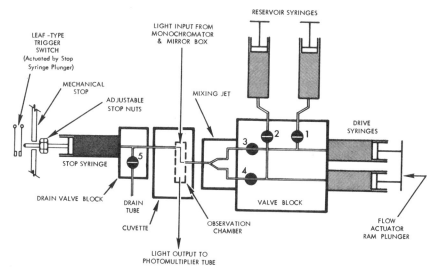

Figure 34. Flow system diagram for the Durrum Stopped-Flow Spectrophotometer.

ment is designed to rapidly mix two solutions (99.5% complete mixing in 0.002 second) and to measure the change in the absorption of light by their reaction product as a function of time. Reactions having a half-time of 0.005 second can be followed. Because of the speed at which measurements are made, the output of the instrument is displayed on an oscilloscope (with storage tube) for visual observation or photographic recording. The stopped-flow mixing chamber in the instrument corresponds to the cell compartment in a conventional spectrophotometer and requires only 0.2 ml of each component per determination. The system is pictured in Figures 34 and 35. In operating the instrument, the investigator first sets the monochromator to the wavelength corresponding to the maximum absorption for the desired compound and adjusts the bandwidth accordingly. He fills two reservoir syringes with the two reaction components. These fill, in turn, two drive syringes. With the flow actuator, he forces a portion of each component rapidly through the mixing jet and into the cuvet (observation chamber). The mixture flows through the cuvet into the stop (or collection) syringe.

Movement of the stop syringe over a fixed distance triggers the beginning of the time-base scan on the oscilloscope. When the plunger reaches the end of its travel, the flow stops abruptly, allowing measurements to be made in a turbulence free environment. Before each new measurement is made, the system is purged automatically.

The oscilloscope readings correspond to changes in photomultiplier voltage, thus recording changes in the transmission of light from the monochromator to the photomultiplier caused by reactions taking place in the cuvet. The oscilloscope storage tube can retain several traces for comparative observation.

The spectrophotometer covers the wavelength region from 250 to 800 mμ. A selector switch on top of the monochromator permits selection of either a prism or grating as the dispersing element. The prism is employed for the UV region as it permits the use of wider slits for maximum energy while retaining a narrow bandwidth, since a prism has better dispersion than a grating for short wavelength

radiation. The grating is employed in the visible and near-IR region where narrow bandwidths are required for precise measurements. A typical tracing is shown in Figure 36. The cost of the instrument is about $10,800.

Figure 35. The Durrum Stopped-Flow Spectrophotometer.

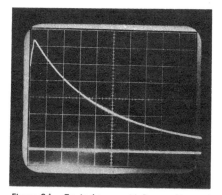

Figure 36. Typical trace with Durrum Stopped-Flow Spectrophotometer. Vertical axis is 0.1 volt/division with 100% transmittance at top of scale and 0% at bottom. Horizontal scale is 0.2 second per division.

Golden Instruments, Inc.

The Spektrokem spectrophotometer (Figure 37) covers the region from 400 to 750 mμ with a bandwidth of 15 mμ. This

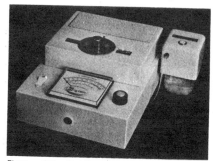

Figure 37. Golden Instruments Spectrokem Spectrophotometer.

transmission grating instrument is unique in that rather than having the sample in a fixed position and rotating the grating to isolate different wavelengths of light, the cuvet, exit slit, and photoconductive detector are moved laterally to intercept the desired portion of the spectrum, a patented feature. In order to stabilize the instrument against line voltage fluctuations, a portion of the current is fed both to the light source and the photoconductive cell in such a manner that fluctuations in the voltage to the light source are compensated by the detector. The photoconductive cell has a linear response so that the current flowing to the meter is directly proportional to the light impinging on the cell. The current is sufficiently large to permit the use of a milliammeter rather than a galvanometer. A flow cell, which employs a peristaltic pump, is offered as an optional accessory. The instrument costs about $330 and with a flow cell system about $395.

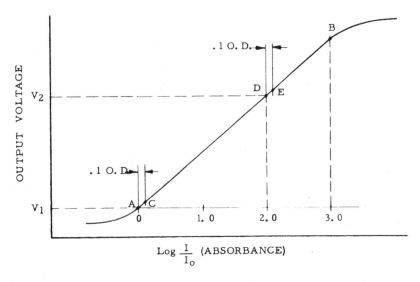

Figure 40. Output voltage versus absorbance for a 1P28 phototube in the Gilford photometer circuit.

Gilford Instrument Laboratories, Inc.

The Gilford model 2000 and model 240 spectrophotometers Figures 38 and 39 are similar in that they incorporate the Gilford photometer unit for both precise absorbance measurements and automatic sequential sampling. The electronic components can also be adapted to other mono-

Figure 38. Gilford Model 240 Spectrophotometer.

Figure 39. Gilford Model 2000 Automatic Spectrophotometer.

ponents can also be adapted to other monochromators such as certain Beckman, Zeiss, or Unicam instruments to convert them into direct reading spectrophotometers for fixed wavelength absorbance measurements. The photometer circuit covers the range from 0 to 3 absorbance units and differs from other circuits by having its output in a direct logarithmic relationship to the radiant energy striking the phototube. Since Beer's Law is also a logarithmic relationship, the output of the photodetector circuit is thus proportional to the concentration of the absorbing material. The photometer system

incorporates a 1P28 photomultiplier tube in a special feedback circuit which adjusts the voltage applied to the dynodes of the tube inversely to the amount of light falling on the cathode. Thus, for high light levels the sensitivity is decreased while with low light levels the sensitivity is increased. The system is designed so that these changes occur automatically and instantaneously which permits operation of the phototube at extremely low anode output currents regardless of the amount of light impinging on the photocathode. This makes it unnecessary to protect the photocell with the customary shutter when the cuvet compartment is opened. Similarly, because the photometer circuit has an inherently constant gain throughout its range, it is unnecessary to employ a separate gain or sensitivity control. Nor is a dark current balance required since the system does not use the complete absence of light as one of its reference points, and the full range of three absorbance units is reached without entering into the region where the dark current is significant. Furthermore, the photometric accuracy of the circuit employed does not depend on the use of high value load resistors, avoiding the instabilities and attendant need for humidity control with desiccants.

Since the feedback signal which controls the phototube sensitivity originates from the anode, an essentially constant anode current is maintained that is immune to power fluctuations and fatigue in the phototube and other components. The output of this constant anode current system is derived from the changing dynode voltage through suitable voltage dividers. Utilization of the dynode voltage as the source of the output signal presents two advantages: (1) Large signals are available for recording without the need for additional amplification. (2) The relationship between the radiant light energy entering the phototube and the voltage is essentially logarithmic over a wide range, as shown in Figure 40, so that in addition to spanning a range of 3 absorbance units, each 0.1 portion of the absorbance range can be expanded to full scale on the recorder.

The photometer system may be used for

conventional manual measurements; it may also be employed in conjunction with the automatic sampling positioner. This unit will automatically position in turn up to four standard spectrophotometer cells or four flow cells into the light path in a continuing sequence and record the changes in absorbance which occur over a period of time, a feature of great interest in studying reaction kinetics. The cost of the illustrated model 2000 system is about $7300. The model 240 monochromator covers the wavelength region from 185 to 1000 mμ and is also designed to function with the sequential positioning system. Additional accessories such as thermostatted cell compartments, flow cells, etc. are available.

Figure 41. Gilford Model 300 Micro-Sample Spectrophotometer.

The Model 300 spectrophotometer is designed for rapid analysis of a series of samples of similar composition. The instrument covers the region from 340 to 700 mμ and features a solid state photometer, a rapid sampling system, and automatic digital absorbance read-out. The rapid sampling system, which is offered as an accessory, permits the analysis of over 100 samples per hour. The cell (which requires a half-ml sample) is filled, the absorbance is measured, and the cell is flushed by a vacuum system operated by a single control level. This instrument costs about $1700 and is pictured in Figure 41.

The Heath Co.

The Heath Co. is introducing a series of optical instruments employing a number of modular units built about their recently introduced model EU-700 monochromator. The monochromator employs a Czerny-Turner mounting with diagonal mirrors to provide entrance and exit beams on a common optical axis. Using a 3 mm slit, the resolving power of this grating monochromator is stated to be better than 0.1 mμ and the line profile half-width less than 0.05 mμ. The wavelength range of the monochromator can be extended to below 180 mμ by nitrogen flushing and the upper limit is 1 μ. However, with the standard 1P28A photomultiplier detector furnished by Heath, the upper limit is restricted to 700 mμ. The monochromator employs a ganged bilateral exit and entrance slit which is symmetric to the optical axis. The slit width is continuously variable from 5 to 2000 microns and can be read directly in microns by means of a 4 digit counter. Provisions are also made for four different slit heights which range from 0.5 to 12 mm.

The wavelength is read directly on a 5

Figure 43. MP-System 1000 Recording Spectrophotometer.

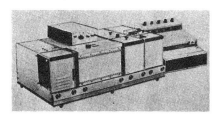

Figure 42. Photograph of Heath Co. single beam spectrophotometer.

digit counter which has a fractional scale of 0.02 mμ per division to permit wavelength readability of 0.01 mμ. An electronic scanning control can be employed for wavelength selection. This control (which employs integrated circuit digital techniques) supplies pulses to a stepping motor which drives a precision lead screw to give scanning speeds which vary from 0.005 to 2.0 mμ per second in nine steps. These pulses also control the drive of the chart recorder motor to permit exact synchronization of the wavelength drive with the chart. In addition, rapid slewing can be done in either direction at approximately 540 mμ per minute.

For single beam spectrophotometric measurements, a light source module (EU-701-50) is attached to one end of the monochromator. A cell compartment module (EU-70-11) and a photomultiplier module (EU-701-30) are attached to the other end of the monochromator. The output from the photomultiplier is then presented on a Heath strip chart recorder (EUW-20M). The cost of the monochromator alone is $1195; the cost of the complete single beam spectrophotometer shown in Figure 42 would be approximately $2300. It is

anticipated that an additional module will be introduced in the near future which will permit conventional double beam recording.

McKee-Pedersen Instruments

MPI also offers a modular approach for spectrophotometric instrumentation which is built around the use of the MP system 1000. This system is essentially an analog computer consisting of a group of operational amplifiers. The signal is presented in this case by the spectrophotometer and the readout is presented on a recorder. The instrument is pictured in Figure 43.

The monochromator employs a 590 line per mm grating which is blazed at 400 mμ and provides a wavelength range from 200 mμ to 1.5 μ with an accuracy of about ± 1 mμ. However, the use of the instrument for spectrophotometry is currently limited to the visible and near-IR region unless the user can supply a deuterium source. The slit width is adjustable and both slit widths and wavelengths are read from turn indicating dials. Single or double beam operation is possible through the use of a pair of matched phototubes. In addition the readout (in transmittance or absorbance units) may also be done by manual null balancing or digitally. By use of appropriate operational amplifier connections, derivative or integrated spectra can be recorded and measured.

The cost of the simple null balance spectrophotometer is about $2150. For about $4400 the user can purchase instrumentation which will enable him to perform all spectrophotometric techniques possible with the MP-system 1000. The system is very flexible and individual units such as a sample compartment can be purchased separately in order to make modifications of the instrument for experimental purposes.

Shimadzu

The Model QV 50 (Figure 44) spectrophotometer both in external appearance and internal optical design, appears to be quite similar to the well-known Beckman

Figure 44. Shimadzu Model QV-50 Spectrophotometer.

DU instrument. The instrument employs a Littrow mounted prism monochromator and a null balance potentiometric system for reading either absorbance or transmittance. An outlet for connecting an external recorder is provided. Although the monochromator wavelength dial is calibrated from 180 to 2000 mμ, the usable range for the instrument is from 183 to 1200 mμ. A semitransistorized power supply is used to operate a conventional type UV-visible light source which also incorporates a mercury lamp for wavelength calibration. The basic instrument costs about $2500 and a large number of accessories can be purchased.

Other Instruments

It must be re-emphasized that many of the instruments described by Lewin in 1960 (1), such as the Zeiss PMQ2, Optica Model CF4, Unicam Model SP 500, etc., are still offered, have been improved to meet current demands, and maintain their positions of excellence.

UV-Visible Spectrophotometry

Supplement

Figure 46. The Bausch and Lomb Spectronic 70 Spectrophotometer.

Advances and changes in instrumentation for spectrophotometry have been made since the early part of 1968, when this article was written. These differences have ranged from a change in nomenclature (what was referred to then as the millimicron, $m\mu$, is now the nanometer, nm) to advancements incorporating the latest achievements in electronics. Many of the instruments encompass virtually totally solid state electronics—from transistor regulated power supplies to linear log amplifiers, a digital display of absorbance or transmittance readings, the linear readout of absorbance, or a direct indication of concentrations through the use of integrated circuit components, as well as the ability to transmit data directly to a digital computer. In addition, the available accessories for instruments have been both expanded and improved; from automatic zero-set to wavelength programming, circular dichroism attachments, and temperature-jump capabilities. Interestingly, it appears that there have been no major breakthroughs in optical design, but rather the adaptation of the improvements in electronics to the previously developed optical systems. The instruments available are more sophisticated, and much attention has been given to make the instruments both more "eye appealing" and functional by changing the shape and colors of the controls, as well as the instrument case.

American Instrument Co.

The dual wavelength spectrophotometer of this manufacturer has been changed to make it possible to operate the instrument in either a split beam or dual wavelength mode. As was previously described, in dual operation, two wavelengths of light simultaneously traverse the sample. In the recently introduced split beam mode, the instrument may be employed for conventional recording of spectra. A beam splitter is employed to sequentially pass the light beam through the reference and sample cell respectively, and allow the instrument to function as a double beam "in time" spectrophotometer. The dual gratings in the double monochromator are locked together to function as a single grating monochromator. The dispersed light is then alternately passed through the sample and reference cells, and appropriately amplified prior to presentation of the data. The Aminco-Chance Dual Wavelength/Split-Beam recording spectrophotometer is shown in Figure 18.

Bausch and Lomb

The single beam Spectronic 100 Spectrophotometer permits the analyst to read photometric results in transmittance, absorbance, or concentration directly from single, multiple or micro flowcells. The instrument has a digital readout through the use of Nixie tubes and the capability to be connected to a variety of recording or printout devices, either digital or analog. The instrument covers the wavelength region from 325–925 nm with an 8.0-nm bandpass, and the readout is to four digits. The maximum absorbance reading is 2.000, minimum 0.000. The instrument encompasses virtually all solid state electronics for stability and linearity in the electronic system. It employs an AC log converter for direct absorbance or concentration readings. Two phototubes are employed to cover the spectral range, a grating is employed as the dispersing device, with a pre-aligned tungsten light source. The instrument sells in the $2000 range and also serves as an integral part of the Bausch and Lomb Spectronic 400 system. This system, Figure 45, incorporates the Spectronic 100 into an automated package with a capability of analyzing 400 samples per hour. The results of the analysis are printed out automatically or on command. The sequence programmer permits adjustment of the

Figure 45. The Bausch and Lomb Spectronic 400 System.

sample and purge times.

The Spectronic 70 instrument is pictured in Figure 46. This instrument employs the same optical system as the Spectronic 100. The readout is on a meter with an 8-in. mirrored scale. An accessory outlet permits the use of a recorder, or the use of a Bausch and Lomb digital readout or concentration computer for three digit display of the readings.

Beckman Instruments

Three double-beam UV-visible spectrophotometers are included in the ACTA series: the models V, III and II. The instruments have four-digit photometric readout, either in absorbance or concentration, the more important measurements in spectrophotometry. Conversion for transmittance readings, which are useful for emission phenomena such as reflectance spectra or fluorescence studies, may be made through a circuit board interchange that is available as an optional accessory. The instruments may be purchased either linear in wavelength or wavenumber and, as with many other manufacturers, are offered in the same basic model to operate from either 60 or 50 Hz. The wavelength scales for the three models of the ACTA spectrophotometers are calibrated from 160 to 1000 nm ($62,500-10,000$ cm^{-1}). However, the operating range for the instruments is from 190 to 800 nm ($52,630-12,500$ cm^{-1}). The ACTA V spectrophotometer employs a double monochromator, and built-in recorder; it has the highest resolution, wavelength and photometric accuracy. The ACTA III employs a single monochromator and also has a built-in recorder. It covers the same wavelength region but with a slight decrease in the wavelength accuracy and resolution, and higher stray light level. The ACTA II spectrophotometer appears identical in optical specifications to the ACTA III. It does not have a built-in recorder, but output terminals that will permit it to be connected to an external 100 mv, 1-megohm internal impedance recorder such as the Beckman 10-in. model. A large number of accessories are offered, all of which will fit all three models. Accessories include a reflectance attachment, microcells, gas, solid, and microcell sampling attachments, temperature regulated cell holders, a gel scanner for electrophoresis densitometry, and a linear programming attachment from ambient to 100°C for thermal denaturation analysis. The ACTA V instrument is illustrated in Figure 47; the optical diagram is shown in Figure 48.

Cary Instruments

The Model 17 Spectrophotometer covering the region from 186 to 2650 nm has been introduced by Cary Instruments. The instrument features solid state electronics, a pen period control, multipots for base line adjustment, a com-

Figure 47. Beckman Instruments ACTA V Spectrophotometer.

bined flatbed-strip chart recorder, and coupling between the wavelength scan and chart drive mechanism. Interesting features include a spectral bandwidth meter that displays the width of the band isolated by the monochromator and a pen fidelity meter that indicates the ability of the recorder to respond to the photometric system output signal. Also included are provisions for connection of an external recorder for plotting a first derivative curve at the same time as the absorption curve. The derivative curve may add considerable information in cases where the sample shows weak sidebands or overlapping bands. The instrument, shown in Figure 49, has a wavelength accuracy of better than ±0.4 nm throughout most of the spectral range, a resolution of 0.1 nm in most of the visible-UV region, and 0.3 nm in the near IR. There are eight absorbance ranges covering combinations of the span from 0 to 2 absorbance units, with an accuracy of ±0.002 absorbance on the 0–1 range, and ±0.005 near an absorbance of 2. A choice of scan and chart speeds are available. A Model 17H is also offered with identical specifications to the Model 17, but with a special watercooled sample compartment and other modifications for high temperature work such as fused salt studies or other work requiring a large sample compartment, as, for example, in high pressure studies, radiation work, etc. The Model 17I, a modification of the basic Model 17, is designed for improved performance in the near infrared, and covers the region from 225 to 3000 nm.

Figure 49. Cary Instruments Model 17 Recording Spectrophotometer.

Coleman Instruments

The well-known Coleman Junior single beam spectrophotometer has been updated in the new Junior IIA version. The instrument has been redesigned electronically to permit the user to read absorbance directly on an 8-in. linear scale. This is done by switching the instrument function control knob to the absorbance position, which places a log amplifier into the circuit. For routine determinations, blank scales can be placed in the readout window, and these scales can be calibrated in concentrations units using known standards for the direct reading of concentrations, a convenience factor for frequently performed analyses. Linear transmittance readings may also be made. The wavelength region from 325 to 825 nm is covered, and the bandpass is 20 nm. The instrument, similar in

appearance to Figure 31, accepts the standard Coleman accessories, such as the Coleman data reader, flow-through cells, etc. The cost of the instrument is about $750.

The Model 44 Spectrophotometer also covers the region from 325 to 825 nm with a bandpass of 20 nm. The instrument also has a linear readout of absorbance. Readings are made on a 7-in. mirrored scale. The instrument costs about $750, and a wide range of accessories is offered.

The Model 46, a single beam spectrophotometer, appears very similar in optical design to the Hitachi Perkin-Elmer illustrated in Figure 33. The Model 46 features digital presentation of transmittance and also digital presentation of absorbance on a linear scale and direct readout of concentration. The instrument covers the region from 195 to 800 nm, has variable slits, and resolution of 0.2 nm. Numerous accessories are offered, including attachments for flame photometry, spectral energy recording, fluorometry, reflectance, and automatic sample changing. The cost of the basic spectrophotometer, shown in Figure 50, is $4800.

The Model 111 single beam spectrophotometer covers the range from 200 to 900 nm with a bandpass of 2 nm. The instrument employs a meter for readout; a data recorder is available as an accessory for linear absorbance readings. In external appearance and optical specifications it appears very similar to the Model 101, previously described (Fig. 32). Several accessories are offered for the instrument, which has auxillary output terminals for making connections to a recorder or the digital readout device. The basic instrument sells for $2550.

Figure 50. Coleman Model 46 UV-Visible Digital Spectrophotometer.

The Model 295 has been recently introduced as a teaching spectrophotometer. This single beam, diffraction grating instrument, selling for less than $300, covers the wavelength region from 400 to 700 nm with a 35 nm constant bandpass. A number of Coleman attachments can be used with the instrument. An educational kit is available that includes a teacher's manual, schematic diagram of operating principles, and demonstration materials. The instrument is shown in Figure 51.

Durrum Instrument Corporation

Durrum has improved the stopped flow spectrophotometer previously de-

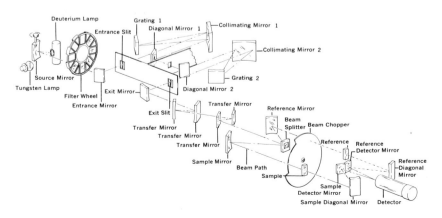

Figure 48. Optical diagram for the ACTA V Spectrophotometer.

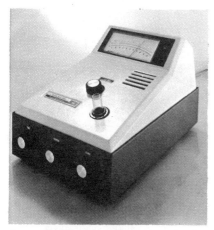

Figure 51. Coleman Model 295 Teaching Spectrophotometer.

scribed and illustrated (Figs. 34–36). Included in the present instrument as standard modifications (which may be purchased as accessories for earlier instruments) are an optical base-plate assembly that serves as an optical platform for the entire instrument and prevents misalignment of the optical path in case the instrument is moved, an air actuator assembly kit to replace the hand actuated hydraulic assembly, and a tungsten-iodine quartz-envelope light source. Also included in the current instruments are a calibrated zero offset, a normalized 10 V output buffer amplifier, and a linear/-logarithmic amplifier. The latter unit provides for a direct reading of the absorbance on the oscilloscope trace and, with an additional log converter unit, provides a log-log output that results in production of a straight line whose slope is proportional to the reaction rate for a first order reaction.

Of particular interest is the temperature-jump system that extends the kinetic capabilities of the stopped flow spectrophotometer. In the temperature jump system, an electrical discharge causes an abrupt, nearly instantaneous temperature increase in the solution initially at equilibrium. A reaction takes place as the concentrations adjust or relax to the new conditions, and this change in concentrations is monitored. Controls are provided to permit operation of the instrument under varying conditions. With this system reactions can be measured that are taking place too fast for the conventional stopped flow technique, such as reactions whose half time is less than 100 μsec.

Gilford Instrument Laboratories, Inc.

Gilford has increased the accessories offered for their instruments and has incorporated the Model 240 spectrophotometer with the 2000 system to develop the new Model 2400 automatic sample changing instrument. Included in the new system are many mechanical and electronic advances; solid state circuitry is employed throughout. Included in the accessories are capabilities for column monitoring, gel electrophoresis scanning, temperature monitoring, and wavelength scanning.

Perkin-Elmer Corp.

The Model 402 spectrophotometer covers the wavelength region from 190 to 850 nm. This double beam, Littrow prism monochromator instrument, with built-in "flow chart" recorder (a recorder sprocket driven and synchronized with the monochromator) is shown in Figure 52. The whole spectrum can be recorded in either, 2, 10 or 40 min with automatic change-over from UV to visible source. Specifications for the instrument include a wavelength accuracy on the chart of ±0.5 nm at 200 nm to ±2 nm at 750 nm; a resolution of 0.2 nm at 210 nm to 1.5 nm at 600 nm; photometric accuracy of ±0.01 absorbance units on the chart for the range of 0–1 absorbance. The instrument can be operated in a number of modes such as repeated scans with either sequential or superimposed recording, single scan recording, and fixed wavelength recording against time. Any 0.3 absorbance unit can be expanded to cover the full ordinate scale giving a fivefold scale expansion. Accessories offered for the instrument include flowthrough cells, a solid sample holder, a microspecular reflectance accessory for measuring film thickness by the interference fringe technique, and a set of beam attenuators to be inserted in the reference beam to allow absorbance values up to 3.5 to be measured. The cost of the instrument is $7400.

Figure 52. Perkin-Elmer Model 402 Spectrophotometer.

Shimadzu

The Shimadzu AQV-50 employs the same monochromator as the QV but automates many of the operations. Optically the two instruments are essentially identical; single beam Littrow prism spectrophotometers for the range from 183 to 1200 nm. The AQV-50 features automatic setting of the zero point by pushing a button when the reference material is placed in the light path, and digital readout, to 3 figures. Two special accessories that have been developed for this instrument are an automatic sampling unit and an automatic sample cell positioner. According to the manufacturer, a wider range of accessories is offered for both the QV-50 and AQV-50, than for any other spectrophotometer, to extend the instrument for atomic absorption, flame emission, circular dichroism, optical rotatory dispersion, reflectance, fluorescence, reaction rate, enzyme kinetics, and column chromatography monitoring work. The

Figure 53. Shimadzu AQV-50 Spectrophotometer.

AQV-50 sells for $3500 and is shown in Figure 53; prices on the accessories vary from $5275 for the optical rotatory dispersion unit to $100 for a chromatogram scanning accessory.

Unicam Instruments, Ltd.

Of particular interest for repetitive work is the Unicam SP 3000 automatic spectrophotometer covering the region from 175 to 750 nm. The instrument is designed for automatic sample introduction, calibration, measurement and digital printout. The sample changer holds up to 50 samples and transfers them sequentially into a special flow cell in the instrument. Two types of sample transfer systems are available. The SP 3002A model flushes each sample to waste after the measurement; the SP 3002P returns the sample to the original container after the measurement, with little cross contamination between samples. Two modes of operation may be employed; each individual sample may be measured automatically at up to 10 wavelengths before the next sample is introduced or the program may be set up to measure the first sample at the first preselected wavelength, the second at the second preselected wavelength, etc., up to 10 sample and wavelength positions. The instrument covers the absorbance range from 0 to 2 with a readout resolution of 0.001 absorbance unit in the range from 0 to 1.200 absorbance units, and 0.01 absorbance unit from 1.200 to 2.000.

An optical diagram for the instrument is shown in Figure 54. A Littrow prism monochromator is employed in a single beam instrument. To prevent drift and improve instrument stability, a second tungsten lamp source is employed as an auxiliary radiation standard. When the reference sample (blank) is introduced into the light beam from the monochromator, an attenuator placed before the sample compartment reduces the light intensity passing through the sample until it equals that of the modulated radiation standard. The sample cell is then moved into the beam. Operating as a null balance system, the transmitted radiation is measured as a function of the intensity of the radiation standard. This function is then converted into the transmittance or absorbance value; calibration is normally carried out automatically at the beginning of each measurement sequence. The automatic digital printer consists of an electric typewriter and associated serializer unit to

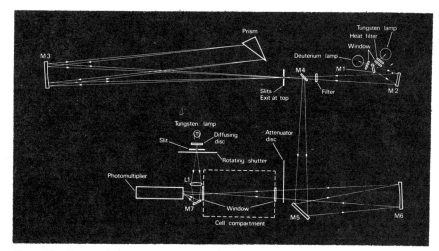

Figure 54. Unicam SP-3000 optical diagram.

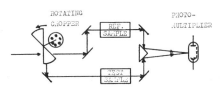

Figure 56. Optical layout of McKee-Pedersen Model MP-1033 Photometer.

print out the absorbance or transmittance of each sample preceded by an index number generated sequentially for the identification of each sample. Also attached to the spindle of the null balance potentiometer is a coded disc that converts the displacement of the potentiometer shaft into a digital result that is displayed on the face of the instrument to four digits for either the transmittance or absorbance. The instrument is illustrated in Figure 55. In addition to the described functions, built into the instrument is a cell corrector control for adjusting the differences in absorbance between the reference and sample cells; among the accessories offered are a fast programmer and constant temperature cell holder.

Other spectrophotometers introduced by Unicam include the Model SP 8000 and the SP 1800. The Model SP 8000 is somewhat similar to the SP 800. This grating monochromator, UV-visible instrument now includes built-in scale expansion and the ability to work with either a strip-chart or flat-bed recorder. Several modular accessories are available, including an automatic sample changer, wavelength programming, digital printout of concentration, as well as sample compartment accessories.

The SP 1800, a newly introduced double beam instrument with a grating monochromator, covers the region from 190 to 700 nm; with a special phototube the range can be extended to 850 nm. The cost of the basic instrument is $5100. Resolution is 0.1 nm, and the readout on the basic instrument is on a meter; four absorbance scale settings are employed. The instrument has the Unicam "second sample" position provision to provide a beam balance control for linearity in the base line over the whole wavelength range and a photometric accuracy of 0.002 absorbance unit. A large number of accessories are offered, such as a recorder, reflectance and fluorescence attachment (the latter uses the instrument for excitation and filters for isolation of the fluorescence wavelengths), automatic sampling, enzyme kinetic and thermostatting accessories.

Other Instrumentation

An interesting accessory recently introduced by McKee-Pedersen Instruments

should be mentioned. This unit, the Model MP-1033 double beam photometer, sells for $1550 and is stated to convert any UV to near IR monochromator into a double beam spectrophotometer. The unit accepts the dispersed light from a monochromator and converts it by means of a rotating chopper into double beam-in-time light pulses, as shown in Figure 56. The light pulses are then measured by the photometer; the readout is on the photometer meter or may be presented on an accessory recorder in either units of transmittance or absorbance. By means of a six position range switch on the photometer, measurements from 0 to 200%T or -0.3 to $+3$ A units may be made with with offset for scale expansion. It can be used directly or, if desired, with MPI accessories. The photometer assembly is shown in Figure 57.

No coverage of instrumentation can be complete, because of the rapidity with which new instrumentation if offered. This review is limited to the commonly offered, routinely used, UV-visible spectrophotometers. An abbreviated coverage of other instruments, particularly reflectance photometers, may be found in the current annual reviews issue of "Analytical Chemistry" (5). Spectrofluorophotome-

ters that may also be used for UV-visible spectral recording have been previously covered (4).

The fact that certain instruments are not treated or that modifications of earlier instrumentation are only briefly mentioned is not to be taken as any implication of poor design or obsolescence, but rather the fact that the basic principles of their design have been previously discussed. Furthermore, because of the vast variety of optional accessories offered for each of the instruments and the fact that these accessories are really quite similar (e.g., flame photometer attachments, reflectance attachments, fluorescence attachments), the discussion is limited to the basic spectrophotometer. Concurrently, the lengthier discussion of certain instruments in this article should not be considered as a bias in their favor, but only an attempt to describe what may be considered more novel instrumentation.

Figure 57. McKee-Pedersen Model MP-1033 Double-beam Photometer.

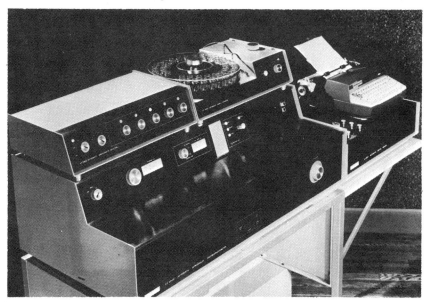

Figure 55. Unicam SP-3000 Automatic Spectrophotometer.

Bibliography

(1) LEWIN, S. Z., THIS JOURNAL **37**, A401, A455, A507, A637 (1960).

(2) BAUMAN, R. P., "Absorption Spectroscopy," John Wiley & Sons, Inc., New York, **1962**.

(3) SHIBATA, K., "Spectrophotometry of Opaque Biological Materials—Reflection Methods," a chapter in "Methods of Biochemical Analysis," Vol. IX, p. 217–234, Interscience Publishers, div. of John Wiley & Sons, Inc. New York, **1962**.

(4) LOTT, P. F., THIS JOURNAL **47**, A327, A421 (1964).

(5) BOLTZ, D. F., and MELLON, M. G., *Anal. Chem.*, **42**, 152R (1970).

Addresses of Manufacturers

An alphabetical listing of the addresses of the manufacturers of these instruments follows; many of the instruments may be purchased from laboratory suppliers.

American Instrument Co., Inc.
8030 Georgia Ave.
Silver Spring, Md. 20910

A. R. F. Products, Inc.
Gardner Road
Raton, New Mexico 87740

Bausch & Lomb, Inc.
22967 Bausch St.
Rochester, N. Y. 14602

Beckman Instruments, Inc.
2500 Harbor Blvd.
Fullerton, Calif. 92634

Burrell Corporation
2223 Fifth Ave.
Pittsburgh, Pa. 15219

Canal Industrial Corporation
5635 Fisher Lane
Rockville, Md. 20852

Cary Instruments, a Varian Subsidiary
2724 So. Peck Rd.
Monrovia, Calif. 91016

Coleman Instruments Corporation,
Division of the Perkin-Elmer Corp.
42 W. Madison St.
Maywood, Ill. 60153

Durrum Instruments Corp.
925 E. Meadow Dr.
Palo Alto, Calif. 94303

General Electric Co.
1 River Rd.
Schenectady, N. Y. 12305

Gilford Instrument Laboratories, Inc.
132 Artino St.
Oberlin, Ohio 44074

Golden Instruments, Inc.
3180 Lakes Shore Dr.
Chicago, Ill. 60657

The Heath Company
Benton Harbor, Mich. 49022

McKee-Pedersen Instruments
P.O. Box 322
Danville, Calif. 94526

Perkin-Elmer Corp., Instrument Div.
721 Main Ave.
Norwalk, Conn. 06852

Phoenix Precision Instrument Co.
3803-05 N. 5th St.
Philadelphia, Pa. 19140

Raytheon Education Company
285 Columbus Ave.
Boston, Mass. 02116

Sage Instruments, Inc.
2 Spring Street
White Plains, N. Y. 10601

Schoeffel Instrument Corp.
15 Douglas St.
Westwood, N. J. 07675

Shimadzu Seisakusho, Ltd., Tokyo and Kyoto,
(Represented in the United States by Cosmos Scientific Co., 921 Bergen Ave., Jersey City, N. J. 07306)

Unicam Instruments, Ltd., Cambridge, England,
(Represented in the United States by Philips Electronic Instruments, 750 So. Fulton Ave. Mount Vernon, N. Y. 10550)

Warner and Swasey Co.
Control Instrument Division
32-16 Downing St.
Flushing, N. Y. 11354

Carl Zeiss, Inc. Oberkochen, W. Germany
(Represented in the United States by Brinkmann Instruments, Cantiague Rd., Westbury, N. Y. 11500)

𝔄ntimony

Topics in...

Chemical Instrumentation

...a *ChemEd* feature

Edited by **GALEN W. EWING**, Seton Hall University, So. Orange, N. J. 07079

The Journal of Chemical Education
Vol. 46, pages A781, A859
November and December, 1969

XLVIII.   Raman Spectroscopy

Bernard J. Bulkin, *Hunter College of the City University of New York, New York, N. Y. 10021*

WHY A RENAISSANCE?

Raman spectroscopy, for many years merely a textbook phenomenon and not a laboratory reality to most chemists, is emerging from the shadow of its complementary technique, infrared spectroscopy. This renaissance, largely stimulated by the development of continuous gas lasers, encouraged by progress in photomultiplier tube technology, has been capped by the revival of some old ideas in optics and some new thoughts about sample handling. The result is that the Raman spectrum of CCl_4 shown in Figure 1—obtained in 3 minutes exposure time with a mercury arc (plus about 25 minutes of film processing) on 30 ml of sample has been replaced with the CCl_4 spectrum shown in Figure 2—taken in 3 minutes with 3 microliters of CCl_4 using He-Ne laser excitation at 632.8 nm.

In this article we wish to discuss the current status of Raman instrumentation from the viewpoint of components—lasers, monochromators, detectors, amplifying systems, and sample handling. The theory of the Raman effect, adequately treated in many monographs (1) will not be discussed here.

To understand the design features of contemporary Raman instruments, it is best to briefly reconsider the problems of the older Raman instruments. One of these, the Cary Model 81, has actually made the transition from mercury arc excitation to laser sources.

It is fair to say that most of the problems of Raman stem from the need for several intense, monochromatic sources of radiation, preferably but not necessarily in the visible region of the spectrum (so that glass optics and cells can be used). These sources should be stable over long periods of time, easily turned on and off, and readily incorporated into an optical system.

The Toronto mercury arc, closest approximation to these goals for many years, had numerous shortcomings. In general the arcs were made of glass and an intense line at 435.8 nm was used. The arcs were difficult to start, usually required continuous pumping to achieve the high vacuums needed, and frequently involved a mass of water lines to cool the electrodes and rheostats. They were not particularly stable sources, making any sort of integration over long time periods such as that required for measurement of depolarization ratios, rather difficult unless a ratio recording method of some sort was employed. Many colored compounds absorb radiation at 435.8 nm and their spectra could not be obtained. This problem severely limited the application of Raman spectroscopy to the problems of inorganic chemistry—problems which it is often uniquely powerful in solving—for many years. Some valiant attempts to overcome this difficulty using Cd, Na, Rb, He, and other arcs, both dc and radio frequency or microwave excited, were made by several groups and the spectra of numerous inorganics obtained this way, but other difficulties of preparation and lifetime thwarted the developments in this area.

Bernard J. Bulkin is assistant professor of chemistry at Hunter College of the City University of New York. Prior to joining the Hunter College staff in 1967, Dr. Bulkin was a postdoctoral fellow at the Swiss Federal Institute of Technology in Zurich, Switzerland. He received the B.S. degree from the Polytechnic Institute of Brooklyn in 1962 and the Ph.D. degree from Purdue University in 1966. Dr. Bulkin's research interests include study of intermolecular forces using infrared and Raman spectroscopy, as well as synthesis and spectra of organometallic compounds.

The mercury 435.8 nm exciting line was also uncomfortably close to another intense line at 404.1 nm, and it was frequently difficult, even with good filtering, to determine which exciting line was responsible for an observed Raman line. For example, studies of the 1640 cm^{-1} vibration of liquid water were not feasible using 435.8 nm excitation because the region of interest was overlapped by the O–H stretching vibration excited by Hg 404.1 nm.

It was impossible, using mercury arcs and anything but the most sophisticated, homemade, high resolution monochromators, to observe Raman lines close to the exciting line itself. This is one of the most interesting regions of the spectrum, however, especially as it is also difficult to look at by infrared techniques.

While all of the instrumental problems of Raman spectroscopy have not been solved, the problems discussed above, which arose from the lack of truly suitable sources of exciting radiation, have been reduced to the point where they are of no concern to the user of a contemporary Raman spectrometer.

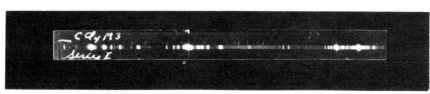

Figure 1.   Raman spectrum of liquid CCl_4, recorded photographically in a 3-min exposure using 30 ml of sample, Hg 435.8 nm excitation.   Excitation by other wavelengths can be seen to left and right of 435.8 nm spectrum.   Calibration spectrum appears above the Raman spectrum.

Chemical Instrumentation

APPLICATIONS

What functions can Raman serve in the modern chemical laboratory which are not already adequately performed by the myriad of infrared, nmr, mass spectrometers, etc.? We will attempt here only the briefest outline of Raman applications, before passing on the details of the instrumentation.

1. *Center of symmetry:* There is a selection rule involving allowed and forbidden infrared and Raman transitions which is known as the rule of mutual exclusion. This states that if a molecule has a center of symmetry (i.e., belongs to a point group which contains the inversion operation, *i*, as one of its symmetry operations) then there are no transitions which are allowed in both the infrared and the Raman spectrum. Frequent use has been made of this rule to distinguish between two possible structures or conformations, one of which has a center of symmetry. Some examples: puckered versus planar cyclobutanes, various types of photodimers, cis-trans isomerism. This rule applies strictly only in the gas phase and holds up fairly well in solution. It should never be applied to solids using the symmetry properties of the isolated molecule.

2. *Totally symmetric vibrations:* There may be a change in polarizability even when there is no change in dipole moment, so that a band which is forbidden in the infrared is allowed in the Raman spectrum. An obvious application comes in studying bond strengths in homonuclear diatomic molecules, which give good, strong, Raman spectra. One of the oldest and most important proofs of the nature of univalent mercury comes from the measurement of the Raman band of Hg_2^{2+} at 169 cm^{-1} by Woodward in 1934. There have been many other recent applications of this type, particularly important ones being those involving metal-metal bonding. There are undoubtedly some important biological applications of this sort as well.

Under highly symmetric vibrations we should also mention the importance of studying the vibrations of carbon-carbon double and triple bonds by Raman spectroscopy. These are quite weak in the infrared, even in relatively unsymmetrical compounds, but are very intense in the Raman spectrum. The vibrational frequencies are also quite sensitive to small changes in the electron density of the pi system and to strain in the molecule.

3. *Solids:* Current Raman instrumentation offers several new avenues regarding the vibrational spectra of solids. It is quite easy to run spectra of powdered solids using Raman instruments, with no sample preparation or matrix necessary. Thus the KBr pellet and the Nujol mull are not needed for Raman. The technique is also nondestructive and needs only a few micrograms of sample.

Intriguing possibilities in the study of single crystals by Raman spectroscopy are now opening up. By taking advantage of the highly polarized nature of the laser beam, and space group theory predictions about the polarization of the scattered Raman light, much information about forces and structure in crystals can be obtained. The first work on ionic crystals has been appearing in the last few years, and initial studies on organic crystals are being undertaken. Not only can internal vibrations be observed in these spectra, lattice vibrations are also seen in many cases; these can be used to study intermolecular forces, solid-solid phase transitions, etc.

4. *Intermolecular vibrations and perturbations:* Because it is possible to observe low frequency vibrations in the Raman spectrum without too much difficulty (at least this appears to be easier in many cases than making comparable measurements in the far infrared spectrum), we should be able to study the vibrations of hydrogen bonds themselves, rather than relying on their perturbation of other bonds in the molecule. Some far infrared work in this area has already appeared, and it is likely that Raman studies are now in progress. This may prove especially useful in biological systems. Other work on biological systems perturbed by intermolecular effects, such as base pairing and stacking, etc., has already shown Raman to be useful as a complementary technique to infrared and nmr.

5. *Aqueous solutions:* It is well known that Raman spectra are obtained in glass or quartz cells, and not subject to any sampling limitations due to water. Further, the Raman spectrum of liquid water is rather weak, and many studies can be carried out on aqueous solutions. This is proving quite useful in current studies on various aqueous equilibria for a large number of metal ion systems.

A further application in water has been the study of the effect of dissolved electrolytes on the spectrum of water itself, giving a measure of the relative strength of hydrogen bonding between anions and water.

6. *Group frequencies:* It is not really known yet whether a comparable number of useful group frequencies will be obtained from Raman spectra such as are now widely used in infrared spectroscopy. However, at the 1969 Pittsburgh Conference on Analytical Chemistry and Applied Spectroscopy there were a few encouraging reports in this direction. There are some interesting group frequencies for substituted benzenes, and certain nuclei such as sulfur provide a very characteristic set of frequencies when bound in different ways.

7. *Polymers:* Several groups have been working on a number of polymer chemistry problems using Raman spectroscopy with very interesting results. These studies have usually made use of the polarization properties of Raman lines. Some work on biopolymers has also been reported, and it seems as if conformational data in polypeptides may be obtainable from Raman spectra.

The above is merely a sketch of some leading Raman applications, which makes no pretense of completeness. It is intended only to show some areas of current interest and to give an idea of the diversity of applications for these instruments.

LASERS

Although pulsed, high-power lasers were used to obtain Raman spectra in the early 1960's, and led to the discovery and development of the laser-Raman effect, this work is primarily of interest to physicists. It should not be confused with what is usually discussed as Raman spectroscopy, and which is best referred to as laser excited Raman spectroscopy.

Helium-Neon: The most commonly used laser exciting line is 632.8 nm from a He-Ne laser. These lasers, now commercially available with powers ranging from less than a milliwatt up to about 100 mw, fulfill most of the requirements of a good source of monochromatic radiation for Raman spectroscopy. A power of about

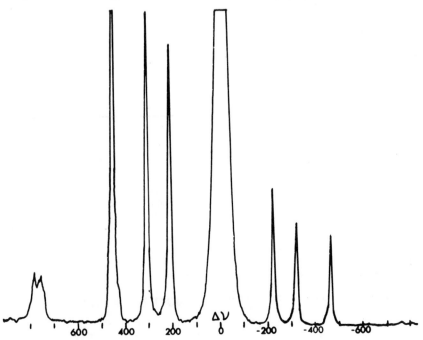

Figure 2. Raman spectrum of liquid CCl₄, scanned photoelectrically at 500 cm⁻¹/min, using 3 μl of sample, He-Ne 632.8 nm excitation.

Chemical Instrumentation

50 mw proves adequate for obtaining spectra of a wide variety of samples. These lasers appear to be extremely stable, with usual short-term peak to peak fluctuations stated by manufacturers as less than 1%. The laser beam, easily visible in a lit room, can be used for a rapid alignment of spectrometer components. We have been able to bring a laser, monochromator, etc., into our laboratory, set them on a big table, and do all the necessary alignments so that spectra could be obtained at near peak performance in a few hours. The technology of He-Ne laser construction is very far advanced, and in our experience few adjustments are necessary on the laser as it arrives from the manufacturer.

There are several nonlasing helium and neon lines emitted. These are of much lower intensity than the laser line and are usually eliminated by use of a spike filter which costs about $200. These are available from many companies such as Corion, Thin Film Products, Oriel, etc. Alternatively, one can make use of the fact that the nonlasing lines diverge much more rapidly than the lasing line; thus if the distance from the laser to the entrance slit of the monochromator is large the nonlasing lines can be eliminated or greatly attenuated by interposing a small hole (e.g., an iris diaphragm which may be opened and closed).

An important property of the laser lines when used as Raman sources is their polarization. The lines are completely polarized, and the axis of polarization can be turned conveniently by mirrors or half wave plates. This greatly facilitates measurements of depolarization ratios to high accuracy. It also makes alignment of the spectrometer, sample, and laser axes quite easy. This is usually accomplished with one or more small pieces of Polaroid and visual observations. The polarization of the laser beam also makes possible the very elegant studies now in progress in several laboratories on oriented single crystals. Much information about lattice vibrations is emerging from these studies, which lie on the border between solid-state physics and chemistry.

Helium-neon lasers have been extremely valuable in the study of colored compounds. Many of the compounds of interest, which absorb strongly in the blue region of the spectrum, have no significant absorption at 632.8 nm and beyond.

There are, nevertheless, certain disadvantages to the use of helium-neon lasers versus other types of continuous gas lasers now commercially available. First, Raman is a scattering phenomenon, and the Raman intensity is a function of the actual frequency of the Raman line, rather than simply of the Raman shift which is usually reported. Scattering intensity is proportional to the fourth power of the frequency, and so the relative intensities of a Raman line excited by 632.8 nm and one excited by an argon ion laser at 488 nm is ca. $(488/632.8)^4$ or 0.35. Secondly, the response of the best photomultiplier tubes avaliable for Raman spectroscopy begins to decline rapidly beyond 632.8 nm, whereas

they are near peak sensitivity at 488 nm. This and other factors such as blue compounds, fluorescence at 632.8, and higher available powers have increased the interest in other laser sources, especially rare gas ion lasers.

Many helium-neon lasers are now commercially available from an ever proliferating group of companies. Most popular has been the Spectra-Physics Model 125 (now called the Model 125A) which is called a 50-mw laser. This means that it is guaranteed to have a power of at least 50 mw for a period of one year. In practice a new model 125 laser has about 75 mw power, and this very slowly declines with use. Formerly a standard feature, radio frequency excitation is now optional at extra cost on the model 125A. Experience with the model 125 indicates that the RF adds about 6% more power, and may also result in a quieter signal from the laser. The laser tube must be replaced when power falls too low or the laser fails to ignite, at a cost of about $800.

The Perkin-Elmer Corp. supplies its own He-Ne laser with its Raman spectrometers, although the Spectra-Physics laser can and has been used. The Perkin-Elmer instruments have used low-power lasers in the past, in the 1–8 mw range. Although satisfactory for some routine samples, such as common solvents, high scattering powders, etc., these lasers do not seem to provide the desired Raman intensity for difficult samples. Relatively few compounds are decomposed by the focused laser beam from the He-Ne sources at 80 mw, and until the laser intensity causes decomposition or other undesirable effects, it probably should be increased as much as possible to maximize Raman intensity. The low-power lasers are considerably cheaper, however.

Rare gas ion lasers, especially Ar+: Almost any commercially available Raman setup is sufficiently modular in approach that lasers can be substituted for one another with no trouble. Using Brewster angle prisms or coated mirrors, many workers have arranged two lasers as part of one spectrometer, changing easily between them. This is desirable in that compounds of different colors can be readily examined, and it has been a general

observation that if there is a severe fluorescence problem with one laser exciting line, changing wavelengths often results in a significant decrease in the background.

An alternative to having two or more lasers is to have one laser that can lase at several wavelengths. The rare gas ion lasers seem to offer the greatest promise in this direction.

Argon ion lasers have their most intense lines at 488 and 514.5 nm. Lasers are available having very high power relative to the lasers we have discussed so far, as much as 900 mw in each line. This appears to be more than is really useful for Raman work, as many samples are destroyed in a focused laser beam of this intensity and frequency. There are several less intense lines throughout the visible spectrum which also offer significant power for Raman excitation.

By returning to the high-frequency end of the visible spectrum (488 nm) from 632.8 nm of He-Ne, one regains much of the Raman intensity lost through both the ν^4 factor in scattering intensity, as well as the decrease in phototube response. When one further considers the high power available from Ar+ lasers the expected increase in Raman intensity becomes approximately a factor of 64 at the exciting line, and increases to a factor of 150 at a 3000 cm^{-1} Raman shift.

The only remaining problem then is the absorption of the exciting light by colored samples. The solution to this problem appears in the offing now, with the development of the krypton ion laser, and the mixed argon-krypton ion lasers. Some problems still exist with commercial versions of the mixed lasers, however, and at least one model has been withdrawn. Kr+ has its strong line at 647.1 nm, well into the red region. Another good laser line appears at 568.2 nm.

Table 1 shows the Ar+ and Kr+ lasers which have found application in Raman spectrometers.

MONOCHROMATORS

If low-frequency Raman lines are to be observed, the level of stray light must be extremely low. This is the most stringent constraint imposed upon Raman mono-

Table 1. Rare Gas Ion Lasers for Raman Spectroscopy

Manufacturer	Gas	Model No.	Power (mw) at	Wavelength (nm)
Spectra-Physics	Ar	140	900	488.0
			900	514.5
Spectra-Physics	Ar	141	100	488.0
			100	514.5
Coherent Radiation	Ar/Kr	52	700	488.0
			700	514.5
			150	647.1
			60	568.2
			60	520.8
	Ar	54	180	488.0
			180	514.5
Carson Labs	Ar	300	1200	488.0
			1200	514.5
	Ar	100	400	488.0
			400	514.5
	Ar (Krypton is also available)	10	40	488.0
			40	514.5
RCA	Ar	LD-2100 (3 models)	10-100-1000	488.0
			10-100-1000	514.5

Chemical Instrumentation

chromators. It is also desirable that the
monochromators be capable of moderately
high resolution, *ca.* 0.5 cm^{-1}, be of the
scanning type with a wide range of scan
speeds, e.g., from 0.5 to 2000 cm^{-1}/min,
and that they maintain peak performance
in all modes over a long period of time with
minimum maintenance.

The best solution to these problems
seems to lie in the use of two coupled
monochromators. A variety of configurations based on this idea have been developed and marketed. Unlike the lasers
discussed previously, which with two exceptions are made by companies not
connected with Raman spectroscopy, all
the monochromators are actually the
heart of total Raman systems devised by
the manufacturers. However, it is possible to mix components and in most cases
to purchase only the monochromators
without the other accessories if this is
desired, so it is reasonable to treat these
separately rather than as part of total
instruments.

The Cary Instruments Model 81 (this
is actually the model number for the entire
Raman system) is probably the most
familiar of any Raman monochromator,
as it was used in the Cary 81 equipped with
mercury arcs for many years. A large
number of these monochromators can thus
be found in labs around the country, some
still operating with mercury arcs. The
Cary 81 optical diagram is shown in
Figure 3, the portion enclosed in the dotted

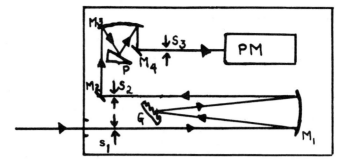

Figure 4. Optical diagram of the CODERG Type CH 1 monochromator. Light enters through slit S$_1$, strikes mirror M$_1$, is reflected to grating G, back to M$_1$, through slit S$_2$, into a prism post-monochromator which acts as a filter, through exit slit S$_3$ to the photomultiplier tube.

line being the monochromator section.
This monochromator is extremely large
compared to the others which will be
described below, in order to make an
efficient match to the image slicing sample
optics, which may be familiar to some
readers from the mercury arc instrument.
The monochromator uses a double Littrow
mount, with a focal length of 1000 mm.
The slits, which are curved, are 10 cm high
but can be masked to 5 or 2.5 cm. In
general, the monochromator uses two
1200 l/mm gratings blazed at 500 nm
in the first order. With the large slit
heights required by the image slicer system, the Cary 81 must use curved rather
than straight slits if it is to retain resolution and maximize light gathering power
for the photomultiplier tube. The only
real disadvantage in using curved slits is
that they inevitably are more expensive to
produce than straight slits. When the

image height is considerably smaller, e.g.,
about 5 mm or less, there is little advantage
to curved slits.

The Societé de Conversion des Energies
(CODERG) has developed two monochromators which are used for Raman
spectroscopy. They are known as the
CH1 and PH1, to indicate that one is for
use by chemists and the other by physicists. The reader should not take these
designations overly seriously, however.
The optical diagram for the CH1 is shown
in Figure 4. The grating monochromator
through which the light passes first is a
600-mm focal length Fastie-Ebert mount.
It uses curved slits, and a 1200 l/mm
grating blazed for 750 nm. Most other
Raman systems use gratings blazed at
shorter wavelengths than 632.8 nm, in
particular either Jarrell-Ash or Bausch and
Lomb gratings blazed at 500 nm. The
reason for this is that the efficiency of most
gratings falls off more rapidly to the short
wavelength side of the blaze (e.g., the
Jarrell-Ash grating efficiency curve shown
in Figure 5. The "chemical" monochromator then sends the light to a second
Fastie-Ebert monochromator, which is
shown as a prism monochromator here but
can be fitted with a grating instead. The
second monochromator is used as a filter,
and does not scan but is set to a particular
position depending upon the region of
interest.

In the CODERG PH1 monochromator
two Fastie-Ebert 600 mm focal length
monochromators identical to the first
monochromator of the CH1 system are
mechanically coupled in a side by side
arrangement, with an intermediate slit.
This arrangement shares the problem of
the Spex Industries Model 1400 and 1401
monochromators, to be discussed presently, of keeping the two gratings at
exactly the same angle at all wavelengths.
The advantage, however, is that these are
effective double dispersing monochromators, whereas other arrangements such as
that used by Jarrell-Ash and Spectra-
Physics in their monochromators are only
single dispersing and thus result in lowering
of stray light but loss of throughput.
This matter has been extensively discussed
in the literature, with a leading reference
being the presentation by Christiansen and
Potter (2).

By contrast, the Model 25-100 double
monochromator designed by Jarrell-Ash
Co., the optical diagram of which is shown
in Figure 6, does not have problems of
grating angle matching discussed above.

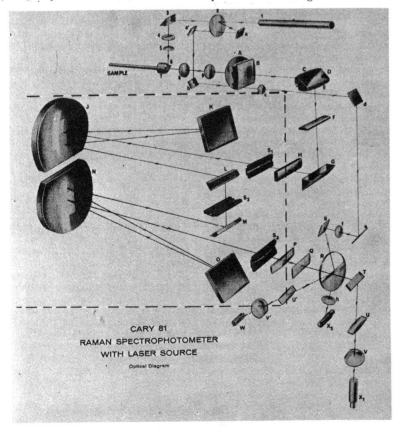

Figure 3. Optical diagram of the Cary 81 Raman spectrophotometer. Dotted line encloses the monochromator portion. Light enters the monochromator through slit S$_1$, strikes collimating mirror, J is reflected to grating K, back to J, then reflected by mirror L through intermediate slit S$_2$ to a second monochromator which is identical with the first. (Courtesy Cary Instruments, a Varian Subsidiary.)

Chemical Instrumentation

This is because the two gratings are attached to a single post, rather than being connected by a bar. This is an extremely stable arrangement. Despite the lack of

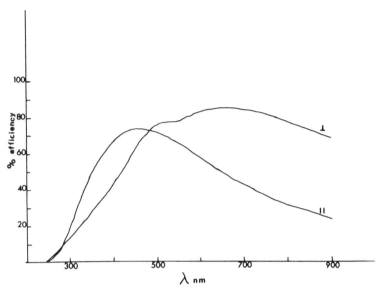

Figure 5. Wavelength dependence of the efficiency of a Jarrell-Ash grating blazed near 500 nm. The efficiency is shown for illumination of the grating with light polarized parallel and perpendicular to the rulings. (Courtesy Jarrell-Ash Div. of Fisher Scientific Co.)

double dispersion, we do not wish to imply that the stacked Jarrell-Ash monochromator automatically has lower throughput than the side-by-side double monochromators such as the CODERG PH1 and the Spex Model 1401. The final balance sheet on luminosity is a result of these factors in instrument design as well as the reflectivity of the surfaces in the monochromators, in particular the grating efficiencies, and these may narrow or widen the gap caused by fundamental design differences. The

Model 25-100 uses two Czerny-Turner 1000 mm focal length monochromators, with Jarrell-Ash 1180 l/mm gratings blazed at 500 nm in the first order. As shown in Figure 6, the light first passes through the top monochromator, then the bottom one.

Still another type of double monochromator arrangement for Raman is found in the Perkin-Elmer model LR-3 Raman spectrometer. The optical diagram in Figure 7 shows that this can even, in some senses, be thought of as a triple monochromator. The light, first dispersed in a double-pass, grating monochromator by a 1440 l/mm grating blazed at 620 nm, then passes into a grating post monochromator which also uses a 1440 l/mm grating. Although the double passing arrangement in the first monochromator results in sig-

nificant improvement in resolution, it does not have the great effect in stray light reduction which is characteristic of double monochromators. The double passing monochromator portion was the building block of earlier Perkin-Elmer infrared instruments, and was also utilized in an earlier version of the Raman spectrometer, the LR-1.

A recently developed small Raman spectrometer was shown in early 1969 by the Spectra-Physics Co. The monochromator for this instrument is a double Ebert monochromator with the two monochromators back to back. The gratings are linked together in a parallelogram arrangement.

Perhaps most popular of all monochromators that have been used for laser excited Raman spectroscopy have been the Spex Industries Model 1400 and 1401. The 1401 is the current version of this monochromator for Raman work, providing linear wavenumber display as opposed to the linear wavelength display of the 1400. An optical diagram of the Spex Model 1401 monochromator, in Figure 8, shows that this monochromator consists of two $^3/_4$ meter Czerny-Turner monochromators in a side-by-side arrangement. Although until recently Bausch and Lomb 1200 l/mm gratings were chosen by most Spex users, the apparent superiority of Jarrell-Ash gratings (1180 l/mm) with respect to efficiency and grating ghosts has made these the choice in several instruments. Other users have selected 600 l/mm gratings for use in the second order at 632.8 nm, with some loss in efficiency but a considerable savings in price.

The problems in the side-by-side arrangement of monochromators have already been mentioned above. It has been pointed out (3) that all double monochromator systems share the problems of equalizing the focal lengths in the two monochromators as well as the angles of incidence and diffraction. It is a simple

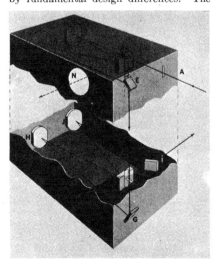

Figure 6. Optical diagram of the Jarrell-Ash Model 25-100 double monochromator. Light enters through slit A, strikes mirror B, is reflected to grating C, thence to mirror D, plane mirror E, and through intermediate slit F to an identical second monochromator. A single shaft connects gratings C and I. Mirror M can be swung into place for use as a single monochromator. (Courtesy Jarrell-Ash Div. of Fisher Scientific Co.)

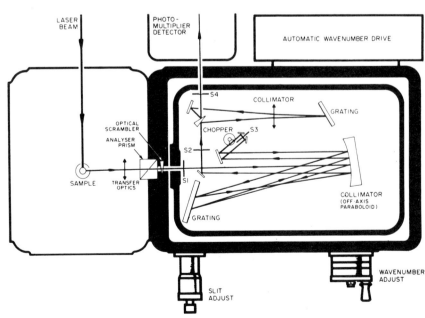

Figure 7. Perkin-Elmer Model LR-3 Raman spectrometer. Light is double passed through a first grating monochromator in which it is also chopped, then passes through intermediate slit S_2 to a second single pass grating monochromator and through exit slit S_4 to the photomultiplier. (Courtesy of the Perkin-Elmer Corp.)

Chemical Instrumentation

matter to test a double monochromator for grating angle matching. Suppose the three slits are set to the same opening, e.g., 100-100-100 microns, and the intensity of a line is observed. If the monochromator is properly adjusted no change in intensity will be observed if the intermediate slit is opened.

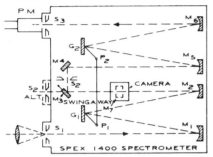

Figure 8. Spex Model 1400 (or 1401) spectrometer optical diagram. Light enters through slit S_1, strikes mirror M_1, is reflected to grating G_1, thence to mirror M_2, and passes through intermediate slit S_2 into an identical second monochromator. Light from the second monochromator exits through slit S_3 to the photomultiplier tube. P_1-P_2 is the tie bar connecting the two gratings. (Courtesy Spex Industries.)

DETECTION AND AMPLIFICATION

The problems of detection and amplification of Raman signals, inherently very weak signals, are considerable. Most older Raman work was done with photographic detection, with exposures as long as 200 hours being common. Several advances have now obsoleted the photographic detection systems, and the inevitable increasing use of computer averaging techniques in Raman spectroscopy should do more in this direction.

Almost all Raman spectrometers now use photoelectric detection with one of three forms of amplification: (1) simple direct current using a picoammeter; (2) lock-in amplifier using a reference signal developed by chopping of the laser beam at some point; (3) photon counting with discrimination circuitry.

To some extent, the choice of which system is to be used depends on the Raman spectrometer selected. In the Cary and Perkin-Elmer systems a phase sensitive detector system is built in to the total Raman system. Probably the Spectra-Physics Raman spectrometer will also have a built-in amplification system, but this new instrument is still in a state of change and it is not yet clear just what system that will be. In the CODERG, Jarrell-Ash, and Spex Raman spectrometers there is really no requirement of a particular amplification system although the brochures describing the CODERG Raman spectrometers usually discuss only DC amplification.

The choice of phototube will depend on which amplification system and which laser lines will be used. Many photomultiplier tubes can offer high sensitivity in the region of Raman scattering near the argon laser line, 488 nm, but by the time one has moved into the red region of the spectrum, e.g., excitation by He-Ne or Kr, it is difficult to find a sensitive photomultiplier tube. The overwhelming choice of Raman users has been for an end-on tube with extended S-20 response (trialkali photocathode). This response is shown in Figure 9, from which it can be seen that the sensitivity is best in the blue region, but is still better than other available tubes well into the red region of the spectrum. Two tubes of this response which have proved popular with Raman spectroscopists and which are used in the majority of commercial instruments are the EMI 9558B and the ITT FW 130. The former was used in earlier instruments, the latter is the one commonly supplied with most current Raman spectrometers. There are indications however, that the

other new photomultiplier tubes just appearing on the market may be quite suitable for Raman work. Among these are the Bendix Channeltron tube, an RCA tube with a gallium phosphide dynode, and tubes made by the Hamamatsu Company in Japan.

The need for cooling of the photomultiplier tube to reduce the dark current is still a topic of some discussion among Raman spectrocopists. With the ITT FW 130 photomultiplier tube, the very small (3-mm diameter) cathode results in a rather low dark current even at room temperature. In the author's laboratory, for example, a tube of this type gives 63 electrons/sec at room temperature, which is a fairly low background. For most Raman work on powders and strongly scattering samples no further reduction in background would be necessary, but if weak bands are to be measured, e.g., dilute aqueous solutions, etc., it is desirable to further reduce the dark current to as low a value as possible. This need results from the fact that the dark current fluctuates statistically rather than being a simple, constant background. Tube cooling is fortunately not nearly so difficult a job as it once was. For the tubes in question, dark current can be reduced to a very low value (less than 5 counts/sec in our experience) by cooling to $-25°C$. This can now be accomplished with thermoelectric coolers (Peltier devices), and some very elegant photomultiplier housings using the thermoelectric effect have been made by Products for Research Corp. Figure 10 shows the effect of cooling on dark current for an ITT FW 130 photomultiplier tube. It seems that these coolers, which are efficient and require little or no maintenance, are ideal for Raman work. Although a dry ice or liquid nitrogen cooler is slightly less expensive, it requires frequent attention and probably involves large changes in tube temperature every day.

The photomultiplier tubes described

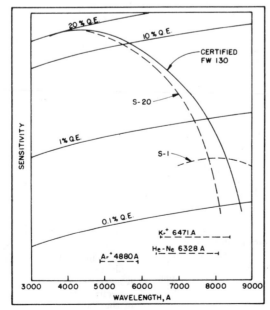

Figure 9. Sensitivity of certified ITT FW-130 photomultipliers compared to S-20 and S-1 standard responses. The dashed horizontal lines represent the range of 3500 cm^{-1} Raman shift from common laser exciting lines. (Courtesy Spex Industries.)

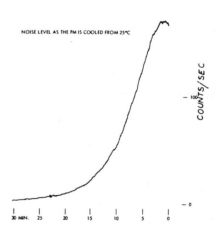

Figure 10. Use of a thermoelectric refrigerator for cooling a photomultiplier tube from room temperature to $-25°C$. The dark-noise level (counts/sec) is plotted as a function of time after the thermoelectric cooler has been turned on. (Courtesy Jarrell-Ash Div. of Fisher Scientific Co.)

here, such as the ITT FW 130 generally run under quite high voltages, in the 2000

Chemical Instrumentation

volt range, although the actual operating voltage must be determined for each tube individually to give optimum signal-to-noise ratios. Considerations in this choice of voltage have been developed in some detail in a paper by Nakamura and Schwartz (4), which is recommended reading for any newcomers to this area. The optimum voltage to be used may change in the early days of tube use, as the characteristics seem to change slightly during the first period of exposure to light. The photomultiplier tubes should be kept in the dark and under full voltage when not being used.

Returning to the discussion of amplification systems, detailed comparison of the lock-in amplifier and photon counting systems has been made by Nakamura and Schwartz (4). It seems that from "theoretical considerations" photon counting is the better system for signals in the range of Raman lines. The main advantage of the lock-in system at the moment appears to be the continuous monitoring of the laser intensity, and the consequent removal of any fluctuations in this intensity from the resulting spectrum. Both dc and photon counting do not provide for any such monitoring, although it could presumably be built into the system, and in the Jarrell-Ash system provision has been made for this.

Although it has been argued in the literature that photon counting should provide up to a factor of three advantage over dc with respect to signal-to-noise ratio for weak signals, it has been found experimentally by many workers that this is not true until one approaches very weak signals indeed. For most Raman bands comparable S/N is found by dc and photon counting. Further, photon counting is considerably more expensive than dc amplification, adding approximately $3100 to the cost of a Raman system.

There are certain advantages, nevertheless, to a photon counting amplification system. These stem from the essentially digital nature of the photon counting system, as opposed to the analog dc current. First it is possible in photon counting to use discrimination circuitry to remove pulses of higher and lower energy than the Raman pulses. Low energy pulses result from electrons originating down the dynode train rather than at the cathode. These produce smaller anode pulses which can be eliminated by discrimination in the amplifier stage of the photon counting system. High energy pulses generally provide little interference in Raman spectroscopy.

The greatest advantage of photon counting comes in digital applications of the data. If the signal is sent from the pulse height analyzer to a digital ratemeter or counter, or in fact directly to a computer which can accept pulses, automation of data collection is a simple matter. In other amplification systems some type of analog to digital conversion system would be needed. Even if the spectrometer is not automated, data collection using a digital ratemeter can rapidly determine depolarization ratios by accumulating all

pulses in a spectral region in both positions of the analyzer. Counting statistics then give information about the uncertainties in the values. The digital nature of the photon counting system, especially as more and more computers are available in laboratories, is certainly its biggest positive feature. For continuous recording of spectra most Raman spectrometers use an analog ratemeter which records counts/sec and sends this information to a recorder

Princeton Applied Research Corp. seems to be the leading source of lock-in amplifier systems which are being used for Raman spectroscopy, aside from those amplifiers built in to the Cary and Perkin-Elmer spectrometers. DC picoammeters are available from several sources, although most in use now come from either Victoreen or Keithley. Photon counting equipment is again available from a very large number of manufacturers, principally those concerned with nuclear equipment and X-Ray detection and amplification. Many of the automation systems based on photon counting for X-Ray diffractometers appear to be applicable to Raman with little modification.

SAMPLE ILLUMINATION AND HANDLING

As with any spectroscopic method, each user seems to develop his own battery of sampling techniques, depending on his own experience and the types of samples he encounters most frequently. In the Raman effect, it seems that the method of illumination of the sample with the laser beam and collection of scattered radiation may be as important as any of the other instrumental factors described thus far.

It is convenient to separate the sampling problems by phase, although in many laboratories the same or similar systems are now used for powdered solids and liquids. A further classification or subheading under solids is that of single crystals.

Gases are difficult to study by Raman spectroscopy, because of their low scattering. It appears that it is only practical to study gases with a He-Ne laser if the sample is placed inside the laser cavity or if photographic detection is used. With an Ar laser it should be possible to obtain good spectra of gases outside the laser cavity using multiple passing through the Raman cell. In general one would have to say that Raman is not a routine technique for study of gases at the moment. Further developments may change this situation, however.

The pioneering effort in sample illumination of liquids was made by the Perkin-Elmer Corp. with a multi-pass cell. This cell is still used by Perkin-Elmer and by Cary Instruments for illumination of liquid samples, although both have other types of liquid cells available as well. In the P-E cell the laser beam enters at one end of the cell, which is silvered and wedge shaped, and is passed down to the other end in a series of reflections. The Raman scattered light is observed at 90° to the incident radiation.

Subsequent to the development of the Perkin-Elmer cell, several workers showed that it is more advantageous to focus the

laser beam to a diffraction limited point in a small sample, rather than try for multiple passing through a larger cell. If the liquid is clear, the focused beam will pass through the liquid and can be reflected back again for another pass. Since the exit of a laser is itself a mirror, further reflection to the sample can be achieved. Thus a considerable gain in Raman intensity can be achieved beyond a single pass. This is now the preferred scheme for examination of liquid samples, with some variations as to the geometry of the small sample cell. It allows one to obtain good Raman spectra on extremely small samples, in the micro-

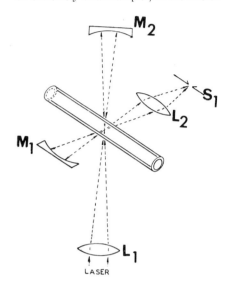

Figure 11. Transverse excitation-transverse viewing optical system for examining Raman spectra of small quantities of sample. Light from the laser is focused on to the sample by L_1, and the scattered light from the sample is focused on the monochromator entrance slit, S_2, by lens L_2. M_1 collects the back scattered light. M_2 allows for multiple passing, which in practice yields a gain of nearly 2 for clear liquid samples. (Courtesy Spex Industries.)

liter or even nanoliter range for liquids. Indeed, since the volume of the focused He-Ne laser beam is only about 8 nl at the diffraction limited point, increasing sample volume does not help. Capillary cells (even inexpensive melting point capillaries are fine) can be used to contain the samples. If the capillary is illuminated along its full length, the rounded bottom of the capillary serves to further focus the radiation. Alternatively, as suggested by Landon and Freeman (5) a transillumination system (Figure 11) is possible.

The Cary 81 spectrometer uses primarily coaxial viewing of the Raman radiation, that is, the Raman lines are observed at 180° to the incident exciting radiation. For small samples a capillary cell is fixed to a hemispherical lens which collects the Raman light. The light is then focused on the entrance slit of the monochromator. If 90° viewing of the Raman light is desired in the Cary 81, the Perkin-Elmer multi-pass cell is used. This is necessary if accurate depolarization ratios are to be measured, because the internal reflections of the Raman radiation from the walls of the capillary in the 180° viewing mode results in substantial depolarization.

For powdered solids, it is best to use a

Chemical Instrumentation

small amount of sample and focus the laser beam directly on to the sample. Although good spectra can be achieved with either 180° or 90° viewing, to observe low-frequency lines it is necessary to maximize the ratio of Raman to Rayleigh scattering, and this is best achieved by 90° viewing. Solid sample handling is considerably simpler for Raman spectroscopy than it is for infrared work, as a small amount of solid sample can be handled with no matrix such as KBr or Nujol. For both powdered solids and liquids it is worthwhile to have a mirror which takes the back scattered light into the monochromator.

For solid single crystals it is a simple matter to install a standard goniometer head in the sample compartment. Considerable information about forces in crystals can be obtained from examination of the Raman spectrum of oriented single crystals, using the high degree of polarization of the laser exciting line. A considerable number of studies have appeared on ionic crystals, and it seems likely that an increasing number of spectra of oriented organic crystals will also be forthcoming.

There are several other important features in a good Raman sample illuminator. First, for single crystal studies and for all depolarization ratio measurements it should be possible to easily turn the axis of polarization of the incident beam and to have an analyzing polarizer in the path of the Raman scattered light. The use of a half wave plate serves the first purpose, and a standard piece of Polaroid such as used in photography is suitable as an analyzer. Second, provision should be made for insertion of filters, both spike filters to isolate the laser line from non-lasing emission lines, and neutral density filters to attenuate the laser beam if necessary. For some samples the high-energy beam will cause decomposition. Third, if accurate depolarization ratios are to be obtained, a quartz wedge or other device for scrambling the polarizations must be placed between the analyzing polarizer and the gratings. This is because gratings respond differently to light polarized in different directions, and the relative efficiencies to the different polarizations is a function of wavelength.

Perhaps the most important requirement for a sample illuminator which is to take full advantage of the microsampling capabilities of Raman spectroscopy is provision for positioning of the laser beam on the small sample and subsequent focusing of the image from this sample into the entrance slit of the monochromator. Different manufacturers have handled this problem in different ways. In the Cary and Perkin-Elmer spectrometers, the sample is always held in the same position, exactly, and no provision is made for moving either sample or laser beam to improve imaging. Further, since the distance from the slit remains unchanged, no adjustment is made in this direction either.

In the Spex and Jarrell-Ash spectrometers this is not the case. The Spex sample illuminator (Fig. 12) keeps the distance between the sample and the entrance slit

fixed, and a lens which focuses the Raman scattered light is usually not adjusted in the course of a measurement. To align laser beam and sample, the Spex illumina-

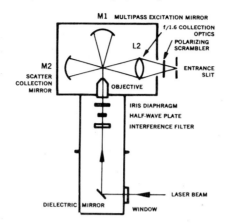

Figure 12. Schematic diagram of the Spex Industries No. 1430 Sample Illuminator. (Courtesy Spex Industries.)

tor has a floating microscope lens mounted just below the sample. This lens focuses the laser beam on the sample and adjustments of the lens in the x and y directions (laser beam proceeding along the z direction) allow the laser beam to be moved several millimeters in each direction. These adjustments can be made with the sample illuminator cover closed. The Jarrell-Ash illuminator is somewhat more complicated, but also is potentially more flexible. The sample is mounted on an optical bench, and as two such benches are provided, one below and one inverted above the entrance slit, illumination can take place from either above or below the sample. Also, the distance between the sample and slit can be changed continuously, though it need not be, of course. For positioning of the sample with respect to the laser beam, Jarrell-Ash cells are mounted in spring clips on a goniometer head, and the goniometer is adjusted for optimum illumination of the sample. The condensing lens between sample and slit is mounted in a sort of floating arrangement, and it is adjusted in all three directions for peak intensity. Presumably, for similar samples if no change in the position of the goniometer head is made, then little adjustment of the lens is needed. To perform all these adjustments, however, the sample compartment doors must be open. This holds a certain disadvantage, as with wide slits it will generally be necessary to darken the room to avoid the high background from room lights. The CODERG spectrometers offer a wide variety of unique sampling cells for liquids and solids, including the only commercially available variable temperature accessories for Raman work.

FUTURE TRENDS

There are two leading areas in which considerable effort is being made at the moment. These are the development of small Raman spectrometers for routine analytical laboratory use and the automation of new and existing instruments.

Several companies are already developing small spectrometers. The new instru-

ment of the Spectra-Physics Corp. has been discussed in this article, and the Perkin-Elmer spectrometer has always been of the small type. In order for such an instrument to be successful, however, it must be able to produce good spectra from a wide variety of samples, including what might be considered difficult samples, such as polymers, powders, colored solutions, etc. As each component of the larger Raman instruments has been optimized so that spectra may be obtained from difficult samples, it is not yet clear that it will be possible to produce an instrument at substantially lower cost which still has the high performance needed. The problems of Raman instrumentation in this regard are more complex than those of infrared spectroscopy, for example. It is not generally possible to trade off resolution for intensity by opening the slits, because the problems of high backgrounds may make it impossible to obtain spectra at the wider slit widths.

There are several design features of a smaller instrument which would be of convenience to many workers, however. These include preprinted charts (using either a strip chart or X-Y recorder), easy interchange of samples, and compatibility with infrared charts. The latter feature has been incorporated in the Perkin-Elmer LR-3, which features charts the same size as those from Perkin-Elmer 21 series infrared instruments.

The automation of Raman instruments is an important area in which the first advances are just now being made. It is, in principle, rather easy to automate a Raman instrument, especially if photon counting amplification is being used. This is a digital output which can be punched almost directly for data collection. To digitally record frequency information a stepping motor can be used as the scanning motor, and the steps either counted from a predetermined reference point or absolutely defined by shaft encoder. The Spex Model 1401 monochromator and the Spectra-Physics instrument already have stepping motors built in as the scanning motors of the instrument. A SloSyn motor made by the Superior Electric Co. and a Responsyn motor made by the United Shoe Co. are suitable for this application.

The problems of collection of frequency and intensity information are thus not difficult ones for Raman. There remains the automation of the scanning of the spectrometer for such things as multiple scanning, solvent spectrum subtraction, radiometric corrections, depolarization ratios, and other signal averaging methods. At least part of this can be done with only slight modification of any one of a number of systems designed for automation of X-ray diffraction equipment. In this application one wants to step through a small angle of rotation and punch position and counts. This is virtually the same requirement as that of Raman. Automation systems of this sort are currently being made by such companies as Canberra Industries, Digital Automation Co., and Ortec, Inc.

We have generally refrained from giving detailed prices of instruments in this article because they have been changing very

Chemical Instrumentation

rapidly (both up and down, strangely enough) and because there are often many component selections to be made by an individual user. Prices may fluctuate greatly even for the same model instrument.

Many useful references on Raman spectroscopy can be found in the biennial review in *Analytical Chemistry*, and these generally cover theory, applications, and instrumentation. There also exists a Raman Newsletter, which is sent free to those who contribute to it regularly. Further information can be obtained from Miss P. Wakeling, Raman Newsletter, 1613 Nineteenth Street, N.W., Washington, D. C. 20009.

Raman Spectroscopy
Supplement

SMALL RAMAN SPECTROMETERS

Many manufacturers have attempted, with some success, to combine the principles of Raman instrumentation described for research instruments to produce a small, tabletop, Raman spectrometer. Several such instruments are now on the market, and these will be discussed briefly.

The *Spectra-Physics* model 700 spectrometer is one of the first attempts in this direction. The back-to-back double monochromator system featuring two 0.4-m Ebert monochromators is additive in dispersion. A particular feature of this instrument is the logic circuitry which automatically selects the time constant for a scan after a slit program (1, 2, 4, or 8 cm^{-1}) and scan-speed have been chosen. The instrument contains both DC and photon-counting electronics, and automatically switches to DC at high signal levels. Photomultiplier tube (ITT FW-130, thermoelectrically cooled) voltage is preset at the factory for optimum signal-to-noise ratio. An X-Y recorder is used, rather than the strip-chart recorder which is employed in all the larger instruments discussed above. The instrument scanning motor is a stepping motor, which is useful for digitizing the frequency shift information; however, no provision has yet been made for doing this as part of the instrument.

The *Spex Industries* Ramalab is, in many ways, a smaller version of the larger model 1401 Ramalog described previously. The spectrometer employs two 0.5-m Czerny-Turner monochromators, a strip chart recorder which is linked to the scanning drive, and DC amplification. A stepping motor drive offers the possibility for digital output of the data, but in this case there is no photon counting equipment so the DC signal would also have to be digitized. The slits are linked so that one control sets all three slits to one of four possible spectral slit widths. The laser is mounted outside the spectrometer cabinet, making it possible for any laser to be selected when the instrument is purchased.

The *Engis Equipment Co.* is offering a very reasonably priced small Raman package which uses Hilger and Watts model D330/331 double monochromator. In this instrument, two 0.3-m Czerny-Turner monochromators are mounted in a back-to-back arrangement, which differs from the Spectra-Physics system in that both gratings are mounted on the same block, rather than linking them in a parallelogram scheme. The instrument uses a low power argon ion laser, and either synchronous detection or photon counting amplification with the photomultiplier tube. It should be mentioned that this system, using as it does a monochromator built for electronic spectroscopy, has a linear wavelength rather than linear wave number scanning drive. All of the other instruments discussed in this article are linear in wavenumber, which saves some trouble in computing the positions of Raman lines.

A small instrument was recently shown in the United States by the *Japan Electron Optics Laboratory Co.*, the JRS-C1. The instrument is similar to the others described above, using two 0.25-m Czerny-Turner monochromators back-to-back, with the gratings mounted on a single bar whose rotation scans the spectrum. A helium-neon laser is mounted inside the instrument, and DC amplification is used (photon counting optional). An attempt has been made to simplify controls, but this has not been done to the extent that is described for some of the other small instruments.

The *Jarrell-Ash* model 500 uses a stacked double Ebert (0.5-m) monochromator system, in what is billed as an all digital, laser-excited, Raman instrument. One control adjusts all three slits, but these are continuously variable over a wide range. Photon counting electronics is used exclusively. The system, however, is considerably faster than the usual analog ratemeters, as it counts up to 1.6 million pulses/sec. A stepping-motor drive is used, and the instrument is set up for digital readout of the frequency shift information. A digital X-Y recorder is used, in connection with a pulse-integrating system of variable time constant. Logic circuitry similar to that used in the Spectra-Physics instrument is employed, with warning lights that indicate if an incorrect choice of operating conditions is made. The instrument usually is supplied with a Coherent Radiation model 54 argon, krypton, or mixed gas ion laser.

The appearance of a large number of tabletop spectrometers seems to be a sign that instrument manufacturers see a future for Raman spectroscopy outside the research lab of the molecular spectroscopist and inside the analytical laboratory. For this to develop, however, a great deal more work must be done on group frequencies, structural changes giving rise to small frequency shifts, and sample handling. In addition, the prices of many of the small spectrometers are rather close to those of the larger instruments described earlier, and these must come down to the $10,000–15,000 range before many instruments will be sold. Infrared instruments are too strong competitors at the moment.

Developments in laser technology will help in the spread of the small Raman instruments. Several laser prices have been decreasing, and the mixed gas (argon-krypton) lasers have become more reliable. Furthermore, the helium-cadmium laser offers the possibility of excitation at 441.6 or 325.0 nm with reasonable power levels at low prices. This laser is now being marketed by Spectra-Physics Co. and RCA. The UV excitation should permit many resonance Raman effect studies not previously accessible.

PULSED LASER RAMAN SPECTROMETER

Early attempts at Raman spectroscopy using lasers employed the pulsed ruby lasers almost exclusively. As continuous gas lasers became available, these displaced the pulsed lasers, and no commercial instruments were made using the pulsed ruby lasers.

Recently, *Block Engineering*, Inc., announced a Raman spectrometer system employing a pulsed nitrogen laser, which operates with high power pulses (up to 130 kW) at 337.1 nm in the ultraviolet. The pulses are of about 5 nsec duration, with a capability of about 1000 pulses/sec. The photon counting system is linked to the pulses, so that counting is only done at the proper time. The instrument uses a Spex model 1401 double monochromator with an RCA gallium phosphide coated photomultiplier. This instrument is particularly useful in the area of gas phase remote Raman spectroscopy. Laser pulses sent out into the atmosphere are sufficiently intense that Raman signals from the light returning at 180° can be detected. The instrument, using the UV excitation, has also found application in resonance Raman effect studies in solution.

RAPID SCANNING RAMAN SPECTROSCOPY

Delhaye (6) has modified the *CODERG* Raman spectrometer to obtain spectra over very short time periods. This might be extremely useful in chemical reaction studies, and several examples along this direction have been presented. Two methods were used. In the more conventional modification, a rapid scanning device was used to get spectra over a 950 cm^{-1} frequency range in time periods on the order of 0.25 sec. Repetitive scanning techniques were used to accumulate spectra.

Chemical Instrumentation

More promising is the use of image intensifier tubes. Such a tube can be viewed as a cross between a photographic plate and a photomultiplier tube. The quantum efficiencies of the photomultiplier are maintained, but there is no loss of information due to scanning. Good spectra are obtained in millisecond time periods, using either continuous or pulsed lasers. Image intensifiers should prove extremely useful for Raman studies in combination with rapid infrared techniques using interferometric methods.

REFERENCES

(1) See, for example, G. Herzberg, *Infrared and Raman Spectra of Polyatomic Molecules* (D. Van Nosstrand, Princeton, 1945); R. P. Bauman, *Absorption Spectroscopy* John Wiley & Sons, New York, 1962).

(2) Christiansen, R. L., and Potter, R. J., *Appl. Optics*, **2**, 1049 (1963).

(3) Landon, D. O., *Appl. Optics*, **6**, 346 (1967).

(4) Nakamura, J. K., and Schwartz, S. E., *Appl. Optics*, **7**, 1073 (1968).

(5) Landon, D. O. and Freeman, S. K., *Anal. Chem.*, **41**, 398 (1969).

(6) Delhaye, M., *Molecular Spectroscopy*, Edited by P. Heppie, Institute of Petroleum (London) 1968.

The octogram, an eight-pointed star drawn with one stroke of the pen.

Topics in...

Chemical Instrumentation

Edited by **GALEN W. EWING**, Seton Hall University, So. Orange, N. J. 07079

The Journal of Chemical Education
Vol. 46, page A9
January, 1969

... a feature

XLII. Infrared Equipment for Teaching

PAUL A. WILKS, JR., President, Wilks Scientific Corporation, South Norwalk, Connecticut 06856

THE VALUE OF INFRARED ANALYSIS

After the original work of W.W. Coblentz at the Carnegie Institution of Washington (1), the development of infrared spectroscopy became a university pursuit. During the late thirties and early forties, infrared spectra were laboriously hand plotted at Michigan, Johns Hopkins, Princeton, and elsewhere. It took the impetus of World War II and the synthetic rubber program to bring the new analytical technique into the industrial laboratory. Since that time, infrared spectroscopy has proliferated until there are in the neighborhood of 20,000 instruments in use today in this country. Although one of the oldest instrumental techniques, it is estimated that more money is still spent annually on infrared equipment and accessories than any other category of instruments (2).

There are three reasons for the continued popularity of infrared analysis: First, the basic information contained in the infrared spectrum provides more clues to the nature of the molecule being examined than that provided by any other analytical technique. High resolution mass and nuclear magnetic spectra may approach infrared in their ability to characterize a molecule but the complexity of the IR spectrum and the shear volume of infrared data in the literature make it unlikely that it will ever be surpassed as a source of qualitative information.

Second, unlike any other spectrographic technique, it is possible to obtain an infrared spectrum on practically any sample, no matter what its phase or condition. Liquids, solutions, gases, solid materials, powders, coatings and finishes, films—any sample that can be gotten into the instrument sampling space—and some that can't—can be made to produce an infrared spectrum. NMR and UV for the most part require liquid samples, while to obtain a mass spectrum the sample must be vaporized into a vacuum. No other

analytical method approaches infrared in versatility.

Finally, the cost of obtaining infrared data has steadily declined both relatively and actually. The recent introduction of instruments in the $3000 range means that it now costs no more to establish an infrared laboratory than one for gas chromatography. Similar reductions in accessory prices bring infrared analysis within the budget of nearly every institution in which chemistry is taught.

THE ROLE OF INFRARED IN TEACHING

Having established the value of infrared analysis to chemistry in general, what should be its role in education and how should the educational institution equip itself to fill this role?

Originally, of course, universities seized upon infrared spectroscopy as a tool of research. Increasingly complex instruments were built since each step forward in resolving power seemed to provide large increases in molecular information. Because the early instruments required close temperature control and freedom from excessive vibration, and made use of exotic, moisture sensitive materials, infrared laboratories were set up in air conditioned areas (when such were an extreme rarity) and often in remote basement locations. Well known examples are "Bryce Canyon" at the University of Minnesota and the spectroscopy laboratory at MIT. As infrared spread into industry the same sort of pattern was followed. The result was that, unlike ultraviolet spectroscopy, which went immediately into the hands of the bench chemist, infrared analysis became the province of the specialist. He demanded instruments with increasingly higher performance and versatility. In addition he built a certain mystique around the technique which protected his domain from mere chemists who "more often than not would draw completely wrong conclusions if left on their own with

Born in Springfield, Mass. (1923) and a graduate of Harvard College, **Paul A. Wilks, Jr.** has been associated with the analytical instrument field, particularly in the areas of Infrared Spectroscopy and Gas Chromatography since joining the Perkin-Elmer Corporation in 1944. He participated in many phases of the company's instrument development program. He was Director of Marketing when he left Perkin-Elmer in 1958 to become a co-founder of Connecticut Instrument Corporation.

While with CIC, he was responsible for the commercial introduction of several new infrared sampling techniques, including equipment for attenuated total reflection. When CIC merged with Barnes Engineering Company in 1961, he became General Manager of the Commercial Instrument Division. In 1963 he left Barnes to form Wilks Scientific Corporation, which company continues the development of new techniques and equipment for spectroscopy and chromatography.

Mr. Wilks has published a number of papers on Infrared Instrumentation and applications, and has been the editor of several company publications. He has had two patents issued and has several pending in the instrumentation field.

an infrared spectrometer!"

The instrument companies, recognizing the great untapped market that existed for infrared analysis began to break down the walls of the infrared domain through the introduction of a new class of instruments that were lower in price and operated with more or less fixed parameters so that the mere organic chemist could not foul up his spectra too badly through improper instrument settings. Though introduced with some trepidation, the new instruments quickly proved themselves invaluable tools in the control laboratory and for routine research applications. As

Table I. Infrared Spectrophotometers*

Manufacturer/ model/distributor	Principle	Range (cm⁻¹)	Chart presentation	Scanning speeds	Adjustable slit program	Ordinate scale expansion	Sizes (inches) length × width × height	Base price
Research Instruments								
Block FTS-14 Dunn Division	Fourier transform interferometer	4,000–250	Flow chart	1 sec–several hours	Yes	Yes	(Three unit console)	$65,000
Beckman IR-11	Filter grating optical null $f/10$	800–33	Flat bed Flow chart	9 min 3 weeks	Yes	25×, built-in	60 × 38 × 17	36,000
Perkin-Elmer 180	Filter grating $f/6$ ratio recording	4,000–180	Flat bed Flow chart and X Y	7 min 8333 hr	Yes	20×, built-in	Optical bench 58 × 30 × 31 Recorder console 42 × 30 × 40	33,000
Perkin-Elmer F1S-3	Filter grating optical null $f/5$	400–30	Flat bed Flow chart	7½ min 64 hr	Yes	4×, built-in	54 × 42 × 49	32,500
Perkin-Elmer 325†	Prism-filter grating $f/5$ optical null	5,000–200	Drum	12 min 140 hr	Yes	Accessory with external recorder	55 × 29 × 16	31,000
Beckman IR-12	Filter grating optical null $f/10$	4,000–200	Flat bed Flow chart	5 min 5 days	Yes	25×, built-in	60 × 38 × 17	23,800
Perkin-Elmer 621	Filter grating $f/6$ optical null	4,000–200	Drum	.12 cm⁻¹/min 60 cm⁻¹/sec	Yes	20×, built-in	40 × 20 × 22	23,250
Beckman IR-9D	Double monochromator prism grating optical null $f/10$	4,000–400	Flat bed Flow chart	5 min 5 days	Yes	25×, built-in	60 × 38 × 17	21,900
Perkin-Elmer 521	Filter grating optical null $f/4.5$	4,000–250	Drum	.44 cm⁻¹/min 150 cm⁻¹/sec	Yes	20×, built-in	40 × 20 × 22	22,250
Beckman IR-4	Double monochromator prism-grating optical null $f/10$	10,000–615	Flat bed Flow chart	3 min 5 hr	Yes	25×, built-in	60 × 38 × 17	17,930
Perkin-Elmer 221	Prism optical null $f/4.5$	10,000–650	Drum	2 min 320 min	Yes	20×, built-in	40 × 20 × 22	17,050
Medium Priced—General Purpose Instruments								
Beckman IR-20A	Filter grating optical null $f/10$	4,000–250	Flat bed Flow chart	5 min 40 hr	Yes	Accessory external recorder	36 × 23 × 16	8,950
Perkin-Elmer 457	Filter grating optical null $f/5$	4,000–250	Flat bed Flow chart	6, 14, 55 min	Yes	Accessory external recorder	28 × 18 × 15	8,900
Unicam SP1200A‡ Phillips	Filter grating optical null $f/6$	4,000–400	Flat bed Flow chart	5, 10, 30 min	Yes	Automatic built-in	36 × 22 × 16	8,700
Beckman IR18A	Filter grating optical null $f/10$	4,000–600	Flat bed Flow chart	4 min 30 hr	Yes	Accessory external recorder	36 × 23 × 16	7,900
Unicam SP 1200‡ Phillips	Filter grating optical null $f/6$	4,000–400	Flat bed Flow chart	5, 10, 30 min	Yes	Accessory external recorder	36 × 22 × 16	7,900
Unicam SP200GA‡ Phillips	Filter grating optical null $f/10$	4,000–650	Flat bed 2 sheets/scan	5, 10, 30 min	Yes	Automatic built-in	36 × 22 × 16	7,700
Perkin-Elmer 257	Filter grating optical null $f/5$	4,000–625	Flat bed Flow chart	5, 12, 48 min	Yes	Accessory external recorder	28 × 18 × 15	7,750
Hilger & Watts‡ H-900 Wilks	Prism grating double monochromator optical null $f/6$	4,000–650	Flat bed X Y 3 chart sizes	4, 8, 16 min	Yes	5×, built-in	45 × 18 × 19	6,900
Unicam SP200G‡	Filter grating optical null $f/6$	4,000–650	Flat bed 2 sheets/scan	5, 10, 30 min	Yes	Accessory external recorder	36 × 22 × 16	6,850
Shimadzu 270-IR§ Bausch & Lomb	Filter grating optical null $f/6$	4,000–400	Drum 3 sizes	10, 20, 45, 90 min	Yes	2×	38 × 20 × 16	6,800
Perkin-Elmer 247§	Filter grating optical null $f/5$	4,000–650	Drum	1.5–40 min	No	No	38 × 20 × 16	6,600
Perkin-Elmer 157‡	Prism optical null $f/5$	4,000–650	Flow chart	1–5 min	Yes	Accessory	28 × 18 × 15	6,450
Hilger & Watts‡ H 901 Wilks	Prism grating double monochromator optical null $f/6$	4,000–650	Flat bed X Y	8 min	Yes	Accessory	45 × 18 × 19	6,200
Unicam SP200‡	Prism optical null $f/6$	5,000–650	Flat bed	5, 30 min	Yes	Accessory external recorder	36 × 22 × 16	5,750
Perkin-Elmer 137B	Prism optical null $f/5$	4,000–650	Drum	3, 12 min	Yes	Accessory external recorder	28 × 18 × 21	5,750
Low Cost Instruments								
Unicam SP100‡	Filter grating optical null $f/4.5$	4,000–625	Flat bed	2, 5 min	No	None	25 × 23 × 12	4,500
Hilger & Watts‡ H-1200 Wilks	Filter grating optical null $f/6$	4,000–650	Flat bed 2 Sizes Sheet or flow chart	1.5 to 20 min	No	None	27½ × 17½ × 8½	3,500
Beckman IR33	Filter grating optical null	4,000–600	Flow chart 2 sizes	2.5, 7.5, 22 min	No	None	26 × 24 × 11	3,300
Perkin-Elmer 700	Filter grating optical null $f/6$	4,000–650	Flat bed	2.5, 8 min	No	None	29 × 19 × 10	3,150

* Note: This table has been revised according to manufacturer's and distributor's list prices in effect, March 1, 1970.
† Made in Germany.
‡ Made in England.
§ Made in Japan.

Chemical Instrumentation

their use spread, the infrared specialist found his position enhanced rather than reduced in stature, since he was called upon to set up analyses and interpret results on a much broader basis than before.

To the academic world, the new class of instruments meant that infrared analysis dropped from the sole realm of faculty members or graduate students engaged in research, to the upper undergraduate level at least. Now that $3000 instruments are available it will not be long before any student taking organic chemistry, from high school on, will expect to have ready access to infrared analytical equipment.

Thus, infrared spectroscopy fulfills a dual role in the educational institution. It is still a basic research instrument, but more and more it is becoming as fundamental a tool in the practice of chemistry as the pH-meter or analytical balance. Since, today, practically every industrial chemist has more or less ready access to an infrared spectrometer, the graduating chemist must be able to use the instrument and be able to make general interpretations of the results. In addition, he must also be able to use the accessory equipment available in industry. Quite naturally, the equipment best suited for each of infrared's roles is quite different.

CHOOSING AN INFRARED SPECTROMETER

Today there are some thirty different models of general purpose infrared spectrophotometers marketed in the United States by five different manufacturers. They fall naturally into three categories (and price ranges!).

At the top of the list are the "research" type instruments. They are characterized by large size, extremely complex controls with resulting versatility. Charts are generally large to take advantage of photometric accuracy and resolution. Prices are in the twenty thousand dollar range. Though well suited for fundamental research in structural analysis, the preparation of standard infrared spectra, and where extreme quantitative accuracy is required, they are not as well adapted to the running of large quantities of general purpose spectra as the generally faster and more compact medium priced instruments. The instruments in this category should

not be purchased solely for their value as status symbols.

Next come the medium priced instruments that have become the "work horses" of the infrared laboratory. Originally conceived as low priced instruments for the non-spectroscopist, their overall performance and versatility have improved to the extent that they can now handle at least 80% of the infrared work load. The medium priced instruments use smaller charts, generally can record routine spectra in less than ten minutes and, with fewer controls, are much easier to operate properly than their larger brothers. Capable of some research use and able to fulfill all routine educational requirements, they are all the instrument required by the great majority of infrared laboratories.

At the bottom of the price scale are the recently introduced "routine" or "educational" instruments. In spite of their low prices, they are still able to turn out remarkably good infrared data—some of grating quality. And, since they have no adjustable controls at all, it is virtually impossible to operate them incorrectly (although correct sample preparation is just as important as in the other categories.) These low-priced instruments (about $3000) will be widely used in colleges and even high schools. A possible drawback is that the low-priced models may not stand up to the rugged use of the classroom as well as the better made medium priced models.

A further study of available instruments reveals that there are not thirty completely different instruments, but rather four families whose members differ in a few characteristics only, plus several individual instruments of foreign manufacture. The four families (and the initial model in each) are the research Beckman (IR-4), the research Perkin-Elmer (Model 21), the mid-range Beckman (IR-5) and the mid-range Perkin-Elmer (Model 137). All currently available models are shown in Table I.

In addition to the general purpose instruments there are a number of specialized spectrometers available. Although probably of most interest to industry, some of them may find their way into universities as equipment for particular research projects. The special purpose models are described in Table II.

With so many makes and models on the market, the choice of just the right equip-

ment for a given educational situation is a difficult one. Even when funds are relatively unlimited, the automatic purchase of a large research instrument may not be the best move—for the same number of dollars two or three mid-range instruments may be purchased that can produce five times as many spectra as the big one and still solve all but the most demanding problems.

It may be of interest at this point to note Table III which shows the plans to buy by instrument category for 1968 by the educational field and industry. The source is the Wilks Scientific Infrared Survey made in the spring of 1968. Approximately two thousand returns were received covering nearly 4000 instruments, representing about 20% of the instruments in use in the United States today. Note interest of the educational field in the low-priced models.

As we have indicated, infrared spectrophotometers are currently offered by five manufacturers: Bausch & Lomb, Beckman, Hilger & Watts, Perkin-Elmer and Philips. The entries of B & L, Hilger & Watts, and Philips are imported, the first from Japan and the latter two from England, while some of the P-E units are made by its English and West German subsidiaries. Baird and Cary have withdrawn from the field, the latter probably only temporarily.

Table III. Plans to Buy Infrared Spectrometers by Class Per 100 Instruments Now In Use

	University	Industry
Low cost	6	2
Mid-range	4	5
Research	3	3

CHARACTERISTICS OF IR SPECTROMETERS

Most manufacturers of infrared instruments are large, financially sound companies with a nationwide network of sales and service offices. All recognize the importance of the infrared field and are making a serious and sustained effort to secure or maintain a position in it. Hence, the prospective purchaser can select an instrument with confidence in the manufacturer.

PRINCIPLE

All general-purpose instruments now being manufactured are of "Double-beam, Optical Null" construction. What is recorded is the difference in energy between the two beams versus wavelength or frequency. Such a system, when properly aligned, can eliminate from the recorder such unwanted information as atmospheric absorption bands, variation in energy due to source fluctuation, most slit effects and any other energy-affecting factor common to both beams.

The optical null servo system drives an energy varying wedge into the reference beam to compensate for energy absorption in the sample beam caused by the sample. The amount of movement of the wedge required to bal-

Table II. Special Purpose Instruments

			Approximate Price
Wilks Model 8	Internal reflectance (ATR)	Solid sampling, reflectance, surface studies	$8–10,000
Wilks Models 111 and 116	GC-IR	The analysis of gas chromatographic fractions	$9–20,000
Warner & Swasey Model 501	Rapid scan	Kinetic studies, GC-IR analyses	$20,000+
Block Engineering	Interferometer	Surface studies, emission GC-IR analyses	$30,000
Beckman FS-720*	Interferometer	10–500 cm^{-1}	$7450
Beckman LR-100*	Interferometer	3–70 cm^{-1}	$10,500
P-E Model 012C (1120-12-G, etc.)	Single beam, building block construction	Physical measurements, special studies of all kinds	$12,500
Wilks Model 44	ATR goniometer	Determination of optical constants, precise reflectance measurements	$20,000

* Manufactured in England.

Chemical Instrumentation

ance or "null" the two beams is recorded on the recorder as the absorption of the sample. In such a system linearity is a function of the mechanical accuracy of the wedge which tends to remain constant, putting no burden on linearity of the electronics which may change with time.

In the "ratio recording" system, which has been used occasionally, the difference is measured electronically and recorded. Though more difficult to manufacture and maintain, this system offers considerably more flexibility in special adjustments such as zero suppression, electronic attenuation, scale expansion, etc.

PERFORMANCE CRITERIA

The three basic criteria of instrument performance are resolution, photometric accuracy and wavelength accuracy. To a certain extent these characteristics are interrelated.

Resolution depends primarily upon the dispersion of the monochromator. Originally, prism and double prism systems were used exclusively. Nearly all the instruments currently in production, however, make use of gratings or a combination of prism and grating. The reason of course is the far greater resolution possible with properly designed grating systems.

Though the dispersion of a grating is large, a complicating factor is the presence of several overlapping spectra or "orders" when the optical beam is reflected from the grating surface. This means that several wavelengths may be present at any given angle and some method must be used to get rid of the unwanted spectral energy. Two common methods are to use a foreprism so that relatively monochromatic radiation strikes the grating, or a series of broadband filters which accomplish the same result. The foreprism system is more complex optically, but generally provides better results in terms of spectral purity and freedom from scattered radiation.

The most obvious advantage of a high resolution system is its ability to provide more spectral detail. A less obvious but equally important result of greater resolving power is a potential increase in quantitative accuracy. As Potts and Smith (3) point out, a high resolution is necessary to precisely locate the bottom of a narrow absorption band. The more accurately this point can be determined, the more accurate will be the quantitative measurement.

A further advantage of the high resolution grating system is its ability to produce a greater energy through-put for a given spectral slit width. The energy passed by the slits increases as the square of the slit opening, while resolution decreases proportionally. A grating instrument with the slits opened to give the same resolution as a prism instrument will have a far greater signal-to-noise ratio. This factor can be important when accessories such as beam condensers or ATR attachments which result in high energy losses are used. (See also the discussion of f-number below.)

Photometric accuracy also depends on a number of factors among which are those which contribute to a high signal to noise ratio (detector sensitivity, amplifier quality, total energy through-put, etc.) and the linearity of the mechanical wedge in a null balance system or of the electronics in a ratio system.

The accuracy with which the location of absorption bands can be determined is primarily the result of the instrument manufacturer's skill in designing cams that relate prism and grating angle to the recorder drum position. Here again the use of a grating simplifies the situation to a certain extent since its dispersion can be converted to linear wavelength or frequency recording by a simpler mathematical relationship than a prism.

These criteria of instrument performance are not independent however, and the infrared spectrometer user should thoroughly understand their interrelationship. For example, to increase the resolution means narrower slits and hence less energy. Less energy requires more filtering in order to preserve the signal-to-noise ratio which in turn reduces the permissible recording speed in order that the more slowly responding pen can follow the increased detail of the spectrum and record true band positions.

In the larger research instruments all possible instrument controls are provided so that the operator can set conditions precisely to obtain the desired results. Likewise, the inexperienced operator can obtain completely fallacious records with improper settings so that considerable education is in order before turning students loose on the large spectrophotometer. On the smaller instrument, the manufacturer has set many of the operating parameters permanently, leaving less margin for error but also reducing overall flexibility.

OPTICAL SPEED

Another design characteristic of the infrared spectrometer that is of considerable importance to the user is the "f-number" or optical speed. This is the ratio of focal length to mirror diameter. The lower the f-number, the greater the total energy flux through the spectrophotometer and the greater the energy loss that can be sustained before the servo system becomes inoperative. Of course, other factors such as detector sensitivity, slit opening, etc. also contribute, but the f-number is inherent in the design and cannot be changed. A low f-number puts a greater demand on optical quality and alignment and also makes the design of focusing accessories somewhat more critical.

WAVELENGTH RANGE

This feature is closely tied to chart presentation since even though a long transmission range may be desirable, if it is all condensed in the last inch of chart, it may not be very valuable. Sometimes the extra range requires somewhat curious breaks in the chart, making for unnatural looking spectra. Further, most libraries of standard spectra cover the rock salt range only (2.5 to 15μ or 5000 to 625cm^{-1}) since the bulk of the characteristic absorption bands occur in this region. Thus, the so-called mid-range instruments are quite

adequate for nearly all work. However, more and more work is being done with longer range instruments and again if financial considerations are not important, the relatively high cost of the few extra wavenumbers of transmission may be worthwhile.

Incidentally, a strong trend exists toware linear wavenumber rather than linear wavelength instruments (despite their less efficient use of chart space!) so that the former presentation should be chosen unless the laboratory is already equipped with wavelength instruments.

CHART PRESENTATION

Here is where the instrument salesman and personal preference have a field day. There are nearly as many different types of recorders as there are instruments:

Drum recorders are somewhat awkward to handle and often produce a crick in the neck as one observes the spectrum being produced. Yet when the chart is laid on a table these minor problems disappear. The small drum instruments (Figures 1 and 2) record on an 8$^1/_2$ × 11 in. chart which is excellent for filing and even where

Figure 1. Perkin-Elmer Model 237-B Grating Infrared Spectrophotometer.

Figure 2. Bausch & Lomb/Shimadzu Spectronic 270-IR Infrared Spectrophotometer.

two charts are required for the complete spectrum, this becomes a minor drawback when it is recognized that more often than not only the fingerprint range need be run (5.5 to 15 μ). Here, however, watch out for awkward spectral breaks.

Flat bed recorders (Figures 3 and 4) are certainly more comfortable to use but they too have their problems. They are a little more complex mechanically, often require more bench space and on some instruments annoying breaks appear on the record. We've spilled many a sample on a freshly recorded spectrum on our flat bed instrument with its chart spread out just in front of the sample space! In some of the more recent instruments the flat recorder is mounted on the back part of the instrument avoiding such problems.

A few instruments offer a choice of chart sizes which can be quite convenient. A large spectrum can be run for precise

Chemical Instrumentation

Figure 3. Beckman Instruments Model IR-20 Infrared Spectrophotometer.

Figure 4. Unicam Instruments Model SP-200 Prism Infrared Spectrophotometer.

quantitative studies and a small one for filing, etc.

ADJUSTABLE SLIT PROGRAM

This is a highly desirable feature present on all the research instruments and on the recently introduced moderately priced models. Energy increases as the square of the slit opening, while resolution goes down proportionally. Hence, a small increase in slits makes up for a large loss in energy (as in ATR studies) with a small change in resolution. Again, unless cost is a problem, only instruments with variable slit programs should be selected.

SCALE EXPANSION

This is a built-in feature of all research instruments and an accessory item on all moderately priced instruments (except the Hilger & Watts Infrascan with built-in scale expansion, no longer available in the U.S.). It can be a very useful feature in trace component measurements (as in GC-IR studies). However, it need be purchased only when required.

SIZE

Here the family resemblance is readily apparent and there is a definite difference between the larger research instruments and the smaller low-priced models. Not only can two or more moderately priced instruments be purchased for the cost of one research model, but they together may occupy no more bench space!

COST

The old adage that you generally get what you pay for holds true with IR instruments as elsewhere. However, several thousand dollars may be laid out for a strip chart recorder or extra transmission range with no fundamental increase in performance. Thus, last year's model which, with some dickering, may some-

times be had at a discount may represent an excellent choice. Also, demonstration instruments, when available, are good bets since their running-in will have uncovered any weak components.

INFRARED ACCESSORIES

To achieve the full capability of the infrared spectrophotometer, a broad range of accessories is necessary. That industry is much more aware of this fact is readily apparent from Table IV. Note, for example, that there are 26 ATR units per 100 industrial IR spectrometers, while there are only 6 units per 100 educational instruments. Similar differences exist in other categories. Apparently the use of IR equipment in industry is quite different from its use in education—a fact that is worthy of investigation by chemistry teachers.

Fortunately for institutions having multiple installations, all of the instruments available in the United States have a more or less standard sampling compartment. They have a 2 × 3 in. cell slide and at least 10 cm of clear sample space. For the most part, infrared accessories are completely interchangeable making it quite practical for a laboratory to mix makes, models and categories.

Table IV. Summary of Accessory Use By Use Category

Accessory	Quantity per 100 spectrometers	
	University	Industry
Heated or cooled cells	9	10
Specular reflectance	2	7
IR polarizer	2	5
Pyrolyzer	1	5
GC-IR attachments	2	8
Long path gas cells	10	14
Micro cells	11	22
ATR attachments	6	26

In order to properly equip an instrument for multiple sampling problems about 20% of the initial purchase price will usually be put into accessory equipment. And 10 to 15% per year should be budgeted for accessory maintenance and improvement. Standard accessory prices have steadily decreased in the past few years and a whole group of standard items have been developed specifically for student use. The academic purchaser should investigate the products of the companies that specialize in such accessories rather than automatically buying all equipment from the instrument manufacturer.

Typically an instrument will require an assortment of sealed and demountable cells for liquids and solutions; a gas cell; KBr pellet equipment and an ATR attachment for solids; and supplies of paper, ink, pens, etc. Special sampling cells such as hot and cold cells, specular reflectance attachments, long path gas cells and the like can be purchased as required.

SPECIAL PURPOSE INSTRUMENTS

There are so many diverse applications for infrared analysis that it is inevitable that special purpose instruments will begin

to be available with improved performance over general purpose spectrophotometers equipped with an accessory (Figure 5). So far, these have generally been associated with special sample handling problems such as GC-IR combinations and attenuated total reflection measurements.

Figure 5. Wilks Scientific Model 116, a modification of the Perkin-Elmer Model 221, shown with the cover removed from the gas-chromatographic sampling system.

Of the special purpose instruments, perhaps of most interest to the academic world are the ATR goniometer, interferometer and rapid scan spectrometers, since these offer the best possibilities for research projects.

The ATR goniometer is designed for extremely precise determination of absorption versus angle of incidence in internal reflection spectroscopy and for other reflection measurements. Crawford and co-workers (4) at the University of Minnesota have demonstrated the value of the ATR approach in determining absolute intensities and indices of refraction with a high degree of accuracy.

The interferometric spectrometer (Figure 6) is unique in that among its special features is the fact that multiple scans can be made over the same spectral region and the information fed into a computer (5). In this manner, weak signals can often be brought out of the background noise level. Among the applications are infrared emission measurements, GC-IR analysis and fundamental studies at extremely long wavelengths.

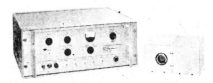

Figure 6. Block Engineering Model 196 Interferometer Spectrometer.

The rapid scan instruments provide IR spectra in milliseconds. Their applications are kinetic studies, explosion monitoring and rapid scan GC-IR analysis.

FUTURE DEVELOPMENTS

There is no question but what infrared spectroscopy will continue to be at the right hand of the chemist for many years to come. In industry the trend is toward greater automation of sample handling and computer readout. In the academic world, the need is for lower cost and simplified equipment on the one hand and more accurate instruments covering a broader range on the other.

Like many other instrumental techniques and perhaps most important of all, a first hand knowledge of infrared spectroscopy is a must for every chemistry student. And, thanks to the new low-cost

Chemical Instrumentation

instruments, infrared is now within the reach of all institutions teaching college level chemistry.

Supplement

Table 1 has been updated to May, 1970, deleting several discontinued models (including one or two mentioned in the text), and adding new. It should be pointed out that one research-grade spectrophotometer, the Perkin-Elmer 180, is now available with ratio-recording rather than optical null.

Manufacturers of Infrared Spectrophotometers

Acton Laboratories, Inc.
  531 Main St., Acton, Mass. 01720
Beckman Instruments, Inc.
  2500 Harbor Blvd., Fullerton, Calif. 92634
Block Engineering, Inc.
  19 Blackstone St., Cambridge, Mass. 02139
Hilger & Watts, Ltd. (England)
  via Engis Equipment Co., 8035 Austin Ave., Morton Grove, Ill. 60053
Perkin-Elmer Corp.
  Main Ave., Norwalk, Conn. 06825
Shimadzu-Seisakusho, Ltd. (Japan)
  via Bausch & Lomb, Inc., 77466 Bausch St., Rochester, N. Y. 14602
Unicam Instruments, Ltd. (England)
  via Philips Electronic Instruments, 750 S. Fulton Ave., Mount Vernon, N. Y. 10550

Warner & Swasey Co.
  32-16 Downing St., Flushing, N. Y. 11354
Wilks Scientific Corp.
  140 Water St., South Norwalk, Conn. 06856

Literature Cited

(1) COBLENTZ, W. W., "Investigations of Infrared Spectra," Carnegie Institution of Washington, 1905.
(2) "1967–69 Spectroscopy Census," (Analytical Chemistry); Reinhold Pub. Corp., N. Y. 1967.
(3) POTTS, W. J., JR., and SMITH, A. L. Applied Optics, 6, 257 (1967).
(4) GILBY, A. C., BURR, J. JR., and CRAWFORD, B. JR., J. Phys. Chem., 70, 1520 (1966).
(5) LOW, M. J. D. THIS JOURNAL, 43, 637 (1966).

This talisman has its origin in the Gnostic conception of the World. It represents the four elements.

Topics in...

Chemical Instrumentation

...a ChemEd *feature*

Edited by **GALEN W. EWING**, Seton Hall University, So. Orange, N. J. 07079

The Journal of Chemical Education
Vol. 47, pages **A163 A255 A349 A415**
March to June, 1970

LI. Fourier Transform Spectrometers

M. J. D. LOW, *Department of Chemistry, New York University, New York, N. Y. 10453*

Dr. Manfred J. D. Low is Associate Professor of Chemistry at New York University. He received his Ph.D. (in physical chemistry) from New York University in 1956 and then worked at Davison Chemical Co., Division of W. R. Grace, and at Texaco, Inc. on fundamental and applied heterogeneous catalysis. He then joined the faculty of the School of Chemistry of Rutgers, The State University and returned to New York University in 1967.

His main research interests lie in the physics and chemistry of adsorption phenomena, with emphasis on infrared studies of surface effects, including measurements of infrared emission spectra of chemisorbed species. The need of highly sensitive instrumentation for infrared emission studies led to work with Fourier Transform spectrometers. The potential of such instruments prompted his interest and work in applying Fourier Transform spectroscopy to a variety of analytical problems.

The choice of the title of this article was prompted by the usual editorial requirements of brevity and conciseness as well as a desire to avoid long, ungainly structures such as "Instrumentation for Recording Spectra by Means of Fourier Transform Methods," or "Spectrometers Using Interferometers to Measure Spectra," and the like. It is unfortunate that the more descriptive titles are so clumsy, because snappier, shorter titles might give the impression that the spectra produced by Fourier Transform spectrometers are different from those produced by the conventional dispersion spectrometers. The final data are identical.

The end results produced by a Fourier Transform spectrometer system and a conventional dispersion spectrometer are the same. Each type of instrument yields a spectrum—a plot of intensity versus frequency. As the ways spectra are used in the various areas of chemistry are well known and widely described, the present article will not consider applications or uses of spectra to any extent. However, there are drastic differences in the way spectra are produced. Entirely different concepts and consequently entirely different hardware are involved, so that a Fourier Transform spectrometer system is thoroughly unconventional with respect to the well-known dispersion instruments, even though conventional spectra in the visible to far-infrared ranges are produced. In considering the current status of spectrometers employing Fourier Transform methods for producing spectra in the visible to far-infrared ranges, and especially in understanding the design features of commercial instrumentation, it is consequently useful to outline the principles of Fourier Transform spectroscopy.

OUTLINE OF FTS

In contrast to the conventional spectrometer, the Fourier Transform spectrometer does not contain a monochromator or dispersing element. In the conventional dispersion spectrometer the incoming polychromatic radiation is separated into bundles of almost monochromatic radiation because the frequencies of the radiation are so high that a detector cannot discriminate among them. The detector can only respond to the intensity of radiation, but cannot discriminate among frequencies in the spectral range in which the detector can operate. The dispersed radiation is consequently swept past a detector in sequence. The detector then produces a simple electrical signal which is proportional to the intensity of the almost monochromatic radiation striking it. The spectrum produced in this fashion is thus the result of a series of radiometric measurements. In contrast, the optical system of the Fourier Transform spectrometer transforms the polychromatic signal in its entirety, so that it may be analyzed for frequency-intensity content without dispersion.

The Fourier Transform method is based on the fact that the signal (termed "interferogram") produced by the detector of a two-beam interferometer and the spectral distribution of the radiation entering the interferometer are Fourier cosine Transforms. Precise descriptions of the processes involved are rather complex, and are given in detail elsewhere (1–3). However, the way interferograms are produced and processed to result in spectra can be readily understood if one considers the simple case of monochromatic radiation or a mixture of just a few wavelengths.

Suppose a beam of monochromatic radiation of wavelength λ enters a Michelson interferometer and is divided into two equal components at a beamsplitter. This is shown schematically in Figure 1 (but only the rays leading to the detector are shown there, and a compensator plate was omitted). The plane, first-surface mirror M_1 is mounted on a suitable carriage and is mechanically or electrically

moved. The second plane mirror is stationary. Each of the two rays is reflected at a mirror and then returned to the beamsplitter, where their amplitudes will add. If the rays r_1 and r_2 arrive at the beamsplitter in phase, they will interfere constructively, and a signal proportional to the sum of their amplitudes will be produced by the detector. The rays r_1 and r_2 will be exactly in phase when the mirrors are equidistant from the beamsplitter. If mirror M_1 is displaced from that position by a distance $\lambda/4$, the path of ray r_1 is changed by a distance $\lambda/2$ so that, on recombination at the beamsplitter, the rays r_1 and r_2 will be 180° out of phase. There will then be destructive interference, and the detector output will be zero. This will happen for all displacements which are precisely odd multiples of $\lambda/4$, or for path differences or retardations of $\pm\lambda/2$, $\pm3\lambda/2$, $\pm5\lambda/2$, and so on. The plus-or-minus signs indicate mirror displacement positions on either side of the zero position. Similarly, constructive interference will occur when the retardation is at even multiples of $\lambda/4$, with partial destructive interference oc-

Chemical Instrumentation

curring at in-between displacements. If M_1 were moved in small even steps, the signal produced by the detector would change incrementally and would approximate the fluctuation of a cosine wave. This is shown schematically by the "stepped" plot in Figure 1. Obviously, the smaller the mirror steps, the smoother will be the detector output, and a continuous cosine wave will be produced if the mirror is moved continuously. This is shown by the lower trace in Figure 1.

Suppose that, instead of being moved stepwise, the mirror M_1 is moved smoothly with a velocity V. As the cosine wave produced by the detector goes through one cycle if M_1 is displaced by a distance $\lambda/2$, the frequency of the detector signal is then,

$$f = V/(\lambda/2) = 2V\nu$$

Or, for a constant mirror velocity, there is a linear relation between the frequency ν in wavenumbers of the incoming mono-

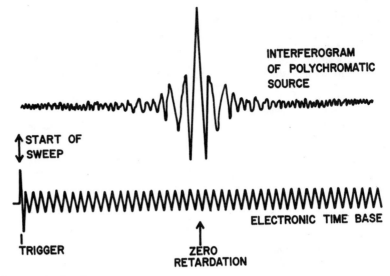

INTERFEROGRAM OF POLYCHROMATIC SOURCE

START OF SWEEP

ELECTRONIC TIME BASE

TRIGGER

ZERO RETARDATION

Figure 2. Interferogram.

mation of all such waves and is a complex signal like that shown in Figure 2. The center peak in the interferogram occurs

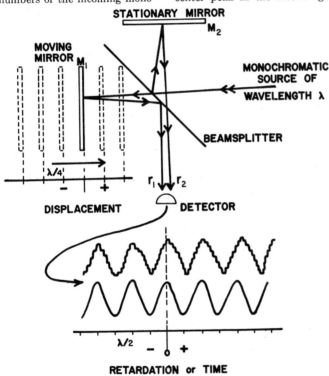

Figure 1. Michelson interferometer.

chromatic radiation and the frequency of the detector signal. For example, with a mirror velocity of 0.5 mm/sec, monochromatic radiation of 10-μ wavelength, equivalent to 1000 cm^{-1} and a frequency of 3×10^{14} Hz, will produce a detector signal of 50 Hz. For 5-μ radiation, $f = 100$ Hz, and so on. The amplitude of the low-frequency signal is proportional to the intensity of the incoming monochromatic radiation.

The extension of this principle to polychromatic radiation follows from the above: each frequency component of polychromatic radiation is made to undergo such a transformation in frequency and produces a detector wave of unique frequency. The signal or interferogram produced by the detector is then the sum-

at zero retardation when the two mirrors M_1 and M_2 are equidistant from the beamsplitter, so that all components reaching the detector have the same phase. The peak amplitude of the interferogram is proportional to the total energy in the incident beam. The peaks of smaller amplitude along each side of the central spike carry intensity and frequency information.

The simple cosine wave can be expressed by,

$$I(X) = I(\nu) \cos 2\pi X\nu \quad (1)$$

where $I(X)$ is the intensity of the output signal as function of the mirror displacement X, and $I(\nu)$ is the intensity of the source as function of optical frequency ν. For polychromatic radiation one must sum all components, and the result is expressed by,

$$I(X) = \int_{-\infty}^{+\infty} I(\nu) \cos(2\pi X\nu) \, d\nu \quad (2)$$

As in eqn. 1, $I(X)$ is the interferogram, and $I(\nu)$ is the spectrum of the source. The complement to eqn. 2 is,

$$I(\nu) = \int_{-\infty}^{+\infty} I(X) \cos(2\pi X\nu) \, dX \quad (3)$$

The Fourier Transform pair, eqns. 2 and 3, provides the relations between the spectrum, $I(\nu)$, and the interferogram, $I(X)$. In practice, interferograms are rarely as symmetrical as that shown in Figure 2 or as implied by eqn. 2. It is difficult to construct interferometers which are perfectly compensated over an extended wavelength range, and imperfect compensation leads to sine as well as cosine components in the interferogram. In general, the interferogram and spectrum are then related by the complex Fourier Transform pair,

$$I(X) = \int_{-\infty}^{+\infty} I(\nu) e^{2\pi i X\nu} \, d\nu \quad (4)$$

$$I(\nu) = \int_{-\infty}^{+\infty} I(X) e^{-2\pi i X\nu} \, dX \quad (5)$$

In general, then, in order to obtain the spectrum of a source, one need only (a) set up a simple optical system like that shown in Figure 1; (b) measure the detector output as a function of mirror displacement; and (c) perform a well-established mathematical operation. There have, of course, been rather severe experimental difficulties, concerned chiefly with defining the mirror position, as well as an enormous problem with the mathematical operation. There is also a number of secondary requirements or "difficulties," many of them changing in severity with the wavelength range which is being examined. For example, the wavelength range which can be examined can be limited by the transparency of the plate on which the beamsplitter material is deposited or by the beamsplitter material itself, the efficiency of the beamsplitter is determined by the nature of the material and how and how well it is deposited, and

Chemical Instrumentation

the planarity of the beamsplitter and compensation plate affects resolution. These and other items, such as the planarity of the mirrors or detector frequency response need not be taken up in detail. However, the two major items concerning the mirror translation and the Fourier Transform should be considered.

Mirror Motion

The detector signal I can be easily measured but, as indicated by the discussion and equations above, I must be known as a function of X, the mirror displacement, in order to make any sense out of the intensity measurement. Or, X has to be known precisely. The precision to which X must be known depends on the spectral range, and it is easy to see that mirror positioning becomes increasingly difficult with decreasing wavelength.

In the far-infrared range, where wavelengths are of the order of hundreds of microns, it is relatively easy to displace the mirror M_1 by fractions of a wavelength and to measure the mirror position (or displacement from zero) with high precision. This can be done with machine shop-type micrometers. Consequently, a mirror drive for a far-infrared instrument is usually a suitable mechanical arrangement by which a mirror or mirror mount is coupled to a micrometer screw operated by an electronically controlled motor. The mirror motion could be continuous, resulting in a smooth signal like the lower trace of Figure 1, or be incremental. In a stepped drive the micrometer screw is advanced by, say, 5 μ at regular intervals, resulting in a trace similar to the upper trace of Figure 1.

Such stepping or mechanical mirror positioning is not feasible at the shorter wavelengths, because the mirror position would have to be measured to fractions of a micron. However, Block Engineering, Inc., partly overcame this problem by using continuous drives. The mirror was mounted on a spring and displaced by a device similar to the voice coil of a loudspeaker, using complicated electronics to assure linearity of mirror motion. An electronic time base and trigger, shown schematically in the lower part of Figure 2, were generated simultaneously. The relations are obvious: if the mirror is moved at constant velocity $V = X/\Delta t$, the displacement X is known because Δt is known by means of the electronic time base. The Block Engineering instruments were successful and useful, but are

now superceded. In order to be able to circumvent the severe requirements of a strictly linear drive and consequently be able to go to longer mirror excursions with attendant higher resolution, Block Engineering developed so-called fringe-referenced interferometers which generate their own time base. How this is done is shown schematically in Figure 3.

One transducer is used to move two mirrors simultaneously through identical displacements. The sample radiation to be processed is passed into one interferometer, and produces the "sample interferogram." Monochromatic light from a laser is processed by the second interferometer and, as in Figure 1, produces a cosine wave. White light from a small source is also passed into the second interferometer and produces a burst when the light paths in the reference cube are symmetric. The two interferometers are aligned so that the white-light burst occurs at the beginning of a sweep or scan. As the two mirrors move in concert, the three signals are "locked" together. Consequently, the requirement of a strictly linear drive is not necessary, and the mirror position can be determined by measuring the laser line interferogram, i.e., counting the laser fringes as the mirror moves from the white-light burst reference position. Such fringe-referencing has made it possible to increase mirror sweep lengths and consequently to increase resolution.

Resolution

The resolution obtainable with an interferometer depends on (a) what may be termed the optical quality of the interferometer, i.e., planarity of mirrors, coplanarity of beamsplitter and compensation plate; (b) the degree to which radiation entering the interferometer diverges or converges—if there are marked deviations from parallelism, there are secondary interference effects; and (c) the length of the mirror sweep. This last item is the most important.

Note the integration limits of eqns. 3 and 5. They mean that, in order to get a

precise reproduction of the spectrum, the mirror displacement must change from $-\infty$ to $+\infty$. It is obviously physically impossible to meet that requirement. In order to see how this limitation affects resolution, let us again consider monochromatic radiation.

An ideal monochromatic line would produce an infinite cosine wave if the mirror travel were infinitely long ($-\infty$ to $+\infty$, shown schematically in Fig. 4, A). The Fourier Transform of such a wave would be a spike of zero width (trace B). As it is possible to bring about only a finite mirror motion, the duration of the cosine wave is finite and the wave form is truncated. The Fourier Transform of such a truncated wave form, which contains only a relatively small number of wavelengths, takes the form of the function sin x/x. The function is centered at the same frequency as the spike in trace B, but now has width, as shown schematically in trace C. The width is inversely related to the length of the truncated wave form, and hence to the length of mirror movement. If the latter is increased, the function narrows, e.g., trace D, and resolution improves. Resolution is thus inversely related to the mirror excursion.

Data Reduction

Fourier Transform spectroscopy owes much of its strength and most of its weakness to the need to perform a Fourier Transform of the observed results in order to obtain a spectrum. In principle, all the spectral information may be extracted by performing the appropriate mathematical operations. In practice, reducing an interferogram has not been accomplished easily until recently. The computations involved in data reduction are so complex, lengthy, and tedious that manual computation is out of the question. This difficulty was one of the main factors which had kept Fourier Transform spectroscopy in obscurity until recently. That problem has now been solved.

Data reduction can be done by analog means. The interferogram is reproduced repeatedly and each harmonic component is measured with a narrow band-pass elec-

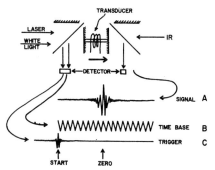

Figure 3. Fringe-referencing.

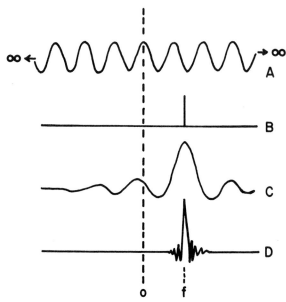

Figure 4. Effects of truncation.

Chemical Instrumentation

trical filter whose frequency can be varied over the region of interest. The device is, essentially, an electrical wave form analyzer or "wave analyzer." The method is mechanized by changing the frequency of the wave analyzer at a constant rate and plotting the output on a strip chart recorder. The result is a graph of the spectral intensity of the source as a function of wavenumber. A number of such analog devices exists and can be used to reduce interferograms. However, digital computing is better.

The interferogram is digitized and expressed as a sequence of numbers. The latter is then transformed by a digital computer which performs the operation accurately and economically. The computer method is much more versatile than analog methods. It is not only more precise, but corrections can be made, spectra can be ratioed and, in general, tricks can be played with the data simply by writing the appropriate software. This capability gives the method a versatility well beyond that of any other spectrometer.

Digital data reduction can be carried out with large computers, but it is necessary to interface the spectrometer and computer. This can be done, for example, by digitizing the interferogram and punching the information on paper tape. The paper tape is then fed to the large computer which carries out the Transform. The spectrum is then plotted out. Unless one has direct access to one's own computer, this method is obviously somewhat cumbersome. However, the advent of the minicomputer will alleviate that problem also. The instruments of two manufacturers now incorporate minicomputers, so that data handling and data reduction are greatly facilitated.

Multiple Scanning

The instruments operating in the visible to infrared ranges have short scan periods (of the order of seconds), so that multiple scanning becomes readily feasible. This leads to an improvement in the signal-to-noise ratio (S/N). The signal proper is always positive *or* negative, while noise is positive *and* negative at random. If the same system is measured repeatedly and the results are added, then S grows linearly with the number of measurements, while N increases only as the square root of the number of scans. S/N consequently increases as a function of the square root of the number of scans. S/N is doubled as the number of scans is quadrupled. With short scan periods it is therefore readily feasible to increase S/N by a factor of 10 or 20. However, provision has to be made to permit the coherent addition and storage of the multiple signals. The minicomputers incorporated in some Block Engineering instruments are used for this purpose.

Advantages

The Fourier Transform spectrometer possesses significant advantages in terms of S/N with respect to conventional dispersion instruments, based on the amount of signal which can be processed, and the time used to process it.

Dispersion or filtering is not required, so that energy-wasting slits are not needed. An interferometer has a relatively large, circular entrance aperture and relatively large mirrors (for example, in the mid-infrared range, a mirror diameter of 15 mm; in the far-infrared, several inches). The throughput, the amount of radiation which can enter the optics of the Fourier Transform spectrometer, is consequently quite large in comparison to that of a conventional spectrometer.

The second major advantage is not quite as obvious as the throughput gain, and also arises from the absence of the need to disperse or filter. In the conventional spectrometer, each radiation bundle or resolution element of the spectrum is scanned across the detector. Consequently, if there are M resolution elements, the intensity of each element is measured for only a fraction T/M of the total scan time, T. The signal proper (the intensity of an element) is directly proportional to the time spent observing it, while noise, being random, is proportional to the square root of the observation time. The S/N is then proportional to $(T/M)^{1/2}$. With the interferometer, however, the entering radiation falls on the detector, so that each resolution element is observed throughout the entire scan period, with the result that S/N is proportional to $T^{1/2}$. The improvement by the factor of $M^{1/2}$ for the case of the interferometer, termed Fellgett's Advantage, can be quite large under conditions of relatively high, or high resolution. Fellgett's Advantage is realized with detectors which are detector noise limited—i.e., as the signal level increases there is no increase in detector noise.

The advantage in S/N can be traded off for rapid response. The scan time can be decreased. A spectrum can be measured with a Fourier Transform spectrometer in the same time as with a conventional spectrometer, but with a better S/N, or in a much shorter time with an equivalent S/N.

In order to compare Fourier Transform and conventional techniques, consider examining a spectrum composed of 2000 resolution elements, and suppose that an observation time of 1 sec per resolution element is required to obtain good S/N. The dispersion spectrometer requires 2000 sec or 33 min to make the measurement; the interferometer completes it in 1 sec. Under the same conditions, improving S/N further, say, by a factor of 2, the dispersion instrument would take 2 hr and 12 min, while the interferometer would require only 4 sec to complete the measurement. Or, on the other hand, improving the resolution by a factor of 2, the entrance and exit slit-widths of the dispersion instrument must be halved. This decreases its throughput by a factor of 4. Therefore, in order to regain the desired S/N, the required measurement time is increased by a factor of 16 to 8 hr and 48 min. By contrast, resolution in an interferometer is doubled by doubling the length of traverse of the moving mirror. The resolution is doubled by going from a 1-sec measurement interval to 2 sec.

INSTRUMENTS

The series of commercially available instruments covers a rather wide spectral range. The range covered by any one particular instrument is governed largely by the beamsplitter and mirror motions, as indicated earlier. Ranges are shown in Figure 5. The lamellar grating used in a far-infrared instrument will be described later. Some flexibility is afforded through the use of interchangeable beamsplitters. However, it is convenient to put the far-infrared instruments in a class by themselves, especially because most of them differ considerably in construction from other instruments, and first consider spectrometers usable in the high-frequency regions. The descriptions given below are based on material supplied by the manufacturers up to January 1, 1970. Prices are given only for major items.

IR-UV SPECTROMETERS

Block Engineering, Inc.

Two commercial systems are available for the high-frequency regions, both manufactured by Digilab, Inc., subsidiary of Block Engineering, Inc. Each instrument consists of a scanning Michelson interferometer and a data-handling system which, together, make up a Fourier Transform spectrometer system which produces spectra and not just interferograms.

The more "advanced" system, the Model FTS-14 spectrometer, is the first of the Digilab series of analytical instruments. (Digilab is a family of individual analytical "heads," integrated with, controlled by, and feeding into a shared, low-cost computer.) The two major items which constitute the FTS-14 spectrometer are the Model 296 Fourier Transform spectrometer unit and the Digilab data-handling system. The entire spectrometer is shown in Figure 6 and 7. The box on the left of Figure 6 contains the interferometer, sampling system, source, and detector, and is continuously purged. The central chassis contains electronics for the interferometer and the Digilab data system.

The Model 296, shown with its cover removed in Figures 8 and 9, is a fringe-referenced Michelson interferometer with 2 in.-diameter mirrors. The mirror drive is electromagnetic. A drag-free mirror drive is produced by floating the mirror shaft on an air cushion within a close-fitting sleeve. The plastic tubing, moisture trap, gas regulator, and gauge are all associated with the air bearing. The black tube at the top left of Figure 8, seen end-on in Figure 9, is the housing of the He-Ne laser used for fringe-referencing. The laser beam is deflected down into the large cube by a small mirror. The base of the second interferometer used for fringe-referencing is the black rectangle fastened by four screws to the top of the cube. The reference interferometer itself is shown with its assembly in Figure 10. The small glass cube is made of two prisms, the joined surfaces of which form the beamsplitter. The moving mirror of this interferometer is a polished flat on the back of the main interferometer's moving mirror.

Chemical Instrumentation

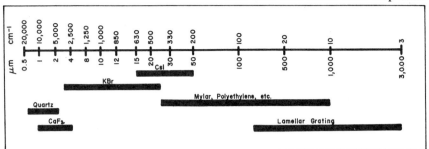

Figure 5.   Wavelength ranges covered by various beamsplitters

cover which houses the overall cube, drive, and reference assembly. A thermistor bead heat sensor servo loop estab-

DOUBLE BEAM SAMPLING OPTICS

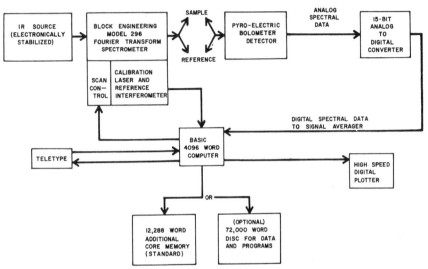

Figure 6.   Schematic of Model FTS-14

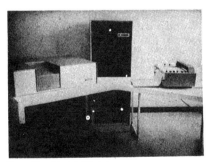

Figure 7.   Model FTS-14

Figure 8.   Model 296

The knurled rings at the back of the stationary mirror are slightly wedged, so that mirror tilt can easily be adjusted. The wires come from two photodiodes which monitor both the laser and a small white-light source built into the assembly. Thermostatically controlled heaters are attached to the inside surface of the dust

Figure 9.   Model 296

Figure 10.   Reference cube of Model 296

lishes temperature control of these optical components to within $\pm 1°C$ to eliminate any long-term tendencies to thermal drift. The infrared source is a glower operated at approximately $1100°C$. This source is stabilized by an electronic servo to limit any infrared intensity fluctuations to within $\pm 0.2\%$ at $2\ \mu$. At longer wavelengths the fluctuations become too small to measure. This servo assures long-term repeatability and consistent performance.

The standard infrared detector used with the FTS-14 is the state-of-the-art triglycine sulfate (TGS) pyroelectric bolometer. The TGS is a room-temperature detector, possessing essentially flat wavelength response ranging from the near-infrared through the far-infrared. This single detector, therefore, can serve both the near- and the far-infrared operating configurations of the FTS-14. The TGS bolometer also can handle signal frequencies of up to several thousand Hertz. This permits the interferometer to be scanned at a moderately rapid velocity, helping to minimize both the noise and the time required to complete a given measurement. A shaping filter is used in the preamplifier electronics, enhancing the spectrometer's overall response at each end of the spectral range relative to the higher-intensity middle region. The spectrometer's response curve is thus relatively flat in comparison to a backbody curve. Another feature about the analog signal electronics is the complete pre-emption of a manual gain control. The gain of the amplifier train is factory-adjusted so that the detector noise level does not exceed the minimum digitization sensitivity of the computer analog-to-digital (A/D) converter. This, by definition, is the maximum gain setting possible without processing pure noise. Thus, an extremely weak signal, as from a sample of high optical density, is always processed at maximum amplification; yet strong signals do not cause overloading because the computer's 15-bit A/D converter can handle the widest dynamic range of signal intensities without distortion. The system, is, therefore, set up to produce ratio-recorded spectra of the highest photometric accuracy automatically, for vastly differing experimental situations, all virtually exclusive of human intervention.

An optical diagram of the FTS-14 is shown in Figure 11. The shaded line traces the path of the beam from the infrared source and collimating mirror through the interferometer and sampling compartment (outlined by the dashed lines) to the TGS detector.

Locating the interferometer between the source and the sample not only protects the sample from intense thermal energy radiating from the source, but it also serves to prevent stray or unwanted radiation from affecting the measured spectra (as the beam passes through the scanning interferometer, it becomes modulated or converted into a beam of fluctuating intensity). The electronics associated with the TGS detector responds only to fluctuating (or ac) signals, discriminating totally against unmodulated (or dc) radiation emitted by the sample or any warm object within the detector's field of view.

Chemical Instrumentation

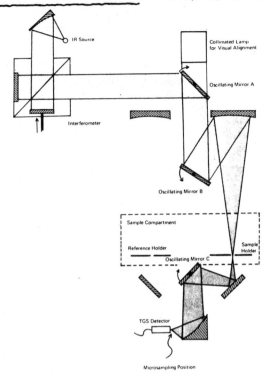

Figure 11. Sampling optics diagram of Model FTS-14

Oscillating mirror A directs the beam from the interferometer into the sampling compartment. When rotated 45° clockwise, this mirror allows a collimated white light to pass through the sampling optics. The dimensions of the alignment beam exactly match those of the infrared beam so that the operator can quickly and accurately optimize the adjustment of any sampling device. This highly visible white-light beam is a valuable aid in setting up and aligning all standard multiple reflection devices. Oscillating mirrors B and C are shown directing the beam through the sample path. During a ratio-recorded measurement they are flipped by computer command to their alternate positions, directing the beam first through the reference path, then through the sample path. The beam comes to a focus within the sample compartment. This 5-mm diameter focus is arranged to accommodate all standard transmission and reflectance sampling devices. The fast foreoptic in front of the detector serves two functions. It assures that all of the available energy (defined essentially by the aperture of the IR source) is focused onto the detector's sensitive area without overfilling it. Secondly, its 1-mm diameter focus makes an excellent microsampling position. To make use of this, a special microsampling stage has been developed, as an accessory.

The double-beam sampling compartment in the FTS-14 accepts all commonly used solid, pellet, and liquid cell holders, short-path gas cells, ATR and specular reflectance devices, so that the instrument can be used to record spectra using all of the conventional techniques. Also, by removing the source assembly and replacing it with a focusing mirror, infrared emission spectra can be recorded. That

is, the self-emitted radiation of a sample at or near ambient temperature can be recorded and, if required, ratioed against the emission of a blackbody. Such infrared spectral emission measurements, which are extremely difficult to carry out with conventional dispersion spectrometers, are easily made with the Fourier Transform spectrometer.

The Digilab/FTS-14 data system is constructed around a dedicated, third-generation Data General Nova minicomputer. The central processor has a 4096-word memory, a 16-bit word length, a cycle time of 2.6 μ, and a fully complemented software package. Two configurations of the Digilab data system are available for the FTS-14. The all-core unit consists of: the central processor with 16k words of memory, a 15-bit A/D converter, a high-speed digital plotter, and a Teletype control unit. Working with the Model 296 spectrometer, this all-core data system can calculate a fully ratio-recorded spectrum (2.8 to 25 μ) at 2-cm^{-1} resolution in 30 sec calculation time. In the half-wavenumber high-resolution mode, selected spectral intervals of the 2.8 to 25-μ region can also be calculated and ratioed in 30 sec. The alternate Digilab/FTS-14 data system utilizes the 4096-word memory in the central processor, supplemented by a 72k-word expandable disc. This configuration is capable of first storing and then performing separate Fourier Transform calculations on a pair of 16,000-word interferograms, yielding half-wavenumber ratio-recorded spectra of the entire 2.8 to 25-μ interval in under 120 sec calculation time. Survey spectra of 2-cm^{-1} resolution are calculated in 30 sec.

During the measurement cycle, while data are being accumulated, the large computational capacity required to carry out the Fourier transformation routines is

not idled. It is used to operate the spectrometer and to monitor its performance throughout each measurement. It controls the mirror drive to take either high-resolution or lower-resolution scans as directed via the Teletype by the operator; it counts off the number of scans requested by the operator and even activates a solenoid which automatically switches the optical system between the sampling mode and the reference mode. At the end of every measurement the computer automatically branches into its transform calculations, ratios the sample and reference spectra (if the operator has selected the double-beam mode), then it activates and drives the digital plotter to trace out a spectrum between the wavenumber limits requested by the operator. The computer also automatically traces out a wavenumber scale calculated from the reference laser calibration information. If the operator so desires, the computer will find the spectral maximum, rescale it to 100%, and plot a scale-expanded spectrum. After plotting out the spectrum and wavenumber scale, the computer types out the maximum and minimum intensity levels which occur in the plotted spectral interval. The precise percent transmission of any given spectral peak can therefore be learned simply by directing the computer to plot out the spectral interval in which it occurs; the percent transmission will then be typed out automatically.

An automatic calibrating program for balancing the sample and reference beams is written into the Digilab software. The operator types CALIB on the Teletype to initiate this program. One scan each of the sample and reference beams is recorded at low resolution (i.e., 2 cm^{-1}). The computer finds the intensity difference between the two by ratioing the interferograms and uses this number as a scaling factor to derive the 100% line. This value is retained and used in the ratio of all double-beam spectra until such time as the operator deliberately changes it by retyping CALIB.

The Model 296 can be used with a Digilab system, or in conjunction with other data-handling systems. Alternate schemes are given in Figure 12. Another spectrometer, the Digilab, Inc. Model 297, is similar to the Model 296 but is especially configured for remote sensing infrared measurements for astronomic, air pollution, and other atmospheric studies. That instrument is equipped with focusing optics to allow easy interchange of various detectors. The Model 297 could be used with a Digilab data system or with any of the alternate data processing methods. As these methods are digital, a number of significant operations becomes possible with these instruments. Derivative spectroscopy, for example, is merely a matter of introducing a simple software routine into the computer and invoking it whenever it is desired to determine the exact wavelength location of a given shoulder or change in slope of a particular band. Again, for those who are interested in setting up their own spectral search programs, the measured spectra are always in digital form and can be readily transferred into the virtually unlimited files of a large batch computer via direct line, magnetic

Chemical Instrumentation

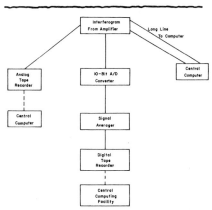

Figure 12.   Other methods of data processing

paper tape, or punched cards. Once a data bank is accumulated in the batch computer, spectral correlations or identifications of unknowns become a matter of simple routine. Also, as the efficiency of spectral search and correlation programs in terms of reliable and unambiguous identifications is heavily dependent upon the frequency of both the spectra on file and the spectrum of the unknown, the automatic and continuous laser-based wavelength calibration built into the FTS-14 system assures the lowest possible error factor. Also, the requirement for repeating a complicated and tedious calibration routine each time a search spectrum is to be measured is obviated.

The Model 296 (and 297) can cover a wide spectral range in steps, depending on the beamsplitter. The fingerprint range is covered by a beamsplitter made by evaporating Ge onto a KBr substrate.

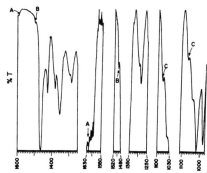

Figure 14.   Scale-expanded spectra. See legend of Figure 13. Portions of the lower trace of that figure were scale-expanded. Spectral resolution is 2 cm^{-1}.

As KBr is hygroscopic, the exposed surfaces are coated with a moisture-resisting film which is devoid of absorption bands in that infrared region. Two additional beamsplitters, calcium fluoride and quartz substrates coated with iron oxide, provide coverage down to the shorter wavelengths. At these wavelengths, the mechanical tolerances governing optical surface flatness and the linarity of the moving mirror transport become fairly critical. It is, therefore, necessary to equip the Model 296 for operation at these shorter wavelengths with optionally available high-

precision optical components and mirror drive mechanism in the interferometer. Plastic films are used in the far-infrared, to be described later. The ring bearing the beamsplitter is seen in Figure 9.

The FTS-14 system as outlined above is completely automatic and, as its performance is computer-maximized, does not require a highly skilled operator. Spectra of optimized quality and highly scale-expanded spectra can be obtained relatively quickly. Some examples are spectra of γ-dodecalactone shown in Figures 13 and 14. The upper full-range spectrum of Figure 13 was obtained with four 2-second scans. The spectral features are quite apparent, even though the spectrum is somewhat nosiy above 2000 cm^{-1}. The lower full-range spectrum resulted from 256 scans. Were it not for the essentially trivial noise near 2400 cm^{-1} and above 3200 cm^{-1}, the spectrum would meet the Class II criteria for research-quality spectra of the Coblentz Society. The spectral resolution for both

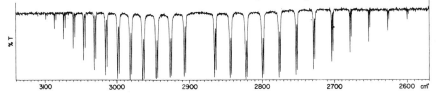

Figure 15.   Spectrum of HCl.   HCl in 10-cm cell, 40 mm Hg; 128 8-sec scans, electronically ratioed, scale-expanded.   Spectral resolution was 0.5 cm^{-1} throughout the entire range of 3900–400 cm^{-1}.

spectra was 2 cm^{-1} throughout the entire range. As the spectrum is stored, scale expansion can be carried out at will without rescanning. On command, by specifying that the plot should be scale-expanded between 600 and 1600 cm^{-1}, the system produced the lower plot of Figure 13. In similar fashion, any segment of the spectrum can be scale-expanded, so that some feature of interest may be examined. Figure 14 shows further expansions of various parts of the spectrum.

The high-resolution capabilities of the FTS-14 system are illustrated by Figure 15 showing the HCl fundamental. The resolution was 0.5 cm^{-1} throughout the entire range of 3900-400 cm^{-1}. The doublets, caused by the presence of the two isotopes Cl^{35} and Cl^{37} in the ratio 3:1 are very well resolved. The measurement time including data reduction for the entire high-resolution spectrum from 3900 to 400 cm^{-1} was about 34 min.

The FTS-14 system shown in Figure 7 and the Model 296 shown in Figure 8 are fairly large. A smaller system is Block Engineering's Model 197 Fourier Transform spectrometer. The Model 197 optics are similar in some aspects to the earlier Models 195 or 196 (these were used in Block Engineering's Model 200 and FTS-12 Fourier Transform spectrometer systems) in that the main interferometer cube is quite small—about 7 cm on edge. However, the Model 197 is fringe-referenced, and employs two separate cubes. These are shown in Figure 16. A reference interferometer, in tandem with the measurement interferometer, monitors a white-light lamp and a monochromatic line source to provide high-accuracy signal sampling information. The reference cube is the left one in Figure 16. The

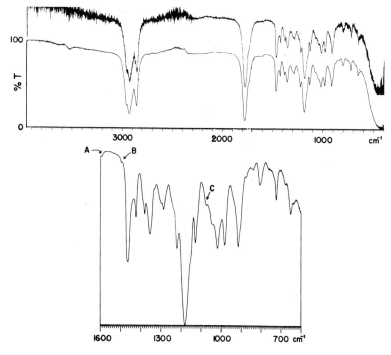

Figure 13.   Spectra of γ-dodecalactone.   Neat sample between NaCl flats, measured with the Model FTS-14.   Upper full-scale plot: 4 scans; lower full-scale plot: 256 scans. Both spectra are electronically ratioed spectra, with resolution 2 cm^{-1} over the entire range.   The lower trace is an ordinate scale-expanded segment of the 256-scan spectrum. On command, the computer placed the highest %T point within the 600–1600 cm^{-1} region to the 100% line, the lowest %T point to the zero line, scaled in-between points, plotted the scale-expanded spectrum, and then printed out the %T values.

Chemical Instrumentation

Figure 16.   Model 197 optical system

stationary mirror (all mirrors are 15-mm diameter) is under the circular plate of the reference cube. The fringe-referencing technique affords improved measurement accuracy, so that the resolution capabilities of the Model 197 are greater than those of earlier, simpler models.

The Model 197 cubes are fitted with various beamsplitters and detectors:

Spectral range, microns	Detector	Resolution, (cm^{-1})
0.3–3	PMT	20
0.8–3	PbS	20
1–4.5	PbSe	20
2.5–16	Bolometer	4
4–40	Bolometer	4

Sweep times are nominally $1/8$, $1/4$, $1/2$, and 1 sec. The entrance aperture is 3 cm, the aperture stop 15 mm. Field of view is 10° full angle. The Model 197 can be used for emission, transmission, and reflectance measurements, and can be used in conjunction with conventional sampling attachments. A special diffuse reflectometer attachment incorporating a large ellipsoidal mirror is available.

The Model 197 is less useful than the Model 296 because of the lower resolution, but makes up for this by reason of its small size and high scan rate. The resolution of the Model 197 is adequate for many purposes, and its small size would lead to flexibility, in that it is easy to incorporate the optical system into an experimental setup.

Several methods of data processing are available for use with the Model 197. These include those shown in Figure 12. Analog data reduction by means of suitable wave analyzers would also be possible. However, a favorable configuration results from combining the Model 197 with the Digilab data system. The latter makes available all of the flexibility of the digital system for data storage, data reduction, and data manipulation outlined for the case of the FTS-14 spectrometer system.

Block Engineering, Inc., also manufactures a variety of other interferometers and Fourier Transform spectrometer systems for use in the UV-IR region. Such special-purpose instruments, e.g., tiny interferometer spectrometers for use in earth-orbiting satellites, or dual-beam spectrometers used for astronomical purposes, need not be described here. Block Engineering, Inc. and Digilab, Inc. will also supply special-purpose attachments for use with the Model FTS-14, Model 197,

and other systems, e.g., for gas chromatography or ultramicrosampling.

Prices: the basic Models 296, 297, and 197 spectrometers are $25,000. The FTS-14 system, incorporating the Model 296 and the basic Digilab data system, is $65,000.

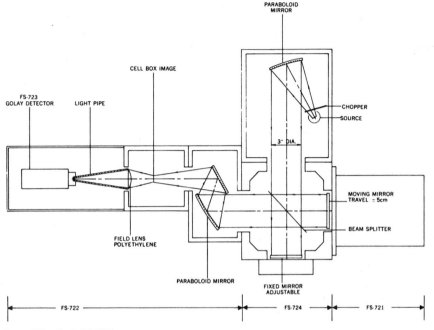

Figure 17.   Model FS-720 spectrometer

Figure 18.   Model FS-720 spectrometer

FAR-IR SPECTROMETERS

Beckman-RIIC, Ltd.

Beckman Instruments, Inc. markets three spectrometers produced by Beck-

man-RIIC, Ltd., a British subsidiary of Beckman Instruments, Inc. The Models FS-720 and FS-820 spectrometers are similar and are made up of compact, light modules. They can be easily changed because of their modular nature. The beamsplitter can also be changed in

seconds without requiring realignment. The location of the source and detector lends itself to both source and detection studies. Both instruments incorporate Michelson interferometers and use 7.5-cm diameter mirrors. The whole optical system can be evacuated to eliminate atmospheric water vapor absorption. The system is capable of holding a pressure of below 0.1 torr, which is necessary to eliminate this absorption. The FS-720 interferometer uses reflecting optics to maximize energy throughput. The FS-820 has a simplified optical system utilizing polyethylene lenses. The addition of a light pipe to both interferometers has also increased the light-collecting efficiencies of the instruments. The FS-720 modular interferometer is designed for use in the

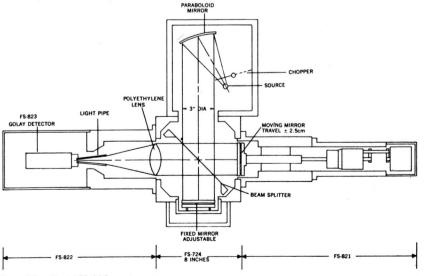

Figure 19.   Model FS-820 spectrometer

Chemical Instrumentation

10–500 cm⁻¹ region and for resolutions of up to 0.1 cm⁻¹. The FS-720 employs off-axis paraboloid reflecting optics throughout, thus maximizing the energy throughout the range of the instrument. There are no Cassegrainian optics to introduce obscuration and reduction of energy. The FS-720 is shown in Figures 17 and 18, and consists of several modules.

The FS-724 module contains the following items. Source: 125-W high-pressure mercury lamp water-cooled. Source aperture: variable in the following steps: 3-, 5-, and 10-mm diameter. Collimator: f:1.5 surface aluminized off-axis paraboloid. Chopper: synchronous motor driven at 15 Hz. Beamsplitter: polyethylene terephthalate. Beamsplitter mount: the most efficient available on the market affording easy replacement of beamsplitter without the necessity for re-aligning. The beamsplitter mount is made of spot-ground stainless steel. Fixed mirror: 3 in.-diameter surface aluminized plane mirror incorporating the RIIC cantilever adjustment principle. Fixed mirror mount: held in position over O-rings; removable for other instrumental configurations, arrangements, and accessories. Dimensions: beamsplitter module, 8 in. aluminum alloy cube.

The FS-721 module contains a low-voltage synchronous motor and multispeed gearbox used to produce mirror path difference speeds of from 0.5 μ/sec to 500 μ/sec. The path difference is ±10 cm. The moving mirror is 7.5 cm in diameter. The mirror motion is monitored through the use of a Moiré grating. The Moiré fringes produced by the system are used in a manner similar to the laser fringe-referencing described earlier.

The FS-722 sample, condenser, and detector modules house surface aluminized off-axis paraboloid and plane mirrors used to form an image of the source in the center of the accessory module, an electroformed copper light pipe used to increase the efficiency of the system, and a Golay detector fitted with a diamond window. The "sampling area" is a 5³/₄ in. cube and will house 10-cm gas cells, liquid cells, and solid sample holders. The source is imaged in the center of the module to allow the best arrangement for positioning the sample and accessories.

The FS-200 electronics module consists of a lock-in type amplifier, a digitizer, and a punch. A 12-bit A/D converter is triggered by the Moiré grating pulses and associated electronics. The Addo punch records the interferogram on paper tape in 12-bit binary code.

The FS-820 is shown schematically in Figure 19. The instrument basically consists of the source and interferometer modules of the FS-720. However, a stepping drive is used in place of the Moiré drive, and polyethylene lenses as the condensing optics. The instrument is for use principally in the range 10–200 cm⁻¹ with resolutions of up to 0.2 cm⁻¹. Although the frequency range is limited to 10–200 cm⁻¹ by the polyethylene lenses and quartz-windowed detector, the range can be increased to 400 cm⁻¹ by substituting a

FS-722 module and diamond-windowed detector for the normal condensing optics and detector. The FS-820 will usually consist of the following modules.

The FS-821 drive module contains the stepping motor and micrometer drive. A lapped ¹/₂ in.-diameter piston carries the moving mirror through a total distance of 5 cm. The mirror is 7.5 cm in diameter.

The FS-822 module contains an f:1.2 polyethylene lens fitted to the exit port of the interferometer. The cell compartment

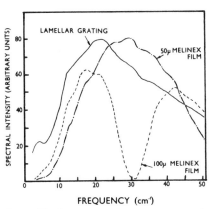

Figure 20. Low-wavenumber energy comparison, between the FS-720 spectrometer employing 50- and 100-μ melinex beamsplitters, and the LR-100 lamellar-grating spectrometer (grating constant 0.95 cm).

incorporates a light-pipe. A Golay detector fitted with a quartz window is used. The FS-200B electronics module is similar to the FS-200 module, and contains a lock-in amplifier, a 12-bit A/D converter triggered by the stepping motor drive

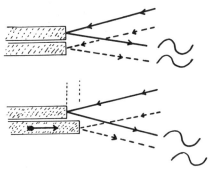

Figure 21. Interference by lamellar grating

circuit, and an Addo punch. The FS-MC1 electronics are used to operate the stepping

drive in 2.5-, 5-, 10-, 20-, 40-, and 80-μ intervals with stop time adjustable from 0.53–34 sec.

The third instrument, the LR-100, differs from the others in that a completely reflective "beamsplitter," a lamellar grating, is used rather than the 50%-transmitting, 50%-reflecting ones described earlier. The reason for this choice becomes apparent in examining Figure 20.

Figure 23. LR-100 spectrometer

In the very far-infrared region the plastic-film beamsplitters become inefficient. Energy is wasted because the films transmit poorly and multiple interference effects impair spectral quality. The lamellar grating, however, uses reflecting surfaces to produce interference effects. How this is done is shown schematically in Figure 21. Suppose monochromatic beams are reflected from the ends of two closely spaced plates, as shown in the upper part of Figure 21. Each beam will travel the same distance after being reflected and will be in phase with the other. However, if one of the plates is moved with respect to the other, then the paths differ and interference effects similar to those described earlier occur. A set of such plates constitutes a lamellar grating.

The LR-100 lamellar grating spectrometer is shown in Figures 22 and 23. The LR-100 is of modular construction and uses modules similar to those designed for the FS-720.

The lamellar grating located in the LR-101 module is comprised of two sets of parallel intermeshing metal plates whose front surfaces have been lapped accurately flat. The plates are mounted so that these faces can be kept parallel to each other within a few seconds of arc. The total area of the grating is 8 cm × 8 cm. The movable set of plates is positioned by a stepping motor operating in conjunction with a micrometer drive.

The standard lamellar grating supplied with the instrument has a "grating constant" of 0.95 cm, and allows interferograms of up to ±5 cm optical path difference to be scanned. This unit is inter-

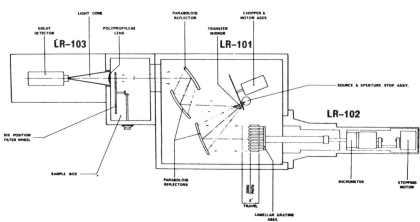

Figure 22. Schematic diagram LR-100

Chemical Instrumentation

changeable, however, and other lamellar grating assemblies can be supplied to order giving either a maximum of 10 cm optical path difference in one direction for higher resolution, or a "grating constant" of 0.64 cm for an increased operating range to ~100 cm^{-1}.

The radiation reflected by the lamellar grating is deflected by the off-axis parabolic mirror, to a transfer mirror positioned directly beneath the source aperture. This transfer mirror directs all the radiant energy toward the sample chamber, via two off-axis parabolic mirrors. The light source is a 125-W high-pressure mercury vapor lamp, water-cooled and fitted with a thermal cut-out safety switch. Source radiation passing through a 1-cm aperture is chopped at 15 Hz. The radiation is collimated by a surface aluminized off-axis parabolic mirror.

Lamellar-grating drive module (LR-102): settings of the movable lamellar-grating plates are effected by a stepping motor triggered by the control unit FS-MC1. The motor is coupled to a micrometer drive consisting of a cylinder and a nonrotating lapped piston to which the grating plates are attached.

Sample, condenser, and detector module (LR-103): the sample and condenser chamber is approximately a 15-cm cube. The incoming light beam is imaged in the center of the chamber in order to permit the best possible positioning of specimens and accessories. The image size is approximately 12 mm. The top plate of the chamber has a large glass observation window, and the module is provided with an externally adjustable rotating wheel which accommodates either five filters or five solid samples. This feature permits filters or samples to be changed without

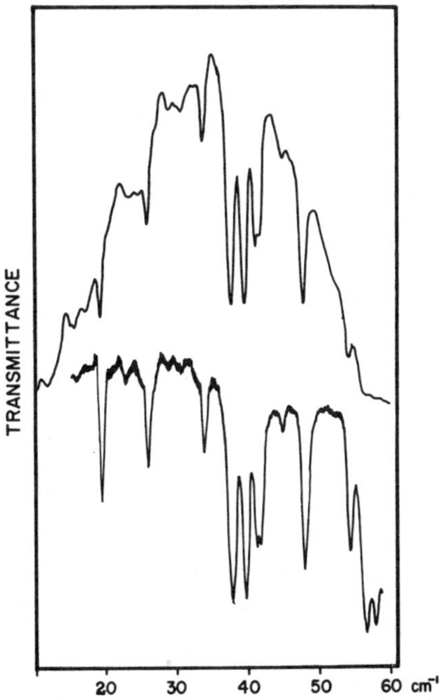

releasing the vacuum. The detector is an improved Golay cell (Unicam) having a 3-mm or a 5-mm diameter quartz window, and a vacuum nozzle.

The FS-MC1 control for the stepping motor and the electronics are like those used for the FS-820.

The three spectrometers produce interferograms. The paper tape thus serves as interface to a digital computer used to perform the data reduction. An alternate approach involves the use of the FTC-100 Fourier Transform computer. While an interferogram is produced, the analog information is processed by an A/D converter and stored serially in one of two ferrite core matrix memories as binary words of 12 bits. Two sections of 1024-word capacity memory are available to enable a ratio of two spectra to be obtained by simultaneous transformation. A paral-

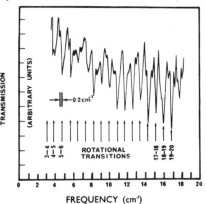

Figure 25. Pure rotational spectrum of N_2^{14}-O^{16}. The spectrum was measured with the LR-100 spectrometer, transformed with the FTC-100 electronics, and is a ratioed spectrum. It was recorded under conditions of 1-meter optical path and about 100 torr pressure. The recording time was 1 hour for each interferogram (±5-cm optical path difference).

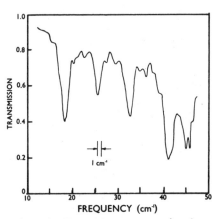

Figure 26. Transmission spectrum of a 2-mm powdered disc of C_6I_6. The spectrum was measured with the LR-100 spectrometer, transformed with the FTC-100 electronics, and is a ratioed spectrum. The recording time was 20 min for each spectrum (±1-cm optical path difference).

lel 12-bit input facility is fitted for paper tape or other already digitized data. (The paper-tape output of the spectrometer can be retained, so that more extensive processing of data by a large digital computer is possible.) Computation consists of cycling the stored information, passing the digitized data through a D/A converter, and then passing the analog output

(the stored interferogram) through a scanning wave analyzer. The output of the wave analyzer is a spectrum and is recorded by an X-Y plotter. The program provides for either section of the memory to be transformed and displayed independently, or for simultaneous transformation to produce the ratioed spectrum. A "background" interferogram can be stored.

A single-beam spectrum and the corresponding ratioed spectrum produced by the FS-720 using the FTC-100 analog data reduction system are shown in Figure 24. Ratioed spectra recorded with the LR-100 spectrometer are shown in Figures 25 and 26.

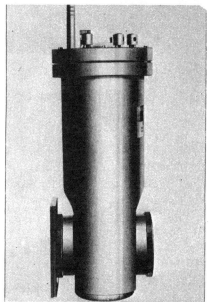

Figure 27. Variable-temperature cell

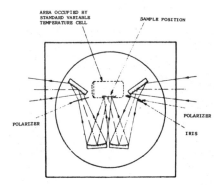

Figure 28. FS-7RF specular reflectance attachment

Beckman Instruments, Inc. supplies a variety of cells for gas, liquid, and solid samples, including a variable-temperature cell (Fig. 27) and a temperature controller for that cell. Also available are the FS-7RF specular reflectance attachment shown schematically in Figure 28, and wire grid polarizers. Spectra recorded with these attachments are shown in Figure 29.

Beckman Instruments, Inc. also markets RIIC's Model FTC-300 control and data-handling system. The FTC-300, shown in Figure 30, is made up of two major parts: (a) the FS-300 unit contains all of the electronics for operating and controlling the FS-720, FS-820, or LR-100

Figure 24. Spectra of water vapor. Upper trace: single-beam spectrum of water vapor. Lower trace: ratioed spectrum of water vapor. Both spectra were recorded with the FS-720; data reduction was analog, carried out with the FTC-100 system.

Chemical Instrumentation

spectrometers, and produces a digitized interferogram. The latter is punched on tape for processing in an off-line computer, or is passed to the FTC-300 core memory; (b) the FDP-300 data processing unit contains the logic control, memory circuitry, wave analyzer, and the flat-bed potentiometric recorder used to produce conventional spectra on preprinted charts.

The FTC-300 system is more sophisticated and more complex than the earlier FTC-100 system, but is also a hybrid system. The digitized interferogram is stored in a 20k-bit memory. When complete, the digitized interferogram is converted by a D/A converter and the now analog signal is processed by a wave

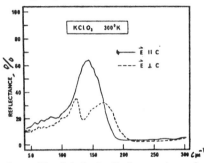

Figure 29. Polarized reflection spectrum of a KClO₃ single crystal measured at room temperature, with the beam electric vector parallel (continuous line) and perpendicular (broken line) to an axis of crystal symmetry. An FS-720 spectrometer was used in conjunction with the FS-7RF reflectance attachment and 4-μ wire grid polarizers.

analyzer. As with the FTC-100, both background and sample interferograms can be stored, reduced individually or simultaneously, to produce single-beam or ratioed spectra. The ratioed spectra are

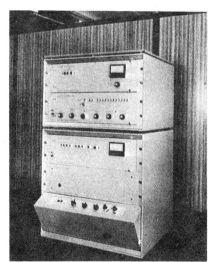

Figure 30. RIIC Fourier Transform electronics FTC-300. The top module of two panels is the FS-300 electronics required for operation of the interferometers, and produces a digitized interferogram on punched tape. The lower module is the FDP-300 data processing unit and produces conventional spectra on preprinted charts. The lowest panel with the slant front can be pulled out; a flat-bed recorder is housed within.

accurate to within ±1% when the background intensity has dropped to 3%. Plotting times vary from 8 to 64 min.

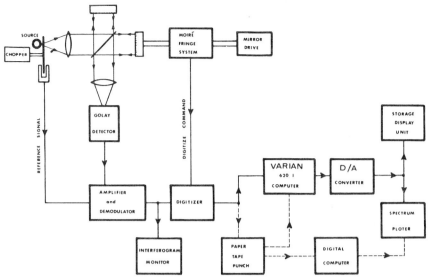

Figure 31.   Block diagram of MIR-2 far-infrared spectrometer

Prices: due to the modular nature of the instruments, many combinations are possible. The FS-720 is $7,550, without electronics. For complete *systems*, the FS-720 + FTC-300 system is $38,550; the FS-820 + FTC-300 + FS-MC1 system is $35,650. The LR-100 is about $11,000.

CODERG

The MIR-2 far-infrared spectrometer manufactured by CODERG (Société de Conversions des Énergies, Clichy, France) is now marketed by Scientific Instrumentation, Inc. A block diagram of the MIR-2 is given in Figure 31. The layout of the optics is shown in Figure 32, and additional sample space configurations are shown in Figure 33. The optics are mounted in two chambers. A conventional Michelson interferometer with f:2 entrance aperture is used. The beamsplitter is polyethylene terephthalate, and various thicknesses ranging between 6 and

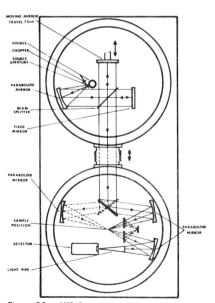

Figure 32.   MIR-2 spectrometer

100 μ are required for complete coverage of the spectral region. The light source is a high-pressure water-cooled mercury vapor lamp, and the light-source unit can

be readily removed so that a laser beam can be used for alignment purposes. The beam is modulated at 12.5 Hz, and a variable iris diaphragm controls the source aperture. The moving mirror is mounted on a high-precision slide bar. Motion is provided by a stepping motor in 10-μ steps, or a continuous motion controlled by Moiré fringe-referencing is used. The beam emerging from the interferometer is parallel and enters a second chamber

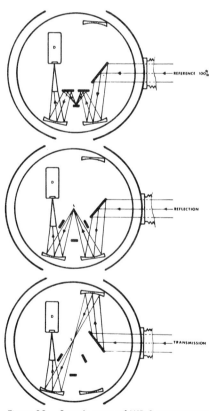

Figure 33.   Sample areas of MIR-2 spectrometer

where, depending on the sampling configuration (Fig. 33), the beam is reshaped. A Golay detector is used. The system can be evacuated to 10⁻⁶ torr.

The instrument covers the range 800–10 cm⁻¹. The maximum resolution is 0.1 cm⁻¹. The maximum resolution is 0.1 cm⁻¹ over a 75-cm⁻¹ region in the single-beam mode, or a 30-cm⁻¹ region in the double-beam mode. Scanning rates vary.

Chemical Instrumentation

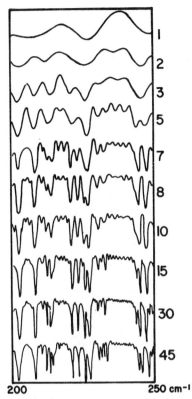

Figure 34. Spectra of water vapor

As shown in Figure 31, the MIR-2 can be used to produce digitized interferograms. These are punched out on tape, which serves to interface with an off-line computer. The more interesting system is the one incorporating a Varian 620/i mini-computer and a Tektronix storage display unit.

As with the Digilab, Inc., FTS-14 system, the computer of the MIR-2 system totally controls the entire interferometer/ data handling and processing systems. However, the data reduction methods differ. With the FTS-14, the entire interferogram produced by a scan is stored, and additional successive interferograms produced by multiple scanning are coherently added until the S/N ratio is acceptable. The entire interferogram is then transformed and the resulting spectrum is plotted out. With the MIR-2 system, computation is continuous. The scan is started and produces a segment of the interferogram. The computer quickly performs a 750-value transform, stores the data, and also displays the rudimentary spectrum on a screen. This procedure is repetitive, and as the mirror continues to move and the computer continuously computes and adds data to those already stored, an increasingly refined spectrum appears on the viewing screen. This procedure is illustrated by the spectra of water shown in Figure 34. Each trace was displayed on the viewing screen and photographed. The time in minutes elapsed since the scan was started is shown beside each trace. When the spectrum has reached an acceptable S/N value, determined by examining the spectrum displayed on the viewing screen, the spectrum

is plotted out. For example, when the water spectrum had reached the quality indicated by the lowest trace in Figure 34 after a 45-min scan period, the plotter was activated and the spectrum of Figure 35 resulted.

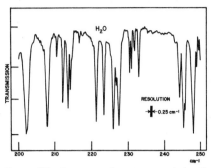

Figure 35. Spectrum of water recorded with MIR-2 far-infrared spectrometer; see Figure 34 and text.

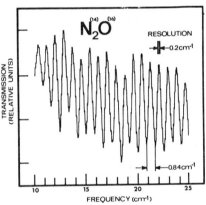

Figure 36. Spectrum of $N_2^{14}O^{16}$ recorded with the MIR-2

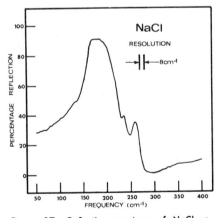

Figure 37. Reflection spectrum of NaCl recorded with the MIR-2

In practice, the operator would insert the sample and, after a vacuum was established, would give the computer the required data about (a) single- or double-beam operation; (b) the spectral range; (c) smallest separation between two points on the spectrum; (d) number of points, being 750 points maximum for double-beam, or 1500 points maximum for single-beam; (e) motor steps or motion; (f) apodising function (a choice of ten is available) for the transform. Operation thereafter is automatic, and the instrument shuts itself off when the required resolution is obtained. Scan times for 1500 points single-beam or 750 points double-beam

spectra vary, e.g., 5 min at 1-cm⁻¹ resolution for a 50-cm⁻¹ range; 2.7 hr at 0.1-cm⁻¹ resolution for a 250-cm⁻¹ range. Some spectra recorded by the MIR-2 are shown in Figures 36 and 37.

Prices: the MIR-2 equipped with electronics and tape readout is $19,600. The data system incorporating the computer is $26,250, so that the computerized system is $45,850.

Block Engineering, Inc.

The Digilab, Inc. subsidiary of Block Engineering, Inc., has two instruments capable of producing far-infrared spectra. One of these, the FTS-14, has already been described. Although principally intended for the fingerprint range, Model 296 of the FTS-14 can be modified.

The design tolerance of a Michelson interferometer set up to measure spectra in the mid-infrared region exceeds that required for the far-infrared. Only a few simple alterations are, therefore, necessary to modify the FTS-14 for far-infrared extension to 1000 μ (10 cm⁻¹). These include replacing the glower source with a high-pressure mercury lamp whose emissivity beyond 100 cm⁻¹ is much superior to that of a glower. The beamsplitters for this region are stretched Mylar polyester films of various thicknesses. The same type of triglycine sulfate (TGS) detector, having good sensitivity and flat response in the far-infrared, can be used. The far-infrared physical conversion is rapidly and simply done, with an equally simple prepackaged software change in the Digilab data system, the changeover is completed in less than an hour. The con-

Figure 38. Kit for conversion of Digilab FTS-14 system for far-infrared operation. A high-pressure mercury lamp, chopper, Mylar film beamsplitter, and TGS detector are used.

version is done by means of a "conversion kit," shown in Figure 38. The result is a computer-controlled far-infrared spectrometer; see the earlier description of the FTS-14 system.

The second Digilab, Inc., instrument is the FTS-16 spectrometer shown schematically in Figure 39. The optical system

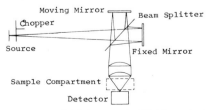

Figure 39. Digilab Model FTS-16 optical diagram

64

Chemical Instrumentation

Figure 40. FTS-16/Digilab far-infrared spectrometer

may be evacuated or purged. Wavelength range: 10 cm⁻¹ to 200 cm⁻¹, or 5 cm⁻¹ to 650 cm⁻¹ optional. Resolution: 0.5 cm⁻¹ (operator variable) is standard, 0.1 cm⁻¹ is optional. Interchangeable, stretched dielectric beamsplitters are used. A precision stepping motor provides 5-μ steps for the mirror motion. Each step can be adjusted to 0.5, 1, 2, or 4 sec nominal dwell time. The source is a high-pressure mercury arc. Detectors are the Golay cell or TGS detector used with the FTS-14 system. Cryogenic detectors can be incorporated. The detector signal is processed by a stable and linear lock-in amplifier tuned to the chopper frequency. The interferogram produced in this manner can be processed in several ways:

The detector signal can be fed to a 10-bit A/D converter and punched out on paper tape. An oscillator provides the required pulses to step the mirror drive and control the A/D converter. The interferogram can then be reduced by an off-line computer. A more favorable if more expensive configuration involves the Digilab data system. The FTS-16 is used with the basic Digilab system, which is used to produce fully ratio-recorded far-infrared spectra. The FTS-16 system itself plus a Teletype, Digilab system, and plotter are shown in Figure 40. A water vapor spectrum measured with the FTS-16 spectrometer, computed by off-line computer, is shown in Figure 41.

Prices: the conversion kit for the FTS-14 is $2500. The Model FTS-16 with electronics and tape readout is about $15,000. A basic Digilab system for use with it is about $40,000.

Grubb Parsons, Ltd.

Grubb Parsons produces two instruments which are marketed in the United

States by Edwin Industries Corporation. The IS3 spectrometer complete with electronics and console is shown in Figure 42. Optical diagrams for the IS3 are shown in Figure 43. The range of the IS3 is 10–675 cm⁻¹ if a diamond detector window is used, and 10–200 cm⁻¹ if a quartz window is used. The resolution is variable to 0.1 cm⁻¹ maximum. The moving mirror is accurately mounted on a glass block and moved in discrete steps of 5 μ path difference. These steps can be arranged by the operator to occur at time intervals of 0.5, 1, 2, 4 sec or longer. Polyethylene terephthalate beamsplitters in a range of thicknesses are used. Different sample

Figure 42. Grubb Parsons IS3 far-infrared spectrometer

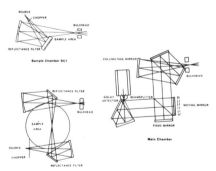

Figure 43. Optical diagram for IS3 spectrometer

chambers are available (see Fig. 43). The small-chamber SC1 will accommodate 10-cm gas cells as well as other cells. A rotary system enables up to six solid samples

to be placed in the sample chamber before evacuating. The large-chamber SC2 accommodates larger attachments and heating or cooling devices.

The IS3 electronics include a lock-in amplifier tuned to the 16-2/3 Hz or 20 Hz chopper, an A/D converter, a pulse drive unit for driving the moving mirror system and A/D converter, and an Addo paper-tape punch. The necessary transforms are carried out by an off-line computer.

Accessories available for the IS3 include a pen recorder, Pirani gauge, a microsampling accessory shown in Figure 44, and cell kits such as that shown in Figure 45.

The second instrument is the Mark II

Figure 44. Microsampling accessory for IS3

"Cube" interferometer designed by the National Physics Laboratory at Teddington, England, and developed by Grubb Parsons. An optical diagram of the Mark II is shown in Figure 46. Figure 47 shows the instrument plus electronics, and an "exploded" view of the interferometer is given in Figure 48.

The Mark II covers the 10–200 cm⁻¹ range, using a Golay detector fitted with a quartz window. An extension of the range to 650 cm⁻¹ is possible with special accessories. The resolution is variable at the operator's discretion to a maximum of 0.5 cm⁻¹, but can be improved to 0.1 cm⁻¹ by means of a special mirror drive unit. The mirror movement is accomplished by means of a high-precision stepping motor

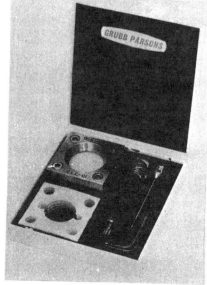

Figure 45. Grubb Parsons liquid cell kit

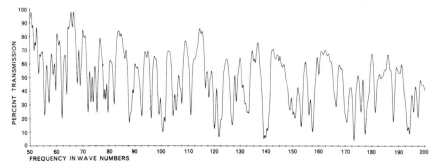

Figure 41. Water vapor spectrum, measured by Digilab FTS-16 spectrometer

Chemical Instrumentation

which produces discrete changes of path differences of 5 μ. These steps can be arranged at intervals of 0.5, 1, 2, or 4 sec, as required. Stretched dielectric beamsplitters such as those shown in Figure 49 are used. The source is a high-pressure mercury vapor lamp. The electronics include power supplies, a high-gain lock-in amplifier, an A/D converter, and an oscillator for operating the mirror drive and triggering the A/D converter. An encoding serializer is then used to drive a paper-tape punch. Data processing is carried out by an off-line computer.

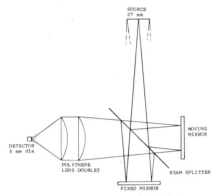

Figure 46. Optical diagram of Mark II cube

Figure 47. Grupp Parsons Mark II interferometer and electronics

A variety of sample cells is available for the Mark II. A dewar attachment bearing a copper cell mount for low-temperature work is shown in Figure 50. Solid and liquid sample holders are shown in Figure 51. The spectrometer can also be

Figure 48. Mark II interferometer

easily adapted for measurements of refractive index. If the sample material is placed in one of the two arms of the interferometer instead of the normal sample position, a phase delay is introduced which is different for each of the spectral components and manifests itself as a displace-

Figure 49. Accessories kit for Mark II spectrometer. The beamsplitters are carried by the large rings.

Figure 50. Dewar attachment for Mark II

Figure 51. Solid and liquid sample holders for Mark II

Figure 52. Mark II spectrometer set up for refractive-index measurements

ment of the central burst of the interferogram as well as asymmetry of the interferogram. The refractive index can be calculated from these data. The Mark II, set up for refractive-index measurements, is shown in Figure 52.

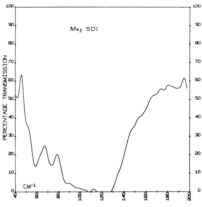

Figure 53. Spectrum measured with Mark II spectrometer

Two examples of spectra measured with the Grubb Parsons Mark II spectrometer are given in Figures 53 and 54.

Prices: the IS3 interferometer including electronics and tape readout is about $20,000. The Mark II interferometer including electronics and tape readout is about $12,000.

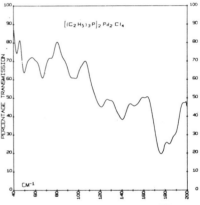

Figure 54. Spectrum measured with Mark II spectrometer

Manufacturers and Distributors

RIIC, Research and Industrial Instruments Company, a division of Beckman-RIIC Ltd.
Worsley Bridge Road
London SE 26, England
U. S. Distributor:
Beckman Instruments, Inc.
2500 Harbor Boulevard
Fullerton, California 92634

Digilab, Inc., a subsidiary of Block Engineering, Inc.
19 Blackstone Street
Cambridge, Massachusetts 02139

Chemical Instrumentation

CODERG, Société de Conversion des Énergies
15 Impasse Barbier
92 Clichy, France

U. S. Distributor:
Scientific Instrumentation, Inc.
P.O. Box 1345
Palo Alto, California 94302

Sir Howard Grubb Parsons & Co., Ltd.
Walkergate, Newcastle upon Tyne 6
England NE6 2YB

U. S. Distributor:
Edwin Industries Corporation
11933 Tech Road
Silver Spring, Maryland 20904

References

(1) VANASSE, G. A., AND SAKAI, H., in "Progress in Optics," Vol. VI, E. Wolf, ed., John Wiley & Sons, Inc., New York, 1967; pp. 261ff.
(2) CONNES, J., *Rev. Opt.*, **40**, 45, 116, 171, 231 (1961). An English translation is available from the Clearing House for Federal and Technical Information, Springfield, Va. 22151, as NAV-WEPS report 8099, NOTSTP 3157, order number AD 409 869; 178 pg., $18.40.
(3) MERTZ, L., "Transformations in Optics," John Wiley & Sons, Inc., New York, 1965.

Acknowledgment

The cooperation of the instrument manufacturers in supplying various photographs of instruments and data is gratefully acknowledged, as is partial support from the National Center of Air Pollution Control.

Topics in...

Chemical Instrumentation

Edited by **GALEN W. EWING**, Seton Hall University, So. Orange, N. J. 07079

The Journal of Chemical Education
Vol. 45, pages A861, A947
November and December, 1968

XLI. Reflectometers, Colorimeters, and Reflectance Attachments

WESLEY W. WENDLANDT, *Department of Chemistry University of Houston, Houston, Texas 77004*

Dr. Wesley W. Wendlandt was born on November 20, 1927, in Galesville, Wisconsin. He attended and graduated from high school in La-Crosse, Wisconsin, in 1945; he received the B.S. degree in chemistry and mathematics at Wisconsin State University (River Falls) in 1950; and his M.S. and Ph.D. degrees in inorganic and analytical chemistry from the State University of Iowa in 1952 and 1954, respectively.

From 1954 to 1966 he was on the faculty at Texas Technological College. In 1966, he became Chairman and Professor of Chemistry at the University of Houston. The summers of 1954 and 1955 were spent at the Argonne National Laboratory and he was a Visiting Professor at New Mexico Highlands University during the summer of 1961.

He has published 163 research papers, 20 book chapters, and 5 books. His research interests are in thermoanalytical techniques, solid-state reactions of coordination compounds, reflectance spectroscopy, photochemistry, and the thermal properties of coordination compounds.

INTRODUCTION

Perhaps unfamiliar to the average chemist is the spectroscopic technique called diffuse reflectance spectroscopy. In this technique, the radiation reflected from the surface of a sample is detected and recorded as a function of wavelength or wavenumber. The technique, contrary to transmittance measurements, may be applied to powdered solids, solids, liquids, or pastes.

Diffuse reflectance measurements of a sample are obtained by use of a reflectometer, spectroreflectometer, or colorimeter. A reflectometer is usually considered to be an instrument which employs a series of filters to obtain approximately monochromatic radiation; the spectroreflectometer is a reflectometer used in conjunction with a grating or prism monochromator; and lastly, the colorimeter is a filter type reflectometer but only three filters are employed, those for the C.I.E. color coordinate values of X, Y, and Z. The terms are used rather loosely in the literature and hence, there is some confusion as to their exact definitions.

Diffuse reflectance measurements can easily be made in the laboratory because reflectance attachments are available for practically all of the well known commercial spectrophotometers. These attachments are basically of three types, as illustrated in Figure 1. In (*a*), the integrating sphere type is illustrated. Monochromatic radiation enters the sphere through a side aperture, strikes the sample, the reflected radiation is collected by the sphere; and then detected by a photo-detector. This is by far the most widely used type of reflectance attachment. In (*b*), the reflected radiation from the sample is collected by an annular, ellipsoidal mirror and then detected by the photo-detector. A

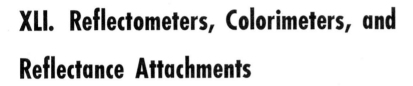

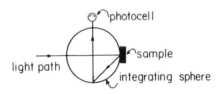

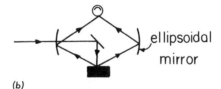

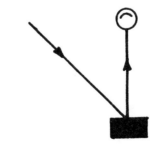

Figure 1. Types of reflectance attachments. (a) Integrating sphere type. (b) Annular, ellipsoidal mirror type. (c) Reflection type.

simple arrangement is shown in (*c*) where the reflected radiation is detected by a photo-detector without any type of collector device being used.

No attempt will be made here to discuss the spectrophotometers which are employed with the reflectance attachments. These instruments have previously been discussed by Lott (1, 2) in this series. Only the attachments, the specific reflec-

tometer, or colorimeter, will be described here.

More and more interest is being shown in the technique of diffuse reflectance spectroscopy. Several books have appeared on the subject (3, 4) as well as a recent symposium (5). With the wide availability of the instrumentation, it is expected that the technique will find even greater uses in the future.

REFLECTANCE ATTACHMENTS

Bausch and Lomb. Two reflectance attachments are available for Bausch and Lomb spectrophotometers. For the Spectronic 505 spectrophotometer, the attachment shown in Figure 2 is employed for reflectance measurements. It consists of a 6 in. in diameter cast aluminum sphere,

Chemical Instrumentation

Figure 2. Reflectance attachment for Bausch and Lomb Spectronic 505 spectrophotometer.

Figure 4. Reflectance attachment for the Beckman DB spectrophotometer (shown mounted on the spectrophotometer).

coated on the interior with barium sulfate and equipped with an end-on window EMI 6255 photomultiplier tube, sample and reference holders, and a lens system. Monochromatic radiation is focused by the lens system onto the sample and reference material surfaces. The incident radiation illumination is circular in area and about 16 mm in diameter. The specular component of the reflected radiation may be rejected by using light traps located at an angle of 90° to the sample and reference materials. A low stray-light level of less than 0.1%, or better, is claimed, with a usable wavelength range of 220 to 650 mμ, or, 220 to 700 mμ with a special photomultiplier tube and cut-off filter.

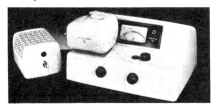

Figure 3. Spectronic color analyzer for Bausch and Lomb Spectronic 20 colorimeter-spectrophotometer.

A somewhat lower accuracy system, the Spectronic 20 Color Analyzer, is available for use with the Spectronic 20 colorimeter-spectrophotometer. The attachment, as shown in Figure 3, consists of a small integrating sphere, a mirror and lens system, shutter, and a photomultiplier tube. The sample and/or reference material is mounted on the top, rather than at the side of the sphere. A rectangular spot of incident radiation, 2 × 8 mm in area, strikes the sample or reference surface. Fixed slits provide a band-pass width of 20 mμ from 400 to 700 mμ. Conversion of percent reflectance values to the C.I.E. trichromatic X, Y, and Z values is accomplished by a chart supplied with the attachment.

Beckman. Perhaps the most widely used attachments for reflectance studies are those for Beckman spectrophotometers. At the present time, four attachments are available.; they are for the Models DU, B, DB, and DK-1A and DK-2A spectrophotometers, respectively.

In the Model B reflectance attachment, monochromatic radiation enters the integrating sphere through a conical tube and is reflected to the surface of the sample or

ment is that designed for the Model DU manual spectrophotometer. The attachment has been employed for a wide variety of applications in both the pure and applied science areas. With the advent of automatic recording spectrophotometers, it is perhaps not as widely used at the present time. The attachment consists of a single metal casting containing an ellipsoidal mirror optical system and a sample-reference drawer. It is attached to the spectrophotometer by a simple interchange with the cell compartment. Diffuse radiation, reflected upward from the sample at angles between 35° and 55°, strikes the ellipsoidal mirror, and passes through an opal glass diffusing screen to the detector.

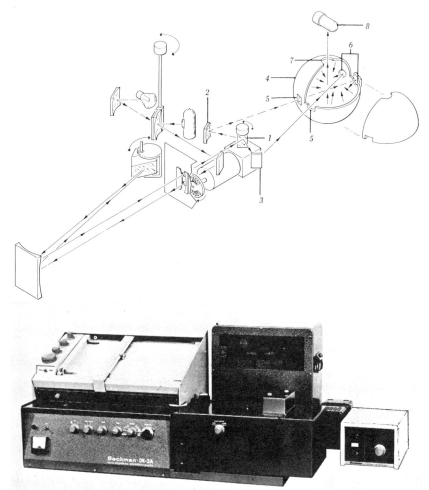

Figure 5. Reflectance attachment for Beckman Models DK-1A and DK-2A spectrophotometers. (a) Schematic diagram. (b) Attached to the Model DK-2A spectrophotometer.

1. oscillating mirror; 2. sample beam mirror; 3. reference beam mirror; 4. integrating sphere; 5, entrance ports (cells are placed here for absorbance and transmittance measurements); 6. exit ports (sample and reference materials are placed here for reflectance and fluorescence measurements; otherwise, ports are normally covered with magnesium oxide plates); 7. integrating sphere opening (filter is placed here for fluorescence measurements); 8. detector.

reference material by a plane mirror set at a 45° angle. Reflected radiation from the sample or reference material, between 0° and 25° to the normal, returns to the mirror, thus excluding the specular component. Normally, a rectangular area, 1/8 × 3/16 in., is exposed to the incident radiation. A beam expander is available to increase the area to 1/4 × 3/8 in. if so desired, for larger, less homogeneous samples. The reflectance attachment may be used for solid, powder, or liquid samples in the wavelength range of 325 to 1000 mμ.

The most widely used reflectance attach-

Radiation striking the detector (either red- or blue-sensitive phototubes or on later models, a photomultiplier tube) produces a voltage across a load resistor which is then amplified and registered on the null meter. The sample drawer accomodates either a solid or a liquid sample and a standard reference material. The sample and the reference material should be approximately the same size, 2.5 in. long × 1.5 in. wide × 1 in. thick; a small cup or beaker (not supplied) is used to retain a liquid or paste-like sample.

The attachment for the Model DB spec-

Chemical Instrumentation

trophotometer is illustrated in Figure 4. This unit consists of a magnesium oxide coated integrating sphere with two side apertures for the mounting of sample and reference materials. A dual plane mirror system directs the monochromatic radiation from the spectrophotometer cell compartment into the integrating sphere. Using the red-sensitive photomultiplier tube, the wavelength range is 380 to 720 mμ.

The reflectance attachment for the Models DK-1A and DK-2A spectrophotometers is shown in Figure 5. This attachment also utilizes a magnesium oxide coated integrating sphere. The sample and reference materials are illuminated by monochromatic radiation from the monochromator by means of an oscillating plane mirror beam splitter. Either the total or diffuse reflectance of a sample may be measured; the former is obtained by mounting the sample at an angle of 5° to the incident radiation beam. Using a 1P28 photomultiplier tube and/or a lead sulfide cell as detectors, the range of the attachment is 210 to 2700 mμ.

Cary. Three diffuse reflectance attachments are available for the Cary Models 14 and 15 spectrophotometers. All three of the attachments are high resolution instruments capable of performing measurements of diffuse reflectance, diffuse plus specular reflectance, and diffuse transmission of transparent or semi-opaque liquid, powdered, or solid samples.

The Model 1411 diffuse reflectance attachment, as illustrated in Figure 6, uses either an integrating sphere or an annular type mirror collector. With the integrating sphere, two beams of chopped monochromatic radiation enter the sphere, via a lens-and-mirror system, at a 90° angle to each other, and strike the sample and reference materials at a 0° angle of incidence.

The reflected radiation is collected by the sphere and detected by a photomultiplier tube mounted over an aperture at the top of the sphere. In another mode of measurement, the sample and reference materials are illuminated by an external lamp source, and the reflected radiation detected by the photomultiplier tube after being passed through the monochromator.

With the ring collector accessory, the sample is illuminated at a 0° angle of incidence with chopped monochromatic radiation. The reflected radiation is collected by an annular mirror mounted above the sample and is detected by a photomultiplier tube. A screen attenuator located in the reference beam permits the reference beam to be standardized to a reference sample.

The specifications and resolution of the two accessories (sphere and ring mirror) for the Model 1411 reflectance attachment are given in Table I.

Figure 7. Cary cell space diffuse reflectance attachment, showing the sample sphere (top) and the reference box (bottom).

The cell-space total diffuse reflectance attachment, Model 1411750 (for the Cary Model 14 spectrophotometer) and 1511000 (for the Cary Model 15 spectrophotometer), is illustrated in Figure 7. The reflectance attachment consists of two integration chambers; a spherical chamber which is used for the sample, and a light integrating box in which is positioned the reference material. Each chamber contains a plane mirror which is used to focus the monochromatic radiation onto the sample and reference materials, respectively. The reflected radiation is then detected by a photomultiplier tube. A compensation system is employed to compensate for any difference in absorption by the sample and reference chamber walls.

The third reflectance attachment, the Model 50-400-000, is shown in Figure 8. This attachment, introduced only recently, has been described in detail by Hedelman and Mitchell (6). The attachment is intended for measurements of reflectance and total diffuse transmittance using a two in. photomultiplier tube and a 25 cm (9 in.) diameter integrating sphere, coated on the inside with magnesium oxide. The sample is mounted in a horizontal plane to facilitate measurements on uncovered liquids, pastes, and powdered samples, as well as solid samples.

Monochromatic radiation from the spectrometer cell compartment is directed into the attachment aperture by means of a lens and mirror arrangement. A flat mirror directs the radiation downward from the sphere aperture to the sample aperture at the bottom of the sphere. The reflected sample radiation is collected by the sphere and detected by either a photomultiplier tube or a lead sulfide detector. The photomultiplier tubes suggested for the attachment are the Dumont 7664, RCA 7326, or RCA 6217. Apparently, no single tube's spectral response curve is ideal for all types of reflectance measurements. Measurements can be made, using any of the above detectors, from 2500 to 20,000 Å (can be extended to 25,000 Å by use of a pen-period control accessory). To make measurements of diffuse transmittance, the sample is placed into the holder "V" block at the entrance aperture of the sphere and the detector at the top aperture.

General Electric. Perhaps the oldest and most widely used recording spectrophotometer for the visible range is the General Electric instrument, developed by A. C. Hardy in the 1920's and first sold commercially in the 1930's. More than 300 of the instruments have been sold to date. Some of the units shipped in the 1930's are still in use today; this is certainly some sort of a record for instrument longevity and use. The instrument, as shown in Figure 9, is used to measure and record the color of samples in the visible range of 380 to 700 mμ. An optical accessory extends the range to 1000 mμ. Also available is an optional automatic tristimulus integrator which simultaneously computes the X, Y, and Z color coordinates.

In principle, the instrument is a double-beam spectrophotometer with a double monochromator. Monochromatic light from a slit is transmitted successively through a balance Rochon, Wollaston prism, rotating polarizing filter, separate double lenses, and into the 8 in. diameter

Figure 6. Cary Model 1411 diffuse reflectance attachment with integrating sphere in position.

Chemical Instrumentation

Table I. Specifications and Resolution of The Cary Model 1411 Reflectance Attachment

	Sphere	Ring Mirror
Wavelength range	2500–7000 Å	2200–7000 Å
Nominal resolution	2.5 to 7 Å	1.3 to 7 Å from 3000–6000 Å; 20 Å at range limits
Detection	Integration, 180° (2π solid angle)	Directional, 45° ±7° angle cone
Sample	With either accessory, semi-opaque or transparent liquid, powdered, or solid samples	

integrating sphere. The reflected light from the sample and reference materials is collected by the sphere and reaches the phototube through a plastic rod light pipe. The diameter of the sample and reference material aperture is $^{13}/_{16}$ in. Smaller areas can be obtained by use of a beam reducing lens accessory.

Perkin-Elmer. A diffuse reflectance attachment is available for use on the Models 350 and 450 spectrophotometers. The attachment, as shown in Figure 10, consists of two precision cast plastic spheres, each coated with magnesium oxide on its interior surface. Incident monochromatic radiation, at an angle of incidence of 16° 15′, is reflected by either the sample or reference surface and then collected by the integrating sphere and detected by the photomultiplier tube. A choice of either black or white aperture covers enables the investigator to reject or accept the specular reflectance component. Each sphere has an individual end-window EMI 9592B photomultiplier tube which is usable in the wavelength range from 220 to 750 mμ. Using a lead sulfide detector, range extension to 2.5μ is possible (special order only).

Precision attainable for reflectance measurements is said to be ±0.25% in the 0–100% reflectance range. For fluorescent samples, the effect of fluorescence is observed by placing the 100% setting of the instrument at half-scale to provide a 0–200% reflectance range. The 100% reference base line is flat to within 2% re-

flectance. Reflecting samples may vary in thickness from thin paper to solids 45 mm thick. The sample support can be removed altogether to accomodate bulky objects. Minimum diameter at sample and reference apertures is 45 mm while the maximum sample area is approximately 180 × 100 mm.

The diffuse reflectance attachment for the Hitachi Perkin-Elmer Model 139 spectrophotometer also employs an integrating sphere. Monochromatic radiation strikes the sample or reference material surface at a 0° angle of incidence. The reflected radiation is collected by the 140 mm diameter sphere, and detected by the photodetector which is mounted at a 90° angle to the sample. Minimum size for solid samples is 25 × 25 mm while the powder cell is 22 mm in diameter by 10 mm in depth. The wavelength range of the instrument is 400 to 750 mμ.

Unicam. Three reflectance attachments are available for the Unicam

Models SP500, SP700, and SP800 spectrophotometers, respectively.

The Model SP540 reflectance attachment (for the Model SP500 spectrophotometer) consists of a concave annular mirror optical system mounted in a rigid frame and containing a sliding sample and reference drawer. Incident monochromatic radiation is reflected down onto the sample by means of a plane mirror. The reflected radiation is then collected by the annular mirror and passed through a glass diffusing screen onto the photocell. Only the sample or reference material is exposed to the incident radiation at any one time by means of the sliding drawer. This attachment is similar to that previously described for the Beckman Model DU spectrophotometer.

The Model SP735 reflectance attachment (for the Model SP700 spectrophotometer), illustrated in Figure 11, is a rather compact unit which is fitted into the cell compartment of the spectrophotometer. Radiation from the cell compartment sample and reference beams is reflected through the apertures in the side of the ellipsoidal mirror by means of four plane mirrors and two lenses. Located inside the annular ellipsoidal mirror are two additional plane mirrors which reflect the incident radiation onto the sample and reference materials at angles near normal incidence. Images of the prism are formed on the sample and reference apertures by means of two additional lenses to ensure even illumination. The area illuminated by each beam is $^1/_4 \times ^1/_2$ in. Reflected radiation is collected by the ellipsoidal mirror and focused onto the photomultiplier tube. Two detectors are employed, an E.M.I. 6255B photomultiplier tube for the 200 to 700 mμ range, and a lead sulfide cell for use in the 700 mμ to 2.5μ range. The standard sample holder has a semicircular aperture. For smaller samples, the sample beam may be stopped down and an expanded scale on the spectrophotometer used, or both beams may be stopped down. Since samples are mounted horizontally on top of the attachment, there is virtually no limit on the size of sample that can be accommodated.

The third reflectance attachment for Unicam spectrophotometers is the Model SP890 diffuse reflectance attachment for the Model SP800 spectrophotometer. The attachment, as illustrated in Figure 12, is similar to the Model SP735 attachment except that the sample is mounted vertically rather than horizontally. The attachment contains an ellipsoidal mirror reflector and an optical system similar to the Model SP735 attachment. A specular reflectance accessory is also available.

Figure 8. Cary Model 50-400-000 reflectance attachment for the Model 14 spectrophotometer.

Chemical Instrumentation

Zeiss. There are three reflectance attachments for Zeiss spectrophotometers; they are the Models RA2, RA3, and RA20 for the Zeiss PMQII and RPQ20A spectrophotometers, repsectively.

The model RA2 attachment, as illustrated in Figure 13, is a nonintegrating sphere type in which monochromatic radiation from the slit and field lens strikes the surface of the sample at an angle of 45° to the perpendicular. The field lens at the exit slit of the monochromator forms an image of the aperture of the monochromator prism on the surface of the sample, so that a change in slit width only affects the brightness of the sample surface. The reflected radiation, at 90° to the surface of the sample, is measured using a phototube. Sample and reference materials are pressed against the measuring aperture by means of springloaded discs. Two samples may be compared against a reflectance reference by use of the sample changer. The RA2 attachment is usable in the wavelength range from 200 to 600 mμ.

The Model RA3 attachment, which also is attached to the PMQII spectrophotometer, is illustrated in Figure 14. Two modes of operation of the attachment are employed; mode A (illustrated in Figure 14), in which the sample is placed over an aperture at the top of the integrating sphere and illuminated by diffuse radiation from the lamp; and in mode B, the sample is illuminated by monochromatic radiation from the monochromator and the

Figure 10. Perkin-Elmer reflectance attachment with cover open showing reflecting spheres.

reflected radiation detected by a phototube. Reflectance measurements using the Model RA3 attachment can be made in the wavelength range from 380 to 2500 mμ.

The Model RA20 reflectance attachment is used in conjunction with the Zeiss Model RPQ20A spectrophotometer. The attachment uses an oscillating mirror to deflect both the sample and reference beams upwards through apertures in the

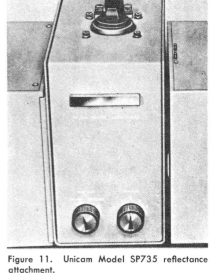

Figure 11. Unicam Model SP735 reflectance attachment.

integrating sphere onto the sample and reference surfaces. Reflected radiation is collected by the sphere and detected by a photomultiplier tube mounted over an aperture on the side of the sphere. A 1P28 photomultiplier tube covers the wavelenth range from 200 to 620 mμ while a PbS detector is used for the 620 to 2500 mμ wavelength range.

REFLECTOMETERS AND COLORIMETERS

Agtron (Magnuson Engineers, Inc.). The Agtron reflectometer is used principally to obtain reflectance measurements of various food products. There are several different models available, the Model M-400-A reflectometer is illustrated in Figure 15. The sample chamber is illuminated by monochromatic light obtained from two concentric discharge tubes—a mercury lamp and a neon lamp. Filters are used in conjunction with the lamps to isolate the 436, 546, and 640 mμ lines with the 585 mμ line available as an optional feature. The reflected light is collimated and focused onto a phototube. Various sample holder configurations are available for solid, paste, or liquid materials.

Colorcord (National Instrument Laboratories, Inc.). The Joyce-Loeble Mark IIA Colorcord is an integrating sphere

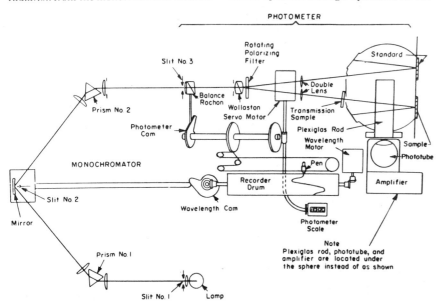

PHOTOMETER

(a) Schematic diagram.

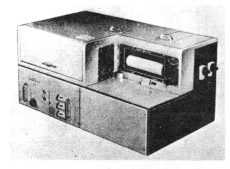

(b) Complete instrument.

Figure 9. The General Electric recording spectrophotometer.

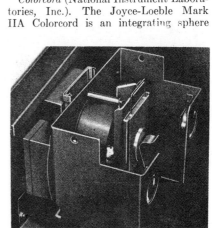

Figure 12. The Unicam Model SP890 diffuse reflectance accessory.

Chemical Instrumentation

type colorimeter. The sample color is read out directly in terms of the X, Y, and Z coordinates on three digital dials located on the bottom front of the instrument.

Gardner. The Gardner Laboratory, Inc., has available multipurpose, precision, and portable reflectometers, tristimulus colorimeters, gloss measurement reflectometers, color-difference meters, goniophotometers, hazemeters, clarity meters, and so on. Due to the lack of space, only the reflectometers will be covered in part here. The reflectometers are classified according to their operation: manual, automatic, precision, and portable instruments. They are used mainly in applied science areas such as paint color, opacity, and hiding power; paper brightness; color of various materials; fluorescence; gloss; and luminous transmission measurements.

The Gardner Model 100M-1 Multipurpose Reflectometer is illustrated in Figure 16. The instrument is used to obtain reflectance, transmission, and color measurements using a 45°, 0° geometry. Light from an incandescent projection lamp, after passing through appropriate filters, is divided into two beams. One beam goes to a fixed comparison photocell, the other, after being reflected by the sample, goes to a movable (test) photocell. The

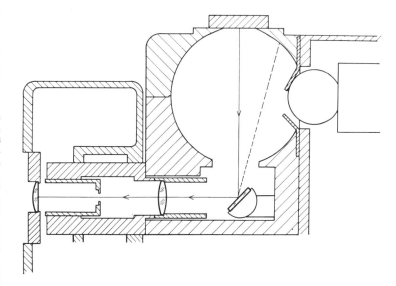

(a) Schematic diagram.

(b) Attached to the PMQII spectrophotometer.

Figure 14. Zeiss Model RA3 reflectance attachment.

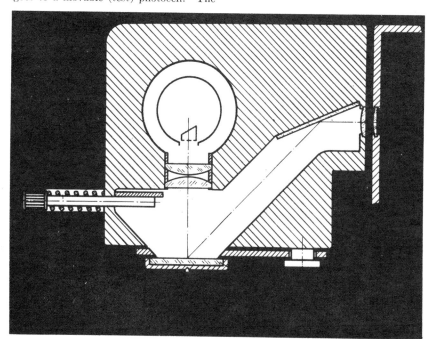

(a) Schematic diagram.

Figure 13. Zeiss Model RA2 reflectance attachment.

(b) Mounted on PMQII spectrophotometer.

distance of the movable photocell from the sample is manually adjusted until the currents in both cells are equal, as shown by the zero indication of the galvanometer. A scale attached to the movable cell is calibrated in percent reflectance. For normal reflectance measurements, the instrument comes equipped with green, amber, and blue filters. Response of the light-filter-photocell combination closely approximates that of the C.I.E. Standard Observer. An attachment for measurements of transmission and 75% gloss and one for measurements of 60% gloss are available.

The Gardner Automatic Multipurpose Reflectometer is similar to the above except that the outputs of the photocells are used to actuate a servo-mechanism which adjusts the position of the movable cell until the two photocell currents are equal.

Figure 15. Agtron Model M-400-A reflectometer with the Model M-30-A wide area viewer.

Hunterlab. The Hunterlab D25 Color Difference Meter, as illustrated in Figure 17, is used to make precise measurements of the color of flat surfaces, as they appear in daylight. Values for the color are read directly from digital dials on the three color scales:

Chemical Instrumentation

L—measures lightness and varies from 100 for perfect white to 0 for black;

a—measures redness when plus, gray when 0, and greenness when minus;

b—measures yellowness when plus, gray when zero, and blueness when minus.

With these three scales, it is possible to represent colors by position in a three dimensional coordinate system.

The instrument consists of two parts; an optical unit containing the light source, window for the sample, and the phototubes; and an electrical unit which converts the phototube currents to scale readings proportional to color. Light from one of the lamps is projected onto the sample from opposite sides at 45° angles. Between these two beams is a vertical light pipe which accepts the light reflected from the sample in and near the perpendicular direction. At the upper end of the light pipe, light is distributed to three 1P39 phototubes, each having a different tristimulus filter in front of them. The phototube currents are converted to voltages proportional to C.I.E. X, Y, and Z values, respectively.

The Hunterlab Model D40 Whiteness Reflectometer, illustrated in Figure 18, is a precision, photoelectric instrument with 0°, 45° geometry for measuring luminous (green) and blue reflectance of white and near-white samples. An attachment is available for measurement of the contribution of fluorescent brightness to whiteness.

The section on the left side of the instrument is the optical unit containing

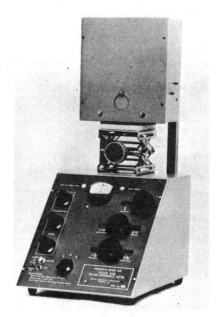

Figure 17. Hunterlab Model D-25 color and color difference meter.

the light source, beam-forming lens, phototubes, light pipes, and light baffles. The right side of the unit contains the power supply, standardizing rheostats, and digital read-out dial. With each pair of phototubes, one tube is exposed to the light source, the other to light reflected from the sample. The beams cross a single blue or green filter, placed in front of the phototubes. The dial is turned in such a direction to null the meter and the corresponding Y and Z values obtained.

Figure 18. Hunterlab Model D40 whiteness reflectometer.

Other reflectometer type instruments available from Hunterlab include the Model D-16 Glossmeter; the Model D25P Sphere Color Difference Meter (similar to the Model D25 but containing an integrating sphere); the Model D10 Recording Goniophotometer; and the Model D44 Color Monitor.

IDL (Instrument Development Laboratories). Two varieties of the IDL Signature Model Color-Eye are available. The two differ only in the size of the attached integrating sphere; the Large Sphere Signature Model Color-Eye incorporates an 18 in. diameter integrating sphere.

Figure 19. The IDL Signature Model Color-Eye

The Signature Model Color-Eye, as shown in Figure 19, is a four filter tristimulus colorimeter and a 10 or 16 filter abridged spectrophotometer. The instrument can measure both fluorescent and non-fluorescent materials and can make both reflectance and transmittance measurements. Using an integrating sphere, the sample and reference materials are illuminated by an incandescent light source. The reflected light beams are focused into a single beam, chopped by a rotating flicker mirror, and alternately focused on the photomultiplier tube through either the tristimulus or spectral filters. The reflected light reaching the photomultiplier tube consists of a steady component representing the mean brightness of the sample and reference, and a "flicker" component representing the difference in brightness between sample and reference. By use of various electronic circuits, the flicker component is converted to percentage difference in the reflectance of sample and reference materials. The illuminated sample area is elliptical, 7/16 × 3/4 in. in center of a 1 in. diameter aperture. Accessories per-

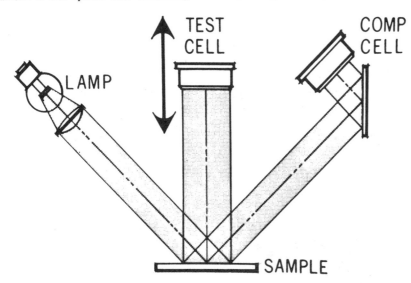

(a) Schematic diagram.

(b) Complete instrument.

Figure 16. Gardner Model 100 M-1 multipurpose reflectometer.

74

Chemical Instrumentation

mit viewing smaller sample areas. Special purpose analog computers are available for automatic data reduction and analysis of the Color-Eye data.

Figure 20. Large sphere Signature Model Color-Eye.

The Large Sphere Signature Model Color-Eye, as illustrated in Figure 20, is designed primarily for measurement of the reflectance of highly directional textile materials or samples with irregular surface characteristics. It incorporates an 18 in. diameter integrating sphere with diffuse illumination and viewing at 8° from normal. Sample viewing size is approximately 3 in. in diameter, reducible to approximately 1 3/4 in. when larger samples are not available. A source of ultraviolet illumination for measuring samples containing optical brighteners or fluoroescent dyes can be obtained as an accessory.

More sophisticated industrial process control instruments are available from IDL, such as the Colorede (Automatic Colorimeter for continuous processes), the Color-Eye Shade Sorting System, and the Color Analysis Computer System. It is beyond the scope of this article to discuss these instruments in detail.

Zeiss Photoelectric Reflection Photometer. The Zeiss Elrepho (*Electric Reflectance Photometer*), illustrated in Figure 21, is used for measuring the directional reflectance of diffusely reflecting materials in the visible wavelength range. It employs an integrating sphere and special filters for the

determination of the tristimulus values in the C.I.E. system. Measurements with other suitably selected filters result in values for the white content of nearly white materials.

The sample is placed at a 35 mm diameter aperture located in the bottom of the sphere and is indirectly illuminated by

Figure 21. The Zeiss Elrepho.

diffuse light from two lamps located at the sides and at 90° to the sample. A standard white plate is placed at an aperture near the sample. Reflected radiation from the sample and the standard white plate is imaged on two photocells, respectively. An adjustable measuring diaphragm is located in front of the standard white plate photocell. Provision is made also for the introduction of the various filters in front of each of the photocells. The photocells are connected together in a differential circuit, the output of which after amplification is connected to a zero indicator. A neutral gray wedge, located in front of the sample photocell, is used to equalize the two photocell currents with a given diaphragm position.

The integrating sphere may also be illuminated with a high pressure xenon lamp to detect the fluorescence of "optical brighteners." This results in a sample reflectance curve which has a fluorescence curve superimposed upon it.

MANUFACTURERS

Bausch & Lomb, Inc., Analytical System Division, Rochester, N. Y. 14602.

Beckman Instruments, Inc., Scientific Instruments Division, Fullerton, Calif.

92634.

Cary Instruments, 2724 South Peck Road, Monrovia, Calif. 91016.

Gardner Laboratory, Inc. P. O. Box 5728, Bethesda, Md. 20014.

General Electric Co., Industrial Process Control Division, 40 Federal St., West Lynn, Mass. 01905.

Hunter Associates Laboratory, Inc., 9529 Lee Highway, Fairfax, Va. 22030

Instrument Development Laboratories, 67 Mechanic St., Attleboro, Mass. 02703.

Magnuson Engineers, Inc., 1010 Timothy Dr., San Jose, Calif. 95133.

National Instrument Laboratories, Inc., 12300 Parklawn Dr., Rockville, Md. 20852.

The Perkin-Elmer Corporation, Norwalk, Conn. 06852.

Unicam Instruments, Ltd., York St., Cambridge, England; in U.S.A.: Philips Electronic Instruments, 750 South Fulton Ave., Mount Vernon, N. Y. 10550.

Carl Zeiss, Inc., 444 Fifth Ave., New York, N. Y. 10018.

BIBLIOGRAPHY

(1) LOTT, P. F., This Journal, **45**, A169 (1968). [Reprinted in this volume]

(2) LOTT, P. F., This Journal, **45**, A273 (1968). [Reprinted in this volume]

(3) WENDLANDT, W. W. and HECHT, H. G., "Reflectance Spectroscopy," Interscience, New York, 1966.

(4) WENDLANDT, W. W. (Ed.), "Modern Aspects of Reflectance Spectroscopy," Plenum Press, New York, 1968.

(5) 154th National Meeting of the American Chemical Society, Chicago, Ill., Sept. 11–12, 1967.

(6) HEDELMAN, S. and MITCHELL, W. N., "Modern Aspects of Reflectance Spectroscopy," W. W. Wendlandt (Ed.), Plenum Press, N. Y. (1968); p. 158.

Topics in...

Chemical Instrumentation

Edited by **GALEN W. EWING,** Seton Hall University, So. Orange, N.J. 07079

...a

feature

The Journal of Chemical Education
Vol. 44, pages A7, A99, A187, A289, A399, A499
January to June, 1967

XXXII. X-Ray Diffraction Analysis[*]

Part One—Safety and Generators

Reuben Rudman, *Department of Chemistry,
Brookhaven National Laboratory, Upton, N.Y. 11973*

Dr. Rudman is presently an Assistant Professor of Chemistry at Adelphi University. He is currently serving as a member of the Education Committee of the American Crystallographic Association and of the Crystallographic Apparatus Commission of the International Union of Crystallography.

—1970—

Reuben Rudman obtained his Ph.D. in X-ray crystallography and inorganic chemistry from the Polytechnic Institute of Brooklyn in 1966. He received his B.A. from Yeshiva University (1957). Dr. Rudman is in the latter half of a two-year appointment as a Research Associate at Brookhaven National Laboratory, after which he intends to pursue an academic career. His principal research interests include low-temperature X-ray diffraction techniques, solid-solid phase transitions, and materials research. He is presently studying crystal structures using single-crystal X-ray and neutron diffraction techniques. Dr. Rudman is a member of the American Crystallographic Association, the American Chemical Society and the American Association for the Advancement of Science. He has also consulted for industry.

This review is limited to a discussion of the principles, features, and availability of X-ray diffraction equipment. X-ray diffraction theory has been kept to a minimum. (For more detailed discussions of theoretical principles the reader is advised to consult the many texts that are available (5–11); an extensive list of books in this and closely related fields is found in the catalog of Polycrystal Book Service.)

During the course of an X-ray diffraction analysis, the investigator often uses apparatus which is not unique to this field. Examples are: materials for sample preparation, radiation detectors, electronic components for detector panels, voltage stabilizers, and film comparators. In such cases representative examples of available equipment are given, with emphasis on the equipment designed for specific applications to the field of X-ray diffraction.

In line with the policy of this series, the discussion has been limited, with few exceptions, to manufacturers with sales outlets in the U.S. The scope of this review is further limited by the fact that apparatus used in the determination of the crystal structure from intensity data, as well as those instruments designed for studying crystal perfection and surfaces, have not been included.

The preparation of this article would not have been possible without the cooperation of the representatives of the many companies whose products are discussed. I would like to thank them for their assistance in furnishing up-to-date technical information and photographs. The scale of price ranges and a complete listing of "Manufacturers and Distributors" are published as an appendix to this paper.

Introduction

A crystal, consisting of atoms arranged in a pattern which is repeated regularly in three dimensions, acts as a three-dimensional diffraction grating for X-rays (with wavelengths from about 0.5–2.5 Å). The X-rays interact with the atoms (more precisely, with their electrons) present in the crystal and the waves of scattered X-rays reinforce one another in certain directions. Bragg has shown that reinforcement will occur when the rays diffracted from parallel planes are in phase with one another, i.e., when the path difference is an integral number of wavelengths. From reference to Figure 1 we see that this can be represented by

$$n\lambda = 2d \sin \theta,$$

where λ is the wavelength, d is the distance between parallel planes, θ is the angle of incidence (and reflection) of the X-rays, and n is an integer. This is the well known Bragg Law.

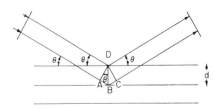

Figure 1. Bragg Law: Reinforcement of the diffracted rays occurs when the path difference is equal to an integral number of wavelengths ($n\lambda$); i.e., $\overline{AB} + \overline{BC} = n\lambda$, where n is an integer. Since $\overline{AB} = \overline{BC} = d \sin \theta$, $n\lambda = 2d \sin \theta$.

[*] Prepared under the auspices of the Atomic Energy Commission.

Chemical Instrumentation

X-rays may be produced when high-energy electrons strike matter of any sort. In practice, X-rays are produced in "X-ray tubes" when electrons, emitted from a hot filament (cathode), are accelerated toward and strike a metallic target (anode). The cathode is generally maintained at a potential of thirty to fifty thousand volts relative to the anode.

The radiation emitted when the electrons strike the target can be classified as: (a) the continuous spectrum; (b) the characteristic or line spectrum.

The continuous spectrum is due to X-rays emitted by electrons interacting with the fields of force of the nuclei in atoms of the target material. The minimum wavelength in the continuous spectrum is a function of the accelerating voltage only; it is independent of the target material. An increase in the applied voltage causes the minimum wavelength to shift toward smaller values (shorter wavelengths correspond to higher frequencies or higher energies).

When the applied voltage is sufficiently high, the emitted radiation contains a group of sharp lines (the line spectrum) in addition to the continuous spectrum. These lines occur at specific wavelengths whose locations are characteristic of the composition of the target. On striking the target, the impinging electrons interact with electrons of the target atoms, raising them to higher energy levels or ejecting them from the atoms. A vacancy then exists in the original electron energy level; it is filled almost instantaneously by an electron from a higher level. When an electronic transition of this sort occurs, from a higher to a lower energy level, a photon is emitted. The wavelength (or frequency) of the photon is a function of the energy difference between the two levels. It is therefore *characteristic* of the target element.

All types of X-ray apparatus have certain characteristics in common: an X-ray unit consisting essentially of a transformer, which converts line voltage to the high voltages needed, together with the necessary controls; an X-ray tube; a system of slits (collimators) to define the x-ray beam; means of holding, orienting, and rotating the specimen in the X-ray beam; and radiation sensitive devices, usually film or counters, to record the scattered radiation.

RADIATION HAZARDS AND SAFETY PRECAUTIONS

The potential danger to the operator of X-ray diffraction equipment, and to others, makes it mandatory for him to be familiar with basic safety precautions.

This discussion of radiation hazards and safety precautions associated with X-ray diffraction equipment has been taken from the following sources: Report of Apparatus and Standards Committee of the American Crystallographic Association (1a); Report of the International Union of Crystallography, Commission of Crystallographic Apparatus (2); and Section 6, Volume III of the International Tables for X-Ray Crystallography (3). Each of these reports contains references to other material describing safety precautions and governmental regulations.

Exposure to ionizing radiation can result in injuries that manifest themselves in the exposed individual and in his descendants; these are called somatic and genetic injuries, respectively. The objectives of radiation protection are to prevent or minimize somatic injuries of persons occupationally exposed to ionizing radiation and to minimize deterioration in the genetic constitution of the population. It should be stressed that recommendations in respect of the installation and operation of the equipment are not in themselves sufficient to guarantee adequate protection. Such protection depends largely on the expert knowledge of the staff and on their cooperation in carrying out the instructions prepared in the interests of radiation protection. When X-ray analysis equipment is used for research purposes, the nonroutine nature of the work greatly enhances the possibility of radiation injuries. It is therefore essential that the personnel should be aware of the hazards and how to guard against them.

In various localities a number of regulations already exist, which usually apply to the use of equipment in factories, educational establishments, etc., and to which individuals are required, or advised, to adhere.

One aspect of the nature of radiation leakage as it applies generally to X-ray diffraction equipment must be stressed: In many cases, this type of leakage is in the form of highly collimated beams of high intensity. Thus the leakage at a given location may be truly negligible while a fraction of an inch from this location it may reach dangerous proportions. This fact should always be kept in mind when checking X-ray diffraction equipment for radiation leakage.

The following comments have been adapted from Reference (2):

"*Sources of Unwanted Radiation.* X-ray crystallographic techniques involve the use of an X-ray generator to which is coupled some form of apparatus such as a photographic camera or counter diffractometer. In operation, dangerous leakage and scattering of radiation can occur in the following ways.

(1) Escape of primary X-rays when a tube-housing window is open and apparatus is not in position at that window.

If, while an X-ray tube is energized, a tube-window is open when apparatus is not in the operating position at that window, there can be an extremely dangerous escape of the primary beam. It is essential that this must not occur and that some safety interlock device between tube-housing and diffraction apparatus be used. Safety measures adopted should (a) have a fail-safe characteristic and (b) not restrict the responsible operator from making those adjustments recognized as an essential part of the procedure associated with the particular apparatus involved.

(2) By scattering of the primary X-ray beam at interfaces between an open window of a tube-housing and the diffraction apparatus being used at the window. Leakage and scattering of X-rays, from the region of coupling between an X-ray tube window and the apparatus in use at that window, is one of the most serious hazards, and is likely to give rise to narrow but intense beams of stray radiation.

(3) Escape of the residual primary X-ray beam from the diffraction apparatus.

(4) Leakage, from openings in the diffraction apparatus or through the walls of the apparatus, of radiation (including unwanted fluorescent radiation) scattered by the air or by mechanical parts of the apparatus.

(5) Leakage of radiation generated by valve rectifiers in the high-voltage power-units of X-ray generators. Thermionic rectifiers used in X-ray generators can act as powerful sources of penetrating X-rays. Causes include under-running of rectifier filaments, and, in a faulty valve, the passage of some inverse current at high voltage. Walls of high-voltage circuit tanks have not always been adequate to prevent this radiation from escaping. Monitoring of equipment should include this region of the generator.

(6) Penetration of radiation through the walls and (closed) window-shutters of an improperly shielded X-ray tube-housing.

"*Detection of Stray Radiation: Personnel Monitoring.* It is essential that any experimental arrangement be surveyed with a counter-tube sensitive to the soft X-rays normally used in X-ray diffraction. The use of portable survey meters employing counter tubes with low inherent filtration (i.e., low-absorption windows and high quantum efficiency in the low-energy X-ray region) is recommended for surveying installations. If a thin sheet of lead long enough to cover the counter tube length and extended about 10 cm beyond the window is wrapped around the tube, the counter will be directionally sensitive and simplify the location of radiation leaks. Large pieces of film placed around the apparatus may be used to locate narrowly defined leaks, particularly if exposed a sufficiently long time (i.e., 1 hr or more).

Film badges worn on the chest, wrist, or finger may be used for personnel monitoring, but it should be noted that they are liable to provide a false sense of security because the stray radiation which may occur in X-ray diffraction apparatus is often in the form of narrow low-energy beams.

The problems involved in complying with local and national safety codes and regulations, insurance, etc., should be taken into consideration in setting up a code of laboratory practice. The appointment of a person responsible for the proper instruction of all personnel, and the keeping of inspection records and checking of X-ray units at regular intervals are, however, most important. No one should be permitted to use X-ray apparatus until they have received proper instruction in safe practices. When new apparatus is being mounted, or existing apparatus modified or re-aligned, it is necessary to survey the apparatus. Regular departmental meetings on safety practices, reviewing the safety of existing apparatus, etc., are useful and often necessary to emphasize the importance of the problem.

"*Factors Responsible for the Hazards.* Each of the various possible sources of radiation hazard listed above results from one or more of the following:

(*a*) Negligence by the user.

(*b*) Failure by the user to recognize and deal with incompatibility of design when coupling diffraction apparatus to X-ray generators of different manufacturing origin.

(*c*) Faulty equipment.

(*d*) Unsatisfactory design of equipment.

The use of X-ray 'warning' signs at the entrance to an X-ray laboratory, small signs attached to the X-ray units, lights indicating what apparatus is in operation, etc., are useful in minimizing the risk of accidental exposure. Wherever possible it is preferable to place X-ray units in a room separate from where personnel do other work."[1]

A number of manually and electrically operated shutters and couplings have been described in the literature. Most of these are not commercially available (as yet), but can be built easily. Details of their construction are found in (*1a*) and the references listed in (*3*). However, the following should be kept in mind (*1a*):

"It is realized that metallic lead may easily be shaped by hand to satisfy most coupling requirements; however, because it is so readily deformable, it is unacceptable as a material for couplings which are to be used over and over again. Even when used on an emergency basis, a lead coupling should be monitored with counters or film each time it is used to assure the operator that excessive radiation leakage does not occur through some subtle defect due to inadvertent deformation of the coupling. The only exception to this recommendation is if lead is used as an X-ray absorbing lining in a coupling made, for example, of steel or brass, where the lead would be protected from contact with other materials. These comments do not pertain to the use of lead as a gasket material. In fact, if properly used,

---

[1] I am thankful to Bob Blumenthal (of Enraf-Nonius) for bringing the following to my attention: The Dutch government is considering revision of the law governing the protection of people working with X-ray equipment. Among the recommended additions are:

1. Persons working with diffraction apparatus must have a medical check-up twice a year.

2. Rooms with X-ray apparatus must have a warning sign on the door.

3. Safety precautions in laboratories for students and/or inexperienced persons must be more rigorous than for experienced personnel.

4. The X-ray tube window may permit X-rays to pass only when a camera is placed in front of it. When a camera is not in place the window shutter must close automatically.

5. When X-rays are on, with a camera in place, a conspicuous light must be illuminated. There shall be no X-rays if the bulb burns out.

6. A radiation baffle, which will prevent leakage of secondary radiation from the shutter, must be provided.

lead is a most satisfactory gasket material for purposes of radiation shielding."

Finally, any user of X-ray diffraction equipment should be aware of the fact that the high-voltage (25–100 kvp) transformers are dangerous. Whenever it is necessary to handle the transformer or cables (e.g., when changing an X-ray tube), always be sure to ground the transformer first.

Monitoring Devices

Dosimeters, Counters, etc. A list of companies marketing radiation monitoring equipment is found in the Directory of Products and Services for Radioactivity Laboratories (*4*) under the following headings: dosimeters and personal and radiation survey equipment.

Film Badges. A pamphlet describing the principles of photographic radiation dosimetry and giving detailed information concerning their line of personal monitoring films is available from Eastman Kodak (Kodak Sales Service Pamphlet No. P-31). Kodak Personal Monitoring Film, Type 3 is used for detecting and monitoring beta-, gamma-, and X-radiation. The packet is of the 2-film type; one film is of exceedingly high sensitivity, the other slow. This emulsion combination permits the recording of ionizing radiation dosage over an extremely wide range (13 milliroentgens to 1800 roentgen of radium-gamma radiation). They are available in packages containing 150 2-film packets at \$8.35/package.

DuPont has a line of dosimeter film packets with one or two films per unit (Types 544, 545, 556, 558); price per package of 150 is approximately \$8.50. They may be supplemented with certain added features such as 5- and 6-digit consecutive numbers indicated on the packet and visible on the film; and lead shield over part of the packet.

Tracerlab offers a commercial film badge service. Their compartmentalized badge not only contains an interchangeable short-term film packet that is replaced and processed by Tracerlab on a weekly, biweekly, or monthly basis, but also a permanent safety back-up film packet. Each year the badge containing the back-up film is replaced. Should the short-term film become lost or damaged, the back-up film will serve as the permanent record. It may also be used to verify dosages recorded on the short-term film. Electronic data processing equipment enables reports to be sent to the subscriber within 24–48 hr after receipt of the film. Typical prices/unit are \$0.65 for 3–10 units and \$0.50 for 61–100 units.

Picker also offers a film-badge service.

Shielding Materials

Commerical sources of lead bricks, sheet, shot, tubing, molded shapes, and lead bonded to concrete and wood are found under "Shielding Materials: Lead and Compounds," in Ref. (*4*). One particularly interesting material is Chemtree 31 (Chemtree Corp.). Chemtree 31 is a leaded mortar which, with the addition of water, may be poured, cast or troweled. If it is self-supporting, easily bonded to most common structural materials and

can be drilled, sawed, and otherwise worked. By weight, it is nearly all lead, and thus has approximately the identical shielding effects against X-rays as does lead. However, its density (6 gm/cc) is about half that of lead and it must be twice as thick as lead for equal attenuation. A number of other formulations are also available; in small quantities the prices are approximately \$1/lb.

Another useful material is the tungsten alloy, Kennertium (T. M. of Kennametal, Inc.). This alloy is available in two formulations; one is 50% more dense than lead, the other 62%. It can be machined (when carbide tooling is employed) and joined to itself or other metals by silver or copper brazing, soldering, press or shrink fitting, or by mechanical attachment. Because this material is quite hard it does not possess the disadvantages of lead that were described above.

3M Lead Scotch Tape No. 420 is an adhesive-backed lead tape which is extremely useful for preventing radiation leaks from tube housings and window shutters.

Warning Signs

Colorful signs displaying the international radiation symbol and containing a warning of specific interest to users of X-ray equipment are available, at no cost, from Siemens America, Inc. (Fig. 2).

CAUTION RADIATION AREA

X-RAY EQUIPMENT PRODUCES RADIATION ((60 Kev) WHEN ENERGIZED AND SHOULD BE OPERATED ONLY BY TRAINED PERSONS.

Before energizing x-ray tube, make sure that diffraction and/or spectroscopy equipment is properly set-up to prevent possible exposure to direct or stray radiation. Likewise, make sure that x-ray tube has been de-energized before making any changes in the equipment arrangement. Do not tamper with safety or protection devices and check their proper function periodically

(*Courtesy of Siemens America, Inc.*)

Figure 2. Warning Sign to be placed near X-ray equipment.

Shutters

X-ray tubes are generally available with two to four beam-exit ports (or windows). This allows for the concurrent use of more than one instrument. It is therefore necessary to furnish each port, or window, with its own shutter so that X-ray beams cannot escape from those ports not in use. The shutters serve other purposes as well:

(*a*) They protect the fragile windows.

(*b*) They can be used to control the exposure time at each window individually.

(*c*) A properly designed shutter will close automatically whenever a camera is pulled away from the port.

Chemical Instrumentation

Not all of the shutters on the market fulfill all these requirements. Some of the more interesting devices are: G.E. manufactures a shutter which is attached to the tube rather than to the tube-housing. As long as the shutter is in place, the tube is protected (even while in storage or in transit). Picker and Siemens manufacture electronically controlled timers which permit each window to be controlled individually. The Picker shutter is attached to the X-ray tube. Philips currently markets a Norelco unit in which a mechanical interlock prevents the shutter from being opened unless the filter wheel is on lead.

Aremco offers a safety shutter ($100) which can be mounted permanently on either Norelco (Cat. No. X-100-P) or G.E. (Cat. No. X-100-G) X-ray tube-housings. The attachment is made of brass and is supplied with lead gaskets and nosepiece for convenient mounting of cameras. The shutter operates on the gravity principle, so that if the camera is removed, it automatically shuts. A filter wheel is an integral part of the shutter (Fig. 3).

(Courtesy of Aremco)

Figure 3. Aremco Safety Shutter (X-100-P).

E & A manufactures a safety shutter with a timer and automatic closing mechanism. The safety feature is a warning light on the shutter which remains lit whenever the shutter is open (if the bulb burns out the shutter closes automatically). It is available for all standard X-ray ports at a cost of $300 (Fig. 4).

(Courtesy of E & A)

Figure 4. E & A Electric Fail-Safe Shutter with timer.

The Super Balanced Filter Assembly (to be described in a later section) can also be used as an automatically controlled shutter.

The following safety considerations are especially valid in those situations where novices are handling X-ray diffraction equipment (e.g., in laboratory classes); consequently, they are of particular interest to the readers of THIS JOURNAL.

(a) Never leave an unused port without a shutter or lead stop, even if the X-rays are off. The shutter protects the tube against accidental breakage; it also protects subsequent investigators who may not be sufficiently knowledgeable or responsible to check *all* the ports before turning on the equipment.

(b) Conversely, before turning on any X-ray diffraction unit check *all* the ports, especially those not in immediate use.

(c) It may be wisest to turn the X-rays completely off before placing or removing cameras or samples.

GENERATORS

The X-ray generator contains a high-voltage transformer and a number of circuits, including the filament and high-voltage circuits.

Filament Circuit. The purpose of the filament circuit is to heat the X-ray tube filament to a point where it will give off electrons. A step-down transformer lowers the incoming voltage to approximately 10 v; further control of this voltage is provided by a variable rheostat in the primary circuit. Output is thus varied and the filament heat is regualated so as to provide the proper number of electrons for a definite current flow across the tube. A filament ammeter enables the operator to read current flow in the filament circuit and to use its figures as a means of presetting filament heat (Fig. 5).

High-Voltage Circuit. The high-voltage circuit contains the high-voltage step-up transformer and its primary and secondary circuits. An autotransformer controls the voltage fed into the primary; changes in this voltage are made by varying the kvp control. The high-voltage transformer secondary is connected to the X-ray tube.

When the primary is energized, high voltage induced in the secondary drives the filament electrons against the tube target producing X-radiation. The potential induced in the secondary will be an alternating current of the same frequency as that in the primary. Since the X-ray tube cathode and anode are connected to the secondary (terminals "X" and "Y") their potentials are also changing (Fig. 5).

Electrons are accelerated to the anode only when it is positively charged. Therefore, as shown in Figure 6, current flows through the tube on the half-cycle when the anode is positive; no current flows through the tube on the half cycle when the anode is negative.

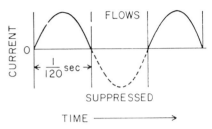

(Courtesy of General Electric Co.)

Figure 6. Representation of current flow in a self-rectified circuit using 60 cycle a.c.

This circuit is called a self-rectified circuit because the X-ray tube itself determines which alternations shall flow through it. A circuit of this type has several limitations imposed upon it by the X-ray tube. One is that the unit must be operated at a rating which will not unduly heat the anode. Should the anode become hot enough to emit electrons, these electrons will be attracted to the cathode and generally result in irreparable damage to the tube filament. A second limitation is the alternating character of the X-ray output caused by the half-wave rectified circuit.

A number of other features are usually built into the control unit (Fig. 7):

(a) A milliameter is also connected in the secondary circuit and indicates the current flow through the X-ray tube. This meter is at ground or zero potential so that it may be mounted on the X-ray control in full view of the operator.

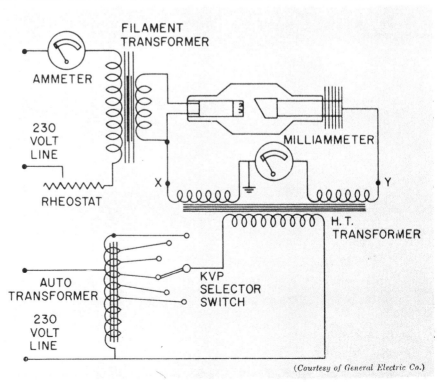

(Courtesy of General Electric Co.)

Figure 5. Diagram of the filament and high-voltage circuits in an X-ray generator.

(b) A warning light, which indicates when X-rays are being produced.

(c) An overload relay set to open the circuits when the tube current exceeds normal operating values.

(d) A pressure switch in the water line to open the circuits when the flow of water to the tube is insufficient to cool the tube.

(e) An electric exposure timer which will shut off the unit at the end of a pre-set time.

(f) A time totalizer which indicates the length of time the unit has been on.

(g) Safety interlocking switches which turn off the high voltage if the panels of the unit are opened.

Self-rectified X-ray equipment is manufactured because of its simplicity and economy.

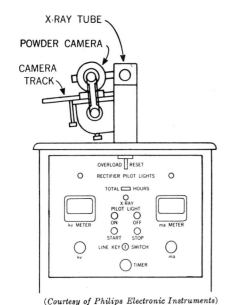

(Courtesy of Philips Electronic Instruments)

Figure 7. Schematic representation of a Norelco X-ray generator control unit.

Full-Wave Rectification. Full-wave rectification is used to smooth the X-ray output by producing X-rays on both halves of each alternating current cycle. Using a circuit that is very similar to the basic or self-rectified circuit, it differs in the addition of rectifying kenotron tubes. These tubes serve the function of maintaining the X-ray tube anode at a positive and the cathode at a negative potential at all times.

A kenotron tube is similar to an X-ray tube in that it contains a heated filament and an anode in an evacuated glass envelope. No attempt is made to control the filament temperature as it is in the X-ray tube; it is allowed to produce electrons far in excess of the number produced by the X-ray tube. These electrons are attracted to the anode whenever this structure is at a positive potential and are repelled by the anode whenever its potential is negative.

Figure 8 illustrates a full-wave rectified X-ray circuit using four kenotron tubes. Only the high-voltage section of the circuit is shown, and this is further simplified by the elimination of the filament transformers used to heat the kenotron tube filaments.

Assuming that a 60-cycle supply is used, terminals "X" and "Y" are changing potentials 120 times a second. However,

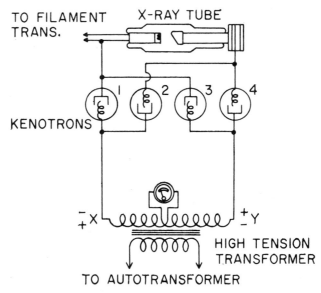

(Courtesy of General Electric Co.)

Figure 8. Diagram of a four Kenotron, full-wave rectified circuit.

these terminals are not directly connected to the X-ray tube as in the self-rectified unit. Rather, the potentials are directed through kenotrons so that they are in the proper polarity at the X-ray tube.

When "Y" is positive and "X" negative, electrons flow from the filament to the anode of kenotron 1, through the X-ray tube, to the filament and then to the anode of kenotron 4, completing the circuit to the secondary of the high-voltage transformer. Kenotrons 2 and 3 are not conducting electrons during this alternation because their anodes are negative with respect to their cathodes.

On the next alternation "X" is positive and "Y" negative. Kenotrons 1 and 4 do not conduct this time as their elements are at the wrong polarity. Electrons flow through kenotron 3, through the X-ray tube, and finally through kenotron 2 completing the circuit. Thus, unlike self-rectified apparatus the second half cycle of the alternating current in a full-wave rectified unit is conducted through the X-ray tube and is utilized in the production of X-rays. (This description has been adapted from Publication 8A-3258B of the G.E. X-Ray Department.)

Constant Potential Systems. Further smoothing of the X-ray output is accomplished by the use of a constant potential assembly. In essence, this consists of a high-voltage capacitor and protective circuits. During the peak part of the transformer pulse, current flows through the X-ray tube and, at the same time, charges the capacitor; between transformer pulses current flows through the tube from the capacitor. In the case of either half-wave rectification or full-wave rectification, although X-rays are intense at the peaks, no X-rays are generated in the intervals between pulses. On the contrary, in the case of the constant potential system distribution of X-ray intensity is averaged, counting-loss of intense X-radiation generated locally by the peak current is reduced and instantaneous overload on the X-ray tube filament is eliminated, hence tube life is extended (Fig. 9).

Voltage Stabilizers. Line voltage variation will lead to corresponding variations in the high voltage applied to the X-ray tube. More serious however is the effect of such fluctuations on the tube current. The tube current is dependent upon the output of electrons from the hot tungsten filament which itself is strongly influenced by slight changes in the filament temperature. Thus relatively small voltage variations causing small filament temperature changes can bring about serious tube current changes which may endanger the life of the tube being operated at or near full rating. When a counter, rather than film, is used as the detecting device the use of a stabilized line voltage is a necessity, for in this case the intensities measured at different times must be comparable or the data are useless. Commercial X-ray generators have built-in stabilizers, or else they are available as optional equipment.

Although commercially available units generally are designed for use with a specific type and make of X-ray tube, in this article the tubes and generators are discussed separately so that the special features of the various models can be compared more easily.

Standard Generators

Norelco X-Ray Generator. The standard X-ray generator is equipped to supply regulated power and cooling water to a Norelco Four-window X-ray Diffraction Tube. The cabinet is made of sheet metal and houses all electrical components and the water circulating system. All panels are hinged for easy access and interlocked for safety. The instrument table is available in circular or rectangular design and provides rigid support for all types of diffraction accessories. Mounted in the center of the table is the heavily loaded bronze tube-housing. The housing is so designed that when an X-ray tube is lowered into position and secured, all connections for tube power and cooling water are made automatically. Each beam port on the housing has a machined face set at a 6° angle from vertical, and

Chemical Instrumentation

normal to the X-ray beam path. Each port is supplied with an Indexing Filter Selector, fitted with five metallic filters and one open aperture. Over each filter disc is a gravity operated shutter, which automatically closes the port when not in use (Figs. 7 and 10).

The unit uses 190–240 v ac, from a 50 or 60-cycle single phase line. It requires a minimum water flow of 30 gal/hr at 35–90 psi pressure. The milliampere output is 2–50 ma, stepless control. Cat. no. 12045 (Price Range G)* has a 10–60 kv peak (Kvp) full-wave rectified output,

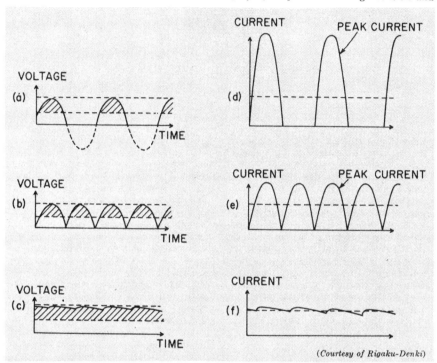

Figure 9. Representation of voltage and current flowing through X-ray tube in half-wave rectified (a and d), full-wave rectified (b and e), and full-wave rectified plus constant potential (c and f) circuits.

stepless control; Cat. No. 12215 has a 10–50 kv constant potential (Kvcp) output, maximum 10% ripple, peak to peak

(Courtesy of Philips Electronic Instruments)

Figure 10. Norelco X-Ray Generator.

*See Appendix A.

under full load, stepless control. It requires the use of the Line Voltage Regulator (Cat. No. 52474) and the Current Stabilizer (Cat. No. 52529).

An Adaptation Kit (Cat. No. 51069) is available which permits attaching an external cable to the generator. This can be attached to an Auxiliary Diffraction Tube Housing (Cat. No. 52571), which allows one to maintain the X-ray tube in a horizontal position.

Philips has recently marketed a high-stability generator (Cat. No. PW 1010) suitable for single-crystal investigations. This constant-potential X-ray generator, in which two stabilizing systems are employed, keeps both the high tension and the tube current constant within narrow limits, even when variations occur as a result of differences in temperature, higher harmonics, etc.

The stabilizers keep the tube current and the high tension constant within 0.1% with maximum mains voltage fluctuations of ±10%. Contrary to the more conventional forms of stabilization, the generator uses a system of rapid feedback from a high-resistance network across the high-voltage secondary. This method eliminates the effects of temperature drift and other fluctuations.

After suitable electronic amplification the feedback signal operates a fast transductor in the high-voltage primary. Since the range of this transductor may be exceeded, an automatic auto-transformer driven by a servomotor returns the voltage to the operating range of the fast transductor. The stabilizers are mounted on separate chassis and provided with plugs, so that they can quite easily be taken out, facilitating inspection and testing of the apparatus.

The transformer is connected to the X-ray tube via a flexible shockproof cable.

Philips also markets a Norelco Table Top Diffraction Generator (Price Range G) which is supplied complete with a

copper target four-window X-ray tube, a camera track assembly and drive motor, a 57.3-mm Debye-Scherrer powder camera and a film punch and cutter. An optional timer is available. This generator is designed for film work and is useful for student instruction (Fig. 11).

(Courtesy of Philips Electronic Instruments)

Figure 11. Norelco Table Top X-Ray Generator.

Picker. The Full-Wave X-Ray Diffraction Unit (Series 6147, Price Range G) is designed for use with photographic X-ray diffraction techniques. It has full-wave rectification and stepless control of the Kv and ma settings, and is equipped with a filament voltage stabilizer. It is used with a three-window tube and has optional timer-controlled, individual electromagnetic shutters and warning lights. Other safety features include individual warning lights to indicate when X-rays are on, if the circuit breaker is operating, and which ports are open. The line switch starts and stops the water flow; safety switches and pressure regulators protect against under- and overpressure.

(Courtesy of Picker X-Ray Corp.)

Figure 12. Picker Ultrastable Solid State D. C. X-Ray Generator (6238 Series) set up for two-tube operation and the Picker Transistorized Radiation Detector Panel.

Picker's newest unit is an ultrastable solid state DC X-ray generator (Series 6238, Price Range H) which can be used with automatic single-crystal diffractometer systems (Fig. 12). The kv stability is 0.1% or a maximum of 60 v including warm-up. The ma stability is 0.02% or a maximum of 10 microamps, including warm-up. A 0.1-mfd capacitor is used to provide constant potential. In addition, a feedback system is provided which results in ripple on the high voltage of less than 50 v, resulting in a nearly straight dc operation.

When two tubes are to be operated simultaneously (Cat. No. 6238E), each tube has its own timers and ma stabilizer chassis. This provides for simultaneous operation of both tubes, each operating constant potential with the same stability and ripple specifications as for a single tube. The total kv-ma product cannot exceed the unit rating of a 2400 w load. Both tubes must be operated at the same kv, but they will tolerate current differences up to 6 ma.

Enraf-Nonius. A number of different X-ray generators are available from Enraf-Nonius; all but one are equipped with a 3-phase 6-selenium-rectifier system and filament current stabilizers. They all have a choice of manual or timer controlled operation, electromagnetic water valve, timer, safety interlocks, and warning lights. The tubes are attached by 225 cm long cables and can be used in either the horizontal or vertical position. The window height is adjustable between 23 and 28 cm. The vertical position is different from most other units in that the target is below the filament and the take-off angle slopes up from the table-top (Price Range G).

The new one-tube Model Diffractis P has been designed for use with camera work and engineered to provide maximum safety. The shutter design is such that the possibility of stray and scattered radiation is reduced as much as possible. Safety interlocks prevent the use of the generator unless a camera and/or collimator is in place. It is equipped with six filter possibilities and can be either electrically timed or manually controlled. The prominent warning light is safety wired so the generator is inoperable if the bulb burns out.

Convenience outlets are provided which are either timer controlled or always on. The table size is 100 × 80 cm and 115 cm high. Magnetic filament stabilization is provided to 1.3 kva.

Model K (Cat. no. Y553) has a 1 kva power rating and is designed for use with one tube. There is continuous regulation of the kv and ma with a choice of two sets of ranges (0–50 kv and 0–20 ma or 0–35 kv and 0–30 ma); a built-in selector switch permits the choice of either one.

Model FG (Cat. no. Y550) has a 2.5 kva rating with a continuously regulated range of 0–50 kv and 0–50 ma.

Model L (Cat. No. Y560) is designed for use with counter diffractometers; it has been designed to maintain optimum stability. The kv can be adjusted from 20 to 60 kv in steps of 1 kv; the ma is adjustable at 13 set values between 3 and 40 ma. The high voltage is stabilized within 0.15 kv for an input fluctuation of ±10%; the tube current is stabilized within 0.2% of the instantaneous value.

Models H2 (Cat. no. Y530) and H3 (Cat. no. Y520) (with power ratings of 3 kva) are built for use with 2 or 3 tubes, respectively.

Siemens. The Crystalloflex line of generators is manufactured by Siemens. The Crystalloflex IV (Price Range H) is designed for use with counter diffractometers and spectrometers. It is a constant potential unit with input voltage and beam current stabilization. The latter circuit checks the high voltage at the output against an internal reference voltage. It is designed to prevent a less than 0.1% change in output voltage and beam current with an input fluctuation of 10% (Fig 13).

(Courtesy of Siemens America, Inc.)

Figure 13. Crystalloflex IV Generator and Detector Panel.

Other Crystalloflex models are available for uses which require less stabilization.

G.E. The XRD-6 Control and High Voltage Power Assembly (Cat. No. A4980A, $5000) has a 3.7 kva rating designed for use with one or two tubes. The output from the transformer for simultaneous two-tube operation is half-wave rectified; the use of separately available constant potential assemblies to each of the transformer outputs will provide a uniform output for both tubes. The kvp meter is of dual range type and allows continuous adjustment from 0 to 30 and 9 to 90 kvp. The ma can be set at any four currents up to 100 ma and is controlled by a four-position switch. A line voltage stabilizer (0.1% deviation) and standard warning lights and safety features are also available (Fig. 14).

Rigaku-Denki Model. D-6C (Price Range G) is a 60 kv, 50 ma, full-wave rectified unit equipped with a constant potential system. It can be used for X-ray diffractometry or X-ray spectrography. There is a one point pre-set system for the tube voltage and a five point pre-set system for the tube current. A safety feature is an automatic high-voltage discharger which automatically grounds the high-voltage cable and the tube when the input voltage is cut off. This unit is designed for use with either Rigaku or Philips X-ray diffraction tubes.

Tem-Pres Research, Inc. has developed a complete X-ray diffraction system designed for routine powder diffraction and high-temperature diffractometry; the system has continuously adjustable slits and an independent sample-rotation axis. The complete system, Model XD-1 (including generator, X-ray tube, horizontal diffractometer, detector, and recorder) is available for $8,950 (see Fig. 72).

Micro-Focus Units

The attainment of a high useful in-

tensity is recognized as the most important factor in the design of X-ray diffraction generators and tubes. Conventional vacuum-sealed X-ray tube configurations permit operation up to a maximum rated intensity of only about 100 w/sq mm over an area of about 10 sq mm. If that intensity rating is exceeded, pitting and contamination of the target area result which adversely affect the purity of the spectrum and eventually destroys the (costly) X-ray tube.

The Microfocus X-ray Tube (Hilger

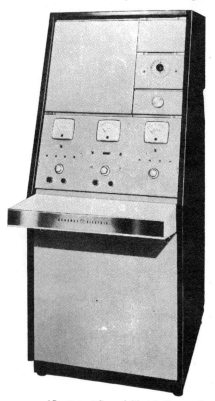

(Courtesy of General Electric Company)

Figure 14. G. E. XRD-6 Control and High Voltage Power Assembly.

Chemical Instrumentation

and Watts) delivers from nine to thirteen times the intensity of the vacuum sealed tube, because of greatly increased electron bombardment at the target by means of a unique electrostatic lens system which serves to focus the electrons into a localized target area (1.4 mm × 100 microns, or optionally, to a 40μ spot). The Microfocus Tube can be quickly demounted so that low-cost replacement targets, lasting from 600 to 1000 hr and more, can be interchanged, in minutes, to prolong indefinitely the life of the tube itself.

Not only are the targets interchangeable, but cathode guns, as well, can be quickly interchanged to yield various spot shapes, sizes, or orientations assuring unlimited applications versatility from one tube.

Jarrell-Ash markets a microfocus generator (Cat. no. 80–000, $7000) and X-ray tube (Fig. 15). The latter is based on the Hilger-Watts design. Two types of X-ray tubes are available. Each form consists of a cylindrical tube $1^3/_8$ in. od, terminating at the top in a block $1^1/_2$ in. square by 2 in. long, mounted either horizontally or vertically. The block is bored to take the target and to provide the X-ray ports, while the cathode guns are mounted on the end of the block. The windows are $10^1/_2$ in. (26.7 cm) above the table-top and the overall height of the vertical tube with a simple electron gun is $11^1/_2$ in. Shutters at the windows provide protection against stray radiation.

The anode is a double copper tube—one tube within another—so that cooling oil can be pumped through it. It is positioned centrally in the vertical evacuation pipe and held rigidly in place near its top end by a spider-like insulator, which permits evacuation. The vacuum seal is demountable and the main insulator at the bottom is so made that the anode can be withdrawn when necessary.

Several interchangeable targets (copper, iron, chromium, cobalt, nickel, silver, molybdenum, tungsten, and lanthanum) are available, all inserted and removed easily with a special tool provided with the generator.

The X-ray tube is evacuated by a large three-stage oil-diffusion pump through a water-cooled pipe which prevents backstreaming of oil vapor. This pump is backed by a suitable two-stage rotary pump. An ion-type vacuum gauge is supplied to monitor continuously the quality of the vacuum. The control system is interlocked to prevent harmful operation; for example, the X-rays are switched off if the vacuum is inadequate and the diffusion pump is turned off if the water cooling supply fails.

Oil is pumped through the double tube of the anode to cool its tip, to which is clamped one of the interchangeable targets. The oil passes through a heat exchanger, which is kept cool by a constant flow of cold water. A safety switch prevents damage to the target by turning off the high-voltage supply if the flow of oil becomes inadequate.

The high-voltage transformer (full-wave rectified), selenium rectifiers, and filter condenser, comprising the constant potential supply, are all mounted in an oil-filled metal tank in the cabinet. This heavy duty supply has a rating of 60 kv at 10 ma assuring long life and trouble-free operation at the X-ray tube ratings. The backing pump, diffusion pump, and high voltage must be switched on in turn, since neither of the latter will work unless conditions are correct; for example, the high voltage cannot be switched on unless the cooling oil is flowing. Any subsequent failure, say of the cooling water, causes the appropriate circuits to be switched off.

There is also available a high-voltage connection which enables this generator to be used as a power supply for a conventional vacuum-sealed tube.

Engis Equipment Co. manufactures a microfocus unit (Y33B, Price Range G) in which the tube is also the Hilger-Watts Microfocus Tube (Fig. 16).

(*Courtesy of Engis Equipment Co.*)

Figure 16. Y-33B Microfocus X-Ray Generator (the replaceable target is mounted in the block at the top of the vertical evacuation pipe).

The vacuum system selection is left to the user. Two ion pump systems offer freedom from oil contamination, minimum maintenance, and increased target and filament life. For those who require the most rapid pump-down and change X-ray tube components often, the ion pump with Sorption pump is suggested. For more typical users, the ion pump with mechanical roughing and oil-diffusion pumps may be specified.

The high-voltage circuit uses full-wave rectification with the complete unit in a sealed container. This circuit is fully interlocked with the vacuum, cooling oil, and water systems. Audio and visual alarms are incorporated to indicate any malfunction. The control circuitry provides fail-safe operation.

Both the low power Microfocus cathode gun (40μ spot focus) and the high power options are available. The high power anode system uses oil to cool the target. Energy of up to 45 kv at 10 ma is possible with copper targets for a beam of 0.1 × 6 mm. Other target materials may be interchanged.

Both the Jarrell-Ash and the Engis microfocus units can be equipped with automatic shutters and various styles of cathode guns.

Rigaku-Denki has a Microflex Microfocus X-ray diffraction unit (Cat. no.

(*Courtesy of Jarrell-Ash*)

Figure 15. Jarrell-Ash Microfocus Generator and Detector Panel (a powder diffractometer is positioned near the tube).

4181, Price Range H) which provides a 10μ diameter point source of X-rays. The tube is horizontally mounted, with copper or platinum targets available. The transformer, vacuum system, and lens current stabilizer are housed in the console, with the input voltage stabilizer kept in another console several yards away from the main unit.

Enraf-Nonius manufactures a microfocus installation (Cat. no. Y-720, Price Range H) complete with microfocus tube, high tension generator, control panel, and vacuum-pumping system. A number of anode materials are available.

Rotating Anode

The Rigaku-Denki Rota-Unit Ru-3H (Price Range I) is a rotating anode X-ray diffraction generator. The target rotates in a vacuum and can be operated at 60 kv-100 ma under constant potential conditions. Extreme care has been taken in the precision machining and precise dynamic balance of the rotating anode so as to eliminate vibration of the X-ray tube and drifting of the focal spot. The tube is available in either the horizontal or vertical position; automatic shutters are supplied for the two X-ray ports and a number of anodes are available (Fig. 17).

(Courtesy of Rigaku-Denki)

Figure 17. Rigaku-Denki Ru-3H Rota Unit with tube in horizontal position.

The Scientific Equipment Department of Westinghouse Corp. has designed a demountable high intensity rotating anode X-ray generator which can be run at 75 kv-100 ma. There are a number of options that are available regarding the size of the beryllium windows, the size of the focal spot, etc. Each unit is essentially "custom-built" to the customers own specifications.

Part Two—X-Ray Tubes and Monochromatization

X-RAY DIFFRACTION TUBES

The design of modern, commercially available, permanently evacuated X-ray diffraction tubes is based upon the first successful high-vacuum X-ray tube built by W. D. Coolidge in 1913. Electrons which are boiled off a coiled tungsten filament are accelerated across a potential

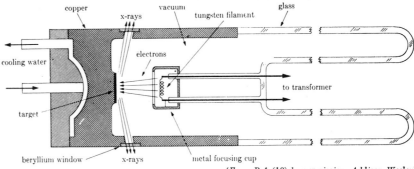

(From Ref. (10) by permission, Addison-Wesley)

Figure 18. Cross section of permanently evacuated X-ray tube (schematic).

drop and bombard a target material which is (*a*) coated on, (*b*) embedded in or (*c*) is a water- or oil-cooled copper block (Fig. 18). Of principal interest to the investigator is the wavelength, shape, and intensity of the characteristic X-ray beam. The wavelength, of course, depends upon the target material; it is necessary to use as pure a target as possible. Contamination of the target by tungsten from the filament or iron from the tube-housing will result in an impure tube X-ray spectrum and if the investigator is not aware of the impurities he may obtain spurious results and arrive at wrong conclusions. The loss of tungsten from the filament will eventually cause the tube to burn out, so that even if the tube is designed to prevent contamination (by the use of an electric grid system) its lifetime is limited. The effective lifetime of a tube will be increased if it is run slightly below the manufacturer's maximum power ratings. For example, if the tube is rated at 50 kvp/20 ma, running it at 40 kvp/15 ma will extend its lifetime considerably; the inconvenience of slightly increased exposure times (required by the decrease in beam intensity due to the lower power ratings) is offset by the longer effective tube life.

As was previously mentioned, the targets must be cooled; 98% of the electron's energy is converted to heat. It is evident that the intense heat which is generated limits the intensity of the beam. If the power is too high the target cannot be cooled sufficiently, resulting in melting or pitting of the target. In addition, electrons may be emitted from the target and the self-rectifying action of the tube would be lost, resulting in irreversible damage to the tube. To overcome the limitations on beam intensity smaller filaments (bombarding smaller portions of the target) can be used; in this manner a more intense beam (albeit with a smaller focal spot) is generated. Since a smaller, more intense beam generates essentially the same heat as a larger, less intense beam, the same cooling capacity can be used. A design of this type has resulted in the evolution of the fine-focus tube and the microfocus tube. Tubes with rotating anodes have also been constructed. As the target (anode) rotates, a relatively cool portion of the target is bombarded. The effective cooling capacity is increased and intense beams with large focal spots can be obtained (Fig. 19). In the past mechanical problems associated with rotating anode tubes prevented their commercial manufacture. Recently, vibrations due to the motor and wandering of

the focal spot have been subdued to the point where such tubes can be built commercially.

Microfocus and rotating anode tubes are generally an integral part of the entire generator assembly and a discussion of the commercially available units has been included in the section on generators.

Focal Spot. The shape of the focal spot depends upon (*a*) the shape of the filament and (*b*) the angle and direction from which the spot is viewed. Generally, the filament is a coiled tungsten wire with the plane of the target parallel to the filament axis. A rectangular portion of the target is bombarded with electrons and X-rays are emitted in all directions generating a hemisphere of X-rays. The tube should be constructed so that X-rays can escape only through the windows (ports); tubes with one to four windows have been constructed. The windows are set slightly below the level of the target and the target can be viewed at any angle from 1 to

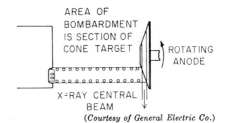

(Courtesy of General Electric Co.)

Figure 19. Principle of rotating anode X-ray tube.

10 degrees. The angle from which the target is viewed is known as the "take-off angle." The windows of the tube are located along the sides of the rectangular focal spot; thus, a four-window tube has two windows at the end of the rectangle and two along the sides.

The shape of the effective focal spot depends upon which window is used and its size depends upon the size of the filament and the take-off angle. The end view results in what is called the "spot focus" and the side view results in the "line focus" (Fig. 20). Under similar viewing conditions (i.e., providing the entire beam is intercepted), the intensity of the spot focus equals that of the line focus. However, the line focus covers a larger specimen area while the spot focus is concentrated on a relatively small area. For this reason, powder diffractometers are set up opposite the line focus (the aim is to irradiate a large sample area) while single-crystal instruments are located at the spot focus.

Chemical Instrumentation

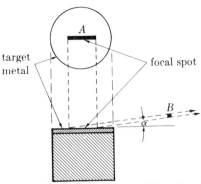

(From Ref. (10) by permission, Addison-Wesley)

Figure 20. Shape and size of effective focal spot depends upon take-off angle (α) and the position from which it is viewed. At point B the "spot focus" is obtained; the "line focus" would be observed from a position perpendicular to the plane of the paper.

An interesting geometrical effect caused by the emission of electrons from the coiled tungsten filament has serious practical consequences. If a pinhole photograph is taken of the focal spot it will show two areas of maximum brightness with a darker area between them. To the crystal it appears as if there were two filaments (or two X-ray sources). A single-crystal specimen will reflect the focal spot profile. If it encounters the beam due to the two intense spots successively, the diffracted beam appears to be doubled, while if it encounters them simultaneously, the profile shows a single, intense peak (Fig. 21). Those instruments using counter detectors are especially sensitive to the beam profile. For this reason the use of a horizontally or vertically mounted tube will depend upon the type of diffractometer being used. If the crystal is rotated about a vertical axis then a horizontally mounted tube must be used; if the rotation axis is horizontal, then a vertically mounted tube is used.

One other potential source of trouble is that caused by wandering of the focal spot. As the tube is used the filament

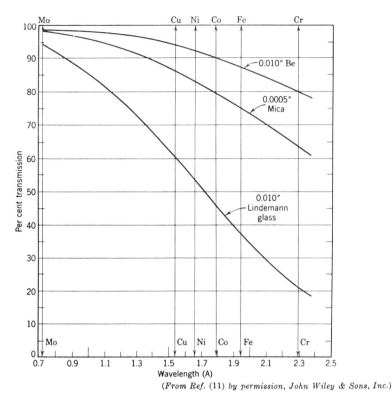

(From Ref. (11) by permission, John Wiley & Sons, Inc.)

Figure 22. Transmission of X-ray tube window materials for K_α radiation.

changes characteristics and the target is eroded, however slightly. A combination of these two effects may cause a long-term wandering of the focal spot; this would necessitate a realignment of the X-ray diffraction instruments. Also, in poorly built tubes, the anode may expand due to thermal expansion effects and the focal spot will wander continuously. In some cases the X-ray beam profile is not uniform. This may be caused by using a horizontally mounted tube where a vertically mounted tube is required (or vice versa), by a poor filament, or by a nonuniform target. These defects are particularly troublesome when sensitive counter detectors are used and occasionally it is necessary to reject two or three tubes before a suitable one is found.

Windows. The number of useful windows per tube depends upon (*a*) the generator and the instruments used with it and (*b*) whether the tube is mounted horizontally or vertically. The maximum practical number of windows is four for a vertically mounted tube. The windows should be set back slightly so as to prevent tungsten from the filament from depositing on them and reducing the beam intensity.

The window must be composed of materials with as low an atomic number (absorption coefficient) as possible. Lindemann glass (containing boron, lithium, and beryllium), mica, and beryllium foil have been commercially used. Lindemann glass (0.010 in. thick) absorbs over 50% of the radiation for wavelengths longer than that of copper; it also tends to devitrify in a moist atmosphere. Nowadays, it is not generally used. Mica (0.005 in. thick) has better transmission properties, but is vulnerable to puncture by stray electrons. For this reason, the mica is protected by a thin beryllium foil. The most satisfactory window material seems to be thin beryllium metal. A 0.010 in. thickness of beryllium transmits about 80% of chromium radiation (which is the longest, most easily absorbed, wavelength commonly used in X-ray diffraction investigations). Furthermore, since beryllium is a fair conductor of heat and electricity, it can be placed close to the focal spot without becoming overheated and charged by secondary bombardment. This results in an additional gain in intensity by permitting equipment to be brought closer to the X-ray source (the focal spot) (Fig. 22).

Choice of Target. The use of long wavelength radiation results in greater dispersion of the diffracted beams and larger absorption effects than does the use of short wavelength radiation. Thus the choice of the target material depends upon a number of factors, among which are the following:

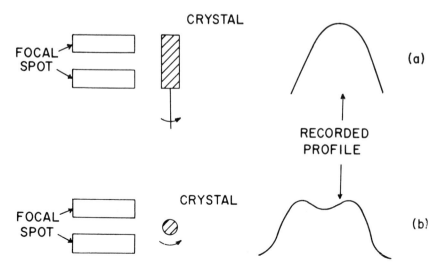

Figure 21. Electrons emitted from a coiled filament cause the focal spot to have two areas of maximum intensity. The recorded profile of a diffracted beam depends upon the orientation of the rotation axis of the crystal relative to direction of electron travel.

(a) Absorption coefficient of sample.

(b) Lattice constants of sample-Large unit cells have many diffraction peaks which will overlap if the radiation wavelength is too short.

(c) Fluorescence Effects—for example, a very high background which obscures the diffraction peaks occurs when an iron-containing sample is investigated with copper radiation. This adverse effect can be prevented by taking special precautions (e.g., placing a monochromator between the sample and detector) or by using radiation of a different wavelength.

The most commonly used targets are copper and molybdenum (the latter is used primarily in single-crystal investigations). A number of other targets are commercially available in sealed, permanently evacuated X-ray tubes. These include: chromium, iron, cobalt, nickel, silver, and tungsten. If many special targets are required, then a demountable tube should be used; if only one or two special targets are required for extensive work, then, in many cases, they can be supplied on special order in permanently evacuated tubes.

A demountable tube, as its name implied, is one which is not permanently evacuated and sealed. The target can be set in place and the system evacuated with roughing and diffusion pumps. The cost and bother of maintaining a vacuum system usually defers people from buying these tubes. However, commercially available demountable tubes are all of the microfocus (or rotating anode) type where the problem of maintaining the vacuum seals and system is offset by the increased intensity of the X-ray beam. Targets for microfocus tubes last about 200 hr and cost $15–20 each.

Some comments concerning the care of tubes are in order.

(a) If the tubes are stored and not used regularly, they tend to become gassy because of the adsorbed gas in the metal and glass parts of the tube. A gassy tube will discharge and not work properly. For this reason every tube should be used regularly. If no regular needs for special target tubes arise, then they should be run a few times each year (every 2–4 months). If a tube is gassy, it can be restored by slowly increasing the operating conditions, while preventing the tube from discharging. Explicit instructions describing how to degas a gassy tube are available from the manufacturer. These instructions should be consulted, so that no permanent damage to the tube is caused.

(b) The tubes must make electrical contact with the high-voltage supply at some point. Occasionally these contact points become corroded or dirty and poor electrical contact results. Careful cleaning of the contact points (with fine sand paper, do not use emery paper because the silicon carbide is conducting) will generally clear up these problems. Some manufacturers recommend specific cleaning procedures, on request.

(c) Some tubes have built-in screens which distribute the cooling water, uniformly, to the target. Although only filtered water should be allowed to enter the tube, some particles may get through and the screens should be cleaned oc-

casionally.

(d) Finally, remember that the windows are very thin and that they support a vacuum. Therefore, they are extremely fragile. Extreme caution is urged whenever the tubes are handled. Never touch the windows with your fingers!

Norelco Four-Window Diffraction Tube. This tube (Price Range E) allows simultaneous use of up to four pieces of apparatus from a single source (Fig. 23). There are two spot and two line foci. The windows are mica-beryllium for minimum X-ray absorption. Another feature of these tubes is that water and electrical connections are made automatically when the tube is secured in position. Furthermore, anode height is carefully controlled, so that minimum realignment is required after changing tubes. There are three types available:

(a) Standard: The focal spot on the anode is 1.2 × 10 mm. The spot focus as seen through the windows at a 6° take-off angle is 1.2 × 1.2 mm; the line focus is 10 × 0.1 mm. Cu, Mo, Fe, W, Cr, Ni, and Co tubes are available.

Figure 23. Norelco Fine Focus X-Ray Tube—window No. 4 is shown.

(b) High Intensity: The focal spot is 3.2 × 12 mm, resulting in a 0.4 × 12 mm line focus at 6° take-off angle and a 1.2 × 3.2 mm spot focus. It is available with copper target only and has a higher power rating than the standard tube.

(c) Fine-Focus: Focal spot on the anode is 0.4 × 8 mm; spot focus is 0.4 × 0.80 mm, and line focus is 0.04 × 8 mm at 6° take-off angle. The unit area loading for the fine focus copper tube is approximately three times that of the standard copper tube. It is available with Cu, Mo, or Co targets.

G.E. Diffraction Tube. This tube (Price Range E) is constructed with three beryllium windows: two spot foci and one

line focus (Fig. 24). Each window can be protected by a spring-operated shutter which automatically snaps shut when not in use; they are available as optional equipment. The focal spot is only 1 in. from the end of the tube, permitting close proximity of diffraction instruments.

Standard (CA-8S): Nominal focal spot dimensions are 0.8 mm wide × 15 mm long. Available with Cu, Ni, Co, Fe, Cr, Mo, W, and Ag targets. High-Intensity (CA-8L): Similar to the standard tube, but with a slightly larger focal spot of approximately 1.3 × 15 mm. Available with the same targets as the CA-8S tube.

(Courtesy of General Electric Co.)

Figure 24. G.E. X-Ray Diffraction Tube with spring-operated window shutters in place.

Rigaku-Denki. This is a four-window X-ray diffraction tube (Cat. No. 9401, Price Range E) with a 1 × 10 mm target area; there are two line and two spot foci. It is advertised as being interchangeable with the Philips PWRD 50/1 or RDF 50/1 tube and is available with Cu, W, Mo, Co, Fe, or Cr targets.

Dunlee. All Dunlee X-ray Diffraction Tubes (Fig. 25) have four beryllium windows ($795). An earlier model three-window tube has been rendered obsolete.

(Courtesy of Dunlee Corp.)

Figure 25. Dunlee X-Ray Diffraction Tube.

The standard beryllium window thickness is 0.75 mm; 0.375 mm thick windows are available on special order. The tubes have been designed with a higher wattage filament to reduce the rate of evaporation and subsequent contamination of the target. Six different types of tubes are available with Cu, Mo, Ag, W, Cr, Ni, Co, or Fe targets:

(a) DZ-1B (Standard): Focal Spot 0.75 × 15 mm

(b) DZ-1BH (High intensity): Focal spot 1.5 × 15 mm

(c) DZ-1BF (Fine focus): Focal spot 0.4 × 10 mm

(d) DZ-2 (Standard, Philips equipment only): Focal Spot 0.75 x 12 mm

(e) DZ-2H (High intensity, Philips equipment only): Focal Spot 1.5 × 12 mm

(f) DZ-2F (Fine focus, Philips equipment only): Focal Spot 0.4 × 10 mm

Dunlee tubes are standard equipment on Picker and Tem-Pres units and can be adapted for use on other generators as well.

Siemens. Siemens produces a line of three- and four-window tubes of various

Chemical Instrumentation

power ratings (Price Range E). The 1 × 10 mm size of the focal spot on the anode is exactly defined by a slit between the filament and the anode. Cr, Cu, Mo, and Ag targets are available. Another line of tubes with a 5.5 × 16.5 mm focal spot and offering Au, Cr, W, and Mo targets is also available for fluorescence spectroscopy. All the tubes are built with oil-filled shields for high-voltage protection. The window material is beryllium (Fig. 26).

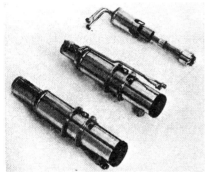

(Courtesy of Siemens America, Inc.)

Figure 26. Siemens X-Ray Diffraction Tubes.

Machlett. Machlett produces a number of X-ray tubes suitable for X-ray diffraction purposes. Models 0-2, 0-2L, A-2, and A-2L each have two beryllium windows and are available with a number of targets (tubes with suffix "L" have the window opposite the line focus rather than the spot focus). They differ in that the shockproof housings of the "O" tubes are filled with an insulating oil and those of the "A" tubes have air as the insulating material. The "A" tubes may also be obtained in nonshockproof models (Fig 27).

The power ratings for all the tubes vary with the type of tube, target material, and generator used (full-wave, half-wave, or

(Courtesy of Machlett Laboratories, Inc.)

Figure 27. Machlett 0–2 X-Ray Tube.

constant potential). Manufacturers' specifications should be consulted.

MONOCHROMATIC RADIATION

The electrons bombarding the surface of a target give rise to both white radiation and characteristic radiation. In the diffraction procedures discussed in this article the wavelength (λ) is known, the diffraction angle (θ) is measured and the interplanar d-spacing is calculated using the Bragg Law: $\lambda = 2d \sin \theta$. In order to obtain reliable data, λ and θ must be known accurately. The experimental determination of θ will be described in those sections in which X-ray diffraction cameras and diffractometers are discussed. The determination of λ depends upon the degree to which the X-radiation emanat-

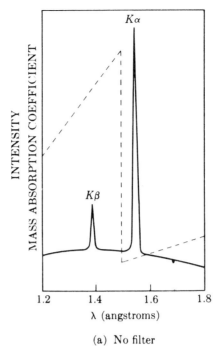

(a) No filter

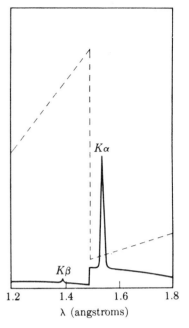

(b) Nickel filter

(From Ref. (10) by permission, Addison-Wesley)

Figure 28.   Comparison of the spectra of copper radiation (a) before and (b) after passage through a nickel filter (schematic).   The dashed line is the mass absorption coefficient of nickel.

ing from a given target can be monochromatized. The following techniques have proved to be successful: beta-filter, crystal monochromator, and pulse-height analysis. The principles, applications, and limitations of these devices will be described.

Beta-Filters

The white radiation spectrum consists of a continuous band of wavelengths; it will not give rise to a specific diffraction peak but will increase the general background or streak associated with each diffraction peak. In general, a reflecting set of planes gives rise to a white radiation streak as well as to characteristic lines. When the K-shell is excited, the $K_{\alpha 1}$ and $K_{\alpha 2}$ are so close that they cannot be separated easily; the K_{β} (which is of shorter wavelength) and some of the white radiation streak can be removed by the use of a beta-filter.

The beta-filter is a material of atomic number one or two less than that of the target whose absorption edge lies between the K_{β} and K_{α} wavelengths (Fig. 28). It is inserted at some point between the incident X-ray beam and the detector and

can be located on either side of the sample. For the targets generally in use, the beta-filter is a thin metallic foil or pressed pellet with a uniform distribution of the metal (or its oxide) in it. The thickness of the foil will determine the final ratio of the intensities of the K_{β} and K_{α} wavelengths. For powder work, a thickness of the beta-filter sufficient to reduce the intensity of the K_{α} to approximately one-half its original value will result in a reduction of the K_{β} from about one-ninth of the K_{α} in the incident beam to about $1/500$ of the K_{α} in the transmitted beam (Table 1).

The placing of the beta-filter depends upon the design of the specific instrument being used. When film is used the filter is nearly always placed between the X-ray source and the sample; when counter detectors are used it is not unusual to find the beta-filter in front of the counter window. If the filter is placed near the counter window, there is a possibility that under certain conditions (i.e., with the use of balanced filters, see below) it will cause an increase in background due to fluorescence. Most manufacturers furnish beta-filters and holders compatible

Target	Filter	Incident beam $\dfrac{I(K\alpha)}{I(K\beta)}$	Filter thickness for $\dfrac{I(K\alpha)}{I(K\beta)} = \dfrac{500}{1}$ in trans. beam		$\dfrac{I(K\alpha) \text{ trans.}}{I(K\alpha) \text{ incident}}$
			mg / cm^2	in.	
Mo	Zr	5.4	70	0.0043	0.30
Cu	Ni	7.5	18	0.0008	0.42
Co	Fe	9.4	13	0.0006	0.47
Fe	Mn	9.0	12	0.0006	0.47
Cr	V	8.5	9	0.0006	0.48

Table 1.   Filters for Suppression of K_{β} radiation

 a From Ref. (10) by permission, Addison-Wesley

with their instruments. Some of these support only one filter and others are small wheels or discs with holes for five or six different filters. A few representative examples of these are described:

Aremco distributes a safety shutter and filter wheel attachment for either Norelco or G.E. equipment (Fig. 3). There are six spaces and filters are also supplied upon request. The wheel can be turned by means of a simple lever-gear system and the operator's hands do not have to approach too close to the tube.

Philips supplies a six-position disc (with filters) which attaches directly to the tube-housing and can be used for all apparatus except the diffractometer. In the latter case a single-filter holder fist over a slit in front of the counter. The filters are designed to reduce the K_β/K_α integrated line intensity ratio to $^1/_{100}$.

The Siemens horizontal tube holder (Price Range E) is equipped with a 7-hole wheel which has five different beta-filters and two open holes for passage of the direct beam.

The material for the filters is generally supplied by the manufacturers; thin metal foils suitable for use as filters can also be obtained from a number of chemical supply houses (e.g., MacKay).

A word of caution: Remember that the correct filter must be used with a given target. If weak patterns or spurious results are obtained, one of the first items to check is the beta-filter.

Balanced Filters. When accurate intensity measurements are made using a counter detector, the background under the peak must be determined. This can be done (a) by taking counts on either side of the peak, extrapolating through the peak and the subtracting the background or (b) by using balanced filters (or Ross filters). Neighboring elements of the periodic system have absorptions which vary with wavelength in a similar manner, except that the absorption edge for the heavier element is shifted to a smaller wavelength. Two filters can be balanced, by adjusting their relative thicknesses, so that they absorb almost exactly the same for all wavelengths except in the gap between their absorption edges. This gap can be chosen so as to bracket the characteristic radiation of the target (Fig. 29). If the peak is scanned twice, once with each filter in place, then the differences between these two scans is that due to the scattering of characteristic radiation from the peak. Pairs of filters suitable for use with various targets are listed in Table 2.

Charles Supper Co. markets a balanced-filter interchanger which can be supplied with a base to fit any commercial X-ray tube-housing ($395). The interchanger proper is only $^3/_4$ in. thick, so the increase in the X-ray beam path is kept at a minimum. The device has three states: (a) X-ray beam blocked, (b) X-ray beam passing through filter A, and (c) X-ray beam passing through filter B. These three states can be controlled manually by means of a switch for use with manually controlled diffractometers, or they can be controlled automatically for use with automated diffractometers. The use of state (a) permits the crystal to be spared

irradiation when a reflection is not being recorded. This is most important if the crystal tends to decompose when irradiated, as with many crystals of biological interest. The effective thickness of each filter can be adjusted by varying the angle it makes with the direction of the X-ray beam. This can be done safely while the intensity of the filtered beam is being measured with the diffractometer.

G.E. manufactures incident-beam attenuators and diffracted-beam balanced filter devices which are incorporated into the collimators. They include a power supply and pushbutton control station

and are suitable for use in automated systems.

Picker offers a balanced-filter attenuator accessory for use with the Goniostat Four-Angle Programmer Automatic System.

Tem-Pres manufactures a slide-type balanced-filter system for G.E., Norelco, Picker and Tem-Pres diffractometers. It can be obtained for manual, semi-automatic or automatic operation (Price Range E).

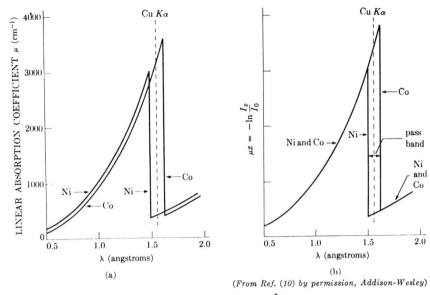

Figure 29. Balanced filters for CuK_α radiation ($\lambda = 1.5418$ Å): (a) absorption coefficients of filter materials; (b) after balancing.

(From Ref. (10) by permission, Addison-Wesley)

			Filter elements					
Radiating element			(a) Component for simple filter or (b) First component of balanced filter			Second component of balanced filter		
Z	Element	K_α wavelength	$Z-2$	Element	Absorption edge	$Z-3$	Element	Absorption edge
47	Ag	0.5608 Å	45	Rh	0.5338 Å	44	Ru	0.5605 Å
46	Pd	0.5869	44	Ru	0.5605	43	Tc	?
45	Rh	0.6147	43	Tc	?	42	Mo	0.6198
44	Ru	0.6445	42	Mo	0.6198	41	Nb	0.6529
43	Tc	?	41	Nb	0.6529	40	Zr	0.6888
42	Mo	0.7107	40	Zr	0.6888	39	Y	0.7276
41	Nb	0.7476	39	Y	0.7276	38	Sr	0.7697
40	Zr	0.7873	38	Sr	0.7697	37	Rb	0.8155
			$Z-1$			$Z-2$		
30	Zn	1.4364	29	Cu	1.3804	28	Ni	1.4880
29	Cu	1.5418	28	Ni	1.4880	27	Co	1.6081
28	Ni	1.6591	27	Co	1.6081	26	Fe	1.7433
27	Co	1.7902	26	Fe	1.7433	25	Mn	1.8964
26	Fe	1.9373	25	Mn	1.8964	24	Cr	2.0701
25	Mn	2.1031	24	Cr	2.0701	23	V	2.2690
24	Cr	2.2909	23	V	2.2690	22	Ti	2.4973

Table 2. Data for Preparing Balanced Filters

[b] From Ref. (6) by permission, John Wiley & Sons

Chemical Instrumentation

Pulse-Height Analysis (PHA)

In systems in which the detectors convert the radiation pulses to electrical impulses, electronic devices can be used to discriminate between the energy of the incoming pulses and to pass only certain ones. High energies produce large amplitude pulses and low energies produce small amplitude pulses. The PHA unit acts as an "electronic filter" and can be employed in conjunction with proportional or scintillation counters. The use of pulse height analysis is far from unique to X-ray diffraction systems and is widely used. A detailed discussion has already appeared in these pages (see THIS JOURNAL 38, A314 (1961)). There are numerous brands of PHA devices on the market and many of these can be adapted for use with X-ray diffraction systems. Certain manufacturers supply PHA units which are designed to be compatible with X-ray diffraction equipment (i.e., fit a standard rack, proper ratings, etc.) and some of these will be discussed briefly. The single channel PHA units usually used deliver an output suitable for driving scalers and ratemeters when the amplitude of an incoming pulse falls within the threshold levels of the discriminator window settings. The settings are made by establishing a minimum level and a window. For example, one would set a base level (E) of 4 volts and pass all pulses in a 10-v window (ΔE), i.e., from 4 to 14 v. Pulse height analysis can be used to remove the "higher-order effects" that a monochromator cannot remove (see below) and/or the white radiation spectrum. However, in order to pass a reasonable intensity of the diffracted beam the window cannot be made too small and so it cannot be used to discriminate between the K_α and K_β radiations. Therefore PHA is generally used in conjunction with a beta-filter or a monochromator.

The PHA unit can also be used to discard all pulses below a minimum value and to pass all pulses above this value.

Hamner Electronics Model NC-11 ($325) can be used with an amplifier having output pulses of 0.5 microsecond or wider; it can be used to drive instruments of 10 v input sensitivity. The baseline may be set between 0.3 and 10 v, with a window of 0–10 v. These settings are made with a 10-turn precision helipot. The PHA accepts positive pulses from the counter amplifier and performs either integral or differential analysis of the pulse. In integral work, only the E dial is used. This sets a baseline above which all pulses will be counted. When in differential mode, a window ΔE wide is set starting at the baseline.

The Picker Amplifier/PHA unit is built into the completely transistorized Radiation Analyzer Panel (Fig. 12). Its features include a choice of manually or automatically driven lower-level scan; window width stability better than 1%/day for 10% window; and overall baseline stability better than 0.5%/day.

The G.E. SPG 6 module (Price Range F) combines a linear amplifier and a PHA unit. The PHA can be set at any base

between 1 and 100 v and the pulse height selector window can be set between 0.1 and 30 v. There is a resolving time of 0.75 μsec and a choice of four modes of operation: (a) lower limit discrimination only, will reject all pulses below a level, E_L; (b) passes all pulses between E_L and $E_L + \Delta E$ (ΔE can be up to 30 v); (c) passes all pulses between $E_L - \Delta E$ and $E_L + \Delta E$ (ΔE can be up to 30 v); (d) passes all pulses above E_L but below E_U (range of 99 v).

The Norelco PHA unit (Price Range F) has a 0–50-v baseline and a 0–50-v window; baseline scanning can be done manually or automatically. There is a choice of differential or integral output and the resolution time is 1 μsec.

Jarrell-Ash Model 80–370 ($1100) combination amplification and PHA unit has a 50-v window, with differential or integral operation and permits automatic baseline scanning.

Crystal Monochromators

A monochromatic beam can be obtained by using a single crystal set to reflect the desired component of the radiation emanating from an X-ray tube. If the crystal is set to reflect the K_α wavelength, the resulting beam is made up (primarily) of the two wavelengths of the K_α doublet and (partly) of harmonics. The harmonics arise because the higher orders of the interplanar spacing reflect $\lambda/2$, $\lambda/3$, etc. at the same Bragg angle. That is λ, $\lambda/2$, $\lambda/3$, etc., are reflected at the same angle (θ) by d, d/2, d/3, etc., respectively. These higher harmonics can be removed by (a) using a crystal with a weak second-order reflection and keeping the kv below the potential needed to excite $\lambda/3$; or (b) using PHA to remove the higher order reflections.

The use of a monochromator results in a lower background as recorded by either film or counter detectors. Furthermore, if the sample fluoresces in the characteristic radiation (e.g., CuK with a sample containing iron atoms) a monochromator placed between the sample and the counter will remove the fluorescent radiation.

A number of monochromator designs have been described in the literature. They employ bent, ground, or straight crystals; monochromatize by transmission or reflection; are focusing or nonfocusing and, if focusing, symmetric or asymmetric. Only one type will be described in detail.

Theoretically perfect focusing can be achieved by using a bent and ground crystal monochromator; i.e., all of the K_α radiation intercepted by the crystal will be focused to a point. The design is based on the following geometric principles: (a) the tangent to a circle is perpendicular to the radius at the point of tangency; and (b) in a circle, all of the inscribed angles intercepting the same arc are equal.

A rectangular crystal with a set of reflecting planes parallel to its surface is bent so that the parallel planes become concentric circles of radius R. In this way all the plane normals are made to pass through the center of the circle of radius R. This point is located on the same circle as the beam source. The

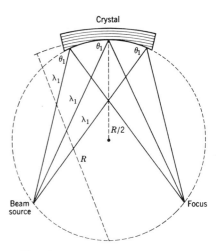

(From Ref. (11) by permission, John Wiley & Sons)

Figure 30. Diagram of accurate focusing which can be obtained by a bent and ground monochromatizing crystal.

crystal is then ground into radius $R/2$, so that all the rays from the beam source will make the same angle with the planes and will be reflected to the same point, i.e., to the focus (Fig. 30).

A properly designed system can resolve the $K_{\alpha 1}$ and $K_{\alpha 2}$ beams.

Those using crystal monochromators should be aware of the fact that monochromatized radiation is polarized. Therefore the polarization factor used in determining the magnitudes of the structure factors from the observed intensities must take into account the fact that monochromatized radiation was used.

Some of the more commonly used crystals are listed in Table 3.

Siemens marketed the first commercial diffracted-beam monochromator (Price Range E). The monochromator is located between the sample and detector of the diffractometer and utilizes a bent and ground quartz crystal. By suitable adjustments, the α_2 beam can be attenuated to such a degree that the diffraction pattern is obtained almost exclusively with α_1 radiation.

E & A offers a monochromator ($1100–1400) that can be adapted to fit G.E., Norelco, Picker, or Siemens units (Fig. 31). The E & A Primary Beam Monochromator (Models CCW (counter clockwise) and CW (clockwise)) has horizontal and vertical slit systems, is fully enclosed for radiation protection and has a graduated dial on the outer shell to allow setting of the required 2θ position. The crystal is mounted on a eucentric goniometer head which can accommodate a singly bent, doubly bent, or flat monochromatizing crystal. A doubly bent LiF crystal is available as optional equipment. Another feature is the removable track which can support an X-ray diffraction camera in the exit beam.

G.E. offers a diffracted-beam monochromator ($1350) for use on their diffractometer with either single-crystal or powder equipment. Built by Manlabs, Inc., it is designed for a cylindrical, doubly bent, spheroidal LiF crystal which is available as optional equipment ($300). The monochromator can be easily adjusted to diffract Cu, Ni, Co, Fe, or Cr K radiation; simple reassembly provides for Ag or

(Courtesy of E & A)

Figure 31. E & A Monochromator (note the removable track which can support an X-ray diffraction camera in the exit beam).

Mo K radiation.

Picker manufactures an incident-beam monochromator for single crystal investigations (Cat. No. 3735, Price Range G).

Picker also distributes a Manlabs Monochromator (Price Range F) designed especially for the Picker 3488 series Bi-plane Diffractometer. It utilizes a singly bent and ground silicon crystal (111) which is supplied and is recommended for use with a high-intensity tube. Doubly bent crystals are available on special order.

The Jarrell-Ash Monochromator (Cat. No. 80–101, $550) uses a curved quartz crystal and has the usual features required for proper adjustment of the crystal. The 80–118 Slit System controls the angular divergence of the monochromatic beam; it is claimed that the $K_{\alpha 2}$ can be eliminated completely.

Enraf-Nonius offers a monochromator similar to that used on their Guinier-DeWolff I Camera. It uses a curved quartz crystal in which the curvature is a logarithmic spiral. The unit is mounted on a stand so it can be used with any tube. Model Y 821a ($320) is for use with Mo radiation and model Y 821 ($320) is for use with Cu, Fe, and Co radiations.

Rigaku-Denki also offers a monochromator (Price Range E) designed for use with its generators; Cat. No. 1705 is for use with a camera and Cat. No. 2722 is for counter work.

The AMR Focusing Monochromator (Price Range F) is designed as an attach-ment for the Norelco Wide Range Goniometer (diffractometer). It is mounted on the detector arm of the diffractometer and employs a curved LiF crystal placed behind the receiving slit so that the rays diffracted by the specimen and passing through the slit impinge upon the curved crystal of the monochromator (Fig. 32). The monochromator is readily adjustable to different wavelengths and has marked settings for the $K\alpha$ radiation of Ag, Mo, Cu, Co, Fe and Cr. By properly adjusting the θ angle for the monochromator crystal, the center of this band is placed between the $K_{\alpha 1}$ and $K_{\alpha 2}$ wavelengths of the radiation being used and these two components are transmitted in their correct relative intensities.

Stoe also offers a monochromator especially constructed for the Philips diffractometer; here, too, the crystal is situated between the specimen and counter.

Flat or bent and ground crystals can be obtained from Harshaw Chemicals.

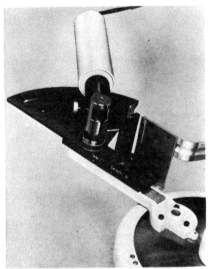

(Courtesy of Philips Electronic Instruments)

Figure 32. AMR Focusing Monochromator mounted on detector arm of Norelco Wide Range Goniometer.

The use of a monochromator between the sample and detector has the following advantages:

(a) Discrimination against unwanted radiation (K_β, continuous radiation, target impurities).

(b) Discrimination against fluorescent radiation from the specimen. Thus monochromatized CuK radiation can be used with iron-containing samples without the need for PHA.

(c) If radioactive specimens are used, the detector is easily shielded against direct radiation from the specimen.

Part Three—Detectors

DETECTORS

Three methods are commonly used to detect X-rays: fluorescent screens, film, and counters.

Fluorescent Screens

Materials which emit visible light when placed in an X-ray beam have been coated on cardboard or plastic to form fluorescent screens. Zinc cadmium sulfide and calcium tungstate are two phosphors which have been widely used. The screens usually have a protective screen coating over the fluorescent material that provides resistance to abrasion, chipping, and peeling (Fig. 33).

Fluorescent screens are used for two purposes:

(a) To locate the primary beam when adjusting apparatus. The screen should be placed at the end of a long rod so that the operator's hands need not be exposed to the direct beam.

(b) To decrease the required exposure times when film is used as the detector. In this case the screen is placed in contact with the film so that the film is exposed to the light emitted by the screen as well as to the diffracted X-rays.

Examples of commercially available screens are Radelin X-ray screens (various types of intensifying (medical and industrial) and fluoroscopic screens) produced by the U.S. Radium Corp. and the DuPont Cronex Xtra Life Intensifying Screens (PAR speed at $6.75 a pair and HI speed at $7.40 a pair).

Film

X-rays affect photographic film in much the same way as does visible light. However, the emulsion on ordinary film is too thin to absorb much of the incident X-radiation. For this reason, X-ray films are made with rather thick layers of emulsion and large grain size. Film techniques are used for recording both the positions and intensities of the reflections. If meaningful intensity measurements are to be made, special care must be taken in the handling and developing of film. The proper technique has been discussed elsewhere (Ref. (6), pp. 78–111; Ref, (11), pp. 364–76). Morimoto and Uyeda (13) reported the results of tests on 43 types of commercially available X-ray films made by eighteen manufacturers. These films were tested for: film speeds for Cu and Mo radiation, fog density, granularity, absorp-

Crystal	Reflection	Spacing d in Å	Properties of Reflection			Properties of Crystal		
			Peak Intensity	Breadth	Multiplicity	Stability	Mechanical Properties	Special Uses
β-Alumina	0002	11.22	Weak			Perfect	Hard, brittle	For long wavelengths
β-Alumina	0004	5.61	Weak-medium	Moderate	Great			
Gypsum	020	7.58	Medium-strong	Very small	Negligible	Poor	Soft, flexible	For low-angle scattering; focusing; long wavelengths
Pentaerythritol	002	4.39	Very strong	Moderate	Great	Poor	Soft, easily deformed	General purposes
Quartz	10$\bar{1}$1	3.34	Weak-medium	Very small	Negligible	Perfect	Can be elastically bent	For low-angle scattering; focusing
Fluorite	111	3.15	Medium-strong	Moderate	Small	Perfect	Moderately hard	For eliminating harmonics; general purposes; short wavelengths
Urea nitrate	002	3.13	Strong	Very large	Very great	Very poor	Very easily deformed	For large specimens
Calcite	200	3.03	Medium	Small	Negligible	Perfect	Moderately soft	For low-angle scattering; isolation of α_1
Rock salt	200	2.81	Medium-strong	Large	Great	Slightly deliquescent	Can be plastically bent in warm water	For focusing; general purposes
Diamond	111	2.05	Weak	Very small	Negligible	Perfect	Very hard	For eliminating harmonics

Table 3. Reflection Characteristics and Properties of Various Monochromator Crystals (12)

Chemical Instrumentation

tion of Ag, Mo, Br, Cu, Fe, and Cr K_α radiation, film homogeneity, sensitivity toward safety lights, aging effects, and density versus exposure curves. The latter test indicates over what range of blackening density the film is linear with respect to exposure time. At low and high degrees of blackening the intensity is not a linear function of exposure time.

The following firms sell industrial X-ray film (in the U.S.): Eastman Kodak, Ilford, DuPont, and General Aniline and Film Corp. (GAF). Representative prices are $15 for a box of seventy-five 5 × 7-in. sheets; $3.50 for a 25-ft roll of 35-mm film.

Polaroid. In recent years the Polaroid Co. has adapted its 4 × 5 Land Film Holder for use with X-ray diffraction equipment, marketing the unit as the Polaroid XR-7 System. An intensifying screen is permanently located in the cassette; after the film is inserted a lever releases an "air pillow" which distributes a 20 psi pressure evenly over the entire 4 × 5-in. picture area, thus keeping the film in contact with the screen. Using Type 57 Land Film Packets (emulsion speed 3000 ASA equivalent), pictures can be taken in minutes (compared to hours for standard film) and a dry print obtained in 15 sec. The film-screen combination is approximately eight times faster than conventional X-ray films of comparable resolution and quality.

The use of this film has greatly increased the speed at which the alignment and preliminary investigation of a crystal can be carried out. Although certain industrial radiography units can dry process their film, the cameras must still be loaded in a darkroom; with the Polaroid system a darkroom is never needed. However, there are a few disadvantages to the Polaroid system. (a) As of the present, Polaroid Land film can only be used with flat cassette cameras (e.g., Laue and precession). (b) So far as is known to the author, methods have not been developed for obtaining accurate relative intensities from Polaroid Land film. Thus it is excellent for routine flat-plate film work but not for cylindrical cameras or for intensity measurements.

Film Readers

Films are read for either of two reasons. The first is to determine the positions of the reflections on the film. (Once the coordinates of a point on the film are known the diffraction angle (2θ) can be determined and the d-spacing calculated from the Bragg Law.) The second is to determine the intensity of the reflection. Therefore film readers are measuring devices or densitometers or both. Film techniques are utilized in many branches of science and technology and there are a large number of commercially available film readers. We shall limit our discussion to those instruments designed specifically for use with X-ray diffraction films.

Measuring Devices. A number of firms market film illuminators equipped with measuring devices. The illuminators have metal cabinets and slanted opal glass tops to permit ease in viewing.

The *Norelco* Measuring Device (Cat. No. 52022, Price Range D), manufactured by Philips, accepts 35 mm film up to 54 cm in length. The scale is 50 cm long with an illuminated cursor, including a vernier and scale magnifying lens. The scale reads directly to 0.5 mm and with the vernier to 0.05 mm (Fig. 34).

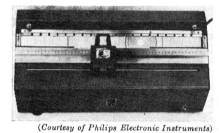

(Courtesy of Philips Electronic Instruments)

Figure 34. Norelco Film Illuminator and Measuring Device.

The *G.E.* scale accepts 35-mm film up to 40 cm long; the scale can be read to 0.10 mm, using the attached vernier.

The *Rigaku-Denki* Measuring Rule (Cat. No. 1861, Price Range D) can be read to 0.05 mm and can accommodate 5 in. wide films. The cursor and film are in close contact so as to eliminate parallax.

In the above three devices, the distances are measured by sliding a mark on the cursor over a line on the film and reading the scale.

Picker and Siemens have illuminators with measuring devices in which the scale itself can be moved. The *Siemens* model (Price Range D) is available with a 20 or 30-cm scale. This consists of a glass rule which is graduated on the underside, and which slides within a frame, driven by means of a finely threaded spindle. The right end of the glass rule rests against the measuring pin of a dial gauge. In operation, the lines to be measured are brought in alignment (coincidence) with the neighboring division lines on the glass scale. The distance the glass scale has to be moved to accomplish this coincidence is indicated on the dial gauge. The scale is then read, millimeters on the glass rule and millimeter fractions on the dial gauge. The scale has a reading error of approximately 0.02 mm; it is designed to hold 35-mm film. The illuminator can be tilted and locked in any position for ease of operation.

The *Picker* Film Scale (Cat. No. 364, Price Range E) works on the same principle. It has a 40-cm glass scale with a full range tolerance of 0.05 mm. Measurements up to 0.01 mm can be made with interpolation to 0.005 mm. Provision is made for holding 5 × 7-in. films as well as 35-mm films.

Enraf-Nonius offers a film reader and measuring device specifically designed for Guinier and Debye-Scherrer films (Cat. No. Y99, $1000). It is based on a projection system which enlarges both the film and scale for viewing on a ground-glass screen. This system reduces the need to measure small distances and permits close lines to be seen separately. Parallax is eliminated as both the film and the scale are projected on the same surface. The stated accuracy is 0.01 mm.

The Nonius Negatoscope (Cat. No. Y885, $150) is an illuminated light box provided with a transparent rule with distance divisions on the underside. Also available for use with the Negatoscope is a plastic triangle for Weissenberg exposures and rulers for Debye-Scherrer exposures at different radiations.

MRC has recently developed a film reader (Price Range E) in which the film is shown on a 8 × 10-in. screen; the precision etched scales are directly readable in millimeters with an accuracy better than 0.03 mm.

Ernest F. Fullam, Inc. sells a film reader which effectively magnifies and records the pattern on a chart recorder. It was initially designed for use with electron diffraction films, but is equally adaptable for some X-ray diffraction films. The apparatus reproduces on a chart the spacing of any selected portion by rotation of the pattern; it may be read to 0.01 mm if the slowest scanning speeds are used. A stabilized light source permits determination from the chart of relative intensities of the lines and spots of the film. The Measuring Device (Cat. No. 1182) falls in Price Range F; the Recording Apparatus (Cat. No. 1182-RDB) in Price Range G (Fig. 35).

The *David W. Mann Co.*, Division of GCA Corp., has developed the Mann Type 1222 (Price Range J), an analytical and data recording instrument for precision measurement of X-ray diffraction films. Type 1222, designed specifi-

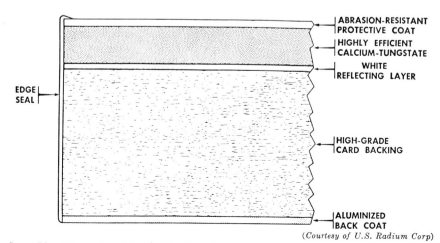

Figure 33. Cross-section of Radelin Intensifying Screen.

ABRASION-RESISTANT PROTECTIVE COAT

HIGHLY EFFICIENT CALCIUM-TUNGSTATE

WHITE REFLECTING LAYER

EDGE SEAL

HIGH-GRADE CARD BACKING

ALUMINIZED BACK COAT

(Courtesy of U.S. Radium Corp)

(Courtesy of Ernest F. Fullam, Inc.)

Figure 35. E. F. Fullam Measuring Device with automatic recorder.

cally for measuring distances and optical density on X-ray diffraction film, consists of a film measuring precision comparator motion, a photoelectric setting device for accurate positioning of the line of interest on the reference axis for measurement, a film density measuring capability, and a digital readout for accurate recording of information by an output writer (Fig. 36).

(Courtesy of David W. Mann Co., Div. of GCA Corp.)

Figure 36. Type 1222 Photoelectric Line Setting Instrument.

In one mode of operation the film is mounted on a stage and aligned manually. Lines of interest can be selected by viewing a 6× magnified image on a rear surface projection screen. Line settings are made photoelectrically.

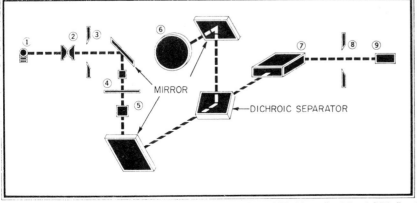

(Courtesy of David W. Mann Co., Div. of GCA Corp.)

Figure 37. Type 1222 Electro-Optical System (see text).

Density readings, taken from the oscilloscope, are manually entered into the memory of the digitizer. The operator then depresses a foot switch to record this readout information, density, and position, in hard copy on an output writer or other optional recorder. If desired, a second mode of operation provides for the power scanning of the film with the density trace being recorded on a strip chart recorder (available at extra cost).

The x-coordinate provides 380 mm of travel with a least count of 0.005 mm. Calibration assures that any position of the stage, when traveling in the measuring direction, will not deviate from the position indicated by the dial reading by more than ±0.005 mm or ±0.003% of stage travel.

Illumination for the optical system is provided by a tungsten lamp (1) (Fig. 37) operated at 6VDC, the light being gathered by a condensing system (2) to illuminate the pre-slit (3). The pre-slit is projected to the film plane (4) at a reduction of 2:1. The pre-slit controls scattering and off-axis light enabling measurement of images having high densities.

The image of pre-slit and film image combined are picked up by substage projection lens (5) after which it transmits a large portion of the visible light to focus at the screen (6). A small portion of the light in the far blue end of the spectrum is used in the photoelectric system exclusively. Magnification of the screen image with respect to film plane is approximately 6×. The photoelectric image is reflected from a mirror at the harmonic scanning device (7), focusing at the exit slit (8). When the scanner is operated the image moves back and forth across the exit slit in a controlled manner. The photomultiplier tube (9) then sees a constantly changing portion of the image field as a function of the mirror position. By triggering the linear sweep of the oscilloscope each time the mirror reaches an extreme of its angular amplitude, an image of the density profile of the line in the scan field is presented on the scope face for each swing of the mirror; first from one direction and then from the other. The profile traces observed are, therefore, mirror images of each other and the criterion for centering is coincidence of the peaks of the line trace. The output of the photomultiplier tube is fed through a compression amplifier so designed that the amplitude of the scope trace is essentially linear in optical density.

Charles Supper Co. produces a circular film-measuring instrument (Cat. No. CFMI, $265). This consists of a cast aluminum frame 12 in. square, with a rotating circular frame mounted in its center. The circular frame contains a glass window upon which the film is mounted for illumination from below. The position of the frame can be determined to 5 min of arc (with a vernier). A large glass rider, with a hairline for linear measurements, slides on a straight millimeter scale and reads to 0.05 mm. It was designed for precession photographs and is useful for any flat-plate photograph.

Densitometers. Enraf-Nonius offers three different microdensitometers. Model 1 (Cat. No. Y870, $2250) is designed for the investigation of integrated Weissenberg exposures; Model 2 (Cat. No. Y874, $4000) is for all types of photographs; Model 3 is a recording microdensitometer for photographic negatives on film or plate.

Model 3 (Cat. No. Y877, $17,450) uses a double-beam servo null balance system; a servo-operated linear-density compensation wedge is balanced against the point being measured (Fig. 38). The position of the wedge is an indication of the density of the specimen. An optical density of 0–3.0 can be measured with 0.5% accuracy.

The film can be positioned by electrically driven translation and rotation motors; it also has options for using punched tape to automatically position the film and for choosing the method of intensity measurement and recording. Models 1 and 2 are manually operated; the degree of blackening of the reflection and the surrounding background are measured with a galvanometer and the intensity is obtained from the calibration chart furnished with the instrument. These two models also provide translations in the base; in addition, Model 2 has a rotational motion. Movements can be made either continuously or in predetermined intervals. A number of different diaphragm sizes are provided.

The Joyce-Deeley Flying Spot Integrating Microdensitometer (distributed by National Instruments Laboratories, Inc.)

Chemical Instrumentation

(Courtesy of Enraf-Nonius)

Figure 38. Nonius Y-877 Microdensitometer.

scans the reflection and records the integrated intensity of the area under investigation (Cat. No. 6-5000, $10,700).

The instrument consists of two major sections (Fig. 39):

(a) The flying spot microscope, one of the two sample film tables, the direct viewing system, and the photo-multiplier receiver are housed in a light-tight case on one side of the instrument.

(b) On the other side of the instrument is the operating section, with all controls and meters, the long persistence monitor tube and the second sample film table.

The sample film to be measured is placed on the table under the flying spot microscope and the copy of the sample film, or any film taken from the pack of films

exposed together in the X-ray camera, is placed on the operating section table. A punch is supplied which enables the two films to be punched together when the detail on both is coincident. The two tables have locating pins to take the punched films.

Direct viewing of the sample to be measured is obtained by an auxiliary light beam through the sample film under the flying spot microscope in a direction opposite to that of the scanning beam by means of a manually retractable mirror. This viewing arrangement serves the combined purposes of ensuring that the microscope field and the light spot in the

(Courtesy of Jarrell-Ash)

Figure 40. Jarrell-Ash Recording Microphotometer.

center of the second or "dummy" table are correctly aligned with respect to one another and also that the microscope is focused. When these two conditions are satisfied, the case containing the flying spot microscope may be closed and all further operations can now take place on the operating side of the instrument.

The purpose of the second table is to provide a simple means for orientation in a complex field and to locate the sample detail to be measured.

The *Jarrell-Ash* Model 24-400 Recording Microphotometer ($8500) is a table model instrument with a film holder that

will support Weissenberg, Laue, precession and powder diffraction patterns (Fig. 40). It has an eleven-step density filter starting at 100% transmission; each additional step reduces the transmission by approximately 50% until the eleventh step, which is 0% transmission. The scanning precision is ±0.05 mm maximum lead error per 25 mm with a traverse motion of 490 mm; it has a 12-speed reversible drive. Apparatus No. 22-260 is a 5 in. plateholder with a circular film measuring rule, longitudinal vernier (to 0.1 mm), and a circular scale (to 5 seconds of arc).

The *Wooster* Mark III microdensitometer (Crystal Structures, Ltd., $7000) has only one optical adjustment. Optical densities from 0 to 4 are recorded on a linear scale without altering the sensitivity. Suitable neutral glass wedges can readily be interchanged to investigate smaller ranges with greater accuracy.

Powder films up to 40 cm long and plates and films up to 20 cm diam can be accommodated. For preferred orientation investigations the stage of 20 cm diam can rotate with a variable speed range. There are two mutually perpendicular automatic traverses recording on digital counters and an automatic scanning over 2, 4, or 6 mm can be invoked perpendicular to the X traverse.

The integral chart recorder with long roll chart uses an electric pen giving instantaneous clean fine lines. The chart movement is directly coupled to the Y traverse and distances on the chart may be measured to 0.05 mm. The chart can move in either direction at one of six ratios selected by turning a knob, magnifying the film traverse from 5 to 200 times. A full-scale deflection takes less than 1/4 sec.

The integrated intensity of a line, a spot, or an area on the specimen can be read on a dial on the front panel. A simple adjustment allows for the background blackening.

Rigaku-Denki (MP-3, Price Range H) and Siemens (Price Range F) also manufacture recording densitometers.

Counters and Electronic Panels

The most accurate intensity measurements are made using Geiger, proportional, or scintillation counters. The pulses received by these counters are amplified and fed into ratemeters, scalers,

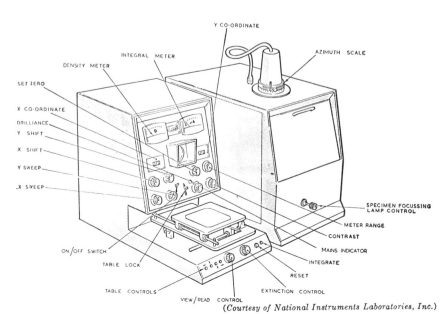

(Courtesy of National Instruments Laboratories, Inc.)

Figure 39. Joyce-Deeley Flying Spot Integrating Microdensitometer.

Detector Selection Chart

(Courtesy of Philips Electronic Instruments)

Figure 41. The efficiency of the counter is a function of the wave-length of the radiation under investigation. This is of particular importance in X-ray fluorescence investigations.

and/or recorders. The principles of most of these devices have been discussed in earlier articles in this series (the following references are to THIS JOURNAL **38** (1961)): counters—A135; power supplies for detectors—A225; amplifiers—A234, A305; scalers—A317; ratemeters—A327. A complete discussion of recorders is found in THIS JOURNAL, **36**, A729 (1959); **37**, A7, A65, and A129 (1960).

In order to obtain the highest accuracy, the various components must be quite stable with linear responses; the line voltage must also be stabilized. All of the major commercial suppliers of X-ray diffraction generators manufacture a line of counters (Fig. 41) and electronic panels. There are also a few companies that manufacture counters or electronic panels designed to be compatible with existing commercial X-ray diffraction equipment. The features of these devices will be discussed.

Counters. The line of Norelco X-ray Detectors (Price Range E) are a good example of the type of detectors which are available for X-ray diffraction studies. The Geiger counter has a linear response up to a maximum of about 1000 counts per second (cps), with quantum efficiency peaked from 1 to 2 Angstroms (about 60% at 1.5 Å); pulse height discrimination cannot be used with Geiger Counters. The proportional counter has a side window tube mounted on a small electronic chassis equipped with a pre-amplifier and a cathode follower. It is linear over an extremely large range (above 20,000 cps). The flow proportional counter is a two-window proportional tube designed to permit a continuous flow of gas within itself. The scintillation counter consists of a thallium-activated sodium iodide crystal mounted on a photomultiplier tube. The detector is equipped with a preamplifier and cathode follower. It is extremely efficient over the entire spectral range (0.5–2.0 Å) and is linear up to approximately 100,000 cps with a resolution time of 1 μsec (Fig. 42).

The G.E. proportional gas counter tube (SPG-6, A4952) has a peak efficiency at 1.9 Å. It is filled with xenon gas and has a 0.025-cm beryllium window. The G.E. flow proportional gas counter tube (SPG-7, A4952EE) is designed for detection of long wavelengths, with maximum sensitivity at 1.54–11.9 Å. The G.E. scintillation counter tube (SPG-4, A4952DH), designed for use from 0.15 to 2.74 Å, uses a NaI (thallium-activated) crystal (Price Range D) (Fig. 43).

(Courtesy of General Electric Co.)

Figure 43. G.E. SPG7 Counter Tube and SPG 4 Scintillation Counter mounted in SPG 4 Preamplifier.

Siemens, too, markets Geiger (krypton or argon filled), proportional, flow proportional, and scintillation counter tubes. The proportional counter is available with the window parallel to the counter wire (Type B) or vertical to the counter wire (Type E). The scintillation counter has a quantum efficiency close to 100% for wavelengths of 0.5–2.0 Å.

LND, Inc. manufactures a line of end window, square window, and round-side window proportional counter tubes (400 Series, Price Range D). These are designed for the detection of low-energy electrons, gamma- and X-radiation. They are available in a choice of windows (mica,

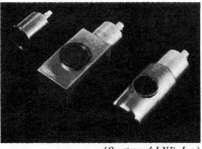

(Courtesy of LND, Inc.)

Figure 44. LND proportional counter tubes.

beryllium, or aluminum), gases (neon, argon, krypton, or xenon, with 10% methane), lengths and sizes of window openings (Fig. 44).

(Courtesy of Philips Electronic Instruments)

Figure 45. Norelco Data Processor and Data Control Electronic Panel.

Proportional and scintillation counters are also available from Hamner electronics (see below).

Electronic Panels. Electronic panels are sold by G.E., Philips (Norelco) (Cat. No. 12206, Price Range H), Siemens (Fig. 13), Picker, Hilger-Watts (Engis), Jarrell-Ash (Fig. 15), and Rigaku-Denki (Cat. No. 5052, Price Range G). Each panel is designed to be compatible with its own system but can be adapted to other systems. Printed and transistorized circuits are becoming more common and the overall reliability of electronic panels is improving.

For example, Philips has developed a computer coordinated, all solid-state Data Processor and Data Control Electronic Panel (Fig. 45). It has controlled-programming capability from a remote source (using optional interfaces). The Data Control unit controls such functions as detector high voltage, pulse shaping, attenuation, linear amplification, pulse height analysis, and determination of pulse rate. The Data Processor unit controls timing, scaling, digital readout (indicating), count rate computing, and provides for printed or punched output.

Picker, too, utilizes completely solid-state circuitry (Cat. No. 6245, Price Range G) (Fig. 12).

In general, the individual modules of the panels can be purchased separately. However, if you wish to combine components from different manufacturers be sure that they are compatible.

The GE XRD-6 detector unit (Fig. 46) can be used as an example of the type of modules that are built into these panels. The XRD-6 detector console (Cat. No. SPG-4, A4984A, Price Range E) is

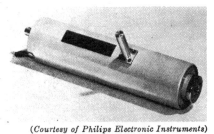

(Courtesy of Philips Electronic Instruments)

Figure 42. Norelco Scintillation Counter.

Chemical Instrumentation

arranged as a floor unit with controls and instruments mounted on front sloping panels above an incorporated writing shelf. Individual chassis and circuit components are mounted on tracks and rollers for front access and inspection.

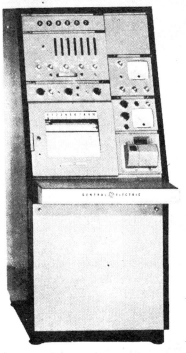

(Courtesy of General Electric Co.)

Figure 46. G.E. XRD-6 Detector Unit.

All service can be performed from the front of the console.

The detector cabinet contains the ratemeter, scaler, time register, power supply, linear amplifier/pulse height selector, and provision for operational choice of a recorder and/or digital printer. The preamplifier and counter tube assembly is included in the basic unit, but mounted separately on the diffractometer. The signal received from the counter and preamplifier as the result of diffracted X-rays is amplified, measured, and displayed by the detector. The linear amplifier feeds the signal to both the scaler and ratemeter; the scaler then accumulates counts over a predetermined time or count and displays the count visually on neon lighted scales. The ratemeter integrates the pulses from the amplifier and displays the rate of count on a meter. The output of the scaler can be registered on a strip chart recorder, and the output of the ratemeter can be registered on a digital printer.

Hamner Electronics Equipment. Hamner Electronics offers a line of instruments that can be used to design an electronic panel which will fit the needs (and budget) of the user. They are designed to be adapted to existing X-ray diffraction apparatus and can be obtained as individual modules to update existing equipment or as entire electronic panels. Detector assembly adapters are available for Siemens, G.E., Norelco, Picker, and other diffractometers. The units are designed to provide for automation of experimental procedures and computer processing of the resulting data. Fully transistorized circuits, front panel accessibility of modules, high stability of the components, and versatility of operating modes are some of the features of this equipment. All Hamner instruments are designed to conform with the specifications of AEC Document TID-20893; all modules conforming to these specifications are mechanically and electrically interchangeable, regardless of manufacturer. Currently Hamner offers the Amperex 300PC Proportional Detector and the Harshaw SHG-X Premium Low Energy NaI (Tl) Integral Scintillation Detector. The former offers 75% detection efficiency with CuK_α; the latter, 97%. These are designed to be used with preamplifiers, power supplies, linear amplifiers, pulse height analyzers, scalers, timers, and rate meters.

Film versus Counter Detectors

Both film and counter detectors have been used since the earliest days of X-ray diffraction investigations. However, because of the overall inefficiency and unreliability of the earliest counter detectors, film was used in nearly all of the reported investigations of the twenties, thirties, and early forties. It is only recently that reliable and efficient counter detectors and electronic panels have been developed.

Some of the major differences between film and counter detection are as follows:

(*a*) With the use of film, entire regions of reciprocal space can be investigated; counters will only record information concerning the particular point at which the counter is located. It is more difficult to investigate *all* of reciprocal space using counter techniques.

(*b*) When film techniques are used, the crystal passes through each reflecting position many times; random variations of the X-ray beam intensity will be averaged out. Counters are set to record each point individually; the X-ray tube output must be properly stabilized if the intensity measurements are to be meaningful.

(*c*) The outlay for equipment (detectors, stabilizers, electronic panels, etc.) is much greater for counter work than it is for film work.

(*d*) The counting efficiency of film is lower than that of counters; the most accurate relative intensity data is collected when counter detectors are used in conjunction with a reliable electronic panel.

Any serious investigation, whether it be of a powder or single crystal specimen, should utilize both techniques. For example, in single crystal investigations, it is recommended that the preliminary investigation of the lattice parameters and symmetry be carried out with film techniques; intensity data can be collected with either film or counter detectors, depending upon the desired accuracy. However, even when counter techniques are used, we have found it most helpful to collect a set of film intensity data. This furnishes a permanent record which is independent of any errors in the electronics or in the recording of the intensity. The recent advances in automatic counter data collection have greatly reduced the tedious portion of data collection and the possibility of errors in recording the intensity; but it is still helpful to have a set of film data as a check on the operation of the instrument.

(Courtesy of MRC)

Figure 47. MRC Collimizer-590.

MRC Collimizer-590

The alignment of X-ray collimation systems generally requires the use of fluorescent screens and a darkened room. The MRC Collimizer-590 ($225) is a novel instrument which simplifies and speeds this process. This device is a small, lightweight, battery-operated unit with a small sensing head or detector (which fits almost all collimation systems) mounted at the end of a flexible cable. The sensitivity of the detector can be adjusted according to the type of radiation and size of the collimator opening. The intensity of the X-ray beam is indicated on the meter and the maximum beam intensity can be obtained easily (Fig. 47).

Part Four — Powder Cameras and Techniques

A sample consisting of a randomly-oriented array of crystallites will, theoretically, have all the possible reflecting planes in diffracting position at all times. In practice, it is difficult to obtain such a sample. However, the larger the diffracting sample and the smaller the crystallites (within certain limits) the greater the chances of approaching random orientation of crystallites. To accomplish this the instruments built for use with powder samples are designed to allow the use of large samples and/or moving samples.

There are two disadvantages inherent in the powder method. The first is that, because of the large number of crystallites present, one does not know the position of the particular crystallite(s) diffracting at a given Bragg angle. If there are a number of planes with different hkl values but the same or nearly the same diffraction angle, overlapping or partially overlapping diffraction lines will be observed. This severely limits the uniqueness of each of the powder data.

The second disadvantage is that the small crystallites have a low diffracting power. Therefore, the intensities of the reflections obtained from a powder sample are generally weak; they are orders of magnitude weaker than the intensities obtained from a single crystal of comparable dimensions.

However, the advantages of rapid sample preparation and data collecting combine to make the powder method a powerful tool for routine sample identification. Also, in special cases it may be possible to obtain sufficient intensity data to allow the crystal structure to be determined. This has been done most often in cases where there are few molecules per unit cell, the crystal system is simple and it is not possible to obtain single crystals.

Two-Theta Angle. Figure 48 shows that the angle between the direct beam and the diffracted beam is equal to 2θ. This is the angle which is most easily obtained experimentally and we shall refer to it often in the following pages.

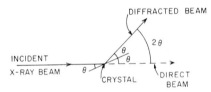

Figure 48. The angle between the diffracted beam and the transmitted beam is equal to 2θ.

Sample Preparation. A properly prepared sample (consisting of randomly-oriented crystallites) is essential to obtaining meaningful powder data. Improperly prepared samples result in spotty lines (on film) or preferred orientation (i.e., incorrect relative intensities, with some reflections predominating and others unusually weak or totally absent). Samples are generally crushed, pulverized or ground and then passed through a -250 to -325 mesh screen. However, care must be taken that the nature of the sample is not changed during the process (e.g., loss or gain of water of hydration or initiation of a solid-state reaction); when in doubt, examine a portion of the untreated sample. Another problem that is encountered is that if the sample is a mixture, one of the components may not grind as finely as the others and will be left behind in the sieve. This latter occurrence is occasionally helpful in isolating the components of a mixture. Other techniques and warnings are found in the recommended texts (*8, 10, 11*).

A number of grinders, crushers, pulverizers, and sieves are commercially available. The products described here are offered only as examples.

Thermovac Industries offers a pulverizer designed for use with frozen tissues. It contains a reservoir which can be filled with dry-ice-acetone or liquid nitrogen. The sample, in a special holder, is inserted in the reservoir and cooled. The pulverizer uses a recoil action pestle; by pressing

the handle the pestle pulverizes the sample. This device would be helpful in powdering organic materials that are soft and tacky at room temperature.
Crescent Dental Manufacturing Co.

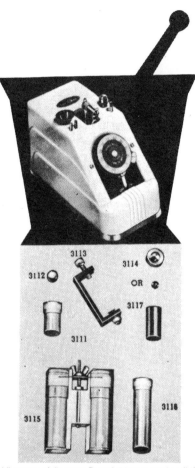

(Courtesy of Crescent Dental Manufacturing Co.)

Figure 49. Wig-L-Bug and accessories; this device can be used to powder small samples.

manufactures and markets the Wig-L-Bug which is a dental tool generally used for triturating amalgams. However it is available with plastic and metal vials and ball pestles in which small samples can be ground to better than -200 mesh (Fig. 49). With this instrument, samples of quartz can be ground to a fine powder in a few minutes. Model No. 3A has the same capacity but has a more durable motor; model No. 6000 is a more powerful machine with a larger capacity.

Spex, Industries sells the Wig-L-Bug as well as a set of nylon sieves and trays in 100, 200, 325, and 400 mesh. You can also obtain more expensive but stronger metal screens (e.g., of stainless steel) from screening material supply houses and use it in a home-made support.

Powder diffractometer sample holders generally have a cavity into which the sample is placed. In certain cases forcing a sample into the cavity causes preferred orientation; in other cases, there is not enough sample available to fill the cavity. In such situations, the sample may be dusted onto a glass slide (cut to the same size as the sample holder) on which a very thin, barely noticeable layer of (amorphous) petroleum or silicone grease has been spread. Aside from those cases where the sample is a very poor diffractor, a satisfactory pattern will be obtained.

Samples for use with Debye-Scherrer or flat-plate cameras can be placed in thin-walled glass capillary tubes made of non-absorbing glass. *Unimex-Caine Corp.* markets imported glass and quartz thin-walled (0.01 mm) capillaries. They are available in a number of diameters between 0.2 and 2.0 mm. The 0.2 and 0.3-mm tubes are best suited for general powder work. These tubes have built-in funnels for easy loading. The sample is placed in the funnel-top and tamped to the tip by vibrating the tube with the flat edge of a file or by dropping it through a

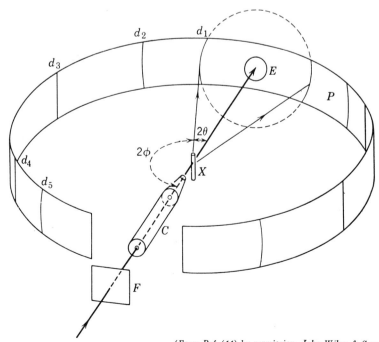

(From Ref. (11) by permission, John Wiley & Sons Inc.)

Figure 50. Geometrical features of the Debye-Scherrer technique: The incident beam passes through beta-filter F and collimator C before hitting the small cylindrical sample X. Sections of the various diffracted cones are intercepted by the cylindrically disposed strip of photographic film P.

Chemical Instrumentation

2-ft length of small diameter glass tubing onto a firm surface.

Debye-Scherrer Powder Camera

The principle of the Debye-Scherrer powder camera is illustrated in Figure 50. A small cylindrical sample (0.1–0.5 mm in diam) is surrounded by a strip of film (usually 35 mm wide). The randomly-oriented crystallites diffract the incident beam in a cone with a semi-apex angle of 2θ in the front-reflection region and $2\phi = \pi - 2\theta$ in the back-reflection region. The intercept of the diffraction cones and the cylindrical film results in curved lines on the film. The 2θ angle can be calculated by measuring the arc(s) from point E to the diffracted line. If the radius (R) is known precisely then 2θ (in radians) $= S/R$ and 2θ (in degrees) $= (S/R)$ (57.3). Many cameras are built with $2R = 57.3$ or 114.6 mm so that 1 mm on the film equals 2 or 1° 2θ (respectively).

The design features of the camera include: a precision-made cylinder for holding the film in a light-proof container; a device to hold the film flush against the cylinder; a sample holder with provision for centering the sample; a device for rotating or oscillating the sample (so as to increase the effective randomness of crystallites); a collimator (which limits the incident beam to the sample); and a beam stop or exit-tube (which prevents the direct beam from exposing the film or escaping into the room) (Fig. 51).

There are three ways of loading the film in the camera (Fig. 52). Method (c) is most commonly used in modern commercial cameras. The location of $2\theta = 0$ and 180° is accomplished by measuring equivalent arcs about these positions. The distance between them can then be calculated to determine the effective radius of the camera; if the diameter is 114.6 mm then the distance should be 180.00 mm.

The lines on the film can be read to 0.05–0.01 mm depending upon the film reader used and the width of the lines. Unless special care is taken the diffraction lines themselves are 0.02 mm or more in width.

Debye-Scherrer powder cameras can be used at either a line or point focus. Powder pattern arcs obtained with the point focus are generally more uniform, but at the center of the arcs (where the film should be read) the pattern obtained using the line focus is quite acceptable.

Philips, Jarrell-Ash, Picker, and *E & A* sell powder cameras based on M. J. Buerger's design (Fig. 51). They are 57.3 or 114.6 mm in diameter. The film and sample are placed inside a light-proof metal cylinder equipped with removable collimator and beam-stop. A fluorescent screen covered by protective lead glass

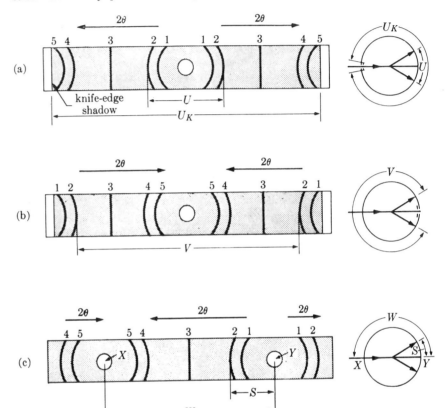

Figure 52. Methods of film loading in Debye-Scherrer cameras. (a) Transmitted beam leaves through hole in film; (b) Incident beam enters through hole in film; (c) Straumanis method contains an internal standard because distance W must correspond to 180° in 2θ. (Corresponding lines have the same number in all films.)

(From Ref. (10) by permission, Addison-Wesley)

and permanently mounted at the outer end of the beam stop facilitates the alignment procedure. The specimen is rotated with a 1-rpm drive motor which is housed in the camera track assembly. The Jarrell-Ash, Picker (57.3 mm: Cat. No. 3592 A, Price Range D; 114.6 mm: Cat. No. 3591A, Price Range E) and E & A cameras are built to fit the standard ACA dovetail track.

Philips offers a number of Norelco accesories (each falls in Price Range D) for use with their camera (Price Range E) (Fig. 53). These include a specimen centering device, a camera track adjusting device, the track assembly, and a mono-

chromator crystal holder. The latter is designed to be attached to the threaded holes in the Norelco X-Ray Generator tube-housing.

A unique feature of the Jarrell-Ash camera (114.6 mm: Cat. No. 80-035, $490; 57.3 mm: Cat. No. 80-400, $390) is that both front and back covers can be removed for ease in sample alignment (Fig. 54).

In addition to its standard powder cameras (57.3 and 114.6 mm diams, $550), E & A manufactures a deluxe stainless steel model (Cat. No. 65-8-DE1, $2300). The deluxe model has the following features:

(a) The film holder is built into the cover. This allows for loading the camera while the specimen is still aligned on the track.

(b) It is built entirely of stainless steel and can be easily decontaminated (if radioactive specimens are used).

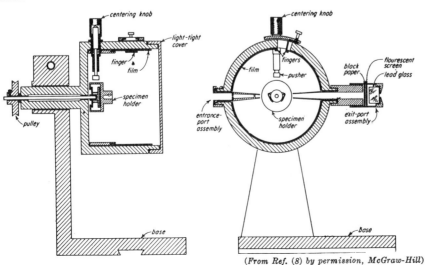

(From Ref. (8) by permission, McGraw-Hill)

Figure 51. Schematic diagram of typical Debye-Scherrer powder camera.

(Courtesy of Philips Electronic Instruments)

Figure 53. Norelco Powder Camera mounted on Camera Track Assembly and attached by a pulley to the 1 rpm drive motor.

(c) The centering device consists of three mutually perpendicular translations and an arc rotation (±8°).

(d) There is a precentered and prefocused microscope.

All cameras can be bought with bases to fit any commercial track (Fig. 55).

(Courtesy of Jarrell-Ash)

Fig. 54. Jarrell-Ash 57.3 and 114.6 mm diam powder cameras.

(Courtesy of E & A)

Figure 55. E & A Deluxe Stainless Steel Powder Camera

The *Siemens* Powder Camera (Price Range D) is designed to fit a Siemens track (Fig. 56). Two models are available —one for 35 mm film and one for 70 mm film. The specimen can be centered either inside or outside of the cameras; the camera can be mounted either horizontally or vertically.

The *G.E.* Powder Camera (Cat. No. A4969, $720) has an effective film diameter of 14.32 cm. This results in the effective film circumference of 45.00 cm (in order to find 2θ, one multiples the distance (in cm.) from $2\theta = 0°$ to the line being

(Courtesy of Siemens America, Inc.)

Figure 56. Siemens 114.6 mm diam powder camera

measured by 8.0). The sample is held in one of four specimen holders; a rotating magnetic chuck (360°) or one of three oscillating (20–30°) holders—wedge, flat-

(Courtesy of General Electric Co.)

Figure 57. G.E. powder camera mounted on camera track set for 6° take-off angle.

ended pedestal, or yoke. The film is held between an inner and outer apron (cassette) and the sample is always visible (Fig. 57).

The *Rigaku*-Denki Powder Camera is

available with one specimen holder which will accommodate any of three film cassettes: 114.6, 57.3, or 90 mm diams (Cat. Nos. 1109, 1111, 1112, Price Range D).

Enraf-Nonius markets a General Purpose Camera (Cat. No. 860, Price Range E) which can be used for Laue, Rotating Crystal, Flat-plate Powder, Sheet Metal, Block Specimen, or Debye-Scherrer techniques. The Debye-Scherrer adapters are available in 57.3, 114.6, and 90 mm diams (Fig. 58).

Flat-Plate Powder Cameras

If a flat-plate film cassette is mounted perpendicular to the direct X-ray beam, the intersection of the film with the cone of diffracted X-rays will result in a series of concentric circles (Fig. 59). In this case $\tan 2\theta = S/R$. It is clear that the maximum value of 2θ that can be recorded is limited by the size of the cassette and the specimen-to-film distance; for a 5 cm distance and a 4×7 in. cassette (with the direct beam aimed at the center of the film) $2\theta_{max}$ is approximately 50°.

The use of a 5.0-cm specimen-to-film distance is recommended for routine investigations because the diameter (D) of the recorded circle is $2S$ and $\tan 2\theta = D/(2)(5)$. Thus $\tan 2\theta$ is equal to the diameter in decimeters.

(Courtesy of Enraf-Nonius

Fig. 58. Nonius Y-860 General Purpose Camera set up for Debye-Scherrer and rotating crystal methods.

A flat-plate cassette with an opening provided for the collimator to pass through

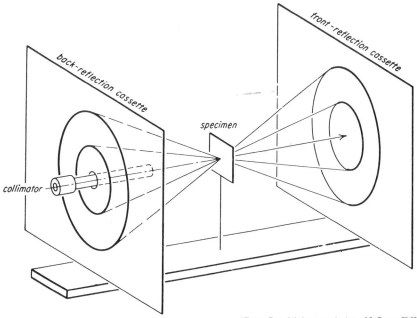

(From Ref. (8) by permission, McGraw-Hill)

Figure 59. Geometry of front- and back-reflection flat-plate cameras.

Chemical Instrumentation

can be used as a back-reflection camera (see following section). In such a case provision must be made to prevent the direct beam from irradiating the room.

The general features of flat-plate cameras are the same whether they are used for powder, single-crystal, or Laue investigations. A good camera should have: (a) a flat-plate cassette mounted *perpendicular* to the direct beam; (b) a specimen holder with provision for centering and rotating (or oscillating) the sample (the latter is not required for Laue studies); and (c) some means of varying the specimen-to-film distance.

The effective random orientation of a polycrystalline sample is increased by rotating or oscillating it. The sample may be a powder contained in a capillary tube or it may be a large block. In the latter case, the beam is directed at an edge of the specimen and the sample is oscillated; in this manner some of the diffracted rays are not absorbed by the specimen. Thus, it is more convenient to study large samples with the flat-plate camera than with the Debye-Scherrer camera.

A variable specimen-to-film distance is useful because (a) the resolution of closely spaced diffraction lines can be improved by increasing the specimen-to-film distance; or (b) the exposure time can be reduced by shortening this distance.

A number of flat-plate cameras are designed specifically for Laue studies and contain no provision for rotating the specimen. These cameras are discussed under single-crystal cameras. Also, most single-crystal cylindrical rotation cameras can be adapted for use as flat-plate cameras by mounting a track suitable for holding a flat-plate cassette on the camera.

At least four companies market flat-plate cameras that are suitable for use with powder samples. The Picker, Enraf-Nonius, and Norelco flat-plate cameras are designed for front and/or back reflection investigations, while the Rigaku-Denki camera (Price Range D) can only be used as a back-reflection flat-plate camera.

The *Picker* camera (designed by M. J. Buerger and built by Technol, Inc., Arlington, Mass.) is available in two styles (each in Price Range E). Style No. 262103 is for polycrystalline samples of varying thickness and Style No. 262104 is for both polycrystalline and

single-crystal techniques. The camera is designed for simultaneous recording of front- and back-reflection patterns. The cassettes take 4 × 5 in. film and are held at a fixed specimen-to-film distance of 5.00 ± 0.05 cm.

The *Enraf-Nonius* General Purpose Camera features a variable specimen-to-film distance in both the front- and back-reflection regions, a film-holder for two exposures, and a sample holder for two samples (Fig. 60). The Rigaku-Denki Back-Reflection camera also has a variable specimen-to-film distance. These cameras use circular films of 78 and 124 mm diam, respectively.

Symmetrical Back-Reflection Focusing Camera

In studies of expansion coefficients, phase diagrams, and solid solutions, slight changes in lattice dimensions must be followed. In such cases one studies reflections in the "back reflection" region, at Bragg angles close to 90°. It can be seen, from Bragg's Law, that when θ is close to 90° small changes in "d" (or $\sin \theta$) correspond to large changes in θ. With reasonable care it is possible, using reflections in the back reflection region, to determine lattice dimensions to better than 1 part in 25,000. Very small relative changes are, therefore, easily detected.

Philips has developed a Norelco camera (Cat. No. 52058, Price Range E) for the precision determination of lattice parameters. Convergent X-rays from the line focus of the tube enter through a narrow slit and irradiate a large portion of the sample which is shaped so as to lie tangent to the focusing circle. A camera radius of 60 mm and the use of focusing geometry provide excellent resolution in the 59–88.74° *theta* range. Its special features include: adaptability for use with powdered specimens and foils, integral film calibrating notches, sealed construction (permitting use of gas or vacuum), specimen oscillating mechanisms, and built-in fluorescent screen for use in aligning the camera.

The focusing principle is described in Fig. 61 and is similar in principle to the focusing principle of the monochromator (Fig. 30).

Guinier Asymmetric Focusing Camera

An asymmetric focusing arrangement, similar to that described for the symmetric back-reflection focusing camera, can be used. Guinier designed a camera which uses a crystal monochromator and the asymmetric focusing condition to allow investigation of the region between 10 and 90° in 2θ.

Characteristic X-radiation from the source (Fig. 62) is focused by the crystal monochromator (A) to point C.

A sample is placed at B and the X-rays passing through the sample are diffracted at the proper Bragg angle ($\angle DBC = 2\theta$).

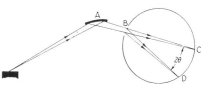

(From Ref. (9) by permission, McGraw-Hill)

Figure 62. Geometric principle of Guinier asymmetric focusing camera.

The beam reflected from the monochromator is rather wide at sample B and covers a large area; all rays diffracted at a given Bragg angle are focused to the same point (D). The focusing conditions are best met if the sample lies along the focusing circle; in practice it is usually tangent to the circle.

By properly adjusting the camera (that is, by rotating it about point B), the diffracted $K\alpha_1$ and $K\alpha_2$ lines can be made to coincide at point D. Generally they are made to coincide at $2\theta = 30°$; this eliminates separation of the doublet between 0 and 60° 2θ. The reduction in background (due to the use of the monochromator) and the superposition of the $K\alpha_1$ and $K\alpha_2$ lines allows one

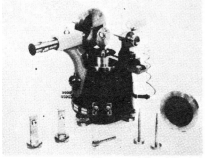

(Courtesy of Enraf-Nonius)

Fig. 60. Nonius Y-860 General Purpose Camera set up for flat-plate powder method with comparison segment cassette and double-specimen holder.

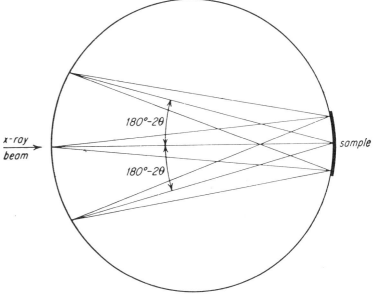

(From Ref. (8) by permission, McGraw-Hill)

Figure 61. Principle of symmetric back-reflection focusing camera for powder samples.

to obtain a good photograph in relatively little time. This camera is good for routine identification and phase transformation studies.

The diffraction angle is easily calculated. Arc CD is measured (S), the radius (R) is known and 2θ (in degrees) $= (S/2R)$ (57.3).

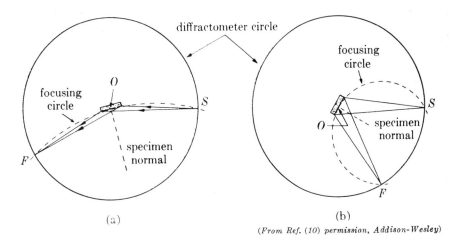

(From Ref. (10) permission, Addison-Wesley)

Figure 65. Diffractometer focusing geometry for flat specimens in (a) forward reflection and (b) back reflection regions.

(Courtesy of Enraf-Nonius)

Figure 63. Nonius Y-919 Guinier-deWolff Focusing Camera II.

P. deWolff designed a Guinier-type camera that allows simultaneous exposure of a number of samples. These cameras are commercially produced by Enraf-Nonius and are available in two styles.

The *Guiner-deWolff* Quadruple Focusing Camera I (Cat. No. Y820, $1850) permits the simultaneous exposure of four different specimens. It is furnished with a Vacuum-tight plastic cover so that the camera and sample space can be evacuated thus eliminating air scattering. It is designed for use at the (vertical) line focus of a horizontal X-ray tube. One monochromator is required for Cu, Fe, and Cr radiations, another for Mo radiation.

The *Guineir-deWolff* Quadruple Focusing Camera II (Cat. No. Y919, $2600) is similar to Model I, but has several added features (Figs. 63, 64). The unit can be mounted on its base in a position suitable for either a vertical or horizontal X-ray tube. The monochromator, as well as the camera portion, is enclosed in the plastic cover; when it is evacuated, air scattering is reduced considerably. Only the film cassette need be removed in order to load the camera. A single monochromator is used for four types of radiation. The sample holder is loaded with a single piece of transparent tape on which the four samples are mounted. Sample loading is simplified and the motion of the sample

during the exposure eliminates the need for precise sample thickness.

Powder Diffractometers

The focusing conditions described previously (Figs. 30, 61, 62) have been utilized in the construction of powder units using counter detector. Such instruments are known as powder diffractometers. The X-ray source (S), sample (O) and the receiving slit of the counter (F) are the three points used to define the focusing circle (Fig. 65). As the sample is rotated through θ-degrees, the counter rotates 2θ-degrees. A focusing circle of infinitely variable radius is obtained by maintaining constant source-to-sample and sample-to-counter distances.

Maximum intensity is obtained by passing the beam from a line focus through Soller slits (a set of parallel, closely spaced metal plates which limit the divergence of the beam) and diffracting from a flat sample which is kept tangent to the focusing circle at all times (Fig. 66). The focusing is not perfect because a flat specimen is used but the aberrations can be kept small. Recently a curved specimen holder has become available (see description of Norelco diffractometer).

Powder diffractometers are available in two general designs: those in which the counter rotates in the horizontal plane and those in which it rotates in the vertical plane. The former must use a horizontally mounted X-ray tube and latter a vertically mounted tube. Picker's diffrac-

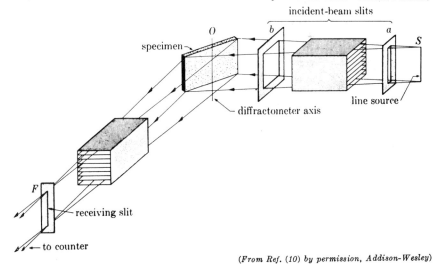

(From Ref. (10) by permission, Addison-Wesley)

Figure 66. Arrangement of slits in diffractometer. The parallel, closely spaced metal plates are known as Soller slits.

(Courtesy of Enraf-Nonius)

Figure 64. Copy of photograph of simultaneous exposure of four specimens in Nonius Y-919 Guinier-deWolff Focusing Camera II.

tometer is "biplanar" in nature and can be mounted either horizontally or vertically.

Philips produces a powder diffractometer (Wide Range Goniometer) in which the plane of rotation of the counter is vertical (Fig. 67). The Norelco Wide Range Goniometer (Price Range G) is a high-precision device with tolerances of ±0.0025°, designed for use with Geiger, proportional, flow proportional, or scintillation counters. It can be read directly to 0.01° 2θ; and, with the vertically mounted Norelco X-ray tube, it has a useful range of 165° 2θ. Other features include a choice of scanning speeds and unidirectional or oscillatory scanning.

Chemical Instrumentation

Accurate intensity measurements can be obtained by using very slow scanning rates or by step-scanning (see below). The Norelco unit has provision for automatic

(*Courtesy of Philips Electronic Instruments*)

Figure 67. Norelco Wide Range Goniometer with scintillation counter and Flat Rotating Specimen Holder.

step-scanning at intervals of 0.01, 0.02, or 0.05°.

The Norelco Wide Range Goniometer is also available in a horizontal model; this unit has a scan range of −100 to +165° 2θ and has independent θ/2θ scanning. It will accept a wide variety of accessories.

The *AMR* Focusing Monochromator which is designed to fit the Norelco diffractometer has been described previously (Fig. 32). AMR also markets an Auto-Focusing Attachment (Cat. No. 52601, Price Range F) for the Norelco diffractometer. This device continuously bends a flexible specimen holder so that the specimen surface conforms to the focusing circle of the diffractometer throughout its angular range. It is bent, mechanically, through the use of a cam and roller linkage.

A number of firms market diffractometers in which the counter rotates in the horizontal plane; they are used with a horizontally mounted X-ray tube.

The *G.E.* SPG Spectrogoniometer (Cat. No. A4954) is a precision θ–2θ device, incorporating an angle bisector, with a range from −25 to +165° 2θ. A number of sample holders can be mounted on this base, including a flat-sample holder, small-sample spinner, and flat-sample spinner for powder work. (The single-crystal accesories will be discussed in a later section.) A conversion kit is available that allows a number of scanning speeds from 4° 2θ/min to 0.002° 2θ/sec. The *Picker* Biplane Diffractometer (Price Range G) is available in two models with (6239 series) and without (6240 series) omega motor drive (i.e., the option of keeping the counter stationary as the sample rotates through θ), for use in single crystal, polycrystalline, and emission

(*Courtesy of Picker X-Ray Corp.*)

Figure 68. Picker Biplane Diffractometer.

techniques (Fig. 68, also Fig. 12). Among its features are:

(*a*) It can operate in either the horizontal or vertical plane.

(*b*) The chassis supports the X-ray tube for precise orientation. The adjustable support at the anode end aligns the focal spot to the diffractometer axis; the adjustable support at the cathode end allows one to vary the take-off angle without disturbing the alignment of the focal spot.

(*c*) The 2θ angular range is from −55° to +165°.

(*d*) A full 360° *omega* motion, independent of detector, is present.

(*e*) It includes the ACA accepted dovetail tracks for the support of specimen holders, goniostats and detectors.

(*f*) The *omega* and 2θ backlash is less than 0.003°.

The *Jarrell-Ash* Diffractometer (Cat. No. 80-301, $2800) has an angular range of 168° 2θ through zero to −130° 2θ (Fig. 15). Other features include five scanning speeds, continuously adjustable receiving slit, independently adjustable sample holder, and direct reading to 0.01° 2θ.

The *Hilger-Watts* Y144 Powder Diffractometer (Price Range G) has several unique features. A telescope is fitted to the diffractometer so that the various components can be aligned optically. This offers the advantages of safety, accuracy, and speed over adjustment by trial and error. Three modes of continuous scanning are available: (*a*) θ–2θ tracking; (*b*) 2θ only (sample stationary); or (*c*) θ only (detector stationary), at speeds of 2, 1, 0.5, 0.4, 0.2, or 0.1° 2θ/-min. Step-scanning at intervals of 0.2, 0.1, 0.05, 0.04, 0.02, or 0.01° can also be accomplished. An important safety feature is that a special shutter, linked to the cover of the sample holder, automatically cuts off the radiation from the X-ray tube when the chamber is opened, e.g., for insertion or removal of sample. (Fig. 69)

Stoe manufactures two horizontal counter diffractometers (Price Range F) for powder and single crystal measurements. The large model DC 40 is mainly designed for automatic use, and the small model DC 20 (Fig. 70) for semiautomatic operation. Both diffractometers have a high positional

(*Courtesy of Engis Equipment Co.*)

Figure 69. Hilger-Watts Y144 Powder Diffractometer.

accuracy of better than 0.005° and a reading accuracy of 0.01° on both circles. The circles can be driven independently at various speeds or they can be linked together. The diffractometers have large stationary plates to permit the mounting of auxiliary equipment. Limiting switches, degree marks, and various collimators for powder and single crystal measurements are supplied with the instrument. Support for Stoe Eulerian cradles and controls for various automatic or semiautomatic operations are available on request.

(*Courtesy of Stoe Instrument, Inc.*)

Figure 70. Stoe DC 20 Diffractometer.

The *Siemens* Omega Diffractometer (Price Range G) is equipped with two separate synchronous motors for either θ/2θ or *omega* drives. The rate of 2θ motion can be one of five angular speeds between 2 and 1/8°/min; the *omega* movement takes place at a constant 1/4°/min. The 2θ range is from −100° to +160° (Fig. 71).

Horizontal diffractometers are also available from *Rigaku-Denki* (Cat. No. 2151, Price Range G) and Tem-Pres (Fig. 72). The complete Tem-Pres system (including generator, X-ray tube, detector, and recorder) is available for $8950.

Special Diffractometers

Commercially available for the first time is the *Picker* theta-theta diffractometer (Cat. No. 6239D Price Range H). The name, theta-theta diffractometer, is derived from the geometric configuration of the relative motions between the X-ray source, the sample, and the detector. In normal diffractometric techniques, the X-ray source remains stationary in space while the detector (2θ motion) rotates about the center axis. At the same time, the sample rotates about the center axis at one-half of the velocity of 2θ, thereby

(Courtesy of Siemens America, Inc.)

Figure 71. Siemens Omega Diffractometer.

maintaining equal angles of incidence and reflection. The relative motions then are 2θ for the detector relative to the source, θ for the sample relative to the source, and θ for the detector relative to the sample. The motions are, therefore 2θ and θ with respect to the stationary X-ray source.

(Courtesy of Tem-Pres Research, Inc.)

Figure 72. Tem-Pres Diffractometer Unit XD-1.

The theta-theta diffractometer, on the other hand, is one in which the sample is held stationary in space while the necessary relative motions are achieved by driving both the X-ray source and the detector through the motion θ with respect to the sample (Fig. 73).

Because the sample remains stationary and horizontal, liquid samples may be investigated without the necessity of covering them to prevent spilling. This feature is particularly applicable to the study of samples of the mesomorphic state, polymers, melted materials, or any other where the sample is presented in the liquid state or in a noncompressible powder form.

Another feature of this instrument particularly suited for liquid sample investigation is the ability to select a stationary source—stationary sample mode of operation. In this mode, a glancing angle of incidence may be chosen and the standard

(Courtesy of Picker X-Ray Corp.)

Figure 73. Picker Theta-Theta Diffractometer.

motion selector will allow scanning of the detector at 2θ relative to both the sample and source, the sample still remaining horizontal and level. A constant depth of penetration can thereby be achieved if required.

Another instrument of interesting design is the *New Brunswick* Double Crystal Diffractometer for polycrystalline materials. A crystal is used to monochromatize the X-ray beam from the tube; this monochromatized beam is used as the incident beam on the sample. Provision is made for taking a Debye-Scherrer photograph using a Transparent Camera. The film is developed and replaced in the camera, after which the counter is moved behind the reflections which are to be studied. The camera and film are then removed and the intensities recorded by the counter directly. This instrument is designed for use with samples having large crystallites which are to be studied individually. Most of its applications would appear to be in metallurgical or mineralogical studies. There is no provision for automatic scanning.

One of the first applications of automatic data recording in the field of powder diffractometry is the use of the *Datex* SDS-1 for recording powder diffractometer output. This system consists of an absolute position encoder mounted in the X-ray system strip-chart recorder (which indicates the intensity) and a second absolute position encoder mounted on the 2θ drive of the diffractometer.

By using a peak-and-valley mode, the peak height (or valley minimum) and its 2θ position are recorded on punched cards or tape. This information can then be used for a computer search of the ASTM Powder Data File (see below).

The price of this system is $6000–10,000, depending on options; it can be utilized on any diffractometer.

Step-Scanning. The production of X-rays is a statistical phenomenon in which the uncertainty of counting is inversely proportional to the total number of recorded counts. In a continuous powder diffractometer scan, the precision of measured intensities can be increased by reducing the scanning rate. Another technique which can be employed is to record measurements obtained at each of a number of increments in 2θ. This is known as step-

scanning. Two methods of step scanning are available;

(a) The total count obtained in a fixed-time period is recorded;

(b) The time necessary to reach a fixed-count total is recorded.

The former method is to be preferred.

A number of diffractometers (e.g., Norelco, Hilger-Watts) are built with provision for automatic step-scanning. *Humphrey Electronics* manufactures a precision Step-Scanning Device (Cat. No. 701 A; Price Range F) which is intended for use with any spectrogoniometer and detector with digital printer. The step angle can be set at any 2θ value between 0.01 and 0.99° in increments of 0.01°. The unit is hooked into the electronic panel so that after the printer records the intensity the drive-control holds all circuits in reset position until the motor advances to the next step. The sequence is repeated until manually stopped.

Humphrey also manufactures a Transistorized Programmer (Price Range G) which provides for automatic positioning of the spectrogoniometer to any or all of six, nine or twelve angles (Models 901, 911, 921, respectively). This is available with or without step-scanning.

Diffractometer Alignment. In a discussion of the alignment procedures for diffractometers the American Crystallographic Association (ACA) Apparatus and Standards Committee (A & SC) has made certain recommendations. These are reprinted here so as to emphasize the points which must be kept in mind when buying or aligning diffractometers.

Recommendations of the A & SC of the ACA (1b):

"The A & SC would like to make certain recommendations based on information discussed here or otherwise available to the A & SC. These recommendations are offered only on an advisory basis and should not be considered mandatory, final, or binding on any individual, group, or organization:

1. The manufacturers should examine their alignment procedures for possible improvements, such as (but not limited to) incorporating additional adjustments to insure parallelism of source, sample axis, and receiver slit.

2. The manufacturers should evaluate the possibilities of providing the user with detailed specifications on critical alignment dimensions or of providing him with physical means and/or techniques for testing alignment characteristics himself.

3. The manufacturers should evaluate the possibilities of providing the user with 2:1 tracking and 2θ calibration data when he buys a diffractometer and/or provide calibration service of this type when needed.

4. The tools, specifications, and procedures supplied by the manufacturer should cover at least the following alignment and calibration features:

(a) Zero degree alignment.
(b) Alignment perpendicular to the plane of rotation of the diffractometer.
(c) Accuracy of the 2:1 ratio setting over the entire angular range of the diffractometer.

Chemical Instrumentation

(d) Accuracy of the 2θ dial reading over the entire angular range of the diffractometer.

(e) Parallelism of the line source, sample axis of rotation, and receiving slit axis.

(f) Equality of sample axis to line source and sample axis to receiving slit distances.

(g) Specimen centering.

5. The user should expect to pay the manufacturer for the accessories, equipment, or services required to carry out the alignment and calibration procedures he desires.

6. The user should not expect to obtain precise and accurate lattice parameter data automatically even if all the alignment and calibration features described above were incorporated in diffractometers. To obtain both precision and accuracy involves a consideration of additional factors, such as a knowledge of the spectral distribution of X-rays used. The factors under consideration in this report are concerned only with the geometry of the diffractometer; i.e., mechanical factors amenable to measurement and checking using an instrument maker's tools. Only after these mechanical factors have been properly accounted for, would it be possible to assess the experimental effects of the remaining factors on the precision and accuracy of lattice parameter data."

Comparison of Powder Methods. A detailed comparison of the uses, advantages, and disadvantages of the various powder methods is beyond the scope of this article. For a complete discussion the reader is referred to one of the standard texts (*8, 11*). However, some major points will be discussed.

The powder diffractometer uses a reflection technique, is not affected by highly absorbing materials, and gives a quantitative measure of the intensities; with a properly aligned instrument accurate diffraction angles are obtained easily.

Film methods have the general advantages discussed at the end of the section dealing with Detectors, require less initial outlay for equipment, are easy to store and allow convenient comparisons of different films. The disadvantages include the need for darkroom equipment, the need for proper measuring techniques, and the possibility of detrimental absorption effects.

The Debye-Scherrer technique can be used to obtain a nearly complete record of the diffraction pattern throughout a large range in 2θ; it requires very small samples. The Guinier method is restricted to a portion of the pattern (usually 0–90° 2θ); however, the patterns, which are extremely sharp, are useful for qualitative analytical work. Back-reflection techniques (both flat-plate and symmetrical focusing) are used for obtaining precision lattice constants. The front-reflection flat-plate method is used to investigate the entire diffraction cone (e.g., in cases of preferred orientation) and a variable specimen-to-film distance allows resolution of closely spaced diffraction lines. (Closely spaced lines can also be resolved by using

X-rays of longer wave-length, but then absorption effects become greater.)

Interpretation of Data

A number of charts and devices have been designed to facilitate the interpretation of powder photographs and diffractometer traces (diffractograms). Some of these are commercially available and will be described.

A set of transparent plastic scales for Debye-Scherrer films are sold by N. P. Nies. These scales are graduated in d-values (Angstrom units). The scale and films are superimposed and the information is read directly. Each set, on one plastic sheet, consists of 5 or 6 scales, all for one nominal diameter and radiation, but of slightly different actual length to cover a 1% variation in film length (this compensates for film shrinkage and other factors). They are available for Co, Cr, Cu, Fe, Mo, or Ni radiations and 57.3, 90, 100, 114.6, 140, 143.2, or 190 mm camera diameters. A set of similar scales which are available in the same diameters, but graduated in $\sin^2\theta$ values are also offered. These are useful in indexing procedures (i.e., assigning the proper Miller indices to a given diffraction line). Each of these charts is $18.

Nies also has a number of transparent plastic charts for use with a diffractogram run at 2°2θ/in. ($18) and at 5°$2\theta$/in. ($20), using CuKα radiation.

Paper charts for use with a diffractogram run at 2°2θ/in. are available from J. Simon. A set consisting of three double scales for the CuKα, FeKα and MoKα radiations, each covering a range of 0–90° 2θ is priced at $12.

These charts are time-savers for routine work; however for the most accurate work the d-spacing is obtained from a carefully measured diffraction angle. Tables for Conversion of X-ray Diffraction Angles to Interplanar Spacings are available as NBS Applied Mathematics Series—No. 10 and can be purchased from the U.S. Government Printing Office. It has tables of d as a function of θ for Mo, Cu, Ni, Co, Fe, and Cr Kα_1 radiations and as a function of 2θ for Cu and Fe Kα_1 radiations. Similar tables are available from Polycrystal Book Service for Cu and Fe Kα_1 radiations; they also have tables of $1/d^2$ versus 2θ for use in indexing procedures.

Polycrystal Book Service also sells charts (PC-35) which are of aid in indexing certain crystal classes; for a description of these charts see Ref. (*11*), p. 351.

Similar charts can be obtained from Enraf-Nonius.

DuPont has built a special-purpose analog computer to resolve rapidly spectra, chromatograms, and other experimental curves into component peaks (X-ray diffraction diffractograms included). The Du Pont 310 Curve Resolver has provision for duplicating an experimentally obtained curve which cannot easily be described by explicit mathematical functions. Unsymmetrical shapes can be generated by the instrument, thereby allowing peaks to be skewed to duplicate tailing. The operator can reproduce the peak and then switch on each component peak individually. This is useful when the height, width, and position of each component peak is desired. It

can be used to resolve completely or partially overlapping peaks caused by the same or different phases in the system. Cost of the unit ranges from $4970 for a three-channel unit to $8350 for a ten-channel unit.

ASTM. The most common use of powder techniques is in the identification of unknown materials. Each compound has a unique powder pattern in which the d-spacings are a function of the unit cell size and the intensities a function of the atomic positions. Thus, for example, a hydrate will have a different pattern than the parent compound, and it is not unusual to find one compound occurring in more than one crystalline form. However, there is no *a priori* method of identifying the compound from the powder pattern. Therefore a set of reliable standard patterns must be available. To this end the American Society of Testing and Materials (ASTM) has established a Powder Diffraction File. This project is operated on an international basis and cooperates closely with the Data Commission of the International Union of Crystallography. Data are collected continually and carefully edited to check the purity of the compound and the reliability and accuracy of the data. A new set of data for approximately 1000 compounds is issued each year. Presently there are about 12,000 compounds on file (Fig. 74). The File is available as individual 3 × 5 in. index cards ($110–200/set) or keysort cards ($170–225/set). A number of indices are available and knowledge of the d-spacings and intensities of the three strongest lines is sufficient to identify most substances. An alternative index uses the eight strongest lines with no indication of their relative intensities. The indices are available in book form or on IBM cards.

As of now, Sets 1–5 are available in book form (Inorganic Compounds: Cat. No. DS1S-5i RB, $45; Organic Compounds: Cat. No. DS1S-5o RB, $25), and Sets 6–10 will be issued shortly; there are 15 sets of cards presently available, with Set 16 soon to be published.

A new optical coincidence (Termatrex) indexing system (Price Range E) has been developed recently (by ASTM and the Jonkers Corp.) for use in retrieving and locating inorganic compounds listed in the ASTM file.

Each Termatrex card represents a single characteristic of data. For example, one card represents the MATTHEWS group of "d" values 1.60–1.69; another card represents materials containing the element Calcium. The index contains cards for positive and negative "d" spacing values, positive and negative element cards plus cards for positive and negative alloys, hydrates, and minerals. Each material in the X-ray Diffraction Data File has the identical position in each Termatrex card. Holes drilled into a given card represent the reference materials in the index which have the characteristic represented by that card. A unique tabbing method allows random selection and filing of cards.

On output, a search is made by selecting those cards which represent the characteristics of the unknown pattern and superimposing these over a light source. The points at which light shines through all of

5-0628 MINOR CORRECTION

d	2.82	1.99	1.63	3.258	NaCl		
I/I₁	100	55	15	13	SODIUM CHLORIDE	(HALITE)	

Rad. CuKα   λ 1.5405   Filter Ni
Dia.   Cut off   Coll.
I/I₁ G. C. DIFFRACTOMETER   d corr. abs.?
Ref. SWANSON AND FUYAT, NBS CIRCULAR 539, VOL. II, 41 (1953)

Sys. CUBIC   S.G. O_H^5 - FM3M
a₀ 5.6402   b₀   c₀   A   C
α   β   γ   Z 4
Ref. IBID.

εα   nωβ 1.542 εγ   Sign
2V   Dₓ2.164 mp   Color COLORLESS
Ref. IBID.

AN ACS REAGENT GRADE SAMPLE RECRYSTALLIZED
TWICE FROM HYDROCHLORIC ACID.
X-RAY PATTERN AT 26°C.

REPLACES 1-0993, 1-0994, 2-0818

d Å	I/I₁	hkl	d Å	I/I₁	hkl
3.258	13	111			
2.821	100	200			
1.994	55	220			
1.701	2	311			
1.628	15	222			
1.410	6	400			
1.294	1	331			
1.261	11	420			
1.1515	7	422			
1.0855	1	511			
0.9969	2	440			
.9533	1	531			
.9401	3	600			
.8917	4	620			
.8601	1	533			
.8503	3	622			
.8141	2	444			

2003

(Courtesy of ASTM and The Joint Committee on Chemical Analysis by Powder Diffraction Methods)

Figure 74. Specimen ASTM Powder Diffraction File card.

the superimposed cards represent those reference materials which have the characteristics of the unknown. Coordinates of these light dots are reference numbers of the appropriate materials in the Powder Data File. A Conversion Manual relating coordinates of the Termatrex index to the name of the material, the card and set number of the material in the ASTM Data File is included with the purchase of this index. This Conversion Manual also contains the eight strongest lines of the material. As new sections are added to the data file, the Termatrex cards are returned for the additional data to be entered.

Part Five—Single Crystal Methods

SINGLE-CRYSTAL TECHNIQUES

Single-crystal techniques are used for orienting crystals, obtaining unit cell dimensions and space group information, and for the collection of intensity data suitable for crystal structure analysis; they are too complex to be used for routine identification problems. The experimental data usually consist of several hundred to several thousand independent reflection intensities. The arrangement of scatterers (i.e., atoms) in a given crystal is responsible for the observed intensities; it is the crystallographer's job to start from these intensities and to determine the atomic arrangement.

Most crystals used in X-ray diffraction investigations have a number of dislocations and consist of mosaic blocks tilted at small angles (of the order of 3°) to each other. (The techniques used for investigating the degree of perfection of a crystal and for measuring the number of dislocations present will not be discussed.)

In order to properly understand the basis of single-crystal methods, use must be made of the "Reciprocal Lattice" concept. The reciprocal lattice is a representation of the crystal in which

each family of parallel planes with indices, hkl, and interplanar spacing, d, is represented by a single point located at the end of a vector whose magnitude is a reciprocal function of the interplanar spacing (k/d, where k is usually λ or 1) and whose direction is determined by the normal to the planes from the unit cell origin.

The origin of the reciprocal lattice is placed at the intersection of the direct beam with the surface of an imaginary sphere (of radius equal to 1 when $d^* = \lambda/d$) which surrounds the crystal. This sphere is called the "Sphere of Reflection."

Whenever a reciprocal-lattice point lies on the surface of the sphere of reflection the family of planes represented by that reciprocal-lattice point is in reflecting position. Figure 75 shows that the rotation of the reciprocal-lattice point through angle *theta* corresponds to the rotation of the crystal through the same angle.

When the reciprocal-lattice point lies on the sphere of reflection the Bragg conditions are fulfilled and *theta* is the Bragg angle. Thus, by examining the orientation of the reciprocal lattice about its origin, one can determine which family of planes is in diffracting position.

From this diagram we can also see that the *maximum* d^* value is 2; i.e., the *minimum* d value is λ/2. (The same information, of course, is obtained directly from the Bragg Law for θ = 90°.) In order to increase the number of observable reflections, one must go to shorter wavelengths.

Mounting of Single-Crystal Specimens

Aside from the large crystals which are aligned using back-reflection Laue techniques, single crystals used in X-ray diffraction investigations are generally no larger than 0.4 mm in cross-section. During these investigations (and this is particularly true when counter methods are employed) the crystal must be held firmly in the desired orientation. For this reason mounting adhesives and goniometer heads (devices for orienting and centering the specimen) are of primary importance.

Mounting Adhesives. A discussion of specimen preparation and mounting is found in Ref. (14c), pp. 21–34. Included in this discussion is a table in which the properties of a number of adhesives are discussed. New adhesives have become available in recent years and one of these, Eastman 910 Adhesive (distributed by Armstrong Cork Co.), has been used quite successfully in a number of laboratories.

Goniometer Heads. Commercially available goniometer heads usually contain two arcs (perpendicular to each other) and two translations (perpendicular to each other and parallel to the arcs). A number of models are also equipped with an elevator (up-down or z-translation). These devices should be equipped with ACA-

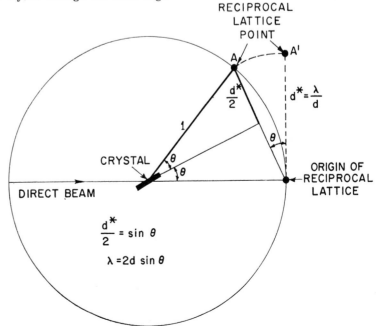

Figure 75. When a set of parallel planes of interplanar spacing d is in reflecting position the reciprocal lattice point representing these planes is located on the surface of the sphere of reflection.

Chemical Instrumentation

threads (15) so as to be interchangeable between various goniometers. The goniometer heads should be designed so as not to creep or slide once the crystal has been adjusted. Oversized gears, dovetails, and spring-loading have been used in attempts to improve the stability of these objects.

There are two common classes of goniometer heads: standard and eucentric.

Standard goniometer heads are generally approximately 65 mm in height from the base to the top of the mounted specimen. The arcs are mounted above the translations; most models have no elevator. Eucentric goniometer heads have a fixed common intersection of the principal axis and the centers of both arcs and provide translations and elevation only above the arcs.

Standard goniometer heads are available from *Supper* (Cat. No G20F, $185) (over 5000 are in use, Fig. 76), *Nonius, Jarrell-Ash* (Cat. No 80–025, $185, Fig. 77), *Stoe* and *Philips* (Cat. No. 52319).

Stoe manufactures an eucentric goniometer head (GH 13) with a center 63.96 mm above the base and a *z*-translation. This is approximately the same height as the standard goniometer head but it is higher than most of the other eucentric

(Courtesy of Charles Supper Co.)

Figure 76. Supper Goniometer Head G20F.

goniometer heads which are available. The Stoe eucentric goniometer head has arcs that can be adjusted up to ±15° rather than the ±25° that are available on most other models, with a corresponding reduction of the cone of obstruction to the diffracted beams.

Stoe also manufactures a micro goniometer head (GH 11), which can be adjusted to ±20°, the translation movements being ±2.5 mm. This head can be used with Stoe and Siemens Eulerian cradles. (Their standard goniometer head (GH 12) can be adjusted to ±30°, the translational movements being ±20 mm.)

All Stoe goniometer heads are in price Range D.

(Courtesy of Jarrell-Ash Co.)

Figure 77. Jarrell-Ash Goniometer Head 80-025.

Eucentric goniometer heads in which the common center is 49 mm (1.929 in.) above the base are available from Supper (Cat. No. G 30E, $220), Nonius, E & A, Jarrell-Ash (Cat. No. 80–026, $210) and W & W. The W & W model ($250) has very large arcs with a range of ±45° (Fig. 78). This is useful on most counter instruments, but is inconvenient on a number of film instruments (the arcs interfere with the film cassette).

Nonius offers a number of different Goniometer Heads. The Y914 Eucentric Goniometer Head ($225) comes with the center at two different heights: Y914-H is 63.96 mm and Y914-L is 49.0 mm.

(Courtesy of W & W)

Figure 78. W & W Eucentric Goniometer Head.

Both have a height adjustment of 2 mm. An adapter is available for the low model, which will raise its center to 63.96 mm. The standard head used on their cameras is the Y814 ($170) with a height of 51.3 mm (Fig. 79); the Y845 ($160), also a standard head, has a height adjustment from 48.4 to 61.4 mm.

The *E & A* model ($300) has an optional removable device ($20) which attaches to the side of the goniometer head and points to the approximate center of the arcs (Fig. 80). This is an aid in placing the crystal on the goniometer head. They also furnish a goniometer key (used for adjusting the arcs and translations) with a handle approximately 3 in. long. The alignment of crystals mounted on counter instruments takes place with the X-ray beam on and the use of a long-handled

(Courtesy of Enraf-Nonius)

Figure 79. Nonius Goniometer Head Y814.

key permits the operator to make adjustments while his hands are safely away from the direct beam.

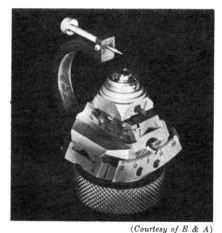

(Courtesy of E & A)

Figure 80. E & A Eucentric Goniometer Head.

Single-Crystal Film Methods

Single-crystal film methods fall into one of three categories:

(a) Stationary Crystal—Stationary Film: Laue
(b) Moving Crystal—Stationary Film: Oscillation-Rotation (cylindrical and flat-plate)
(c) Moving Crystal—Moving Film: Weissenberg and Precession

A number of charts have been devised to aid in the interpretation of the diffraction photographs. Some of the more useful ones are commercially available on transparent plastic from N. P. Nies and on transparent paper from Polycrystal Book Service.

Stationary Crystal-Stationary Film Techniques

The photograph obtained from a stationary crystal irradiated by an unfiltered, polychromatic X-ray beam is known as a Laue photograph. Under these conditions the Bragg Law is satisfied by various combinations of wavelength (λ) and interplanar spacing (d). The symmetry of the Laue photograph is quite sensitive to crystal alignment, but the identification of each reflection spot is complicated by the lack of knowledge of the wavelength of the diffracted beam. Nowadays, the Laue method is generally limited to crystal alignment procedures. Most often the crystals are rather large and back-reflection Laue techniques must be employed. The Laue method usually employs a flat-plate cassette, a crystal holding device and a collimator-beam-stop arrangement. In addition to the flat-plate powder cameras which have already been discussed, a number of cameras specifically designed for Laue investigations can be used.

A back-reflection camera in which the film cassette has provision for an intensifying screen and the specimen holder is designed to accommodate samples up to 1-in. in diameter and 10 in. in length is available from Philips (Cat. No. 52486, Price Range E), GE (Date Sheet A7016B) and MRC (Models XR-507 and XR-508, $675). The crystal can be moved along an arc of 60°, either vertically or horizontally; once the crystal is aligned the entire assembly can be removed and mounted on the bed of a grinding machine without disturbing the alignment.

Polaroid has modified a film cassette for use in taking back-reflection Laue photographs. The full assembly consists of the No. 57-1 Cassette with a track adapter (for Norelco or G.E. units), beam collimator, radiation cap, and 5 cm extension ($325). The advantages of Polaroid film can be fully utilized in the Polaroid Laue system. The Polaroid cassette can be used in place of the film cassette in the unit described above (Fig. 81). A second track adapter is needed for transmission Laue investigation.

(Courtesy of Polaroid Corp.)

Figure 81. Back-Reflection Laue Camera with Polaroid No. 57-1 Cassette mounted on Norelco track.

MRC sells the Polaroid system and has also adapted it for Siemens tracks.

(Courtesy of Jarrell-Ash Co.)

Figure 82. Jarrell-Ash Laue Camera set up for back-reflection studies.

The *Jarrell-Ash* Laue camera (Fig. 82) is suitable for both back-reflection and transmission Laue studies (Cat. No. 80–045, $600).

(Courtesy of E & A)

Figure 83. E & A Laue Camera with Polaroid No. 57-1 Cassette.

E & A has also adapted the Polaroid cassette for use in its Laue camera (Fig. 83). Unit-1 is a back-reflection Laue camera ($700) with a built-in electric shutter (up to 5 min). Unit-2 is for both back-reflection and transmission Laue photographs ($1150).

(It should be noted that transmission Laue photographs can also be taken on the precession camera.)

Moving Crystal—Stationary Film Techniques

Oscillation-Rotation Cameras. From our brief discussion of the reciprocal lattice it should be evident that the reciprocal-lattice points lie on planes perpendicular to real directions in the crystal. Therefore, the reciprocal lattice of a crystal rotating or oscillating about a crystallographic direction consists of planes of reciprocal-lattice points perpendicular to the rotation (or oscillation) axis of the crystal. The diffracted beams lie on the surfaces of co-axial cones. If the crystal is surrounded by a cylindrical film, the reflections will be recorded on circles which appear as straight lines when the film is laid out flat (Fig. 84). If a flat cassette is placed perpendicular to the

X-ray beam, rows of reflections are recorded on a series of hyperbolas (Fig. 85).

A crystal can be aligned by interpreting the deviation of the recorded diffraction pattern from that of an aligned crystal. Once the crystal is aligned, the repeat distance of the crystallographic direction along which the crystal is aligned can be calculated from the spacings between the rows of reflection (these rows, which are the projections of reciprocal-lattice planes onto the film, are known as layer lines, with the line through which the direct

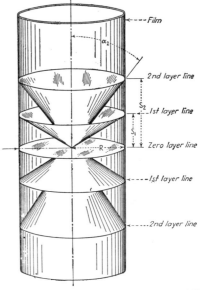

(From Ref. (9) by permission, McGraw-Hill)

Figure 84. Diagram showing the formation of layer lines from the intersection of cones of diffracted X-rays with cylindrical film.

beam passes designated as the zero-layer line, Fig. 84). In a rotation photograph, because each layer line consists of an entire plane of reflections projected onto the film, one loses information concerning the

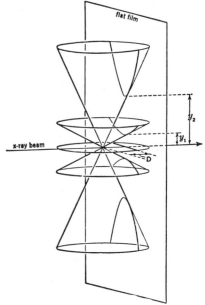

(From Ref. (5) by permission, John Wiley & Sons, Inc.)

Figure 85. Diagram showing the formation of layer lines from the intersection of cones of diffracted X-rays with flat film.

Chemical Instrumentation

two-dimensional arrangement of the points on the plane. This information can be obtained from oscillation photographs over small angular ranges (5–15°) or by using either the Weissenberg or precession technique (see below). If the reflections can be indexed (i.e., assigned the correct *hkl* values) the intensities of the spots can be used to determine the crystal structure. However, because of the difficulties involved in properly indexing all the reflections, rotation and oscillation photographs are rarely used in the collection of intensity data.

A camera suitable for oscillation-rotation photographs should have a cylindrical or flat-plate film cassette, a support for the goniometer head and a rotation-oscillation mechanism. The oscillation mechanism consists of either a heart-shaped cam arrangement or a pair of microswitches. When the former method is used a different cam is required for each angular setting; when microswitches are used the angular settings can be changed continuously. In both cases provision is made for disengaging the oscillatory motion so that a full-rotation photograph may be obtained.

The interpretation of the x-y film coordinates is in terms of the projection onto the film of the intersections of reciprocal-lattice points with the sphere of reflection. A number of charts for this purpose are available for cameras of various diameters (cylindrical) and specimen-to-film distances (flat-plate).

The commercially available flat-plate cameras have been discussed under flat-plate powder cameras.

A few companies manufacture cylindrical cameras in which the crystal is rotated about a vertical axis. For example, Rigaku-Denki offers a cylindrical film cassette (Cat. No. 1141) which may be used on the Rigaku-Denki Universal Mount (Cat No. 1301, Price Range E). The Mount may also be used with a flat-plate film cassette and has an oscillation range which is adjustable to ±20°. The *Enraf-Nonius* General Pam-era has a cylindrical, as well as a flat-plate, film cassette (Fig. 58). Furthermore, all commercially available Weissenberg goniometers can also be used as cylindrical oscillation-rotation cameras.

Moving Crystal—Moving Film Techniques

Weissenberg Goniometer. The limitation of the oscillation-rotation method (i.e., the loss of information caused by the projection of planes of reciprocal-lattice points onto lines) is overcome by the use of the Weissenberg Goniometer (Fig. 86). The heart of the instrument consists of a mechanism for translating the cassette which is coupled to the oscillation

mechanism such that a given section of the film is in the same position each time a given reflection occurs. A layer line screen (i.e., a slotted cylinder which fits over the goniometer head and crystal, but inside the film cassette) is adjusted so that the reflections from only one plane of reciprocal-lattice points (corresponding to a single layer line on the rotation photograph) can pass through the slot. In this manner a distorted reciprocal-lattice plane is recorded on the film (Fig. 87). Charts used to interpret the photographs are available from Nies (plastic) and Polycrystal Book Service (paper).

The intensity of any reflection obtained from a given crystal is usually not evenly distributed over the entire spot as it

(Courtesy of J. Ladell, Philips Laboratories and The Norelco **Reporter 12,** *48 (1965))*

Figure 87. Positive print of a Weissenberg photograph of 0k ℓ zone of topaz taken with Zr-filtered Mo radiation. Note that the spots lie at the intersections of a series of festoons. The b^* and c^* reciprocal-lattice repeat distances and the angle between these axes can be measured on this film.

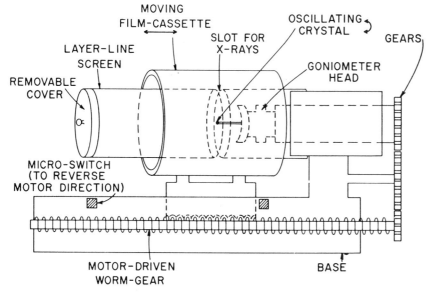

Figure 86. Schematic diagram of Weissenberg Goniometer

appears on the film. However, in order to obtain reliable intensity data one must determine the intensity of the complete spot (integrated intensity), not just that of the darkest portion of the spot (peak height). It is often possible to approximate the integrated intensity if the blackening of a spot is compared with the spots on a calibrated intensity film. In those cases in which the shape of the spot or some other factor prevents an approximation of the integrated intensity two instrumental methods can be used: (a) measure the intensities with an integrating microdensitometer (discussed in an earlier article); or, (b) take the photographs with an integrating Weissenberg goniometer. In the latter instrument the film cassette is translated in a direction normal to the beam as in the standard Weissenberg goniometer, but it is supported by a special mechanism. When it reaches the end of its translation it trips a lever which causes the film cassette to change its position on the support.

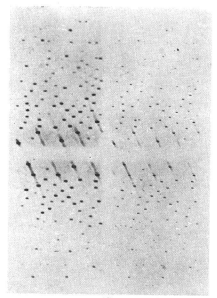

Figure 88. Right: Standard Weissenberg Photograph; Left: Integrated Weissenberg photograph. Both were taken over the same oscillation range.

The mechanism is arranged so that all the portions of a reflection are superimposed on one small section of the film. This section contains the integrated intensity and by measuring the optical density (peak height) of this spot one determines the integrated intensity of the reflection (Fig. 88).

Upper-Level Weissenberg photographs. For investigations of layers other than the zero layer, the interpretation of the film is simplified if the axis of rotation of the crystal is tilted at an angle to the direct beam such that the origin of the reciprocal lattice plane under investigation lies on the sphere of reflection (Fig. 89). This is known as the equi-inclination setting (Fig. 90).

All commercial Weissenberg goniometers can be used as oscillation-rotation or Weissenberg goniometers and they can be arranged in the equi-inclination setting as well as in the normal-beam setting.

The *Charles Supper Co.* manufactures a

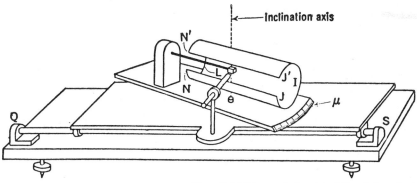

Figure 90. Schematic of Equi-Inclination Weissenberg Goniometer set up for upper level.

number of different Weissenberg goniometers:

(*a*) Standard Model W-20, $1710 and Model W-21, $1805: These models differ in one respect only—Model W-20 has a 57.3 mm diam film cassette and can accommodate only one film; Model W-21 has a film cassette with a larger diameter which can hold up to eight films for use with multiple film techniques (Fig. 91).

(*b*) Integrating Model WI-24, $2595: This model has a special film cassette containing an integrating mechanism, a power box which activates the integrating mechanism, and a microswitch which advances the integrating mechanism. The integrating mechanism differs from that found in many other integrating Weissenberg goniometers. The translational and rotational motions are achieved by means of special steel cams which are advanced electrically, rather than mechanically, to prevent wear in the mechanism. The cams, which are precision machined to impart precisely the correct increment of motion to the film holder, are of the "heart" variety. Thus, there is no disruptive discontinuity in any portion of the path followed by the film.

The integrating mechanism has been incorporated as part of the film cassette. The film-holding portion of the cassette, which may be detached from the integrating mechanism, can accommodate up

to eight films. The unit may be operated in either the normal or integrating modes. (Models W-20 and W-21 can be converted to an integrating device by factory installation of the power box and switch, and purchase of the integrating type cassette.)

(*c*) Model W22 ($2095)—Standard unit with a 77-mm diam film cassette.

(*d*) Back-Reflection Model W23, $2685: This unit is based on Buerger's original design (Ref. (5), p. 435). It has a 114.6-mm diam film cassette and is used for precision lattice constant determinations.

The Supper units feature extensive use of high-precision ball bearings, considerably reducing wear on the motor, lead screws, and cassette carriage.

The *Stoe* Weissenberg Goniometer (Price Range F) can be obtained in either an integrating WI 20 or standard WN 22 model. The inclination angle can be set up to ±55°. The limiting contacts can be set anywhere between 2° and 230°

Figure 91. Supper W-20 Weissenberg Goniometer.

of crystal oscillation for Weissenberg photographs and between 2° and 359° for oscillation photographs. The ratio of rotation-to-translation can be set at 1° to 8 mm for precision back-reflection Weissenberg photographs or 1° to 2 mm (for normal use) or for rotation but no translation (for counter attachments).

The following attachments are available:

(*a*) Counter attachment for manual operation.

(*b*) Straumanis-type film holder for precision lattice parameter measurements from rotation and oscillation photographs.

(*c*) Double radian attachment (114.6 mm diam) for precision back-reflection Weissenberg photographs.

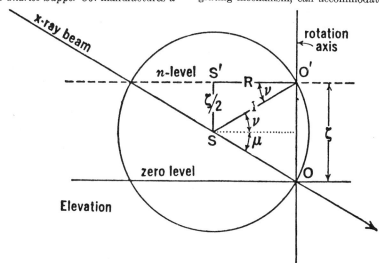

Figure 89. Diagram of Equi-inclination Weissenberg setting. O is the origin of the zero-level of the reciprocal lattice; O′ is the origin of the nth level; μ is the angle of inclination. The diagram shows the X-ray beam inclined to the reciprocal-lattice; in practice the crystal, not the X-ray tube, is tilted (Fig. 90).

Chemical Instrumentation

(*d*) Low and high temperature attachment for photographs or counter measurements between −170 and +250°C.

Enraf-Nonius offers four different models of the Weissenberg Goniometer ($2500–3500). There are integrating and nonintegrating models, and each of these can be obtained either for normal temperature work or for normal, high, and low temperature work (the temperature range is −150° to +300°C) (Fig. 92). The integrating mechanism, which is typical of most integrating cameras, operates as follows:

The apparatus is provided with a mechanical device by which the camera performs the following two additional movements at the end of every translation. (*a*) A small translation parallel to its axis, always in the same sense. (*b*) A

(Courtesy of Enraf-Nonius)

Figure 92. Nonius Integrating Weissenberg Goniometer with low-temperature attachment.

small rotation around its axis. When the total displacements in the two directions exceed the dimensions of the spot a plateau of uniform density is found in the center of the parallelogram.

The integrating mechanism is operated by a ratchet wheel provided with pins (Fig. 93). Toward the end of every ordinary camera translation one of the pins is caught by a stud causing the wheel to rotate 360/14 degrees. This rotation of the wheel is converted by the integrating mechanism into the additional translation and rotation of the camera. The magnitude of the additional camera movements can be varied easily so that they may be adapted to the dimensions of the reflection spots. Even very small spots can thus be integrated easily.

(Courtesy of Enraf-Nonius)

Figure 93. Close-up of integrating mechanism on Nonius Integrating Weissenberg Goniometer.

A Weissenberg goniometer with inclination angle of ±40° is obtainable from Jarrell-Ash (Cat. No 80-020, $1675). It allows a translation of up to 110 mm

(crystal oscillation of 220°) and is designed to eliminate wear and stress on the lead screw and motor (Fig. 94).

(Courtesy of Jarrell-Ash Co.)

Figure 94. Jarrell-Ash Weissenberg Goniometer.

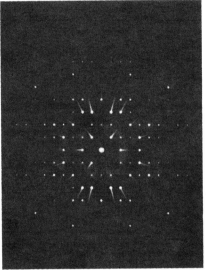

(Courtesy of Robert L. Snyder)

Figure 95. Precession photograph of the hk0 reflections of ammonium oxalate monohydrate taken with Polaroid film.

Precession Camera. In the early 1940's M. J. Buerger developed a camera in which the plane of the film is maintained parallel to a plane of reciprocal-lattice points. The crystal is adjusted so that a reciprocal-lattice plane is perpendicular to the direct beam. The projection of the reflections onto the film (as the reciprocal-lattice points pass through the sphere of reflection) is undistorted and easy to index (Fig. 95). A layer line screen is used to eliminate reflections due to other planes. The principle is outlined in Figure 96.

This method is a front-reflection method in which the mechanical arrangement of the instrument limits the number of reflections that can be investigated to those planes having λ/d less than 1.4. Thus, approximately as many reflections can be measured on the precession camera using molybdenum radiation ($\lambda = 0.710$ Å) as can be measured on the Weissenberg goniometer using copper radiation ($\lambda = 1.542$ Å). The advantage of being able to investigate all of the reciprocal lattice with only one mounting of the crystal has caused the precession camera to become one of the most popular single crystal units in use. The usefulness of the precession camera has been enhanced with the adaptation of the Polaroid cassette to the Supper precession camera (#57-1 Cassette, $325; #57-1 Adapter only, $35). The Polaroid method can also be adapted to other precession cameras.

Both standard and integrating precession cameras are available.

The most commonly used crystal-to-film distance is 6.00 cm. The use of a shorter distance will decrease the required exposure time, while the use of longer distance will improve the resolution of the spots on the film (this is useful for crystals with large unit cell dimensions, i.e., small reciprocal lattice dimensions). Precession cameras with variable as well as fixed crystal-to-film distances are available.

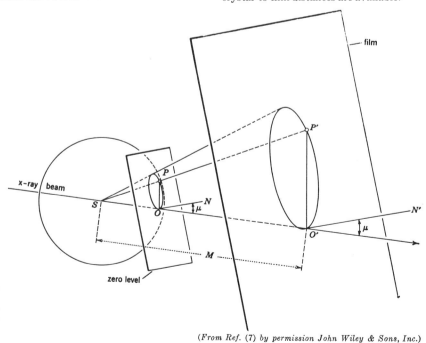

(From Ref. (7) by permission John Wiley & Sons, Inc.)

Figure 96. Principle of Precession Camera. This diagram shows the diffraction by a point P in a zero-level plane of the reciprocal lattice. The place is precessed about line $\overline{SO}$ with a precession angle of μ.

Figure 97. Supper Precession Camera with layer-line screen in place (the counter-weights are on the right).

Supper, which built the first working model of the precession camera, markets a number of models: Model P20 ($2095) is the standard unit (Fig. 97), with crystal-to-film distance of 6.00 cm (for zero-level photographs). Model P21 ($2695) is an integrating precession camera; and

(Courtesy of Jarrell-Ash Co.)

Figure 98. Jarrell-Ash Precession Camera. The Autocollimator and layer-line screens are on the table top; no counterweights are used.

Models PV-2 ($2390), PV-3 ($2510), and PV-4 ($2650) have 2, 3, and 4 variable crystal-to-film distances, respectively. Standard equipment includes an optical autocollimator, as well as layer line screens, beamstop, and X-ray collimator. A cone axis film cassette is available as an extra.

Stoe markets a standard (PS 30) and an integrating version (PI 31) of the Precession Goniometer (Price Range F) with a single set distance of 6.00 cm. A combined microscope-telescope with auto-collimator simplifies the optical adjustments. A low- and high-temperature attachment for photographs between −170° and +250° is available. The Polaroid XR-7 system can be used on the instrument and is available on request.

Jarrell-Ash has two models available. Model 80–010 ($1950) is a standard unit (Fig. 98). Model 80–011 ($2600) is a variable precession camera in which the crystal-to-film distance can be set to 5, 6, 7.5, or 10 cm. Both units are provided with autocollimators. It is claimed that the use of a 1-rpm direct drive motor and the

anti-backlash clutch eliminate the need for gear devices and counterweights. Alignment of the unit is facilitated because a hollow drive shaft provides a direct line-of-sight through the length of the instrument.

Enraf-Nonius offers an Integrating Precession Camera (Cat. No. Y925, $3790) with a choice of a fixed crystal-to-film distance of 5, 6, 7.5, or 10 cm from stock and any distance between 5 and 10 cm on special request. The counterweight is eliminated in this unit also (Fig. 99). The integrating mechanism is designed so that the film describes a Lissajous figure and each reflection is smeared into a blot, the density of which is proportional (at the center) to the integrated intensity of the reflection. Optional equipment includes a telescope, autocollimator, and Polaroid Adapter.

(Courtesy Enraf-Nonius)

Figure 99. Nonius Integrating Precession Camera.

Other Uses of Precession Camera. If a slow-drive motor is attached to the spindle handle by a belt it can be used to rotate the sample. The precession camera can then be used as a flat-plate rotation camera for single-crystal or powder specimens. It can also be used for taking transmission Laue photographs of accurately aligned crystals. Needless to say, we do not recommend purchasing a precession camera solely for these purposes, but if one is available it can readily be adapted to these other uses.

Summary. Let us consider a crystal aligned along the crystallographic c-direction; the reciprocal-lattice planes parallel to the plane defined by reciprocal-lattice axes a^*b^* are perpendicular to this direction.

The repeat distance along the c-direction can be obtained from the measurement of the layer-line separation on an oscillation or rotation photograph. The layer lines themselves are projections of the a^*b^* planes, and the Weissenberg goniometer would be used to photograph each a^*b^* plane (hk0, hk1, hk2, etc). Information concerning the 00l reflections (i.e., those along c^*) cannot be obtained from either of the above two methods, unless the crystal is realigned.

On the other hand, if the crystal is aligned along c^*, use of the precession camera allows one to investigate any of the planes defined by c^* and any direction in the a^*b^* plane. The major limitations of the precession camera are: (a) reflections at large Bragg angles cannot be

recorded and (b) the symmetry of the a^*b^* plane is not indicated on any one photograph. In practice the two methods complement each other.

A number of other single-crystal film methods have been devised, but are not in common use. For example, the retigraph (based on the deJong-Boumann method) is used to obtain undistorted photographs of reciprocal lattice planes perpendicular to the axis of alignment. Retigraphs are available from Enraf-Nonius and Crystal Structures, Ltd. However, because they are rarely used in the U.S. they will not be discussed.

Part Six—Single Crystal Methods (continued) and Miscellaneous Methods

Single-Crystal Counter Methods

The development of sensitive counters, reliable electronic panels, and stabilized X-ray generators over the past fifteen years has been accompanied by the development of sophisticated X-ray diffraction apparatus.

The apparatus used in conjunction with a counter detector is, basically, a crystal orienting device or goniometer. While it is true that film devices must be constructed so that the crystal remains in the center of the X-ray beam as it is rotated or precessed about some axis, greater accuracy is required of counter devices. For this reason, counter instruments are elaborate, rather expensive items.

Two basic geometries have been utilized in the construction of counter instruments. The first is that of the Weissenberg goniometer, with the film replaced by a counter mounted on a moveable arm. This instrument is known as the Counter Weissenberg and data is collected exactly as on a Weissenberg camera.

The second differs from the first in the following manner: Upper-level reflections are measured by tilting the crystal rather than the detector. The detector remains in one plane (e.g., the horizontal plane, when a horizontal diffractometer is used) and the crystal is carried in a two-circle assembly which is mounted on the *omega* (ω) axis (the ω axis lies parallel to the 2θ axis). The two-circle assembly consists of a χ-circle, which rotates about a horizontal axis and a ϕ-circle, mounted on the χ-circle so that the ϕ-axis is parallel to a radius of the χ-circle. The crystal is mounted on a goniometer head carried on the ϕ axis (Fig. 100). A reflection is obtained by setting the 2θ, χ and ϕ angles so that the reflection is in the zero-level plane (i.e., the plane perpendicular to the ω axis which contains the direct beam).

The two-circle assembly must be designed so that a properly aligned crystal will remain centered in the X-ray beam regardless of the angular settings. These devices, known as Eulerian cradles, have been built with a ϕ-circle capable of rotating 360° and a 90, 180 or 360° χ-circle.

Chemical Instrumentation

Some of the more popular terms used to describe these instruments are goniostat and full-circle or quarter-circle diffractometer; they are also described as having Eulerian or zero-level geometry. The term diffractometer has been applied to all counter units (e.g., powder diffractometer, Weissenberg diffractometer). In this article we shall refer to those instruments employing zero-level geometry

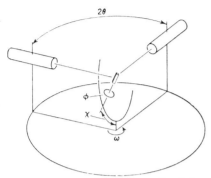

(From Ref. (6) by permission, John Wiley & Sons, Inc.)

Figure 100. Schematic diagram showing relationship of 2θ, ω, ϕ, and χ axes.

as goniostats, unless the manufacturer's term differs.

One major difference between these two types of diffractometers is that, under similar conditions (same crystal, same radiation, and using only one mounting of the crystal), practically the whole of the reflecting sphere (up to the maximum 2θ value) can be explored with the goniostat, while a smaller volume can be explored with the counter Weissenberg. The range of the counter Weissenberg is limited by the maximum attainable value of the equi-inclination angle.

A minor difference between these two instruments is that data processing (e.g., Lorentz-polarization corrections) is simpler for goniostat data than for counter Weissenberg data. The reason for this is that upper-level Weissenberg corrections are more difficult to apply than are zero-level corrections and only zero-level corrections are needed for the goniostat data. However, with the present availability of computer programs for data-processing, this objection to the use of counter Weissenberg instruments is not very serious.

Goniostats have been designed for use with horizontally or vertically mounted X-ray tubes; because of the splitting of the focal spot discussed in a previous article, the *omega* axis must be perpendicular to the tube mounting (e.g., a vertical *omega* axis is used with a horizontally mounted tube).

Once a crystal of known unit cell dimensions is mounted, it is a routine, and rather tedious, matter to adjust the crystal to the angles necessary to bring a given set of planes into reflecting position and to measure the intensity. The most recent advance in X-ray instrumentation has been the automation of counter units. The angles are automatically adjusted for a given reflection, the intensity is measured and the instrument moves on to the next reflection. Computer, punched-card, and

punched-tape operated units are on the market. These instruments are presently the most sophisticated and costly X-ray diffraction units available.

Manually Operated Counter Diffractometers

Counter Weissenberg Instruments. Both manually and automatically operated counter Weissenberg goniometers are available from Supper, Stoe, and Crystal Structures Lmtd. and a manual device is marketed by Rigaku-Denki.

The *Rigaku-Denki* apparatus (Cat. No. 2522, Price Range G) allows one to investigate Bragg angles up to a value of $164.5°$ 2θ (Fig. 101). An added feature is that the instrument has provision for the mounting of a cylindrical camera so

(Courtesy of Rigaku-Denki)

Figure 101. Rigaku-Denki Counter Weissenberg.

that an oscillation photograph can be taken. Although it is entirely possible to align a crystal on the diffractometer, the rough alignment can usually be carried out more efficiently with film methods. Furthermore, crystal anomalies (e.g., twinning) can be easily recognized by examining the film. The availability of a film cassette negates the need for transferring the crystal from one unit to another.

Goniostats. G. E. manufactures a quarter-circle goniostat ($2700) designed for use with a horizontally mounted tube; angles of -25 to $+160°$ in 2θ, a full $360°$ in θ, and -10 to $+90°$ in χ can be set on this apparatus (Fig. 102).

(Courtesy of General Electric Co.)

Figure 102. G.E. Quarter-Circle Single Crystal Orienter on SPG-2 spectrogoniometer.

G.E. also offers the E & A Full-Circle Goniostat for manual operation (see below) or automatic operation (with the Datex XDC-II system, see next section).

Crystal Structures, Ltd. makes three-, four- and five-circle goniometers. The

three-circle model is of the Weissenberg design, the four-circle device is a goniostat with a χ-circle which can be rotated through $180°$, and the five-circle unit utilizes both motions (of the three- and four-circle units). These units may also be automated.

E & A offers a full-circle goniometer ($5000) which can be mounted on a number of standard diffractometer bases (G.E. Norelco, etc) (Fig. 103). A full-circle unit is the most efficient instrument for examining all regions of reciprocal space. For example, all equivalent reflections can be compared; this is quite useful when measuring intensity differences due to anomalous dispersion or

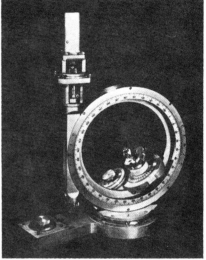

(Courtesy of E & A)

Figure 103. E & A Full-Circle Goniometer.

absorption effects. Cumulative angular errors are claimed to be less than 1 min of arc; cumulative linear errors to 0.001 in. These units can also be completely automated and pre-programmed.

The *Picker* Full-Circle Goniostat (Cat. No. 6239C, Price Range I, sold only with diffractometer) has an inside diameter of 10 in.; the manual controls are interchangeable with automatic high-speed motors and angle encoders. The χ and ϕ drives are disengageable to facilitate alignment and rapid preliminary examinations (Fig. 104). The instrument accepts either the 63.96 or 49.00 mm goniometer heads with the use of an auxiliary base

(Courtesy of Picker X-Ray Corp.)

Figure 104. Picker Full-Circle Goniostat.

that allows repeated removals and replacements with an accuracy and precision of better than ±0.01° and 0.001 in. The angular range of the circles is as follows: ϕ (360°); χ (360°); ω (360°, less obstruction); and 2θ (360°, less obstruction). In the latter two cases the obstruction depends on (1) how the instrument is placed relative to the tube and (2) the type of collimators used.

Stoe (EC 62) and *Siemens* (Fig. 105) market a full-circle Eulerian Cradle (Price Range F) of outer diameter 150 mm which provides the χ, ϕ, and ω motions and can be mounted on the counter diffractometer. The crystal is mounted

(Courtesy of Siemens American, Inc.)

Figure 105. Siemens Full-Circle Goniometer.

on a special goniometer head (the Stoe microgoniometer head) which can be made to comply with international standards (ACA-thread) by the use of an adapter. Stoe has developed a large full-circle Eulerian cradle EC 63 of outer diameter 330 mm (Fig 106); Stoe eucentric goniometer heads can be used. The angular accuracy is better than ±0.01°. The large Eulerian cradle can be operated either manually or automatically (with encoders).

(Courtesy of Stoe Instrument, Inc.)

Figure 106. Stoe Full-Circle Goniometer EC63.

Automated Counter Diffractometers

The instrumentation for automatic diffractometers, based on both the Counter Weissenberg and the Goniostat designs has been developed at a number of levels. For example, automatic diffractometers can be controlled by computers, punched-card, or punched-tape units. In the latter

two cases the angular settings are all pre-calculated; the instrument reads the settings for a given reflection, sets the angles, measures and records the intensity (according to pre-determined instructions), and moves on to the next reflection. On the other hand, the use of a computer permits a much larger choice of operational modes. The computer is controlled by the user's own program. The simplest programs involve calculating the angles from lattice parameters and crystal orientation, setting the angles and measuring and recording the intensities; more complex programs contain provision for searching the area near the peak, choosing the angular settings which give the maximum intensity, measuring the intensity and correcting for Lorentz-polarization factors.

As a second example, we can differentiate between those manual units which have been automated (by the addition of encoders) and those diffractometers that have been specifically designed for automation. A number of manufacturers have products which are designed for use as interfaces between existing units and punched-tape or computer control units; also, some computer manufacturers have connected their computers to standard diffractometers. For the most part, the design of each instrument depends upon the equipment available to the user and on his specific needs. Much of the equipment is, therefore, custom designed. Automated single-crystal X-ray diffractometers have been commercially available for but a few years and advances and improvements are constantly being made. The instruments described in this section are representative examples of the type of equipment that is commercially available; it is not a thorough coverage of all existing equipment.

(Courtesy of Datex Corp.)

Figure 107. G.E. Single Crystal Orienter and Datex Encoderdynes.

The *G.E.* goniostat with either the G.E. Quarter-Circle or the E & A Full-Circle Goniometer can be automated with the use of the Datex XDC-II, Automatic Single-Crystal Digital Control System utilizing Datex Encoderdynes (motor-absolute encoder in a unitized package) (Fig. 107).

A number of options are available, with the completely automatic system allowing axis positioning, fixed angle counting, constant speed scanning, and transfer of data to output recorder (punched tape, IBM cards, or digital printer). It also allows for control of a balanced-filter pair and a primary beam shutter; paper-tape, punched-card, and direct computer options are all available.

The *Picker* Full-Circle Goniostat can be combined with the Picker Four Angle Programmer to produce an automated single-crystal diffractometer (Fig. 108).

The Picker Four Angle Programmer's function is to control the four angular movements of the single crystal diffractometer; to coordinate the operation of the electronic measuring equipment, and to process and record the accumulated data. (Data entry and readout are in the form of punched cards. An IBM-026 card punch is utilized for this purpose.) The programmer consists of a combination of relay and transistor logic and contains no vacuum tubes. The system is capable of three different measuring modes: fixed position, constant velocity scan, and step-scan.

Also included are a balanced filter and automatic attenuator mechanism. An automatic shutter operation, permitting the shutter to be closed during crystal reorientation, and a crystal monochromator are optionally available.

Charles Supper Co. and Picker X-Ray Corp. jointly market a Buerger Automatic Single-Crystal X-Ray Diffractometer, SPAD (Model D50, approx. $20,000), in which the basic unit is the Supper diffractometer. The control system is completely solid-state and completely digital (manufactured by Pace Controls Corp.). The unit performs two types of functions: (1) the Tape Functions, which are controlled by the punched input tape, determine such operations as angular positioning of detector and crystal, scanning direction and rate, control of background counting time, use of balanced filters, and automatic interruption of routine. (2) The Fixed Functions which are built into the design system and cannot be varied. They control the counting pulses, opening and closing of fail-safe X-ray shutter, recording stored count in output tape, and automatic stop-start control of strip-chart recorder. A manually operated model (D-30, $4000) is also available.

The *Nonius* Three-Circle Automatic Single-Crystal Diffractometer (Cat. No. ACD, Price Range J) contains a position controller for the automatic positioning of the three circles, a range selector, digitizer, and built-in measuring program. The printer and reader unit are standard 8-channel units. With one setting of the crystal, one hemisphere of reciprocal space can be measured out to $\theta = 58°$ and the other hemisphere to $\theta = 15°$.

The counter moves in the equatorial (zero-level) plane; reciprocal lattice points are brought into this plane by positioning ϕ, χ, and θ through the use of incremental angular settings. The input tape contains the minimum of information, i.e., the angular increments and the indices of *hkl*. The possible methods of measurement are (1) θ-2θ scan (moving crystal-

Chemical Instrumentation

(Courtesy of Picker X-Ray Corp.)

Figure 108. Picker Four-Angle Programmer System for Four-Circle Automatic Diffractometer.

moving counter); (2) ω scan (moving crystal-stationary counter); (3) peak height (stationary crystal-stationary counter); or a combination of these. The method of measuring can be chosen through the use of the patch board. The output is printed and punched on tape.

The choice of the measuring sequence is up to the user; thus reference reflections can be put in where wanted. Because of the incremental character of the angle-setting system, the input tape can be prepared before the crystal has been mounted and adjusted. Flexibility is assured because each type of intensity measurement can be performed with the same input tape; the method of measuring the intensities is determined by the patch board.

An automatic slit-width and scan-angle control is built-in; by inserting the proper instructions on the input tape, the same band pass can be used for all reflections.

A manual model can also be obtained with the automation added at a later date if desired.

Humphrey Electronics specializes in designing interfaces for the automation of existing diffractometers for use with punched-tape or computer control. A number of models which can be made compatible with the user's equipment are available. Some examples are: tape punch, single-axis, and coordinated multi-axis positioning controls and data interfacing equipment which can convert almost any presentations of digital data to another format that might be required in order to use particular data-handling equipment.

At least three computer manufacturers have developed computer-controlled diffractometers. In all three cases commercial diffractometers were automated by designing an interface system between the diffractometer and the computer. Each system is custom designed and for more information the potential customer is referred to the specific companies. They are IBM (Fig. 109), 3M (Model 2018 Control Computer, exclusive sales rights to Picker Corp.), and SDS. (Also see the following description of a PDP-8 hook-up to a Hilger-Watts diffractometer.)

Hilger-Watts offers two different diffractometers. The more conventional Four-Circle Single-Crystal X-Ray Diffractometer (Model Y230, Price Range J) has the four circles arranged so that the crystal can rotate about the ϕ, χ, and ω axes, while the detector rotates about the 2θ axis which is coincident with the ω axis. All four circles are set by tape control; a flexible order code allows a wide variety of measuring sequences to be programmed.

The measuring system of this diffractometer is rather unique. To provide a measuring system entirely free from the mechanical effects of wear and backlash, the movement and positioning of the circles is controlled by a Moiré fringe technique.

Three of the circles are engraved with radial gratings having 3600 lines, while the 2θ circle is engraved with 1800 lines and the measuring heads are fitted with a small segment of a similar grating as a reference. Light is projected through the two gratings and, as the circle revolves and the gratings move past one another, a Moiré fringe pattern travels across the field at right angles to the direction of movement.

The measuring heads, each of which contains a lamp and photocell assembly, examine the fringe movements through four separated silicon photocells whose output takes the shape of approximately sinusoidal waveforms in quadrature, thus

providing complete and continuous information of the direction of movement and the position of the circle.

The sinusoidal output from the measuring heads is converted electronically into $^1/_{10}$ fringe increments, and these represent circle displacements of 0.01° for the ϕ, χ, and ω circles (which have radial gratings with 3600 lines), and 0.02° for the 2θ circle (which has a radial grating of 1800 lines). This is the accuracy to which the circles can be set, and it is entirely unaffected by the gearing or by backlash in the drive.

Each radial grating is marked with a datum point which is detected by a separate measuring head to the same accuracy; so an automatic check can be made on the setting accuracy at any time. The absolute position of the circles can be read from scales and micrometer heads.

All information for controlling the axes movements, measuring sequences, and special functions is provided by a suitable coded punched paper tape.

Recently, an interface system has been designed which connects the Four-Circle Diffractometer to a PDP-8 computer (produced by Digital Equipment Corp., Maynard Massachusetts 01754). This system, Model Y290, operates in a manner similar to that of Model Y230, with the added advantages of direct computer-control.

The *Hilger-Watts* Linear Diffractometer (Model Y190, Price Range I) is another instrument for the automatic measurement of X-ray reflections from single crystals. The linear diffractometer incorporates a novel analog system for orienting the counter and crystal, the settings of which are linearly related to the lattice dimensions of the crystal under investigation.

During the measurement of a reflection the crystal is oscillated through the reflecting range by a cam-controlled mechanism which can be set for any oscillation angle up to 5°. In operation a background is measured on one side of the reflecting position, the integrated reflection intensity is then measured as the crystal turns and a second background intensity is measured on the other side. Each background measurement takes half the time of the total intensity measurement, so that by subtracting the sum of the backgrounds the total intensity above background is determined. The time for

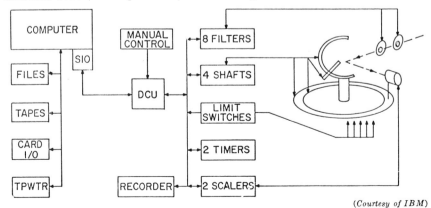

(Courtesy of IBM)

Figure 109. Block Diagram of Computer-Controlled Automatic Diffractometer. The Diffractometer Control Unit (DCU) is connected to the computer through a cable attached to the Serial Input-Output (SIO). The SIO permits the connection of special input-output devices to the computer.

a measuring cycle may be 60, 30, or 15 secs, and for each reflection the number of measuring cycles may be pre-set to a maximum of 10.

The intensities are measured along one row in the reciprocal lattice, back along the next row, and so on, with limit switches determining the areas scanned. A single adjustment of constant magnitude is required between consecutive layers. The maximum Bragg angle is 35° corresponding to a spacing of 1.35 Å with copper radiation, or 0.62 Å with molybdenum radiation.

A comparison of the linear and four-circle diffractometers has been published by U. W. Arndt in the *Hilger Journal* (**7**, 69 (1963), reprints are available from Hilger and Watts.

The *Stoe-Güttinger* Automatic Weissenberg Goniometer System (AW 31, Price Range J) is a semiautomated Stoe Weissenberg Goniometer. Before starting with counter measurements, Weissenberg photographs can be taken on the same instrument. The control system was designed and manufactured by Güttinger (Switzerland) and Addo tape-punching units are used. Continuous or balanced-filter scanning can be performed with moving crystal and either moving or fixed counter. The angular range 2θ of the counter is 0°–158° and the maximum equi-inclination angle is 45° (limiting switches are provided). The crystal and counter positional accuracy is 0.01°; the travel rate is 5°/sec; and scanning rates can be selected between 0.25 and 4°/min. A modified incremental technique is used which is based on absolute angular values. An automated absolute angular check is provided. Contacts for automatic selection of scanning rates and filters are available on request. A monochromator which is mounted in front of the X-ray tube and a low- and high-temperature attachment are available as optional accessories.

Stoe-Güttinger also manufactures an automatic 4-circle diffractometer (ADS 1) which is based on the large Stoe diffractometer DC 40 and the large Eulerian cradle EC 63. Optical converters are used and the four-circles are positioned simultaneously; both X-ray and neutron versions are available.

One of the few instruments specifically designed for automated-data-collecting purposes is the *Norelco* "Philips Automatic Indexing Linear Reciprocal-Space Exploring Diffractometer" or PAILRED (Cat. No. 12229, Price Range J). The design concepts of the PAILRED system, from the positioning of the tube to the automation of the instrumentation, were modified to fit the requirements of an automated single-crystal unit. The use of a modular design concept allows assembly of the complete PAILRED system starting from the basic high-stability generator, according to the needs and means of the purchaser. Thus, although it was designed for use with a punched-tape controller it can also be obtained for manual use.

Some of its more interesting features are the following (Fig. 110);

(a) *Incident-Beam Assembly.* In addition to providing a facile and reproducible instrument alignment when filtered radiation is employed, the use of a tilting tube

(Courtesy of Philips Electronic Instruments)

Figure 110. Norelco PAILRED Precision Diffractometer.

tower permits, with equal facility, the control of spectral bandpass through crystal monochromatization. A wide choice of characteristic wavelengths are available from silver, molybdenum, tungsten, copper, or other target tubes.

(b) *Eulerian-Cradle Assembly.* The Eulerian Cradle provides a wide angular range for orienting and aligning the "study" crystal as well as the development of equi-inclination data up to $\bar{\mu} = 45°$, about any crystal axis, with a cone of 130°. Alternatively, this assembly can be employed in the acquisition of intensity data according to the flat-cone technique. Thus with one mounting of the "study" crystal, the sphere of reflection within a solid cone of 310° can be investigated by using a combination of both techniques.

(c) *Precision-Diffractometer Assembly.* This assembly provides traverse of the detector on a spherical surface. The crystal shaft, supported on precision bearings, provides for 360° crystal ω rotation in steps of 0.001°. The ω and collinear γ (*upsilon*) axes are rotatable up to 45° about a vertical $\bar{\mu}$ axis as in Weissenberg geometry.

The Precision-Diffractometer Assembly together with the Eulerian-Cradle Assembly provides four rotational degrees of freedom for the crystal and two for the detector. PAILRED is thus a "six-circle" instrument. This versatility provides a facile means for accurate crystal alignment, rapid ascertainment of crystal symmetry, the feasibility of simultaneously diffracting from two sets of reflecting planes, etc.

(d) *Linear Reciprocal-Lattice Tracking Assembly.* Provides for coordinating detector and crystal motion so as to track along a line in reciprocal space. With this assembly, the diffractometer is set by dialing the Cartesian reciprocal-lattice coordinates of a given reflection; the investigator is relieved of the need to know or compute crystal and detector setting angles.

(e) *Automation Assembly.* Permits the automatic collection and indexing of intensity data within a zone or level in the reciprocal lattice. After initialization, consisting of supplying reciprocal repeats for a zone or level (two or four numbers) and directing the instrument assembly to

an initial reflection, the diffractometer proceeds in accordance with a self-contained program to each reflection in the zone or level, executes an intensity scan, and reports the results by means of a digital printout tape, a punched paper tape or a strip chart (profile) monitor. A wide range of scanning rate and slewing speeds are pre-adjustable. The automatic sequence can be interrupted at any time by manual intervention without loss of sequence or extensive re-initialization.

MISCELLANEOUS TECHNIQUES

Low-Angle Scattering (Small-Angle Scattering) Cameras

Materials such as proteins, viruses, glasses, catalysts, resins, and clays very often have interplanar spacings greater than 200 Å. Extremely small scattering angles are encountered, even when copper or chromium radiation is used. Generally, X-rays diffracted at such small angles are obscured by the direct beam and the average X-ray diffraction camera or diffractometer cannot be used to record these interplanar spacings. Instruments which have been successfully used for obtaining measurements at very low Bragg angles employ either slits or blocks to limit the direct X-ray beam.

(Courtesy of Jarrell-Ash Co.)

Figure 111. Luzzati-Baro Low-Angle Scattering Camera.

The *Luzzati-Baro* Low-Angle Scattering Camera (manufactured for Jarrell-Ash by Establishments Beaudouin, Paris, France) uses a two-slit system (Cat. No. 80–065, $3500). The first slit limits the width of the X-ray beam and the second eliminates secondary scattering of X-rays (Fig. 111). The slits have a symmetrical aperture adjustable to 1/100 mm between 0 and 2 mm. A Guinier-type monochromator (not supplied) is utilized to focus the X-rays; the instrument operates under vacuum, with detection by photographic film. Under ideal conditions the camera can measure diffused scattering to 2×10^{-3} radians, giving a spacing of 800 Å using CuK radiation. Optional accessories include a sample holder for low- and high-temperature operation.

Rigaku-Denki manufactures two Low-Angle Scattering Goniometers (Price Range G) suitable for use with either film or counter detectors. Model 1 has a 5° (2θ) scanning range and Model 2 has a 20° (2θ) range (Fig. 112). The systems use two defining slits and a scatter slit. Under proper conditions spacings as high as 1000 Å can be measured with CuK radiation.

Chemical Instrumentation

Figure 112. Rigaku-Denki Low-Angle Scattering Goniometer, Model 2.

Siemens offers a Kratky Small-Angle Scattering Camera (Price Range G). In this instrument the X-ray beam is defined by three finely polished metal blocks (Fig. 113). This collimation system is said to eliminate the scattering effects normally encountered when slits with knife edges are used. A number of collimation systems are available. The choice of the proper system depends upon whether intensity or resolution is of primary importance. An evacuable tube is located between the sample and the detector (film or counter). Spacings of better than 20,000 Å, corresponding to a diffraction angle of 10^{-4} radians (16 secs of arc), can be measured. Other optional equipment includes curved crystal monochromator, balanced filters, programmed step-scanner, and temperature cell for liquid samples (variable between -10 and $+90°C$).

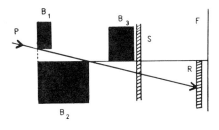

B_1, B_2, B_3. — BEAM LIMITING BLOCKS
S — SPECIMEN
F — FILM
R — PRIMARY BEAM STOP
P — PRIMARY BEAM

Figure 113. Collimation system of Kratky Small-Angle Scattering Camera.

Humphrey Electronics has developed a step-scanner (Series 721, Price Range F) for use with the Kratky small-angle cameras, which allows one to program different step widths and the number of steps throughout sequential portions of a total scan. In other respects it is similar to the Humphrey step scanner described previously.

Micro Cameras

A number of cameras have been designed for studies of extremely small specimens or limited portions of larger samples.

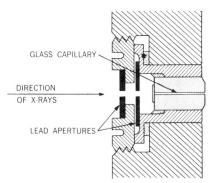

Figure 114. Collimation system of Norelco Micro Camera.

The *Norelco* Micro Camera (Cat. No. 52055, Price Range E) uses a glass capillary-tube collimator (available with either 50 or 100 micron bore) (Fig. 114) which permits examination of a specimen area of 10^{-3} or 10^{-2} mm², with a short film-to-specimen distance (10 to 15 mm). Holders are available for fibers or flat microtomed sections.

Rigaku-Denki Micro Camera Model B-3 (Price Range G) permits use of either transmission or back-reflection methods.

Figure 115. Rigaku-Denki Micro Camera, Model B-3.

A micrometer adjustment allows a fine control of the specimen position (Fig. 115). Model B is a Micro Laue Camera in which a telescope and a Geiger-Müller counter are used to align the sample. Either transmission or back-reflection photographs can be obtained.

Environmental Changes

It is often useful, and occasionally necessary, to change conditions of pressure and temperature. The study of phase transitions and solidified liquids and gases are two of the most important applications of these specialized techniques. Many of the cameras and accessories that are used in high-pressure and in high- and low-temperature studies have been custom-built by and/or for individuals

and it is only recently that commercial manufacturers have turned to the sale of these devices.

Low-Temperature Equipment. Commercially available low-temperature X-ray diffraction apparatus is used to cool the sample by one of two methods: (1) conduction through a copper block and (2) blowing cold gas over the sample. The former is usually used for powder diffractometers, while the latter has been utilized most often in single-crystal investigations (however, there are important exceptions to each). The I.U.Cr. Low-Temperature Bibliography (*16*) contains a summary of low-temperature techniques and applications. A simple, but very effective, gas-blowing apparatus can be constructed by referring to the references indicated in the apparatus section of the Bibliography.

Figure 116. MRC Cryogenic Diffractometer Attachment, Model X-86-NC.

MRC manufactures a Cryogenic High-Vacuum Diffractometer Attachment. Model X-86-GC ($5000) is designed for use with horizontal units (e.g., Norelco) (Fig. 116). The samples, which are powders or solids, are held on a copper specimen stage which is cooled by conduction through contact with a gold-plated liquid helium, vacuum-jacketed, dewar. The sample can be maintained at any temperature between 4.2 and 310°K. High temperature conversion kits are available which can be used to heat the sample to 2500°C.

Figure 117. E & A Cryoex System mounted on G.E. Diffractometer.

E & A markets its Cryoex-Cryogenic Specimen Holder (Cat. No. 65-7-C11, $12,000) for use with horizontal diffractometers (Fig. 117). Cooling (to 4.2°K) is achieved by throttling a flow of helium through a tube connecting the liquid helium dewar and the specimen holder. The helium vent valve acts as a coarse control over the amount of cooling applied, and an electric heater is used to permit rapid and accurate attainment of the desired temperature. Calibrated platinum and germanium sensing units permit temperatures to be read accurately to 0.1°K.

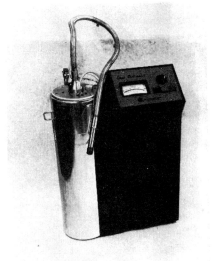

(Courtesy of MRC)

Figure 118. MRC Cryo-Collimator.

Rigaku-Denki has an attachment for its horizontal diffractometer that can be used to attain liquid nitrogen temperatures (Cat. No. 235, Price Range F). The sample is maintained in an evacuated chamber and is cooled by conduction.

MRC has developed a cold-nitrogen-gas generator, the Cryo-Collimator (Cat. No. XCC-107, $2190), which operates on the same principles as the glass device pioneered by Post, Schwartz, and Fankuchen (17). Liquid nitrogen in a dewar is transferred to gas under pressure and then cooled by passing it back through the liquid nitrogen. The cold gas passes through a flexible vacuum-insulated transfer tube and special collimation nozzle onto the sample (Fig. 118). The 15-liter capacity of the dewar allows 6 hr of operation at the −180°C level. Optional automatic refilling equipment is available.

Air Products and Chemicals is currently developing a device for cooling single crystal specimens mounted on a metal pin attached to the cold surface of their Cryo-Tip Refrigerator (approximately $3000). The Cryo-Tip Refrigerator is an open-cycle Joule-Thompson device which works on compressed nitrogen. The sample is mounted on a goniometer inside an evacuated chamber (Fig. 119). The vacuum chamber is provided with an X-ray window of beryllium or any of a variety of amorphous substances chosen to match the characteristics of different sources of radiation.

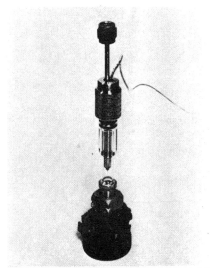

(Courtesy of Air Products and Chemicals)

Figure 119. Nitrogen Cryo-Tip Refrigerator mounted on modified eucentric goniometer head.

Stoe and *Enraf-Nonius* both offer heating and cooling attachments for their Weissenberg goniometers, which permit one to attain any temperature between that of liquid nitrogen and +300°C. These devices use heat exchangers to cool or heat gas which is then blown over the sample.

High-Temperature Equipment. High-temperature X-ray diffraction techniques are discussed in a special I.U.Cr. bibliography (18); research and commercial furnaces for photographic and diffractometer uses are described.

A number of commercially available furnaces are designed to heat powder or solid samples by conduction. This is true of the MRC X-86-G (for horizontal diffractometers) and X-86N-11 (for vertical diffractometers) models ($4600 and $4250, respectively) and of the R. L. Stone Co. Model XR-6 for vertical diffractometers. The Tem-Pres Furnaces (Fig. 120), Models HX ($4400–5800) and SX ($5800–7300) support the sample on a refractory pedestal at the geometric center of the heating chamber (both models can be adapted to either horizontal or vertical diffractometers).

(Courtesy of Tem-Pres Research, Inc.)

Figure 120. Tem-Pres Furnace, Model SX.

All these furnaces have provision for maintaining the sample in vacuum or in an inert-gas atmosphere. In the MRC and R. L. Stone units the samples are supported on resistance heaters and the outsides of the furnaces are water-cooled.

With the MRC furnace, temperatures up to 2500°C can be attained; up to at least 1300°C with the R. L. Stone unit; and up to 1400°C for the Tem-Pres Model HX and 1850°C for Model SX.

The MRC units are now manufactured with provision for radiant heating in addition to resistance heating; the heaters can be used individually or together. Temperature control is by an auto-proportional controller; maximum temperature with the radiant heater is approximately 1500°C. Both MRC units are supplied with a cold stage for cryogenic work to −120°C and liquid helium cryogenic (4.2°K) converters are available with the X-86-G model.

Tem-Pres also supplies five highly purified powders as standards (Cat. No. XRS-665, $10) for calibrating the furnace (the standards have transition temperatures between 146 and 583°C).

E & A uses radiant heating with platinum elements in the two units it manufactures: Refractex I ($3600) for diffractometers (Fig. 121) and Model 65-4 HTC-1 for photographic work. The latter can attain temperatures up to 950°C; the former can reach over 1000°C and also has an adapter ($10,400) for low-temperature work (down to 4.2°K).

(Courtesy of E & A)

Figure 121. E & A Refractex I.

Enraf-Nonius has a unique High-Temperature Camera (Cat. No. Y909, $7000) of a modified Guinier type, which uses a Pt/Rh spiral heater with a water-cooled cover and beryllium window. It has a movable film carriage with 8 different speeds (Fig. 122). The temperature can be increased continuously and synchronized to the film movement; in this manner the pattern at each temperature can be obtained and splittings due to phase changes or thermal expansion of the lattice constants can be followed.

Rigaku-Denki manufactures a similar instrument. The film cassette is 90 mm in diam and it can move automatically and continually; the specimen can be rotated (Cat. No. 1211, Price Range F).

Another Rigaku-Denki accessory, the High-Temperature X-Ray Diffractometer Attachment (Cat. No. 2321, Price Range F) can heat the sample to 1500°C. A

Chemical Instrumentation

(Courtesy of Enraf-Nonius)

Figure 122. Nonius High-Temperature Camera.

Programme System Automatic Temperature Controller (Cat. No. 5181, Price Range G) is available which will oscillate the diffractometer over a pre-set angular region and automatically change and record the temperature. In this manner a phase diagram can be obtained.

High-Pressure Equipment. To the best of my knowledge, only one company manufactures high-pressure X-ray diffraction cameras.

The *MRC* XKB-100 camera ($4950) uses a pair of optically flat diamond crystals in a piston and anvil arrangement with the X-rays passing through the sample and diamonds (Fig. 123). The piston diamond is attached to a gas-pressurized piston, which provides a mechanical advantage of approximately 2500 times the applied gas pressure. Dry nitrogen from a standard gas cylinder is normally used. A surrounding copper black contains heaters to raise specimen temperatures up to several hundred degrees centigrade.

(Courtesy of MRC)

Figure 123. MRC High-Pressure Camera and Controller.

Pressure limits are about 100 kilobars at the specimen. The diamonds are set in removable supports which can be re-used many times without danger of cracking. The supports contain the collimators (2 sizes are supplied).

The *MRC* XVB-200 ($3000) uses an externally loaded pressure cell. The X-rays pass through the sample perpendicular to the line of force; the film covers 10 to 170° 2θ.

Note

Shimadzu X-Ray Equipment—a newcomer to the U.S. market, Shimadzu offers a complete line of X-ray diffraction equipment, including generators, diffractometer, powder cameras (Debye-Scherrer, Back-Reflection, Micro-, High-Temperature), monochromator, film comparators, etc. Unfortunately, the descriptions of their instruments were not received early enough to be included in the proper places in the text. A photograph of their latest generator (Type VD-1) is shown in Fig. 124.

(Courtesy of Shimadzu)

Figure 124. Shimadzu X-Ray Generator and Accessories.

Editor's Note

Although it has not been the purpose of this review to discuss or recommend texts in the field of X-Ray Diffraction, a recent publication merits special attention: *Single Crystal Diffractometry* by U. W. Arndt and B. T. M. Willis (Cambridge University Press, 1966). There are a number of excellent texts dealing with the theoretical and practical aspects of X-ray film units. However the treatment of single-crystal diffractometers has been neglected and in the past one's introduction to the use of these instruments has been through manufacturers' instruction books and articles scattered throughout the scientific literature. *Single Crystal Diffractometry* is the first book devoted solely to a discussion of single crystal diffractometer techniques (neutron, as well as X-ray diffractometers are discussed). After a short introductory description of diffraction geometry, the remainder of the book is evenly divided between a description of instrumentation (design principles, detectors, electronic circuits, and primary-beam production) and techniques for obtaining meaningful intensity measurements (background counting, systematic errors, integrated intensities, and accuracy). A final short chapter discusses computer programs and on-line control. The level of the book assumes that familiarity with basic crystallographic concepts which is to be expected of anyone contemplating the use of these instruments.

X-Ray Diffraction
Supplement

The preceding review of X-ray diffraction instrumentation was completed in October, 1966; the present supplement was written in January, 1970. During the intervening time, over twenty additional firms have offered equipment of interest to the X-ray diffractionist. As was true in the first article, the emphasis will be on suppliers with active outlets in the U.S. (The author is presently compiling the *Index of Crystallographic Supplies* under the auspices of the Commission of Crystallographic Apparatus of the International Union of Crystallography. This Index will contain a list of suppliers and manufacturers on a global basis; it will be available for distribution through Polycrystal Book Service in 1971.)

In general, products that have recently become available are improved versions of existing equipment. Extremely stable generators and advances in the precision, accuracy, and computer interfacing of single-crystal diffractometers are two notable examples. However, a few new products have also become available. The present review will be restricted to a discussion of new and/or unusual products or services and a brief description of products offered by manufacturers not mentioned in the main review. The reader can obtain information concerning general improvements direct from the manufacturer.

Monochromators

Union Carbide has developed the "Ucar" Graphite monochromator, a high purity, hot-pressed, self-bonded form in which the basal planes of the crystallites have a high degree of alignment with respect to each other and the surface. The $2d$ interplanar spacing is 6.708 Å. Singly bent and flat pieces in various sizes are available in a range of mosaic spreads. The top grade has a mosaic spread between 0.3 and 0.5°.

These monochromators provide 2 to 3 times more X-ray intensity than lithium fluoride crystals with no compromise in other characteristics. They are currently offered as standard equipment by several manufacturers or can be purchased directly from Union Carbide.

Detectors and Film Scanners

Optronics International, Inc., manufactures a number of microdensitometers capable of rapidly converting photographic data into digital form. The output of their Photoscan scanner can be interfaced directly with any small computer or tape unit. The photographic record is scanned in a raster with a line separation of 50, 100, or 200 μ in the x and y directions. The photographic density is said to be measured with an accuracy of 0.02 between $D = 0$ and $D = 3$ and is presented at the output in 7-bit binary form together with the x and y data. The scanning window is adjustable.

Photometrics, Inc., offers an isodensitometry data reduction service, which provides for extracting continuous tone data from photographic records using a two-dimensional scanning microdensitometer.

Powder Technique

Philips offers a new horizontal powder diffractometer in addition to its vertical model. They also have designed a number of new diffractometer accessories including a curved crystal focalizer for transmission diffractometry, a vacuum attachment useful for improving measured intensities, a texture attachment and an automatic sample changer that can handle up to 35 samples.

The ASTM Powder Diffraction File has developed a computer search system using data input from the user's laboratory via teletype.

Single Crystal Techniques

Goniometer Heads

The diffractometer settings for any accessible reflection can be calculated for a single crystal once the orientation of the crystal on the unit is known. It is not absolutely necessary to align a crystallographic axis parallel to the goniometer head axis. For this reason, the use of arcs (a major source of undesired crystal misalignment) is not mandatory. A few distributors (e.g., Blake Industries, Enraf-Nonius) presently offer goniometer heads with x, y, z translations only. These have a large range of travel and are very stable.

Retigraph

Enraf-Nonius has introduced an integrating retigraph. This instrument, which is in common use in Europe, is not too often seen in the U.S. It offers a number of advantages over the Weissenberg goniometer, including an inclination angle range of -50 to $+60°$. In addition, it records the reciprocal lattice without distortion and with constant spot sizes. It can be obtained with a Polaroid adapter; several accessories are available for conversion to other uses, such as Laue photographs.

SCMF Camera

The Toshiba Stationary-Crystal Moving-Film (SCMF) Camera has both the function of a precession camera and of an equi-inclination Weissenberg goniometer. Conversion from one to the other requires exchange of layer-line screen and cassette but no readjustment of the crystal. The specimen remains stationary during exposure so that the crystal environment can be controlled easily. The novel feature of this instrument is that the X-ray tube moves while the crystal remains stationary (Fig. 125). In the Weissenberg version, the tube rotates around the crystal, while in the precession version, the collimator travels on a conical surface. Layer-line-screens and film motions are properly coordinated with the desired mode of operation.

Figure 125. Toshiba stationary-crystal moving-film (SCMF) camera. (Courtesy of Toshiba International)

Diffractometer

Philips offers a full-circle computer or manually controlled unit, as does Syntex.

Enraf-Nonius has introduced a new design in their CAD-4 computer-controlled 4-circle diffractometer (Fig. 126). The specific characteristic of this new goniostat is the introduction of an axis, called the kappa axis, which intersects the omega axis at an angle smaller than 90°. In this design the angle is 50°. The phi support is attached to the kappa axis. The phi axis also intersects the kappa axis at an angle of 50°. As a result, by moving kappa, the phi axis describes a cone with a top angle of 100°.

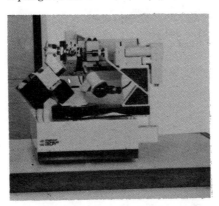

Figure 126. Enraf-Nonius CAD-4 computer-controlled four-circle diffractometer. (Courtesy of Enraf-Nonius)

Since a chi circle, or Eulerian cradle, is no longer present in this design, a full omega rotation can be carried out without obstructions between goniostat and collimators.

The only obstructed region is almost equal to the small cone-shaped shadow of the goniometer head. A full omega rotation is possible, and a reflection can be observed with constant azimuth in two different ways. Some reflections can be observed in four different ways, because the phi axis can be tilted 10° above the equatorial plane.

The advantages of the new design include:

(1) Ample space above the equatorial plane for accessories, such as a low temperature device.
(2) The bearing-mounted axis construction is basically more accurate than an arc/cradle design.

Environmental Changes

High-Pressure

High pressure cameras and attachments for single-crystal and powder studies are now also sold by Enraf-Nonius (Fig. 127) and by High Pressure Diamond Optics. Both devices consist of high-pressure cells which can be removed from the (modified precession) camera so that the crystal can be grown under observation through a microscope. A tetrahedral press with X-ray attachment for studies up to 2000 tons is manufactured by Bliss-Barogenics, Inc.

Figure 127. Enraf-Nonius high-pressure cell, mounted on specially adapted precession camera. (Courtesy of Enraf-Nonius)

Low-Temperature

Andonian Associates has developed a variable temperature liquid helium sample holder for use with horizontal diffractometers. Available with both a holder for powdered specimens and a one-arc goniometer head, it permits X-ray studies at temperatures from 1.5°K to 100°C.

Enraf-Nonius recently introduced a gas-stream low-temperature apparatus with adapters for Weissenberg, precession, diffractometer, retigraph, and Guinier-Lenne powder camera. It has a continuously variable temperature range down to $-170°C$, with a maximum consumption rate of approximately 2 l liquid nitrogen/hour. The unique design of this instrument provides a stable temperature ($\pm1.5°C$) over long periods of operation, including the refill cycle. Special adapters, Dewar tubes, mounting brackets and a copper-constantan thermocouple are furnished with the instrument.

118

Supplement

---

[2] The International Union of Crystallography (I.U.Cr.) publishes an Index of Cyrstallographic Supplies which is revised periodically. The next edition, scheduled to appear late in 1967, will be available through Polycrystal Book Service (the U.S. Clearinghouse on all publications of the I.U.Cr.)

---

* Manufacturers only.

Digital Automation Co., Inc.
68 Highway 31
Pennington, N. J. 08534
(General line of electronic counting circuitry, and automation equipment utilizing solid-state components and designed for X-ray diffractometry)

Espi, Inc.
854 So. Robertson Blvd.
Los Angeles, Calif. 90035
(Thin metal foils for use as beta-filters)

General Electric Co.
Analytical Measurement Section
25 Federal Street
West Lynn, Mass. 01905
(In addition to their own line, G. E. now distributes Rigaku-Denki products in the U.S.)

Harshaw Chemical Co.
Crystal and Electronics Products
  Department
6801 Cochran Road
Solon, Ohio 44139

High Pressure Diamond Optics, Inc.
929 MacKall Avenue
McLean, Virginia 22101
(High pressure beryllium diamond cell for use in single-crystal and polycrystalline X-ray diffraction work, adapted for precession camera; other high pressure accessories)

JEOL (Japan Electron Optics Laboratory)
828 Mahler Road
Burlingame, Calif. 94010
(Complete line of diffraction equipment)

Lemont Scientific, Inc.
Pike Street
Lemont, Penn. 16851
(X-ray safety shutter; small goniometer head furnace capable of heating single crystals to 1000°C)

Millipore Corporation
Bedford, Mass. 01730
(Manufacturers of filters and filter holders; useful in preparation of polycrystalline samples for diffractometer study)

Nuclear Equipment
931 Terminal Way
San Carlos, California 94070
(Si(Li) detectors and associated equipment)

Optronics International, Inc.
7 Stuart Road
Chelmsford, Mass. 01824
(Microdensitometers capable of converting photographic data into digital form at rates compatible with standard computer peripheral equipment)

Ortec, Inc.
1000 Midland Road
Oak Ridge, Tenn. 37830

(A complete line of solid state electronic instrumentation for detection, amplification, signal analysis and data acquisition, including cooled semiconductor detectors)

PhotoMetrics, Inc.
Lexington, Mass. 02173
(Isodensitometry data reduction service on work order or complete program basis)

Syntex Analytical Instruments
3401 Hillview
Palo Alto, Calif. 94304
(Distributor of general line of X-ray instrumentation including high-temperature powder camera, Guinier camera, and automatic four-circle single-crystal diffractometer)

Technical Operations, Inc.
South Avenue
Burlington, Mass. 01803
(Present distributor of Joyce, Loebl double-beam automatic recording microdensitometer)

Toshiba International Corp.
465 California St.
San Francisco, Calif. 94104
(Complete line of X-ray equipment, generators and cameras; powder diffractometer; distributed in the U.S. by Beckman Instruments).

Union Carbide Corp.
Carbon Products Division
270 Park Avenue
New York, N. Y. 10017
(Highly oriented graphite monochromators)

Literature Cited

(1) Beu, K. E., Chairman, Apparatus and Standards Committee of the American Crystallographic Association, 1962, (a) Report No. 1; (b) Report No. 2.

(2) Commission on Crystallographic Apparatus, *Acta Cryst.* **16**, 324 (1963).

(3) Cook, J. E., and Oosterkamp, W. J., *International Tables for Crystallography*, Vol. III, Section 6, Birmingham, England, Kynoch Press, 1962, p. 333.

(4) Directory of Products and Services for Radioactivity Laboratories, *Nucleonics* **23**, No. 6, 93–112, June 1965.

(5) Buerger, M. J., *X-Ray Crystallography*, John Wiley & Sons, New York, 1942.

(6) Buerger, M. J., *Crystal Structure Analysis*, John Wiley & Sons, New York, 1960.

(7) Buerger, M. J., *The Precession Method*, John Wiley & Sons, New York, 1964.

(8) Azaroff, L. V., and Buerger, M. J., *The Powder Method*, McGraw-Hill, New York, 1958.

(9) Barrett, C. S., *Structure of Metals*, McGraw-Hill, New York, 1952.

10) B. D. Cullity, *X-Ray Diffraction*, Addison-Wesley, Reading, Mass., **1956**.

11) Klug, H. P., and Alexander, L. E., *X-Ray Diffraction Procedures*, John Wiley & Sons, New York, 1954.

(12) Lipson, H., Nelson, J. B., Riley D. P., *J. Sci. Instr.* **22**, 184 (1945)

(13) Morimoto, H., and Uyeda, R., *Acta Cryst.* **16**, 1107 (1963).

(14) *International Tables for Crystallography*, Birmingham, England. (a) Vol. I. Symmetry Groups, 1952; (b) Vol. II. Mathematical Tables, 1959; (c) Vol. III. Physical and Chemical Tables, 1962.

(15) Commission on Crystallographic Apparatus, *Acta Cryst.* **9**, 976 (1956).

(16) Post, B., Low-Temperature X-Ray Diffraction, Bibliography 2, I.U.-Cr., Commission on Crystallographic Apparatus, 1964. (Polycrystal Book Service.)

(17) Post, B., Schwartz, R. S., and Fankuchen, I., *Rev. Sci. Instr.* **22**, 218 (1951).

(18) Goldschmidt, H. J., High-Temperature X-Ray Diffraction Techniques, Bibliography 1, I.U.Cr. Commission on Crystallographic Apparatus, **1964**. (Polycrystal Book Service.)

APPENDIX A

Price Ranges

In many instances it was not feasible for the distributors to furnish exact prices. However, for the convenience of the reader, they did indicate the *price range* according to the following scale: A = under $10; B = $10–50; C = $50–100; D = $100–500; E = $500–1000; F = $1000–3000; G = $3000–7000; H = $7000–12,000; I = $12,000–25,000; J = above $25,000. The letter symbols are used in the text.

Topics in...

Chemical Instrumentation

...a

feature

Edited by **S. Z. LEWIN,** New York University, New York, N.Y. 10003.

The Journal of Chemical Education
Vol. 43, page A683
September, 1966

XXX. Microwave Absorption Spectroscopy

Galen W. Ewing, *Department of Chemistry, Seton Hall University*
South Orange, New Jersey   07079

Microwave absorption is primarily a tool for observing and measuring rotational transitions in molecules which possess a permanent electrical dipole moment. Absorptions can also be observed under appropriate conditions in molecules possessing magnetic moments, such as O_2, NO, NO_2, ClO_2, and in free radicals. It thus supplements on the one hand far-infrared and Raman spectroscopy which can provide information about molecular rotations, and on the other hand conventional magnetic susceptibility measurements.

The region of the electromagnetic spectrum included in the term "microwave" has no sharp boundaries, and can be considered to overlap the far infrared. It is often assigned the rough limits of 1 mm to 100 cm wavelength, corresponding to a frequency range of 300 to 0.3 Ghz (1 gigahertz, Ghz, = 10^9 hz; 1 hz = 1 cycle per second). The region which has proved most fruitful for absorption spectroscopy lies between approximately 8 and 40 Ghz (7.5 mm to 3.75 cm). It should be noted that this includes the frequencies useful in electron-spin magnetic resonance (ESR), so that the two methods share some features in common. The phenomena underlying them, however, are very different, and their fields of utility therefore also differ. The microwave region is distinguished from the infrared primarily by the very different apparatus and experimental techniques required. Before discussion of the techniques, we will consider briefly the theoretical implications.

Theory of Microwave Absorption

There are no microwave absorptions which are characteristic of specific bond types or functional groups, such as we often find in optical and infrared spectra. Qualitative analysis can only be applied by means of comparison with known spectra. The microwave region is excellent as a complement to the infrared in the identification of covalent compounds by the "fingerprint" approach. This is partly due to the very great number of frequencies available. If absorption bands can be resolved at a separation of 200 khz, a not unreasonable figure, this gives a total of 160,000 spaces or "channels" in the 8 to 40 Ghz region [$(40 - 8) \times 10^9 / 2 \times 10^5 = 1.6 \times 10^5$]. By comparison, if we assume with Gordy (6) that an average resolution of 1 cm^{-1} is attainable over the entire infrared region from the visible up to 50μ, only 10,000 spaces are available. Of course there are many frequencies which do not appear in any known spectra, so these figures may be somewhat misleading; nevertheless, an extremely large number of compounds can in principle be examined in a microwave absorptiometer with little probability of overlapping. The chief restriction is that only gaseous samples can be studied, though the vapor pressure need not be high; 10^{-1} to 10^{-3} torr is the usual pressure range.

The quantitative absorption law (Lambert's law) can be expressed in the form,

$$P = P_0 \cdot 10^{-\alpha b} \text{ or } \log (P_0/P) = \alpha b$$

where P_0 and P represent respectively the radiant (microwave) power incident upon and passing through the absorption cell, which has an effective length b. The absorption coefficient, α, corresponds to the absorbance, A, as usually employed in the optical region, taken per unit length of the absorption cell. Ideally it is a function only of the number of absorbing molecules per unit path length, and hence should be related to the partial pressure of the substance in a mixture of gases.

The degree to which such a relation is valid depends on the way in which the absorption coefficient is measured and utilized. At low pressures and power levels, where saturation effects are not evident, the pressure is proportional to α_{max}, the height of the absorption maximum. However at higher pressures, the value of α_{max} becomes constant, and an increase in pressure results in the broadening of the line as measured at half-height. Above this point it can be shown that the integrated absorption is very nearly proportional to the pressure:

$$\int \alpha(\nu) d\nu = Kp$$

where ν is the frequency, p is the pressure, and K is a proportionality constant. Figure 1 shows the absorption curves of a typical band at a series of pressures.

If a nonreacting gas which does not absorb at the same frequency is added to the sample, the value of the integrated absorption is found to be proportional to the mole fraction of the absorbing gas. Hence this integral (the area beneath the curve) is a valid quantitative analytical tool. Its measurement is rather tricky, however. It is necessary to be certain

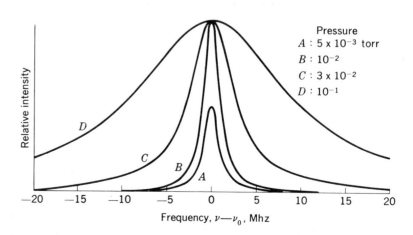

Pressure
A : 5 x 10^{-3} torr
B : 10^{-2}
C : 3 x 10^{-2}
D : 10^{-1}

Relative intensity

Frequency, $\nu - \nu_0$, Mhz

[*From Ref. (7) by permission, John Wiley & Sons.*]

Figure 1. Variation of microwave absorption curves with pressure change. Note that the peak intensity remains constant over a wide pressure range.

Chemical Instrumentation

that the power level is not too high, so that neither the absorbing sample nor the crystal rectifier detector becomes saturated. The temperature also must be controlled. With due care standard deviations of the order of 3–5% are to be expected (but remember that the sample may be only a few micromoles).

In this connection it must be emphasized that both theory and techniques in microwave absorptiometry are still young. There is every reason to believe that the signal-to-noise ratio can be increased drastically over the best now available, so that the precision as an analytical tool should be improved by at least a factor of ten.

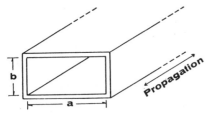

Figure 2. Geometry of a rectangular waveguide.

Microwave Techniques

Microwave technology in its modern form originated with the development of radar during and immediately following World War II. It was recognized early in the game that the most convenient and effective way to direct the radiation along desired paths within a piece of laboratory apparatus is by means of *waveguides*.

In current practice waveguides most commonly consist of straight or curved metallic pipes or conduits of rectangular cross section. It can be shown, both theoretically and by experiment, that a waveguide of dimensions a and b (Fig. 2), where the ratio a/b is between 2.0 and 2.5 (the usual proportions) is capable of propagating an electromagnetic wave

of wavelength not greater than $2a$. The relative orientations of the electric and magnetic fields in the waveguide at any instant in time take the form suggested in Figure 3. The wave is moving along the guide, so that at an instant one-half period later the fields will have assumed a pattern identical to that of Figure 3 except that all arrow heads will be reversed. This pattern is referred to as the *dominant mode* of propagation, and it is

the only mode possible for wavelengths greater than a but less than $2a$. Waves of lengths less than a will be propagated but various more complex field configurations (modes) will be possible and the measurements and interpretations of results may no longer be unambiguous.

In practice the optimum wavelength for a particular guide is specified as approximately two-thirds of the maximum wavelength, or about $1.3a$. The useful range is commonly taken as between $1.1a$ and $1.7a$. This means, of course, that it is impossible to cover the whole microwave region of interest with a single system of guides. Hence the region is divided into five bands with the properties given in Table 1. Many other bands have been established but are not included in the region of chemical interest.

For absorption measurements, a spectrophotometer must be assembled from microwave components, and as in any

---

[1] To equal the microwave output of a klystron, the usual equations indicate that a black body would have to be operated at 10^{14}°C!

---

spectrophotometer must include a variable monochromatic source, an absorption cell (the equivalent of a cuvet), and a detector.

Sources

Incandescent or black-body sources as used in the infrared emit far too little power in the microwave region to be useful.[1] A radiofrequency arc is a white

Table 1. Conventional Waveguide Bands

Designation EIA[a]	Band[b]	Inner dimensions (in.)	Wavelength range (cm)	Frequency range (Ghz)
WR-137	J	1.372 × 0.622	3.66 –5.66	5.30 –8.20
WR-90	X	0.900 × 0.400	2.42 –3.66	8.20–12.4
WR-62	P	0.622 × 0.311	1.67 –2.42	12.4 –18.0
WR-42	K	0.420 × 0.170	1.13 –1.67	18.0 –26.5
WR-28	R	0.280 × 0.140	0.749–1.13	26.5 –40.0

[a] EIA = standards adopted internationally by the Electronics Industries Association.
[b] The letter designations vary somewhat between manufacturers and in various fields of application; those given are as used by Hewlett-Packard.

source in this area, but it also does not have sufficient power in any small wavelength interval. Fortunately we do not need to depend on monochromatizing a white source, as we have available a number of electron devices which can be made to oscillate monochromatically, and which can be adjusted to any wavelength over a considerable range.

The earliest of these to be used was the *magnetron*, a vacuum tube in which the trajectories of electrons are affected by a constant magnetic field in such a way as to cause the electrons to give up part of their kinetic energy to oscillating electromagnetic fields in a cavity resonator. The magnetron is capable of generating high power at frequencies up to about 25 Ghz, but it is difficult to tune, and inconvenient to use because of the heavy magnet required.

Better suited to microwave spectroscopy is the *reflex klystron*, the basic structure of which is shown in Figure 4. Its action is best explained by assuming that oscillating fields already exist in the metallic resonator cavity (which also acts as the anode).

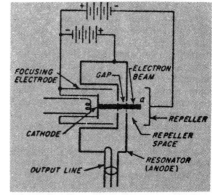

[From Ref. (13) by permission, McGraw-Hill Book Co.]

Figure 4. Schematic representation of a reflex klystron oscillator. As shown here it is connected to a coaxial transmission line ("output line"), but it could equally well connect to a waveguide.

Electrons emitted by the cathode pass through a central hole in the walls of the resonator. The rapidly alternating electric field in the cavity causes a potential between the opposite walls which alternately accelerates and decelerates the

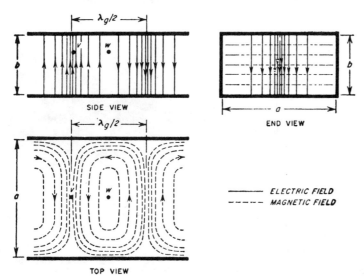

ELECTRIC FIELD
MAGNETIC FIELD

[From Ref. (13) by permission, McGraw-Hill Book Co.]

Figure 3. Field configuration of the dominant mode in a rectangular waveguide. The symbol λ_g refers to the wavelength in the guide, which is slightly different from the wavelength in free space.

electrons passing through. Hence some electrons move out of the gap with a higher velocity than their immediate predecessors, and soon catch up with them, forming a bunch. All the electrons are repelled or reflected back through the gap again by the negatively charged repeller electrode. If the anode and repeller potentials are correctly chosen, each "bunch" of electrons will arrive back at the gap at just the right phase point in the cycle that they are slowed down by the field, and hence deliver energy to maintain the cavity in oscillation. The dc voltages required by the klystron are rather critical, and must be carefully regulated.

To be more versatile, a klystron may be designed so that part of the resonating cavity is outside the vacuum envelope. It can then be tuned by means of a plunger which changes one dimension of the cavity. The repeller potential must be changed simultaneously to maintain proper phase relations. The widest tuning range attainable is about 2:1, but in most commercial versions it is less than this, which constitutes the major limitation of the klystron in spectroscopy. Figure 5 shows a commercial klystron.

The most recently developed microwave source is called the *backward-wave oscil-*

(*Courtesy of Varian Associates.*)
Figure 5. A typical klystron. The waveguide connects on the back. Note the threaded plunger for mechanical tuning.

lator (BWO). This is a highly specialized vacuum tube which can be made in a variety of physical forms, all depending on the same operating principle.

Consider a rectangular waveguide repeatedly folded back on itself, as in Figure 6, with a beam of electrons passing from left to right through a series of holes in the guide. Assume that a wave appropriate to the dimensions of the guide is passing through it in the right-to-left net direction ("backward" with respect to the electron beam). The electrons passing through the first gap are alternately aided and retarded by the electric field, which it will be recalled is always directed *across* the narrow dimension of the waveguide. So just as in the klystron, the electrons tend to become bunched. Since it takes the wave a longer time to get from one gap to the next than it does the electron, when the electron gets to the second gap, it finds a wave slightly lagging in

phase with respect to the wave it encountered at the first gap. Because of the fold in the waveguide, the direction of the field relative to the motion of the electron is effectively just short of one *half* cycle rather than of a whole cycle. The result is that the electrons experience a slight deceleration in crossing the second gap, but the bunching effect is increased.

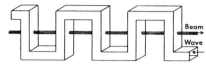

[*Redrawn from Ref. (13) by permission, McGraw-Hill Book Co.*]
Figure 6. Schematic diagram of a BWO employing a folded waveguide structure.

The process is repeated at each gap (perhaps 50 in all), the electrons become more highly bunched, and the phase lag becomes greater. The resulting repeated deceleration of the electrons causes the transfer of a considerable fraction of their kinetic energy to the oscillating fields. This power feedback is the source of the

oscillations. The frequency generated can be shown to be

$$\nu \cong \frac{[1 - (1/n)]}{2(T_e + T_w)}$$

where ν is the frequency of the microwave (in Ghz), n is the number of gaps in the folded waveguide, T_e is the average time of transit of an electron from one gap to the next (in secs), and T_w the corresponding time of propagation of the wave. If $n \gg 1$, then

$$\nu \cong \frac{1}{2(T_e + T_w)}$$

The values of n and T_w are determined by the geometry of the tube, and hence are constant. T_e can be varied by changing the accelerating potential applied to the electron gun. This means that the BWO can be tuned by control of a dc voltage, a major advantage. One manufacturer reports that he can cover the 8 to 40 Ghz span with four interchangeable BWO

tubes, where ten klystrons were formerly required.

Figures 7 and 8 show a modern BWO tube, schematically and photographically. In this tube the folded waveguide is replaced by a helix (index number 5 in Figure 7) of metallic tape. A microwave is conducted along the helix from right to left, and, provided that the circumference is between one-quarter and one-half the wavelength, the electrical potential between successive turns will alternate so as to play the same role as the successive gap potentials in the folded-guide structure. For most efficient operation the electron beam must be annular in cross section, passing as close to the turns of the helix as possible. The electron gun, consisting of a cathode (1) with its internal heater, a control grid (3), and an anode (4), is specially shaped to produce such an annular beam. The microwave energy from the helix is transferred to waveguide fields, and passes out through the guide window shown at the top (9).

There are several semiconductor devices known which can be made to produce sustained oscillations in the microwave region.

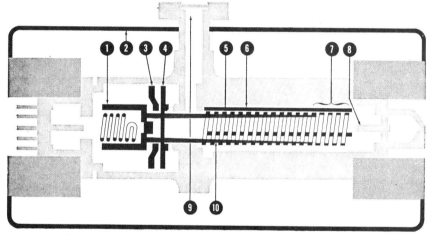

(*Courtesy of Varian Associates.*)
Figure 7. Schematic representation of a helical BWO; (1) is the cathode, (2) a magnetic shield, (3) the control grid, (4) a pierced anode, (5) low-voltage helix, (6) ceramic support rods, (7) the termination region, (8) a grounded element to collect stray ions, (9) waveguide flange, extending through the vacuum envelope, (10) hollow (annular) electron beam; electrical connections are made through the prongs to the left.

(*Courtesy of Varian Associates.*)
Figure 8. Outward appearance of the BWO diagrammed in Figure 7. Note the waveguide flange.

These include the tunnel (Esaki) diode, the Read diode, and the Gunn-effect device. These have the same advantages compared to klystrons and BWO's that transistors have relative to their vacuum-tube counterparts: more efficient use of power, lack of appreciable heat dis-

Chemical Instrumentation

sipation, small size, and no need for vac-uum-tight construction. They can be made to produce as much microwave power as a small klystron. So far as is known to the writer, none of these devices has yet been applied to spectroscopy, but judging by the history of other semicon-ductor devices, they probably will become the sources of preference for this purpose. For an introductory discussion, particu-larly describing the Gunn device, see the article by Bowers cited in the bibliog-raphy (2).

The output from either a klystron or a BWO can readily be introduced into a waveguide system. Since the signal is essentially monochromatic, no component is required analogous to the monochro-mator (entrance slit, prism or grating, and exit slit) of an optical spectrophotometer.

Detectors

By far the most common type of detec-tor for microwaves is a simple silicon crys-tal diode. It is conveniently mounted on an insulating stud, with a short, stiff wire projecting into the waveguide like an antenna, about one quarter-wavelength from a closed end. The diode is inherently a nonlinear device, and is selected to give an output proportional to the square of the electric field which it senses in the waveguide. Hence its indication is di-rectly proportional to power.

Occasionally a bolometer, also called a barretter, is employed as detector. This measures the heating effect of the field, which is proportional to the square of the current, hence also a measure of power. The silicon diode is much less expensive, and has a smaller time constant, so that it can respond quicker to changes in micro-wave level.

Spectrometers

There are on the American market two fairly sophisticated microwave spectro-meters. In addition, many investigators have built their own instruments, largely from commercially available components. Thus the field is in a somewhat similar situation to that of infrared absorption a generation ago.

Before describing the two commercial spectrometers, we will lead up to them gradually by discussing some simpler instruments which have been found useful. One of the simplest is diagrammed in Figure 9. It consists of a section of wave-guide acting as an absorption cell, con-nected through mica windows to a kly-stron or BWO source at one end and to a crystal detector at the other. Linked to the source through a special waveguide section which withdraws some of the energy from the main guide is a precision cavity wavemeter (frequency meter) and a second crystal detector. The power sup-ply for the microwave generator includes a sawtooth sweep oscillator. The sweep voltage is applied to the appropriate

² Actually it will be somewhat curved because of the inequality of power output of the oscillator at different frequencies.

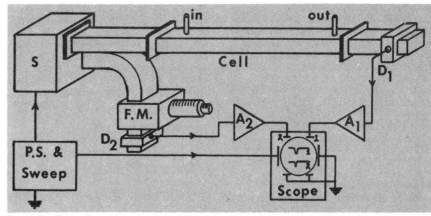

Figure 9. "Video" spectrometer, schematic. S represents the source; F.M., a frequency meter; P.S., the power supply which incorporates a sweep generator; D₁ and D₂, two crystal detectors; A₁ and A₂, two wide-band amplifiers, each connected to one set of vertical deflection plates in the dual oscillo-scope.

electrode of the oscillator tube to vary its output frequency over as wide a range as desired or as feasible, and at the same time to the horizontal deflection plate of a dual-beam oscilloscope. The rectified signal from the primary crystal, after amplification, is impressed upon one pair of vertical deflection plates. If no absorb-ing gas were present in the cell, the trace produced by this signal would be a hori-zontal straight line,² but if absorption oc-curs at one or more particular frequencies, a corresponding number of sharp dips will appear in the trace, as shown in the figure. The signal from the secondary crystal, similarly amplified, is applied to the other pair of vertical deflection plates of the dual oscilloscope, to produce a trace which shows a single sharp dip, usually called a "wavemeter pip." Manual adjustment of the wavemeter cavity by means of its calibrated micrometer screw will cause the pip to move across the screen. When it lines up exactly with a dip in the primary trace, a reading of the wavemeter indicates the frequency of absorption in the sample. This simple system can give excellent frequency precision, provided the absorption maximum is a strong one, but it is not very sensitive.

Modulation

The sensitivity of a microwave spec-trometer can be greatly increased by modulating the wave, preferably some-where in the range of 10–100 khz. The advantages of this procedure are com-parable to the advantages of chopping the beam of radiation in an optical spectro-photometer.

Perhaps the most obvious way to modulate the wave is by applying an alternating potential in the khz region to the control element of the oscillator tube. Then the output of the crystal de-tectors can be amplified by means of ampli-fiers tuned at the modulating frequency. This has been done, with a resulting hundredfold improvement in sensitivity. There are disadvantages which make it quite inconvenient to search for absorp-tion lines, and as this method is little used, we will not describe it further.³

Another method of modulation, which has nearly pushed all others out of the picture, is based on the *Stark effect*. This is the effect upon a rotational absorption

line produced by the application of an in-tense, uniform, dc, electric field parallel to the electric vectors of the microwave being absorbed. The line is shifted or split into a number of component lines related in a known manner to the rotational and magnetic quantum numbers, J and M. The number of lines produced by the Stark effect may be equal to J, to $J + 1$, or to $2J + 1$, depending on the symmetry type of the molecule. The spacing of the Stark lines, or the shift if only one is produced, is dependent on the magnitude of the applied field.

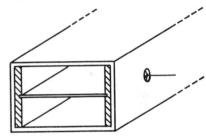

Figure 10. Geometry of a waveguide, showing septum for Stark modulation. The septum is energized through the wire connection to the right.

The application of the Stark effect to modulation is accomplished by the use of an absorption cell waveguide with a central metallic septum parallel to its broader side (Fig. 10). The septum is electrically insulated from the walls of the waveguide by inert spacers. Provision is made (not evident in Figure 10) for the circulation of the sample gas both above and below the septum. The septum is electrically con-nected to a square-wave generator in the 10–100 khz frequency range, at a potential high enough to produce a field of several hundred volts per cm. The square-wave is grounded through a diode ("clamped to ground," the electronic expert would say) so that the waveguide septum is alter-nately at ground and at a high positive potential. When the septum is grounded, the normal absorption peak appears,

³ A similar, though not identical, system was reported recently by Johansson (9) who found it valuable as a detector in gas chromatography. He used a klystron at 9.7 Ghz, modulated at 50 hz.

Figure 11. Possible appearance of oscilloscope screen: at left, an absorption band, un-modulated; center, with Stark modulation but simple detection; at right, Stark modulation with phase-sensitive detector.

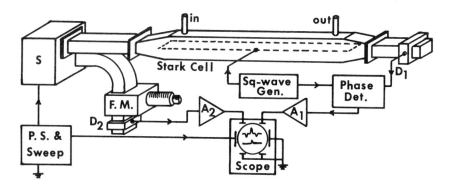

Figure 12. Elementary spectrometer with Stark modulation and phase detection. Symbols as in Figure 9.

while during the half-cycles when the Stark field is ON, the unmodified absorption is replaced by the Stark components. If the output of the absorption cell were detected and impressed directly on an oscilloscope screen, as in Figure 9, all components would be visible simultaneously, as neither the screen nor the observer's eye could follow the rapid ON-OFF changes.

This confusing situation can be simplified by means of a phase-sensitive detector. This is a network of diodes, resistors, and capacitors arranged to combine the signal from the waveguide crystal with a reference signal taken from the modulating generator, so that the signal coming from the guide while the Stark voltage is OFF will be displayed upward on the 'scope, and that arriving when the Stark voltage is ON will be displayed downward. Figure 11 shows the type of response to be expected for an absorbing molecule with $J = 1$.

The elements of a simple Stark modulated instrument are shown in Figure 12. Note that the Stark absorption cell is of somewhat larger size waveguide than the rest of the system, to reduce attenuation losses.

The great advantage of modulation, by whatever technique, is the reduction in noise (or enhancement of signal-to-noise ratio) which it makes possible. The key is the use of amplifiers responsive only to signals at the modulation frequency. This increases the sensitivity greatly, but for intense absorptions, the selectivity of the unmodulated spectrometer is still superior.

Double-Beam Spectrometers

It is possible to construct a microwave spectrometer utilizing the double-beam principle so well known in optical spectrophotomers. The power from the source is split between two identical sections of waveguide, one of which contains the sample, then recombined. The recombination is carried out in a microwave device called a "magic T" because of its unexpected and unusual properties (not all of which are utilized here). The magic T accepts the two waves through two of its arms, and emits through a third arm a wave with amplitude given by the *difference* in amplitude of the two entering waves. This means that a signal will appear *only* at a point where an absorption is taking place in the sample. The signal is best measured by the superheterodyne method. It is combined with a wave from a secondary oscillator to give an intermediate beat frequency, for example at 30 Mhz, which can be amplified in a tuned amplifier, then rectified and applied to an oscilloscope or a recorder.

Zeeman Modulation

The Zeeman effect is the magnetic analog of the (electric) Stark effect, and it can be utilized in an analogous manner to modulate a microwave absorption process. It is easily applicable to substances with a marked magnetic dipole, particularly the paramagnetic gases and free radicals. For other molecules, much stronger magnetic fields are necessary.

Zeeman modulation is readily produced in the paramagnetic samples by winding a helix around a waveguide. The waveguide must be constructed of a nonmagnetic metal, and if free radicals are to be studied. must be lined with glass or other material which will not catalyze recombination, The helix is energized from an oscillator, just as in the case of Stark modulation. If a square wave is required, the frequency must be comparatively low, because the inductive effect of the helix makes difficult the production of a sharp square wave. The other parts of the spectrometer can be identical with the Stark instruments.

When strong magnetic fields are required, the waveguide can be coiled between the poles of an electromagnet. Large magnetic fields cannot be made to alternate at frequencies high enough to be effective as modulation, so the instrument becomes, in effect, an electron-spin resonance spectrometer.

A novel spectrometer has been described by Radford (*11*), based on a form of Zeeman modulation. Radford has eliminated the use of a waveguide as a sample cell by transmitting the microwaves as a beam in free space from a transmitting antenna to a receiving antenna, some 30 cm or more distant. The sample to be studied is held in, or flows through, a glass container in the space between the antennas.

Modulation is achieved by a technique which Radford calls "field spinning." The sample cell is placed within two pairs of orthogonal Helmholtz coils. The two pairs of coils are fed 90° out of phase at 50 khz, which produces a magnetic field rotating at that frequency. So the *direction* of the field is varied periodically rather than its *magnitude* as in conventional modulation techniques. The instrument was tested by observing the absorption bands of the OH free radical. For further details, the original paper should be consulted.

The Hewlett-Packard Microwave Spectrometer

The H.-P. Model 8400[4] is a completely integrated spectrometer system covering the 8–40 Ghz region with four inter-

[4] Hewlett-Packard, Palo Alto, Calif. 94303.

(*Courtesy of Hewlett-Packard.*)

Figure 13. H.-P. Model 8400 Microwave Spectrometer.

Chemical Instrumentation

changeable BWO sources. A photograph of the instrument is shown in Figure 13 and a block diagram in Figure 14. The right-hand panel in the photograph contains the controls and inlet ports for the vacuum pumps and sample-handling system. Three glass stopcocks facilitate introduction of samples or flushing gas. In the section directly in front of the operator's table can be seen several sections of waveguide with panel controls for power level and calibration. To the operator's left are the controls for the microwave source and monitoring equipment together with the read-out oscilloscope and recorder.

The operation of the instrument can be followed with the aid of the simplified block diagram of Figure 14. Microwave power is taken from the BWO, through a manually adjustable attenuator, A-1, to the Stark cell and detector. As in any single-beam spectrophotometer, it is desirable to maintain the source at a constant power level. In the present instrument, this is accomplished by a leveling loop which consists of a thermistor detector feeding a power meter which in turn regulates the power output of the oscillator.

The 6-ft Stark cell is modulated by a variable-voltage square wave at a frequency of 33.333 khz, which also serves as reference for the phase-sensitive detector. A unique feature of this spectrometer is the "signal calibrator." This unit, shown at the top of the diagram, takes some of the power from the BWO, modifies its form, and returns it to the main line just ahead of the Stark cell. The side path includes an attenuator, A-2, identical with A-1, a modulator operating at the same 33.333 khz frequency, and finally an adjustable phase shifter. The modulator in this path does not operate on the Stark principle, but rather through a special silicon diode called a PIN diode, because its p- and n-type regions are separated by a layer of intrinsic silicon. The PIN diode acts as a normal diode to frequencies below about 100 Mhz, but above this, including the microwave region, it does not rectify, but acts as a high resistance. When the 33.333-khz modulating signal is applied, the diode (which is shunted across the waveguide) alternately acts as though it were a short circuit and an open circuit, and the microwave is thereby completely modulated. To use the calibrator, the variable attenuators and phase shifter are adjusted so that the simulated signal which is produced exactly matches the signal produced by Stark modulation of an absorbing sample. This amounts to adding just enough power to the line to exactly offset the absorption by the sample. One result of this is that the crystal detector operates at a relatively high power level during a measurement, which improves the signal-to-noise ratio and hence the sensitivity.

The operating frequency of the microwave source is continuously displayed on an electronic counter, the output of which can be used to place frequency markers directly on the recorder chart.

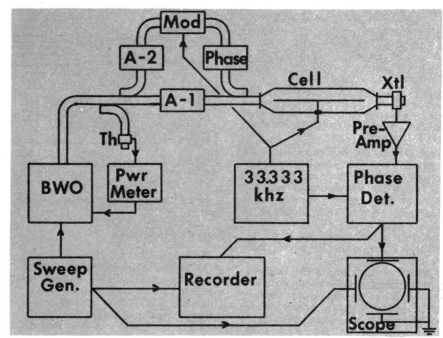

(*Redrawn from complete diagram, courtesy of Hewlett-Packard.*)

Figure 14. H.-P. Model 8400 Spectrometer, simplified block diagram. "Mod" refers to a PIN modulator in the calibration section; A-1 and A-2 are identical attenuators; "Phase" is a manual phase shifter; Xtl is the detecting crystal; Th is a thermistor detector in the power-leveling loop. For details of operation, see text.

The Tracerlab Microwave Spectrometer

Tracerlab[5] offers three spectrometers with designated model numbers, 1001, 3001K, and 4001K, of successively greater degrees of sophistication. The respective frequency stabilities, for example, are given as 10^{-3}, 2×10^{-5}, and 1×10^{-6} per hour (all are better if measured over short terms); attainable sensitivities are $\alpha = 10^{-9}$, 10^{-9}, and 5×10^{-10} cm^{-1}.

In addition they offer an extensive line of components which can be assembled into any of a number of configurations, to meet the requirements of a particular customer's application. Figure 15 shows the appearance of one such assembly. Some of the available features are given in Table 2.

[5] Tracerlab Division of the Laboratory for Electronics, Inc., Waltham, Mass. 02154.

(*Courtesy of Tracerlab Division of Laboratory for Electronics.*)

Figure 15. Tracerlab Model 4001KKA Microwave Spectrometer.

Table 2. Tracerlab Options[a]

Sources: (1) BWO's, 8–90 Ghz, (2) klystrons, 8–185 Ghz, (3) klystron with diode frequency multipliers, (4) quartz crystal oscillator with frequency multipliers for fixed frequency operation in the range 8–60 Ghz.

Stabilization of Source: (1) unstabilized sweep, (2) high-Q cavity, (3) phase-locking to a high-stability reference oscillator.

Frequency Read-out: (1) Wavemeter, (2) digital counter, (3) selectable marker tied to display unit, (4) frequency programming.

Display: (1) oscilloscope, (2) strip-chart recorder, (3) digital voltmeter.

Receivers: (1) video crystal with short time-constant, (2) long time-constant crystal, (3) superheterodyne system.

Cells: (1) Simple, (2) Stark, with modulating field parallel to or perpendicular to the electric vector, (3) single or dual beam, (4) in-line reference cell, (5) temperature controllable from $-80°$ to $+300°C$, (6) material: copper, steel, brass, aluminum, silver, (7) length up to 8 ft.

Vacuum and Sampling System: (1) pressure stabilized, (2) static sampling, (3) dynamic (flow) sampling.

Modulation: (1) Stark, 100 khz square wave or sine wave up to 2000 volts, (2) dual-frequency Stark, (3) frequency-modulation.

[a] Taken from the manufacturer's bulletin, with omission of a few sub-categories, which seem to the author irrelevant to the present treatment.

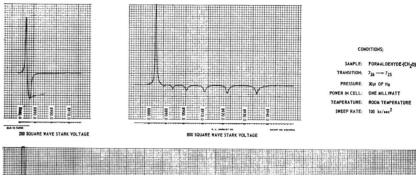

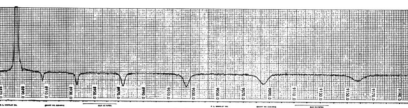

CONDITIONS:
SAMPLE: FORMALDEHYDE (CH_2O)
TRANSITION: $7_{26} \rightarrow 7_{25}$
PRESSURE: 30μ OF Hg
POWER IN CELL: ONE MILLIWATT
TEMPERATURE: ROOM TEMPERATURE
SWEEP RATE: 100 kc/sec^2

(Courtesy of Hewlett-Packard.)

Figure 16. Stark-modulated spectra of formaldehyde vapor, showing the effect of varying the Stark voltage.

Other Commercial Instruments

The only other commercial microwave spectrometer known to the author is a product of Japan Electron Optics. Their representative[6] states that this has not

[6] JEOLCO (U. S. A.), Inc., Medford, Mass. 02155; The parent company is Japan Electron Optics Laboratory Co., Inc., Tokyo.

yet been made available in the United States, and that they have no technical information concerning it.

Typical Spectra

A few representative microwave absorption spectra are shown in Figures 16–18. Details are given in the captions. No atlas of absorption spectra in the microwave region is at all up-to-date, but references to many spectra may be found, especially in references (*6*), (*12*), and (*14*).

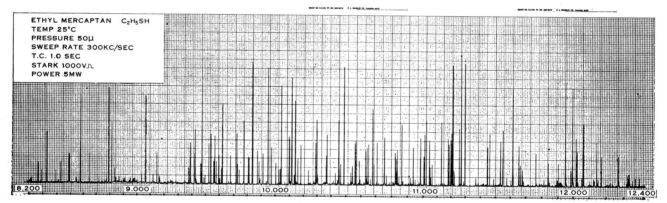

ETHYL MERCAPTAN C_2H_5SH
TEMP 25°C
PRESSURE 50μ
SWEEP RATE 300KC/SEC
T.C. 1.0 SEC
STARK 1000V⊓
POWER 5MW

(Courtesy of Hewlett-Packard.)

Figure 17. Stark-modulated spectrum of ethyl mercaptan.

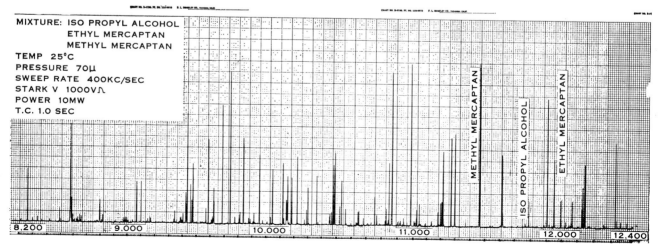

MIXTURE: ISO PROPYL ALCOHOL
ETHYL MERCAPTAN
METHYL MERCAPTAN
TEMP 25°C
PRESSURE 70μ
SWEEP RATE 400KC/SEC
STARK V 1000V⊓
POWER 10MW
T.C. 1.0 SEC

(Courtesy of Hewlett-Packard.)

Figure 18. Stark-Modulated spectrum of a mixture of three substances.

Microwave Absorption

Supplement

A number of reviews have appeared recently on microwave absorption or "molecular rotational resonance" (MRR) spectroscopy (*16–18*), as has one monograph (*19*). Extensive tables of microwave data are made available by the Bureau of Standards (*20*).

Applications

Microwave absorption has been found useful in several new directions. Since the resonant frequencies depend on the exact structure of the absorber, any slight change in the latter shows up in the spectrum. This provides a very sensitive tool for the study of isotopic abundances. Figure 19 illustrates this with a portion of the spectrum of *p*-chlorotoluene; each line appears as a doublet corresponding to the ^{35}Cl and ^{37}Cl isotopes. The relative proportions of the two species can be estimated with good precision.

Another area where highly precise measurement of dipole moments is fruitful is the study of conformational isomerism. The trans and gauche forms of *n*-propyl halides, for example, can be distinguished, and the height of the energy barriers separating the isomers estimated (*22*).

Microwave techniques provide nearly the only way to measure barriers to internal rotation of methyl and other small atomic groups in organic compounds. To date, more than 100 such barriers have been measured to an accuracy within a few percent (*16*).

Harrington (*23*) has introduced a new parameter with which to measure the intensity of an absorption line. His coefficient is defined as

$$\Gamma = \frac{P_0 - P}{b\sqrt{P_0}}$$

in which the symbols have the same meanings previously used in this article. This parameter is convenient for several

reasons. The spectrometer can be so adjusted that the signal is directly proportional to Γ. Furthermore, Γ can always be resolved into two factors, one of which is a linear function of the molecular density of the absorbing species. The other factor depends on the broadening of the absorption line, which we have seen to be pressure-dependent, and on the value of P_0.

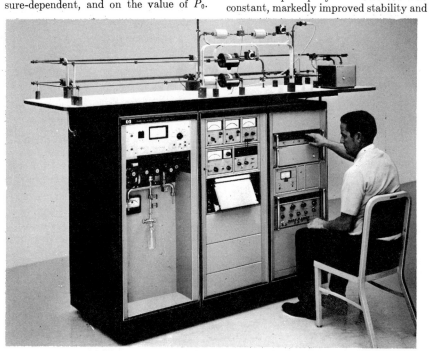

Figure 20.   Hewlett-Packard Model 8460A MRR spectrometer.

With the new parameter, it becomes possible to measure absorption at the maximum spectrometer signal, whereas previously it was necessary to reduce the power to a low level to prevent saturation. This development goes far toward increasing the usefulness of microwave absorption as an analytical tool.

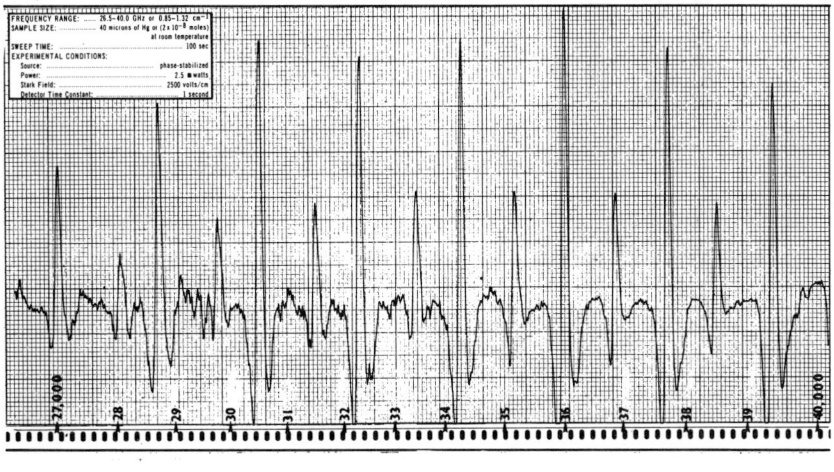

Instrumentation

The Hewlett-Packard Model 8400 spectrometer has been replaced by the 8460A, shown in Figure 20. Comparison with Figure 13 suggests the considerable simplification which has been incorporated. The waveguide system is now readily accessible on the table top. Frequency band switching is accomplished in a matter of minutes simply by interchanging RF units, with the sample retained in the Stark cell. The microwave circuitry of the new model is only slightly different from the old. The performance specifications differ principally in a decreased time constant, markedly improved stability and resolution.

Abridged microwave spectrometers, suitable for student experiments, have been built and operated successfully at Princeton and Michigan State Universities (*24*). Commercial instruments for this service are offered by Research Systems, Inc., Lexington, Mass., and by Hewlett-Packard. The Research Systems spectrometer makes use of a Gunn diode as source. It can be mechanically tuned over a 1 GHz range within the X-band, and then swept ±90 MHz around the selected point. This instrument is of modular construction, and sells for about $3000 without oscilloscope or recorder.

A home-built spectrometer with a Gunn diode source has also been reported (*25*).

Bibliography

The works listed below are those which have been most useful to the author in preparing this brief review. The list is not intended to be exhaustive, but will provide an entry into the field.

Figure 19.   Microwave absorption spectrum of *p*-chlorotoluene.   In each pair of peaks, the stronger is due to the ^{35}Cl isomer, the weaker to ^{37}Cl.   (*Courtesy of Hewlett-Packard.*)

128

(1) Barrow, G. M., "The Structure of Molecules," Benjamin, New York, 1964. (Paperback; an elementary account which places microwave spectroscopy in a good perspective compared to other tools.)

(2) Bowers, R., "A Solid-State Source of Microwaves," *Scientific American*, **215**, (No. 2), 22 (August 1966). (The Gunn effect and other semiconductor devices compared to the klystron.)

(3) Dailey, B. P., "Microwave Spectroscopy," in A. Weissberger, ed., "Physical Methods of Organic Chemistry," being Vol. I, Part IV, of A. Weissberger, ed., "Technique of Organic Chemistry," chapter XL, p. 2635; Interscience Div. of Wiley, New York, 1960. (A short discussion of the experimental and instrumental aspects of the subject.)

(4) Ginzton, E. L., "Microwaves," *Science*, **127**, 841 (1958). (An excellent introductory survey of the field and of its applications, primarily nonchemical.)

(5) Goldstein, J. H., "Microwave Spectrophotometry," in I. M. Kolthoff and P. J. Elving, eds., "Treatise on Analytical Chemistry," Part I, Vol. 5, chapter 62, p. 3233; Interscience Div. of Wiley, New York, 1964. (A good, brief, discussion of theory; the experimental part reflects largely the author's personal experience with partly home-built equipment.)

(6) Gordy, W., "Microwave and Radiofrequency Spectroscopy," in W. West, ed., "Chemical Applications of Spectroscopy," being Vol. IX of A. Weissberger, ed., "Technique of Organic Chemistry," chapter II, p. 71; Interscience Div. of Wiley, New York, 1956. (An excellent over-all treatment of theory, with no discussion of apparatus.)

(7) Gordy, W., Smith, W. V., and Trambarulo, R. F., "Microwave Spectroscopy," Wiley, New York, 1953. (Good background material, in spite of its age.)

(8) Harvey, A. F., "Microwave Engineering," Academic Press, London and New York, 1963. (Extensive, modern treatment of microwave devices and their operation; required reading for anyone planning to build his own spectrometer.)

(9) Johansson, G., *Anal. Chem.* **34**, 914 (1962). (Use of a source-modulated spectrometer as detector in gas chromatography.)

(10) Lide, Jr., D. R., "Microwave Spectroscopy," *Annual Revs. Phys. Chem.* **15**, 225 (1964). (A "state-of-the-art" discussion, primarily concerning applications, very little on instrumentation; bibliography of 171 references.)

(11) Radford, H. E., "Free Radical Microwave Absorption Meter," *Rev. Sci. Instr.* **37**, 790 (1966). (Describes a home-built Zeeman-modulated spectrometer.)

(12) Sugden, T. M. and Kenney, C. N., "Microwave Spectroscopy of Gases," Van Nostrand, London, Princeton and Toronto, 1965. (A good, up-to-date treatment of both theory and practice; contains a bibliography of published microwave spectra of 346 compounds.)

(13) Terman, F. E., "Electronic and Radio Engineering," McGraw-Hill, New York, 4th edition, 1955. (Good treatment of waveguides, microwave tubes, etc., but not of spectroscopy.)

(14) Townes, C. H., and Schawlow, A. L., "Microwave Spectroscopy," McGraw-Hill, New York, 1955. (This is a classic, if that word can be applied to so new a field, and is still the most comprehensive source, particularly in terms of theoretical background.)

(15) von Hippel, A. R., "Dielectrics and Waves," Wiley, New York, 1954. (A unified treatment of the interaction of radiant energy, including microwaves, with matter.)

(16) Wilson, E. B., Jr., *Science*, **162**, 59 (1968).

(17) Kirchhoff, W. H., *Chem. Eng. News*, **47**, #13, 88 (1969).

(18) Armstrong, S., *Appl. Spectrosc.*, **23**, 575 (1969).

(19) Wollrab, J. E., "Rotation Spectra and Molecular Structure," Academic Press, New York, 1967.

(20) "Microwave Spectral Tables," Natl. Bur. Stds. Monograph 70; several volumes are now in print.

(21) Trambarulo, R., and Gordy, W., *J. Chem. Phys.*, **18**, 1613 (1950).

(22) Hirota, E., *J. Chem. Phys.*, **37**, 283 (1962); Sarachman, T. N., *J. Chem. Phys.*, **39**, 469 (1963).

(23) Harrington, H. W., *J. Chem. Phys.*, **46**, 3698 (1967); **49**, 3023 (1968).

(24) Schwendeman, R. H., Volltrauer, H. N., Laurie, V. W., and Thomas E. C., *J. Chem. Educ.*, **47**, 526 (1970).

(25) Hrubesh, L. W., Anderson, R. E., and Rinehart, E. A., *Rev. Sci. Instrum.*, **41**, 595 (1970).

Three arrows bound together;
the sign of unity.

Topics in...

Chemical Instrumentation

Edited by **GALEN W. EWING**, Seton Hall University, So. Orange, N. J. 07079

...a

Chem Ed

feature

The Journal of Chemical Education
Vol. 45, page A467
June, 1968

XXXVIII. Refractometers

L. E. MALEY, *Anacon, Inc., Ashland, Mass. 01721*

Refractometry is a well-known but little understood technique of analytical measurement. A determination of the refractive index of a liquid is often made as a check on its composition if it is a solution, or its purity if it is a single compound. Likewise, the continuous measurement of a liquid flowing through a tube whether it be a process stream or the eluate from a chromatographic column can provide a direct indication of a change in composition or quality.

Relatively few of the refractometers being used today in the laboratory differ significantly from the original designs of Abbé (1874) and of Pulfrich (1887). In fact, these names have almost become generic terms to describe certain types of refractometers.

Theory

A ray of light which passes obliquely from one medium into another of different density is changed in direction when it passes through the surface. This change in direction is called refraction. When the second medium is optically denser than the first, the ray will become more nearly perpendicular to the dividing surface. The angle between the ray in the first medium and the perpendicular to the dividing surface is called the angle of incidence, i, while the corresponding angle in the second medium is called the angle of refraction (see Fig. 1). The refractive index, n, expresses the ratio of the velocity of light in the two mediums which form the boundary through which the light is passing. Snell's law expresses n as the ratio of the sine of the angle of incidence to the sine of the angle of refraction:

$$\text{Refractive index} = n = \frac{\sin i}{\sin r}$$

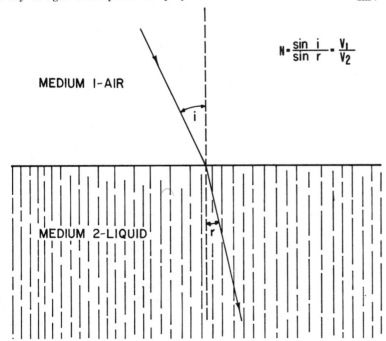

$$N = \frac{\sin i}{\sin r} = \frac{V_1}{V_2}$$

MEDIUM 1-AIR

MEDIUM 2-LIQUID

Figure 1. Basic principle of refractive index.

L. E. Maley is a graduate of Carnegie Institute of Technology with a B.S. in Electrical Engineering and of Duquesne University with a Doctorate of Laws.

In 1948, Mr. Maley joined Minneapolis-Honeywell, Inc. as a production and application engineer. He left Minneapolis-Honeywell in 1950 to join the Instrument Division of Mine Safety Appliances Co. From 1950 until June, 1960, Mr. Maley was in charge of the sales and application engineering of analytical instruments for Mine Safety Appliances Co.

Mr. Maley left Mine Safety Appliances Co. in 1960 to join Water Associates, Inc. Waters Associates was a new company embarking in the field of continuous refractometers. Mr. Maley was Vice President, Secretary and a Director of Waters Associates, Inc. until 1967 when he left Waters Associates to form Anacon, Inc.

Mr. Maley has been actively engaged in the design and application of analytical instruments and sampling systems for over 16 years. He has authored numerous recognized technical articles such as: "Analysis of Fractionation of Polymers," Journal of Polymer Science, No. 8, pp. 253–268; "Liquid Chromatography Monitoring with a New Refractometer," Pittsburgh Conference on Analytical Chemistry, March, 1961; "Design of Multiple Stream Sampling Systems," Control Engineering, November, 1961.

Chemical Instrumentation

It is normal practice to refer the index of refraction to that of a vacuum which is arbitrarily defined as having a refractive index of unity; reference to air introduces less than 3 parts in 10^4 error. The refractive index n, therefore, is a dimensionless constant whose value for light of a given wave length is determined by the character and state of the liquid or solid medium and of the reference medium, air. If the refractive index of various liquids or liquid solutions is to be compared with each other, then it is necessary to specify the state of the reference medium as well as to control other variables which affect the velocity of light in the sample to be measured. The symbol n_D^{20} for example, means the refractive index for the D lines of sodium measured at 20°C. Most refractive index values given in the handbooks have been measured under these conditions and are so referenced.

Types of Refractometers

Two types of refractometers are commercially available, the *differential* and the *critical-angle*. In the differential refractometer, a light beam is transmitted through a partitioned cell which refracts the beam at an angle which depends on the difference in refractive index between the sample liquid in one part and a standard liquid in the other. Differential refractometers can be accurate to 1×10^{-7} units of refractive index difference between the two liquids. In the critical-angle refractometer the light incident on the surface of the solution changes sharply from reflected to transmitted light at a critical angle. Several different versions of the differential and critical-angle refractometers are commercially available for the laboratory and for continuous process monitoring.

A. Critical-Angle Refractometers

Critical-angle refractometers measure the refractive index of liquids at an interface with air or, more commonly, with a glass prism. As shown in Fig. 2, a light beam is directed at the interface at various angles in the vicinity of the critical angle. The critical angle is the angle from the perpendicular at which the beam changes from light transmitted into the liquid to light totally reflected at the liquid surface. At angles smaller than the critical angle the light is transmitted into the liquid. The critical angle depends not only on the solution composition but also on the prism material. The refractive index can be calculated from:

$$i_c = \arcsin (n_g/n_l)$$
where

i_c = critical angle in radians within the glass

n_g = index of refraction of the glass

n_l = index of refraction of the liquid

The significant feature of a critical angle refractometer is that it measures the refractive index at the surface of a solution. Since surface reflection requires no penetration of the light beam into the

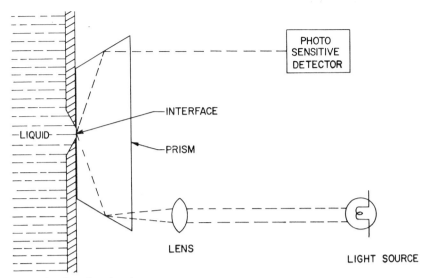

Figure 2. Critical angle refractometry.

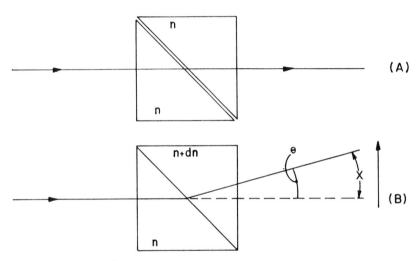

Figure 3. Differential refractometer.

solution, this instrument may be used for highly opaque samples and various murky solutions and suspensions, as well as transparent samples.

B. Differential Refractometers

Differential refractometers are intended primarily for the analysis of liquid mixtures. They are applicable to any mixture whose refractive index is a single-valued function of the composition; this includes nearly all simple binary systems. In refractive index measurements, the temperature control has always been a major problem. The designs of new instruments indicate that this problem is recognized and precautions are taken to minimize the temperature effect.

Each of the instruments employs a sample cell through which a light beam is transmitted. The beam is refracted through an angle whose size depends on the refractive index difference between the sample and a standard which constitutes a part of the cell. The angle of refraction is measured with a photoelectric pickup.

The applicability of a refractometer to the analysis of liquids is usually determined empirically. Samples of known chemical composition are passed through the instrument, and a calibration curve of composition versus refractometer reading is plotted. Various ranges and nominal values of refractive index can be accommodated by the various instruments, and some of them will provide any range down to $\Delta n = 10^{-5}$ at any nominal value.

The minimum detectable index difference for sustained periods of operation is $1/100$ of the range or 10^{-7} units of refractive index for the best of the new instruments. In the most precise instruments sensitivity is limited to this value mainly by mechanical deformation of the instrument caused by changes in ambient and sample temperature.

The principle of the differential refractometer is very simple. First consider the passage of a beam of light through a simple right-angled prism of refractive index n, and a second prism of identical composition in close proximity to the first but with the refracting surfaces reversed (Fig. 3a); the beam will pass through the combination without net refraction. If, however, the second prism is made of material having a very slightly different refractive index from the first, $n + dn$, then the beam will be deflected through the angle θ (Fig. 3b). If dn is very small, θ will also be very small and the distance x, measured at some arbitrary point along the beam, will be proportional to *dn*. Now replace the solid prisms with hollow ones filled with liquids having refractive indices *n* and *n + dn* and we have a simple differential refractometer.

The principle of the differential refractometer is not new and such instruments have been in use for plant control for a number of years, but the literature shows that there is a steadily increasing awareness of the potentialities of the technique, both for laboratory and plant use.

Operation of a typical differential refractometer may be better understood by referring to the schematic diagram (Fig. 4). A light beam generated by an incandescent lamp passes through the slit and mask that confine the beam to the region of the cell. The lens renders the light rays from the slit parallel. This parallel beam passes through the cell to the mirror. The mirror reflects the beam back through the sample cell to the lens which focuses it upon the edge of the beam-splitter mirror, M_2.

At this point, the location of the light beam depends upon the relative refractive indices of the sample and the standard.

As the index of refraction of the sample changes with respect to the standard liquid, the beam emanating from the lens shifts laterally with respect to the beam splitter. When this occurs, the amount of light reaching one photocell changes with respect to the other photocell and an electrical signal is generated in proportion to this difference. The beam deflector driven by the servo motor and its controlling amplifier-photocell circuits keeps the light beam falling equally on the two photocells. This occurs in the following manner. When light falls equally on the two photocells, there is no signal input to the amplifier. A shift in the beam, due to refractive index change of the sample, puts more light on one photocell, impressing a signal on the amplifier. The amplified signal drives the motor and the beam deflector in a direction that equalizes the light falling on the two photocells. The more the refractive index of the sample deviates from the reference, the greater the deviation of the light beam and the more the deflector must turn to balance the split light beams. The position of the deflector, therefore, is a measure of the refractive index differential.

Process Refractometers

Since the early days of World War II, recording refractometers have been used to monitor the purity of butadiene and styrene streams. In addition, process refractometers have been used to control the blending of these two feed components in the manufacture of GR-S rubber. The refractive indices of styrene and butadiene are approximately 1.5434 and 1.4120, respectively. A change of 0.1% is easily detectable by continuous refractometry.

Applications in the food industry involve materials such as soya bean and cottonseed oils having refractive indices of approximately 1.47. The finished products are shortenings and margarine with indices of approximately 1.43. Close control of the end point is required to obtain a product of the desired quality. A process refractometer makes this con-

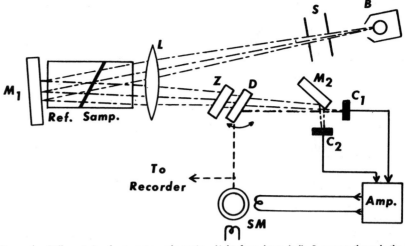

Figure 4. Differential refractometer, schematic. Light from lamp bulb B passes through the slit and mask assembly S, is collimated by lens L, then twice traverses the sample and reference prismatic cells, being reflected by mirror M. The exiting beam is focused by the same lens, and is displaced laterally as necessary by the glass blocks Z, the zero adjustment, and D, the balancing deflector. Part of the light is reflected by the oblique mirror M_2 onto cadmium sulfide photocell C_2, while part passes the edge of M_2 to impinge on cell C_1. The signals from C_1 and C_2 control a servomotor SM, via the differential amplifier. The motor acts to turn the deflector D so as to maintain equal illumination on C_1 and C_2, and at the same time operates a strip-chart recorder.

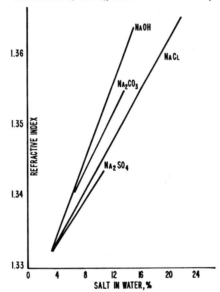

Figure 5. Salt in water.

trol possible without the delays required in making spot checks with laboratory instruments.

Many applications involve following the concentration of solutions. In the concentration of sugar solutions, salt brines and acids, in fact, almost any chemical solutions, the degree of concentration can be easily followed by the change in refractive index (Fig. 5). For example, the refractive index of a sucrose solution increases by about 0.0002 units per 0.1% increase of sugar concentration. It is a simple matter to detect such changes by process refractometry. It is also possible to detect changes as little as 0.02% in aqueous solutions of nitric acid. An even higher ultimate sensitivity is possible with sulfuric acid solutions (Fig. 6).

The separation of aromatics from saturates is an outstanding example of the possibility of using refractive index to follow a process, since the change in refractive index between aromatics and saturates is quite marked. The aro-

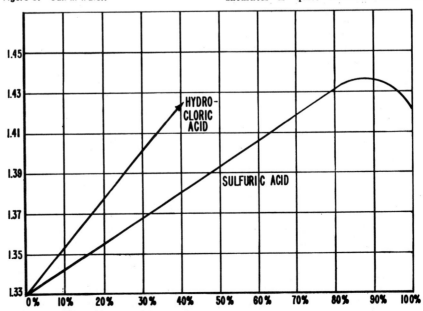

Figure 6. Sulfuric acid, hydrochloric acid.

Chemical Instrumentation

matics have a refractive index of about 1.50, and the saturates about 1.40. Since these measurements can be made to one in the fourth place, a precision of about 0.1% is easily obtained. Alcohol-water analyses are also carried out by this method. Water has a refractive index of 1.33299, while methanol is 1.32920, ethanol 1.36176, and isopropanol 1.37757 (Fig. 7). It is possible to detect changes of a few tenths to a few hundredths of 1% in the concentrations of these various alcohols in water.

Many applications of continuous refractometry are carried out in fractionating tower control. One example of such an application is the separation of cyclohexane and n-hexane. The respective refractive indices of these components are 1.42623 and 1.37486. This difference is sufficient to give measurements of composition changes of less than 0.1%. Another typical fractionating column application is the determination of normal butane in isobutane. These two components in the liquid phase have a refractive index difference of approximately 0.0100. It is possible to detect changes of the order of 0.1% in this case.

While refractometry has been applied to many process streams there was a limitation in the scope of these applications until the development of the in-line critical-angle refractometers. Many processes are amenable to monitoring and control by refractive index, but these streams may often be opaque or contain a large concentration of solid materials such as catalyst fines, pipe scale, undissolved solid product, polymers, etc. For example, in the manufacturing of high boiling petroleum products a relation between refractive index and quality exists. These streams are too dark to be analyzed by the differential refractometer but they can be easily handled with the critical-angle reflectance type refractometer.

The Anacon in-line refractometer (schematic diagram, Fig. 8) has the prism mounted in the pipeline itself. The prism is in intimate contact with the process fluid because its surface is flush with the wall of the pipeline. The entire optical system other than the prism is mounted in an explosion proof housing, clamped to the pipe. In operation, a light beam from an incandescent light bulb is directed through a lens and out the back of the steel tubing or pipe, the process fluid does not have to be removed from the pipeline and transmitted to some remote instrument. Product is measured under actual process conditions, i.e., at the same temperature and pressure conditions prevailing in the process. The short section of pipe in which the prism is mounted is provided by the manufacturer as a part of the instrument, and installation is made simply by flanging the section of pipe into the process line.

The light beam directed from the instrument housing is refracted at the interface between the prism and the process fluid and directed back through a lens and restorer glass on to a detector assembly consisting of two cadmium sulfide photocells. One photocell is located in the

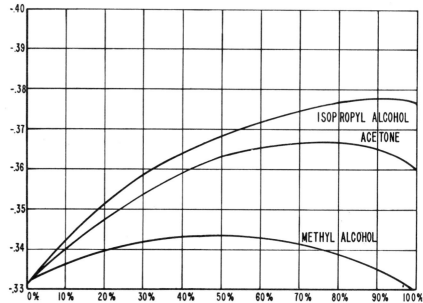

Figure 7. Alcohols and acetone.

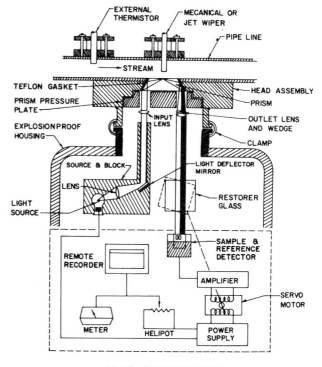

INLINE REFRACTOMETER
BLOCK DIAGRAM

Figure 8. In-line refractometer.

permanently illuminated section, and the other is mounted at the critical angle point, where the beam changes from light to dark. As the refractive index of the process fluid changes, the critical angle changes and causes more or less light to fall on the sample photocell. The other cell, of course, always remains in the full intensity portion of the beam.

Sample Conditions: Viscosity, Coating Action

Since a refractometer operating on the critical angle principle measures the index only at the interface, care must be exercised to make the material at the interface representative of the bulk solution. Agitation or turbulence in the process stream, of course, accomplishes

the required cleansing of the interface quite easily, but when agitation is small or when the solution is highly viscous, a film can build up on the prism surface and produce an erroneous reading. To eliminate this possibility, an automatic jet washing system is provided in the pipeline head.

Temperature Compensation

Another important factor that must be taken into consideration in critical angle measurements is the effect of a temperature change in the process stream. The refractive index of fluids, and of the prism itself, are sensitive to temperature changes. In a well-agitated line, window temperature quickly comes to equilibrium with solution temperature, and measurements can be very accurate and reliable. If the temperature is changing, however, some compensation must be made. A thermistor probe inserted in the pipe can measure the temperature of the process line and compensate for any temperature changes over a reasonable range. It is not possible to compensate over a range as large as 100°F (50°C), but it is possible to compensate for temperature changes of approximately 20°F (10°C). If the process stream is varying radically in temperature, then the head must be provided with thermostatic control.

Laboratory Refractometers

There are three common types of laboratory refractometers—the Abbé, the immersion or dipping, and the Pulfrich instruments. The Pulfrich uses monochromatic light, requires more of the sample than the Abbé type, and is therefore less commonly employed. These instruments have been used extensively in the laboratory for many years. Accurate measurements can be made over a fairly wide range of refractive indices. Some typical accuracies and ranges are shown in Table I.

Table 1. Maximum Accuracy and Range of Critical-Angle Refractometers

Refractometer	Maximum accuracy n_D	Typical ranges of n
Pulfrich[a,b,c]	$\pm 1 \times 10^{-4}$	1.33–1.61 1.47–1.74 1.64–1.86
Abbé[a–f]	$\pm 1 \times 10^{-4}$	1.30–1.70 1.45–1.84
Precision Abbé[d,f]	$\pm 2 \times 10^{-5}$	1.33–1.64 1.36–1.50 1.40–1.70
Dipping (immersion)[b,d]	$\pm 4 \times 10^{-5}$	1.32–1.54 (6 prisms)[d]

[a] Hilger, [b] Zeiss, [c] Bellingham & Stanley, [d] Bausch and Lomb [e] AO Inst. Co., [f] Valentine.

The Abbé Refractometer

The range of the Abbé is normally

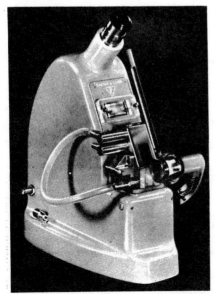

Figure 9. Abbé refractometer.

1.3000 to 1.7000, the maximum precision attainable being 0.0001. It reads the refractive index directly, and requires only a drop of sample. White light is used, and to prevent a colored, indistinct boundary between the light and dark fields two direct-vision prisms, called Amici prisms, are placed one above the other between the objective and eyepiece of the telescope. These are constructed of different varieties of glass and are so designed as not to deviate a ray of light of the wavelength of the sodium D line. Rays of other wavelengths are, however, deviated, and by rotating the Amici prisms it is possible to counteract the dispersion of light at the liquid interface. A typical Abbé refractometer is shown in Figure 9.

The Immersion Refractometer

This refractometer is the simplest to use, but requires 10–15 ml of sample. A single prism is mounted rigidly in the telescope containing the compensator and eyepiece as shown (Fig. 10). The scale is

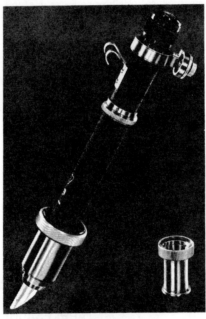

Figure 10. Dipping refractometer.

mounted below the eyepiece inside the tube. The lower surface of the prism is immersed in a small beaker containing the sample, with a mirror below to reflect light up through the liquid.

Pulfrich Refractometer

The Pulfrich refractometer is particularly useful for precise measurements with solutions or liquids that are volatile, reactive, or hygroscopic, with solid plates, for differential studies, and for dispersion measurements. Absolute values obtained with the Pulfrich refractometer are usually not reliable in the fifth decimal place, but a comparison at two wavelengths or of two substances can be made to $\pm 1 \times 10^{-5}$. Solids and liquids of index between 1.33 and 1.86 can be investigated over a wide temperature range. The refractive index n cannot be read directly, but must be determined from a set of observations.

Once the apparatus has been set up, a value of n can be obtained in about 5 min.

The critical boundary is particularly sharp in this type of refractometer even with liquid samples, since it is free from shielding of grazing rays and the full area of the refracting prism may be illuminated.

Direct Reading Critical-Angle Refractometer

A new direct reading laboratory critical-angle refractometer was recently introduced by Anacon. This instrument makes use of the same basic principle, but the change in the light beam is measured by a photosensitive detector and the refractive index value indicated directly on a meter. The Anacon direct reading refractometer is shown in Figure 11. The sample liquid is placed in a cavity where a synthetic diamond prism is mounted. The cavity is covered with a spring loaded, Teflon covered, cap which ensures that the liquid wets the prism surface. When the surface is wetted, the light-to-dark ratio of the critical angle image changes, and the photosensitive detector on which the image is focused changes its resistance. This change in resistance is electronically measured and a signal directed to the indicating meter.

Figure 11. Laboratory direct reading refractometer.

Refractometers in Liquid Chromatography

The chromatography of colorless liquid compounds has long been burdened with tedious methods of fraction collection and estimation. Various analytical techniques such as ultraviolet and infrared absorption, colorimetry, and refractometry have been adopted in an effort to obtain continuous flowing chromatographs which would greatly simplify fraction collection and analysis. An instrument which would continually monitor the refractive index of the column effluent has always appeared to be the most desirable for this application (Fig. 12). Various instruments along these lines have been described in the literature since Tiselius, Claesson and their collaborators first described the use of an optical system registering changes in refractive index based on the "Schlieren" method.

Chemical Instrumentation

Refractometry is appealing as a continuous monitor for liquid chromatography, and one can say that refractive index is to liquid chromatography as thermal conductivity is to gas chromatography. Refractometers will respond to any change in the liquid effluent and therefore

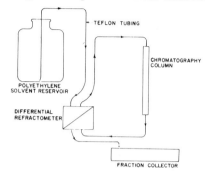

Figure 12. Liquid chromatography monitoring.

can be used as a universal detector. The other analytical techniques mentioned above are specific rather than universal and consequently are limited to particular classes of material. With such instruments it is necessary to know what material is being eluted from the column and at what wavelength it absorbs radiation or what reagent must be used to cause a color reaction. If the material does not absorb radiation at the frequency for which the detector is set the peak will be missed entirely or if the reagent is wrong a color change will not occur and again a peak will be missed. A refractometer responds to any change in the liquid effluent and is therefore a universal, highly sensitive detector.

When to Use a Refractometer

Whenever the sample to be analyzed is a simple binary mixture, such as alcohol in water, the first choice analyzer is a refractometer. Density-measuring analyzers are applicable when the range of compositions is broad. But when the range is narrow, down to about 0.4%, and certainly when an analysis of the liquid phase of a suspension or slurry is required, a refractometer again is first choice.

Batch Analysis: Ketchup, Jams, Jellies

There are also a variety of applications for batch analysis wherein very viscous food products are being cooked to a desired soluble solids content. A typical example is continuous measurement of soluble solids in ketchup (Fig. 13). Since this is a batch type operation, a sample is pumped from the bottom of the kettle through the in-line refractometer and back into the finishing kettle. When a product reaches the desired concentration the kettle can be drained and refilled with another batch. With this type of installation a continuous monitoring is made of the product as it is being prepared.

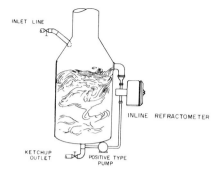

Figure 13. Soluble solids in ketchup.

Carbonated Beverages

Figure 14 illustrates a typical installation of an in-line refractometer at a carbonated beverage plant. Here the unit measures the amount of sugar in the beverage after it has been completely carbonated and just prior to bottling.

Some other applications of interest are measurement of glycerine-water mixtures, formaldehyde-methanol mixtures, synthetic rubber processes, and a variety of instant food products, including coffee and tea. An extremely important applica-

Figure 14. Carbonated beverage plant.

tion is in the hydrogenation of fats and oils. The analysis of these materials as well as standard lubricating oils in refineries has been done for many years with laboratory refractometers. It was felt that it would be impossible to analyze these products continuously, because of large quantities of entrained metal catalysts, such as nickel, platinum, etc. The maintenance problems involved in filtering and cleansing these sample streams for measurement by a differential refractometer or other analytical instruments presented such severe problems that analysis could seldom be made on a continuous basis. The in-line refractometer has been applied to these hydrogenation problems as well as to measurement of waxes in oils, measurement of catalytic cracking-unit feed stock, measurement of octane numbers of gasoline, and detection of interfaces in pipelines. It provides an immediate, simple analysis without a very elaborate, highly complicated, sampling system. In a number of cases the elimination of a sampling system in itself more than justifies installation of the refractometer.

The field of in-line measurements is becoming more and more important with the advent of completely controlled automatic processes. Figure 15 shows an automatic control system for applesauce processing and Figure 16 a monitor for a paper mill recovery boiler. To obtain true automatic control in many cases, analysis must be extremely fast and extremely reliable. The more important factors in analysis instrumentation are reliability and simplicity. In order to simplify the complicated analysis measurements it is virtually essential in many cases that the unit become an in-line measuring device. By measuring within the actual process, many of the problems of analysis are eliminated and the unreliability of complicated sampling systems is re-

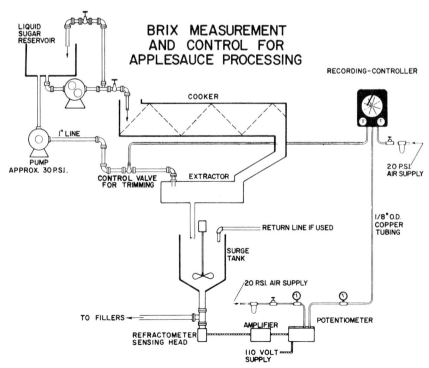

Figure 15. Applesauce control system.

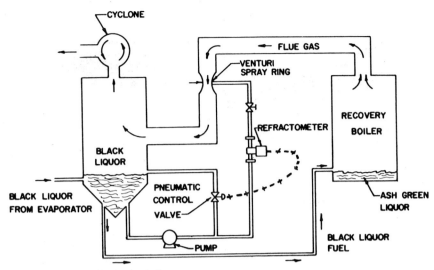

Figure 16. Paper mill recovery boilers.

moved. The in-line measurement provides a very high degree of reliability and a very high speed of response. It is anticipated that in the next few years we will see more and more analytical devices applied to direct in-line measurement.

In summary, refractometers have been used for almost 100 years for the measurement of liquid solutions. Continual improvements are being made, particularly in the continuous process type refractometers. No significant changes have been made in laboratory refractometers in recent years but the advent of solid state circuitry probably will cause some major changes in these instruments in the very near future. Refractometers are reliable, simple, analytical tools for the laboratory, continuous monitoring, and liquid chromatography.

MANUFACTURERS

Anacon, Inc.
62 Union St.,
Ashland, Mass. 01721

AO Instrument Co.
Eggert and Sugar Rds.,
Buffalo, N. Y.
14215

Bausch & Lomb, Inc.
77466 Bausch St.,
Rochester, N. Y.
14602

Bellingham & Stanley, Ltd. (England)
sold by Epic, Inc.,
150 Nassau St.,
New York, N. Y. 10038

Hilger & Watts, Ltd. (England)
sold by Engis Equipment Co.,
8035 Austin Ave.,
Morton Grove, Ill. 60053

Nester/Faust Mfg. Corp.
2401 Ogletown Rd.,
Newark, Del. 19711

Phoenix Precision Instrument Co.
3803–05 N. 5th St.,
Philadelphia, Pa.
19140

Valentine Instrument Co.
2184 Primrose Ave.,
Vista, Calf. 92083

Carl Zeiss, Inc.
444 Fifth Ave.,
New York, N. Y. 10018

Bibliography

BAUER, N., FAJANS, K., and LEWIN, S. Z., "Refractometry," being chapter 18 in Weissberger, A. (editor), "Physical Methods of Organic Chemistry," Pt. II, (3rd edition), Interscience Div. of John Wiley & Sons, Inc., New York, 1960.

WILLARD, H. H., MERRITT, L. L., JR., and DEAN, J. A., "Instrumental Methods of Analysis," chapter 14 (4th edition), D. Van Nostrand Co., Inc., Princeton, 1965.

JONES, A. G., "Analytical Chemistry: Some New Techniques," chapter 7, Butterworth, London and Washington, 1959.

Topics in...
Chemical Instrumentation

Edited by **S. Z. LEWIN,** New York University, New York 3, N. Y.

...a  feature

The Journal of Chemical Education
Vol. 42, pages A565, A619, A775, A853, A945
August to December, 1965

XXIV. Instrumentation for Micrometry and Microscopy—Part One

S. Z. Lewin, *Dept. of Chemistry, New York University, N. Y. 3, N. Y.*

Simple magnifiers of various kinds have probably been made purposeful use of for more than three thousand years. In the ancient near east, glass globes filled with water may have been used by engravers as an aid in their carving of cylinder seals and similar artifacts. Spheres of rock crystal were made by the ancients and may have served a similar purpose. The use of magnifying lenses for reading had already become common in the time of the Roman empire, and by the end of the sixteenth century in Europe, lens grinding was a well established and advanced art. The compound microscope was invented before the beginning of the seventeenth century, and its extensive use during that century was capped by the explanation of its theory given by Huygens in 1703.

Yet, despite this venerable record of utilization and investigation of optical magnifying devices, fundamentally new advances continue to be made in the design and application of this class of instrumentation. It is sobering to reflect that although the role of diffraction and interference phenomena in microscopy was well understood as early as 1873 (and employed then by Abbe to develop the concept of resolving power and numerical aperture), it was more than 60 years later that Zernike hit upon the idea of inserting a simple shield over the light source, and a thin film over the objective, to exploit interference in a manner that so improves the microscopic image as to give birth to a new field of microscopy (viz., phase microscopy) and to win for himself a Nobel Prize (in 1953). The very considerable time lag between the instrumental developments of Abbe and Zernike is a consequence of the remarkable subtlety and complexity of optical phenomena. The nature of light is still but dimly understood, and many dramatic and powerful new applications are doubtless in the offing.

Types of Magnifiers

The purpose of optical magnifiers is to present to the eye an image that is more readily perceived and amenable to measurement or study than the image which the unaided eye is capable of forming. The magnifier may be designed to produce an image upon the retinal surface of the observer's eye, in which case it is a *visual* instrument, or it may be devised to create an image on a photographic plate, a cathode ray tube, or a viewing screen, in which case it is a *projection* instrument.

The simplest magnifiers provide a *single stage of amplification*, i.e., there is a single lens system which forms an image of the object under study. Examples of such devices are the so-called reading glass, hand lens, pocket magnifier, loupe, and the "simple" microscope (as differentiated from the compound, or two stages of amplification, microscope). The most common form of these magnifiers consists of a biconvex lens in a suitable mounting, and may be found in a variety of sizes and shapes, as illustrated in Figure 1, A–C.

The Biconvex Lens

These lenses have spherical shapes, i.e., their surfaces are portions of spheres, as shown in Figure 2. The relationships between the locations and relative sizes of the object and image are shown in the usual idealized way in this figure. That is, the corresponding points of object and image can be located by drawing (a) a ray from the object parallel to the principal axis which is refracted by the lens through the focal point in the image space, and (b) a ray from the object through the center of the lens, which continues on without deviation. If R_1 and R_2 are the radii of curvature of the two sides of the lens, S_1 is the distance of the object from the center of the lens (XO), S_2 is the distance of the image from the center (OY), and n

is the refractive index of the lens, then the equation describing the respective distances is:

$$\frac{1}{S_1} + \frac{1}{S_2} = (n - 1)\left(\frac{1}{R_1} + \frac{1}{R_2}\right)$$

Thus, it can be seen that if the object is small, and is on the principal axis at infinity (i.e., so far to the left of the lens center that the rays coming from it can all be assumed to be parallel to the principal axis) the rays will come to a focus at $S_2 = 1/[(n - 1)(1/R_1 + 1/R_2)]$. If, as in the usual case, both radii are equal and $n = 1.5$, this reduces to $S_2 = R$. The closer the object comes to the lens, i.e., the smaller S_1 becomes, the larger must S_2 be; i.e., the farther away from the lens is the image formed. If the object is located at $S_1 = R$, the image is formed at infinity, i.e., no image is formed. Thus, a real (and inverted) image is formed only for objects located between the limits R and infinity from the lens center.

Magnification

The *lateral magnification* is defined as the ratio of corresponding lateral (i.e., perpendicular to the principal axis) dimensions in the image and object respectively. From a comparison of similar triangles in Figure 2, it will be seen that:

$$M = \frac{YY}{XX} = \frac{S_2}{S_1}$$

The relationship between S_1, S_2 and R given above permits us to rewrite this equation in the form:

$$M = \frac{R}{S_1 - R}$$

which shows that the lateral magnification approaches infinity as the image approaches R (while, as seen above, the image also approaches infinity). When $S_1 = 2R$, $M = 1$. Hence, for lateral magnifications in excess of $1\times$, the object must be closer to the lens than twice the radius of curvature. If S_1 is greater than $2R$, the image is reduced.

If the object is placed closer to the lens than R, a virtual, erect and enlarged image is formed, as shown in Figure 3. This image is *virtual* because the rays are divergent, and the image is seen only as a result of the mental extrapolation of these rays by the observer back toward the point from which they appear to come.

The general relationships between the location of the object vis-a-vis the lens center and the type of image formed are summarized graphically in Figure 4.

When a biconvex lens is used as a reading glass, hand lens, or loupe, it is usually this virtual image that is viewed. The

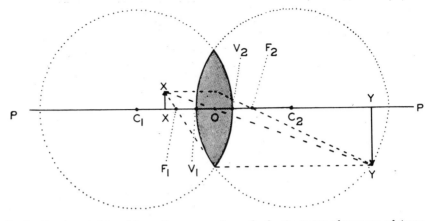

Fig. 2. The characteristics of the ordinary convex lens. C_1, C_2: the centers of curvature of the two spherical surfaces. PP: principal axis of the lens. O: Optical center of the lens. V_1, V_2: vertices of the lens. F_1, F_2: focal points. X,X: object. Y,Y: image (magnified and inverted).

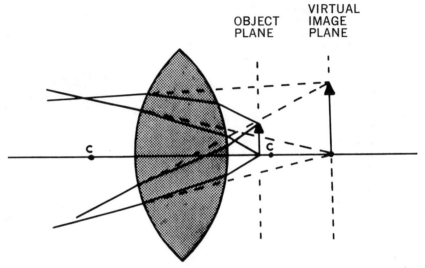

Figure 3. Formation of virtual, erect, enlarged image when object is closer to biconvex lens than the focal distance.

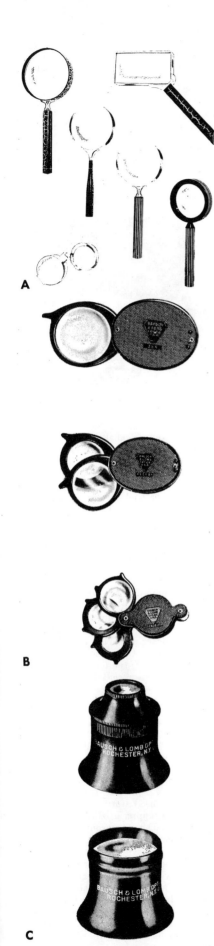

Fig. 1. A. Typical biconvex lens reading glasses (Edroy Products Co., N. Y.). B. Pocket magnifiers of the simple biconvex lens type. C. Watchmakers' loupes of the simple biconvex lens type.

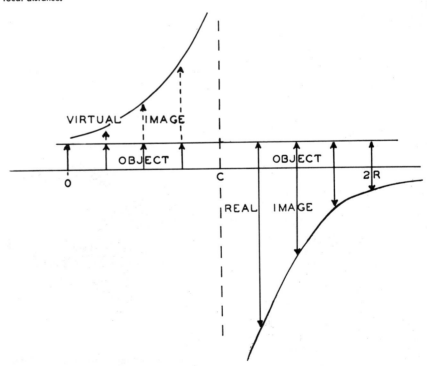

Figure 4. Schematic representation of the *relative dimensions* of image and object as a function of the *location of the object* with respect to the lens. The location of the image is not represented in the diagram. When the object is beyond the focal point, the image is real and inverted. When the object is before the focal point, the image is virtual and erect.

distance from lens to object is small, the magnification is modest, and there is considerable distortion of the lateral dimensions. In the case of reading glasses (Fig. 1 *A*), the focal distance is typically between 6 and 12 inches, the magnification 2–3×, and the price $4–$10. Sample watchmaker's loupes (Fig. 1 *C*) have focal distances in the range of 1–5 inches, with corresponding magnifications between 20× and 2×, and the cost ranges between $6 and $2. These lenses are available from most optical stores, as well as from Bausch and Lomb, Rochester 2, N. Y.; Ealing Corporation, Cambridge, Mass., and laboratory supply houses.

Depth of Focus

An important consideration in the use of these lenses is the *depth of focus* characteristic of the image. If the depth of focus is small, only those points of the object that lie on a given object plane can be seen sharply and in their proper relationship to each other at the image plane; points above and below this object plant are distorted or unsharp. The depth of focus of a lens may be expressed in terms of its *longitudinal magnification*. This is defined as the ratio of the longitudinal displacement (along the principal axis) of a point in the image resulting from a given longitudinal displacement of the corresponding point of the object. As illustrated in Figure 5, the longitudinal magnification is YY'/XX'. By means of the comparison of similar triangles, as in Figures 2, it can be shown that the longitudinal magnification, L_M, is

Figure 5. Longitudinal magnification is defined as the ratio of the longitudinal displacement produced in the image, YY', as a result of an arbitrary displacement of the object, XX'. $L_M = YY'/XX' = M^2$ where M is the lateral magnification.

equal to the square of the lateral magnification, M.

$$L_M = M^2$$

Thus, the longitudinal magnification increases faster than the lateral magnification. If the object is very flat and smooth, one can increase the lateral magnification as far as the lens, lighting, and working distance permit without excessive loss of clarity. However, if the object has relief, irregularities in elevation, etc., the image rapidly becomes fuzzy and indistinct as the lateral magnification increases. In such cases, the best compromise must be sought between enlargement and sharpness.

The focal length of a lens is often specified in terms of its reciprocal, designated *power*, or *dioptry* of the lens. The dioptry is equal to $1/f$, where f is the focal length in meters. For example, a lens of 7⅞ inches focal length has a power of:

$$\frac{1}{(7\frac{1}{8})\left(\frac{2.54}{100}\right)} = \frac{1}{0.20 \text{ meters}} =$$

5 diopters.

An increase in power is equivalent to a decrease in focal length (a smaller radius of curvature means a more highly curved lens surface). The dioptry of a converging lens is positive; that of a diverging lens is negative.

Applications of the Simple Biconvex Lens

The simple biconvex lens is very widely used in laboratory, plant, and field work. For example, this type of lens, with a focal length of 2–4 inches, giving magnifications of $2\times - 4\times$, is mounted at the end of a metal cylinder which is clamped onto thermometers, burettes, etc., as illustrated in Figure 6, to provide a parallax-free arrangement for precise reading of an interface or meniscus.

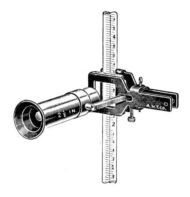

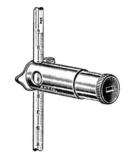

Figure 6. Mounted biconvex lenses for meniscus location in thermometers and burettes.

Two of these biconvex lenses, mounted one in front of each eye, to permit binocular viewing, as shown in Figure 7, represent a convenient extension of normal

Figure 7. The Magni-Focuser binocular magnifier.

stereoscopic visual range without interfering with the free movement of the user. These devices are available with focal lengths from 4 inches to 20 inches, the

corresponding lateral magnifications being from 3.5× to 1.5×. The example shown in the figure is the Magni-Focuser distributed by Edroy Products Co., N. Y. 17, N.Y.; the price ranges from $12.50 to $15, depending upon the focal length. A similar device is the Magna-Sighter of Fairchild Optical Co., Chicago 26, Illinois; magnifications of 1.75× to 2.75× are available at working distances of 14″ to 8″; a model giving 3.5× at 4″ is also produced; the prices are $9–$11.

An elaboration of this type of device is

Figure 8. The ZoomScope adjustable binocular magnifier.

the ZoomScope (available from Allcraft Tool and Supply Co., N. Y. 36, N. Y.; $9–$11). The binocular lens mount is fastened to the head-band by a pair of slotted side-members, so that it can be slid forward or backward relative to the plane of the eyes. In this way, the working distance can be varied, and consequently the magnification can be adjusted within a limited range. The instrument is illustrated in Figure 8.

A biconvex lens of large enough area that both eyes may view the work piece through it, is frequently used in place of the binocular headgear described above when only small magnifications are required. Such a unit is the Clear-Lite, made by Stocker and Yale Inc., Marblehead, Massachusetts. The lens dimensions are 3.5 × 6″, the focal length is 11 inches, and a fluorescent lamp is an integral part of the lens support.

Limitations of Simple Lenses

The application of simple lenses to precise measurements, or to magnifications in excess of about 10×, is severely limited because of the difficulties created by the phenomena of spherical and chromatic aberration. The nature of these difficulties may be appreciated by considering how a lens manages to produce an image of some object.

Spherical and cylindrical surfaces can be manufactured with precision and without excessive cost. Consequently, almost all of the precision optical equipment commercially available is based upon the utilization of such optical surfaces, and we will limit the following discussion to lenses the surfaces of which are segments of spheres.

If light of a single wave length, diverging out from a point source, as in Figure 9, impinges upon a spherical surface bound-

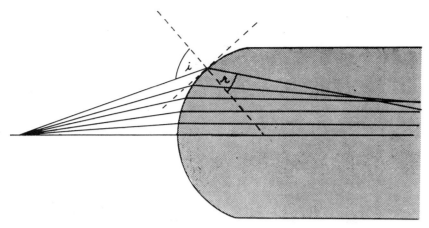

Figure 9. Illustration of the fact that the direction taken by light rays upon entering a medium is determined by the angle of incidence, *i*, and the refractive index of the second medium relative to the first. If the second medium is bounded by a spherical surface, the refracted rays are not brought to a focus.

ing a medium having a different (e.g., higher) refractive index, a substantial amount of the light is refracted into the second medium. Each ray strikes the spherical surface at a definite angle of incidence, i, measured from the normal to the interface and is refracted at an angle, r, determined by the refractive index law:

$$n_1 \sin i = n_2 \sin r$$

where n_1 is the index of refraction of the first medium (e.g., $n_1 = 1$ if the point source is in air), and n_2 is the index of the second medium (e.g., $n_2 = 1.5$ if the light is entering glass). In Figure 9, the direction of each ray in the second medium has been determined by measuring the angle of incidence and calculating sin r, taking $n_2 = 1.50$. It is evident that the light rays do not all intersect in a common point in the second medium; i.e., there is no focussing of the light beam. This is a general result; an *aspheric* (i.e., not spherical) surface would be required to bring the beam to a focus.

If a second spherical surface is added, as shown in Figure 10 *A*, a closer approach to the focussing of the refracted beam results, although the beam is nowhere in true focus. A screen placed at various positions beyond the lens (*A* to *E*) would show light spots of various diameters surrounded by halos, as indicated in Figure 10 *B*. In this case, the best approximation to a focus is at location *D*. This result is also general; the use of spherical lens surfaces is associated with a lack of sharpness in the focussed image.

This spherical aberration can be markedly reduced by decreasing the aperture of the lens; i.e., by limiting the illuminated portion to a small area at the center, as shown in Figure 11. However, this severely reduces the amount of light entering the lens, thus diminishing the visibility of the image. Reducing the aperture also results in loss of resolution in the image. Thus, "stopping-down" a lens is an expedient that can often greatly improve image clarity if the light intensity is high and the magnification is not great. This is commonly done in photography, for example. However, when high resolution is needed at high magnification, other means of removing the aberration must be sought.

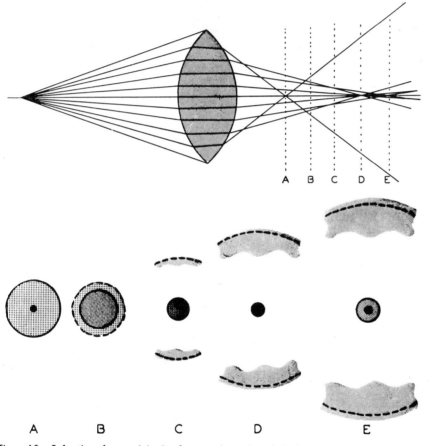

Figure 10. Refraction of rays originating from a point on the principal axis by a convex lens having two spherical surfaces. The rays refracted from the central region of the lens tend to focus at a greater distance from the lens than rays refracted by the marginal region. A, B, C, D, E represent screens placed at the indicated locations to reveal the size and light distribution in the beam.

Spherical Correction

A reasonably effective expedient consists in the construction of a composite system, made up of a converging lens mated to a diverging lens. Since spherical aberration results from the fact that the focal length for central rays is greater than it is for the marginal rays (Fig. 10), it is necessary to produce some divergence for the latter, without much affecting the convergence of the former. How this may be achieved is illustrated in Figure 12.

It will be noted that the combination of lenses can be designed to bring the marginal rays to the same focus, P_2, as the central rays, P_0, but then the intermediate rays come to a focus at a somewhat shorter distance, P_1. The distance P_0P_1 is called

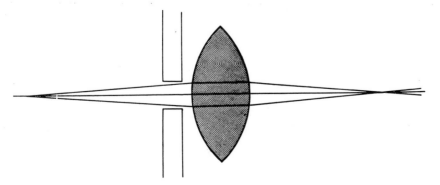

Figure 11. Stopping-down, or reducing the aperture of the lens of Figure 10, improves the sharpness of the focus.

the residual longitudinal spherical aberration; the best image is located between P_0 and P_1.

Off-Axis Spherical Aberration

The spherical aberration discussed so far has been limited to that which afflicts rays parallel to, or from points lying on, the principal axis of the lens. If one considers the problem of bringing to a focus the rays from a point that lies appreciably off the principal axis, then new difficulties arise. As Figure 13 shows, the lens which, when stopped down as in Figure 11, satisfactorily focussed the rays from an on-axis point, no longer yields even an approximate focus for an off-axis source of rays.

This form of distortion can be diminished by proper selection of the curvature and refractive index of the lens system. Abbe showed that a plane area in the object will be seen as a plane area in the image if the ratio of the sines of the angles made by the incident and refracted rays with the principal axis is constant. This condition is illustrated in Figure 14. Lenses that conform to this requirement are said to be *aplanatic*.

Rendering a system aplanatic in the fashion described above does not, however, eliminate other forms of distortion of the image. If the object has appreciable size, so that different parts of it are imaged by bundles of rays that make various

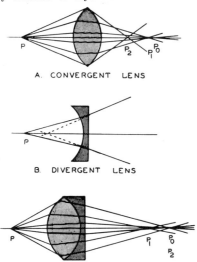

Figure 12. The spherical aberration shown in Figure 10 can be corrected for by combining a diverging lens with the converging lens. The central region of the lens combination now brings the rays to a focus at P_0, which coincides with the focus of the rays from the marginal region of the lens, P_2. However, now the rays from the intermediate region of the lens come to a focus at P_1.

angles of inclination with the center of the lens, *astigmatism, curvature,* and *coma* are produced.

Astigmatism arises from off-axis distortion of the focus, as illustrated in Figure 15. The rays, after passage through the lens, are focussed not to a point, but in two lines in perpendicular directions. The closest approach to an image is the so-called *circle of least confusion* midway between these lines. The greater the angle

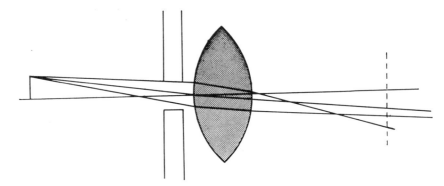

Figure 13. Astigmatic distortions are produced by off-axis points. The same lens and aperture arrangement as shown in Figure 11, which brought on-axis points to a good focus, do not focus an off-axis point. The dashed line is the plane of the focal point in Figure 11.

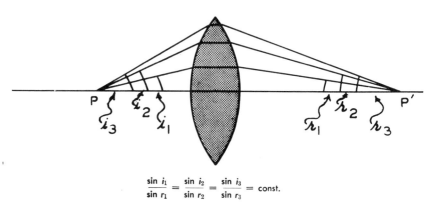

$$\frac{\sin i_1}{\sin r_1} = \frac{\sin i_2}{\sin r_2} = \frac{\sin i_3}{\sin r_3} = \text{const.}$$

Figure 14. The Abbe sine condition for an aplanatic system.

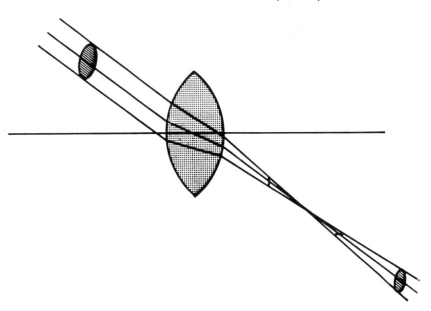

Figure 15. The appearance of astigmatic distortion due to refraction of an inclined beam.

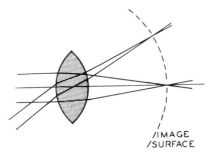

Figure 16. Illustration of astigmatism leading to the formation of a curved image.

between the axis of the lens and the beam of light, the more pronounced is the astig-

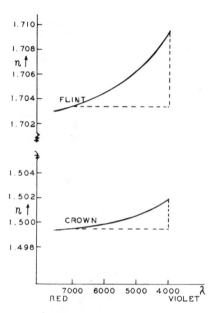

A. PINCUSHION EFFECT: SCREEN NEAR LENS.

B. BARREL EFFECT: SCREEN FARTHER AWAY.

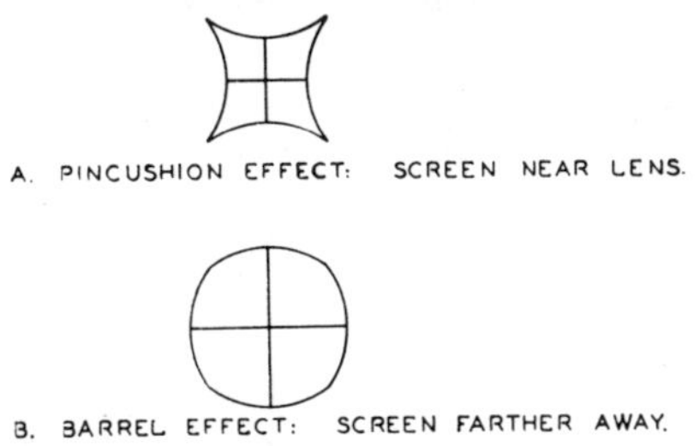

Figure 17. Types of distortion resulting from the projection of a curved image onto a flat surface.

matism. The astigmatic effect can be decreased to some extent by empirical alterations of the shape of the lens surface. Lenses corrected in this fashion are termed *anastigmats*.

If the image of a plane object does not lie in a plane, but on a curved surface, as shown in Figure 16, the system is said to exhibit *curvature of the image*. When the image is projected onto a plane surface, distortions such as are shown in Figure 17 result. A system which is free of such distortions, i.e., one in which angles are correctly reproduced, is called *orthoscopic*. This is achieved if the ratio of the tangents of the angles made by incident and refracted rays with the principal axis is a constant.

It is evident that the formation of an image by a lens system is a complex matter if the object has appreciable size, and the design of lenses involves in large part the selection of the best compromise for the conditions of use.

Complex as the situation is for spherical aberration, it must now be recalled that all of the preceding discussion has been limited to the special, simple case of monochromatic light. If the light illuminating the object contains several wave lengths, or a continuous distribution as in the case of white light, then *chromatic aberrations* arise, and the problem of producing an image that is a faithful replica of the object becomes correspondingly more difficult.

Dispersion

The refractive index of the transparent medium constituting a lens is different for each individual wavelength of light: this is the property known as *dispersion of refraction*. To "disperse" means to spread or separate, and dispersion phenomena are manifested whenever a medium interacts with a disturbance in a manner that is not independent of the energy, wavelength, or other characteristic parameter of the latter. This specificity of interaction can serve to differentiate the individual energies, wavelengths, etc., and hence to sort them out, or "disperse" them. The spreading-out effect implied in the dispersion of refraction is clearly evident, e.g., in the separation in space of the paths followed by rays of different wavelengths when a single beam containing those wavelengths is incident on a prism.

The rate of change of the refractive index as a function of the wavelength is quite specific and characteristic of the medium. Figure 18 shows the refractive dispersions of crown and flint glass, the

Figure 18. Comparison of the dispersion of refraction of crown glass and of flint glass over the visible region of the spectrum.

most commonly employed types of optical glass. Crown glass is a potassium calcium silicate glass with refractive index in the vicinity of 1.50; the name comes from the crown-shaped drop this glass forms at the end of a blowpipe during one phase of the manufacturing process. If some or all of the calcium is replaced by barium in the glass formulation, the product is called a barium crown glass. Flint

glass, so-called after the lead mining town of Flint in Wales, contains sufficient lead to give it a much higher refractive index than crown glass, as well as a faster rate of change of the index with wavelength.

In the visible region of the spectrum, the dispersion of flint glass is about twice that of crown glass. Thus, a flint glass prism of a given angle spreads out the component wavelengths of white light over twice the distance of a crown glass prism of the same angle, as illustrated in Figure 19.

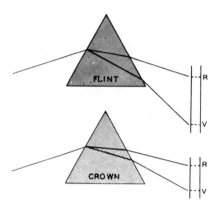

Figure 19. Equal prisms of flint and crown glass yield spectra the spreads of which are in the ratio of 2:1.

Chromatic Aberration

If we consider a lens of crown glass with a suitable aperture that will yield a reasonably good focus for light of a given wavelength, as shown, e.g., in Figure 11, the fact that its refractive index is different for other wavelengths means that it will not bring rays of these other wavelengths to the same image point. Since the refractive index increases with decreasing wavelength, the short wavelength rays will be brought to a focus closer to the lens than will the longer wavelength rays. This is illustrated in Figure 20. If a screen is placed at the image point for the violet rays, a spot will be seen that is violet at the center and red at its extremities; at the image point for the red rays, the spot is red at the center and fringed with violet. As a consequence, the image is indistinct, and the best that can be achieved is a *circle of least confusion*. The circle of least confusion due

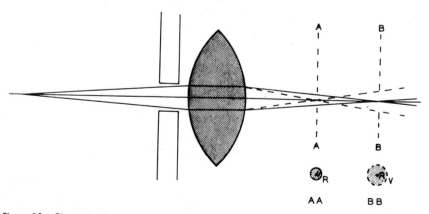

Figure 20. Chromatic aberration arises from the fact that the shorter wavelength rays are brought to a focus closer to the lens than are the longer wavelength rays. Screens placed at *AA* and *BB* show spots that are violet fringed in red, and red fringed in violet, respectively.

to chromatic aberration is about one hundred times larger than the circle of least confusion due to the spherical aberrations previously discussed. Hence, the imperfect behavior of simple lenses stems predominantly from chromatic aberration.

Chromatic Correction

Any two different wavelengths can be brought to a focus at the same image point by a suitable combination of a converging lens with a diverging lens having a different dispersion. For example, Figure 21B shows an *achromat* of the type developed by Fraunhofer, in which a biconvex converging lens of crown glass is united with a plano-concave diverging lens of flint glass. The converging lens tends to focus the violet rays in front of the red rays; the diverging lens applies a negative correction to all these rays, with the effect being greater for the shorter wavelengths (*cf.* Fig. 18). Because the dispersion of flint glass is much greater than that of crown glass, it is possible to choose a focal length for the diverging lens that causes the violet and red images to coincide, while only partially cancelling the convergence produced by the first lens.

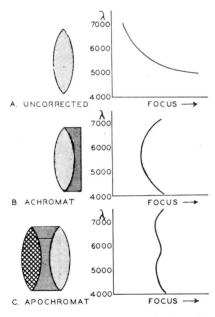

Figure 21. Correction of chromatic aberration. A. Simple, uncorrected lens. Violet rays are brought to a focus closer to the lens than red rays. B. Doublet lens containing two different glasses of unequal dispersions. Two different wavelengths can be brought to equal focus. C. Triplet lens of three different dispersive media. Three wavelengths can be brought to equal focus.

When the focal lengths of the two lenses are so chosen as to make the red and violet rays exactly coincide in the image plane, other wavelengths are not brought to this same focus. That is, a two-piece, or *doublet* lens can correct only for two colors. The two colors for which the lens is corrected can be chosen at will, and if they are properly spaced in the visible range, the amount of residual chromatic aberration observable visually is small.

The correction can be further improved by the use of three lenses having different dispersions, as indicated in Figure 21C. A *triplet* can be color-corrected for three wavelengths; such lenses were first devised by Abbe and are called *apochromats* (the prefix "apo" here means "free from"—hence the lenses are free from color effects; this is meant to carry a stronger implication of success than that of the prefix "a," meaning "without" in the term "achromat"). The substance fluorite is often employed as one of the

a

b

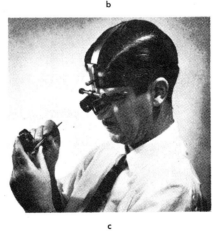

c

Figure 22. Multi-element magnifiers corrected for spherical and chromatic aberration. A. Model 3D3 binocular loupe of Edroy Products Co., New York, 17, N. Y. B. Spectacle magnifiers of Carl Zeiss, Oberkochen. C. Stereomagnifier of Tropel, Inc., Fairport, New York.

components of apochromatic lens systems.

In the process of achieving color correction, both achromats and apochromats correct at the same time for spherical aberration, by the principle outlined previously and demonstrated in Figure 12.

The effect of chromatic and spherical correction can also be achieved if the converging and diverging lenses are not cemented together, as in the examples given above, but are separated from each other (though the required focal lengths would be different in the two cases). Thus, the diverging element may be part of a separate eyepiece or ocular, which is

used in conjunction with an objective lens containing the converging element(s).

Highly Corrected Magnifiers

When doing fine work at close quarters for extended periods of time, a moderate degree of object enlargement that is completely free of distortion is often essential for precision and comfort. Examples of a type of binocular magnifier employing spherically and chromatically corrected lenses are shown in Figure 22. The linear magnification is only $2\times$ to $4\times$. In the case of the Zeiss spectacle magnifiers (Fig. 22B) the field of view is 35 mm at a distance of 200 mm, and distortion is acceptable for continuous work. The Tropel unit (Fig. 22C, $149) gives $4\times$ with a field of view of 30° and a working distance of 5.25 in.; the lens cells can be rotated upward over the forehead when normal vision is required.

A

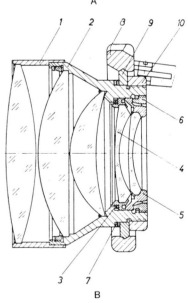

B

Figure 23. Zeiss telescopic spectacles. A. Actual appearance. B. Optical schematic: 1, slip-on reading lens; 2, objective lenses; 3, intermediate ring; 4, 5, ocular lenses; 6, clamping ring; 7, spacer ring; 8, spectacle frame; 9, holding discs; 10, clamping ring.

Zeiss also makes telescopic spectacles that can be used as binocular magnifiers. The design of these spectacles is shown in Figure 23. The basic lens system consists of a two-element converging lens (2, in the figure) combined with a two-element diverging lens (4 and 5). This gives a linear magnification of 1.8$\times$ with a field of view of 23°. For close-range work, an auxiliary lens (1) is slipped over one of the magnifiers (binocular vision is not possible now, because of the short distance necessary between object and lens), yielding magnifications, depending upon the lens selected, ranging from 2× to 6× at working distances of 223 mm to 72 mm, respectively.

Magnifiers used for inspection and dimensional gaging of small parts must give images that faithfully reproduce the lengths, angles, and curves of the object. An inexpensive and imperfect approach that is frequently used is based on the Coddington lens design, illustrated in Figure 24A. The two spherical surfaces of the lens enclose a diaphragm cut into the center of the unit; this stop limits the aperture to approximately the size of the eye pupil, and in this way the aberrations are kept within reasonable bounds. An illuminated magnifier based on the Coddington lens is shown in Figure 24B; in this case, the center cut serves as a reflecting and diffusing surface for the

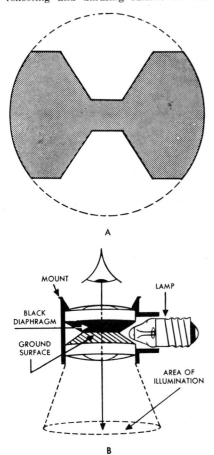

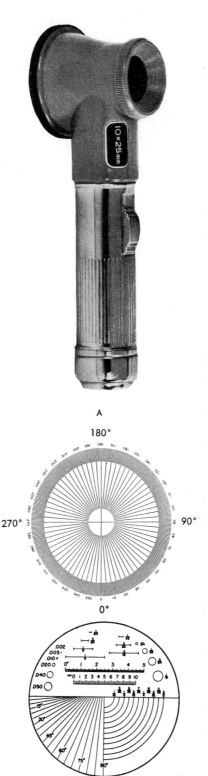

A

Figure 24. The Coddington lens. A. The basic element has two spherical surfaces with a diaphragm cut into the glass between these surfaces to limit the aberrations. B. The Bausch and Lomb Illuminated Coddington Magnifier, in which the diaphragm surfaces are utilized to illuminate the object.

light source as well as a diaphragm to limit the aperture of the system. This is the design of the Bausch and Lomb Illuminated Coddington Magnifier ($7.50); its magnification is 10×, and its light source is powered by flashlight batteries.

A more highly corrected magnifier incorporated in a flashlight illuminator is made by the Titan Tool Supply Co., Buffalo 16, N. Y., and is shown in Figure 25A. The lens is a three-element achro-

B

Figure 25. A. The Titan Tool Supply Co. Flashlight Comparator. B. Reticles for incorporation in the viewing system.

mat, and the magnification is 10×. Various reticles, shown in Figure 25B, may be attached at the object plane of the lens for measurement of the dimensions of the object. The price of this device is $17 complete with one reticle; the reticles are $3 each. A similar instrument is the Bausch and Lomb Illuminated Hastings Triple Aplanatic Magnifier (5×, $25). A 7× triplet lens mounted in a plastic reticle holder with a selection

of reticles, shown in Figure 26, is also available from this manufacturer (designated the Measuring Magnifier, $14; with a set of four reticles, $36).

A modification of the above illuminated magnifiers that is available from Bausch and Lomb is specifically designed for the comparison of two surfaces (the Surface Comparator, $30). The optical scheme is shown in Figure 27; light from a flashlight-type source is split into two beams by a 45°, partially-silvered mirror, and the reflected beams from both paths are viewed simultaneously through the triplet lens at a magnification of 10×.

An illuminated magnifier that provides a wide enough field of view to permit the object to be seen with both eyes, and hence with stereoscopic effect, is the MacroScope made by Ednalite Research Corp., Peekskill, N. Y., and shown in Figure 28 ($213 for magnifier, $75 for illuminated work platform shown). If the top achromat lens is used alone, the magnification is 2.5×, and the field of view is a 5-in. diameter circle. A second achromat may be inserted in a bayonet mount beneath the first lens to increase the magnification to 4× at the same viewing aperture. For still higher magnifications, a third lens may be swung into position beneath these two. The object can now be viewed through one eye, i.e., it is monocular, since the last lens has a

Figure 26. The Bausch and Lomb Measuring Magnifier with case, with one reticle in the holder, and three additional reticles.

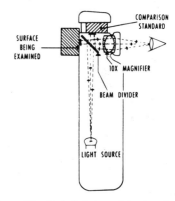

Figure 27. Optical design of the Bausch and Lomb Surface Comparator for simultaneous viewing of light reflected from two specimens.

small aperture. If the latter is used in combination with only the first lens, the magnification is 5×; if used in conjunction with both lenses, the total magnification is 8×. The specimen can be viewed by transmitted light, or by reflected light from a circular fluorescent lamp in the lens housing. Accessories are available for photomacrography, polarized light, dark field, and color contrast work.

A similar large field magnifier employing an aplanatic and achromatic lens with an aperture of 6″×4″, permitting binocular, stereoscopic viewing is available from Vision Engineering, Inc., Danbury, Conn. (Model Stereo-X-Two, magnification 2×, mounted on a universal stand with fluorescent lamps in the lens housing, $118).

For the inspection and measurement of cavities, drilled holes, and recessed areas, an instrument is available from Vision Engineering (the Bore Magnifier, $258)

Figure 28. Cut-away view of the Ednalite MacroScope with its illuminated work platform.

that embodies a 1¼″ diameter achromatic lens below which is mounted a

Figure 29. Bore Magnifier of Vision Engineering Inc. The lamp in the housing at the rear is cooled by an air blower powered by a motor, both of which are contained in the housing. The stand base contains the transformer for the lamp.

semisilvered mirror. Light from a small, high intensity lamp is reflected down along the axis of the lens, and reflected back by the object. Interchangeable lenses are available with magnifications from 1× to 5×; the corresponding depths of field range from 3″ to ⅝″. The lamp is transformer-controlled, and the intensity of illumination is adjustable as required. The instrument is shown in Figure 29.

Figure 30. Model 239 Ramsden Hand Lens of Pacific Transducer Corp.

A hand lens containing a highly corrected optical system designed for 6× viewing of objects without distortion of dimensions or angles is the Model 239 Ramsden Hand Lens ($4) of Pacific Transducer Corp., Los Angeles 64, Calif., shown in Figure 30. The field width is ¾-in., and the distance from eye to object is 3″ (2¼″ of which is occupied by the lens). This lens is optically equivalent to the eyepiece or ocular of a compound microscope, and, in fact, microscope eyepieces can generally be used as hand magnifiers for low-power inspection of objects. However, many microscope eyepiece assemblies are specifically designed to correct for some of the residual aberrations of the objective, and hence

may not give distortionless magnification when used alone.

Limitations of Single Lenses

In order to gain high magnifications with a single lens system, it is necessary to bring the object close to the focal point of the lens, and the lens must have a highly curved surface, i.e., a short focal length, to yield a high power (which, as shown above, is given by the reciprocal of the focal length). If the lens has any appreciable size, the effects of spherical and chromatic aberration will be very pronounced. These effects can be reduced by limiting the light from the object to the small central area of the thick, highly curved lens. Consequently, the higher the magnification desired, the closer the object must be to the lens surface, and the smaller must the aperture of the lens be. However, this severely limits the amount of light that can be collected by the lens, and the contrast and resolution, as well as the field of view, are greatly curtailed.

In the seventeenth and eighteenth centuries, high magnifications were achieved with single lenses by employing tiny beads of glass or quartz as the magnifier. This was the basis of the "flea glass" of Descartes, and of the simple microscopes of van Leeuwenhoek and Robert Hooke. Globules of glass as tiny as 0.5 mm in diameter were used, which were capable of giving magnifications over 400×, but the quality of the

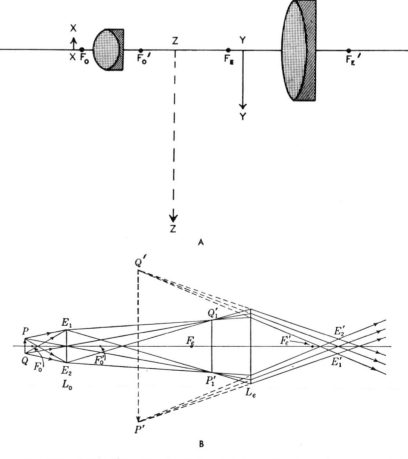

Figure 31. Basic principle of the optical path in a compound microscope. A. Locations of the object, real image, and virtual image with respect to the focal points of the objective and eyepiece. B. Ray diagram showing the formation of the real image, $P_1'Q_1'$, and the virtual image, $P'Q'$, by the objective, L_o, and eyepiece, L_e, respectively. The length E_1E_2 is the entrance pupil, and $E_1'E_2'$ the exit pupil of the microscope.

image was extremely poor, and the difficulty of observing it enormous.

The only way to achieve a high magnification combined with a wide aperture and field of view is to resort to a compound optical system composed of at least two separate and distinct lens systems.

The Compound Microscope

The basic design of a compound microscope consists of (1) an *objective* lens or combination of lenses positioned close to the object, and (2) on *eyepiece*, or *ocular* lens or combination of lenses located a substantial distance away from the objective and close to the eye of the observer. A very simple compound microscope may be composed of an achromat of short focal length as the objective, and another achromat at the eyepiece, as illustrated in Figure 31.

The object is positioned *before* the first focal point, F_o, of the objective, hence this lens forms a real, inverted, enlarged image *beyond* its second focal point, F_i' (compare with the relationships summarized in Fig. 4). That is, the object, XX, is imaged by the objective at YY. The eyepiece is positioned so that this image falls *inside* the lens first focal point, F_e, and therefore the rays emerging from the eyepiece are divergent, and ex-

trapolate back to a virtual, upright, magnified image at ZZ. A more detailed schematic diagram of the components of a modern compound microscope and the paths of the rays is given in Figure 32.

In most microscopes, the distance between the objective and eyepiece is fixed by the length of the tube in which they are mounted. Therefore, the microscope is focussed by bringing the objective down toward the object (or, in actual practice, more properly up away from it, to avoid overshooting the proper position and driving the lens into the object) until the distance between them is such that the real image formed by the objective is at the proper location with respect to the first focal point of the eyepiece (Fig. 31). The total linear magnification is therefore also fixed.

In some cases, it is desirable to make small adjustments in the total magnification of the microscope. This is true, for example, when the image is to be compared with a standard, or measured against a reticle scale, as in toolmakers' and measuring microscopes. Such microscopes are designed with extra lengths of the screw thread on either the objective or the eyepiece housings. If the separation between objective and eyepiece is increased, by screwing either one outwards relative to the other, the objective must be brought closer to the object in

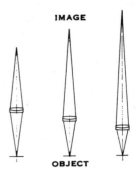

Figure 33. The overall magnification of a compound microscope can be varied by adjusting the distance between objective and eyepiece. Increasing this spacing requires that the objective be brought closer to the object in order to bring it into focus, and results in increased magnification.

order to bring it into focus, and the magnification will consequently be increased. The principle of this method of adjusting the magnification is illustrated in Figure 33. It can only be employed with low-power objectives. Because of the critical nature of the aberration corrections in high-power objectives, the precise objective-to-eyepiece distance for which the optical system has been optimized must be rigidly preserved.

Optical Principles of the Compound Microscope

The purpose of the compound microscope is to produce a bright, clear, undistorted, and greatly enlarged image of the object under study. It has been shown in the preceding that in the case of individual lenses these requirements are mutually incompatible. For the image to be bright and clear, the lens must gather a large amount of light from the object, i.e., it must have a large diameter, or aperture. But a large aperture inevitably results in distortion, and the distortion increases rapidly as the power of the lens is raised. The special role of the compound microscope in providing a way around this impasse arises from the fact that the properties of a system composed of two lenses separated by a considerable distance (i.e., a distance equal to, or greater than the sum of the focal lengths) are strikingly different from those evidenced by these lenses when close together.

For thin lenses that are in contact, or close together, the power (= dioptry; $D = 1/f$) of the combination is the algebraic sum of the powers of the individual components. Thus, as illustrated in Figure 34, a converging lens of $+3$ diopters, in contact with a diverging lens of -3 diopters combine to yield a system of 0 diopters. However, if the centers of these lenses are separated by a distance equal to the sum of their focal lengths (in this case, 66.6 cm) plus an *optical interval*, Δ, this combination can be shown to have an effective focal length, $f_{comb.}$, given by:

$$f_{comb.} = \frac{f_1 f_2}{\Delta}$$

In the example shown in Figure 34, if $\Delta = 33.3$ cm, then $f_{comb.} = 33.3$, and $D_{comb.} = +3$ diopters. If, e.g., the optical interval is reduced to one-half of the preceding

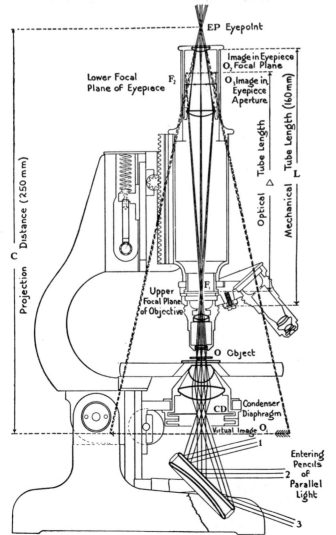

Figure 32. Schematic diagram of a modern compound microscope and the ray paths through it.

Labels on figure:
EP Eyepoint
Image in Eyepiece O_2 Focal Plane
Lower Focal Plane of Eyepiece F_2
O_1 Image in Eyepiece Aperture
Optical Tube Length Δ
Mechanical Tube Length (160 mm) L
Projection Distance (250 mm) C
F_1
Upper Focal Plane of Objective
O Object
CD Condenser Diaphragm
Virtual Image O_3
1
2
Entering Pencils of Parallel Light
3

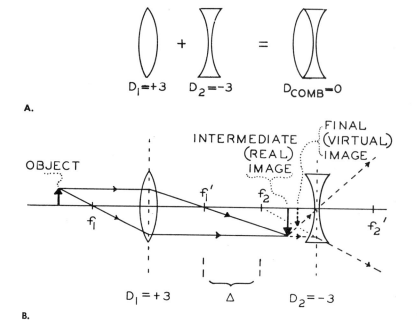

A.

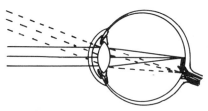

Figure 36. Bundles of parallel rays entering the eye are focussed by the eye lens onto the retina at the rear of the eye. The eye muscles automatically adjust the curvature of the lens to whatever value is required for this focussing.

B.

Figure 34. A. The powers of lenses in contact with each other are additive. Thus, a converging lens of +3 diopters placed close to a diverging lens of −3 diopters gives a resultant power of zero.

B. If the lenses are separated by a distance larger than their focal lengths, the resultant effect is quite different. The image formed by the first lens serves as the object for the second lens. The power of the combination depends upon the optical interval, Δ:

$$D_{comb.} = \frac{\Delta}{f_1 \times f_2} = \Delta \times D_1 \times D_2.$$

value, the resultant power will be +6 diopters. [In the standard compound microscope, the optical interval, Δ, is generally between 16 and 18 cm.]

Role of the Eye

In visual microscopy, the eye of the observer is an integral part of the total optical design. The locations and focal lengths of the objective and ocular lens systems are so chosen that the light rays emerging from the latter enter the eye at such angles that the refraction produced by the latter's lens system causes an image to be formed on the retinal surface at the rear of the eye. The essential part played by the eye lens in determining the apparent size of the image and its position in space must be understood before the detailed design of microscopes can be appreciated.

It has been pointed out before that if an object is positioned at the focal plane of a converging lens, the light rays emerging into the image space have been rendered parallel by the lens, and an image is not formed. Thus, the diverging rays emanating from each point on the object are converted into bundles of parallel rays, as shown in Figure 35. If these

bundles of rays enter the eye, its lens will bring each bundle to a focus at the rear focal plane, i.e., on the retina, as illustrated in Figure 36. The lens of the eye in this case will adopt (automatically and unconsciously) the curvature necessary to focus the rays on the retina. This is the same curvature as is adopted when viewing a distant object, for if the object is sufficiently far away, all the rays coming from it are essentially parallel. Hence, the eye sees the object at the focal plane of the converging lens *as if it were at infinity*. In the "mind's eye," a magnified image viewed at infinity gives the illusion of being located in space at the normal distance for comfortable close viewing, which is about 25 cm from the eye.

Visual Magnification

The apparent size of this image is determined by the size of the retinal image, and is equal to the size of an object which, if located 25 cm from the eye, would give a retinal image of the same magnitude. This is illustrated in Figure 37. The actual object of height x, if 25 cm from the eye, would subtend a visual angle, θ, approximately equal to x/d_0, where d_0 is the close viewing distance. The "mind's eye" image, at this same apparent distance, subtends the visual angle y/d_0, where y is the apparent height seen by the

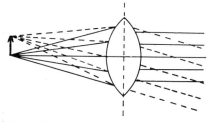

Figure 35. If an object is located at the focal plane of a converging lens, the diverging rays from each point of the object are refracted into bundles of parallel rays.

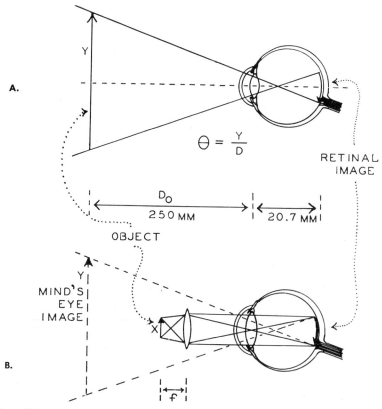

Figure 37. A. An object of height y a distance d_0 from the eye subtends an angle $\theta = y/d_0$ at the eye.

B. An object of height x in the focal plane of a converging lens is seen as a virtual image of height y a distance d_0 away.

eye. Hence, the *angular magnification*, γ, achieved by viewing the object through the lens when the former is in the focal plane of the latter, is:

$$\gamma = \frac{\text{apparent visual angle of image}}{\text{visual angle of object at normal distance of vision}} = \frac{y/d_0}{x/d_0} = \frac{y}{x}$$

which will be recognized as equal to the linear, or lateral magnification. It can further be shown that this ratio is equal to d_0/f. Thus, a 25-mm focal length, biconvex lens ($D = +40$ diopters) used as a magnifier in the sense shown in Figure 37, yields a (visual) magnification of $d_0/25$ mm $= 250$ mm/25 mm $= 10\times$. If this same lens were used to cast a real image on a screen, and the object were located 25 mm beyond the lens focal plane, the lateral magnification would be $f/x = 1\times$ (cf. Figure 4).

In the compound microscope, the distance of the object from the front focal plane of the objective is adjusted so that the real image formed by this lens system occurs at the front focal plane of the ocular, as shown in Figure 38. The ocular, therefore, produces bundles of parallel rays, which the observer's eye lens focuses to a real image on the retina, as in Figure 37, by accommodating for viewing at infinity, and thus the visual image appears to be located at the normal close viewing distance, d_0.

The total magnification of the compound microscope is the product of the magnifications due to the objective and ocular, respectively. The lateral magnification, m, of the objective is:

$$m = \frac{x}{f_1}$$

where x is the distance of the image from the second (rear) focal plane of the lens,

power objective of 4-mm focal length would give about $45\times$ magnification. This magnified image then serves as the object for the ocular. The magnification produced by the latter is $\gamma = 250/f_2$, where 250 is the normal close viewing distance in millimeters, and f_2 is the ocular focal length in the same units. Thus a $10\times$ ocular has an effective focal length of 25 mm. The total magnification, M, is therefore given by the product:

$$M = m\gamma = \frac{(180)(250)}{f_1 f_2}$$

For example, a 4-mm objective used with a 25-mm ocular provides a total magnification equal to $450\times$.

Image Quality

Until this point, the formation of an image by a lens or combination of lenses has been discussed simply in terms of the bringing together of bundles of light rays emanating from (i.e., scattered by) each point of the object. This is, however, a grossly inadequate approach, for the light rays that reach a given place on the image by way of different paths from the same object point, or from nearby points on the object, may as a consequence of the unequal path lengths traversed, differ in phase, and hence interference or diffraction phenomena must play an essential role in determining the variation of intensity from point to point in the image. The principle alluded to in the preceding is indicated graphically in Figure 39.

If a valid image is to be formed by a lens system, *all* of the light rays from *all*

Every lens system has a limited aperture, and this restricts the process of image formation to the combination of only those rays which approach the lens at angles falling within the solid cone defined by the angular aperture (cf. Figure 40). The construction of the lens

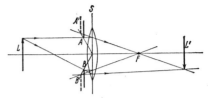

Figure 40. A diaphragm *AB* limits the amount of light entering the lens; the effective aperture of the system is the virtual image of the diaphragm, viz., the opening *A'B'*.

system, e.g., the mounting of the objective and ocular, impose geometrical limitations on the aperture for acceptance of light from the object, as well as for the transmittal of light to the observer's eye. These limitations are often described in terms of the *entrance pupil* and *exit pupil* of the instrument, as illustrated in Figure 41.

The refractive index and curvature of the lens affect the angular range of the light that is passed into the instrument, because of the change in direction caused by the refraction, as shown in Figures 42 and 43.

Similarly, the refractive index of the medium between the object and the lens

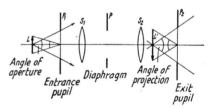

Figure 41. A diaphragm positioned between two lenses determines the size of both the entrance and exit pupils, since these are the images of the diaphragm formed by the respective lenses.

affects the angular spread of the light which can enter the lens, as shown in Figure 44. If the medium has a high refractive index, as is the case when an immersion liquid is used to fill the space between the object and the objective, then rays emanating from the object at what would otherwise be too large an angle, are bent inwards by refraction and can pass into the objective. The total amount of light admitted by an objective is proportional to the product $n \sin \alpha$, where n is the refractive index in the object space, and α is the angle of aperture (cf. Figure 44). The quantity $n \sin \alpha$ is termed the *numerical aperture*, generally abbreviated as N.A.

The numerical aperture is of fundamental importance in determining the quality of the image formed by an objective. A light wave scattered by two points in the object which are close together will form an image similar to that observed in the classical case of Fraunhofer diffraction by two slits, Figure 45.

It is the overlapping and mutual interference, both constructively and de-

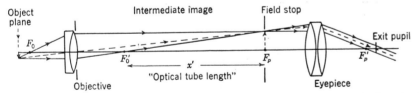

Figure 38. In a compound microscope, the objective lens produces a real image at the front focal plane of the ocular, which therefore refracts the divergent rays from this image into bundles of parallel rays that are directed into the observer's eye.

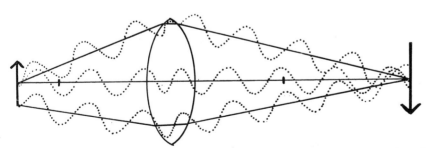

Figure 39. The light rays combined at a given point in the image may differ in phase by virtue of having traversed different total paths from the same or different points in the object.

and f_1 is the focal length. In the standard compound microscope, x is approximately equal to the optical interval, Δ, i.e., about 180 mm. Therefore, a typical low-power objective of 16-mm focal length would yield a lateral magnification of $180/16 \cong 11$, or approximately $10\times$; a high-

parts of the object should be combined in a proper way. If any of the scattered or diffracted rays fail to be included in the image, information inherent in these rays has been lost, and the image cannot then truly disclose all of the features of the object.

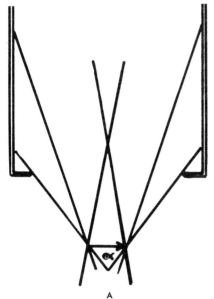

Figure 42. A. Light scattered into a hollow tube from an object completely fills the tube.

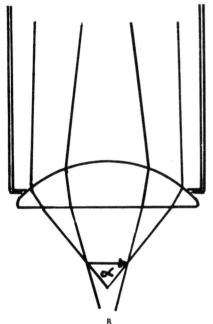

Figure 42. B. A hemispherical lens concentrates, or condenses, the light.

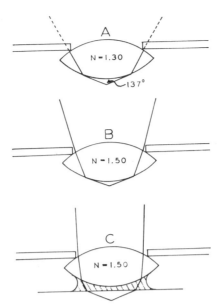

Figure 43. A. A cone of light impinging on a biconvex lens of refractive index 1.30 with a central angle of 137° is partially blocked by the diaphragm.

B. When the refractive index is increased to 1.50, the greater refraction serves to bend all of the light into the system.

C. If the object space is filled with a liquid of appreciable refractive index, the transmitted light is condensed into a narrower beam; hence, a cone of substantially larger central angle than 137° could be used to illuminate the lens.

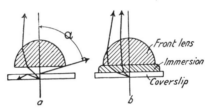

Figure 44. The use of an immersion liquid of high refractive index to fill the object space increases the amount of light from the object that can enter the objective.

n is the refractive index of the medium. For example, this relation shows that if light of wave length 5500 A is employed, the first order peak in air due to scattering centers 1.0 microns apart will be seen at an angle of about 33.5° from the axis. Hence, the numerical aperture of the lens viewing this specimen must exceed 0.55 if the image is to contain the information inherent in the first order of diffraction from these points. If the second order of diffraction is also to be incorporated in the image, the numerical aperture would have to be twice that value, viz., 1.10.

It will be noted that the quantity $n \sin \alpha$, which is the numerical aperture, has turned up now in a fundamentally different way from the preceding considerations of the total light intensity entering the instrument. Increasing the intensity of the light source illuminating the object will result in an increase in the amount of light entering the optical system of the instrument, but this cannot affect the number of orders of diffracted intensities employed in forming the image.

Thus, the quality of an image formed by an objective lens depends upon the numerical aperture. It is generally found that for two points separated by the distance b to be recognized in the image as separate and distinct features of the object, the N.A. must have at least the value given by:

$$\text{N.A.} = \frac{0.61\lambda}{b}$$

Conversely, any given N.A. is capable of yielding an image in which the limit of resolution corresponds to features which are separated by the distance b in the above equation. For example, using white light a lens of N.A. = 0.1 cannot ever resolve features closer together than about 3.4 microns; if the N.A. = 1.2, the limit of resolution is lowered to about 0.3 microns.

structively, of these Fraunhofer diffraction patterns produced by all possible combinations of the rays from all of the points in the object that results in the formation of the image. For a good quality image, it is essential that as many as possible of the orders of diffraction be included in the combination. If the zero order maxima enter the objective along its axis, then the various orders of maximum intensities will enter at angles, α, determined by the wave length of the light, λ, and the *grating constant*, or distance apart of the diffracting centers, b. In this case, the fundamental relation between these quantities is the grating equation for normal incidence:

$$n \sin \alpha = \frac{m\lambda}{b}$$

where m is the order of the maximum and

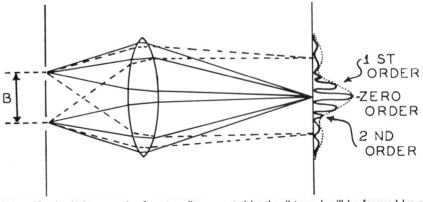

Figure 45. A. Light emanating from two slits separated by the distance b will be focussed by a lens to yield a series of diffraction orders.

B. Photograph of the Fraunhofer diffraction pattern formed by two slits.

Resolution Versus Magnification

It must always be borne in mind that the amount of detail which can be seen in an image depends upon both the magnification and the resolution, and that these factors are independent of each other.

The photosensitive nerve cells composing the retina of the eye have finite dimensions, and in the most sensitive region their centers are spaced about 2 microns apart. Therefore, if two points in an image are to be recognized by the observer as separate and discrete, they must fall on different retinal sensors, i.e., they must be at least 2 microns apart at the retina. The angular separation of the bundles of rays entering the eye from two points on an object would have to be about 0.0012 radians (= 0.062°) if the image points are to be 2 microns apart on the retinal surface. Thus, in an object viewed at the normal close viewing distance of 25 cm, the minimum separation that can be resolved by the eye is given by 0.0012 = $x/250$; $x = 0.3$ mm.

If it is desired to see details in an object that are, e.g., as small as 1.0 microns (= 0.001 mm), then in the image these details must be magnified until they are at least 0.3 mm apart; i.e., the minimum magnification, $M_{min.}$, necessary to reveal details of dimensions L millimeters is:

$$M_{min.} = \frac{0.3}{L}$$

The requisite magnification can be achieved in a variety of ways, e.g., by the use of a high power ocular, by projecting the image onto a distant screen, by photographing it and printing an enlargement, etc. However, even if the magnification is sufficient to cast the image points on separate nerve centers in the retina, the details in the object will not be *recognized* as such unless the image *quality* is adequate, and this is governed not by the magnification, but by the numerical aperture.

Thus, as has already been shown, if details in the object of 1.0 microns dimensions are to be reproduced, the numerical aperture must be great enough to admit at least the first order diffraction maxima, viz., N.A. $\cong 0.3$ with white light.

In the example given, it follows that to see details in the object 1.0 microns apart, two independent conditions must be satisfied: (*1*) the objective lens must have a numerical aperture of 0.3 or greater, and (*2*) the powers of the objective-ocular combination, or the degree of projection or photographic enlargement must combine to yield a minimum magnification of 300×.

In general, the degree of useful magnification that can be employed with an objective lens of given N.A. is equal to 1000 times the N.A. Any greater magnification than this merely enlarges the image without revealing any more detail, i.e., it is *empty magnification.*

Microscope Objectives

The design of an objective for microscopy must seek to achieve a high magnification coupled with a large numerical aperture, while keeping spherical and chromatic aberrations small. These are exacting requirements, and become more so the higher the desired magnification; objective design is a correspondingly complex and sophisticated subject. For N.A.'s between 0.15 and 0.30, two achromats separated by a small distance serve adequately; this is the Lister objective shown schematically in Figure 46A For greater numerical apertures (0.30 to 0.80), a hemispherical front lens is added, as in the Amici objective of Figure 46B. Oil immersion objectives designed for the highest N.A.'s (up to 1.3) have additional component lenses, as indicated in Figure 46C.

A typical low-power objective may have a focal length of 16 mm and a numerical aperture of 0.28. This lens will give a magnification of approximately 10× alone; when used in combination with an ocular, the maximum useful (or, real) magnification it can contribute to is about 280×. The working distance, or space between the cover glass and the front surface of the objective, is about 5 mm. The depth of focus is about 0.007 mm.

A typical high-power dry objective may have a focal length of 4 mm and N.A. = 0.65. This corresponds to a magnification of 45×, and a maximum real magnification of up to 650×. The working distance is only 0.7 mm, and the depth of focus is about 0.001 mm.

Some commercial apochromats showing the method of designating focal length,

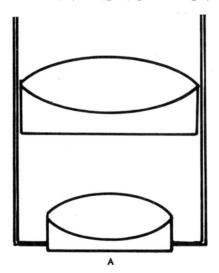

A

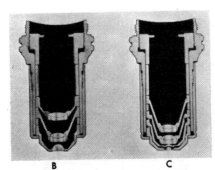

B   **C**

Figure 46. **A. The Lister objective for low power and low N.A. magnification.
B. The Amici objective for medium power.
C. The oil immersion objective.**

N.A., and linear magnification are shown in Figure 47A. A series of achromats is reproduced in Figure 47B. Apochromats

Figure 47. **A. A series of Bausch and Lomb apochromat objectives.**

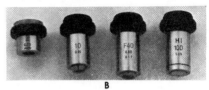

B. A series of Elgeet achromat objectives.

range in price from $94 for a 16-mm, 10×, 0.30 N.A. objective (Bausch and Lomb No. 31–11–71–01) to $326 for a 2-mm, 90×, 1.40 N.A. objective (B. and L. No. 31–11–93–02). Achromats are about one-fifth as expensive for corresponding focal lengths and N.A.'s.

In some applications it is desired to employ ultraviolet light for microscopy, since the shorter wave length permits better resolution to be achieved. A special objective is available for this purpose based upon reflection optics, as shown in Figure 48. Ordinary transmission optics cannot be employed in this instance, for glass absorbs the UV too completely, and quartz, which passes the UV, cannot be readily fabricated into objective lens components. Bausch and Lomb produces two UV objectives: a 53×, 0.72 N.A. dry ($588), and a 94×, 1.00 N.A. water-immersion objective ($700).

A special objective designed to give high magnification while maintaining a relatively large working distance between the object and the front surface of the lens system is made by Unitron Instrument Co., Newton Highlands 61, Mass., for use in their metallographic microscope in conjunction with a heated stage. The Model FF objective has N.A. = 0.45, and gives 40× magnification; the working distance is 5.5 mm, and the price is $149.

A novel type of objective that has come into recent widespread use is the so-called "zoom" objective, which permits the effective focal length and, consequently, magnification to be varied continuously over a prescribed range. This is accomplished by the introduction of a system of three additional lenses behind the regular objective lenses. The zoom lenses have the effect of moving the image due to the objective lenses farther away from the objective, thereby increasing the magnification. By moving the components of the zoom lens relative to each other through an arrangement of cams, the image plane can be shifted over a sufficient range to cause the overall magnification to be variable up to a factor of 2× the normal objective magnification. The zoom principle as applied in the design of the Bausch and Lomb variable objective is illustrated schematically in Figure 49. If a 10× ob-

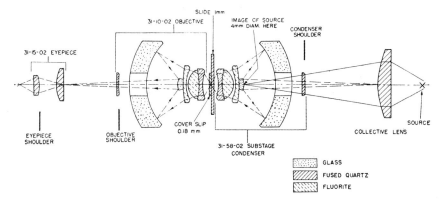

Figure 48. Optical design of the Bausch and Lomb UV objective and its relation to the condenser and ocular.

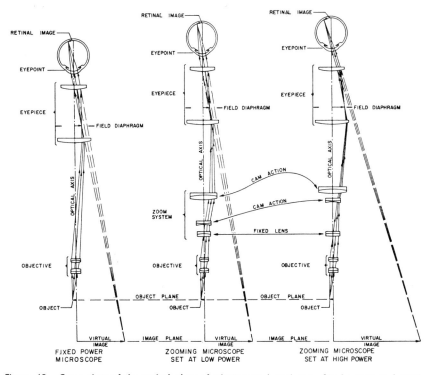

Figure 49. Comparison of the optical plans of microscopes based upon fixed power and zoom objectives, respectively.

jective and a 10× ocular are used, incorporation of the zoom lens results in the overall magnification being continuously variable between 100× and 200×.

The correction of lens aberrations in the objective can be improved by the use of additional lens elements. The objective systems described previously (cf. Figure 46) are rather well corrected, but there is a residual curvature of the image plane (cf. Figures 16 and 17). The Bausch and Lomb Company now offers a line of objectives in which extra lenses are employed to improve the flatness of the visual field. The difference in optical design between the standard achromats and the new "flat field" objectives is illustrated diagrammatically in Figure 50. Other optical manufacturers also offer highly corrected objectives containing large numbers of individual lenses. The novel feature of the Bausch and Lomb "flat field" objectives is the use of a 5× negative doublet lens located in the nosepiece. The upper doublet lens of Figure 50B is common to

all the objectives and stays fixed in the nosepiece as the various objectives are interchanged. In the higher powers, the front hemisphere lens is replaced by a thick meniscus lens.

Oculars

An ocular serves the dual function of: (1) gathering the divergent rays coming from the objective and thus increasing the brightness and the field of view of the final image, and (2) magnifying the image formed by the objective. The standard microscope ocular is itself a compound lens system, consisting of two lenses separated by a distance that is significant relative to their focal lengths.

In the case of the Huygens ocular, shown in Figure 51A, two plano-convex crown lenses are mounted a distance apart equal to one-half the sum of their focal lengths. This geometry is such that no lateral chromatic aberration is produced by the passage of the light rays through the ocular. In the case of the Ramsden ocular, Figure 51B, this condition is only approximated; the two lenses are separated by about two-thirds of the proper distance for lateral achromatism. The Ramsden ocular is thus not as free from lateral chromatic aberrations as is the Huygens ocular, but its design leads to a considerably smaller degree of longitudinal chromatic aberration. The lateral chromatic correction of the Ramsden ocular can be improved by using an achromatic doublet for the eye lens; this design is known as the Kellner, or achromatized Ramsden ocular.

The Huygens ocular consists of two lenses having different focal lengths, with the curved surfaces both turned toward the objective. The lens closest to the objective is called the *field lens*, and it redirects the rays coming from the objective so as to condense them into a smaller cross-section, thus increasing the field of view over what would be seen under the same conditions if the field lens were absent. It forms a real image in the focal plane of the second, or *eye lens*. Since this image is at the focus of this lens, the emergent light consists of bundles of parallel rays. The eye thus sees the image

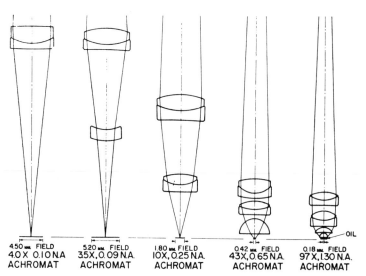

Figure 50. A. Scale diagrams of the standard series of achromat objectives.

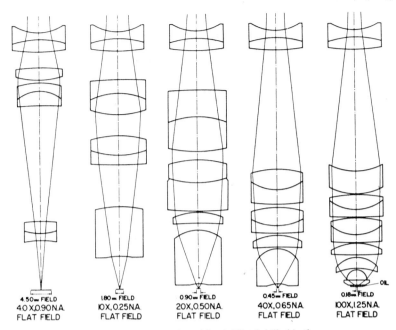

4.50 mm FIELD
40X 0.90 N.A.
FLAT FIELD

1.80 mm FIELD
10X 0.25 N.A.
FLAT FIELD

0.90 mm FIELD
20X 0.50 N.A.
FLAT FIELD

0.45 mm FIELD
40X 0.65 N.A.
FLAT FIELD

0.18 mm FIELD
100X 1.25 N.A.
FLAT FIELD

Figure 50. *B. Scale diagrams of the Bausch and Lomb "flat field" objectives.*

Figure 53. The Unitron zoom ocular.

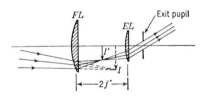

A

Figure 51. *A. The Huygens ocular, consisting of a field lens, FL, having three times the focal length of the eye lens, EL.*

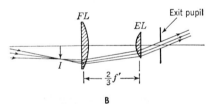

B

B. The Ramsden ocular, consisting of field lens and eye lens having equal focal lengths.

"at infinity," i.e., the normal close viewing distance of 25 cm.

The Ramsden ocular is positioned so that the real image from the objective falls at the front focal plane of the ocular. The combination of the field and eye lenses acting on the rays from this image causes bundles of parallel rays to be presented to the eye of the observer.

Cross-hairs, or a scale or reticle, may be located in the plane of the real image in either ocular. The light rays forming the image serve to illuminate the cross-hairs or scale, and the light scattered therefrom is focussed by the lens (or lenses) of the ocular and eye to yield an image that is superimposed upon the microscope image. In the Huygens ocular, only the eye lens is involved in the formation of the image of the cross-hairs or scale (since the latter must be placed between the field and eye lenses), and hence aberrations are produced. In the Ramsden ocular, both field and eye lenses focus the rays from the cross-hair or scale, and the image is con-

sequently much sharper and clearer. It is easy to recognize whether an ocular is of the Huygens or Ramsden type, since the latter, if removed from the microscope, can be used as a hand magnifier, whereas the former is not a true magnifier.

An additional doublet lens inserted between the field and eye lenses is used in some currently available oculars to provide a wider field of view without introducing excessive aberrations. This optical design is illustrated in Figure 52.

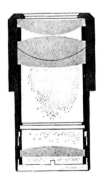

Figure 52. Design of the wide field ocular.

The so-called "high point" oculars are standard Huygens or Ramsden designs in which the focal lengths of the field and eye lenses are selected so that the exit pupil of the ocular is at a greater-than-normal distance from the ocular. This is preferred by some persons who have long eyelashes, or wear glasses, and who find it inconvenient or uncomfortable to bring the head as close to the ocular as the normal design requires. The usual ocular has its eye-point, or exit pupil about 4–8 mm from the eye lens. The high-point ocular may increase this distance by a factor of three.

A zoom ocular, in which the magnification can be continuously varied between 10× and 20× is available from Unitron Instrument Co. ($24.50), and is shown in Figure 53.

Types of Compound Microscopes

The objective and ocular lenses are the *sine qua non* of a compound microscope, but other components must be provided if a practical and useful instrument is to be constructed. There is a variety of ways in which the object under examination may be illuminated, and in which the optics may be positioned relative to it, as well as a number of means of modulating the light rays before and after passing through the magnifying lenses. Certain arrangements and accessories have proved to be particularly useful for specific types of work; these correspond to instrumental designs which are commonly designated by the following special names:

1. Measuring, Universal and Toolmakers' Microscopes; Comparators, Cathetometers. The basic microscope optics are combined with special devices for precise positioning and movement of the specimen or of scales, crosshairs, reticles, or graticles for the estimation of distances or angles. Projection optics may be employed to permit the measurement of dimensions and profiles, and interference phenomena may be exploited for the quantitative estimation of roughness, elevation, and contour.

2. Standard Microscopes. The stage and illumination system are designed primarily for the observation of transparent specimens mounted on standard microscope slides.

3. Phase Microscopes. The contrast and resolution with which transparent specimens are seen in the standard microscope are greatly improved by the incorporation of accessories that make visible the phase shifts which occur when light passes through media of differing refractive indices.

4. Stereo and Operations Microscopes. The optics are designed for a relatively large objective-to-specimen working distance, and for maximum flexibility in the positioning of the microscope with respect to the specimen. Stereoscopic viewing and inversion of the image are employed to facilitate manipulations performed on the specimen while it is under observation.

5. Chemical and Petrographic Microscopes. Accessories are incorporated that polarize the light before it passes through the specimen, and then make visible the effect of the specimen on the constituent wavelengths of the polarized beam. The specimen holder (microscope stage) may be specially designed to permit heating or cooling of the sample while it is under observation.

6. Biological Microscopes. The standard microscope is inverted, so that the underside of a flat, transparent specimen (e.g., a Petrie dish or culture flask) can be placed above the objective, and the contents of the container viewed from below.

7. Metallographic Microscopes. The illumination system and optics are modified to optimize the image quality when viewing the surface of an opaque specimen by reflected light.

8. Photomicrographic Instruments. The optics and illumination are modified to permit substitution of a camera for the observer's eye, and thus to permit the photographic recording of the image, whether for still, time-lapse, or cinematographic presentation.

9. Projection Microscopes. The microscope image is projected onto a screen for simultaneous viewing by a group of persons.

1. Measuring Microscopes

A. Shop Microscopes

A variety of so-called "shop microscopes" is available for the examination of machined parts, tool edges, cracks, scratches, bore holes, fine details of design

and engraving, etc. An example of the simplest of these is the Model 233 Pocket Microscope of Pacific Transducer Corporation, Los Angeles, California 90064 ($20), shown in diagram in Figure 54. It consists simply of an objective that is positioned a specific distance from the surface under examination by means of a surrounding mounting tube (the "chassis" in Fig. 54). The depth of focus of this objective lens is sufficiently broad that the front focal plane of the ocular (eyepiece) can be varied over a range of about 2 in. without excessive loss of clarity in the final image. When the eyepiece is at the lower limit of its range, the image it magnifies is smallest, and the over-all magnification seen by the observer is 20×. When the eyepiece is extended to its upper limit, the objective image at its front focal plane is about 3× larger than in the previous case, and the over-all visual magnification is 60×.

A somewhat more elaborate version of this type of design is the PTC Model 232B Industrial Microscope ($43), the optical diagram of which is given in Figure 55. Two magnifications are provided; 20× when the eyepiece is at its closest position to the objective, and 40× when it is at its far position. A spacer is provided to insure accurate positioning of the eyepiece in these two locations, and a rack-and-pinion gear system allows adjustment of the optical train in its chassis in order to focus on the specimen surface. This device is also available without the rack-and-pinion system (Model 232, $33).

A shop microscope of similar design to the above is the Swiss-made OMAG microscope, distributed in the U.S. by Paul N. Gardner Co., Washington 14, D. C. ($30; magnification 10×–25×, 18×–35×, or 26×–50×, depending upon the eyepiece; additional eyepieces permitting changeover from one of these ranges to another are $10.)

Better image quality can be achieved if the objective-to-eyepiece distance is

maintained constant at that value which places the clearest objective image in the front focal plane of the eyepiece. That is, in these very simple microscopes the convenience of a wide range of magnifications must be sacrificed if the lens system is to yield the best image of which it is capable. This is the approach taken in a number of shop microscopes, of which the Model 65 of Titan Tool Supply Co., Buffalo, N. Y. 14216 ($17) is representative. The magnification is 50×, a built-in, pen-sized, battery-powered flashlight provides the illumination of the specimen, and a reticle with a scale marked in 0.001-in. divisions is incorporated in the eyepiece. This is shown in Fig. 56.

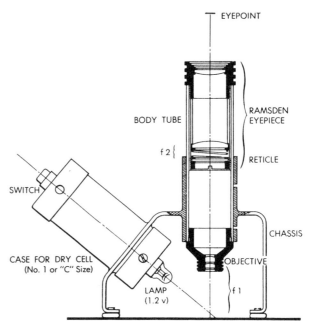

Figure 55. PTC Industrial Microscope, Model 232.

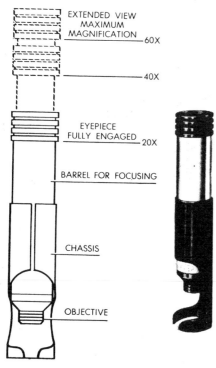

Figure 54. Pocket Microscope, Model 233, of Pacific Transducer Corp. Magnification is varied by adjusting the distance between the objective and the eyepiece.

Figure 56. Shop Microscope, Model 65, of Titan Tool Supply Co.

The Bausch and Lomb Optical Co., Rochester, N. Y. 14602 also supplies microscopes of this general type. One such is their Shop Microscope of 40×, with a reticle graduated in 0.001-in. divisions (No. 31-29-33-34, $100). This manufacturer also offers a series of Wide Field Tube Microscopes, consisting simply of an objective and eyepiece mounted at either end of a cylindrical tube. Magnifications are available of 10×, 20×, and 40× ($41 each); a tripod stand is available for mounting the microscope over the specimen ($15), and reticles can be inserted in the eyepiece ($15). The same basic design is also offered in the form of the B. and L. Wide Field Macroscope, shown in Figure 57. A prism, mounted between the objective and the eyepiece, permits the latter to be inclined for more convenient positioning of the observer's eye, and also serves to erect the inverted, real image produced by the objective.

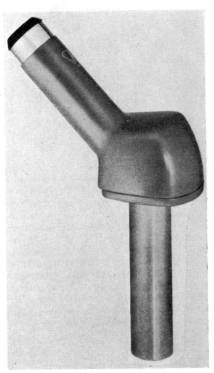

Figure 57. The Wide Field Macroscope of Bausch and Lomb Optical Co.

The Macroscope tube component is about $30 more expensive than the Tube Microscope.

Other models of simple shop microscopes are available from the above-mentioned manufacturers, as well as from the Ealing Corporation, Cambridge 38, Mass., and the Gaertner Scientific Corp., Chicago, Illinois 60614.

B. Micrometer-Equipped Microscopes

The measurement of dimensions and angles in the specimen by direct visual comparison of a microscope image with the simultaneously formed image of a reticle or scale is a technique of inherently low precision and accuracy. This stems in part from the difficulty of mentally aligning the features of the specimen with the markings of the reticle, and in part

from the distortions produced in both images, (specimen, and reticle), by the microscope optics. In addition, a reticle can only serve to give information about the dimensions of features that are completely encompassed within the limited field of view (which is 0.2–0.3 in. wide in a 20× shop microscope; about 0.07 in. at 100×), and cannot be used at all for depth measurements.

(It may be pointed out at this place that the term reticle, or reticule as it is sometimes spelled, has been used in the preceding paragraphs in the sense in which it is currently commonly encountered in manufacturer's literature, viz., as designating any insert in an eyepiece which bears fiducial or scale markings. However, this is a loose usage. More properly, a reticle consists of a reference mark or marks, such as crosshairs. A graticle, or graticule, is the term for any glass disc bearing graduations, scales, or other designs intended for mounting at

Figure 58. Micrometer eyepieces of Gaertner Scientific Corp. A. With external micrometer drum only. B. With revolutions counter mounted on micrometer drum.

the focus of an eyepiece. If the glass contains a linear scale that is ruled and numbered, it is often called a scale micrometer, or simply a micrometer.)

The term "micrometer" has through usage acquired the more general meaning of any displacement device based upon the operation of a precision screw mechanism, and in this sense one speaks of the addition to the basic microscope of micrometer accessories. Such accessories can greatly improve the precision and accuracy of dimensional gaging with the microscope.

1. Horizontal Displacement Micrometer in Eyepiece. One common approach is the micrometer eyepiece, shown in Figure 58A and B. A crosshair or other form of reticle mounted in the focal plane of the eyepiece is attached to the shaft of a precision screw of small pitch (e.g., 0.25 mm). Rotation of the drum fixed to this shaft can thus serve to position the reticle at any desired point in the field of view. For example, a 10× eyepiece used in conjunction with a 5× objective may be fitted with a screw that advances the crosshair image by 0.05 mm in the field of view for each complete revolution of the micrometer drum. The number of complete revolutions needed to move the reticle from one place in the field of view to another can be counted by means of an external revolutions counter, or by reference to a special scale imaged in the field of view by means of a stationary graticle in the eyepiece (cf. Fig. 59). The fractional part of a revolution is read on the external micrometer drum.

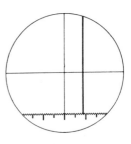

Figure 59. If there is no external revolutions counter, the number of complete drum revolutions can be determined by reference to an internal scale. The crosshairs and notched scale shown are stationary markings on a graticle in the eyepiece. The heavy vertical line is the filar reticle controlled by the micrometer screw.

If the movable reticle consists of a single strand, the device is called a filar micrometer. If a pair of strands, parallel and close together, is employed, the eyepiece is termed a bifilar micrometer. The latter permits more accurate positioning of the reticle on fiducial mark than the former. When the settings are to be made on a profile or boundary that divides the field of view into dark and light areas, the best precision of setting is obtained by the use of a reticle in the form of a 90° crosshairs (cf. Fig. 60).

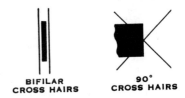

Figure 60. Bifilar crosshairs are best for precise positioning on fiducial marks. Ninety-degree crosshairs are best when setting must be made on a boundary between dark and light areas in the field of view.

A representative filar micrometer microscope is the Gaertner Scientific Corp.

Model 113, shown in Figure 61. The magnification is 50×; the range of measurement on the object is 2 mm, with each complete drum revolution corresponding to a displacement of 0.05 mm on the object ($182 without rack-and-pinion gears for focusing; $30 additional for these).

Filar micrometer eyepieces are available from a number of manufacturers, and can be employed in place of the regular eyepiece in any standard microscope to adapt that instrument for dimensional measurements. Examples of this type of component are the No. 425, 10× eyepiece ($120) of American Optical Co., Buffalo 15, N. Y.; the Micrometer Eyepiece, 10× ($88) of Rolyn Corp., Arcadia, Calif.; the Standard Filar Micrometer, 10× ($70) of Unitron Instrument Co., Newton Highlands 61, Mass.; and the Filar Micrometer, 10× ($150) of the Ealing Corp., Cambridge 38, Mass.

Figure 61. Filar micrometer microscope of Gaertner Scientific Corp.

A micrometer eyepiece equipped not only for linear displacement of the filar reticle, but for its angular rotation as well, is the Unitron Widefield Filar-Goniometer, 10× ($155), shown in Figure 62. The goniometer scale is marked every 2°, and the angular position can be read by means of a vernier to 20' of arc.

2. Vertical Displacement Micrometer in Body Tube. The micrometers described in the preceding section are only capable of yielding dimensions in a plane perpendicular to the optical axis, and if the points of interest on the specimen do not lie on a plane which is also perpendicular to this axis, one can only determine the projection onto this plane of their separation. It is, however, possible to use a micrometer screw for measuring distances along the axis of the optical train if an objective lens of very small depth of focus is used, and the micrometer is employed to displace the eyepiece

relative to the objective. This is the design principle of the Ace Optical Microm-

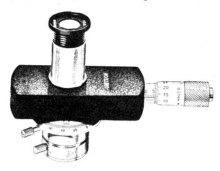

Figure 62. The Widefield Filar-Goniometer eyepiece of Unitron Instrument Co. The scale on the annular ring at the base of the eyepiece is the protractor, or goniometer scale.

eter, available from Ana-Tec/Air Cargo Division, Monogram Industries, Culver City, California 90230 ($273; plus $30 for a support base); the instrument is shown in Figure 63. To make a measurement, the objective is positioned over the desired area of the specimen, and the micrometer thimble is rotated until the first points of interest are brought into sharp focus. The micrometer and vernier are read, and the focus is then changed up or down until the next points of interest come into sharp focus. The barrel is read again, permitting the vertical displacement to be derived from the difference of these determinations (corrected for the refractive index of the medium at the object plane).

3. Micrometer Displacement of Specimen. The maximum distance over which

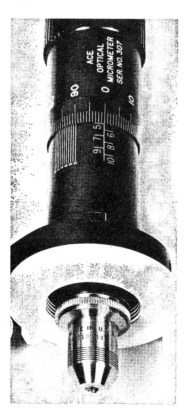

Figure 63. The Ace Optical Micrometer, for depth measurements through vertical displacement of the eyepiece relative to the objective.

an eyepiece micrometer can be employed is limited by the size of its visual image, and is only of the order of magnitude of a millimeter. If larger distances are to be measured, either the specimen or the microscope must be caused to suffer the displacement. Precision slides on which the microscope or the specimen may be mounted, and which are controlled by a micrometer screw, are available from several sources. A micrometer slide for linear displacement is shown in Figure 64; it is the Model 300P of the Gaertner Scientific Corp. ($470) and can traverse a total displacement of 25 mm with direct reading of position to 0.001 mm.

An illustration of the incorporation of the slide micrometer and a microscope in a support designed for the measurement of the positions of, e.g., spectral lines on photographic plates or films is shown in

Figure 64. A slide micrometer made by Gaertner Scientific Corp. for precision displacement of the relative positions of microscope and specimen.

Figure 65. The instrument shown is available from Gaertner as a complete unit; it is their M1160-342 Micrometer Slide Comparator ($1130), with a linear range of 100 mm and direct-reading to 0.001 mm. If a range only half as long, and direct-reading to only 0.01 mm are adequate, the cost of the components is approximately one-half of the preceding.

Figure 65. A Gaertner micrometer slide comparator suitable for the analysis of spectral plates or films.

[The term *comparator* is commonly used to designate an instrument, such as these, in which the positions of various points or lines are compared to each other, or to some common reference, fiducial mark. The term *cathetometer* may frequently be employed interchangeably with comparator. though the former is generally reserved for measuring instruments in which the viewing optics (microscope or telescope) are mounted with the optical axis horizontal, and the displacement that is measured is in the vertical sense.]

Coordinate comparators are constructed by mounting two micrometer slides so that their directions of travel are mutually perpendicular. A simple attachment that can be added to many standard microscopes to adapt them for coordinate measurements, direct-reading to 0.0001-in., is available from Automation Gages, Inc., Rochester, N. Y. 14621 (AG Hi-Precision X-Y Measuring Stage, $240). A Bausch and Lomb microscope equipped with this type of stage is shown in Figure 66 (Toolmakers' Measuring Microscope,

Figure 66. The Bausch and Lomb Toolmakers' Measuring Microscope, with 1-inch travel in both coordinate directions.

$1250, with built-in bright field and surface illumination; 1-in. motion of each micrometer, direct-reading to 0.0001 in.).

An example of a standard microscope mounted on a coordinate comparator with a range of 4 × 3 in. is shown in Figure 67 (available from the Ealing Corporation; Model 11-350, $935). This instrument can be placed on its side and used as a cathetometer; in this instance it is possible to replace the 0.67-in. focal length microscope objective by a 6-in. focal length objective, thereby converting the optics into a telescope (objective lens 11-352, $20). Other coordinate compara-

Figure 67. Coordinate Research Measuring Microscope available from the Ealing Corporation. The longitudinal motion is 4 inches, direct reading to 0.002 mm, and the cross motion is 3 inches, reading to 0.1 mm by means of a vernier.

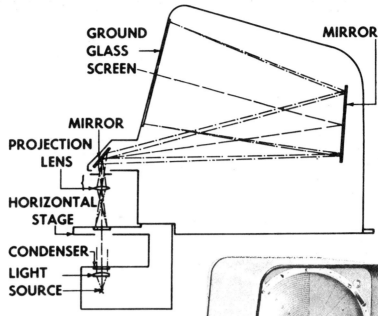

Figure 68. Optical plan of the Bausch and Lomb Ten-Inch Optical Comparator.

tors are available from Gaertner; Ealing; Bausch and Lomb; Titan Tool Supply; Opto-Metric Tools, Inc.; and Unitron.

The measurement of dimensions can be accomplished with considerable ease, at the cost of some precision, with the aid of a *projection comparator*. An optical projection system, such as that shown in Figure 68, is employed to form an enlarged image of the object on a screen bearing a coordinate grid or other scale markings. Thus, the Bausch and Lomb 10-in. Optical Comparator ($900) projects onto a 10-in. diameter screen a 10× to 50× magnified image, depending upon the objective used. The specimen is positioned by means of micrometer slides.

A similar instrument, made by Hydra-Electric Co., Burbank, California 91503, is shown in Figure 69. It is designated the 600 VistaScope, and may be used with objectives giving magnifications from 5× to 62.5× ($995 with one objective lens, but without specimen stage). Other projection comparators are available from Reichert, Vienna (distributed in the U.S. by W. J. Hacker and Co., West Caldwell, New Jersey); Stocker and Yale, Inc., Marblehead, Mass.; and Micro-Vu, Inc., Burbank, California.

C. Toolmakers' or Universal Microscopes

A *universal* microscope is one with which the specimen can be positioned with high precision in any coordinate direction relative to the microscope objective, and with which objects of a great range of sizes and shapes can be examined. Since these are the requirements generally specified by the operators of tool and inspection departments in industrial precision manufacturing establishments, these instruments are also commonly known as *toolmakers'* microscopes. An example of this type of versatile, precision measuring microscope system is the Elgeet-Olympus STM Toolmakers' Microscope made by

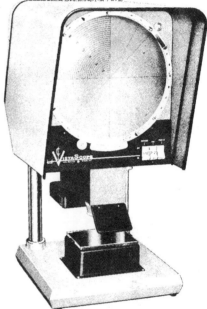

Figure 69. The VistaScope projection comparator of the Hydra-Electric Co.

Figure 70. The Elgeet-Olympus STM Toolmakers' Microscope.

Elgeet Optical Co., Rochester 11, N. Y., and shown in Figure 70. It has a protractor, 10× eyepiece, 3× and 15× objectives, built-in illumination systems for vertical, surface, and substage lighting, a stage with 75 mm lateral movement, 25 mm cross movement, and 360° rotation. The cost is about $1400.

Other toolmakers' microscopes are available from Unitron; Opto-Metric Tools, Inc. (distributor of the Leitz line); and Westwood Engineering Co., Los Angeles 64, California. The most elaborate instrument of this class is made by Zeiss, West Germany (distributed in the U.S. by Carl Zeiss, Inc., New York 18, N. Y.); the cost of the basic unit is about $22,000.

D. Double-Imaging Microscopes

The precise measurement of dimension, shape, symmetry, and orientation can be greatly facilitated by the use of double imaging techniques. A beam splitter is inserted into the optical system, usually behind the objective lens, so that two beams of light are produced which are acted upon by the ocular to yield a pair of images of the original object. These images may be displaced with respect to each other by the adjustment of reflecting surfaces within the body tube of the micro-

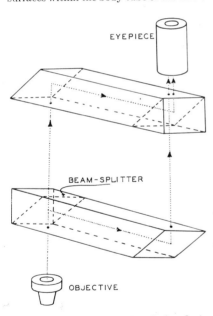

Figure 71. Design principle of the Cooke-A.E.I. Image Splitting Measuring Eyepiece. A semi-silvered mirror in the lower prism splits the light from the objective into two beams, which are brought to a common viewing position by a similar mirror located in the upper prism.

scope, and the amount of adjustment required to produce a given displacement constitutes the measured parameter.

This design principle is illustrated in Figure 71, which shows the component parts of an image splitting eyepiece available from Cooke, Troughton, and Simms Inc., Malden, Massachusetts 02148 (the Cooke-A.E.I. Measuring Eyepiece, $490). Light passed into the lower prism by the objective is divided into two beams by a 45° semi-silvered mirror. These two beams follow separate paths into the upper prism, where they are brought to a common viewing plane by a second half-

silvered mirror. Rotation of one of these prisms relative to the other will cause the two images to be mutually displaced in the viewing plane (or, they are optically sheared) by an amount proportional to the angle of rotation.

Figure 72 shows how this double-imaging principle can be exploited for the measurement of the dimensions of particles. If the corresponding reflecting surfaces of the lower and upper prisms are exactly parallel, the two images coincide in the viewing plane, and only one image is seen, as in Figure 72 A. As the prisms are rotated, the images diverge. When the edges of the two images just touch, as in Figure 72 C, the amount of optical shear that has been introduced is equal to the lateral dimension of the object, and a micrometer drum linked to the prism rotator can be employed to read out the particle size.

This approach to particle size determination has the advantage of convenience and speed, and the results are free of those errors that may arise in estimations based on reticles, graticles, or filar micrometers which stem from parallax, vibration of the microscope optics, or motion of the particles. However, the range of dimensions that can be measured is limited by the field of view at the magnification being employed, and the clarity of the image is reduced because of the division of the light gathered by the objective into two separated beams.

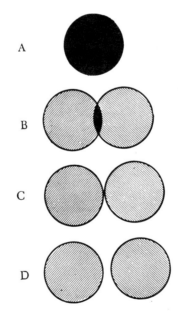

Figure 72. Visual appearance of the field of view when a circular object is seen with the image splitting eyepiece as a function of the amount of optical shear (degree of prism rotation) introduced.
  A. No shear, prism faces exactly parallel; the two images coincide.
  B. Optical shear slightly less than the diameter of the object.
  C. Optical shear exactly equal to object diameter.
  D. Optical shear greater than object size.

When numbers of particles are simultaneously in the field of view, some confusion may arise as to which are the cor-

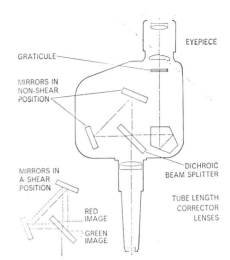

Figure 73. Optical plan of the Watson Image-Shearing Eyepiece, in which a dichroic beam splitter is employed. Light reflected from this component is green, whereas light transmitted by it is red. If the deflecting mirrors are in a position relative to the beam splitter that will cause the two beams to form separated images, the latter will be tinted green or red, respectively.

responding members of the various double images. This difficulty is obviated by causing the two images of a given particle to differ in color or intensity. Such differentiation can be achieved by the use of color, neutral density, or polarizing filters in appropriate positions in the optical train.

The Watson Image-Shearing Eyepiece, made by W. Watson and Sons, Ltd., Barnet, Herts, England (available in the U. S. through W. J. Hacker and Co., West Caldwell, New Jersey 07007; $425) employs a dichroic beam splitter to yield a red and a green image, respectively. The optical design is illustrated in Figure 73. A dichroic component has the property that the spectral distribution of intensities in the light reflected by it is significantly different from that of the light transmitted through it. In this case, the reflected

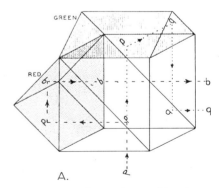

Figure 74. A. Beam-splitter yielding images that are laterally and vertically reversed, respectively, with respect to the object. Only the center point is common.

light is rich in those wavelengths which combine to give the sensation of green, whereas the transmitted light appears red in color. Relative displacement of the two images is accomplished by the micrometer-controlled movement of a slide on which the two reflecting mirrors are mounted.

An interesting application of the double-imaging principle is in the study of shapes and symmetry, and is exemplified in the Zeiss Universal Measuring Microscope, as well as in the Zeiss Toolmakers' Measuring Microscope. Figure 74 A shows a beam-splitter on which are mounted two roof prisms—one of red-tinted glass, the other of green glass. The image formed by the beam which has been reflected into the red prism is a *vertically reversed* image (i.e., turned upside down), whereas that formed by the beam which has passed into the green prism is *laterally reversed* (i.e., turned around sidewise). Only the centers of the images coincide with the center of the original object. If the two images are superimposed, any deviations from central symmetry will become clearly evident.

A beam-splitting arrangement in which the two images are reversed only with respect to a vertical axis is shown in Figure 74 B. This type of component is useful in measuring relative distances of features located on opposite sides of a reference line in the object, as well as for the precise location of the center of a symmetrical object, such as a drill hole.

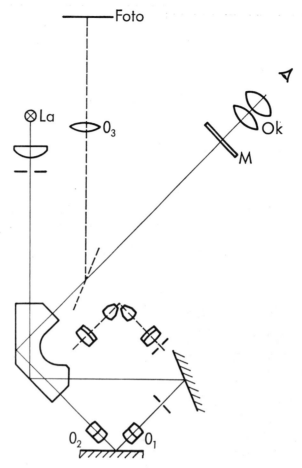

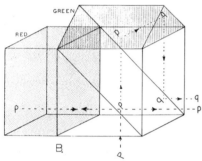

Figure 74. *B. Beam-splitter yielding images that are laterally reversed about a vertical axis.*

Figure 75. *B. Optical diagram. A fine line of light defined by a lens and slit is focused by objective O_1 onto the specimen at a 45° angle to its surface. The reflected light is gathered by objective O_2 and directed to the viewing ocular, Ok. A camera attachment is represented by O_3 and Foto, and M is a reticle. A second pair of microscope objectives is shown above O_1 and O_2 these are rotated into position in place of the first pair when higher magnification is desired.*

E. Surface Profile Microscopes

Details of the surface of an opaque specimen are often best revealed by the use of a light beam that makes a 45° angle with the surface. If the reflected beam is viewed at an equal angle, viz. 45°, then any surface irregularities can be easily seen, for under these conditions a perfectly smooth surface would act like a mirror, reflecting substantially all of the light into the objective lens, whereas local deviations from planarity would deflect parts of the light beam away from the objective. This principle is employed in the so-called *light-section* microscope, the Zeiss version of which is shown in Figure 75. A thin line of light is focused onto the surface by means of one objective, and is viewed at the specular reflection angle by a second objective.

If the surface under examination has a coating of a sufficiently translucent material, multiple images may be formed, as illustrated in Figure 76. These may be employed to yield the thickness and smoothness of the coating, in addition to giving the profile of the substrate and the coating. Some typical fields of view seen through the light-section microscope are reproduced in Figure 77.

The Zeiss instrument shown in Figure 75 has objectives giving 200× and 400× magnification, a mechanical stage with micrometer-controlled displacement of 25 mm × 25 mm in perpendicular directions, an eyepiece micrometer, and a camera; it is priced at about $2500.

A Leitz Light-Section Microscope is available that is based on a similar optical principle, but designed for coarser measurements. The total magnification is 30×, and the diameter of the field of view is ¹/₄-inch. A protractor eyepiece is provided that permits reading of both angular deviations and linear displacements. This

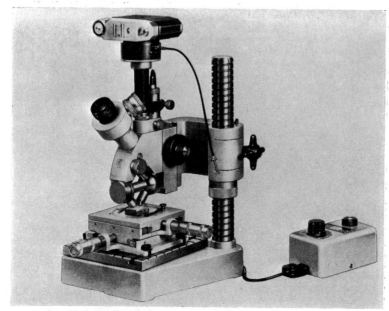

Figure 75. *A. The Zeiss Light-Section Microscope.*

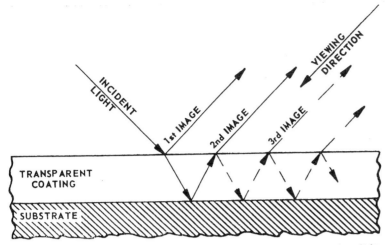

Figure 76. If an opaque specimen bearing a translucent coating is examined with a light-section microscope, multiple images are formed as a result of reflections from the top of the coating as well as from the substrate surface.

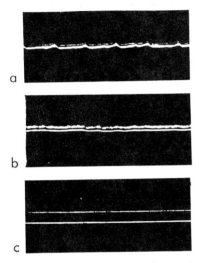

Figure 77. A. Aluminum metal surface coating with a 13-micron thick layer of oxide. The lower light band is from the uppermost surface. Magnification 200×.

B. A metal substrate coated with an 8-micron thick film of a transparent lacquer. Magnification 400×.

C. A 70-micron thick specimen of transparent plastic sheet. The reflections are produced by the front and back surfaces. Magnification 200×.

instrument is available in the U. S. through Opto-Metric Tools, Inc., New York, N. Y. 10013 and costs about $1100.

The approach to the detailed examination of elevations, depressions, slopes, and other aspects of the topography of a specimen's surface which is capable of the greatest degree of resolution using light as the illuminant consists in the application of interferometry. Light from a suitable source is divided into two beams by means of a beam-splitter; one of these beams (the reference beam) is reflected by a flat mirror and combined with the other (sample) beam which has traveled to, and been reflected by, the specimen surface. The total path lengths of the reference and sample beams are adjusted to be equal within one or several wavelengths of light. Variations in the topography of the specimen surface will then result in variations of the path length of different parts of the sample beam, and this effect will be manifested in the form of displacements of the interference fringes which arise when the reference and sample beams are combined in the image plane.

A microscope objective that makes use of this principle is manufactured by W. Watson and Sons, Ltd. The optical design is shown in Figure 78. Light enters through the long, flanged tube and impinges upon the beam splitter prism. A portion of the beam travels straight through and strikes a flat reference mirror, where it is reflected and returned to the beam splitter, which deflects it up into the objective lens. The rest of the original beam is sent down toward the specimen,

which reflects it back up and into the objective lens. The operator adjusts the position of the reference mirror by means of a fine screw mechanism so that the total reference path length is about equal to the sample path length. An example of the interference fringe pattern seen through this objective is shown in Figure 79.

The Watson Interference Objective is

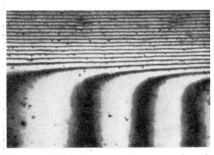

Figure 79. Interference pattern seen through the interference objective. Sample was a germanium wafer, the upper part of which is beveled at a 2° angle.

designed to screw into a standard microscope nosepiece. It has a focal length of 16mm and N.A. = 0.2; the recommended magnification is 100×. The diameter of the field of view is 1.3 mm, and surface irregularities $1/10$th of a wavelength of the light used (i.e., about 0.06 microns for yellow light) can be resolved.

An interference microscope based upon the principles described above is manufactured by A. Kohaut, Bamberg, West Germany (distributed in the U. S. by H. A. H. Behrens, Inc., New York, N. Y. 10017; complete with photomicrographic accessories, about $2500). It provides magnifications of 180× and 270× with a field of view of 1 mm × 1 mm.

The precision with which small displacements of interference fringes can be measured is limited by the gradual nature of the intensity transition at the boundary between the light and dark portions of the fringe system. This can be markedly sharpened, and the precision of measurement correspondingly improved, by means of a photographic technique due to Sabatier. The development of a print is interrupted when the process is about half-complete, the print is strongly illuminated, and the developing process is resumed and

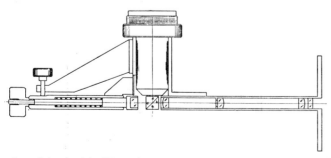

Figure 78. Optical design of the Watson Interference Objective. Light enters through the flanged tube at the right. The beam splitter sends part of the light down to the specimen (not shown), and the rest to the reference mirror mounted on the micrometer-driven shaft at the left. Both beams return to the beam-splitter, and from thence proceed to the objective lens. The latter combines the beams to yield an interference pattern.

Figure 80. At the left is a photograph of a fringe pattern such as is seen through the interference microscope. At the right is a print that has been developed by the Sabatier method, i.e, interrupting the development, strongly illuminating the print, then completing the development. The thin dark lines remaining mark the boundaries between the light and dark regions in the original set of fringes.

completed. This results in a bleaching of the darkest regions of the print, whereas the partially darkened areas remain relatively unaffected. The extent to which this technique can improve the measurability of an interferogram is illustrated in Figure 80.

If the surface irregularities are large compared to the wavelength of the illuminating light, the interference fringes may be so numerous and closely spaced as to make quantitative determination of elevations or depressions difficult. The interpretability of such interferograms is greatly facilitated by the use of a replica technique developed by Zehender. A thin film of transparent plastic is applied to the surface of the specimen, so that an impression of the surface topography is fixed in the plastic. This replica is then removed and transferred to a special chamber for viewing. The chamber has a mirror at its bottom, and the plastic film is placed on the mirror with the replica surface down. An immersion liquid is employed to fill the space between the mirror and the replica.

It will be recalled that the *optical path length* of light traversing the chamber is equal to the sum of all the $n_i \times l_i$'s, where n_i is the refractive index of the ith medium, and l_i is the distance traveled in that medium. Hence, the closer the refractive index of the immersion fluid is to that of the replica, the smaller will be any optical path length differences due to the irregularities in the thickness of the film, and the fewer will be the number of interference fringes formed. The principle is illustrated in Figure 81, and some typical results are reproduced in Figure 82. A kit containing supplies for replica preparation and Zehender chambers is available from H. A. H. Behrens, Inc., for about $300.

Another approach to the use of interference fringes in the study of surface topography is exemplified in the Interference Microscope Objective made by Bausch and Lomb (No. 33-16-02-04,

16-mm focal length, $175), and shown in Figure 83. A small glass test plate which is an integral part of the objective assembly is brought into direct contact with the specimen to be tested. A vertical illuminator attached to the body tube above the objective sends light rays down to the specimen. Rays are reflected from the surface of the specimen, and from the under-surface of the test plate; if these surfaces are sufficiently close to parallel, an interference fringe pattern is seen through the microscope. The spacing of successive fringes with this objective corresponds to a depth variation of 0.000011 inch.

2. Standard Microscopes

The standard microscope, which was historically the first type of compound microscope to come into use, and which is still the most widely-used instrument among students and workers in the life sciences, is primarily intended for use with thin, transparent specimens. Simple instruments of this type, often designated as *students' microscopes*, are made by most of the microscope manufacturers.

The basic components of the standard microscope are: (1) ocular, (2) body tube, (3) objective, (4) stage, and (5) substage illumination system. Examples of the simplest and most inexpensive combinations of these components are given in Figure 84, which shows the Bausch and Lomb series of elementary school and science teaching microscopes, ranging in price from $12 to $50.

Numerous types of more elaborate standard microscopes are available that incorporate combinations of design fea-

Figure 83. The Bausch and Lomb interference microscope objective, in position on a microscope equipped with a vertical illuminator.

tures providing for greater convenience, range of magnifications, image quality, and flexibility of application—all at a correspondingly increased price. These design features can best be described in terms of the component parts of the generalized standard microscope.

OCULAR

Most standard microscopes are provided with one or more oculars that range from 5× to 20× in magnification. Since the real magnification attainable with a given microscope is generally limited by the numerical aperture of the objective, it is not practical to seek to attain very high total magnifications by excessively increasing the power of the ocular. However, in the case of stereomicroscopes, high power objectives cannot be employed, since an appreciable specimen-to-objective distance must be maintained to create a three-dimensional effect in the observer. Consequently, when high total magnifications are required with such a microscope, it is necessary to achieve that result mainly through the power of the ocular. Thus, Bausch and Lomb makes a pair of 33× stereo oculars which will yield a total magnification of 200× in their StereoZoom microscope equipped with the 3× zoom lens.

Microscopes and their oculars and objectives may be designed so as to allow a variety of different lens systems to be used interchangeably in the same body tube. A *parfocal* series of lenses is a set which satisfies this criterion. The manner in which the constituent field and eye lenses of various Huygens and Kellner oculars must be positioned to keep the focal plane of the unit at a fixed location in the body tube is shown in Figure 85. The objectives are also parfocal; each one is designed so that the image it forms lies in the ocular's front focal plane.

Special oculars are available that permit: (A) two observers to look through the same microscope simultaneously, (B) two different microscopes to be viewed at the same time by a single observer, (C) the microscope image to be projected onto a screen, and (D) visual superposition of the microscope image on a drawing board.

An example of an ocular designed for joint viewing is the Unitron Demonstration Eyepiece (Model DE, 10×, $40), shown in Figure 86. The beam from the objective is split into two parts, one for each ocular. The unit may be rotated around the axis of the body tube, and a

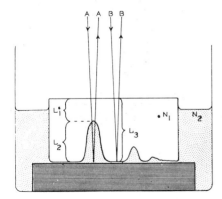

Figure 81. The Zehender chamber employs a mirror, immersion liquid, and replica of the surface under study. Optical path differences due to the topography of the replica surface are reduced in proportion to the closeness of matching of the refractive indices of the replica and the immersion fluid. The following equations apply to the diagram:

Optical Path Length A—A: $2[n_1 l_1 + n_2 l_2]$

Optical Path Length B—B:
$$2[n_1 l_3] = 2[n_1 l_1 + n_1 l_2]$$

Optical Path Difference:
$$\Delta = 2l_2(n_1 - n_2)$$

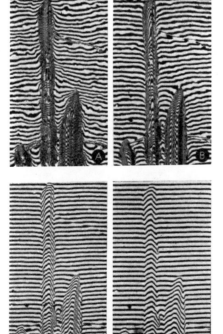

Figure 82. Photographs of the fringe pattern obtained from a polished steel surface with three scratches. *A.* The part is viewed directly. *B.* Replica viewed in the Zehender chamber through air. *C.* Viewed in the chamber through water. *D.* Viewed in the chamber through oil.

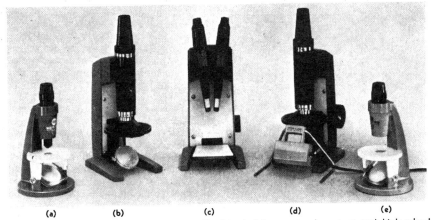

Figure 84. The Bausch and Lomb microscope line intended for use in elementary and high school science courses. (a) Elementary science microscope × 100. (b) Zoomscope × 25–100. (c) Stereomicroscope. (d) Zoomscope 50–200×. (e) Elementary science microscope ×10.

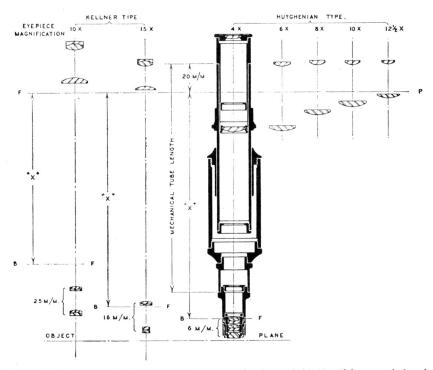

Figure 85. The same body tube may accept a variety of oculars and objectives if these are designed to be parfocal. The spacing of the component parts of the lens systems is such as to cause the focal plane of the ocular to lie on the plane *FP*. The back focal plane of each objective is at such a location, *BF*, that the real image is formed at *FP*.

paper or a template. The principle is illustrated in Figure 89. A prism containing a 45° mirrored surface with a central hole in it is positioned over the ocular. The light beam from the microscope ob-

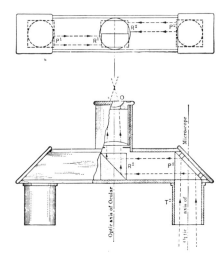

Figure 87. The comparison of two specimens is facilitated by combining the fields of view of two microscopes into a single view consisting of one-half of each field. The plan view at top shows how two prisms, *R1* and *R2*, produce a composite view of the two fields.

Figure 88. Nikon projection screen attachment on the Model S-KE microscope.

built-in, movable pointer located in the plane of the real image of the objective permits one observer to call the attention of the other to specific points in the field of view.

The principle of the comparison ocular is illustrated in Figure 87. Two reflecting prisms are employed to combine the fields of view of two microscopes and place them in side-by-side juxtaposition along a common center line. This technique is particularly useful when an unknown is to be identified by comparison of its structural features with those of standard specimens. An example of this type of ocular is the Comparison Eyepiece of Nikon, Inc., Instrument Division, New York 3, N. Y. ($190).

A reflecting prism or 45° mirror placed over a standard ocular can be employed to project the microscope field of view onto a screen for group observation. An example of such an ocular accessory is the Nikon

Figure 86. Model DE Demonstration Eyepiece of Unitron Instrument Co. for simultaneous viewing by two persons.

projection screen attachment ($170), shown in place on their Model S microscope in Figure 88. Because of the large area of the image on the screen (17 cm diameter for the unit shown in Fig. 88), a very bright source of light must be used if easy visibility of the image by a group of persons is desired.

The *camera lucida* is an optical accessory placed over the microscope ocular that permits the observer to see the microscopic field superimposed upon a sheet of drawing

jective passes up to the observer's eye through this hole, while light from the drawing surface is reflected into his eye by an adjustable mirror plus that mirrored surface in the prism. The Unitron Camera Lucida, Model A ($40) is shown in Figure 90. In addition to the features described, it includes a pair of rotatable Polaroid filters which allow the relative intensities of the two images to be varied as desired.

Protracted observation through a monocular microscope can be a great strain, because of the different images being received by the observer's two eyes, or

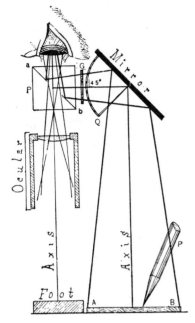

Figure 89. The microscopic field of view can be superimposed on a drawing by means of the camera lucida, the heart of which is a prism containing a 45° mirrored surface with an un-silvered central hole, a—b. G is a glass diffuser plate, Q is a protractor for adjusting the external mirror angle, and AB is the drawing surface.

Figure 90. The Camera Lucida, Model A, of Unitron Instrument Co.

received by the observer's two eyes, or because of the muscular effort involved in keeping one eye closed. Consequently, most standard microscopes are produced in versions that include beam-splitting arrangements, so that the same field of view can be presented to both eyes. Two types of beam splitting used for this purpose are shown in Figure 91, and a third type is shown in relation to the entire microscope optics in Figure 92.

In order for a binocular microscope to achieve its goal of greater comfort and clarity in viewing the image, it is essential that the two ocular images be precisely equal in magnification. Thus, the two optical paths must be equal in effective length (viz., $\Sigma\ n_i \times l_i$), and any extra thickness of glass through which the light travels in one tube must be compensated for in the other. The two images must enter the respective eyes at equivalent angles, and hence the separation of the binocular eyepiece tubes should be adjustable to match the observer's interpupillary distance. This can be achieved most simply, without altering any of the optical

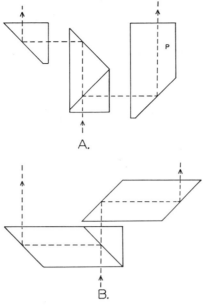

Figure 91. A. Beam splitting arrangement of Jentsch. The prism, P, is elongated to equalize the optical paths traversed by the two beams.

B. The Siedentopf type of beam splitter.

path lengths, by rotating one of the eyepiece tubes about an axis through the center of the beam splitter if the arrangement of Figure 91 is being used, or by a symmetrical movement of both tubes about the center of the beam splitter if an arrangement such as that of Figure 92 is involved.

The simple binocular arrangement described above should not be confused with the true stereoscopic microscope design, in which two *different* angles of view of the object are presented to the eye. In the binocular design, the *same* angle of view of the object is presented to each eye. The former gives depth perception; the latter does not, but does permit more comfortable viewing of a flat field.

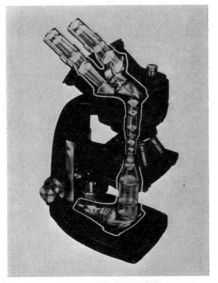

Figure 92. Binocular arrangement employed in the Bausch and Lomb laboratory microscope. The model shown incorporates this company's Dynazoom and Flat Field Achromatic systems.

BODY TUBE

The body tube of a standard microscope provides the means of rigidly supporting and aligning the objective and ocular lens systems, and it is fitted with mechanical arrangements for precise positioning of this optical train relative to the object under study.

For comfort in viewing, some microscopes have the upper section of the body tube inclined at an angle to the vertical. This involves the inclusion of a deviating prism or mirror at a convenient point between the objective and ocular, as shown in Figure 93. Such a component also allows the direction of viewing to be rotated as desired about the axis of the objective; an observer may thus have the option of placing the massive, curved limb of the support stand either toward or away from himself when working at the microscope. The flexibility of mechanical arrangement and ready interchangeability of components characteristic of this design are illustrated in Figure 94.

Some standard microscopes are equipped with a graduated, adjustable draw tube by means of which the mechanical tube length of the microscope can be varied, as shown in Figure 95. This permits the total magnification to be adjusted to a precisely predetermined value, which may be important in work requiring estimations of sizes or distances. Small draw tube adjustments are also helpful in optimizing the microscope optics to give maximum image clarity when cover-glasses of non-standard thickness are present on the specimen.

Standard microscopes intended for low and medium magnifications may have only a simple rack-and-pinion device for moving the body tube; this is the *coarse* focus control. However, when high magnifications are employed, the depth of focus is extremely short (e.g., 0.001 mm for a dry 4-mm objective of N.A. = 0.65), and a *fine* focus motion is essential. This is generally achieved by employing a fine screw to drive a hardened steel plate or ball bearing, the displacement of which is transmitted to the body tube by means of a lever working against a spring-loaded slider, as shown in Figure 96.

A recent microscope design by Reichert, Vienna (the Monopan, available in the U. S. through W. J. Hacker and Co.; shown in Fig. 97), combines the coarse and fine focus adjustments in a single control. Rotation of this control racks the body tube up or down in the conventional coarse focus motion. Shifting this control to or fro in a slotted groove constitutes the fine focus motion.

In some microscope designs, the focusing motion is limited to displacement of the nosepiece assembly, i.e., the objective lens and its mounting. This is much lighter in weight than the entire body tube with its fittings, and better control of the motion is possible. However, this involves variation in the objective-to-ocular distance, and if the objects examined are not all of the same thickness, or cannot be laid flat on the stage, the total magnification will differ from one specimen to another, if indeed the image can be brought into focus at all. If only standard microscope slides are to be examined, the system

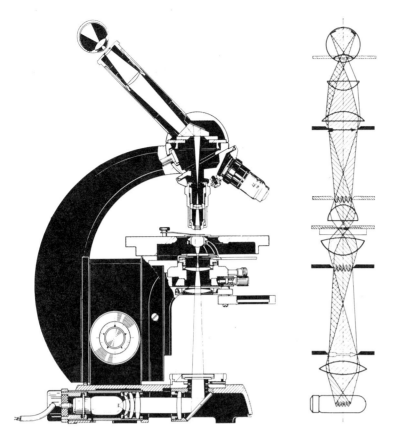

Figure 93. Optical design of the Zeiss Standard Microscope, Model GFL, showing the use of a deflecting prism to allow the body tube to be inclined for more comfortable viewing.

is quite effective. Variable objective-to-ocular distance causes no problems if special objectives are employed which are corrected for infinite tube length, i.e., in which the light rays proceeding from the

Figure 95. A Bausch and Lomb body tube with a graduated, adjustable draw tube.

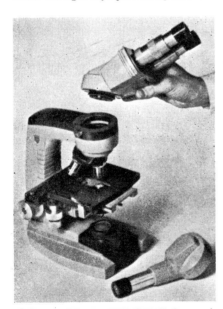

Figure 94. The American Optical Company's Series 10 Microstar microscopes incorporate rotatable and interchangeable body tube sections.

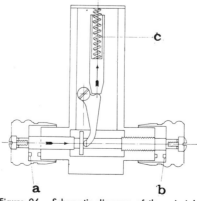

Figure 96. Schematic diagram of the principle of one type of fine focus control. Stops *a* and *b* limit the motion of the micrometer screw-driven spindle. The spring-loaded slider, *c*, communicates its displacement to the body tube.

back surface of the objective are parallel, and hence the ocular can be placed any distance away. A telescope-type optical system must then be used for the ocular. Such objectives cannot be interchanged with the more common type which are designed for a standard optical tube length.

The mechanical design of the American Optical Company's Spencer Series 60 microscope, shown in Figure 98, is of the type described in the preceding. The coarse and fine controls are mounted as concentric knobs; the coarse motion knob rotates a cam, whereas turning the fine motion knob shifts this cam to or fro. Both motions of the cam act on a lever that is attached to the microscope nosepiece.

In some microscopes, the focusing motion is applied to the stage, rather than to the body tube. This is most desirable when it is essential to maintain a precise relative positioning of the light source and objective, or when it is awkward to disturb this disposition, as in the case of certain metallographic microscopes.

OBJECTIVE

The change from one total magnification to another in examining a specimen is so frequent in the type of work for which standard microscopes are designed that it is practically indispensable to have several objectives mounted on the body tube in such a way that one can be switched for another without necessitating any major readjustment of the focus. This is accomplished by employing parfocal lens systems (*cf.* Fig. 85), and by mounting the objectives on a rotatable nosepiece. Most standard microscopes have nosepiece assemblies carrying 2, 3, or 4 objectives; the largest number currently available is 6. Such sextuple nosepieces are available with the Reichert "Biozet" laboratory microscope, and the Wild M20 Research Microscope. The latter is shown in Figure 99

Figure 97. The Monopan microscope of Reichert Optical Works. The black knob at bottom rear is the combined coarse and fine focus control. Rotation of the knob gives the coarse motion; shifting it in the horizontal slot gives the fine motion.

(Model KdGS, $600 without objectives, oculars, or photo accessories; available in the U.S. through Eric Sobotka Co., Inc., New York, N. Y.).

A recent trend in objective design for standard microscopes is in the direction of increasing the area of the field of view, and improving the "flatness" of the field, i.e., the ability to maintain a large part of the field of view in uniform focus. The so-called "flat-field" objectives differ from ordinary objectives by the addition of some

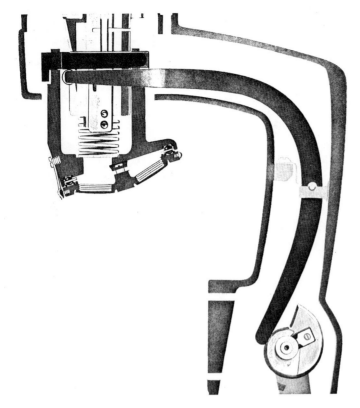

Figure 98. Focusing mechanism of the American Optical Co. Spencer Series 60 microscope. The coarse focus control rotates the cam, whereas the fine focus control shifts it. Both motions are transmitted by a lever to the nosepiece assembly.

the Bausch and Lomb Flat Field Optical System illustrated in Figure 92.

The depth of focus characteristic of high power objectives is so small that difficulty is often encountered in finding the image of the specimen. To minimize the risk, while hunting for that elusive image, of driving the objective into the specimen with sufficient force to ruin a valuable preparation, objectives are available with built-in "specimen protectors." The objective lens fits fairly loosely in a cylindrical tube, and is held in place by a weak spring, as shown in Figure 101. If the proper focal position is overshot, and the objective touches the specimen, the spring gives enough to prevent excessive pressure from being applied to the specimen. The spring-loaded nose assembly shown in Figure 98 functions in a similar fashion to protect the specimen.

STAGE

The stage serves as a support for the specimen, and is commonly either a circular or square platform fastened to the microscope stand. If high magnification is to be used, it is essential that a mechanism be provided for moving the specimen in a smooth and controlled fashion in two mutually perpendicular coordinate directions. Such mechanisms are generally

negative correction (concave curvature) to one or more of the lens elements, as illustrated for a low power objective in Figure 100. A similar result is achieved by the introduction of additional negative lenses in the optical train, as in the case of

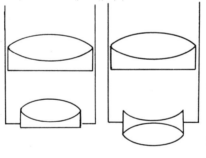

A. ORDINARY   B. FLAT–FIELD

Figure 100. Comparison of the ordinary Lister objective design and a corresponding flat-field objective of similar power.

Figure 102. A circular mechanical stage by Bausch and Lomb.

termed *mechanical stages*. The circular stage is particularly valuable when it is necessary to adjust and/or measure the orientation of the specimen relative to some unique axis, such as that of a camera

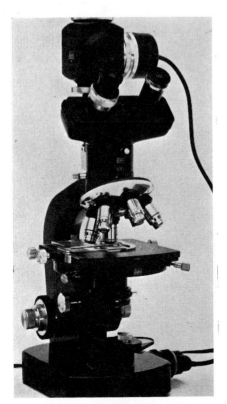

Figure 99. Wild M20 Research Microscope with sextuple nosepiece assembly.

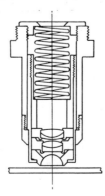

Figure 101. A microscope objective with built-in specimen protection. If the lens is pressed down against the microscope slide, the spring takes up most of the force by being compressed, and the danger of smashing the specimen is reduced.

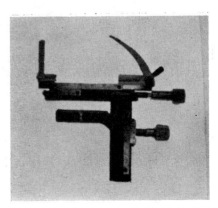

Figure 103. The Nikon Attachable Mechanical Stage.

in photomicrography, or the plane of polarized light in phase, interference, or polarization microscopy. A research-type circular stage by Bausch and Lomb that can be set and read to 6 minutes of arc, and that has a micrometer-controlled travel of 75 mm with front-to-back travel of 50 mm is shown in Figure 102.

Mechanical stages are available as accessories that can be attached to an ordinary microscope stage; an example is the Nikon Attachable Mechanical Stage ($40) shown in Figure 103. Similar accessories are available from most microscope manufacturers, as well as from: Rolyn Corp., Arcadia, California; Graf-Apsco Co., Chicago 26, Illinois; and Edmund Scientific Co., Barrington, New Jersey 08007.

In some investigations, it is necessary to observe a specimen while it is being subjected to special conditions, such as low or high temperatures, vacuum, etc. Accessories are available for attachment to ordinary microscope stages that accomplish those ends.

A cooling device designed to be mounted on a microscope is the Thermoelectric Microscope Cold Stage made by Industrial Instruments, Inc., Cedar Grove, New Jersey (Model CY3-A, $182), shown in Figure 104. It is based on the thermoelectric cooling principle, viz., the fact that if a direct current flows through junctions between dissimilar metals, one junction will heat up and the other will cool down. By using tap water at 15°C

to remove the heat from the first junction, this unit can lower the temperature in the vicinity of the other junction to −60°C. The temperature of the cold junction can be sensed by a thermistor probe thermometer (Model 42SL, $165). If the direction of current flow is reversed, the device can be used as a heater to raise specimen temperatures as high as +80°C.

The cold vapors from a cylinder of liquid CO_2 are used to lower the temperature of a specimen to −20°C in the Leitz Heating and Cooling Stage, Model 80 ($380), the component parts of which are shown in Figure 105. A resistance heater built into the stage permits the temperature to be raised to +80°C. Nitrogen flushing is provided to prevent frost from forming on the specimen; water may be circulated through the stage for moderate cooling. A more elaborate stage that provides a range of −20° to +350°C is also available from this manufacturer (Model 350, $365). A camera lucida arrangement permits the scale of a thermometer mounted in the stage to be seen juxtaposed to the microscopic image of the specimen.

Other specialized heating stages are made by Leitz for work up to 1350°C (Model 1350, $1400), and to 1750°C in vacuum (Model 1750, $1350). The latter can only be used with vertical illumination optics.

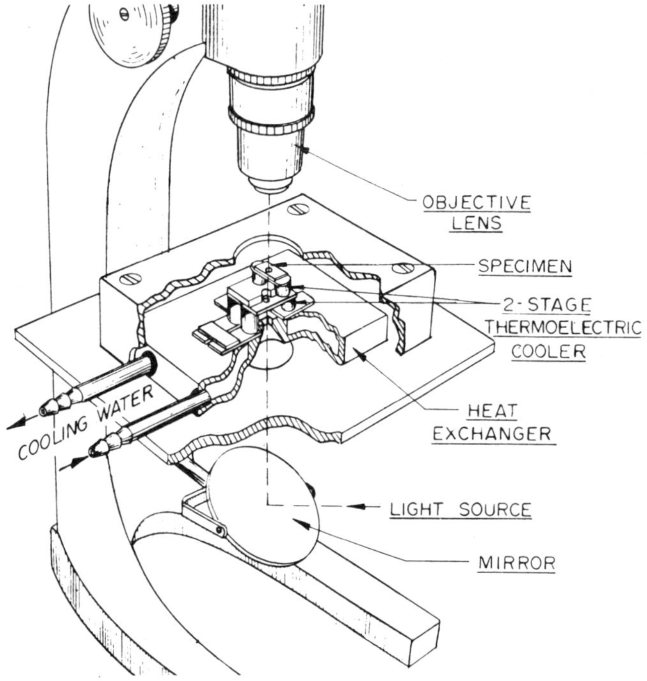

BIBLIOGRAPHY

AVERY, D. G., "Two Optical Instruments Used in Semiconductor Research," *J. Sci. Instr.*, **34**, 16–7 (1957). Describes a light spot microscope.

BARER, R., "Combined Phase-Contrast and Interference-Contrast Microscopy," *Nature*, **169**, 108 (1952).

BARER, R., "Interference Microscopy and Mass Determination," *Nature*, **169**, 366 (1952).

BARER, R., "Lecture Notes on the Use of the Microscope," Blackwell Pub., Oxford, 1953.

BARER, R., "A New Micrometer Microscope," *Nature*, **188**, No. 4748 (1960).

BENFORD, J. R., "Improvements in Microscope Illumination," *J. Opt. Soc. Amer.*, **37**, 642 (1947).

BLAISIE, B. S., "Polarized Beam Interferometer," *J. Opt. Soc. Amer.*, **46**, 950 (1956).

BRACEY, R. J., "The Aberrations of Microscope Objectives and the Variations with Small Departures from Optimum Working Conditions," *J. Roy. Micr. Soc.*, **72**, 1 (1952).

BROCK, T. W., "A Hot Stage Petrographic Microscope for Glass Research," *J. Amer. Chem. Soc.*, **45**, 5 (1962).

BURCH, C. R., "Reflecting Microscopes," *Proc. Phys. Soc.*, **59A**, 41, 47 (1947).

CHERONIS, N. D., Stein, H., and Sharefkin, D. M., "The Use of Small-Scale Experimentation in Teaching Chemistry," *Microchem. J.*, **3**, 191–203 (1959).

CONN, G. K. T., and Bradshaw, F. J., "Polarized Light in Metallography," Butterworth, London, 1952.

CUCKOW, F. W., "The Phase-Contrast Incident Light Microscope," *J. Iron and Steel Inst.*, **161**, 1 (1949).

DAVIS, W. C., "Reticle-Projecting Microscope," *Rev. Sci. Instr.*, **28**, 577–9 (1957).

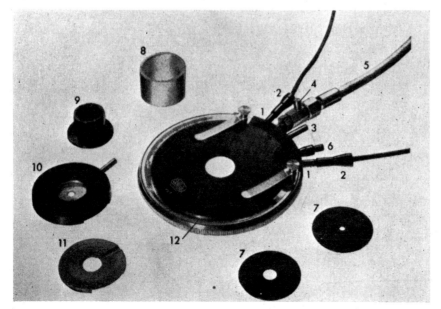

Figure 104. Thermoelectric Microscope Cold Stage of Industrial Instruments, Inc.

OBJECTIVE LENS

SPECIMEN

2-STAGE THERMOELECTRIC COOLER

HEAT EXCHANGER

COOLING WATER

LIGHT SOURCE

MIRROR

Figure 105. Heating and Cooling Microscope Stage, Model 80, of Ernst Leitz/Wetzlar. *1* and *2*: electrical input. *3*: drainage pipe. *4*: safety valve. *5*: flexible metal tube. *6*: temperature regulator. *7*: stage apertures. *8*: keys for apertures. *9*: cover tube. *10*: nitrogen chamber. *11*: temperature-stabilizing chamber. *12*: thermometer.

DYSON, J., "An Interferometer Microscope," *Proc. Roy. Soc.,* **204A,** 170 (1950); **216A,** 493 (1953).

DYSON, J., "Precise Measurement by Image Splitting," *J. Opt. Soc. Amer.,* **50,** 754 (1960).

FAUST, R. C., "Multiple Beam Interference Microscopy," *Proc. Roy. Soc.,* **211A,** 240 (1952).

FOSTER, L. V., "Polarizing Vertical Illuminator," *J. Opt. Soc. Amer.,* **28,** 124 (1938).

FREDERICKSON, A. F., "A Method of Effectively Increasing the Resolving Power of a Microscope to Reveal Unsuspected Detail in Thin Sections," *Am. Mineralogist,* **38,** 815–26 (1953). Use of narrow pencils of highly inclined light effectively doubles the resolving power.

GABOR, D., "A New Microscopic Principle," *Nature,* **161,** 777 (1948).

GAGE, S. H., "The Microscope," Comstock Pub. Co., Ithaca, N. Y., 17th Ed., 1947.

GRIGG, F. C., "Colour Contrast Phase Microscopy," *Nature,* **165,** 368 (1950).

HAITINGER, H., "Fluoreszenz-Mikroskopie," Akad. Verlagsges., Leipzig, 1959.

HALLIMOND, A. F., "Use of Polaroid for the Microscope," *Nature,* **154,** 369 (1944).

HALLIMOND, A. F., "The Polarizing Microscope," Cooke, Troughton and Simms, Ltd., York, 2nd Ed., 1953.

HEYN, A. N. J., "The Interference Microscope in Fiber Research," *Text. Res. J.,* **449** (1957).

HOPKINS, H. H., and BARHAM, P. M., "The Influence of the Condenser on Microscopic Resolution," *Proc. Roy. Soc.,* **63B,** 737 (1950).

INGELSTAM, E., "Merits of Interference Microscopes," *Exptl. Cell Research,* Suppl. **4,** pp. 150–7 (1957).

INOUE, S., AND HYDE, W. L., "Depolarization of Light at Microscope Lens Surfaces. II. Simultaneous Realization of High Resolution and High Sensitivity with the Polarizing Microscope," *J. Biophys. Biochem. Cytol.,* **3,** 831–8 (1957). *Cf. Exptl. Cell Research,* **3,** 199 (1951).

KRUG, W. AND LAU, E., "An Interference Microscope for Observations in Transmitted and Vertical Illumination," *Ann. d. Physik,* **8,** 329 (1951).

LAU, E., AND KRUG, W., "Die Aquidensitometrie," Akademie-Verlag, Berlin, 1957. Includes treatment of Sabatier process.

LOCQUIN, M., "Microscope with an Elastic Focusing Train," *Compt. rend.,* **244,** 2378–80 (1957).

McLEAN, J. D., JR., "Comparison Microscope—a Tool with Unique Possibility," *Bull. Am. Assoc. Petrol. Geol.,* **35,** 96–101 (1951).

MANSOUR, T. M., "A Nondestructive Method of Measuring Thickness of Transparent Coatings," *Mat. Res. and Stds.,* **3,** 29 (1963). Deals with Zeiss light-section microscope.

MARSHALL, C. R., AND GRIFFITH, H. D., "An Introduction to the Theory and Use of the Microscope," G. Routledge and Sons, Ltd., London, 1928.

MARTIN, L. C., "Optical Measuring Instruments," Blackie and Son, London, 1924.

MARTIN, L. C., "The Light Microscope," Chap. 6 in "Physical Techniques in Biological Research," G. Oster and A. W. Pollister, editors, Academic Press, N. Y., 1955, Vol. 1.

MARTIN, L. C., AND JOHNSON, B. K., "Practical Microscopy," Blackie and Son, London, 2nd Ed., 1949.

MERTON, T., "On a Method of Increasing Contrast in Microscopy," *Proc. Roy. Soc.* **189A,** 309 (1947).

MERTON, T., "On Interference Microscopy," *Proc. Roy. Soc.,* **191A,** 1 (1948).

MOTT, B. W., "Progress in Metallurgical Microscopy," *Endeavour,* **12,** 154 (1953).

NOTZOLD, E., "The Microscope in the Preparation of Coal," *Rev. ind. minerale,* **31,** 192–8 (1950). Describes the Leitz "Dialux" stereomicroscope.

OSTERBERG, H., "The Polanret Microscope," *J. Opt. Soc. Amer.,* **37,** 726 (1947).

PAYNE, B. O., "Microscope Design and Construction," Cooke, Troughton and Simms, Ltd., York, 1954.

POLICARD, A. M. B., AND LOCQUIN, M., "Traite de Microscopie. Instruments et Techniques," The Hague: M. Nÿhoff, 1957.

POLLARD, A. F. C., "The Kinematical Design of Couplings in Instrument Mechanisms," Hilger and Watts, London, 1951.

ROBERTSON, J. K., "Introduction to Optics," Van Nostrand, N. Y., 4th Ed., 1954.

ROHR, M. VON, "Abbe's Apochromats," Carl Zeiss, Jena, 1936. Includes reprints of two 1886 papers on the then new objectives in J. Roy. Micr. Soc.

SAYLOR, C. P., BRICE, A. T., AND ZERNIKE, F., "Colour Phase-Contrast Microscopy, Requirements and Applications," *J. Opt. Soc. Amer.,* **40,** 329 (1950).

SCHULZE, R., "Optical Devices: Microscopes," *Z. Ver. dtsch. Ing.,* **100,** 1107–9 (1958). A review of measuring microscopes.

SHILLABER, C. P., "Photomicrography in Theory and Practice," Wiley, N. Y., 1944.

SOUTHALL, J. P. C., "Mirrors, Prisms and Lenses," Macmillan, N. Y., 3rd Ed., 1933.

STEEL, W. H., "The Design of Reflecting Microscope Objectives," *Austral. J. Sci. Res.,* **4,** 1 (1951).

STOCK, E., "The Leitz Ortholux Microscope and its Applications in the Coatings Industry," *Farben, Lacke, Anstrichstoffe,* **4,** 42–5 (1950).

STOLL, S., "A Precious Auxiliary of Microchemical Analysis: the Microscope," *Chim. anal.,* **37,** 409–15 (1955).

STOVES, J. L., "Fibre Microscopy," National Trade Press, London, 1957.

SWANN, M. M., and MITCHISON, J. M., "Refinements in Polarized Light Microscopy," *J. Exptl. Biol.,* **27,** 226–37 (1950).

TAYLOR, E. W., "The Phase-Contrast Microscope with Particular Reference to Vertical Incident Illumination," *J. Roy. Micr. Soc.,* **69,** 49 (1949).

TOLANSKY, S., "Multiple Beam Interferometry," Clarendon Press, Oxford, 1948.

WILKENS, M. H. F., "Precautions in the Use of Immersion Objectives," *J. Roy. Micr. Soc.,* **70,** 266 (1950).

ZERNICKE, F., "Phase Contrast, a New Method for Microscopic Observation of Transparent Objects," *Physica,* **9,** 686–98, 974–86 (1942).

Topics in...

Chemical Instrumentation

... a  feature

Edited by **S. Z. LEWIN,** New York University, New York 3, N. Y.

The Journal of Chemical Education
Vol. 42, page A83
February, 1965

XX. Flash Photolysis—A Technique for Studying Fast Reactions

David N. Bailey and **David M. Hercules,** *Department of Chemistry and Laboratory for Nuclear Science, Massachusetts Institute of Technology, Cambridge, Mass. 02139*

One of the most interesting developments in chemical instrumentation has been the use of flash photolysis for studying excited-state processes in organic molecules. Flash photolysis produces a high power of radiant energy, as compared with steady state sources, by dissipating a moderate amount of energy during a short period of time. Powers of 50×10^6 W for a few microseconds are not uncommon with flash equipment. Such high powers are capable of producing high concentrations of short-lived intermediates in chemical reactions. Some published uses of flash equipment include flame studies *(1),* kinetic studies *(2–5),* obtaining absorption spectra of intermediates *(6, 7)* and triplets *(8),* determining radiative lifetimes of excited states *(9),* stimulating laser emission *(10),* and studying reaction mechanisms *(11).*

Theory

Because the electronic processes that occur within a molecule are independent of the mode of excitation, flash photolysis can be used to study the mechanisms of photochemical reactions. Figure 1 shows the electronic energy-level diagram of a typical organic molecule containing a π-electron system.[1] S_0 is the ground state (a singlet state), S^* is the first excited singlet state, and T_1 and T_2 are the first and second triplet states, respectively. Photon absorption (process 1) raises the molecule to an excited singlet state, in which state it can undergo one of several processes. The most rapid process *(ca.* 10^{-12} to 10^{-13} sec) is the *radiationless* loss of excess vibrational energy called *internal conversion* (process 2). This puts the molecule in the lowest vibrational level of the first excited singlet state where it remains for the lifetime of the state *(ca.* 10^{-8} sec). A molecule in an

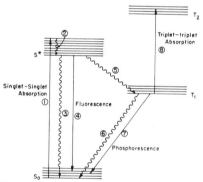

Figure 1. The energy level diagram of a typical organic molecule. Singlet states—S_0 and S^*; Triplet states—T_1 and T_2; Absorption processes—1 and 8; Internal conversion—2, 3, and 6; Fluorescence—4; Intersystem crossing—5; Phosphorescence—7.

excited singlet state may undergo *radiationless* deactivation to the ground state by internal conversion (process 3); it may undergo a radiative transition to the ground state called *fluorescence* (process 4); or it may cross to a triplet state via *intersystem crossing* (process 5), finally reaching the lowest vibrational level of the first triplet state. Because spectroscopic selection rules formally forbid transitions between states of unlike multiplicity, a triplet state has a long intrinsic lifetime *(ca.* 10^{-3} sec or longer) relative to that of a singlet state *(ca.* 10^{-8} sec). Deactivation of the triplet state may be accomplished by internal conversion (process 6) or by the emission of a photon giving a longlived luminescence called *phosphorescence* (process 7). Because of the long lifetime of the triplet state, it has been possible to observe *triplet-triplet absorption* (process 8) for a number of molecules by flash photolysis.

Besides losing its excitation energy by the processes described above, a mole-

David M. Hercules was graduated from Juniata College with a B.S. degree in chemistry in 1954 and received his Ph.D. from M.I.T. in analytical chemistry in 1957. From 1957–1960 he was a member of the faculty of Lehigh University, and from 1960–1963 he served on the faculty of Juniata College. Since 1963 he has been Assistant Professor of Chemistry at M.I.T. His industrial experience in analytical chemistry includes summers spent with U. S. Steel (1954), Sun-Oil Co. (1956), and the Polychemicals Department of DuPont (1959, 1960).

Dr. Hercules' research interests include trace analysis by fluorescence and phosphorescence, relationships between luminescence and molecular structure, chemiluminescence, electroluminescence, and the chemistry of molecules in electronically excited states.

Dr. Hercules is a member of the A.C.S., the AAAS, the Society for Applied Spectroscopy, the Pennsylvania Academy of Sciences, the Photoelectronic Spectrometry Group and Sigma Xi. He is a member of the Publications Advisory Committee for Applied Spectroscopy, and a member of the Advisory Board of Analytical Chemistry.

David N. Bailey, currently a graduate student in Analytical Chemistry at M.I.T. received his B.S. degree in chemistry from Juniata College in 1963. He spent one summer with Gulf Research and Development Company and for the past two years he has been an NSF Fellow. His research interests are in the fields of photochemistry and molecular spectroscopy.

[1] For a review of some essential features of organic photochemistry, see LEERMAKERS, P. A. and VESLEY, G. F. THIS JOURNAL **41**, 535 (1964).

Dr. Bailey is now Assistant Professor of Chemistry, Gustavus Adolphus College, St. Peters, Minnesota 56082. Dr. Hercules is Associate Professor of Chemistry at the University of Georgia, Athens 30601. **1970**

Chemical Instrumentation

cule in an excited state may enter into a photochemical reaction from either singlet or triplet levels. A valuable feature of flash photolysis is that it can be used to detect and monitor triplets and/or free radicals as intermediates in photochemical reactions. Often such observations are impossible by other techniques because generally the concentrations of intermediate species are too low to allow detection under even high-intensity steady-state illumination.

Experimental Design

Apparatus for flash work ranges from simple designs, such as setting up a flash tube and reflector next to a reaction vessel, to highly complex, well baffled systems, such as those used for obtaining spectra of intermediates. Porter has described the latter in considerable detail (12).

Figure 2. An experimental design useful for spectroscopic monitoring of a flash photolysis experiment. A. Sample Cell, B. Water jacket and/or filter, C. Flash tubes, D. Reflector, E. Spectroscopic source, F. Monochromator, G. Monochromator, H. Detector.

Figure 2 shows a generalized design of the apparatus used for spectroscopic flash studies. A is the sample cell; B is a water jacket for thermostatting the cell and/or filtering the output of the flash tube(s); C is the flash tube(s); D is a reflector; E is a spectroscopic source, either a flash tube or a steady-state lamp; F is a monochromator for selecting radiation of a particular wavelength from the spectroscopic source; G is another monochromator for analyzing the light emerging from the cell; and H is a detector, the exact nature of which depends on the experiment. By rearranging the various components and choosing the proper detector, this basic design can be modified for use in a wide variety of experiments.

To determine the complete absorption spectrum of an intermediate, one would use a flash tube at E, no monochromator at F, and a spectrograph with a photographic plate as a detector (combining G and H). A suitable period of time (ca. 10 to 1000 μsec) after the intermediate is produced by the main flash (C), a delay circuit would fire a flash tube at E and the spectrum would be recorded. This may be done as many times as necessary to obtain the proper density on the photographic plate. Figure 3 shows spectra of transients produced in the flash photolysis of duroquinone (14). The change in the absorption spectrum can be seen as a difference in darkening of the plate as a function of the time delay between the photolysis flash and the analyzing flash.

For spectrophotometrically monitoring

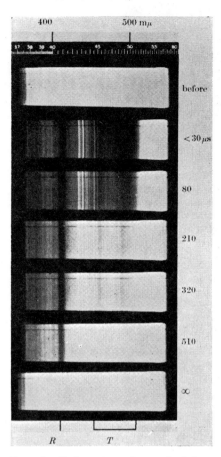

Figure 3. Flash spectroscopic records of duroquinone transients in viscous paraffin. (Adapted from Bridge and Porter (14). Wavelength is recorded along the top of the photograph. Time after the flash is recorded along the vertical right-hand column. The R band is due to the radical and the T band is due to the triplet.

the concentration of an intermediate with a known absorption spectrum, one would use a tungsten or hydrogen lamp as the spectroscopic source at E and a photomultiplier tube with an oscilloscope read-out at H. The monochromator at F would be set to the proper monitoring wavelength while the monochromator at G would be omitted. The study of duroquinone by Bridge and Porter (13) provides an excellent example of this type of experiment. Figure 4 shows the oscilloscope trace obtained by monitoring the absorption due to the duroquinone triplet and the durosemiquinone radical at their respective absorption maxima. This technique also may be used to obtain the complete absorption spectrum of an intermediate by varying the monitoring wavelength to cover the spectral region of interest.

Radiative triplet-state lifetimes (phosphorescence lifetimes) may be determined by setting the analyzing monochromator (G) at the phosphorescence maximum and eliminating the spectral flash (E) and preliminary monochromator (F). Also excited singlet-state lifetimes (fluorescence lifetimes) have been determined using a flash of extremely short duration (ca. 10 nanoseconds) and electronics which respond rapidly (9, 19). The oscilloscope trace obtained for either of these experiments would be similar to

those shown in Figure 3 except the ordinate would record phosphorescence or fluorescence intensity rather than light absorption.

Circuitry

Figure 5 shows a block diagram of a flash type circuit. The high voltage power supply must provide from 200 to ca. 20 kv dc depending on the particular application. The storage section is usually a capacitor bank but may also contain inductors to control the flash duration. The energy and duration at a flash are given by $E = \frac{1}{2} CV^2$ and $t \simeq \sqrt{LC}$ where V is the voltage, and C and L are the total capacitance and inductance, respectively, of the discharge circuit. This indicates that the condition of high voltage and low capacitance and inductance produces the highest power. The charging resistor isolates the storage section from the power supply in order to prevent the capacitor bank from discharging through the power supply when the flash is triggered, and to prevent the flash tube from continuing to fire after the capacitor bank has been discharged. The flash tube is triggered by a pulse of ca. 5 to 25 kv from the trigger section. This pulse is applied either to a trigger wire wrapped around the flash tube or to a spark gap connected in series with the flash tube.

A typical circuit for the operation of flash equipment is shown in Figure 6. Specifically, this circuit is designed to fire two 500 j flash tubes producing a flash of 20 μsec duration (15). The high voltage dc supply charges each of the 10 μf, 10 kv capacitors to 10 kv, a voltage at which the lamps would fire spontaneously without the spark gap in the circuit. The spark gap is a type of high voltage, high current switch which is activated by a triggering pulse (ca. 5 to 25 kv) obtained by discharging a small capacitor through the primary of the trigger transformer. When the spark gap is triggered, the full 10 kv of the main storage capacitors appear across the lamps and they fire. Then the storage capacitors automatically begin to charge for the next flash. The high frequency choke coils in the circuit permit the use of one power supply to charge both capacitors while isolating them from one another during the discharge (7).

Also included in the circuit of Figure 6 are safety features which are strongly recommended due to the combination of high voltages and large capacitances encountered in flash circuits. When the power is off, both the main capacitor bank and the trigger capacitor are shorted through appropriate resistances, by use of relays. Another relay may be inserted in the primary of the trigger transformer to prevent accidental firing of the lamps. All safety features should be of the type which will automatically return to, and remain in, the safety position when the power is shut off.

Should a flash of even shorter duration be desired, special components and circuits must be used. For example, a very short flash has been reported, producing

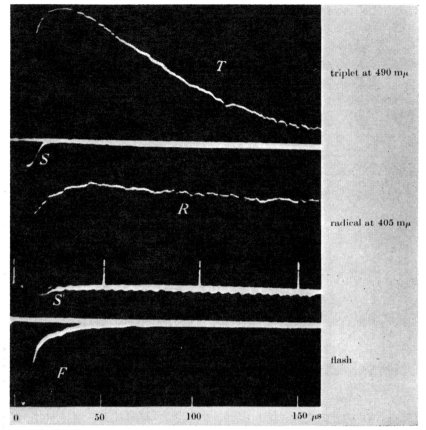

Figure 4. Oscilloscopic traces of duroquinone transients in viscous paraffin solvent. *T*, triplet; *R* durosemiquinone radical; *F*, flash profile; *S* and *S'*, scattered light from photolysis flash Adapted from Bridge and Porter (13).

only ca. 40 μj/flash and having a full width at half-height flash duration of less than 10 nanoseconds (16). A special flash tube was made by opening a General Electric NE-2 neon bulb and bending the electrodes until they were parallel and within 1/64 in. of one another. The bulb was

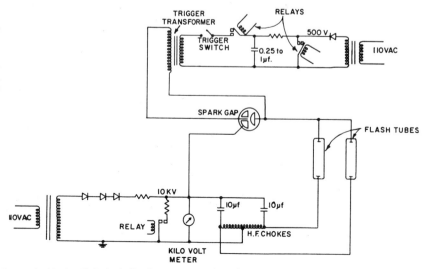

Figure 5. A block diagram of a basic flash lamp circuit.

then connected to a diffusion pump and gently flamed. Pure hydrogen gas was admitted to a pressure of *ca.* 100 mm and the bulb was resealed.

Figure 7 shows the circuit used with this flash tube. Only the distributed capacitance (*C'*) of the circuit was used, unless higher energy per flash was desired, in which case a special low inductance ceramic capacitor (*C*) was used. In place of a spark gap, a 2D21 thyratron tube was used to trigger the flash. The extremely high plate voltage used required a high variable bias to prevent the thyratron from firing because of the plate voltage alone. Although these

Figure 6. A detailed circuit for the operation of two flash tubes.

conditions exceeded the recommended operating conditions for the 2D21, the authors reported that most 2D21 thyratron tubes performed satisfactorily. The 20 Meg resistor between the power supply and the flash tube was large enough to cut off both the flash tube and the thyratron after firing. The 500 k resistor across the lamp prevented the flash tube from breaking down prematurely under the high supply voltage

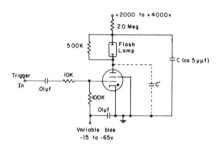

Figure 7. A circuit designed to produce flashes in the nanosecond region.

and prevented it from firing at all unless it fired within a few microseconds of the thyratron. Such a circuit must be wired carefully to eliminate all stray inductance which would prolong the duration of the flash.

Recently an ultraviolet lamp for generating intense flashes, of constant shape, in the subnanosecond region has been reported (19). This lamp was made from a commercially available mercury contact relay, using the spark emission from the contact as the source of radiation. The rise time of the pulse was 0.4 μsec and the half-width of the pulse was 0.5 μsec. An interesting feature is that the lamp can be operated between 3 and 5×10^4 pulses per second.

Currently there are several commercially available complete flash units for flash photolysis studies. Some of these units were developed for high speed photographic (17) use and provide *ca.* 1.5 j/flash with a duration of *ca.* 0.5 sec. Some units have high repeatability rates (up to *ca.* 6000/sec) while others may be flashed only once in 5 sec. Other units were developed for exciting laser action in crystals. Because they provide powers up to 20,000 j/flash (18), these units should have application in flash photolysis studies requiring moderate to high energies.

The construction of flash lamps has been described in detail by Porter (12). Electrodes, spaced appropriately, are sealed into each end of a quartz tube of the proper size and shape. The tube is flamed gently while under vacuum and then filled to a pressure of *ca.* 100 mm. with an inert gas (argon, xenon, or krypton are usually used) and fired several times to outgas the electrodes. It is evacuated again and refilled, then sealed off. Such a tube will have a satisfactory life if the outgassing of the electrodes has been complete and the operating conditions are moderate. It cannot be emphasized too strongly that the quartz to metal seals must be constructed to withstand the violent electrical and thermal shock of firing the

Table 1. Summary of Operating Characteristics of Some Typical Commercially Available Flash Tubes

Mfr	Model	Shape	Energy per flash (j)	Flash dura-time (μsec)	Flash rate	Arc Length, in.	Envelope
EG&G	FX-1	Straight	400	250	1 per 10 sec	6	Quartz
	FX-3	Straight	1.25	2	800/sec	$3^5/_8$	Quartz-inner Corning #7740 outer
	FX-12	Straight	5	6	6000/sec	$1/_4$	Quartz
	FX-38A	Straight	400	1000	1 per 20 sec	3	Quartz
	FX-38A	Straight	200	400	1 per 10 sec	3	Quartz
	FX-38A	Straight	100	80	1 per 5 sec	3	Quartz
	FX-42	Straight	600	600	1 per 10 sec	3	Quartz
	FX-47	Straight	10,000	2200	1 per 4 min	$6^1/_2$	Quartz
	FX-51	U	600	600	1 per 10 sec	$3^1/_2$	Quartz
	FX-100	U	100	150	1 per 10 sec	$2^3/_8$	Quartz
GE	FT-91	Straight	125	....	30/min	3	Quartz
	FT-151	Helix	125	....	2.5/sec	....	
	FT-218	Helix	200	....	....	....	....
EPP	S-13-138	Straight	600	....	6/min	3	Quartz
	S-13-139	Straight	2000	....	2/min	6	Quartz
	S-13-140	Straight	10,000	....	$1/_4$ min	$6^1/_2$	Quartz

Manufacturer's Addresses for Table 1:

E G & G   Edgerton, Germeshausen & Grier, Inc.
160 Brookline Ave.
Boston, Mass. 02114

GE   General Electric
Photo Lamp Department #281
Nela Park
Cleveland 12, Ohio

EPP   Electro-Power-Pacs, Inc.
5 Hadley St.
Cambridge, Mass. 02140

Flash Photolysis

Supplement

Figure 8. The Xenon Corp's Model 710C flash photolysis apparatus.

lamp. It is recommended that sturdy safety shields be used around any flash apparatus, especially if high energies are being dissipated in a short time, because lamp failures are often quite violent.

A large variety of flash tubes is available from several manufacturers at the present time. Normally the tubes are made of quartz, although some are made of Pyrex. Xenon seems to be preferred although on special order the flash tube may be filled with another gas or vapor. There are two basic designs in use, the straight tube and the helix. Table 1 summarizes representative flash tubes and their operating characteristics under "normal" conditions.

Spark gaps consist of two heavy tungsten electrodes separated by a few millimeters with a sharply pointed trigger wire terminated a short distance from one of them. The electrodes are enclosed in a container to reduce noise. The trigger probe provides the first spark which breaks down the gap to allow the main current surge to pass. Table 2 gives the operating characteristics of a large number of triggered spark gaps currently available.

For providing the triggering pulse, ordinary automobile spark coils or tesla coils have been used, but now trigger transformers are common. A wide range of trigger transformers is available for most applications.

Except for flash circuits designed for very short duration flashes, any commercially available capacitor having the appropriate capacitance and voltage ratings may be used in the energy storage section. For flashes of short duration it may be necessary to use special low inductance capacitors. If it is necessary to prolong the flash duration to prevent failure of the flash tube, special inductors may be used in the circuit. These inductors must be capable of handling high currents and must be mounted securely and away from metal objects which may be attracted to the inductor by the high magnetic field generated during the discharge.

The high voltage power supply needs to provide only the appropriate voltage and current necessary to charge the storage capacitors in a resonable time. It need not be filtered since it is effectively disconnected from the circuit during discharge.

Acknowledgment

The authors wish to thank D. K. Roe and V. R. Landi for their helpful comments and discussions. This work is supported in part through funds provided by the U. S. Atomic Energy Commission under contract AT(30-1)-905 and by USPHS grant GM 11766-01 from the National Institutes of Health. We are indebted to the National Science Foundation for a graduate fellowship awarded to David N. Bailey.

The most significant advancement in instrumentation for flash photolysis to occur since the writing of the above article has been the marketing of a commercial flash photolysis unit by Xenon Corporation, along with a series of accessories such as cells and flash tubes.

The availability of an integrated flash instrument should prove to be valuable for most chemists or biochemists doing flash photolysis studies. It eliminates the necessity for matching components— often the source of extended-duration light pulses. The Xenon Corp. unit comes in one of three modifications, the 705C, the 710C or the 720C. The 705C is a unit for obtaining spectra of transients at various delay times after firing of the main flash. The 710C is an instrument designed for kinetic studies utilizing photoelectric detection. A picture of the 710C is shown in Figure 8. The 720C combines flash spectroscopy with the capability for kinetic measurements by use of a combination monochromator-spectrograph. Each unit is capable of pumping up to 2000 j in the main flash. Pulse durations are on the order of 30 μsec at 2000 j and 12 μsec at 250 j. (Figures quoted are one-half amplitude).

Table 2. Summary of Characteristics of Some Commercially Available Spark Gaps

EG&G Model no.	Operating range (kv) min	Operating range (kv) max	Static break-down (kv)	Peak current (amp)	Peak current duration (μsec)	Energy discharge (j)	Trigger potential needed (kv)	Typical delay time (μsec)
GP-11	1.8	3.5	4.2	5,000	20	25	5.5	0.10
GP-12	10.0	24	30	100,000	10	2500	15	0.05
GP-14	12	36.5	42	100,000	10	2500	20	0.05
GP-15	25	70	86	100,000	10	4000	25	0.10
GP-16	1	2	2.6	5,000	20	25	5	0.20
GP-17	4.4	10	12.5	5,000	20	25	7	0.02
GP-19	2.5	5	7	75,000	10	150	5	2.0
GP-20	3.5	11	14	15,000	20	200	10	0.06
GP-22	6	16	19	100,000	10	2500	15	0.04
GP-26	2	3.7	4.8	5,000	20	25	6	0.10
GP-27	2	3.7	4.8	5,000	20	25	6	0.10
GP-30	2	5	7.5	100,000	10	2500	15	0.08
GP-31	2	6	7.5	15,000	20	200	10	0.10
GP-32	20	50	70	100,000	10	4000	25	0.10

Because Xenon Corporation is now a major supplier of flash tubes, flash components and complete flash systems, their address should be added to those of Table 1:

Xenon Corporation
39 Commercial Street
Medford, Massachusetts 02155

Literature Cited

(1) NORRISH, R. G. W., PORTER, G. AND THRUSH, B. A., *Proc. Roy. Soc.* (London), **216A**, 165 (1953).
(2) LINSCHITZ, H. AND PEKKARINEN, L., *J. Am. Chem. Soc.*, **82**, 2411 (1960).
(3) LINSCHITZ, H., STEEL, C., AND BELL, J. A., *J. Phys. Chem.*, **66**, 2574 (1962).
(4) CHISOV, A. K. AND KAPYAKIN, A. V., *Dokl. Akad. Nauk S.S.S.R.*, **153**, 1132 (1963).
(5) STEEL, C. AND LINSCHITZ, H. *J. Phys. Chem.*, **66**, 2577 (1962).
(6) BERRY, R. S., SPOKES, G. N., AND STILES, M., *J. Am. Chem. Soc.*, **84**, 3570 (1962).
(7) PORTER, G., *Spectrochim. Acta*, **12**, 299 (1958).
(8) PORTER, G. AND WINDSOR, M. W., *Proc. Roy. Soc.* **245A**, 238 (1958).
(9) RABINOWITCH, E., *J. Chim. Phys.*, **55**, 927 (1958).
(10) MAIMAN, T. H., *Nature*, **187**, 493 (1960).
(11) BRIDGE, N. K. AND PORTER, G. *Proc. Roy. Soc.* (London), **244A**, 259, 276 (1958).
(12) PORTER, G., in Weissberger, (ed.) "Techniques of Organic Chemistry," Vol. VIII, Part II, Interscience, New York, 1963, pp. 1062–1072.
(13) BRIDGE, N. K. AND PORTER, G., *Proc. Roy. Soc.*, **244A**, 276 (1958).
(14) BRIDGE, N. K. AND PORTER, G., *Proc. Roy. Soc.*, **244A**, 259 (1958).
(15) Adapted from PORTER, G., ref. 12, p. 1067.
(16) MALMBERG, J. H., *Rev. Sci. Instr.*, **28**, 1027 (1957).
(17) Edgerton, Germeshausen and Grier, Inc., Data sheets 5490, 2307, 5021, 501.
(18) Electro Powerpacs, Inc., Model 322, 325, and 330 Optical Maser Power Supply data sheets.
(19) D'AUSSIO, J. T., LUDWIG, P. K., AND BURTON, M., *Rev. Sci. Instr.*, **35**, 1015 (1964).

Topics in...

Chemical Instrumentation

... a *ChemEd* feature

The Journal of Chemical Education
Vol. 46, page A717
October, 1969

XLVII. Multipurpose Electroanalytical Instruments[1]

Galen W. Ewing

Dual trends are discernable in the development of laboratory instrumentation in the last decade or two. On the one hand are specialized instruments, useful for a single purpose, and on the other, versatile instruments applicable to many distinct analytical techniques.

The first general instrument primarily for electrical methods was described by D. D. DeFord of Northwestern University at the 133rd national ACS meeting, San Francisco, 1958. DeFord's instrument was built around an assemblage of Philbrick vacuum-tube operational amplifiers. "Its modular construction permitted it to be used for a wide variety of electroanalytical studies, including voltage-scan voltammetry, current-scan voltammetry, chronopotentiometry, coulometric analysis, controlled-potential electroanalysis, and potentiometric, amperometric, conductometric, and photometric titrimetry. With additional modules still other functions could be performed readily." (1) The accuracy was of the order of $\pm 0.1\%$; the cost of its component parts (including a recorder) was about $2000.

[1] This paper was originally presented at the Center for Professional Advancement, Hopatcong, N. J., August, 1968.

The next entries in the field were reported simultaneously by Unterkofler and Shain (2), by Lauer, Schlein, and Osteryoung (3), by Morrison (4), and by Ewing and Brayden (5). These papers formed part of a symposium on operational amplifiers in analytical instrumentation held at the 144th ACS meeting, Los Angeles, 1963; appropriately, Professor DeFord was symposium chairman.

Unterkofler and Shain (2) saw disadvantages in a patch-cord system such as that employed by DeFord, because of the inconvenience of changing from one experimental set-up to another, and because the many cables with their possibly erratic contacts may constitute a source of noise. They also discarded multiple switching as impractical. Instead they developed a system of plug-in panels. The main chassis of their instrument held eight Philbrick amplifiers with their associated power supplies. Each plug-in unit provided the signal generators, feed-back networks, range-switching, and other circuitry needed to program the amplifiers for the particular requirements of one electrochemical method. A very similar instrument, developed independently, but following a suggestion by Shain, was described in detail by Propst (6) in 1964.

Lauer, Schlein, and Osteryoung (3) chose to program their amplifiers by multiple switching. Their instrument,

with seven amplifiers, incorporated twelve switches and six relays. It was designed particularly for fast, controlled-potential sweeps with simultaneous integration of current, hence is somewhat more specialized than some experimenters might wish.

Morrison (4) constructed what amounted to a versatile analog computer with all connections to its seven amplifiers (except for power supply) brought out to a logically arranged patch panel. This apparatus, which Morrison called an "Instrument Synthesizer," has subsequently been described in full detail (7). The approach adopted by Ewing and Brayden (5) was rather similar in concept; they investigated the applicability of the Heath analog computers to electroanalytical work.

Beckman Instruments, Inc.

The first general purpose electroanalytical instrument to be made commercially available was announced in 1965: the Beckman Electroscan-30 (8). This instrument, shown in Figure 1, is built around a single, high-performance, operational amplifier which can be utilized in three configurations: (1) as a potential-measuring device, particularly for glass-electrode pH, (2) in a galvanostat mode, to maintain controlled current for such applications as coulometric titrimetry and

Figure 1. Beckman Electroscan-30. The center panel is the recorder bed.

chronopotentiometry, and (3) as a potentiostat for polarography, cyclic voltammetry and related methods.

The Electroscan-30 has a built-in, high-impedance strip-chart recorder capable of half-second pen response even with 2 megohms source resistance. The chart drive can be reversed for cyclic voltammetry.

Princeton Applied Research Corp.

P.A.R., in 1968, introduced a multiple purpose instrument designated the Model 170 Electrochemistry System (Fig. 2) (9). The designers point out that the instrumentation for nearly all electrochemical techniques can be resolved conceptually into three portions: firstly, a function generator and control element which applies the required potential or current to the cell; secondly, a circuit to measure the desired quantity; and thirdly, a data-processing system to provide a suitable signal to the read-out device. The P.A.R. instrument panel is laid out with this logic in mind, conditioning controls grouped to the left, measuring controls in the center, and controls for data processing to the right. The P.A.R. instrument employs sophisticated operational amplifier circuitry as well as digital logic, the latter particularly in connection with pulse polarography.

Four innovations are especially deserving of comment. (1) An electromechanical timer is arranged to dislodge drops of mercury from a polarographic capillary at 0.5, 1, 2, or 5 sec intervals. Added circuitry then makes it possible to measure the drop current during a predetermined and exactly reproducible period just preceding fall-off.

(2) A summing amplifier is utilized to

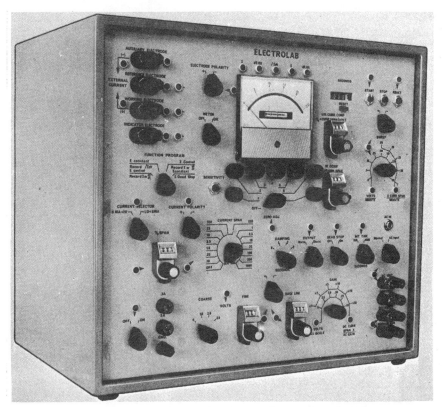

Figure 3. N.I.L. Electrolab. The numerous black dots are the road-map pilot lights mentioned in the text.

apply excitation to the cell without itself generating the excitation signals. The significance of this is that it permits the amplifier to be optimized for one function—delivering power. Other amplifiers generate appropriate signals, such as dc levels, ramp voltage, ac of selected frequency, and a train of pulses for pulse polarography. These various signals are summed as required at the input to the power amplifier. This amplifier is capable of supplying up to 1 A at 100 V or 5 A at 20 V.

(3) A lock-in amplifier (10) is provided for synchronous demodulation in ac polarography. This is inherently phase-sensitive, and can thus provide, on demand, output signals proportional to either the faradaic current or the capacitive current passing through the cell, as these differ in phase by 45°.

(4) Pulse polarography is instrumented with the aid of an analog shift register. This is a series of capacitive memories which store information about currents flowing during successive pulses. The memorized signals are then presented in sequence to a differential amplifier which always produces an indication of the *difference* between current flow from one pulse to the next, a true derivative pulse polarogram.

The readout for the P.A.R. instrument is a built-in X-Y recorder. Fast-changing signals can be observed with an external cathode-ray oscilloscope.

National Instrument Laboratories, Inc.

The only other comparable instrument on the market is the N.I.L. Electrolab (Fig. 3) (11). A unique feature is the "function program" switch with four positions bearing the following legends:

E control, Record I or dI/dt
E constant, Record $\int$Idt
I control, Record E or dE/dt
I constant, Record E (dead stop)

These choices provide the necessary dc conditions for any electroanalytical

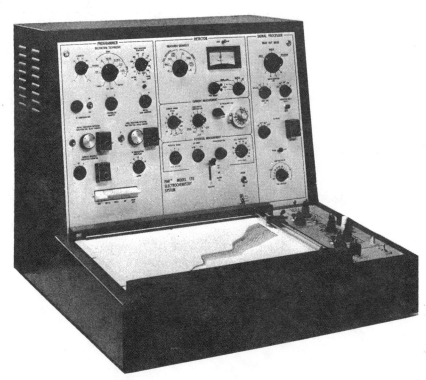

Figure 2. P.A.R. Model 170 Electrochemistry System. The recorder controls are grouped in the right foreground. Connections to the cell are located on the right-hand side panel, out of sight in this photograph.

method. For ac methods an ac signal generator must be connected as an accessory. At each position of the function program switch a series of small pilot lamps adjacent to the various panel controls light up to act as a road map, so that the operator sees at a glance which of the many connections and controls he must attend to. An external recorder is required.

Among the features of interest are the following: compensation for ohmic voltage drop in the solution, linear compensation for capacitive current, internal seconds timer for coulometry, pulse-polarographic capability. As many as four electrodes utilized simultaneously are protected from mutual interaction by buffer amplifiers.

Related Instrument Assemblies

McKee-Pedersen Instruments (12) manu-

Fig. 4. Console of McKee-Pedersen analytical system. The main chassis contains necessary power supplies for external use via the lower front panel, and for internal use to activate the electronic circuits. The various panel segments correspond to plug-in amplifiers, electrometers, reference voltage sources, and other units.

factures a multipurpose electronic console which combines patch-cord and plug-in assembly features. Only those plug-in units likely to be needed in any particular laboratory need be purchased; these are permanently installed in the console and patched into various configurations as needed. This makes it easy to set up nonstandard methods as well as conventional ones, (See Figure 4.)

A.R.F. Products (13) produces an instrument assemblage which should be included here. It consists of a series of monofunctional modules, each with its own power supply, which can be interconnected with patch cords to form a variety of electrochemical instruments.

The Heath Company (14) also offers a series of interconnecting modules with which complete instruments can be assembled. Many of these, including a polarographic unit, are built around a central module containing four operational amplifiers.

The A.R.F. and Heath systems are primarily designed for teaching purposes, where complete visibility of circuit connections is more important than convenience.

Addresses of Manufacturers

1. A.R.F. Products, Inc.
   Gardner Rd., Raton, N. M. 87740

2. Beckman Instruments, Inc.
   2500 Harbor Blvd., Fullerton, Calif. 92634

3. The Heath Company
   Hilltop Rd., Benton Harbor, Mich. 49022

4. McKee-Pedersen Instruments
   P. O. Box 322, Danville, Calif. 94526

5. National Instrument Laboratories, Inc.
   12300 Parklawn Dr., Rockville, Md. 20852

6. Princeton Applied Research Corp.
   P. O. Box 565, Princeton, N. J. 08540

Literature Cited

(1) DeFord, D. D., Abstracts of Papers, 133rd Meeting, ACS, San Francisco, 1958, p. 16B.

(2) Unterkofler, W. L., and I. Shain, *Anal. Chem.,* **35,** 1778 (1963).

(3) Lauer, G., H. Schlein, and R. A. Osteryoung, *Anal. Chem.,* **35,** 1789 (1963).

(4) Morrison, C. F., Jr., *Anal. Chem.,* **35,** 1820 (1963).

(5) Ewing, G. W., and T. H. Brayden, *Anal. Chem.,* **35,** 1826 (1963).

(6) Propst, R. C., "A Multipurpose Instrument for Electrochemical Studies," Report DP-903, E. I. du Pont de Nemours & Co., Savannah River Laboratory, 1964.

(7) Morrison, C. F., Jr., "Generalized Instrumentation for Research and Teaching," Washington State University, Pullman, 1964.

(8) Bulletins 7076B and 7079, Beckman Instruments, Inc., Fullerton, Calif.

(9) Flato, J. B., *Amer. Lab.,* February, 1969, p. 10.

(10) Coor, T., J. CHEM. EDUC., **45,** A533 A585 (1968). [Reprinted in this volume]

(11) Literature of National Instrument Laboratories, Inc., Rockville, Md.

(12) Literature of McKee-Pedersen Instruments, Danville, Calif.

(13) Ewing, G. W., J. CHEM. EDUC., **42,** 32 (1965) and literature of A.R.F. Products, Inc., Raton, N. M.

(14) Literature of the Heath Co., Benton Harbor, Mich.

𝕿𝖍𝖊 𝖘𝖍𝖊𝖕𝖍𝖊𝖗𝖉'𝖘 𝖆𝖓𝖉 𝕭𝖎𝖘𝖍𝖔𝖕'𝖘
𝖘𝖞𝖒𝖇𝖔𝖑, 𝖙𝖍𝖊 𝖈𝖗𝖔𝖟𝖎𝖊𝖗.

Topics in...

Chemical Instrumentation

... a  feature

Edited by **S. Z. LEWIN,** New York University, New York 3, N. Y.

The Journal of Chemical Education
Vol. 42, pages A261, A361
April and May, 1965

XXII. Instrumentation for Electrodeposition and Coulometry

Peter F. Lott, *Department of Chemistry, University of Missouri at Kansas City, Kansas City, Missouri.*

An electric current may be passed through a solution and used as an analytical tool without causing any appreciable change in the composition of the solution as is the case for potentiometry, conductometric measurements, polarography or chronopotentiometry. Under different conditions, the flow of electrons through the solution is used to change the composition of the solution, and this change in solution composition is used as the means of chemical analysis. For example, in the electrodeposition of copper, the electrons could be considered as a gravimetric precipitating reagent for copper. Or, under proper conditions, the total flow of electrons—that is, the number of equivalents—to cause this change could also be measured, and the procedure could be classified as a volumetric process and would be known as a coulometric titration. It is interesting to note that the basic laws for these procedures were developed by Faraday in 1834 and that the determination of copper by electrodeposition was developed around 1860. In contrast, not until about 25 years ago were coulometric titration methods firmly established in analytical chemistry. This came about particularly through the work of Lingane and Swift in the United States.

Electrodeposition and coulometry may both be used for macro and micro quantities of material. Both offer the possibility of very high precision in determinations and, compared to conventional analysis, they offer the possibility of a high degree of automation so as not to require the constant attention of the chemist. Furthermore, their much greater simplicity greatly decreases the danger of mechanical loss and, in both procedures, the change in the solution composition takes place at the two electrodes.

Electrodeposition

Electrodeposition, as well as coulome-

try, may be performed under different conditions. The methods may be classified according to the electrical quantity to which primary control is centered throughout the course of the electrolysis.

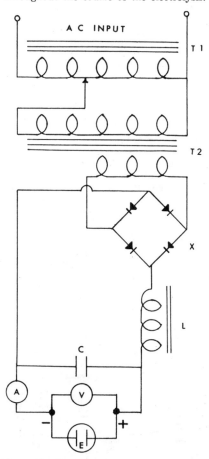

Figure 1. Circuit for "constant current" electrolysis: T 1, Variable voltage transformer; T 2, Stepdown transformer; X, Full wave rectifier; L, Choke; C, Condensor; A, Ammeter; V, Voltmeter; E, Electrolysis cell.

This may be either the initial voltage which is applied across the two electrodes of the electrolysis cell (constant voltage electrolysis), or the current which flows through the cell (constant current electrolysis), or the variation of the potential of the electrode at which the primary electrochemical reaction occurs (controlled potential analysis). The methods are very closely interrelated. However the diversity of their analytical applications, as well as the complexity of the concordant instrumentation, differs greatly from one to the other.

The simplest instrumentation is that for electrodeposition at constant current. The circuitry for such apparatus is shown in Figure 1 and is probably immediately recognized as equipment which is used for the electrodeposition of copper in quantitative analysis courses. A variable voltage transformer (Variac) is used to control the voltage applied to the primary winding of a step-down transformer. In turn, this changes the voltage developed by the transformer secondary. The secondary of the transformer is connected to a full wave rectifier, with a simple choke and condensor to filter the rectifier output, although not much attention is paid to reducing the voltage ripple. A voltmeter is connected across the terminals of the electrodes and an ammeter is used to indicate the current flow through the electrolysis cell. Initially the current is set to a value well above the decomposition potential of both copper and some other ion species in the solution which might be hydronium ions or perhaps a depolarizer such as nitrate ions. As the electrolysis proceeds, the copper ion concentration is depleted and the cathode assumes a more negative potential. Consequently, if there were only one possible method of conduction through the solution, then according to Ohm's Law, the current should decrease. However, as the potential of the cathode is well above that of the other ions in the solution, these continue to conduct more and more of the current. Although the current may have dropped considerably from its initial value, since the other ions offer an alternative path for the conduction of the current through the solution, the current tends to remain at a "constant" value, and accordingly the process is known as "constant current electrolysis." No attempt is made to control the current. The "constant current" nomenclature is here really a misnomer, and only serves as a means of describing the type of electrodeposition. Obviously, the primary reaction—the deposition of copper—does not proceed with 100% current efficiency.

Chemical Instrumentation

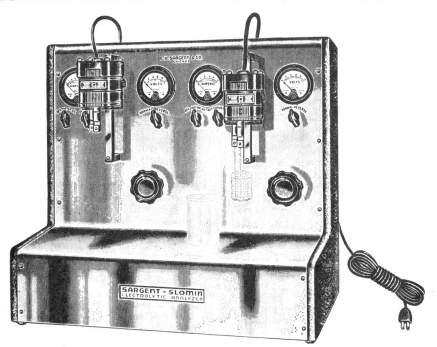

Figure 2. E. H. Sargent Co., Electrolytic Analyzer

The usual commercial units offer a maximum current setting of about 7.5 amp at 10 volts DC. The electrolysis would proceed quite slowly (in time) if migration were the only means of transferring ions from the bulk phase of the solution to the electrodes. Consequently, stirring is necessary. Very often, to further increase the rate of migration of the ions, provision is made for heating the solution. In conventional equipment, as indicated in Figures 2 and 3, the solution is stirred by connecting one of the electrodes to a stirring motor. The electrolysis is usually carried out at a current

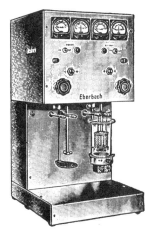

Figure 3. Eberbach Corp., Electrolysis apparatus, model 1000

setting of about 2 to 3 amp.

Equipment is also manufactured to provide a higher current of approximately 25 amp through the solution. This will permit the deposition of 1 g of copper in about 8 minutes compared to about 45 minutes with the conventional apparatus. To prevent undue heating, the cell is

usually water cooled during the electrolysis. A novel arrangement is used to stir the solution. A large permanent magnet is fastened below the sample. The magnetic field of the magnet is at a right angle relative to the field produced by the current flow between the two cell electrodes. Accordingly, the interaction of the two fields causes the solution to rotate and rapid stirring takes place.

These units, as well as conventional electrodeposition apparatus, may be used to separate ions from a solution by making the cathode connection to a mercury pool. With commercial high-current carrying units, it is possible to separate 0.5 g of iron from a steel sample in 10 minutes at the mercury cathode, as might be required as a preliminary to the determination of trace metals such as aluminum or titanium. In addition, the permanent magnet below the mercury pool draws the

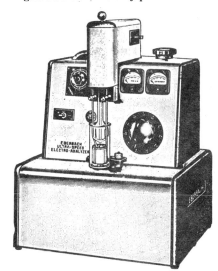

Figure 4. Eberbach Corp., Ultra-Speed Electro-Analyzer

ferromagnetic materials below the mercury surface and provides a continuously clean

mercury surface as well as preventing the resolution of the deposited metals. A picture of such apparatus is shown in Figures 4 and 5.

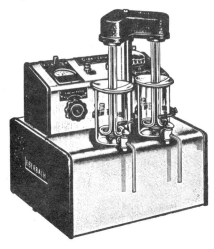

Figure 5. Eberbach Corp., Dyna-Cath Mercury Cathode Apparatus

The disadvantage of this conventional "constant current" electrolysis apparatus is that it is very difficult to plate out one metal selectively in the presence of another unless hydrogen evolution or a similar secondary electrolytic process (e.g., a potential buffer) takes place to limit the potential of the electrode below that of the decomposition of the second ion. Thus, copper can be plated out from nickel in an acid solution because hydrogen evolution takes place before the cathode potential becomes negative enough to permit the deposition of nickel. In a basic solution however, both copper and nickel deposit simultaneously since now the decomposition potential of nickel is reached before hydrogen evolution takes place. This latter separation and similar separations could still be done if somehow it were possible to limit or control the potential of the cathode.

In theory, the initial voltage applied to the two electrodes could be set to a potential necessary only for the deposition of one constituent, and the applied voltage is not changed throughout the electrolysis. This approach is carried out in "constant voltage" electrolysis. Because of the inability to compensate for resistance changes in the solution, as well as the difficulty in assigning values to overvoltages at the electrodes, "constant voltage" electrolysis is more of an academic calculation than a practical means of analysis. The electrolysis in principle could be carried out with the same equipment as is used for "constant current" electrolysis.

Controlled Potential Analysis

Electrodeposition phenomena tend to follow the Nernst equation; thus in the case of copper deposition, the potential of the copper electrode formed as soon as copper is plated on to the cathode would be given by the expression:

$$E_{Cu} = E_{Cu}° + 0.059/2 \log Cu^{++}$$

Accordingly, the potential of the copper electrode would be 0.31 v in a 0.1 M cupric ion solution, and would decrease to

0.22 v when the copper concentration has been reduced 1000 fold to 10^{-4} M. Should it be desired to verify this exponential change of cathode potential as the electrolysis proceeds, we would simply add an independent reference electrode such as the calomel electrode to the circuit and use it to independently measure the potential of the cathode throughout the electrolysis. The addition of this third electrode is necessary as measurements of the cathode potential are meaningful only with regard to the potential of some reference electrode. In the described case, as shown in Figure 6, where the primary interest centers on the cathode, the cathode would be designated as the working electrode and the anode, since it serves only to carry the electrolysis current, would be designated as the counter electrode. Obviously, for anodic processes, the reference electrode would be connected to the anode in order to measure anode potentials. In this case, the cathode would be designated as the counter electrode and the anode, the working electrode.

Conversely, if it were desired, the electrolysis could be terminated as soon as the cathode reached a set potential, for example 0.22 v as in the electrolysis of copper. The analytical chemist, however, would be more interested in an instrument in which the voltage across the anode and cathode is varied as needed to automatically maintain this desired potential between the reference electrode and cathode throughout the electrolysis. Such a procedure would allow the maximum possible current flow between the working and counter electrodes. It would prohibit the deposition of a second ion if its deposition potential were more negative than the limiting value of the cathode potential. This could be an ion which might have a decomposition potential of

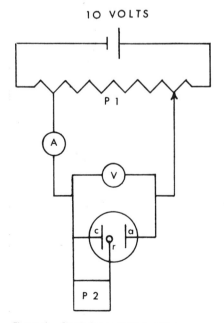

Figure 6. Circuit for controlled cathode potential analysis; P 1, Potentiometer for adjusting voltage applied between anode and cathode; c, Cathode; a, Anode; r, Reference electrode; P 2, Potentiometer or vacuum tube voltmeter for measuring cathode potential; A, Ammeter; V, Voltmeter.

0.20 v in the described case for copper electrolysis. Also, since the only potential measured is the cathode potential, the measurement is independent of secondary changes which occur between the electrolysis electrodes such as the overvoltage changes at the anode, changes in the resistance of the solution as the electrolysis proceeds and line voltage fluctuations. Manual systems to do such controlled electrodeposition were promulgated by Sand (1), and a system of this nature is really shown in Figure 6. The disadvantage of this manual system for controlled potential analysis is that it requires continuous supervision in order to manually change the voltage applied across the anode and cathode, in order to keep the cathode at the desired potential. A graph of the current flow through the system as the electrolysis proceeds is shown in Figure 7.

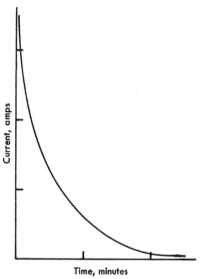

Figure 7. Variation of current with time in controlled potential analysis.

Potentiostats

To make such controlled potential analysis feasible for routine work requires devices which will do this potential adjustment automatically. Such instruments are called potentiostats. Until recently potentiostats were "handmade" in the individual laboratories. Currently, they may be purchased commercially, at a cost ranging from approximately $400 to $3,000. In addition to offering potential control for electrodepositions, potentiostats are also used for polarization studies, electrolytic metal polishing for the preparation of metallurgical samples for microscopic examination, and the selective stripping of metals in alloys in order to determine tie lines in phase rule studies of alloys.

The operation of these potentiostats may be based either on electromechanical or totally electronic principles. Rechnitz (2) has listed some guidelines in the selection of potentiostats. As criteria in the selection of potentiostats, the following considerations are of importance:

(1) High current capacity
(2) High output voltage
(3) Fast response time

(4) Wide operating range for potential control
(5) Highly sensitive and stable potential control
(6) Low control current between the reference and working electrode
(7) Low ripple voltage in the output of the potentiostat

These criteria apply to the selection of all potentiostats, but depending upon the purpose for which the instrument is used, some criteria are of greater importance than others.

For normal analytical work, a prime consideration would be the current capacity of the potentiostat. Based upon Fick's law of diffusion and the Nernst equation, to limit the reduction of the ion to diffusion-controlled conditions alone would require a current of about 2 amp for a 0.01M solution of an ion of +1 charge. For more concentrated solutions as may often be encountered in practice, a current capacity of around 10 amp would be desirable. Since the solution may have a resistance of about 50 ohms between the electrodes, should diaphragm cells be employed, then based on Ohm's law, the voltage required to handle a current of only 5 amp would be 250 v. The response time of the potentiostat may also be an important consideration if the ions to be separated lie closely together. In the beginning of the electrolysis the potential changes very rapidly, and the potentiostat would have to follow this potential change quite closely to prevent the second ion from depositing simultaneously. Possibly, however, the criterion of response time may be of greater importance in polarization studies.

Generally, if the unit supplies a controlled potential range from +2.5 to −2.5 v, this would be sufficient for most purposes. From a practical standpoint this is a minor factor as a "bucking circuit" can be easily constructed by the user to extend the control range. In certain electrolytic procedures taking place in fused salt and nonaqueous media, control voltages of ±5 v may be desirable, and certain commercial potentiostats cover this control range.

The last criteria are interrelated. The potentiostat should control the potential of the working electrode to within ±10 mv. This, consequently, implies also a ripple current of less than 10 mv. The instrument must be stable within this same voltage range, and this operating stability would be inclusive for such interrelated effects as line voltage fluctuations, electronic drift, and the response time of the potentiostat.

All instruments electronically amplify the current flowing between the reference electrode and working electrode. Because of the high input impedance of this amplifier, a negligible current flows between the electrodes; thus the electrode does not become polarized. If the input impedance is sufficiently high, it is possible to use commercial reference electrodes as are found in pH meters. These electrodes generally have a much higher resistance than individually constructed electrodes which incorporate a low resistance salt bridge.

Chemical Instrumentation

Electromechanical Devices

The Analytical Instruments, Inc., Potentiostat would be classified as an electromechanical type potentiostat. By means of a servo mechanism, it automates the operations which were done manually by the Sand system of controlled cathode potential analysis. The operation of this potentiostat for an electrolytic reduction (where the potential of the working electrode is + with respect to the saturated calomel electrode) is illustrated in Figure 8. The motor-driven auto-transformer (Variac) T 201 supplies an AC voltage to the rectifier and filter through a stepdown transformer T 202. The DC output of the filter is applied across the electrolysis cell to the working electrode (cathode) and counter electrode (anode). In the figure, the working electrode is represented as a mercury pool and illustrates the applicability of controlled cathode potential analysis to separations with a mercury cathode. The potential difference between the working and saturated calomel electrodes is connected in series opposition with the known DC output (which represents the preselected control potential of a ten-turn helical potentiometer, R 202), and also in series with the servo amplifier. The amplifier senses any difference between the control potential and the potential of the cell arising between the working electrode and the saturated calomel electrode. This potential difference (error signal) is amplified and is used to cause the control motor to turn the auto-transformer in such a direction that the error signal is reduced to zero.

For quantitative work, the electrode may be weighed if the metal is deposited on the working electrode. Or, if the electrolysis reaction proceeds with 100% current efficiency, the current flowing during the electrolysis may be integrated.

A current integrator is supplied for the Analytical Instruments, Inc., potentiostat and used to obtain an accurate measure of the amount of the substance electrolyzed. The operating mode of the current integrator, which is plugged directly into the potentiostat, is illustrated schematically in Figure 9. The current to be integrated, that is the current flowing between the cathode and anode of the electrolysis cell, is passed through a precision resistor. Since the resistance of this resistor does not change and is accurately known, the voltage drop is proportional to the current. This voltage drop is opposed by the voltage produced by a DC generator (a permanent magnet DC motor, operated as a generator).

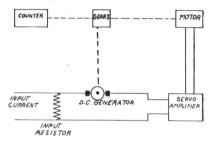

Figure 9. Block diagram of Analytical Instruments, Inc. current integrator

If there is any difference between the voltage drop across the resistor and the generator output, the amplifier converts this DC voltage difference to AC, and amplifies it to operate a reversible two phase AC motor which drives the DC generator. This results in the generator being driven at a speed sufficient to maintain its output very nearly but not quite equal to the opposing voltage drop developed in the standard resistor, R. As the output voltage of the generator is proportional to the speed of the rotation of its shaft, the counting rate of a revolution counter mechanically connected to the generator shaft by appropriate gearing is directly proportional to the generator output voltage, and thus to the in-

stantaneous current in the cell circuit. Thus, by proper choice of the standard resistors and gearing, the revolution counter sums the total current which has passed through the cell.

A similar mechanical type potentiostat has been recently introduced by Fisher Scientific Co. The unit is very suitable for electrodepositions or electroseparations; a current integrator is not incorporated into the instrument by the manufacturer. [Current integration may also be done by connecting conventional current integrators such as a ball disc integrator, a printing integrator, or chemical coulometer into the circuit. Only the commercially offered units are described. Information on adapting equipment as well as a description of laboratory constructed potentiostats, and coulometers may be found in J. J. Lingane, "Electroanalytical Chemistry," John Wiley, New York, 1958.] The unit incorporates two control channels and permits two determinations to be made simultaneously for either the same or two

Figure 10. Fisher Scientific Co., Model 40 potentiostat

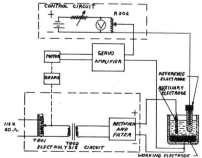

Figure 8. Analytical Instruments, Inc. a) Potentiostat with Current Integrator b) Block diagram of potentiostat

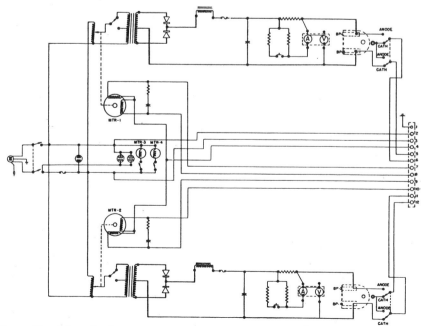

Figure 11. Schematic diagram of analyzer unit, Fisher Scientific Co., Model 40 potentiostat

different metals. As shown in Figure 10, the instrument consists of two units mounted permanently together. The lower unit contains the power unit, combination heaters, beaker supports, beaker cover plates, stirring spindles, meters, fuses and master power switch. The upper unit is a self-contained electronically controlled and regulated amplifier control, designed to control either anode or cathode potentials. Figure 11 indicates the circuit diagram for the analyzer unit (lower unit) and shows the two servo-motors, MTR-1 and MTR-2, which regulate two separate DC rectifier circuits; Motors MTR-3 and MTR-4 are used to stir the solution. The unit incorporates a servo amplifier system which keeps the voltage of the working electrode from exceeding a preset value or from drifting more than a few millivolts from the preset value.

As a group, electromechanical units have the ability to handle large currents. Compared to totally electronic units, they could be considered to suffer from a relatively slow response time. Because of their larger current capacity the electrolysis can be done more rapidly which is an advantage for routine type determinations, as might be found in control work.

Electrodeposition techniques have always played a prominent role in chemical analysis, and the great improvement in precision, convenience, and speed that characterized the introduction of the electromechanical controlled-potential devices stimulated efforts directed toward the elaboration of still more sophisticated instrumentation.

All-Electronic Potentiostats

"Total electronic" potentiostats are offered by several manufacturers. They vary in detail from instrument to instrument but generally operate on similar principles. Transistor circuitry is incorporated in some models to approach the higher circuit capacity of the electromechanical type units.

Typical of the electronic units is the NUMEC Electronic Controlled-Potential Coulometric Titrator which incorporates both electrode potential control and means of integrating the cell current throughout the electrolysis. It is based on an Oak Ridge National Laboratories instrument designed by Kelley, Jones, and Fisher (3). A similar instrument based upon the same electronic circuit is also offered by the Indiana Instrument Corp. The NUMEC instrument is shown in Figure 12. This instrument is totally line-operated, uses a Zener diode constant potential source to eliminate batteries and incorporates a transistor current amplifier stage to simplify the attainment of relatively high electrolysis currents. The unit can be used either for controlled potential reductions or oxidations. The major constituents are the electrolysis cell, control potential source, control amplifier, cell power supply, operational amplifiers power supply, current integrator, and integral read-out device.

A simplified block circuit diagram, in which the unit is operated for reduction of the sample, is shown in Figure 13. In the block diagram, the large triangles represent Philbrick chopper-stabilized difference amplifiers. The Zener diode supplies a constant reference voltage with which to compare the potential of the working electrode in the electrolysis cell. The current amplifier compares and amplifies any voltage difference, which acts as the command signal. This output voltage from the difference amplifier is fed into the base of the transistor current amplifier. The transistor acts essentially as an appropriate variable resistance and consequently controls the current delivered to the cell from the cell power supply in order to maintain the potential of the working electrode. The transistor amplifier in turn is stabilized through a feed-back loop that is completed through the cell.

Figure 12. NUMEC electronic controlled-potential coulometric titrator.

A conventional analog computer type integrator circuit is also included. A separate operational amplifier is used for this purpose. In this operational amplifier the output voltage, which is taken across the 10 μf capacitor, is directly proportional to the time integral of the cell current. A fraction of the output voltage is selected by a voltage divider and presented on the read-out device which might be a digital voltmeter, recorder, etc.

The Wenking Electronic Potentiostat is also a fully electronic, partially transistorized unit which is line-operated except for one battery that serves as the source of the operating potential. The unit is pictured in Figure 14. It operates by comparing a reference input with the actual potential between the working and reference electrodes in a voltage compara-

Figure 14. Wenking potentiostat.

tor. The error signal is amplified about 3000 times in a voltage amplifier. The amplified signal is used to regulate a power amplifier which furnishes the cell current between the working and auxiliary electrodes. No current integrator is supplied with the instrument by the manufacturer, although as accessories a "motor potentiostat" is offered which permits automatic programming of the operating potential of 0–2 v, or superimposing upon the normal operating potential a second potential within this range. This would be quite useful for repetitive or serial tests which involve current-voltage measurements in which the operating potential must be raised at a predetermined rate within a certain critical range. Also offered by Wenking is a multichannel potentiostat, Model MKTR, which permits the operation of six potentiostat channels simultaneously.

Duffers Associates, Inc., potentiostat with power supply and null detector, Figure 15, provides for either direct or external adjustment of the control potential. Hence by using an external potentiometer and the null detector, it is possible to set the desired control potential with great accuracy within the potential control range of the instrument, or to program the control potential during the electrolysis by means of a separate external programmer.

A very interesting potentiostatic system is also offered by A. R. F. Products, Inc. Although these instruments were originally designed as a modular system, primarily for teaching instrumental analysis and consequently somewhat limited in their operating range, the units would also seem to serve well in certain applied work where it is desired to test controlled potential techniques without "major" capital expenditures. The potentiostat is shown in

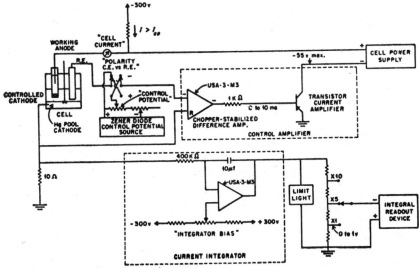

Figure 13. Block diagram for NUMEC potentiostat.

Chemical Instrumentation

Figure 16. It is built into a transparent plastic case and constructed, where applicable, with solid state components. The control signal is transformer-isolated from the power output circuit and thus the control signal can be taken from any point in the output. It also has a manual mode of operation in which the output is controlled by the operator.

Current integrators that are applicable to potentiostatic techniques can be found in the instrument lines of manufacturers who specialize in devices unrelated to wet chemical analysis. For example, current integrators are often employed in radiation detection instrumentation; a number of integrators applicable to the present purpose are made by Eldorado Electronics, Berkeley 10, California.

trolled potential coulometry this is not necessary since the current flow ceases when all the material has been titrated.

Much simpler equipment is required for constant current electrolysis. The commercial instruments are particularly suited for micro and semi-micro analyses and readily lend themselves to incorporation into automatically controlled titration devices. Basically the coulometer consists of a constant current source and an electric timer so connected to the circuit that the timer is in operation only when the electrolysis is taking place. The current ranges are so selected that the coulometer will read out directly as microequivalents of generated titrant. Such coulometric analyzers may be used for a number of precise titrations as, for example, the titration of acids and bases, oxidizing or reducing agents, halides, and

from the ac power supply to the cell of the coulometric apparatus, and ac power to operate a clock motor coupled to a mechanical counter. In series with the electrolysis cell (the load on the instrument) is a calibrated resistor. The current which is delivered to the load is held at a constant value by a control action in which the IR drop across the resistor is continuously compared with the EMF of a standard cell. Any voltage difference which may come about by changes in the resistance of the load as the electrolysis proceeds, or by line voltage fluctuations, is converted in a chopper to positive or negative pulses. The sign of the pulses depends upon the sign of the correction to be applied. After ac amplification and

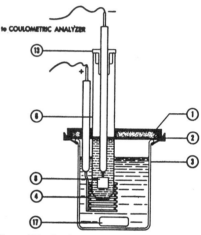

Figure 17. Leeds and Northrup Co., coulometric titration cell.

1   Cell cover
2   Ring
3   Beaker
4   Platinum generating electrode
6   Large tube shield
8   Auxillary platinum electrode
13   Electrode holder
17   Magnetic stirring bar

Figure 15. Duffers, Associates, Inc. potentiostat with power supply and null detector.

Constant Current Coulometry

Constant current coulometry and controlled potential coulometry are related in the respect that in each case the numbers of equivalents consumed in the reaction are calculated by means of Faraday's law. In both cases, for this summation of the current-time flow to be meaningful, the reaction must proceed with 100% current efficiency. In coulometry at controlled current, the current is kept constant throughout the electrolysis so that the total quantity of current passed through the solution is attained simply by the product of the current and the time, $i \times t$. In contrast to coulometry at controlled potential, a means for the detection of the end-point must be available. In con-

certain organic compounds. The organic compounds, such as phenols, may be determined by the amount of bromine (which is electrolytically generated) required to react with the substance. The titrant may be generated in the same reaction vessel (Fig. 17) or it may be generated externally (Fig. 18). In both cases provision may be made so that the products of the cathode and anode reactions are separated and do not intermingle.

The constant current coulometers manufactured by E. H. Sargent Co. (Fig. 19) and Leeds and Northrup Co. (Fig. 20) are similar in basic electronic circuitry. A block diagram of the Leeds and Northrup Co. coulometric analyzer is shown in Figure 21. Closing of the output switch simultaneously applies high voltage dc

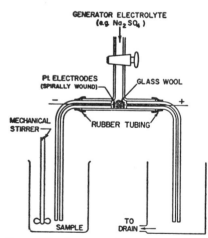

Figure 18. Apparatus for coulometric titration. with externally generated reagents. DeFord, Pitts and Johns, Anal. Chem., 23, 938 (1951).

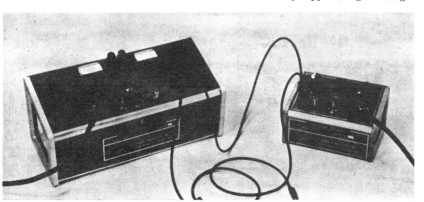

Figure 16. A.R.F. Products, Inc. potentiostat and reference voltage source.

limiting, the signals are fed to a power amplifier tube and reproduced in the current output as unidirectional voltage pulses above or below a prevailing dc level. Thus any changes in the output current are instantaneously corrected by means of the voltage applied to the electrodes of the cell so that the average current output corresponds to the initially selected current levels.

Galvanostats

Potentiostats may also be operated as constant current generators, i.e., galvanostats. The operation of the Analytical Instruments, Inc., potentiostat as a constant current generator is indicated in Figure 22. A resistor, R, is connected in series with the generator electrodes, and this resistor is set to a fixed value which is not changed throughout the electrolysis.

Figure 19. E. H. Sargent Co., coulometric current source.

The voltage drop across this resistor is the command signal for the potentiostat. The control potential of the potentiostat is set to a fixed value which in turn sets the IR-drop across the resistor. Changes in the solution composition throughout the electrolysis would normally cause the

Figure 20. Leeds and Northrup Co., coulometric analyzer model 7960.

current flowing between the electrodes to vary. Concurrently, this would also change the IR drop across the resistor. As this voltage drop is the command signal for the potentiostat, the potentiostat in turn varies the voltage across the electrodes in the solution in order to maintain a constant IR-drop across the resistor. Because R is fixed, the current, I, flowing through the system remains constant.

The magnitude of the IR-drop across the resistor limits the current flowing through the system. If the potentiostat were set to maintain a control potential of 0.50 v, and if the resistance of the resistor were 10,000 ohms, the current flowing between the electrodes would be 0.50/10,000 amp or 50 microamp. In

turn if the magnitude of the resistor were 1 ohm, and the control potential of the potentiosat was set to 1 v, a constant current flow of 1 amp would be maintained throughout the electrolysis.

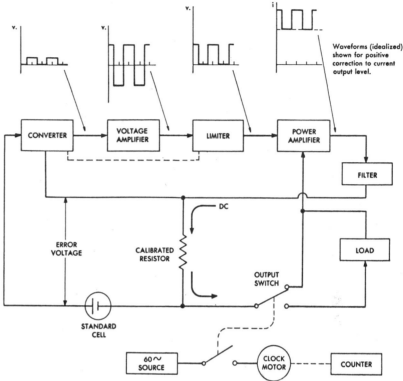

Figure 21. Functional diagram of Leeds and Northrup Co., coulometric analyzer.

This general approach may be used to convert any potentiostat into a galvanostat. Coulometric analyzers are specifically designed to carry small currents in order to be able to perform microanalyses. The higher current-carrying capacities, e.g., one amp or more, of the potentiostats could be very useful in extending coulometric techniques to macroanalyses.

Microcoulometric Titration System

The Dohrman Instrument Co., 990 Varian St., San Carlos, California, has developed a very sensitive instrument for the determination of submicrogram and microgram amounts of certain materials, particularly halides and sulfur in organic compounds such as are found in petroleum materials or pesticide residues. The underlying principle of the unit is to employ a coulometric generator as an additional "detector" with a gas chromatography unit. The microcoulometric titrator may be operated several ways, depending upon the composition of the sample. In a typical chloride analysis, the effluent materials coming from a gas chromatography unit would first be passed into the Dohrman combustion unit. Here the sample is oxidized at about 800°C in an excess of oxygen. The effluent gases from the combustion unit, in the form of carbon dioxide, hydrogen chloride and water vapor, would then pass into the Dohrman titration cell. The titration cell consists of four electrodes, two of which serve as the generator electrodes of the coulometric analyzer and two as the detector

electrodes. The generator electrodes consist of a platinum cathode and silver anode. The sensor electrode pair consists of a silver indicator electrode and a silver in saturated silver acetate reference

electrode. The supporting electrolyte in the titration cell is a 70% acetic acid solution.

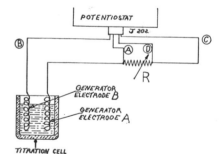

Figure 22. Block diagram for the operation of the Analytical Instruments, Inc., potentiostat as a galvanostat.

As soon as hydrogen chloride enters into the titration cell and silver chloride precipitates, the composition of the electrolyte in the titration cell is altered. This change in the concentration of the electrolyte in turn alters the half cell

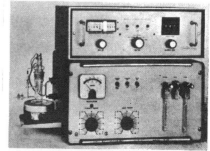

Figure 23. Dohrmann Instrument Co., microcoulometric system.

Chemical Instrumentation

potential of the indicator electrode and gives an error signal to a high gain null-balance servo amplifier in the coulometer. The servo system then applies a voltage to the generator electrodes, whereby an equivalent quantity of silver is dissolved from the silver anode to replace the amount consumed by the sample and, in turn, restore the silver concentration back to its initial value. The current which passes through the generator electrodes flows through a precision resistor network. A

Table I. Manufacturers of Equipment

A. Electrodeposition Units, 110 V, AC

Manufacturer	Model	Sample Capacity	Maximum Output (Amp)	Maximum Output (V)	Comments
Eberbach Corporation P. O. Box 1024 Ann Arbor, Michigan	100 electroanalysis apparatus	2	10/cell	±7.5	Heater available
	1250 ultra speed electroanalyzer	1	25	10	High speed unit
	1500 dyna-cath mercury cathode apparatus	2	25/cell		Mercury cathode unit
	1400 single position electrolysis apparatus	1	5		Model 1410 available for use with external dc source, e.g., batteries
Carlo Erba Via Carlo Imbonati, 24 Milan, Italy	AE 130/°F	1	10	±12	
Prolabo 12 Rue Pelee Paris XI, France	4404	1	10		
E. H. Sargent Company 4647 W. Foster Avenue Chicago, Illinois 60630	S-29464 electrolytic analyzer	2	5/cell	±12	Integral heater available in Model S-29465

B. Potentiostats

Manufacturer	Model	Type	Maximum Output Current (amp)	Output Voltage (v)	Potential Range (v)	Potential Control	Response Time	Comments
Analytical Inst. Inc. North Street Woolcott 16, Connecticut	....	Electro-mechanical	±12.5	35	+3 to −3	±20 mv	<5 sec	Integrator available
ARF Products, Inc. P. O. Box 57 Raton, New Mexico	Model APP-6	Electronic	±5	10	0–3	5 mv or 1%	....	
Duffers, Assoc., Inc. P. O. Box 296 Troy, New York	Model 600	Electronic	±2	8	+4 to −4	<1 mv	<400 μsec	
Fisher Scientific Co. 711 Forbes St. Pittsburgh, Pennsylvania 15219	Model 40	Electro-mechanical	±10	10	0–3	10 mv	<5 seconds	2 channels
Indiana Instrument and Chemical Corp. 916 S. Highland Avenue Bloomington, Indiana	O.R.N.L. Model Q-2005-X50	Electronic	0.25	55	+5 to −5	±3 mv	<20 μsec for 0.1 v step increase	Integrator available
Magna Corp. 11808 S. Bloomfield Avenue Santa Fe Springs, California	Model NP-10	Electronic	±1	8	+3 to −3	±2 mv		
NUMEC Inst. & Control Corp. Apollo, Pennsylvania 15613	O.R.N.L. Model Q-2005-X50	Electronic	0.55	55	+5 to −5	±3 mv	<20 μsec for 0.1 v step increase	Integrator available
Prolabo 12, Rue Pelee Paris XI, France	Model 4411	Electro-mechanical	±10	6	0 to 2	±5 mv	5 sec	
Societa Italiana di Technologia Via F. LLi Gabba 8, Milan, Italy	Model V-2 Model V-3	Electronic Electronic	5 1	12 30		1.5 mv 1.5 mv	100 μsec 100 μsec	
Wenking U. S. Distributor Brinkmann Instruments, Inc. Cantiague Road Westbury, New York	Model 61-TR Model 61-R	Electronic Electronic	±0.3 A-1A ±0.3	30 40	+5 to −5 +5 to −5	1 mv 1 mv	10 μsec 1 μsec	6 channels available

Table I.   Manufacturers of Equipment (*Continued*)

C.   Constant Current Coulometers

Manufacturer	Model	Voltage Standard	Output Range	Output Voltage (v)	Accuracy (%)	Comments
ARF Products, Inc. P. O. Box 57 Raton, New Mexico	APP-7 constant current source	Zener Diode	0.25–25 ma in 5 steps	25	±3	Modular unit with solid state components designed for teaching purposes
Canal Industrial Corp. 4935 Cordell Avenue Bethesda, Maryland 20014	Coulachron					Specifications not available at time of writing
Leeds and Northrup Company 4907 Stenton Avenue Philadelphia 44, Pa.	7960	Standard cell	0.643, 6.43, 64.3 ma or 0.4, 4, 40 μequiv/min	25	±0.05	
E. H. Sargent Co. 4647 West Foster Ave. Chicago 30, Illinois	S-30974	Standard cell	4.825 to 193.0 ma in 6 steps or 0.005 to 0.2 μequiv/timer unit	300	±0.1	

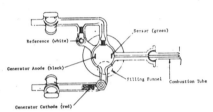

Figure 24. Titration cell for Dohrmann Instrument Co. microcoulometric system.

recorder with a ball disc integrator monitors the IR drop of this resistor network and accordingly displays the total nuber of coulombs used in the titration of the sample.

The Dohrman unit is illustrated in Figure 23, and the titration cell is shown in Figure 24. An additional cell is offered for the analysis of substances which may be titrated with iodine, such as sulfur dioxide. In this cell, all the electrodes are made of platinum; the reference electrode is platinum in saturated triiodide. The supporting electrolyte is a dilute solution of potassium iodide in acetic acid. Any sulfur which has been oxidized to sulfur dioxide in the combustion unit reacts with triiodide in the titration cell to form sulfur trioxide. In turn, this changes the composition of the electrolyte and similarly produces an error signal for the coulometric analyzer which then regenerates sufficient triiodide to return the composition of the electrolyte to its initial concentration.

Figure 25. ACHEM Model 1 electrodeposition unit.

Electrodeposition and Coulometry

Supplement

A number of new instruments have appeared in this field during the last five years.

ACHEM has recently introduced a very compact, portable unit, shown in Figure 25, that is specifically designed for use by the student at his assigned bench space, for the determination of copper in quantitative analysis courses. The unit operates from the 117-V ac line, and delivers a current of about 3 amp under the usual conditions of electrolysis, so that no further current control is required. A second model, of similar size, is also offered that has a filtering unit and variable voltage control, so as to allow a selection of operating conditions, and to extend simple electrolytic procedures to other elements such as chromium.

The Magna Corp.'s Model 4700M potentiostat is shown in Figure 26. This unit features the usual potentiostatic control functions as well as a dual channel potential control, to permit an instantaneous shift to a second potential, or,

by means of their Model 4510 linear scan unit, to a scan or sweep of the applied potential by either a ramp, sawtooth, or triangular sweep function.

The Vari-Tech Co. now offers a constant-current coulometer that covers, in four ranges, a current span from 0.01

Figure 26. Magna Model 4700M potentiostat.

to 10 amp/sec. This is shown in Figure 27.

LKB's Coulometric Analyzer (Fig. 28) permits the rapid recording of the total current flow for a coulometric titration. The unit compares the potential difference between indicator electrodes to the voltage from a potentiometer calibrated in both millivolts and *p*H units. This voltage difference is employed as the control signal for operation of the generating electrodes in the coulometric cell, and as this potential difference decreases throughout the titration, the current decreases asymptotically to zero. An integrator sums the total current flow, which can also be recorded.

McKee-Pedersen offers a potentiostat, their Model MP-1026 (Fig. 29). This is essentially a programmable power supply which can be operated as a potentiostat, a source of constant current, or of constant voltage, or as an operational amplifier.

In addition to these specialized units, a number of multi-purpose electroanalytical

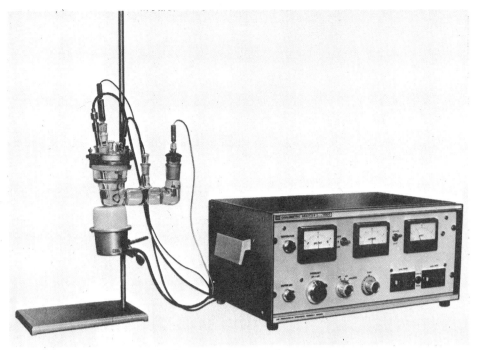

Figure 28. LKB instruments coulometric analyzer.

instruments may be employed for controlled potential electrolysis or for coulometric titrations. Such units include the Beckman *Electroscan-30*, the Princeton Applied Research Model 170 *Electrochemistry System*, the National Instrument Laboratories *Electrolab*, and the McKee-Pedersen *System 1000* multipurpose electronic console. These instruments have been described by Ewing (*4*).

Figure 27. Vari-Tech Model VT-1187 coulometric power supply.

Bibliography

(1) SAND, H. J. S., "Electrochemistry and Electrochemical Analysis," Blackie and Son, Ltd., London, Vol. II, 1940.
(2) RECHNITZ, G. A., "Controlled-Potential Analysis," MacMillan, New York, 1963.
(3) KELLEY, M. T., JONES, H. C., AND FISHER, D. J., *Anal. Chem.*, **31**, 488 (1959); **31**, 956 (1959).
(4) EWING, G. W., J. CHEM. EDUC., **46**, A717 (1969); [Reprinted in this volume.]

Prior research leading to this article was supported in part by Public Health Service Research Grant HEW GM 12830-01 from the National Institutes of Health and the University of Missouri Faculty Research Fund.

Summary of Manufacturers

Table I lists manufacturers of equipment and summarizes the salient information. Some of the instruments are also sold by laboratory supply houses. For those instruments of foreign manufacture, the name of a U. S. distributor is also included where possible.

Supplement

ACHEM, 11830 West Ripley Ave., Milwaukee, Wis. 53226; Model 1, single sample, 5 amp at 10 V; uses magnetic stirrer or stirrer-hot plate. Model 2, single sample, 5 amp at 15 V.

Magna Corp., Model 4700M, electronic type; 10 amp at 10 or 20 V; +6 to −6 V; control to ±5 mV; response time <2 μsec; linear scan and high-voltage power supply available.

LKB Instruments, Inc., 12221 Parklawn Drive, Rockville, Md. 20852.

Vari-Tech Co., 546 Leonard St., Grand Rapids, Mich. 49504; Model VT-1187; output 0.01–10 A, in 4 steps; 0–15 V; accuracy ±1%.

Figure 29. McKee-Pedersen Potentiostat and Regulated Power Supply

Topics in...

Chemical Instrumentation

Edited by GALEN W. EWING, Seton Hall University, So. Orange, N. J. 07079

The Journal of Chemical Education
Vol. 47, page A81
February, 1970

David E. Burge received his B.S. in chemistry from the University of California, Berkeley and his Ph.D. from Oregon State University. He was employed by the U. S. Geological Survey, Shell Development Company and Melabs, and is currently acting as a private consultant. His research interests include electrochemistry, polymer characterization, environmental pollution and analytical instrumentation.

L. Pulse Polarography

David E. Burge, P. O. Box 313, Cupertino, Calif. 95014

Polarography is a well established method of chemical analysis and research. Although it has been known for about forty years, it is still in a state of active development. This discussion will cover one area in which real progress has been made in the last ten years; the use of pulse techniques to improve the analytical usefulness of polarography.

As anyone familiar with polarography is well aware, the usual technique involves using the dropping mercury electrode (DME), applying a continuous and linearly changing voltage to the electrode, and measuring the current passing through the cell. The other electrode of the cell is a standard reference electrode such as calomel or silver-silver chloride. If the reference electrode is constructed properly, its potential relative to the solution will not change with current. Therefore, the total potential applied to the cell will consist of the constant potential of the reference electrode, a potential between the solution and the working electrode—that is the DME—and a potential through the solution due to its resistance.

Under these conditions the mercury electrode is completely polarized and, hence, determines the total current flowing through the cell. The current flowing will depend on reductions taking place at the electrode at a particular potential, as well as on the capacitive charging current required to charge the double layer around each new mercury drop. These two currents are the only ones of significance passing through the cell. Furthermore, since the Faradaic reduction current is the one

of interest, it follows that the capacitive charging current should be minimized in order to increase the sensitivity or improve the measurement limit of the Faradaic current. Suppression of the effects of charging current is the major reason for the use of pulse techniques.

Pulse polarography was first developed by Barker (1) as an outgrowth of his work with square wave methods. His published work has been summarized by Schmidt and von Stackelberg (2).

Pulse Polarography

Let us consider in more detail the use of pulse measurement techniques in polarography, with particular emphasis on the reasons that these techniques permit greater sensitivity than is obtainable with conventional dc polarography. In order to understand why pulse methods are better, it is first necessary to review the principles of polarography, and particularly those factors which ultimately limit its usefulness. Normally, polarography consists of applying a linearly changing voltage to the cell and measuring the resulting current. If the dropping mercury electrode is used, and if the solution contains at least 100 times as much inert electrolyte as there is reducible material, then typical results are as shown in Figure 1. Figure 1a shows the voltage change as a function of time. Figure 1b shows the current as a function of voltage. The current varies drastically during the life of each mercury drop. When the drop first starts to form its area is extremely small, and so is the current. The current then increases with increasing drop area. Let us examine the nature of this current change in more detail. Figure 2 shows the current during two complete drop lifetimes for four distinct conditions in the cell. Figure 2a shows the current observed when Faradaic reduction is the predominant process. It has been shown theoretically and verified experimentally that this current increases as the $^1/_6$ power of the time. Figure 2b shows the current when no reducible species is

present in the solution, and when the potential is more negative than that of the electrocapillary maximum. (The "electrocapillary maximum" is the term used to designate the potential which the mercury electrode assumes spontaneously in contact with the solution. In 1 M KCl it is about 0.5 v negative with respect to the saturated calomel electrode. The name arises from the fact that the mercury–solution interfacial tension is a maximum at this potential.) In this case, due to the growth in size of the capacitance around the mercury drop, the current is a charging current, and is proportional to the rate of growth of the drop area. Since this rate of growth is very large when the drop size is small, the charging current increases sharply just after the previous drop falls off. This current then decays as the $-^1/_3$ power of the time.

In Figure 2c we show the current resulting from conditions similar to those in Figure 2b, except that the cell potential is more positive than that corresponding to the

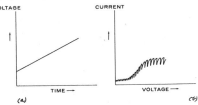

Figure 1. Voltage and current relations in conventional polarography: (a) Sweep voltage as a function of time; (b) Current as a function of voltage, the "polarogram."

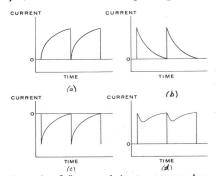

Figure 2. Cell current during two mercury drop lives: (a) Faradaic reduction current; (b) Charging current when cell potential is more negative than the electrocapillary maximum; (c) Charging current, more positive than the electrocapillary maximum; (d) A mixture of a and b.

Chemical Instrumentation

point of the electrocapillary maximum. In this case the current is quite large at the start of the drop and again decays as the $-1/3$ power of the time. However, its polarity has changed. That is, if one considers the polarity of the current in Figure 2b to be positive, then in Figure 2c it is negative. When the concentration of reducible species is small, the average Faradaic current may be comparable to, or even smaller than, the average charging current. In this case, curves as shown in Figure 2d are sometimes observed. Here current increases very sharply at the beginning of the drop, due to the charging current. But before the drop is complete, the Faradaic current becomes predominant and current increases again. Thus, the curve is clearly a mixture of the charging-current type and reduction-current type as shown in Figures 2a and 2b. This type of curve will only be observed when the voltage is more negative than the electrocapillary maximum. This curve makes it quite clear why the charging current eventually limits the sensitivity of determination of Faradaic current.

Another way of looking at this is to smooth the variations due to drop growth and look only at the average current over the complete voltage scan. When this is done the result shows a sloping baseline, due to a change in the average charging current as a function of voltage. Superimposed on this are the reduction waves, as shown diagrammatically in Figure 3. Here the average current is shown. It can be seen that, as current sensitivity is increased, eventually the slope of the background due to charging current will be so steep as to limit the ability to measure Faradaic current. These are well known observations, repeated here to point out a weakness of conventional polarography which to a large extent can be overcome by pulse measurement techniques.

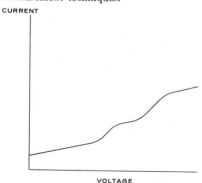

Figure 3. Smoothed or averaged polarogram, at high sensitivity.

Pulse measurements are carried out in two distinctly different ways. In one, a slowly changing dc voltage is applied to the cell as in conventional polarography, but once during the life of each mercury drop a small-amplitude voltage pulse, in addition to the dc ramp, is applied to the cell. This small voltage pulse results in a current pulse made up of two major components, the charging current and the Faradaic current, both caused by the

voltage pulse. The charging current is a sharp spike which decays to zero quickly (a few milliseconds) whereas the reduction current, if any, does not decay to zero but only decays to the diffusion limited value. Therefore, if the current pulse measurement is made only during the last part of the pulse, the charging current pulse will have decayed to zero, and the only part measured will be due to a Faradaic reduction taking place at this voltage. This is illustrated in Figure 4, where the time intervals are shown for application of the voltage pulse, the resulting current pulse, and current measurement. The measurement technique is required to compensate for any current flowing, up to the time the pulse is applied, and to establish a reference level against which the pulse current can be measured. Once this is done, the pulse current is measured during the last part of the pulse when the charging current has decayed. The resulting current value is stored and held until the next pulse is applied.

The second method of measurement makes use of similar timing intervals and current measurement methods but the voltage applied consists of a series of pulses of steadily increasing amplitude superimposed on a constant, preset voltage. This method gives a current-voltage curve similar to that obtained in dc polarography whereas the first method gives a curve similar to the first derivative of the usual polarogram. For clearer understanding of these two methods, the operation of one particular instrument will be considered in more detail. A discussion of the different commercially available instruments will be given in a subsequent section.

Figure 4 shows the sequence of events occurring during two complete mercury drops when measurements are made in the first mode. We see that the reference level against which the pulse current is measured is the level existing just before the voltage pulse is applied. (The true reference current will increase slightly during this period due to the growth of drop area, so there is a slight error in this reference level.) Time delay between the

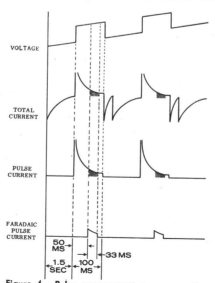

Figure 4. Pulse measurement sequence. The unmarked small peak in the total current curve corresponds to the knocking off of the mercury drop.

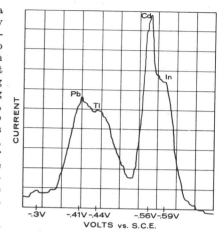

Figure 5. Pulse polarogram of solution containing lead, thallium, cadmium, and indium. Approx. $10^{-5}M$ in $0.01M$ NH_4Cl; the labels "Cd" and "In" were inadvertently interchanged in the figure.

beginning of the pulse and current measurement is 50 msec, which permits the charging current spike to decay to a very small value. The remaining current, assumed to be due to any Faradaic reduction taking place, is integrated for $33 1/3$ msec, or a time equivalent to two complete cycles of the line frequency. This averages out any modulation of the signal caused by line frequency pick-up, and averages out random noise to some extent. The resulting integral is then stored in a conventional pulse stretching circuit for two seconds until the next pulse is applied. When the next pulse is applied, the integrator is zeroed during the 50-msec time interval. Then the signal is integrated during the following $33 1/3$ msec, and the cycle continues to repeat itself.

It can be seen that this approach results in an output which is very nearly the first derivative of the conventional polarographic curve because the pulse current will be largest at the half-wave potential where the current-voltage curve is steepest. Improved sensitivity is achieved through compensation for the charging current, which permits the use of greater electrical amplification of the signal. The first derivative output provides better resolution and better compensation for prior reductions than is achieved by conventional polarography. Therefore, this technique has three distinct advantages; improved sensitivity, improved resolution, and improved compensation for prior reduction. In Figure 5 a differential pulse polarogram is shown for a mixture containing lead, thallium, cadmium and indium. Lead and thallium are reduced at similar potentials, and hence are not separated in conventional polarography. The same is true of cadmium and indium. It can be seen in this figure that at least partial separation is achieved when the concentrations are approximately equal. This result corresponds to the best theoretically attainable resolution when using 30-mv pulse amplitude and is very close to the theoretical limit, which is determined by the half-width of the individual polarographic waves (3).

Another example is given in Figure 6, where the determination of maleic acid in the presence of a large concentration of fumaric acid is illustrated. This is another particularly difficult problem by conventional polarography, because of the lack of resolution between the two reductions; but it can be seen that by careful technique and careful superposition of the curves, a quantity down to 0.1% of maleic acid in fumaric acid can be determined by the differential pulse technique.

Another example is the determination of nickel and vanadium in a single sample, which has been discussed in a paper by Gilbert (5). These reductions are both sufficiently irreversible so that the waves in conventional polarography tend to become very broad, spread out, and ill defined. However, with pulse polarography,

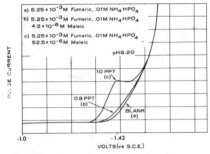

a) 5.25 × 10⁻³M Fumaric, .01M NH₄HPO₄
b) 5.25 × 10⁻³M Fumaric, .01M NH₄HPO₄
4.2 × 10⁻⁶ M Maleic
c) 5.25 × 10⁻³M Fumaric, .01M NH₄HPO₄
52.5 × 10⁻⁶M Maleic

pH8.20

10 PPT (c)
0.8 PPT (b)
BLANK (a)

-1.0   -1.42
VOLTS (vs S.C.E.)

PULSE CURRENT

Figure 6. Pulse polarograms of maleic acid in the presence of large amounts of fumaric acid.

peaks are obtained which are clearly useful for the simultaneous determination of these two elements.

When the voltage pulses applied to the cell have an amplitude which increases linearly with time, the situation is closely related to that of DC polarography. The current pulses contain a capacitive and a Faradaic component just as in the previous technique and the capacitive contribution is minimized by similar methods. Thus the Faradaic component of pulse current is measured and yields a current-voltage curve similar to that of dc polarography. This technique does not provide the sensitivity and resolution obtained with the differential method but it has value for the study of reaction kinetics and for monitoring concentration using solid metal or carbon electrodes.

For our purposes we will call this method integral pulse polarography and the technique using constant amplitude pulses differential pulse polarography. The terms "differential" and "derivative" are both used to refer to the latter case but in fact all pulse methods measure finite, first differences rather than a true first derivative so the term differential appears preferable. Likewise, the terms "normal" and "integral" are used to refer to the first method. The use of the word "normal" arose because the shape of the curve obtained resembled that of the conventional or normal dc polarogram. On the other hand, the word "integral" is analogous to the use of differential or derivative. It seems that the choice in this case is somewhat arbitrary. Due to this variation in usage the exact operation of each mode will be described for each commerical instrument.

Commercially Available Instruments

At the time of writing this article, three commercially available instruments provide the capability to do pulse polarography. Each of them has more than one mode of operation and the details of the pulse measurement differ somewhat between them. In those cases where the instrument can be used in several modes, only the pulse modes will be described in detail.

Southern Analytical, Limited

Pulse polarographic techniques were first developed by Barker at the Harwell Atomic Energy Establishment in England. This work led to the first commercial pulse polarograph which was manufactured by Southern Analytical Limited and sold as the Southern-Harwell Model A1700 Pulse Polarograph. The Model A3100 now sold by Southern is an updated version of the first instrument. It is a two electrode system with two modes of operation, "normal" pulse and "derivative" pulse polarography. In the normal mode the pulses increase in amplitude with time whereas in the derivative mode constant amplitude pulses are superimposed on a ramp voltage.

The operation of this instrument is basically as described in the previous section. A voltage pulse is applied to the cell and at some later time a current measurement is made. The current is integrated for a fixed period and the resulting integral held and presented to the output until the next pulse is applied. The capacitance decay time can be varied from 2 to 40 msec and the current measurement time over the same range. In the derivative mode the pulse amplitude can be varied from 2 to 100 mv.

A mechanical drop synchronizer provides drop times variable from 0.5 to 4 sec. In addition, the natural drop time of the capillary can be used and the timing circuits activated by high frequency signals generated when the drop falls off. When using the drop synchronizer, the voltage pulse is applied just before the drop is knocked off.

A number of minor cell current components can be important when operating at maximum sensitivity. Front panel controls are provided to compensate for residual current and capillary response. The residual current is caused by the growth of drop area during the pulse and the compensator is designed to minimize this effect. Capillary response is due to wetting of the inside wall of the capillary tip. It is roughly proportional to pulse amplitude but varies from drop to drop so the compensation is never perfect. These controls are mainly useful for optimizing performance for maximum sensitivity.

Melabs

The Melabs Model CPA-3 Polarographic Analyzer shown in Figure 7 is a three electrode polarographic instrument which can be operated in four different modes. These include conventional dc polarography, fast sweep polarography in which the complete voltage scan is applied to a single mercury drop, "differential" pulse polarography, and "integral" pulse polarography. The differential pulse mode refers to the case in which a

Figure 7. Melabs Pulse Polarographic Analyzer and Electrode Stand. (Photograph courtesy of Melabs)

constant amplitude pulse is superimposed on a slowly changing voltage ramp whereas in the integral mode pulses of increasing amplitude are applied.

The methods of pulse measurement and the timing sequence for the Melabs instrument correspond to that described in the general discussion of theory. An electrode stand with built-in mechanical drop synchronizer is used in conjunction with the analyzer. Each time a drop is knocked from the capillary, a pulse to the analyzer starts the timing sequence. One and one half seconds later a voltage pulse is applied to the mercury drop. This pulse has a duration of 100 msec and an amplitude, selected on the panel, of 10, 30 or 100 mv. Current measurement starts 50 msec after the pulse is applied. The current is integrated for 33¹/₃ msec (two complete cycles of the 60 Hz line frequency). This integration removes the effect of line frequency modulation of the signal and also smooths out the effect of random noise.

The differential mode is the most generally useful since it provides better resolution and sensitivity than conventional polarography with comparable accuracy.

Princeton Applied Research

Princeton Applied Research Model 170 Electrochemistry System and the Model 171 Polarographic Analyzer shown in Figure 8 both provide for operation in the pulse mode. The Model 170 is a sophisticated system which can be used to perform a wide range of electrochemical measurements (4). Polarographically it performs in a conventional dc mode, in a normal pulse mode, a derivative pulse mode, and a phase-locked ac mode. The Model 171 is primarily a polarographic instrument, and within this limitation, has the same modes of operation as the 170. Both of them can be operated as two electrode or three electrode systems.

In the normal pulse mode, voltage pulses of 48 msec duration are applied just before the end of the drop life. The current flow is measured for 8 msec after a 40 msec delay. Pulse amplitude increases linearly with time. Drop life, determined by a mechanical synchronizer, can be varied from 0.5 to 5 sec with the pulse always applied just before the drop is

Chemical Instrumentation

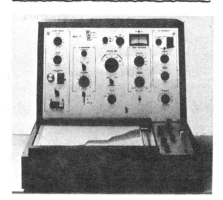

Figure 8. PAR Model 171 Polarography Analyzer and Recorder. (*Photography courtesy of Princeton Applied Research*)

mechanically removed from the capillary. Since the pulse amplitude increases linearly with time, this mode corresponds to the normal or integral modes of the previous instruments.

The derivative pulse mode in this instrument is quite different than that described for the differential or derivative modes of the other instruments. In this case, voltage pulses of increasing amplitude are used just as in the normal mode, but a differential current measurement is used. The current measured for a given pulse is subtracted from that for the succeeding pulse and the resulting difference presented at the output. Therefore, this output resembles a first derivative curve. In contrast to the other difference methods where sensitivity and resolution are determined by selection of a given pulse amplitude, in this case they are determined by the rate of change of pulse amplitude and by the time between pulses.

A comparison of some of the operating features of the three commercial instruments is given in Table 1.

Conclusion

At the time of its development, pulse polarography was the most sensitive method available for analysis of trace constituents in aqueous solution. Since then, anodic stripping methods have been developed which are more sensitive for some elements, but not so convenient or rapid as pulse polarography. For elemental analysis, atomic absorption has comparable sensitivity for many elements and is somewhat faster if only one or two elements are to be determined. There

is no doubt that the advent of atomic absorption methods has been the major deterrent to use of pulse polarography even though many problems could be better solved by the polarographic method.

The most promising area for further development of pulse polarography is in the determination of electroreducible organic compounds. It could be adapted as a relatively specific detector for liquid chromatography if the solvent is sufficiently polar to ionize a salt such as tetramethylammonium chloride. This whole field has been neglected, probably because of a limited knowledge of polarographic methods among most chemists. Polarography has been widely used in Europe, but American chemists have generally found another method to be preferable. This has lead to underdevelopment of an analytical method of great potential.

Manufacturers

Melabs
   3300 Hillview Ave.
   Stanford Industrial Park
   Palo Alto, Calif. 94304
Princeton Applied Research Corp.
   P.O. Box 565
   Princeton, N. J. 08540
Southern Analytical Ltd. (England)
   via The Bendix Corporation
   Scientific Instruments Div.
   1775 Mt. Read Blvd.
   Rochester, N. Y. 14603

Literature Cited

(1) BARKER, G. C., Proc. Congr. on Modern Anal. Chem. in Ind., St. Andrews, 1957, p. 199. BARKER, G. C. and GARDNER, A. W., *Z. Anal. Chem.*, **173**, 79 (1960).
(2) SCHMIDT, H. and VON STACKELBERG, M., "Modern Polarographic Methods," Academic Press, New York, 1963.
(3) PARRY, E. P. and OSTERYOUNG, R. A., *Anal. Chem.*, **37**, 1634 (1965).
(4) EWING, G. W., J. CHEM. EDUC., **46**, A717 (1969). [Reprinted in this volume]
(5) GILBERT, D. D., *Anal. Chem.*, **37**, 1102 (1965).

Supplement

Figure 9 illustrates the A-3100 Pulse Polarograph of Southern Analytical. The photograph was not available in time for the article proper. There have been no further developments to report.

Figure 9. Model A-3100 Pulse Polarograph, Southern Analytical, Ltd.

Table 1. Comparison of Commercially Available Instruments

	Southern	Melabs	PAR
Starting Potential Range (volts)	±2.5	±2.5	±5
Sweep Voltage Polarity	±	±	±
Pulse Height Range (mV) (differential mode)	2–100	10–100	N.A.
Capacitive Decay Time (mS)	2–40	50	40
Current Measuring Time (mS)	2–40	$33^{1}/_{3}$	8
Synchronization	Mechanical and drop-fall	Mechanical	Mechanical
Sweep rate (v/sec)	0.0003 to 0.016	0.0005 to 15	0.0001 to 0.5
Number of electrodes	2	2 or 3	2 or 3
Drop life (seconds)	0.5 to 4 Mechanical or Natural	2	0.5 to 5

Topics in...

Chemical Instrumentation

Edited by **GALEN W. EWING**, Seton Hall University, So. Orange, N. J. 07079

...a *ChemEd* *feature*

The Journal of Chemical Education
Vol. 45, pages A635, A787
September and October, 1968

XL. Oscilloscopes in Chemistry

Joseph E. Nelson, *Chemtrix, Inc., Beaverton, Ore. 97005*

INTRODUCTION

One may say that the oscilloscope is a logical extension of the strip-chart recorder with the X-Y recorder as a natural transition. By comparison the horizontal X-axis of the strip-chart recorder is produced by the paper movement and the vertical Y-axis by the pen movement across the paper. The resultant display on the paper is a chart or graph. The X–Y recorder is similar except that the paper remains stationary while the pen moves across the paper in both coordinates.

If one now imagines that the oscilloscope screen is a piece of graph paper and the electron beam is a pen the relationship is immediately apparent. One might suggest that the comparison is not complete since the waveform or graph drawn by the oscilloscope beam disappears almost immediately. Even this shortcoming has been overcome by the recently introduced storage tube oscilloscopes that have the ability to retain the written image for indefinite lengths of time. Most non-storage type oscilloscopes can present a visual fixed display if the signal is repetitive.

The block diagram of Figure 1 shows the essential working parts of an oscilloscope. The high and low voltage power supplies have been omitted since they will not be discussed here. The working components of importance to the chemist are: (1) the cathode-ray tube, abbreviated CRT, (2) the amplifier, (3) the time-base, and (4) the trigger circuit. Of these, the one most likely to be unfamiliar is the trigger circuit. It is this circuit which has been largely

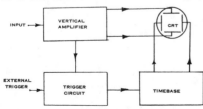

Figure 1. The major component sections of a modern oscilloscope.

responsible for the vast improvements that have occurred in oscilloscope design in the past twenty years. The trigger is discussed in detail below.

THE CATHODE-RAY TUBE

The cathode-ray tube can be considered the end result of a series of transductions. Thus, suppose one starts with a sudden change in temperature and uses a thermistor to convert this to a change in voltage. Within the oscilloscope the voltage change is amplified and applied to the cathode-ray tube where it is converted to a visual display on the screen of the tube. This action of the CRT starts with the *electron gun* where a stream of electrons is produced (See Fig. 2). This gun, located within the narrow neck of the tube, contains an indirectly heated cathode which produces a cloud of electrons. These electrons are then attracted down the neck of the tube to the screen by a potential of several thousand volts between the screen and the electron gun. By use of various elements within the gun it is possible to focus the beam to a small dot where it strikes the screen.

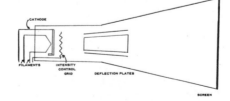

Figure 2. The internal structure of an electro-static deflection cathode-ray tube.

The *screen* consists of a thin layer of cadmium or zinc sulfide which fluoresces momentarily when struck by an electron. Thus, a steady stream of electrons striking the screen will cause a dot of fluorescence. If the construction of the tube is symmetrical, this dot will occur directly in the center of the tube. Later in the text the subject of different screen materials will be discussed in detail.

Joseph E. Nelson received his early training in biochemistry at MIT, and worked in this field for the U. S. Army during World War II. Since then he has had more than twenty years experience in electronic instrumentation, including most recently six years with Tektronix, Inc., on oscilloscope development. During this period he developed a single-sweep polarographic analyzer using a storage oscilloscope. He is now President and Director of Research at Chemtrix, Inc.

At this point the dot on the screen is similar to a resting pen on a recorder, waiting for commands. These commands are applied to the electron beam by two pairs of *deflection plates* contained within the tube neck and located about midway between the screen and the electron gun. When voltages of sufficient amplitude are applied to the pair of deflection plates an electrostatic field is developed between them. If the applied voltages are unequal the steady stream of electrons will be deflected toward the more positive plate. Because of the field between the plates, this type of deflection is called electrostatic.

In Figure 3 the voltages shown represent the potentials applied to the two horizontal deflection plates of the cathode-ray tube. Note that in position 1 the voltage applied to both plates is the same and the electron dot remains at center screen. Position 2 shows 200 v on the left plate and 160 v on the right plate, a difference of 40 v which resulted in a dot movement of 2 cm toward the more positive plate. These data provide the deflection factor of the CRT. In this instance it required 40 v of difference between the plates to move the beam 2 cm or 20 v to deflect 1 cm. This

Chemical Instrumentation

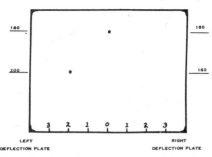

Figure 3. A graphic representation of the voltage applied to the horizontal deflection plates, and the resulting dot position on the CRT screen; the deflection sensitivity is 20 V/cm.

then, is the deflection factor of the horizontal plates: 20 v/cm. It is very possible that the deflection factor of the vertical plates in this same tube will be considerably different than that of the horizontal plates.

Aside from the general interest in knowing how things function, there is a more important reason why the chemist should be aware of deflection plate characteristics. This concerns the application of signals directly to the plates.

Most modern day CRT's contain deflection plate connections directly accessible on the neck of the tube. To use this capability, one simply removes the internal connections and carefully tapes them to prevent electrical shorts, then clips on the new signal connections. In many applications only the vertical plates will be connected to external signals with the internal horizontal time base amplifier used for horizontal deflection.

When electrical signals are passed through attenuators and amplifiers on their way to the CRT screen they frequently suffer slight distortion and time (phase) shift. In most cases, especially where high frequency oscilloscopes are used, these effects are negligible. However, if only a low frequency instrument were available it is possible that the observed signal would be a poor replica of the original. In this case one might benefit by connecting directly to the oscilloscope plates. Or, if some other type of higher-frequency amplifier is available this could be used to precede connection to the plates.

The objection to this method concerns the high voltages required to deflect the beam. Typically for a 6 cm sine wave the change would need to be 120 v with a CRT having a 20 v per cm deflection factor.

One further aspect of CRT operation that can be useful is called *Z-axis modulation*. The Z-axis in this case refers to a change in the intensity of the beam produced by varying the potential between the cathode and a control grid. Usually, the beam is either on or off; a short pulse applied to the cathode can momentarily turn off the beam and serve as a time mark on the screen.

Conversely, one can intensify the trace at precise points by using a pulse of the correct polarity. This can be especially useful to coordinate some exterior event with the waveform shown on the oscilloscopes. For example, if a rising voltage is causing a heating effect in some circuit

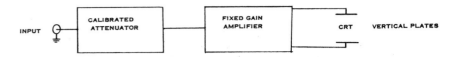

Figure 4. Components of the vertical deflection system.

component that is being followed on the oscilloscope screen, the trace could be intensified at a point in its scan that corresponds to some preset voltage level. Any temperature lag at that point would be immediately apparent. Trace intensification is also used on digital readout oscilloscopes to identify the point at which the measurement is being made.

To close the discussion on CRT's, the chemist should also be aware of screen *persistency*. Different compounds or phosphors continue to fluoresce or persist for different lengths of time, after being struck by an electron, that extend from several milliseconds to as long as 30 sec. The choice of persistency should be determined by the type of experiments planned. Slow-scan work should use a long persistence CRT phosphor.

A long persistence CRT used in a repetitive scanning test can give the impression of a stationary image. Two examples are: (1) a cathode-ray polarographic analyzer in which the scan is synchronized to repeat every 7 sec. If a screen with 10 sec of persistence is used the resulting polarogram will almost appear stationary. (2) where a spectrofluorometer is monitored with an oscilloscope, the X-axis corresponds to wavelength, and the Y-axis fluorescence, so that a continuous repetitive scan will portray a plot of a fluorescence spectrum.

ELECTRONICS

The *vertical amplifier*, also called the Y-axis amplifier, receives the signals to be measured and converts them to a form that can be displayed by vertical deflection of the electron beam of the CRT. In its simplest form it consists of a fixed gain amplifier with a calibrated variable attenuator at its input and the vertical plates of the CRT at its output. This is shown in Fig. 4.

The signal to be displayed is applied to the vertical amplifier through appropriate probes or cables. Generally this signal must be in the form of a voltage or current that has been transduced or changed from another form such as temperature, pressure or vibration. For example, light intensity cannot be applied directly to a vertical amplifier but must first be converted to a voltage signal through a photomultiplier or photosensitive cell. A second example is the change in current through a thermistor as a function of temperature change. In polarography, of course, the quantity of interest is current through the cell which can be applied directly to a current amplifier and does not require a transducer.

Of particular importance to the chemist who requires an accurate replica of a varying quantity is the faithfulness of reproduction of the instrument being used. Most chart recorders and even X-Y recorders are quite limited in this respect. The sensitivity control of the vertical oscilloscope amplifier is calibrated in volts per division of deflection. That is, a 10-v attenuator

setting will require that a 10-v signal must be connected to the input to produce one division of vertical deflection as measured on the CRT graticule. (A CRT screen is usually divided into 8 or 10 vertical divisions and 10 horizontal divisions by a scribed plastic graticule). Because of this, the calibrated attenuator is usually called the volts per division switch and contains switch positions that extend from the most sensitive to the least sensitive in a repeated sequence, i.e., 10, 5, 2, 1 v/division. In many amplifiers the most sensitive position of the volts per division switch also represents the sensitivity of the fixed gain amplifier and the actual attenuation in this case is unity.

When the applied signal is considerably larger than the least sensitive position of the volts per division switch an external attenuator probe can be used to reduce the signal by $1/10$ or $1/100$ th of its original value. Conversely, if the signal level is considerably below the most sensitive position additional amplification is required.

The chemist who uses an oscilloscope would, no doubt, like the displayed waveform on the CRT to be an exact representation of the applied signal. Yet, unless he is familiar with his instrument's *bandwidth* he may be seeing a poor reproduction. Bandwidth, as applied to an oscilloscope vertical amplifier, means the sine-wave frequency at which the displayed signal is 70% (−3db) of the applied signal. This is measured by connecting a variable frequency oscillator of constant amplitude to the vertical amplifier input of the oscilloscope. The volts per division switch is set to 1 and the oscillator amplitude adjusted to produce six divisions of vertical deflection at a sine-wave frequency of 1 kilohertz. (Note: the designation hertz has now replaced cycles per second.) The oscillator frequency is then increased until the displayed waveform reduces to 4.2 divisions (70% of 6 divisions). The frequency at which this occurs is the bandwidth of the amplifier. In the case of an ac coupled amplifier the value described above would be the upper bandwidth. Since the amplifier does not pass dc levels it also has a lower bandwidth which is measured in the same way. A good rule of thumb is to use

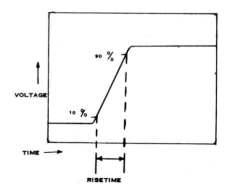

Figure 5. The two points on a waveform that are used to measure risetime.

Chemical Instrumentation

an amplifier that has a bandwidth that is at least ten times the frequency of the signal of interest.

Of even greater interest to many chemists is the response of the oscilloscope amplifier to a step function or square wave. Despite the square-sided appearance of a step function it still requires a finite time to change from one voltage level to another. This time, designated *risetime*, is the time interval between the 10% and 90% amplitude points on the rise of the step function. Figure 5 shows a step function with the period of risetime outlined.

A step function can also be used to measure amplifier bandwidth through the following relationship: The bandwidth times the risetime equals a constant which is typically 0.35.

For example, an oscilloscope amplifier with a bandwidth of 500 kilohertz would have a rise time of 0.35/500 kH = 0.7 μsec.

This means that a fast-rise step function of 0.1 μsec would appear to have a 0.7 μsec risetime when applied to this amplifier. In the same manner fast risetime signals of any kind would be considerably damped due to the low bandwidth of the amplifier. Thus, the displayed waveform may only be a very poor representation of the actual applied signal.

From the foregoing one might conclude that a wide bandwidth oscilloscope would be desirable in all applications. Unfortunately this is not true. High-frequency noise often obscures the signals of interest and the penalty of wide bandwidth becomes apparent. A good example is the amplifier used with a dropping mercury electrode in polarography. To amplify the small currents requires high sensitivity. But the column of mercury acts like an antenna and considerable noise pickup will be evident with a wideband oscilloscope amplifier. Because of this it is necessary to reduce the bandwidth to a point where even small polarographic currents can be measured.

Consequently, some amplifiers have a bandwidth control that permits adjustments of the bandwidth over a considerable range. When using operational amplifiers bandwidth can be reduced by a small capacitor placed across the feedback network.

The *time base* (X-axis) establishes the time value of each horizontal division on the oscilloscope CRT screen. This is accomplished by setting a Time per Division switch to some discrete value. For example, with this switch set to 1 sec, it requires 10 sec for the beam to scan across the full screen (10 divisions).

Internally, the voltage change produced by the time base is a linear ramp that rises to some value and then rapidly falls to zero. Due to the appearance of this wave it is called a sawtooth. The sawtooth is usually developed by integrating a constant current to produce the linear rise. Since the integrator rate of rise is related to both current and capacity, it can be changed to different values by changing the value of either the resistors or the capacitors involved in the integrator. This is actually what takes place when the Time per Division switch is changed.

Time base integrators have been designed to extend from as slow as 10 sec per division to as fast as 1 nanosecond per division. In order to relate these times to some better known reference, consider the elapsed time of 1 cycle of a 60 Hz sinewave signal. The time period of 1 cycle is equal to the reciprocal of the frequency, 1/F which equals 1/60 or 16.6 milliseconds. similarly, one can see that a time base of 1 nanosecond per division would be required to display 1 cycle of 100 megahertz sinewave signal on a 10-division screen.

The integrator output is applied to a *horizontal amplifier* from which it is then applied to the horizontal deflection plates of the CRT.

THE TRIGGERED OSCILLOSCOPE

When a repetitive signal such as a sinewave is applied to an oscilloscope and displayed on the screen it appears as a stationary waveform. A new waveform is superimposed on each scan and the net effect is a stationary-appearing display. This ability to exactly superimpose repetitive signals is one of the most important capabilities of the modern oscilloscope. It is possible because each scan of the time base can be initiated by a trigger impulse that is derived from the signal of interest.

A triggered oscilloscope is one in which each horizontal sweep is initiated or triggered by a signal that is external to the time base. Usually, the trigger signal is received from the vertical amplifier. However, it can also be applied directly to the time base from an external source. In the case of a single sweep, the trigger signal may be applied by actuating a front panel switch.

Almost all triggered oscilloscopes can also operate in a free-run, or nontriggered mode. In this case the sweep is continuous, one cycle after another, with no synchronized relationship to the signal applied to the vertical amplifier. This mode does produce a trace on the CRT

reaches point A.

Once started, the sweep cannot be influenced by triggering signals. When the sweep is completed and retraced to the left side of the screen it will then wait until the triangular wave again crosses point A before scanning. Notice in Figure 6 that the display in each case is the same and when these successive traces are superimposed they appear as a stable display.

The chemist will probably use single events more than repetitive signals and a good illustration of triggering concerns the capture of a fleeting transient. In a flash photolysis experiment the sudden application of a high energy flash produces chemical changes that can be followed electrochemically on the oscilloscope. By using the pulse of light to start or trigger the sweep, the events that immediately follow in the electrolysis cell can be recorded from the screen by camera, or preserved intact by use of a storage tube. The Time per Division switch set on the time base will determine the time period on which events will be recorded following the flash triggering.

In chronopotentiometry a constant current is suddenly applied to the electrolysis cell and the resultant potential is monitored. The change in potential upon application of the constant current can be used to trigger the time base.

Since a trigger signal can be applied to the time base from an external source it is possible to couple electrical signals from mechanical devices. For example, in a dropping mercury assembly one may wish to trigger an oscilloscope when a drop is dislodged. A mechanical striker can knock off the drop and simultaneously close a microswitch to provide a short pulse to the oscilloscope trigger circuit.

Two additional types of triggering are of interest: (1) delay sweep after triggering and, (2) trigger after delay. In the first a timing circuit (delay circuit) is started by the normal trigger. At the end of the preset time, the sweep scans at

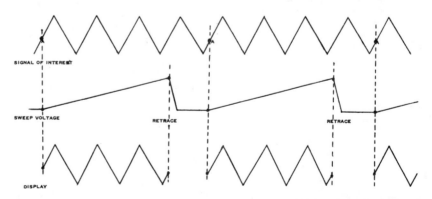

Figure 6. The time relationship between the input signal, the sweep voltage, and the display seen on the CRT screen; each successive display is superimposed over the previous one, and the result appears as a steady waveform.

screen and is useful for amplitude or dc measurement. However, the free-run mode will not produce a stable waveform.

To produce a stable repetitive waveform the sweep voltage applied to the horizontal deflection plates of the CRT must start when the signal of interest reaches a preset point in its rise or fall. In Figure 6 the triangular wave is the signal of interest. The sweep starts when the triangular wave

the rate set by the Time per Division switch. To illustrate consider a sudden change applied to a chemical system and use this change to trigger an oscilloscope. The sweep starts and three seconds later the oscilloscope trace shows a group of small fluctuations that are of interest. However, since the scan must be quite long, the fluctuations of interest are not too well defined. If the region between 3 and

Chemical Instrumentation

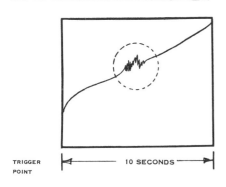

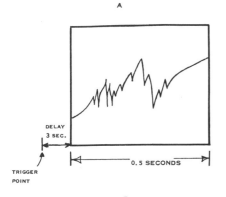

Figure 7. A triggered display with a 10-sec scan. (a) Conventional display, the area of interest is encircled. (b) The central region expanded to fill the screen by means of a delayed sweep; the scan is triggered at the same point, but delayed 4 sec followed by a 0.5-sec scan.

3.5 sec could be spread across the entire screen it should produce considerably more detail. This can be accomplished with an oscilloscope that contains a delayed sweep feature. Thus, the delay would be set for 3 seconds and the sweep Time per Division to 0.05 sec. (0.5 seconds per 10 divisions). The two different cases are shown in figure 7.

The second type is similar except that at the end of the delay period the trigger circuit will accept the next eligible trigger signal. This is useful in cases where some initial turbulence must be ignored before the signal of interest is produced.

To sum up triggering, each scan is started by a triggering signal which can be internally derived from the vertical amplifier or applied directly from some external source.

PLUG-IN AMPLIFIERS

Since many of our modern oscilloscopes accept plug-in units for the X and Y axes, a large variety of vertical amplifier and time base combinations are possible. These units permit one to apply the latest developments available to a particular scientific problem.

Dual Trace Amplifier

This vertical amplifier can accept two signals which are then displayed simultaneously on the CRT. The time relation between events is immediately apparent

and one can think in terms of cause and effect. For example, let channel 1 show an impulse signal applied to an electrolytic cell and let channel 2 show the current through the cell. With a slow scan of both traces moving across the screen, apply the impulse signal and see the response immediately in exact time relationship.

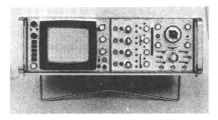

Figure 8. A four-trace amplifier (Type 1804A) used with a variable-persistence oscilloscope (Type 181A). (Hewlett-Packard Co.)

Dual trace is accomplished by internally switching each channel alternately (at 100 kHz) into the output amplifier. It should not be confused with dual beam in which two separate CRT's are contained in one glass envelope although the results are similar. There are also four-trace units that accept and display four signals simultaneously, see Figure 8.

Differential Amplifier

This unit, also called a difference amplifier, has the ability to amplify and display the difference between two signals. Because of this, many chemical instruments have been designed to use differential amplifier oscilloscopes as the readout system. By applying a blank signal to one input and the unknown to the second, only the difference signal is displayed. Thus, background or residual currents including noise and line frequency hum can often be cancelled or considerably reduced.

This ability to cancel similar signals is called *common-mode rejection* measured by the common-mode rejection ratio (CMRR), an important parameter to consider when selecting a differential amplifier. It is measured by applying identical sine-wave signals (from the same source) to each input and increasing the amplitude until some measureable waveform is seen on the screen of the CRT. For example, assume that the Volts per Division switch is set to 0.001 v per division and that the input amplitude of the applied sine-wave must be increased to 10.0 v to produce a 1 division trace on the screen. The ratio between the applied signal to the observed waveform is 10.0 v/0.001 v or 10,000 to 1. Thus, the CMRR of this amplifier would be 10,000 to 1. Amplifiers are available with up to 100,000 to 1 common-mode rejection. One additional function that can be accomplished with a differential amplifier is the application of a variable dc voltage to channel 2 in order to offset a large signal in channel 1. For example, to examine the peak of a waveform in detail would normally require a high sensitivity which would drive the trace well off the screen. By using a variable dc voltage in the opposite channel one can move the peak of the trace back on screen and examine it in detail. Plug-in units that have this ability are called differential comparators and they provide an accurate tool

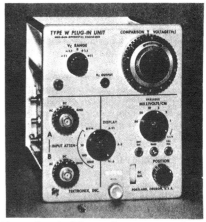

Figure 9. A differential comparator plug-in module (Type W) that permits precise amplitude measurements on displayed waveforms through use of a calibrated dc voltage applied to one channel. (Tektronix, Inc.)

for waveform amplitude measurements (Fig. 9).

Operational Amplifiers

Two special plug-in operational amplifiers are available (Fig. 10). These units each contain two programmable amplifier units as well as a common deflection amplifier. Thus, one can develop a variety of operational systems by using external

Figure 10. An operational-amplifier plug-in module (Type 3A8) that permits components to be applied externally for a variety of analog operations. (Tektronix, Inc.)

components in the input and feedback positions. These units also contain internal components that can be switched into input and feedback positions.

This type of plug-in can frequently occupy a dual role in an experiment. For example, one amplifier can be arranged to generate a staircase function which can be applied to an external system. The response of the system can then be impressed upon the second amplifier, connected as a differentiator. The output of this amplifier is then internally applied to the deflection amplifier and displayed as the derivative waveform on the CRT.

Current measurements are often made by introducing a sampling resistor into the external circuit and then measuring the

Chemical Instrumentation

voltage drop with a voltage amplifier. Since the introduction of the sampling resistor may be objectionable, an operational amplifier can be used to measure small currents without introducing additional components when it is connected as a current to voltage converter.

SPECIAL OSCILLOSCOPES

Under this heading one finds the storage oscilloscope, digital-readout, sampling, and signal-averaging oscilloscopes. Each of these oscilloscopes has special capabilities not possessed by conventional instruments. The following brief description is intended to introduce these instruments and point out their most important features.

The Storage Oscilloscope

In the past, the chemist has been somewhat hampered in using oscilloscopes since many of his experiments were not repetitive. He therefore was obliged to use a considerable quantity of photographic film to record his single transients on an oscillograph. The storage oscilloscope solves this problem in that images can be retained on the screen for indefinite periods. When operated in the storage mode every trace is stored on the screen until erased by the operator. This includes all traces regardless of scan speed up to the limit of the CRT voltage writing rate. A recent variation of the storage oscilloscope is termed variable persistence. This system permits the operator to select the fading time for a trace and extends from almost immediate erasure to complete storage without fade. This technique can be useful in following slowly changing repetitive signals since the previous waveform will still be visible although partially faded with respect to the new waveform. With regard to the storage oscilloscope as a working tool, it is the writers opinion that eventually all low and medium frequency laboratory oscilloscopes will be of the storage type.

The Digital Readout Oscilloscope

Each new development responds to the need for greater accuracy, ease of operation or new areas of measurement. The digital readout oscilloscope was created for all three reasons.

The trace on this oscilloscope contains two small brightened spots called intensified zones that can be moved across the trace to any selected point. The readout will then provide the amplitude between those two points. For example, if one zone is set at the baseline of a square wave and the second zone on the top of the wave, the readout will indicate digitally the amplitude of the squarewave. Thus, the possible ambiguity of reading the position of the trace with respect to graticule divisions is resolved. One significant aspect of this instrument may appeal to those chemists who enjoy automating a series of measurements. For example, since the digital readout is well suited for peak amplitude measurements, a series of peak measurements of a multiple peak waveform could be made in sequence by design-

ing a device that would automatically move the zones to each successive peak sequentially. In addition, most digital readout instruments have auxiliary output signals that can be used to actuate a printer for a permanent record of the peak values.

The Sampling Oscilloscope

Although most chemists rarely deal with signals at frequencies of a gigahertz or higher, they should be aware of the existence of oscilloscopes that can display these microwave frequencies. In the sampling oscilloscope 'real time' (our usual concept of time) is converted to equivalent time. A good analogy here is the moving picture photography of a speed event such as a horse race that is viewed as it occurred (real time) and also viewed in slow motion (equivalent time). Thus, although the input signal to the oscilloscope may have a frequency of one gigahertz the real time frequency of the displayed waveform may be only ten kilohertz.

Sampling is based on taking voltage samples from many input signals with each sample taken at a slightly later point on subsequent signals. As each sample is taken the beam moves horizontally a short distance and to a vertical position dictated by the amplitude of the sample. In this way a replica of the input signal is constructed point by point and displayed on the CRT. In real time the replicas are formed repetitively at perhaps up to 10,000/sec. The oscilloscope Time per Division switch is marked in microseconds, nanoseconds, or even picoseconds.

The Signal-Averaging Oscilloscope

This oscilloscope has as its prime application to extract meaningful repetitive signals that are buried in noise. This latter condition can be visualized by considering a one volt sinewave signal completely covered with 5 v of atmospheric random noise. The conventional oscilloscope display would be simply a five volt band of noise with no visible evidence of the one volt sinewave signal. When this repetitive signal is introduced to the signal-averaging oscilloscope one finds that the sinewave signal can be displayed with almost no evidence of noise.

This is accomplished by averaging the signal waveform at more than 1,000 points in a memory system during successive scans of the sweep system of the oscilloscope. Since the noise is completely random, the average noise level when summed over many scans is practically zero. The signal, on the other hand, is repetitive and each scan adds to the previous one and presents an average that is a true replica of the original. The signal must be repetitive but it can be of a relatively slow nature since the previous scans are held in the memory bank and will wait for subsequent signals to arrive. Due to the mathematical nature of noise averaging, the greater the number of sweeps, the closer the net average of the noise is to zero. Much like tossing a coin in which a 50% heads figure will be reached if enough tosses are made. Accordingly one might consider the averaging oscilloscope in situations where data might be submerged

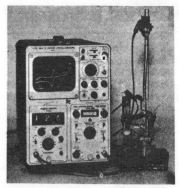

Figure 11. A single-sweep polarographic analyzer (Type SSP-3) that uses a storage tube for analog display, coupled with digital read-out of polarographic peak height. (Chemtrix, Inc.)

in a high noise level and a repetitive scan is possible. Thus, new levels of sensitivity in analytical techniques may be possible.

APPLICATIONS

Within the chemists' own domain one finds an increasing use of the oscilloscope as a part of major instrumentation. Some examples are explored below.

Single Sweep Polarography

The instrument shown in Figure 11 uses a storage tube for visual readout of the polarographic waveform as well as a digital readout of individual peak heights. High speed polarography at 1 millisecond per polarogram is possible with the oscilloscope storage readout. In this application, the time base waveform that drives the CRT horizontal scan also serves, after attenuation, to drive the polarographic current through the cell. The cell current is applied to the vertical amplifier which in turn drives the vertical plates of the CRT. The resultant polarogram is a current-voltage plot where the horizontal axis of the CRT represents the potential applied to the cell. In the example shown the time base waveform was a triangular wave; the combined cathodic and anodic scan is called a cyclic polarogram (also called a cyclic voltammogram).

In the past some objection had been raised due to the cathode-ray tube's limited vertical area for peak height measurements (8 centimeters). It appeared difficult to resolve small differences in peak height. This has been overcome through use of the digital readout since each vertical centimeter can be read to one part in a hundred.

Spectrophotometry

In instruments of this type the major function of the oscilloscope concerns a display of radiant power versus wavelength. The spectrum can be scanned at a variety of scan rates from as little as one millisecond up to several seconds. The plot in this case is wavelength on the horizontal axis with either absorbance or emission on the vertical axis. Since the horizontal scan follows exactly the wavelength scan, precise wavelength measurements are possible through pilot-ion markers within the spectrum. Several commercial instruments that use oscilloscopes in their sys-

Chemical Instrumentation

tem are described by Lott, THIS JOURNAL, **45,** A89, A169, A273(1968). [Reprinted in this volume]

In view of the oscilloscope's ability to be triggered, and also to be triggered after a delay, one might wish to try to display a small segment of the spectral scan. For example, assume a spectral scan of one millisecond that covers 10 mμ. To examine a portion of this scan in detail, one would trigger the oscilloscope delay circuit at the start and then set the delay to end just before the occurance of the peak of interest. The actual oscilloscope scan of the peak could be adjusted in the microsecond range to provide various degrees of magnification.

Waveform Monitor

For classroom instruction the usual five-inch cathode-ray tube is often difficult to see, especially in large lecture halls. To overcome this problem Welch Scientific Company has produced a 12 in. lecture table oscilloscope. The electron beam of the CRT can be brightened to give a well-defined trace that will be visible from any part of the classroom. And as one further improvement, even the instructor is removed from the front of the display tube, since all of the controls plus a small 3 in. CRT monitor are located on the rear of the instrument. In this way the instructor sees the same waveform as the students but does not block any portion of the large screen display.

Additional information on modern oscilloscopes can be obtained from the manufacturers. In the case of Tektronix and Hewlett-Packard their product catalogs contain much tutorial material as well as descriptions of their products. In addition, Tektronix has two publications that will be especially helpful to those not acquainted with oscilloscope terminology. They are (1) *A Primer of Waveforms and their oscilloscope displays*, and (2) *Understanding Operational Amplifiers*. For oscilloscope applications in polarography write for *30 Questions on Single Sweep Polarography* from Chemtrix.

Manufacturers of Oscilloscopes

There are at least 30 manufacturers of oscilloscopes. The following is a selection which seems particularly significant in the present context. Code: ST = storage 'scopes, SP = sampling, SA = signal averaging. Most have various other models besides those coded here.

Analab (Div. of Benrus) [ST, SP]
  Waterbury, Conn. 06720
Fabri-Tek, Inc. [SA]
  Madison, Wis. 53711
Hewlett-Packard Co. [ST, SA, SP]
  Palo Alto, Calif. 94304
Hughes Instruments (Div. of Hughes Aircraft Co.) [ST, SP]
  Culver City, Calif. 90230
Measurement Control Devices, Inc. [ST]
  Philadelphia, Pa. 19125
Northern Scientific, Inc. [ST, SA]
  Madison, Wis. 53711
Nuclear Data, Inc. [SA]
  Palatine, Ill. 60067
Technical Instruments, Inc. [SA]
  North Haven, Conn. 06473
Tektronix, Inc. [ST, SP]
  Beaverton, Oregon 97005
Welch Scientific Co.
  Skokie, Ill. 60076

The triceps, an old Nordic sign. A symbol of heavenly power. By tracing its perimeter from the apex back to the apex we realize the meaning of the words: "The Will of God, descending upon the world, sways to and fro over the Earth and returns again on High."

Topics in...

Chemical Instrumentation

...a  feature

Edited by **S. Z. LEWIN,** New York University, New York 3, N. Y.

The Journal of Chemical Education
Vol. 40, page A5
January, 1963

V. Thermometric Titrators

Joseph Jordan, *Department of Chemistry, The Pennsylvania State University*

Thermometric *Enthalpy Titrations* represent the newest art in the field of instrumental analysis. Their methodological principle is based on the recording of a plot of temperature versus volume of titrant. The titrant is added to an isothermal (*titrate* = solution titrated) in an adiabatic *titration calorimeter*. In most instances, a change in enthalpy occurs concomitantly, yielding a corresponding heat of reaction. As a result, titration curves (called *enthalpograms*), of the shapes shown in Figure 1, are generally obtained.

Curves I and II illustrate an exothermic and an endothermic *titration reaction*, respectively. On both enthalpograms the *base lines* AB represent *temperature-time blanks*, recorded prior to the start of the actual titration. B corresponds to the beginning of addition of titrant, C is the end point and CD the excess reagent line. BC is the *titration branch* proper of the enthalpograms: It has an ascending slope when $\Delta H < 0$, and a descending slope when $\Delta H > 0$. In Figure 1 the *excess reagent branches* of the enthalpograms (CD) were drawn with ascending slopes, because the dilution of most titrants is an exothermic process. In order to minimize variations in heat capacity during titrations, it is customary to use titrants 50–100 times more concentrated than the "unknowns" titrated. Thus the volume of the titrate solutions is maintained virtually constant, but the titrants are diluted appreciably. The heats of dilution can be corrected for conveniently by the linear back extrapolation CB' (see Figure 1). Under these conditions, the *extrapolated ordinate heights*, BB', represent a measure of the change of temperature, ΔT, due to the titration reaction, as well as of the integral heat, Q, evolved or absorbed, viz.:

$$\text{BB' } prop \ \Delta T = \frac{Q}{k} = \frac{-N \cdot \Delta H}{k} \quad (1)$$

where k denotes the effective heat capacity and N is the number of moles reacted in

the titration. It is evident from equation (1) that the quantity BB' (appropriately calibrated) can be used for obtaining the following types of information:

a. The heat of the titration reaction, ΔH, whenever a known amount of titrate is used.

b. The concentration of an unknown titrate, whenever ΔH is known (ΔT is

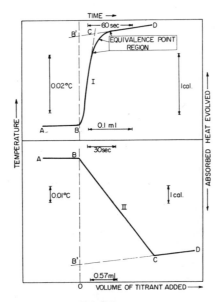

Figure 1. Typical Enthalpograms.

I. Example of exothermic titration reaction: 0.0009 molal potassium chloride titrated with 1.4 molal silver nitrate in a molten $LiNO_3$-KNO_3 eutectic solvent at 158°C, yielding sparingly soluble AgCl(s).

II. Example of Endothermic Titration Reaction: 0.014M magnesium chloride titrated with 1M ethylenediaminetetraacetate (Y^{4-}) in aqueous solution at 25°C, yielding the chelate $[MgY]^{--}$.

AB. Temperature-time blank (base line), recorded prior to start of titration
B. Start of titration
C. End point
CD. Excess reagent line

proportional to concentration and independent of titrate sample size, because—in a given titrate solution—both N and k are proportional to the sample volume).

c. The quantity of an unknown in a known volume of titrate.

Approach *a* is ideally suited for the rapid and convenient determination of heats of reaction in dilute solutions. *B* and *c* can be applied to a unique type of *direct enthalpimetric analysis* which does not require a standardized titrant. A more precise application of enthalpograms to quantitative analysis is inherent in the conventional *end point determinative technique*, to calculate the amount of an unknown in the titrate from the volume of an accurately standardized titrant required to attain the equivalence point. Naturally, the stoichiometry of the titration reaction must be known for this. Conversely, unknown reaction stoichiometries can be determined using known amounts and/or concentrations of both titrant and titrate. In instances, when the titration reaction is appreciably incomplete in the equivalence point region (e.g., Figure 1, Curve I), the enthalpograms exhibit a curvature from which the equilibrium constant (and the corresponding standard free energy) can be estimated.

Joseph Jordan obtained his Ph.D. degree in 1945 at the Hebrew University, Jerusalem, Israel. He came to the U.S. in 1950, served as Research Fellow and Instructor at Harvard and at the University of Minnesota, and in 1954 joined the faculty of The Pennsylvania State University where he is now Professor of Chemistry. Dr. Jordan teaches analytical chemistry on the undergraduate and graduate levels. He conducts research in the electrochemistry and thermochemistry of aqueous solutions and fused salts. He is a member of the International Commission on Electrochemical Thermodynamics and Kinetics, affiliated with the International Union of Pure and Applied Chemistry.

Table I. Synopsis of Applications of Thermometric Titrations

Information obtainable		Procedure for calculating data from enthalpograms
Analytical applications	Concentration of an unknown	Direct enthalpimetric
	Amount of an unknown	End point determinative
Fundamental applications	Heat of reaction	From extrapolated ordinate heights
	Free energy of reaction	From curvature in equivalence point region
	Entropy of reaction	From equation 2
	Reaction stoichiometry	End point determinative

Combining this with the reaction heat which is independently accessible from any enthalpogram, the entropy can readily be calculated from the well-known equation:

$$\Delta S^\circ = \frac{\Delta H^\circ - \Delta F^\circ}{T} = \frac{\Delta H^\circ + RT \ln K}{T} \quad (2)$$

Thus all three important thermodynamic parameters of the titration process (viz. the heat, ΔH°; the free energy, ΔF°; and the entropy, ΔS°) can be obtained *from one single experiment*, which is an accomplishment unmatched by any other method. The nature of the information accessible from the judicious mathematical analysis of thermometric titration curves is summarized in Table I.

Nomenclature

Operationally, enthalpograms are "thermometric" titration curves, the experimental variable monitored being temperature. The basic property which determines their shape is enthalpy change. This accounts for the designations "thermometric titration" and "enthalpy titration" used by various authors. Other terms encountered in the literature for naming similar procedures include: "calorimetric titration," "thermochemical titration," "thermal titration." Growing interest in thermoanalytical titrimetry led in 1957 to the scheduling of a symposium at the Fall Meeting of the American Chemical Society in New York City. The program included a round-table discussion on terminology. The consensus was that the designation "thermometric titration" be preferred in keeping with accepted usage for potentiometric, amperometric, etc., techniques.

Instrumentation

A thermometric titration has been reported as early as 1913: It was carried out by a discontinuous procedure, adding increments of 0.1 M ammonium hydroxide from a conventional volumetric buret to a 0.1 M solution of citric acid in a Dewar flask. After each addition of titrant a corresponding temperature reading was taken from a Beckman thermometer, immersed in the titration cell. The shape of such manual enthalpograms was poorly reproducible. Temperature changes were relatively large (up to several degrees C)

and it was necessary to wait 5–10 minutes before taking each temperature reading, due to the slow response and appreciable heat capacity of the mercury thermometer. A complete titration required about an hour. Heat exchange with the laboratory environment was inevitable, with the result that the titration curves obtained depended somewhat on the patience of each experimenter. This unsatisfactory situation was remedied in the fifties with the invention of automatic thermometric titrators using semiconductor instrumentation and electrically powered continuous constant flow burets. With these devices it has become possible to titrate extremely dilute solutions (10^{-4} to 10^{-2} M), to minimize temperature changes to $<0.1°C.$, to reduce the time required for an entire titration to five minutes or less and to eliminate human error entirely. With the advent of automatic instrumentation, thermometric titrimetry has been transformed into an important quantitative method: This is an interesting instance where *instrumentation represents not merely a convenience, but a methodological necessity.* Use of an instantaneous electric signal for monitoring temperature is an absolute requirement for eliminating the irreproducibility which has haunted the manual thermometric titration techniques of yesteryear.

The major breakthrough which has made possible automation of thermometric titrations was accomplished in 1953 by Linde, Rogers, and Hume at MIT. They adopted as the temperature sensing device semiconductors known in the USA as "thermistors," and in Western Europe as *NTC-s* (= *Negative Temperature Coefficient*).[1] In 1956 an automatic titrator equipped with a motor driven syringe-buret was developed at The Pennsylvania State University. Finally, in February, 1962, the first commercially manufactured automatic thermochemical titration instrument, trade-named the *Titra-Thermo-Mat*, was made available by the American Instrument Company (AMINCO), Silver Spring, Maryland.

The heart of all modern automatic thermometric titration instruments is the thermistor. This is a solid state device, made of sintered metallic oxides. In

[1] Suppliers from whom thermistors can be obtained include: Victory Engineering Corporation (VECO), P.O. Box 373, Union, N.J.; Fenwal Electronics, Inc., Framingham, Mass.; The General Electric Co., Metallurgical Products Department, Detroit 32, Mich.; and Yellow Springs Instrument Co., Inc., Yellow Springs, O.

thermometric titrators, it is used in the shape of a small bead (<1 mm in diameter), covered by a thin glass envelope, and is wired as one arm of a DC Wheatstone bridge. A block-diagram of a typical home-made thermistor-titrator is shown in Figure 2. Thermistors function as highly sensitive resistance thermometers:

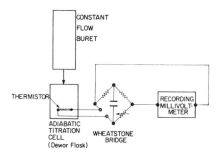

Figure 2. Block diagram of "home-made" automatic thermometric titrator.

In contradistinction to metals, thermistors possess large *negative* temperature coefficients of resistance, as is illustrated in Figure 3.

As can be seen in the figure, the resistance of platinum increases slightly with temperature: This is due to the fact that in metals, the "effective concentration" of electrons available for conduction is the same irrespective of temperature, but their mobility is decreased by enhanced thermal

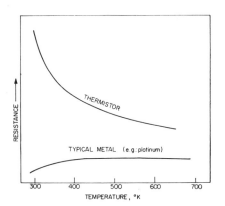

Figure 3. Plots of resistance as a function temperature.

fluctuations of the atomic nuclei. The resistance of thermistors decreases drastically with temperature, because these semiconductors possess a conduction band, separated from the valence band by an energy gap of the order of κT: As the temperature increases electrons are transferred to the conduction band and resistance decreases accordingly.

Within small temperature intervals (up to 1°C) the unbalance potential of a thermistor bridge represents a linear measure of temperature. Typical performance characteristics of thermistors include: Negligible heat capacity; negligible power dissipation; response lag <1 sec; resistance at 25°C ~ 2000 ohm; temperature coefficient −0.04 ohm/ohm/degree. A thermistor, wired as illustrated in Figure 2, produces an unbalance potential on the order of 40 millivolt per degree C, which is equivalent to the response of a two thousand junction thermocouple.

Most thermometric titrators are designed to record automatically "normal" enthalpograms, i.e., plots of temperature *versus* volume of titrant. Their operating principles and performance characteristics are discussed in detail below as exemplified in the Titra-Thermo-mat. Specialized instruments have been constructed in various laboratories for recording differential and derivative thermometric titration curves. Readers interested in these are referred to the appropriate references in the bibliography.

The Titra-Thermo-Mat

A self-explanatory photograph of the Titra-Thermo-Mat is shown in Figure 4.

It consists essentially of: (a) The *Menisco-Matic* Buret; (b) An "adiabatic titration tower"; (c) A thermistor bridge assembly; (d) A recording DC millivoltmeter.

Components *a*, *b*, *c* are assembled in one vertical unit, seen on the left of photograph (*a* is at the top, *c* is in the base of the instrument). *D* is at the right and displays two titration curves of hydrochloric acid with standard sodium hydroxide.

Figure 4. The Aminco TITRA-THERMO-MAT; adiabatic titration tower in loading position.

The Menisco-Matic Buret is equipped with a precision machined Vycor plunger, coupled with a screw which is driven by a synchronous motor at constant speed. As the plunger advances, it displaces a volume of titrant at a rate of 600 microliters/minute. The titrant reservoir is made of Pyrex: Embedded in it is a thermistor which can be connected to the Wheatstone bridge for measuring the temperature of the titrant, if desired. Details of the buret are shown in Figure 5, including a digital counter which provides direct readout of the volume of titrant dispensed.

The "adiabatic titration tower" is shown in a loading position in Figure 4, and in its "titration position" in the cross-sectional line drawing in Figure 6. The tower is made of Styrofoam plastic which serves as thermal insulator. Its lower section contains a 30 ml sample beaker into which the titrate solution is placed. In the "titration position" the beaker is raised to immerse into it the buret tip, a small L-shaped heater, a stirrer, and a "sample" thermistor. The heater can be

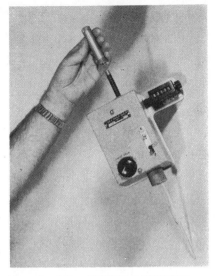

Figure 5. Aminco MENISCO-MATIC BURET.

used as a calibrating device for determining heats of reaction.

Either the sample thermistor or the titrant thermistor can be connected as an arm of the Wheatstone bridge, by a selector switch: If the sample solution is colder than the titrant, it may be warmed up with the aid of the heater. Generally titrant and titrate temperatures should be within 0.2°C. When performing a titration, a "base line" is first obtained on the recorder by adjusting the two balance dials on the instrument panel. The sensitivity dial, which controls the bridge input voltage, is adjusted and the titration is carried out with the thermistor selector switch in the "beaker (= sample) position." On the recorder chart, the ordinate deflection is proportional to change in temperature. The abscissa is proportional to volume of titrant, because the chart is driven by a synchronous motor at a constant rate, as is the Menisco-Matic Buret.

Unique Features and Significance of Thermometric Titrations

Most analytical procedures are ultimately "free energy methods," i.e., they depend on a property related solely to equilibrium constants. For instance, the

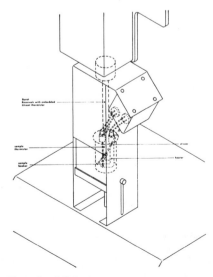

Figure 6. Adiabatic tower in titration position.

*p*H change in a conventional acid-base titration in aqueous solution depends on the relevant ionization constants. As a result, analysis of a very weak acid is not feasible by any titrimetric procedure, when the end point is determined by a conventional free-energy method (as is the case when visual color indicators or potentiometric titration curves are used). However, enthalpograms depend on the heat of the reaction *as a whole*, viz.:

$$\Delta H = \Delta F + T\Delta S \qquad (3)$$

Consequently, thermometric titration curves may yield a well-defined end point when all free energy methods fail, if *the entropy term in eq. 3 is favorable.*

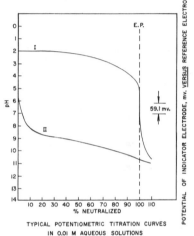

TYPICAL POTENTIOMETRIC TITRATION CURVES
IN 0.01 M AQUEOUS SOLUTIONS

I. Hydrochloric acid titrated with sodium hydroxide.

II. Boric acid titrated with sodium hydroxide.

E.P. Equivalence point.

Figure 7. Potentiometric titration curves of strong and weak acids.

This is indeed the case in many alkalimetric titrations of weak acids: As an illustration, Figures 7 and 8 show potentiometric (*p*H) and thermometric titration curves of hydrochloric and boric acids with sodium hydroxide in dilute aqueous solutions. The potentiometric curves of the two acids (Figure 7) are very different yielding a large "end-point inflexion" for hydrochloric acid ($K_a \approx \infty$) and virtually none for boric acid ($K_a \approx 10^{-10}$). The corresponding enthalpograms (Figure 8) are strikingly similar, because the heats of neutralization of hydrochloric and boric acid are comparable, viz.: -13.5 and -10.2 kcal/mole, respectively. As a result the two acids can be determined with comparable ease, precision, and accuracy by thermometric titration, while any direct free energy method is bound to fail with boric acid.

Since heat of reaction is a most general property of chemical processes, thermometric titrations have a very wide range of applicability in quantitative analysis: They have been applied successfully to all important categories of "titration processes," including proton transfer (acid-base), oxidation-reduction and complexation reactions, in media ranging from water at room temperature to molten salts at 350°C. The main limitation of thermometric titrations is inherent in their very virtue: Because of its general applicability, the method is non-selective.

From a fundamental point of view, thermometric titrations possess extraordinary potentialities for determining rapidly and conveniently heats of reaction in solutions so dilute that the determined ΔH values can be set virtually equal to the ideal thermodynamic parameters corresponding to infinite dilution. The temperature change during a titration ($\pm 0.1°$ or less) is so small that it can be neglected with respect to the variability of thermodynamic parameters: This is of significance in controversial situations, where an actually endothermic reaction may appear spuriously exothermic when the heat is determined by a conventional application of the Van't Hoff Isochore and averaged over temperature ranges of 50°C. Such an ambiguous situation has been recently elucidated with respect to the heat of chelation of magnesium with ethylenediaminetetraacetate, which was believed to be exothermic, but yielded the typically endothermic enthalpogram shown in Figure 1, Curve II.

Reliable thermochemical data on many important reactions occurring in solution are conspicuous by their absence, due mainly to the laboriousness of classical calorimetric procedures. Thermometric titrations can rapidly yield data on heats of reaction under specified experimental conditions. Since the free energies are readily available, this has resulted in making accessible hitherto unknown illuminating information on entropies, which in turn can serve to elucidate associative phenomena, such as ion-association and solvation, and the *detailed nature of chemical species actually prevailing in conventional solvents and molten salts.*

Thus, the analytical and fundamental applications of thermometric titrations

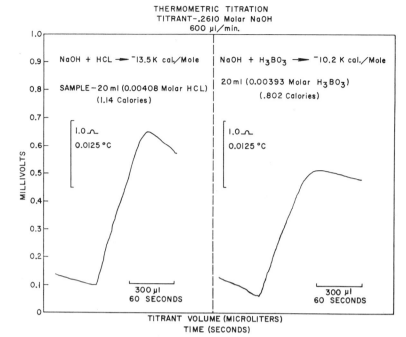

THERMOMETRIC TITRATION
TITRANT—.2610 Molar NaOH
600 µl/min.

Figure 8. Thermometric titration curves of strong and weak acids.

represent two sides of the same coin. It is anticipated that the next few decades may witness advances in thermochemical methods of analysis concomitantly with progress in the understanding of the nature of "entropy compounds." This would parallel the spectacular development of electroanalytical chemistry during the past several decades which has lead to tremendous advances in the knowledge of kinetics and mechanisms of electrode reactions.

Thermometric Titrators

Supplement

Thermometric Enthalpy Titration (TET) was the sole methodology of its type known in 1963. Since that time several "second generation methods" have been developed, based on similar principles. Their common feature is reliance on the accurate measurement of minute temperature increments and decrements in an *adiabatic cell*, under conditions of virtually invariant heat capacity: These "Enthalpimetric Methods of Analysis" are outlined below.

Direct Injection Enthalpimetry (DIE) is a quasi-instantaneous thermometric titration. It involves the rapid injection of a reagent (e.g., with the aid of a syringe) into the solution of an "unknown" in an adiabatic cell, yielding a "temperature pulse," ΔT, which is recorded automatically as the unbalance potential of a thermistor bridge. The reagent is added in sufficient excess to drive the relevant reaction thermodynamically and kinetically to virtual completion. Under these conditions, a reaction of the type

$$xA + yB = zP \qquad (4)$$

proceeds from the "initial" to the "final" situation, *viz.*:

$$\{n_A^i \, A + n_B^i B\} \rightarrow$$

Initial Situation

$$\left\{ \frac{z}{x} \, n_A^i \, P + \left[n_B^i - \frac{y}{x} \, n_A^i \right] B \right\} =$$

Final Situation

$$\left\{ n_P \, P + \left[n_B^i - \frac{y}{z} \, n_P \right] B \right\} \qquad (5)$$

where n_A^i denotes the number of moles of unknown initially present.

$$n_B^i > \frac{y}{x} \, n_A^i \qquad (6)$$

is the number of moles of excess reagent added, while $n_P = N$ is the number of moles of product formed (*cf.* eqn. 1). The stipulated virtual "completeness" implies

$$n_P = \frac{z}{x} \, n_A^i \qquad (7)$$

Consequently:

$$Q = - \frac{z}{x} \, n_A^i \cdot \Delta H = (\text{const}) \cdot n_A^i \qquad (8)$$

$$\Delta T = \frac{Q}{k} = (\text{const})' \cdot n_A^i \qquad (9)$$

Equation 9 clearly shows the proportionality between the temperature pulse ΔT and concentration, which is the basis of DIE. Expressed in terms of the actually measured variable which is the unbalance potential of the thermistor bridge, ΔE prop ΔT, the applicable analog of eqn. (9) is:

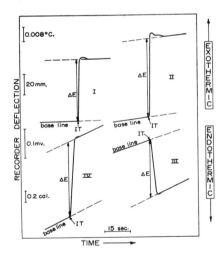

Figure 9. Typical injection enthalpograms.
I. 83.1 µm of HCl plus excess NaOH
II. 101 µm of H_3BO_3 plus excess NaOH
III. 100 µm of Mg^{+2} plus excess EDTA
IV. 104 µm of Pb^{+2} plus excess EDTA
IT. Injection time of 300-µl 1M reagent into 25.0-ml volume of unknown solution
ΔE. Unbalance potential of thermistor bridge (eqn. 10)

$$\Delta E = (\text{const})'' \, n_A i \qquad (10)$$

Equations (9) and (10) are the "working relations" on which DIE relies. Typical "injection enthalpograms" are illustrated in Figure 9.

Gas Enthalpimetry is a variant of TET and/or where a *gaseous unknown* (e.g., a sample of air polluted with sulfur dioxide or carbon dioxide) is added to an appropriate reagent solution (e.g., a large excess of aqueous sodium hydroxide). Gases can thus be determined rapidly and accurately, utilizing reactions of the type

$$SO_2 + 2OH^- = SO_3^{--} + H_2O;$$
$$\Delta H° = -38.6 \text{ kcal/mole} \quad (11)$$

$$CO_2 + 2OH^- = CO_3^{--} + H_2O;$$
$$\Delta H° = -27.0 \text{ kcal/mole} \quad (12)$$

Apparatus used in gas enthalpimetry is illustrated in Figure 10.

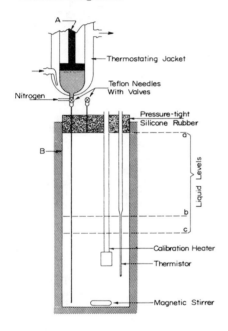

Figure 10. Enthalpimetric gas analyzer.
A. Gas syringe
B. Adiabatic barrier: dewar wall or platinum-lined Styrofoam cavity

Thermokinetic Analysis (TKA) relies on temperature-time curves. Mixtures can be resolved, taking advantage of differences in rates of reactions. Conversely, kinetic parameters can be determined conveniently and rapidly by TKA. The basic principle of TKA is apparent from Figure 11.

Competitive Kinetic Titrimetry (CKT) combines principles of TET and TKA. The rate of addition of titrant is adjusted so that it is comparable to the rate of the relevant reaction. TKA and CKT are powerful methods for studying rates of chemical reactions. The pertinent bibliography on enthalpimetric analysis (given at the end of this write-up) should be consulted for details, which involve complex mathematics.

New Applications of enthalpimetric analysis are described in several hundred recent publications. They range from clay minerals and ion exchange resins to pharmaceuticals, detergent analysis and environmental pollution control: An ex-

Table II. Functional group analysis by thermochemical methods

Functional group	Reagent	Reactions	Precision and accuracy attainable (%)	Method
—COOH	Strong base	Acid-base neutralization	1	TET
—COO⁻ or —NH₂	Strong acid	Acid-base neutralization	1	TET
—C=O or (aldehyde)	NH₂OH·HCl	Oximation	0.6	TET
(aldehyde —CHO)	H₂SO₄-Na₂SO₃	Bisulfite addition	0.6	TET
Ar—NH₂	NaNO₂	Diazotization	0.2	TET
Ar—SO₂NH₂	NaOCl	Oxidation	3–5	TET
Ar—OH	Br₂ + HCl	Bromination	3–5	DIE
R₁—NO₂ plus R₂—NO₂	Strong base	Proton abstraction	7	TKA

Source: "Analytical Calorimetry," R. S. Porter and J. F. Johnson, eds., Plenum, New York, 1968, p. 207.

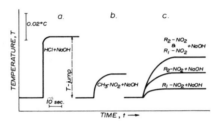

Figure 11. Thermokinetic Analysis (TKA).
Plots recorded upon addition of excess sodium hydroxide to:
a. HCl (reaction instantaneous)
b. A single nitroparaffin (reaction CH₃NO₂ + OH⁻ → CH₃NO₂⁻ + H₂O, slow)
c. Two nitroparaffins singly (reacting at different rates) and to a mixture of the two. Standard methods of "kinetic analysis" are available for treating curves b and c (see bibliography)

haustive synopsis is available in the book by Tyrrell and Beezer (see bibliography). A few examples are summarized below, selected by criteria of significance and novelty.

Quantitative Organic Functional Group Determinations may well be revolutionized by the use of enthalpimetric analysis. This is an area which has resisted instrumental approaches (because even in this day and age of specific electrodes it is difficult to envision, e.g., a "methoxy-electrode"!). and has been haunted by awkward "back titrations" (due to slow kinetics which require excess reagents). Fortunately the classical organic functional group reactions are highly specific *per se*: Consequently they can be readily monitored by even as nonspecific a property as the heat of reaction. Recent functional group determinations by thermochemical methods are listed in Table II.

Water in Organic Liquids has been determined by an imaginative use of an "inverse DIE technique." Moist samples (or pure water for calibration and standardization) were added to an excess of the well-known Karl Fischer reagent, utiliz-

ing the heat of the reaction:

$$C_6H_5N\cdot I_2 + C_6H_5N\cdot SO_2 + C_6H_5N + CH_3OH + H_2O =$$
$$2C_6H_5N\cdot HI + C_6H_5NH\cdot SO_4CH_3;$$
$$\Delta H = -16.1 \text{ kcal/mole of } H_2O \quad (13)$$

Enthalpimetric temperature pulses, linearly correlated with the relevant amounts of water, were recorded and are illustrated in Figure 12.

Protein Analysis and Immunological Reactions. Recent studies have revealed that enthalpimetric analysis may have signal potentialities in applications of biological and clinical relevance. Thus the various types ("sets") of prototropic groups of ovalbumin have been determined. The primary structure of the relevant protein (MW 43,800) consisted of 277 peptide bonds. Accessible side chain proton-donor acceptor groups included 46 carboxyls, 19 primary amines, and 7 imidazoles. This altogether "72-valent Brønsted acid-base moiety" was quantitatively characterized with the aid of one single TET curve, which is illustrated in Figure 13. A conventional potentiometric (pH) titration curve is shown for comparison in the same figure.

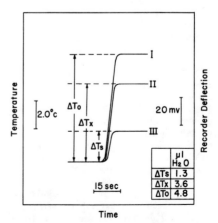

Figure 12. Moisture analysis by inverse DIE. "Enthalpimetric temperature pulses" of water reacting with Karl Fischer reagent.

The enhanced methodological capability of TET is evident. The reason for this is twofold, *viz.*:

a. pH is a logarithmic function (which "dulls its ability to discriminate" between close pK's), while enthalpimetric methods of analysis rely on a linear measure of the progress of the reaction;

b. The various "prototropic sets" vary appreciably in their heats of protonation; they are Brønsted entities whose charge-types are different (e.g., —COOH, —COO$^-$ versus NH$_3^+$, NH$_2$).

TET may also have promise towards instrumenting serological analysis in clinical practice: Well-defined end-points were obtained in thermometric titrations of antigens with the corresponding antibodies.

Microanalysis. It is apparent from Eqn. (1) that enthalpimetric analysis has a remarkable *intrinsic* potential for miniaturization. As has been pointed out in connection with eqn. (1), both the quantity N (in the numerator) and k (in the denominator) are proportional to volume, as long as a solution of a *given concentration* is involved. Consequently, the same "temperature pulse" ought to be obtained (e.g., in DIE) regardless of the volume taken for analysis. Naturally, there are practical limitations inherent in the finite heat capacity of the adiabatic cell and ancillary devices. Microequipment has been designed which is suitable for the determination of micromolal amounts (at millimolal concentration levels) with a precision and accuracy of 1%.

Instrumentation has been diversified during the last several years. An automatic thermometric titrator with digital readout has been described in the literature. Sophisticated "isothermal-adiabatic" titration calorimeters have been developed, where reaction heats evolved (or absorbed) are continuously compensated in an adiabatic cell, with the aid of a Peltier thermoelectric device and feedback circuits. Enthalpimetric process control instrumentation has also been designed for industrial applications. Two broad categories of commercial equipment are offered:

1. sophisticated "titration calorimeters" where ambient temperatures (i.e., the "environment" of the adiabatic cell) are controlled within 0.001°C.

2. thermometric titrators of simpler design with emphasis on rapidity and convenience, primarily for analytical use.

Suppliers of thermometric titrators and related enthalpimetric equipment include: American Instrument Co., Silver Spring, Md.; LKB Instruments, Inc., Bromma, Sweden, and Rockville, Md.; SKC, Inc., Pittsburgh, Pa.; Tronac, Inc., Orem, Utah. Thermistors have remained the preferred temperature sensor used in enthalpimetric analysis. An imaginative novel quartz resonator thermometer (whose frequency is a function of temperature) has been developed (Hewlett-Packard Co., Palo Alto, Calif.). This device has a sensitivity threshold of 0.0001° which is comparable to thermistor circuits. However, the piezo-quartz thermometer has the added advantage that its linearity extends over a range of several hundred degrees Kelvin, while a given thermistor responds linearly with tempera-

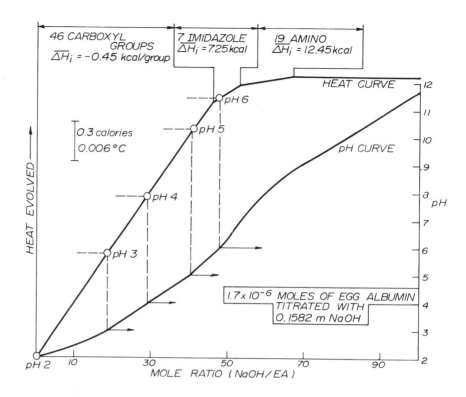

Figure 13. Alkalimetric titration curves of ovalbumin (Egg Albumin, EA). ΔH$_i$: heat of ionization per mole of H$^+$.

ture only within a one-degree span. Unfortunately, handicaps of relative bulkiness, appreciable heat capacity and slow response (at maximum sensitivity) have thus far precluded the use of the quartz thermometer in enthalpimetric analysis.

BIBLIOGRAPHY

General Reviews

American Chemical Society, "Symposium on Thermoanalytical Titrimetry," *Abstracts of Papers, 132nd National Meeting*, New York, N. Y., 1957, p. 5B–13B.

Bark, L. S., and Bark, S. M., "Thermometric Titrimetry," Pergamon, Oxford, 1969.

Christensen, J. J., and Izatt, R. M., "Thermochemistry in Inorganic Solution Chemistry," in "Physical Methods in Advanced Inorganic Chemistry," H. A. O. Hill and P. Day, eds., Interscience, New York, 1968, p. 538–598.

Ewing, G. W., "Instrumental Methods of Chemical Analysis," 2nd ed., McGraw-Hill, New York, N. Y., 1960, p. 347–349.

Hume, D. N., and Jordan, J., "The Nomenclature of Thermochemical Titrimetry," *Anal. Chem.*, **30**, 2064 (1958).

Jordan, J., "Thermochemical Titrations," *Record of Chemical Progress*, **19**, 193, December 1958.

Jordan, J., "Thermometric Enthalpy Titrations," in "Treatise on Analytical Chemistry," I. M. Kolthoff and P. J. Elving, eds., Pt. I, Vol. 8, Interscience, New York, 1968, p. 5175–5242.

Jordan, J., and Alleman, T. G., "Thermochemical Titrations, Enthalpy Titrations," *Anal. Chem.*, **29**, 9 (1957).

Jordan, J., and Ewing, G. J., "Thermometric Titrations" in "Handbook of Analytical Chemistry," L. Meites, Editor-in-Chief, McGraw-Hill, New York. N. Y., In Print (1963).

Tyrrell, H. J. V., and Beezer, A. E., "Thermometric Titrimetry," Chapman and Hall, London, 1968.

Tyson, B. C., Jr., McCurdy, W. H., Jr., and Bricker, C. E., "Differential Thermometric Titrations and Determination of Heats of Reaction," *Anal. Chem.*, **33**, 1640 (1961).

Willard, H. H., Merritt, L. L., and Dean, J. A., "Enthalpy Titrations," in "Instrumental Methods of Analysis," 3rd ed., Van Nostrand, Princeton, N. J., 1958, p. 594–598.

Zenchelsky, S. T., "Thermometric Titrations," *Anal. Chem.*, **32**, 289R (1960).

Zenchelsky, S. T., and Segatto, P. R., "Derivative Thermometric Titrations," *Anal. Chem.*, **29**, 1856 (1957).

Instrumentation

American Instrument Co., Inc. (8030 Georgia Ave., Silver Spring, Md.), "Titra-Thermo-Mat, Automatic Thermometric Titrator," *Bulletin 2375*, 1962.

Becker, D. F., *Chemie Ingenieur Technik*, **41**, 1060, 1105 (1969).

Christensen, J. J., Izatt, R. M., and Hansen, L. D., *Rev. Sci. Instr.*, **36**, 779 (1965).

Phillips, J. P., "Automatic Titrators," Academic Press, New York, N. Y., 1959, p. 110.

Priestley, P. T., *Analyst*, **88**, 194 (1963).

Priestley, P. T., Sebborn, W. S., and Selman, R. F., *Analyst*, **90**, 589 (1965).

Methodological Developments and Selected Applications

Bell, J. M., and Cowell, C. F., "Methods for the Preparation of Neutral Solutions of Ammonium Citrate," *J. Am. Chem. Soc.*, **35**, 49 (1913).

Jespersen, N. D., and Jordan, J., *Anal. Letters*, **3**, 323 (1970).

Jordan, J., and Carr, P. W., "Enthalpimetric Analysis," in "Analytical Calorimetry," R. S. Porter and J. F. Johnson, eds., Plenum, New York, 1968, p. 203–208.

Jordan, J., Henry, R. A., and Wasilewski, J. C., *Microchem. J.*, **10**, 260 (1966).

Jordan, J., Jespersen, N. D., Reich, R. M., Cullis, H. M., and Miller, C. D., Abstracts, Pittsburgh Conference on Analytical Chemistry and Applied Spectroscopy, Cleveland, Ohio, 1970, p. 141–142.

Jordan, J., Meier, J., Billingham, E. J., Jr., and Pendergrast, J., "Thermochemical Titrations in Fused Salts," *Anal. Chem.*, **32**, 651 (1960).

Wasilewski, J. C., and Miller, C. D., *Anal. Chem.*, **38**, 1751 (1966).

Wasilewski, J., Pei, P. T-S., and Jordan, J., *Anal. Chem.*, **36**, 2131 (1964).

Zambonin, P. G., and Jordan, J., *Anal. Chem.*, **41**, 437 (1969).

Ancillary Reference

Siggia, S., "Quantitative Organic Analysis via Functional Groups," 3rd ed., Wiley, New York, 1963, p. 655–682.

ACKNOWLEDGMENT

The work leading to this article has been supported in part by the United States Atomic Energy Commission under contract AT (30-1)-2133 with the Pennsylvania State University, and by a grant from the Research Corporation. Figures 4–8 were made available by courtesy of American Instrument Company.

The work on which this supplement is based has been supported in part by Research Grant GP 11386 from the National Science Foundation.

Bene Valete. Good Health. A form of greeting often found at the conclusion of old papal documents.

Topics in...

Chemical Instrumentation

Edited by **GALEN W. EWING**, Seton Hall University, So. Orange, N. J. 07079

...a ChemEd feature

The Journal of Chemical Education
Vol. 44, pages A571, A629, A685, A853
July to October, 1967

XXXIII. Recent Developments in Calorimetry

Part One—Introductory Survey of Calorimetry

RANDOLPH C. WILHOIT, *Thermodynamics Research Center,
Department of Chemistry, Texas A & M University,
College Station, Texas 77843*

A native Texan, **Dr. Randolph C. Wilhoit** received his formal education at Trinity University (San Antonio), the University of Kansas, and Northwestern University, where he received the Ph.D. working with Professor Malcolm Dole. He spent a year's postdoctoral at Indiana University, then joined the faculty of Texas Technological College, which he left in favor of New Mexico Highlands University. Since 1964 he has been Associate Professor of Chemistry and Assistant Director of the Thermodynamics Research Center at Texas A & M University. His research interests are centered around calorimetry, with particular emphasis on the thermochemistry and thermodynamic properties of organic compounds, and applications of thermodynamics to biochemistry. He is a member of the ACS, AAAS, Sigma Xi, and the Calorimetry Conference.

SCOPE AND PURPOSE OF CALORIMETRIC INVESTIGATION

Calorimeters are used to measure the change in internal energy or enthalpy which occurs when a system changes from an initial state to a final state. Such data are required for many thermodynamic and thermochemical calculations.

Calorimeters are also suitable for analytical tools. They can be used, qualitatively, to detect the presence of exothermic or endothermic processes, and quantitatively, to determine the extent to which processes occur. They can be applied to the study of the equilibrium properties of matter as well as to the study of rates of change of such properties in nonequilibrium states. Careful calorimetric studies of phase changes in condensed systems have produced some of the most accurate data available on the equilibrium temperature for these changes, as well as quantitative estimates of the purity of the samples.

Calorimeters have been used in the study of properties of matter for well over a century and modern calorimetric instrumentation and techniques are the result of a long evolution and extensive refinement. In spite of this long history, calorimetry is still an active area of research which utilizes many of the most sophisticated of modern instruments. Many major laboratories throughout the world are devoted partly, or completely, to calorimetric research, and the demand for these data far exceeds the supply. Because of the wide range of systems, phenomena, and conditions of interest to thermochemists, the choice of standard ready-made calorimeters is limited to a relatively few specific areas. Although several valuable new calorimeters have appeared on the market within the past few years, the thermochemist, more often than not, still finds it

necessary to modify existing instruments or to construct them entirely to his specifications, in order to carry out his objectives.

The principal purpose of this review is to give a brief introduction to calorimetry in general and to describe recent developments, especially in manufactured instruments, in a few areas of calorimetric research. The literature on calorimetry is very extensive. More detailed information can be found in several monographs (1–7), which should be consulted for references to the original literature.

In 1946, at the urging of Dr. Hugh H. Huffman, a group of scientists interested in experimental calorimetry organized a series of informal meetings to discuss common problems. These meetings have continued on an annual basis under the sponsorship of the Calorimetry Conference and have grown to include some 150 to 200 participants, who present fifty or more reports each year on current research in calorimetry. These meetings have had great influence in improving the standards of calorimetric techniques and of reporting data in publications. Similar national conferences have been organized in the Soviet Union, England, and Japan. International conferences on calorimetry are conducted bi-annually by the Subcommittee on Thermodynamics and Thermochemistry of the International Union of Pure and Applied Chemistry.

The design and operation of a calorimeter is simple in principle and much useful data can be generated with very simple equipment. However, precision measurements, compatible with the best modern practice, require exacting technique and meticulous attention to detail. Figure 1 shows, in a schematic way, the basic parts of a typical calorimeter. The calorimeter contains the system under study, one or more thermometers, and perhaps, an elec-

trical heater and devices for mixing or stirring. The calorimeter is surrounded by the jacket, which also contains one or more thermometers and usually some provision for controlling its temperature. Heat may be transferred between the calorimeter and the jacket by conduction through solid materials connecting them, by conduction and convection through any gas which may be present, and by thermal radiation. Work energy may also be transferred to the calorimeter by mechanical motion used in stirring or mixing, by compression or expansion against an external pressure, or by an electric current used to operate a heater. Following the

Chemical Instrumentation

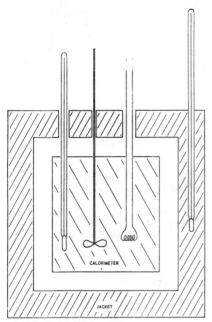

Figure 1. Schematic diagram of a calorimeter.

usual practice in thermodynamics in which the calorimeter plus its contents is regarded as the system, a calorimetric experiment can be described in terms of the heat and work transferred to or from the calorimeter and the changes in the variables of state of the calorimeter and its contents when they change from an initial to a final state. According to the first law of thermodynamics,

$$q + w = E_f - E_i = \Delta E_t \qquad (1)$$

where the subscript, t, designates the calorimeter plus its contents. If the volume of the calorimeter is constant, no pressure-volume work is done and w, in eqn. (1), represents other types of work. If the pressure is kept constant, then,

$$q + w' = H_f - H_i = \Delta H_t \qquad (2)$$

where w' is work other than pressure-volume work. The objective of a calorimetric experiment is to measure the heat absorbed, q, and the work done, w or w', as accurately as possible and to relate them to changes in internal energy or enthalpy of the contents of the calorimeter. Since the measurement of the heat exchange, q, usually presents the most difficult part of the experiment, calorimeters are usually designed to keep the heat exchange as small as possible. However, the heat exchange is deliberately made large in some types of experiments, and the evaluation of ΔE_t or ΔH_t, then depends upon an accurate measurement of q.

Calorimetric studies are now made on a large variety of systems under a wide range of conditions. Systems include the gas, liquid, and solid states, either as single phases or as two or more phases. Nearly all types and classes of chemical compounds are included. Temperatures range from around 1 to over 1500°K, and pressures range from the micron region to several hundred atmospheres. The quantity

of energy measured ranges from around 100 kcal for the study of explosives to around 10 kcal for heats of combustion and down to microcalories for certain studies. The time scale varies from fractions of a second to hours, or even days. In addition to the study of the familiar kinds of phase changes and chemical reactions, calorimeters are used to measure heats of adsorption from both liquid solutions and gases, energies which accompany the relaxation of strains in solids, energies of nuclear transformations, heat generated by living organisms ranging from bacteria to man, and the energy present in electromagnetic or ionizing radiation.

TYPES OF CALORIMETERS

Although the wide range of systems and conditions studied in calorimeters requires a great variety of specific instruments and techniques, certain features common to nearly all calorimeters have evolved into a few specific types. These features are found in a great many combinations in specific calorimeters.

Static calorimeters operate between definite equilibrium (or metastable) initial and final states at constant, or nearly constant, temperatures. *Dynamic* calorimeters are run with a continuous change in temperature, so that the contents never attain a state of true equilibrium. Dynamic calorimeters are usually of small size so that the temperature can be maintained uniform throughout the calorimeter. In *flow* calorimeters a constant flow of material, usually liquid or gas, is maintained into and out of the calorimeter, usually operated in a steady state condition. Calorimeters in which a constant flow of material is introduced, with the products of the reaction being accumulated in the calorimeter, have been called *thermometric titration* calorimeters.

With respect to the interaction between the calorimeter and the jacket, three types can be recognized. In calorimeters having *isothermal* or *adiabatic* jackets, the heat exchange, q, is kept small compared with the total change of energy of the calorimeter and its contents. In *conduction* calorimeters, however, the heat exchange is approximately equal to the change in energy of the calorimeter and contents. The temperature of an isothermal jacket is kept constant during the measurement, while the temperature of an adiabatic jacket is adjusted to match that of the calorimeter at all times. Both adiabatic and isothermal jackets have been used for many years. The principal advantage of adiabatic control is that, ideally, there is no heat exchange between jacket and calorimeter, and thus the temperature of the calorimeter is very nearly constant in the initial and final states. This is very important in some investigations where the system requires an appreciable time to reach equilibrium, and, in itself, may be sufficient to decide in favor of the adiabatic jacket. Besides, it is much easier to measure accurately a constant temperature than a changing one. On the other hand, it is more difficult to construct an adiabatic than an isothermal jacket. Provision must be made for rapid and uniform heating and cooling. Control of the temperature is very tedious and demanding if done

manually, or requires complex instrumentation if done automatically.

In spite of the advantages of adiabatic calorimetry, many workers prefer an isothermal jacket for certain types of highly accurate measurements. The use of an isothermal jacket requires that the heat exchanged between calorimeter and jacket be carefully evaluated. However, it is seldom possible to attain ideal temperature control in an adiabatic jacket, and it is frequently necessary to calculate a correction for heat exchange for the adiabatic jacket also. The principal source of error in the calculation of the heat transfer is the lack of uniformity of temperature on the surfaces of the calorimeter and the jacket. Since the temperature of an isothermal jacket is more likely to be uniform than that of an adiabatic jacket, it is preferable in this respect. Thus, even though the heat exchange may be larger for an isothermal jacket, the error in calculating the heat exchange may be less.

In many instances the choice between an isothermal or an adiabatic jacket of the same overall accuracy is a matter of personal preference. The adiabatic jacket is more difficult to construct and to operate unless elaborate automatic controls are installed. The isothermal jacket requires more thorough data logging and a more complex calculation procedure. The isothermal jacket is to be preferred for those calorimeters which change temperature rapidly.

In the *conduction calorimeter*, all or nearly all of the heat evolved or absorbed in the calorimeter is conducted to or from the jacket. The measurement of ΔE_t or ΔH_t depends entirely on the calculation of the heat exchange. This depends on a knowledge of the temperatures of the jacket and calorimeter at all times during the measurement and on the knowledge of the rate of heat transfer as a function of these temperatures. The best accuracy which can be attained with a conduction calorimeter is less than that which can be attained with isothermal or adiabatic jacket calorimeters. Conduction calorimeters may be of the static type, in which case the jacket is kept at a constant temperature and the initial and final temperatures of the calorimeter are the same as the jacket temperature, or they may be of the dynamic type, in which case the difference between the temperatures of the jacket and calorimeter is kept constant so that the rate of heat transfer is also constant. An adiabatic jacket calorimeter can be easily converted to a dynamic conduction calorimeter. Adiabatic calorimeters which utilize a stirrer are frequently run with the jacket temperature slightly below the calorimeter temperature in order to remove the heat generated by the stirring. This can be called pseudo-adiabatic control.

The *labyrinth flow* calorimeter is a type of conduction calorimeter which is located in a series of coaxial tubes so that any heat which is evolved or absorbed is transmitted to or from water which is circulated through the tubes. The rate of evolution or absorption of heat can be calculated from the rate of flow of water and the difference between the temperature of the water at the inlet and the outlet.

Another variation, which takes many specific forms, is the *twin* or *differential*

Chemical Instrumentation

calorimeter. In this modification two matched calorimeters are placed in either the same or in different jackets. They are operated simultaneously, with a standard system in one calorimeter and the system being studied in the other. The two calorimeters are kept at the same temperature by supplying electrical energy to one of them. Their use rests on the assumption that, if the two calorimeters are at the same temperature, there is no heat transfer between them and that the rate of heat exchange between each calorimeter and the jacket is the same. If this assumption is valid, the electrical energy is equal to the difference between the energy change for the standard system and the energy change for the system being studied. Thus, neither the temperatures of the calorimeters or the heat exchange need be known. The advantage is obvious. If this condition is not met exactly a small correction can be applied. In some designs the two calorimeters are separate and insulated from each other, while in others they are thermally connected. Twin calorimeters may be either static or dynamic, and they may have either adiabatic or isothermal jackets, or be of the conduction type. Twin calorimeters have the greatest advantage when the system being studied is very similar to the reference system used for comparison, for example, a pure solvent and a dilute solution.

A *phase-change* calorimeter contains a relatively large quantity of an auxiliary substance in good thermal contact with the system under study. The auxiliary substance is in the form of two phases in equilibrium at constant temperature. The calorimeter is placed in a jacket at the same temperature, and thus it is both adiabatic and isothermal. The heat evolved or absorbed by the system results in a change, not of temperature, but of the relative amounts of the two phases present. The change in enthalpy of the system is determined by measuring the amount of phase change. The most familiar example of this type of calorimeter is the Bunsen ice calorimeter. The system is placed in a chamber which is surrounded by a mantle of ice, which, in turn, is surrounded by liquid water. The calorimeter is completely filled with the ice and liquid water. A change in the relative amounts of liquid and solid produces a change in volume which is detected by means of a capillary tube filled with mercury connected from the calorimeter to the outside. The quantity of mercury may be determined volumetrically or gravimetrically. Careful measurements have shown that 1 g of mercury corresponds to 64.64 calories of heat. In the presence of the ice–water system, the measurement is restricted to 0°C; however, other substances may be used to establish a solid–liquid equilibrium at other temperatures. Substances used for this purpose should be stable, obtainable in a state of high purity, and should attain a reproducible solid–liquid equilibrium readily. Diphenyl ether is a suitable material; it melts at 26.91°C and gives a calibration constant of 18.91 calories per gram of mercury. Acetic acid, naphthalene, and phenol are other substances

applicable to this type of calorimeter. Phase-change calorimeters based on a liquid-vapor equilibrium have also been used. Evolution of heat generates additional vapor, which can be withdrawn from the calorimeter and condensed to determine the amount. This principle has been used to study reactions which take place in boiling solutions of ammonia and of carbon tetrachloride. The solvent serves as the auxiliary material in these cases.

Drop calorimeters are designed to measure the change in enthalpy which results when the temperature of the sample is changed by a relatively large amount. The initial temperature of the sample, which may be either above or below the initial temperature of the calorimeter, is established in a suitable environment before it is placed in the calorimeter. The sample is then quickly dropped into the calorimeter and the final temperature of the calorimeter and sample determined after they have reached equilibrium. The calorimeter may consist of a block of highly conducting metal, such as copper or silver, or a container of liquid, such as water. From a knowledge of the heat capacity of the calorimeter, the initial temperatures of the sample and the calorimeter, and the final temperature of calorimeter plus sample, the change in the enthalpy of the sample may be calculated. The sample is often placed in a protecting capsule. If this is done, the heat evolved by the capsule must be subtracted from the total. A series of such measurements, with the sample at different initial temperatures, will serve to determine heat capacities, heats of transition, and heats of fusion. The calorimeter may have either an adiabatic or an isothermal jacket.

Figure 2 illustrates a drop calorimeter manufactured by the Dynatech Corp., Cambridge, Mass. They offer a series of such calorimeters which consist of either a copper block or a water bath. These are designed to operate in the range of −150° to 1600°C. They are furnished with various accessories such as furnaces, refrigerators, and temperature indicators, recorders, and controllers. Accuracies in the measurement of specific heat are in the range of 3–5%, depending on the model

Figure 2. Drop Calorimeter (Courtesy of Dynatech Corp.)

and the initial temperature of the sample. The Thermo-Physics Corp., Cambridge,

Mass., markets a similar drop calorimeter consisting of a furnace designed for temperatures up to 1600°C and a Bunsen ice calorimeter to receive the sample. The change in quantity of mercury in the system is determined by weighing. Accuracies of 1% or better in the measurement of heat delivered to the calorimeter are specified. They also will furnish optional equipment which permits initial cooling of the sample down to liquid nitrogen temperatures, and also permits keeping the sample in a vacuum or in an inert atmosphere. The availability of high-temperature drop calorimeters in ready-made form seems to be a rather recent development.

AREAS OF CALORIMETRIC INVESTIGATION

Calorimetric research can be classified into several well recognized areas according to the kind of measurement or system being studied. Each such area is identified by certain common problems and objectives.

Low-temperature calorimetry usually refers to the measurement of heat capacity, heats of phase transitions, and derived properties in condensed systems below room temperature. The objective is frequently to obtain the absolute entropy at room temperature through use of the third law of thermodynamics. Such data also play an important role in theories of the solid state. An excellent review of low-temperature calorimetry has been written by Edgar Westrum (8). In the early years the lower limit for these measurements was usually about 77.35°K, the boiling point of liquid nitrogen. However, absolute entropy based on measurements to this temperature are unreliable, and measurements down to around 20.77°K, the boiling point of hydrogen, or to 4.21°K, the boiling point of helium, have become routine for third law studies. Calorimetric studies at a few tenths of a degree on the Kelvin scale have been made by using the technique of adiabatic demagnetization. Modern low-temperature calorimeters are nearly always of the static type with an isothermal, or, more frequently, an adiabatic jacket. Low-temperature calorimetry has become an advanced and highly developed art. The equipment is custom built and requires a number of years to develop. The Sulfrian Cryogenics Inc., Rahway, N. J., can supply a number of special purpose cryostats, among which is one designed for use with low-temperature calorimeters. The Linde Co., New York, and Cryogenic Engineering Co., Denver, supply storage Dewars and equipment for handling cryogenic fluids such as liquid hydrogen and helium.

Low-temperature calorimeters are also highly automated. The one in use at the Bureau of Mines Thermochemistry Laboratory in Bartlesville, Okla., is an outstanding example. Accuracies of one part per thousand in the measurement of heat capacities in the range from 20° to 300°K are now considered routine, and some published data approach accuracies of one part in five thousand. The limiting factor in many such measurements is in the irregularities inherent in the definition of the temperature scales in this range.

Chemical Instrumentation

Measurements above room temperature fall in the province of *high temperature* calorimetry. The kind of properties measured are the same as in low-temperature calorimetry, but the major difficulties encountered and the hardware are quite different. According to the Stefan-Boltzman law, the rate of radiation of thermal energy from a surface is proportional to the fourth power of the absolute temperature. As a consequence the control of radiative heat transfer, especially above a few hundred degrees Celsius, is the overriding problem in high-temperature calorimetry. The measurement of temperature and the location of suitable construction materials present other difficulties. Static calorimeters with adiabatic jackets, differential dynamic calorimeters, and drop calorimeters are useful up to temperatures of around 1000°C. So far, only drop calorimeters have been found to be practicable much above this temperature. When using drop calorimeters for measurements in the higher temperature range, the sample must be transferred to the calorimeter very quickly to prevent undue loss of heat by radiation. Elaborate mechanisms have been devised for this purpose. The initial temperature of the calorimeter in these measurements is usually around room temperature. However, sometimes when it is desirable to avoid transitions which may take place in the sample at lower temperatures, the calorimeter also may be kept at an elevated temperature.

The heats of formation of most organic compounds, inorganic oxides, and a few other inorganic compounds are derived from measurements of the heats of combustion in oxygen. For compounds in the liquid or solid state, these measurements are made with a *bomb calorimeter* in which the sample is burned in an atmosphere of oxygen at a pressure of 15–40 atms. Gaseous compounds are burned in a flame calorimeter at atmospheric pressure. Heats of combustion have been measured for about 90 years, and the technique has become highly developed and standardized. Some laboratories routinely obtain data accurate to one part in ten thousand, and occasionally, to one part in twenty thousand. These accuracies are needed since heats of formation are obtained as relatively small differences between the heats of combustion and the heats of formation of the products of combustion. The major limitation lies not in the measurement of temperature or energy but in the preparation of samples of sufficient purity, in the determination of the quantity of material burned, and in the identification and application of corrections for the numerous side reactions. In addition to the highly precise measurements of the heat of combustion of pure compounds, many measurements of lower accuracy are made to determine the heat value of fuels. Several manufacturers supply combustion calorimeters. These will be described in Part III of this review.

The standard combustion calorimeter consists of a bomb, capable of withstanding high internal pressure, of about 300 ml capacity which is placed in a bucket containing around 2 liters of water. One to three grams of sample are burned, producing a temperature rise of one or two degrees. Isothermal jackets are usually used although adiabatic jackets are becoming more popular. Aneroid calorimeters consist of the bomb only, or bomb imbedded in a block of metal. These also have been applied to high precision measurements in recent years, especially where only a limited quantity of sample can be obtained. Organic compounds which contain only the elements carbon, hydrogen, oxygen, or nitrogen can be satisfactorily burned in a stationary bomb. However, compounds which contain sulfur or halogens must be burned in a rotating bomb to obtain reliable values. The rotation is needed to mix the liquid products of combustion so that they are homogeneous. Combustions in other oxidizing atmospheres may also be studied in a calorimeter. Some organic and inorganic compounds can be burned in chlorine or fluorine. Fluorine bomb calorimetry has been a valuable new development for the study of many refractory inorganic materials which will not burn in oxygen. Some reactive metals can be burned in nitrogen.

The heat capacities of gases are usually measured in a *vapor flow* calorimeter. The gas is passed through the calorimeter at a constant rate. It first flows around a thermometer placed near the inlet, then over a carefully designed electrical heater, and finally past another thermometer near the exit. Thermocouples or resistance thermometers are usually employed. The heat capacity of the gas can be calculated from its mass flow rate, the electrical power dissipated by the heater, and the temperatures at the inlet and exit. Corrections must be applied for the heat exchange between the calorimeter and the jacket, transfer of heat from the heater to the thermometers by radiation, the Joule-Thompson effect, and the departure from a steady state condition. The success of this method depends upon having a controlled, stable flow of vapor. This can often be accomplished by vaporizing the liquid form of the compound in a boiler. The liquid is kept at the boiling temperature by surrounding the boiler with the refluxing vapor of the same material. Vaporization is accomplished by means of an electric heater immersed in the liquid. The flow of vapor is proportional to the heating rate and may be calculated from a knowledge of the heat of vaporization. Or, by measuring the flow rate independently, the heat of vaporization may be calculated. Thus it is very common to combine the vapor flow calorimeter with the measurement of heat of vaporization. The rate of generation of vapor can be found by condensing the vapor and measuring the amount collected in a given time. The measurement of heat of vaporization and vapor heat capacity may be carried out in separate experiments, or they may be determined simultaneously by collecting the vapor after it has passed through the flow calorimeter. These measurements, of course, may also be conducted over a range of temperature and pressure.

A large variety of calorimeters have been used for measuring heats of solution, heats of mixing of liquids, and heats of reactions in solution. These measurements all have the common characteristic of studying the effect of mixing two samples, either both liquid, a solid and a liquid, or occasionally a gas and a liquid. Thus, there must be some provision for keeping the two samples separate until the proper time and then for mixing them to form a homogenous solution. In order to insure thorough mixing, these calorimeters either contain a motor driven stirrer, or the whole calorimeter is rocked back and forth or rotated.

Calorimetric studies can be carried out, not only on aqueous and nonaqueous solutions in the vicinity of room temperature, but also on solutions at low temperature, such as those in liquid ammonia, and solutions at high temperature, such as in molten salts or liquid metal systems. The heat effect which accompanies the vaporization or condensation of the volatile components of the solution must always be considered as an important source of error in *solution calorimetry*. In fact, the volatility of the materials being mixed influences to a large extent the design of the calorimeter. Highly volatile materials must be mixed in a calorimeter which is completely filled both before and after mixing so that there is no vapor space at any time. Although a vapor space can be tolerated when mixing components whose vapor pressure is less than around 50–100 torr, the vapor space should be kept small in careful work, and the calorimeter should be nearly vapor tight to prevent appreciable loss of vapor. Reactions which produce a gas are always troublesome since the gas tends to carry away some vapor.

Most solution calorimeters are of the static variety, although thermometric titration types are becoming more popular in recent years. A simple calorimeter suitable for measurements of low or intermediate accuracy on relatively nonvolatile systems may be constructed from a Dewar flask fitted with a cover, a thermometer, a stirrer, and a pipet. Measurements of high accuracy are conducted in elaborate calorimeters with isothermal or adiabatic jackets. They may be of the single or of the twin type. Thermometric titration calorimeters are easy to build and generate a large amount of data in a short time. They are particularly advantageous when studying complex reactions which involve several intermediate steps.

Reaction calorimetry includes the study of many chemical reactions other than combustion. To be suitable for calorimetric study, a chemical reaction should proceed quantitatively at a convenient rate to produce products of definite and reproducible composition. This includes many reactions which occur in liquid solution. Heats of hydrogenation, halogenation, and hydrohalogenation of unsaturated organic compounds in liquid and gas phases have been measured in flow calorimeters. A fairly extensive literature exists on heats of polymerization.

Microcalorimetry includes the study of systems in which the total temperature change, the total amount of heat produced, or the rate of production of heat is small. Solution calorimeters often operate in this range. Heats of dilution or of mixing of nonpolar liquids often require the measurement of very small temperature changes. Calorimeters which can detect temperature changes of 10^{-6} deg C have been constructed by using multijunction thermo-

Chemical Instrumentation

couples, and ones which can detect changes of 10^{-5} deg C or a little less by using thermistors. The studies of some biochemical reactions or of the heat produced by radioactive isotopes are examples of situations where the available quantity of material may be limited. Calorimeters which can measure fractions of a millicalorie have been designed for these applications. Although the total amount of heat generated by a slow reaction may be large, a microcalorimeter may be required to detect it. The measurement of the heat produced by the setting of cement or by the germination of seeds are examples.

A major problem in microcalorimetry is the identification and removal of sources of electrical or thermal interference. The heat effects resulting from evaporation or condensation, from friction in stirring or mixing devices, from adsorption, and from temperature gradients in the environment are very troublesome. Twin calorimeters have been used effectively to reduce these sources of interference. Electrical noise always sets a lower limit of sensitivity when using thermocouples or resistance thermometers to detect changes in temperature. The commercially available microcalorimeters are of the conduction type. They will be described in Part III of this review.

The following outline identifies, in a general way, the types of calorimeters which have been used for various kinds of measurements. It does not include all combinations or certain calorimeters designed for specific purposes.

Specific Heats and Heats of Transition in Condensed Phases

Low Temperature: Static calorimeters with either adiabatic or isothermal jackets.

High Temperature: (a) Static calorimeters with an adiabatic jacket: (b) Drop calorimeters with the receiver having either an isothermal or an adiabatic jacket, or a phase-change calorimeter; (c) Dynamic twin conduction calorimeters.

Specific Heats of Gases and Heats of Vaporization

Flow calorimeters (about 50 torr pressure minimum).

Heats of Combustion

Gas Flame Calorimeters: Flow calorimeters.

Bomb Calorimeters: (a) Static "water bucket" calorimeters with either isothermal or adiabatic jackets; (b) Static aneroid calorimeters with either isothermal or adiabatic jackets.

Heats of Solution, Dilution, Mixing, and Reactions in Solution

(a) Dewar flask type with either isothermal or adiabatic jacket; (b) Static calorimeter with either adiabatic or isothermal vacuum jacket; (c) Twin calorimeters with adiabatic, isothermal, or conduction jackets; (d) Flow calorimeters; (e) Thermometric titration calorimeters; (f) Phase-change calorimeters.

Heats of Hydrogenation, Halogenation, and Other Catalyzed Reactions

(a) Flow calorimeters with isothermal jackets; (b) Flow calorimeters using phase-change principle.

PROCEDURES AND CALCULATIONS

Most calorimetric experiments fall into one of two types. These differ not in the design or operation of the calorimeter, but in the objectives of the measurement and in the treatment of the data. The objective of a Type I experiment is to determine the enthalpy of the contents of the calorimeter as a function of temperature. Its derivative with respect to temperature is the heat capacity. Electrical energy, represented by w' in eqn. (2), is added to the calorimeter and its temperature changes from T_i to T_f. For this type of experiment eqn. (2) may be expanded as,

$$q + w' = \Delta H_0 + \Delta H_s = (C_0 + C_s)(T_f - T_i) \quad (3)$$

where ΔH_0 and ΔH_s represent the changes in enthalpy of the calorimeter and its contents, respectively, and C_0 and C_s represent the corresponding average heat capacities over this temperature interval. Sometimes it is convenient to combine part, or all, of q and/or w' with ΔH_0. If this is done, ΔH_0 does not represent the true change in enthalpy of the calorimeter, but is more appropriately called the "energy equivalent" of the calorimeter.

The purpose of a Type II experiment is to determine the value of ΔH (or ΔE) for a transformation which occurs in the contents of the calorimeter at a constant temperature. Such transformations include phase changes, mixing of components, and chemical reactions. These changes are accompanied by abrupt changes in the properties of the system. Let $\Delta H_s(T_r)$ represent the change in enthalpy for the transformation being studied at the reference temperature, T_r, and let C_i and C_f represent the average heat capacities of the contents before and after this transformation. Then, for a Type II experiment, eqn. (2) becomes,

$$q + w' = C_0(T_f - T_i) + C_i(T_r - T_i) + \Delta H_s(T_r) + C_f(T_f - T_r) \quad (4)$$

An accurate knowledge of the heat capacity of the empty calorimeter, C_0, or of the energy equivalent of the calorimeter, is needed for either a Type I or a Type II experiment. This should be obtained in a separate calibration experiment in which a calorimetric standard is placed in the calorimeter. Calorimetric standards may be either substances of accurately known and reproducible heat capacity, or systems which undergo a transformation of known enthalpy change. Specific standards suitable for various applications will be discussed in Part II. The calibration may either be a Type I or Type II experiment, but should, if possible, be similar in all respects to the measurement of the unknown quantity. By making the calibration experiment similar to the measure-

ment, systematic errors in the evaluation of q, w', T_i, and T_f will largely cancel out in the calculation of C_s or $\Delta H_s(T_r)$.

Frequently a calorimeter which is used for a Type II experiment will be calibrated by a Type I experiment where a standard of known heat capacity is placed in the calorimeter, or, if the heat capacity of the sample is small compared to that of the calorimeter, the empty calorimeter alone can be used. The properties of calorimetric standards must be based on calibration experiments in which electrical energy is added to the empty calorimeter. It is difficult to remove systematic errors from such measurements, and they must be carried out in specially designed calorimeters in standardizing laboratories. Thus all modern calorimetric data are based, either directly or indirectly, on calibrations using electrical heating in Type I experiments. The unit of energy used in reporting calorimetric data is either the (absolute) joule or the thermochemical calorie, which is defined as 4.184 (exactly) joules. Calorimetric standards may also be used to test the overall accuracy of the measurement. For this purpose, it is useful to have more than one standard. One can be used for the calibration and the others for testing.

When studying endothermic transformations in Type II experiments, it may be possible to adjust w' so that little or no change in temperature occurs. This makes the term containing the heat capacities, C_0, C_i, and C_f, in eqn. (4) small, so that only approximate values are needed. In a sense, the procedure amounts to conducting the measurement and the electrical calibration simultaneously. A procedure, similar in principle, can be applied to exothermic transitions also, by removing the heat quantitatively. This is done automatically in the labyrinth flow and phase-change calorimeters which have already been described. Heat can also be removed quantitatively in a controlled manner by using the Peltier effect in thermocouples. This is done in the Calvet microcalorimeter to be described in Part III. The availability of new thermoelectric materials in recent years which have large Peltier effects introduces the possibility of removing relatively large amounts of heat quantitatively. Devices based on this effect are now available for use as small refrigerators. Equations corresponding to (3) or (4) may be written for each calorimeter of a twin calorimeter system. If it is assumed that the heat exchange, q, the heat capacities of the two calorimeters, and the temperatures of the two calorimeters are the same, then the differences in the heat capacities of the two calorimeters may be readily related to the differences in w' for the two calorimeters for a Type I experiment, or the differences in the $\Delta H_s(T_r)$ may be related to the difference w' for a Type II experiment.

It is convenient to separate the experimental data and their associated uncertainties into two groups. One group consists of the quantities such as q, w', T_i, T_f, and C_0, and uncertainties in these quantities give rise to the calorimetric errors. The other group consists of data describing the quantity of sample, the extent and kind of transformation which has occurred, the extent of side reactions, and other aux-

Chemical Instrumentation

iliary data needed to convert the enthalpy change to a thermodynamic standard state. Modern calorimetry has developed to the point where the uncertainties connected with the second group limit the overall accuracy of the measurement.

The rate of transfer of heat from the jacket to the calorimeter may be represented as,

$$dq/dt = \iint \kappa(\theta_j - \theta_c)dS + \sum_i \kappa_i(\theta_{ij} - \theta_{ic}) \quad (5)$$

θ_j is the temperature at a point on the inside surface of the jacket and θ_c is the temperature at a point on the outside surface of the calorimeter, opposite to the point corresponding to θ_j. κ is an effective thermal conductivity constant for the transfer of heat by radiation and by conduction and convection through the gas between the calorimeter and jacket. κ, θ_j, and θ_c, in general, are functions of position and time. dS is the surface differential, and the integration is carried out over the surface of the calorimeter. The terms in the summation of eqn. (5) represent conduction through solid connections, such as electrical wires, stirrer shafts, and supports between calorimeter and jacket. The κ_i's are the effective thermal conductivity of each such connection, and θ_{ij} and θ_{ic} are the temperatures at the two ends of such links. For practical calculations, eqn. (5) is nearly always replaced by,

$$\frac{dq}{dt} = K(\Theta_j - \Theta_c) \quad (6)$$

where Θ_j and Θ_c are appropriate average temperatures of jacket and calorimeter and K an overall average thermal conductivity. The rate of change of temperature of the calorimeter, when no electrical energy is being added and no transformation is taking place inside the calorimeter, is

$$\frac{dT}{dt} = \frac{K(\Theta_j - \Theta_c) + S}{C_0 + C_s} \quad (7)$$

where S represents the transfer of energy by stirring, evaporation, or other processes not accounted for by eqn. (6). The quantity K/C_0 is an important characteristic of a calorimeter. It can be calculated from the measurement of dT/dt for two different values of $\Theta_j - \Theta_c$ by use of eq (7). The constancy of K/C_0, or lack of it, at a given temperature is a good indication of the magnitude of the calorimetric errors. For calorimeters designed for highly accurate work K/C_0 should be no larger than around 10^{-3} min^{-1}.

The total heat exchange which occurs when the calorimeter changes from an initial temperature, T_i at time t_i, to a final temperature, T_f at time t_f, is

$$q = K \int_{t_i}^{t_f} (\Theta_j - \Theta_c)dt \quad (8)$$

If the calorimeter and jacket temperatures are recorded graphically or digitally throughout the experiment, the integration in eqn. (8) can be carried out either graphically or by use of a digital computer. Numerous approximations of eqn (8) have been developed for specific types of calorimeters. The general reviews listed previously should be consulted for details of these procedures.

Part Two. Some Associated Measurements

THERMOMETRY

The calorimetrist is likely to spend more of his time in measuring and controlling temperature than in any other single activity. Progress in calorimetry is dependent on the availability of suitable thermometers of sufficient sensitivity and stability in the temperature range of interest. The demands of calorimetry have, in turn, given impetus to the development of better thermometers. Ideally, temperature should be measured on the thermodynamic scale, which is defined through the use of the second law of thermodynamics and is independent of the properties of any particular substance. However, the direct use of the thermodynamic scale is too difficult for practical measurements. The International Practical Temperature Scale (IPTS) was first defined by the Seventh General Conference on Weights and Measures in 1927 and has been revised slightly several times since. It is based on six fixed points ranging from the boiling point of oxygen to the freezing point of gold. Interpolation between these points is made by a platinum resistance thermometer from $-182°C$ to $630.5°C$, and by a platinum, platinum-rhodium thermocouple from 630.5 to $1063°C$. Higher temperatures are measured with an optical pyrometer. The IPTS is not defined below $-182°C$, and several scales, which differ slightly from each other, are now in use. According to the best current estimates, the IPTS differs from the thermodynamic temperature by about $0.04°$ at $200°K$, is identical at $273.15°K$, and differs by about $0.14°$ at $700°K$ and by about $1.1°$ at $1200°K$. Some extensive revisions of the IPTS have been recommended, and it seems very probable that it will be extended down to liquid helium temperatures within two or three years.

The measurement of temperature now forms a major branch of science and technology; but only a few procedures, which have some application to calorimetry, will be discussed here. Several recent references to thermometry in general are listed at the end of this section, and some additional references to specific procedures are given in the text. A list of companies which supply thermometric equipment, and their addresses, is also given at the end of this section. To be suitable for use in a calorimeter, a thermometer, itself, must not introduce appreciable disturbing influences on the measurement. This usually means that it be small in size and in heat capacity and not conduct excessive heat to or from the calorimeter. The use of remote indicating or recording, the ability to with-

stand corrosive conditions, and rapid response to changes in temperature may be important in some applications.

Several kinds of temperature measurement may be made during the course of a calorimetric experiment. The same thermometer may be used for all these measurements; but since the requirements are somewhat different it is more common to find two or more different thermometers in a single apparatus. The temperature at which the measurement is made must be established on a recognized scale. In the vicinity of room temperature, accuracies of around $0.1°$ are usually sufficient. This result does not enter directly into the calculation of the enthalpy change and, except at very high or very low temperatures, presents little difficulty. This will be called an "absolute" temperature measurement. A second kind of measurement is needed to determine the difference between initial and final temperatures. This will be called a "relative" temperature measurement. Since it enters directly into the calculations, it must be measured with care; and errors in this measurement often constitute the principal source of calorimetric error. This difference ranges from a few microdegrees, in microcalorimetry, to as much as ten or twenty degrees, in measurements of heat capacity. In a Type II experiment (see Part I), the relative temperature may be expressed in terms of an arbitrary scale, which is the same for the calibration and for the measurement experiment. However, in a Type I experiment, such as the measurement of heat capacity, the relative temperature must be expressed on an accepted scale, since it enters into the final result.

The differences between the temperature of the surfaces of the calorimeter and the jacket are determined at each stage of the experiment by "differential" temperature measurements. These data are used to calculate the quantity of heat exchanged between the calorimeter and the jacket. Since the heat exchange depends on surface temperatures, thermometers for this purpose may be placed directly on the surface. When this is done it should be established that the thermometer, itself, does not introduce undesirable disturbances in the temperature distribution. High accuracies are not usually required in differential measurements, but these temperatures may change rapidly, and some form of automatic recording is desirable. High sensitivity may be needed since the differences may be small.

Although *mercury-in-glass* thermometers have played an important role in the measurement of calorimeter temperatures in the past, they are not often used at present except in some routine work such as measurement of heats of combustion of fuels. However, they should not be entirely overlooked. In the range from about 0 to 100°C they can, with proper calibration and precautions, permit absolute measurements to an accuracy of $0.05°C$ and relative measurements to an accuracy of about $0.001°C$. Mercury thermometers are simpler, more economical, and more trouble-free than any other kind of thermometer of similar performance in this range. Their size and shape restricts their use in calorimetry.

Chemical Instrumentation

The difficulty in adapting them to remote indicating, recording, and controlling is also a disadvantage. The Brooklyn Thermometer Co. sells a large variety of special purpose mercury thermometers which includes standard and differential types. Several of these are designed for use with combustion calorimeters. The Parr Instrument Co. also sells mercury thermometers for combustion calorimeters.

Thermocouples

A *thermocouple* generates an electric potential which is roughly proportional to the difference in temperature between the hot and cold junctions (Seebeck effect). It is thus well suited for differential temperature measurements. It may also be used for absolute and relative measurements by keeping one junction, the reference junction, at constant temperature. The potential developed by a difference in temperature of one degree is the thermoelectric power of the thermocouple. If current flows in the direction dictated by the Seebeck effect, heat is absorbed at the cold junction and evolved at the hot junction (Peltier effect). Heat is also absorbed when current flows along a temperature gradient in a homogeneous wire in the same direction as the heat flow, and evolved when it flows in the opposite direction (Thompson effect). The Peltier and Thompson effects operate independently and in addition to the ordinary thermal conduction and Joule heating generated by electric current flow.

The use of thermocouples requires the measurement of low, d.c. voltages. The reduction of noise and extraneous sources of potentials below the level of a few microvolts requires the exercise of some care. Potentials of this magnitude are produced by instability in the ground, by electrostatic effects, by changing magnetic fields, by insulation leakage, and by thermoelectric effects. Thermoelectric potentials are developed in any circuit where a temperature gradient exists in a region of inhomogeneity. Circuit components, such as resistors and terminals, which are made from metals other than copper represent inhomogeneities. They can also result from soldered connections and from mechanical strains and the presence of impurities in the wire. A thermocouple composed of copper and the ordinary tin-lead solder generates three microvolts per degree. A special solder composed of 70% cadmium and 30% tin generates only about 0.1 microvolt per degree with copper. For the most precise work however, it is better to avoid the use of solder altogether. A good review of methods of eliminating noise from low-level circuits has been given by D. Nolle (9). Some alloys used for thermocouple wire are more prone to develop inhomogeneities than is copper. Although modern thermocouple wire is superior in this respect to older wire, inhomogeneities are still the major source of error in thermocouples. Methods of testing thermocouple wire are described by Roeser and Lonberger (10), and by Fuschillo (11).

At zero current the potential developed by a multi-junction thermocouple is proportional to the number of pairs of junctions connected in series. The advantage gained by use of more junctions is somewhat offset by the increase in resistance of the circuit and by the increase in thermal conductivity through the wires from the hot to the cold junctions. The total resistance of a multijunction thermocouple consisting of n pairs of junctions between wires of resistivity ρ_1 and ρ_2 and cross-sectional area of a_1 and a_2, with length l from the hot to the cold junctions is,

$$R = \left(\frac{\rho_1}{a_1} + \frac{\rho_2}{a_2}\right)nl \qquad (1)$$

The total rate of conduction of heat from the hot to the cold junctions, for a difference of one degree in temperature, is,

$$dq/dt = (k_1a_1 + k_2a_2)n/l \qquad (2)$$

k_1 and k_2 are the thermal conductivity of the material used to make the wires. It is desirable to reduce the total resistance, R and the conductivity, dq/dt. By combining eqs. (1) and (2) it is easy to show that Rdq/dt is a minimum when the relative sizes of the two kinds of wires correspond to,

$$\frac{a_2}{a_1} = \left(\frac{k_1\rho_2}{k_2\rho_1}\right)^{1/2} \qquad (3)$$

When the thermocouple is connected to a measuring instrument of internal resistance, N, the current generated by a one degree temperature difference is,

$$I = \frac{ea}{R + N} \qquad (4)$$

where e is the thermoelectric power for one pair of junctions. If the number of junctions is increased and the size of the wires varied in such a way that the ratio given by eqn. (3) is maintained and the total thermal conductivity given by eqn. (2) is kept constant, the current will increase at first, pass through a maximum. and then decrease. The maximum current will be attained when $R = N$, and when

$$n = \frac{(Ndq/dt)^{1/2}}{(k_1\rho_1)^{1/2} + (k_2\rho_2)^{1/2}} \qquad (5)$$

The magnitude of the maximum current is,

$$I_m = \frac{1}{2}\left(\frac{dq/dt}{N}\right)^{1/2}\frac{e}{(k_1\rho_1)^{1/2} + (k_2\rho_2)^{1/2}} \qquad (6)$$

Very high sensitivity in measurement of differential temperatures in calorimeters may be achieved by the use of carefully designed multijunction thermocouples. Some microcalorimeters have been built with as many as several thousand junctions. Equations (5) and (6) show that the optimum use of multijunction thermocouples can be most easily attained with a measuring instrument of relatively low internal resistance and high current sensitivity. For example, the data presented in Tables 2 and 3 show that for a copper-constantan thermocouple having a total conductivity of 1 cal sec^{-1} deg^{-1}, the maximum current is produced by three hundred thousand junctions when measured with an instrument whose

internal resistance is 1 megohm. Maximum current would be attained with about 1100 junctions with an instrument of 10 ohms resistance and would be equal to 2.2 milliamps per degree of temperature difference.

Thermocouples have been used for temperature measurements for nearly eighty years. They are applicable throughout the range from 5 to 2500°K. Although the principles of thermoelectric thermometry have not changed during this time, many new materials have been introduced in recent years, so that a rather bewildering array of combinations is now offered. The practice by different manufacturers of selling the same, or nearly the same, alloy under different trade names adds to the confusion. A thorough review of the properties of many of these materials has been written by Caldwell (12) and additional information will be found in reference (13). These materials may be grouped into three principal classes. The noble metals include platinum, rhodium, silver, gold, palladium, and their alloys. The base metals include copper, iron, and alloys of nickel, chromium, magnesium, silicon, and trace amounts of other elements. Refractory metals include rhenium, tungsten, molybdenum, and their alloys, carbon and the carbides of boron, silicon, molybdenum, and tungsten. A list of the more common alloys, and some suppliers of bulk wire is given in Table 1. (The addresses of the suppliers are given in an appendix.)

Table 1. Sources of Thermocouple Wire

Noble metals	
Platinum and Platinum-Rhodium alloys	Sigmund Cohn, Engelhard, Omega
Iridium	Sigmund Cohn, Engelhard
Platinum-iridium alloys	General Electric
Platinel R	Engelhard
Base metals	
Chromel R and Alumel R	Hoskins
Tophel, Nial, and Cupron	Wm. B. Driver
Kanthal PR and Kanthal NR	Kanthal
Advance R, Geminol PR, and Geminol NR	Driver-Harris
Copnic	General Electric
Refractory metals	
Tungsten and Tungsten-Rhenium alloys	Engelhard, Omega
Rhenium	Omega

The alloys Advance, Copel, Copnic, and Cupron are all essentially the same alloy of copper and nickel which is commonly designated as "constantan." Extension leads, which have a low thermoelectric potential for the corresponding thermocouple wire, may be used to connect the thermocouple to the measuring instrument to save the use of the more expensive thermocouple wire. Table 2 lists some properties of metals and alloys used for thermocouples, as well as other materials of interest in the construction of resistance thermometers and calorimeters. Properties such as thermal conductivity and electrical resistivity are highly sensitive to the presence of impurities and to mechanical and heat treatment of metal alloys. Therefore the values of these properties may deviate considerably from

Chemical Instrumentation

the nominal values given in this table. Table 3 lists some characteristics of thermocouples made from pairs of wires.

The upper temperature limit permitted for a given thermocouple depends not only on the composition of the wires but also on the length of time they are exposed to this temperature, the size of the wires, the kind of atmosphere present, and the accuracy desired. The limits listed in Table 3 are for relatively accurate measurements with 20 gauge wire in air. Also shown in this table are the optimum ratio of diameters of the two wires, as calculated from eqn. (3) and the factor $e/[(k_1\rho_1)^{1/2} + (k_2\rho_2)^{1/2}]$ which appears in eqn. (6). All these data apply at 25°C. Some data at other temperatures may be found in the references listed at the bottom of Table 2.

Prepared thermocouples in a great variety of styles, sizes, and protecting tubes may be obtained from many sources. Most of these are designed for industrial use and are too large for most calorimeters. Leeds and Northrup, Omega, Conax, and others supply a series of small thermocouples in protecting tubes down to 0.04 in. in outside diameter. Omega furnishes pairs of welded thermocouple wires in sizes down to 0.001 in. diam.

The copper-constantan thermocouple is the most popular combination for precision measurements in the range from 15°K to around 250°C. It is highly stable and reproducible in this range and has a moderately high thermoelectric power. The copper-gold, 2.11 atomic percent cobalt, thermocouple has been used for low-temperature measurements in recent years. Its principal advantage is a large thermoelectric power at low temperatures. For example, at 13°K the thermoelectric power of a copper-constantan thermocouple is 4.1 microvolts per degree, while it is 11.3 microvolts per degree for the copper-gold, cobalt thermocouple. The advantage of the greater sensitivity is reduced by uncertainties which are produced by inhomogeneities in the gold-cobalt alloy. Base metal thermocouples have a high thermoelectric power and are most frequently used for moderately high temperatures in industrial applications. Noble metal thermocouples, especially the platinum-platinum, rhodium combination, are preferred for applications requiring high accuracy in the range of 500–1200°C. Refractory metals are applicable in the extremely high-temperature range.

The tables given in the NBS Circular 561 (14) are usually considered as the standard for converting thermoelectric potentials to temperature for many of the commonly used pairs of metals. Additional tables will be found in reference (13) and in the catalogs of the manufacturers. For precision work, however, each thermocouple should be periodically calibrated at fixed points or by comparison with a standard thermometer. The potentials developed by thermocouples may be measured with any instrument suitable for measuring low d.c. voltages. The current should be kept small for

Table 2. Properties of some Metals and Alloys at 25°C

Name of Metal	Density (gm cm⁻³)	Heat Capacity (cal deg⁻¹ gm⁻¹)	Heat Capacity (cal deg⁻¹ cm⁻¹)	Thermal Conductivity (cal deg⁻¹ sec⁻¹ cm⁻¹)	Thermal Diffusivity (cm sec⁻¹)	Resistivity (10^6 ohm cm)	Temperature Coefficient of Resistance (deg⁻¹)
Platinum	21.4	0.0316	0.677	0.170	0.254	10.6	0.0037
Platinum– 10% rhodium	20.0	—	—	0.070	—	30.3	0.0013
Copper	8.89	0.092	0.816	0.91	1.11	1.73	0.0039
Iron	7.87	0.107	0.84	0.176	0.21	10.0	0.0050
Nickel	8.9	0.106	0.94	0.21	0.22	7.0	0.0050
Gold	19.3	0.032	0.62	0.75	1.20	2.44	0.0034
Silver	10.5	0.056	0.59	0.98	1.65	1.63	0.0038
Chromel	8.73	0.107	0.94	0.046	0.049	105.	—
Alumel	8.60	0.125	1.08	0.071	0.066	47.	—
Constantan	8.90	0.094	0.84	0.051	0.061	49.0	0.0000
Manganin	8.4	0.094	0.79	0.053	0.067	48.2	0.0000
Stainless steel (304)	7.9	—	—	0.036	—	103.	—
Brass	8.9	0.092	0.82	0.28	0.34	7.0	0.0017
Germanium	5.38	0.081	0.44	0.14	0.32	40.	0.05

Most of the data in this table have been extracted from references (33), (36), (45), (46), and from "Data Book," Thermophysical Properties Research Center, Purdue Research Foundation, Lafayette, Ind.; A. Goldsmith, H. J. Hirschhorn, and T. E. Waterman, "Thermophysical Properties of Solid Materials, Volume 1, Elements, Volume 2, Alloys," WADC Technical Report 58-476, 1960.

Table 3. Characteristics of Some Typical Thermocouples 25°C

ISA Type	Metal No. 1 (Positive)	Metal No. 2 (Negative)	Upper Limit (deg. C)	Thermoelectric Power (microvolt deg⁻¹)	Optimum Radius Ratio (r_2/r_1)ᵃ	$2I(N/-dq/dt)^{1/2}$ (ma cal⁻¹/₂ deg⁻¹/₂ sec⁻¹/₂ cm¹/₂ ohm¹/₂)
S	Platinum	Platinum–10% Rhodium	1600	5.5	1.38	1.97
—	Copper	Gold–2.11 At.%Cobalt	—	40	—	—
T	Copper	Constantan	250	40	2.67	14.1
Y, J	Iron	Constantan	450	51	2.03	17.5
E	Chromel	Constantan	1000	59	0.80	12.4
K	Chromel	Alumel	1000	41	0.72	12.6
—	Tungsten	Tungsten–26% Rhenium	2200	3.3	—	—

The subscripts refer to metals No. 1 and No. 2.

precise quantitative measurements in order to reduce the effect of potential drop in the lead wires. Precision potentiometers can be purchased from Leeds and Northrup Co., Honeywell, Inc., Princeton Applied Research Corp., Sensitive Research Instrument Corp., and others. Digital voltmeters are useful either for indicating or for recording of thermocouple potentials. United Systems Corp. furnishes a voltmeter which converts thermocouple potentials to temperature and presents a direct digital reading.

Resistance Thermometers

The measurement of temperature in terms of the electrical resistance of a sensitive element is a familiar method. *Resistance thermometers* may be made from a coil of metallic wire such as copper, nickel, or platinum; a semi-conducting element such as carbon or germanium, or from a semiconducting metallic oxide. Because of their considerable versatility, reliability, and sensitivity, resistance thermometers have a wide application to calorimetry.

The principal electrical and thermal properties of a resistance thermometer are

its resistance, X, its temperature coefficient of resistance,

$$S = \frac{1}{X}\frac{dX}{dT} \qquad (7)$$

its dissipation factor, D, and its time constant τ. The time constant is a measure of the speed of response of the thermometer to a change in temperature. The dissipation factor is the rate of heat exchange between the thermometer and the surrounding medium, for a difference of one degree in temperature. The time constant is related to D and to the heat capacity of the thermometer, C, by

$$\tau = C/D \qquad (8)$$

The electrical energy which is dissipated by the current through the thermometer raises its temperature above that of the surroundings. The amount of this self heating is

$$\Delta T = XI^2/D \qquad (9)$$

If this rise in temperature is constant, it may be ignored when measuring relative or differential temperatures, but it may be necessary to consider it when measuring absolute temperatures. The dissipation

Chemical Instrumentation

factor may be easily determined by measuring the resistance of the thermometer using several different currents. Generally the temperature rise should be no more than about five to ten times the smallest temperature change which is to be detected, although it may be larger if enough care is taken to keep it constant. However, this self heating sets a practical upper limit to the current which can be allowed to flow when measuring the resistance.

The accuracy of measurement of temperature with a resistance thermometer is determined by the stability and reproducibility of its resistance, the accuracy of the measuring instrument, and the smallest change in temperature which can be detected. This smallest temperature change which can be detected with the combination of thermometer and measuring instrument is the "resolution." When a calorimetric measurement is based on the measurement of a small temperature change, the resolution may determine the overall accuracy of the measurement. In a simple Wheatstone bridge, as in Figure 3, the resolution is set by the amount of self heating which can be tolerated, the characteristics of the thermometer, the values of the resistances in the ratio arms, and the characteristics of the null detector. The current in the null detector, of internal resistance, N, which results from a change in the resistance of the thermometer, δX, from a balanced condition is

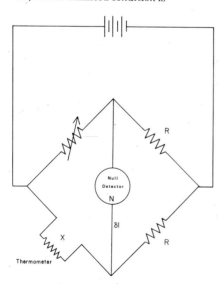

Figure 3. Wheatstone bridge.

$$\delta I = \frac{IR\delta X}{2RX + N(R + X)} \quad (10)$$

If δI now represents the smallest current which can be detected by the null detector, the resolution may be calculated as,

$$\delta T = \left(\frac{4X}{DKS^2}\right)^{1/3}\left[\left(1 + \frac{N(R + X)}{2RX}\right)(\delta I + I_n)\right]^{2/3} \quad (11)$$

where K is the ratio of the temperature rise due to self heating to δT, and I_n, represents the random noise current through the null detector. Electrical noise is produced by many types of phenomena, but, in the ultimate limit, can be no smaller than the thermal (Johnson) noise present in any resistor. The magnitude of I_n is also determined by the band width, or response time, of the null detector. Equation (11) shows that δT is decreased by increasing D, K, and S, as would be expected. The effects of R, X, and N are more complicated, but further study shows that the quantity, $N(\delta I)^2$, which is the smallest power which can be detected by the null detector, is closely related to the temperature resolution which can be achieved. The most sensitive moving coil galvanometers can detect a power of about 10^{-18} watt, but dc electronic null detectors which detect 10^{-21} watt can now be obtained. In general it is best to have the resistance of the ratio arms, R, considerably larger than the thermometer resistance, X. The most effective internal resistance of the null detector is about $2X$ for instruments which can detect a minimum of 10^{-20} watt or greater, and about 3 to 4 times X for instruments which can detect lower levels.

The resistance of resistance thermometers can be measured by means of a simple Wheatstone bridge of suitable accuracy and sensitivity, or by means of some modification designed to achieve a particular objective. C. H. Miller, (15), has discussed the limits of accuracy in the measurement of resistance. He discusses the effects of circuit and detector noise, power dissipation by the resistor, and detector bandwidth. He concludes that, with the present technology, an error of less than about one part in 10^8 is unlikely.

The Mueller bridge is designed for accurate measurements of absolute temperature with a platinum resistance thermometer. It is generally useful in the range of about 1–400 ohms. It employs a circuit which reduces the effects of switch contact resistances in the measuring arm, and it can be used with three- or four-lead thermometers to cancel the resistances of the lead wires. The principles and circuit details are described in references (3) and (13) and also in the manuals furnished by the manufacturers. The Leeds and Northrup Co. supplies several models of the Mueller bridge. The G1 bridge is the most economical one and is suitable for measurements of moderate accuracy (0.01°C). The G2 bridge has six decades, with the lowest in steps of 10^{-4} ohm. Over the past few years the G2 bridge has become well established in many areas of calorimetry. The G3 bridge has an additional decade, giving increments of 10^{-5} ohm. They also offer an automatic recording Mueller bridge with a resolution of 0.001 ohm. In cooperation with the National Bureau of Standards, Leeds and Northrup has developed a very elaborate automatic balancing and digital recording bridge. It has been described by Williams and Mergener (16). Honeywell, Inc., and the J. G. Biddle Co. also sell Mueller bridges, which are about equivalent to the Leeds and Northrup G1.

A modification of the Wheatstone bridge, originally described by C. G. Maier (17), and further described by Wm. F. O'Hara, Ching-Heien Wu, and L. G. Helpler (18), has twice the sensitivity of the circuit shown in Fig. 3. It uses two thermometers, placed together physically, but connected in opposite arms of the bridge. The bridge circuit is easily adapted for differential measurements by connecting the two matched thermometers in adjacent arms, with a fixed resistor in series with one thermometer and a variable resistor in series with the other. A. Michels and J. Strijiland (19), describe a more elaborate bridge which they used to measure very precisely the difference in temperature between two twin calorimeters. The two platinum coils functioned both as thermometers and as heaters.

Any of these modifications may be operated in the usual manner by adjusting the variable resistance until balance, as indicated by the null detector, is found. Or, the signal which results from deviations from balance may be read by an appropriate meter, recorded on a strip chart recorder, or used for control purposes. Equation (10) shows that, for small deviations, the current is approximately proportional to the change in resistance of the thermometer. This procedure is useful for measuring small changes in relative or differential measurements. A. B. Kaufman (20), describes the use of a half bridge, with a constant current supply, for measurement of resistances. This has a larger power output and better linearity, than does the usual full bridge, for a given change in resistance.

The Sensitive Research Instrument Corp. manufactures a bridge specially designed for use with the standard 25.5 ohm platinum resistance thermometer. It has been described by T. H. Dauphinee and H. Preston-Thomas (21). When balanced, it reads temperature directly, thus eliminating the troublesome step of converting resistance to temperature. An accuracy of 0.001°C in the range from −50 to 711.11°C is claimed. Although it is called a bridge, it actually works on the potentiometric principle, described below.

In recent years several resistance thermometers have been placed on the market which read temperature directly. J. M. Janicke (22), describes such an instrument which is manufactured by Radio Frequency Laboratories Inc. It uses a platinum resistance thermometer and displays temperatures from −100 to 500°C with an accuracy of 0.5°. The Yellow Springs Instrument Co. manufactures several "Thermilinear" systems which produce either a voltage proportional to temperature, a strip chart record, or a digital display with printed or punched tape record of the temperature. It uses a thermistor as the sensitive element. United Systems Corp. sells instruments which can employ either a platinum resistance thermometer, a thermistor, or a thermocouple, and gives a digital display of temperature. A variety of probes, of different sizes, shapes, and temperature ranges are available. The model which uses a platinum resistance thermometer has an accuracy of about 0.1°C in the range from −140 to 440°C. Princeton

Chemical Instrumentation

Applied Research Corp. supplies a digital reading thermometer and temperature controller which uses a platinum resistance element as sensor. It operates in the range of −192 to 1000°C with an accuracy of 0.1°C. It uses an automatic balancing Kelvin bridge circuit which is described by E. W. Jones (23). The Parr Instrument Co. has recently developed a temperature measuring system for use with combustion calorimeters in the range of 20–34°C. It consists of a thermistor probe and digital read-out device which indicates relative temperatures. Two models are available; one reading to 0.001° and the other to 0.002°C. A digital printer which permits recording of temperatures at selected times, can also be purchased.

Both direct and alternating current bridges are used for measurements of low and moderate accuracy. Measurements of high accuracy are usually made with direct current. When using alternating current in the bridge, allowance must be made for the effect of inductance and capacitance in the circuit. On the other hand the use of alternating current eliminates some sources of direct current interference, such as thermoelectric potentials. Alternating current electronic amplifiers are easier to construct than direct current amplifiers. Methods of optimizing the signal to noise ratio in resistance thermometer measurements using ac bridges have been described by J. M. Diamond (24). Alternating current bridges for precision measurements have been described by D. L. H. Gibbings (25), and J. J. Hill and A. P. Miller (26). The one discussed by Hill and Miller is designed specifically for platinum resistance thermometry, and is manufactured by the Automatic Systems Laboratories Ltd. Two models are available. The manually operated model has a resolution of 1 part in 10^8 and an accuracy of 4 parts in 10^7 in the measurement of resistance ratios. The automatic balancing model gives a direct digital indication of temperature with a resolution of 1 part in 10^7 and an accuracy of 4 parts in 10^7. An energizing current of 380 Hz is used in these bridges. The effect of lead resistance on the measurement may be made very small by use of four-lead thermometers. These instruments appear to be very versatile and suitable for use in a wide variety of conditions found in calorimetry.

The potentiometric method of measuring resistance offers some advantages in thermometry. A constant current is passed through the thermometer and an external standard resistance which is connected in series. The resistance of the thermometer may be calculated from the measured ratio of the potential drops across the thermometer and the standard resistance. If a potentiometer is used for this measurement the procedure may be simplified by connecting the standard resistance in place of the standard cell. A potentiometer is somewhat more economical than a bridge of the same accuracy, although the potentiometric method requires a more sensitive null

detector and a source of accurately controlled current. It has a very definite advantage for the measurement of low resistances (below 1 ohm). Since no current flows in the potential leads, their resistance has very little effect on the measurement. Finally, automatic recording is easier to achieve with the potentiometric method than with the bridge method. Princeton Applied Research Corp. manufactures a completely self-contained instrument for measurement of resistance ratios by this method. It consists of a constant current power supply to use for the energizing current, and a variable potential source for the determination of the voltage ratios. Accuracies of one part in 10^6 are specified for the measurement of resistance ratios.

Platinum resistance thermometers made to the specifications of the International Practical Temperature Scale consist of a coil of carefully purified strain-free wire sealed in an inert atmosphere. They have a resistance of about 25.5 ohms at 25°C. Such thermometers, with appropriate measuring instruments, furnish the most accurate means of measuring temperature in the range of −182 to 630°C. It seems likely that the platinum resistance thermometer will become standard throughout the range from 15°K to 1000°C. The National Bureau of Standards will now certify these thermometers in various ranges from 15°K to 630°C. Leeds and Northrup makes several types of platinum resistance thermometers to these specifications; some with a glass protecting case and some with metal. They also make a thermometer specially designed for use in calorimeters. The resistance coil is sealed in a platinum capsule about 0.55 by 6 cm. Engelhard sells a series of platinum wire coils in various sizes, shapes, and resistances, which can be used to construct resistance thermometers to particular specifications. Minco Products sells a series of small resistance thermometers of platinum and other metals. These are supplied in metal cases; and some are designed for particular applications, such as measurement of surface temperatures. Thermometers suitable for relative or differential measurements can be constructed from nickel or copper wire. It is possible to use these elements in calorimeters both as a heater and as a thermometer, not, of course simultaneously. The temperature coefficient of resistance of nickel is about 50% greater than that of platinum. The Driver-Harris Co. sells nickel wire, and Sigmund Cohn sells both nickel and platinum wire, designed for construction of thermometers.

Thermistors

The resistance of certain semiconductors made from impurity doped metallic oxides is strongly dependent on temperature. A thermistor is made by connecting two lead wires to a small bead of this material. The assembly is usually fused into a glass bead having an overall diameter of one millimeter or less. In contrast to metals, which have a positive temperature coefficient of resistance, thermistors have large negative temperature coefficient in a limited temperature range. This is typi-

cally in the order of 3–5% per degree. They are usually made to have a resistance of a few hundred to a few thousand ohms in the range over which they are useful. The relation between resistance and temperature is highly nonlinear and difficult to represent accurately with a simple mathematical function. The equation,

$$\frac{R}{R_0} = \exp\left[B\left(\frac{1}{T} - \frac{1}{T_0}\right)\right] \quad (12)$$

is applicable over a limited temperature range. R is the resistance at temperature T, and R_0 the resistance at T_0. B is an empirical parameter.

Since the stability and reproducibility of the resistance of thermistors are less than for metallic resistance thermometers, they are not suitable for accurate measurement of absolute temperature. It is difficult to state, in general, the magnitude of this instability. Experiences of different people vary considerably. This is due partly to the use of thermistors supplied by different manufacturers, although the quality of thermistors is continually improving. The resistance of a thermistor fluctuates with time. Both the resistance and noise generated by thermistors at a given temperature are dependent on their previous treatment. The following phenomena may exert an appreciable effect: temperature cycling, mechanical shock, exposure to light, and passage of an electric current. When using thermistors for precise measurement, the best plan is to examine the effect of these, and any other, plausible factors on the behavior of the particular thermistor being used, and to control those which have significant effect.

The high resistance usually found for thermistors means that variations in lead resistance are much less important than they are with most metallic thermometers. Their small size and rapid response to temperature changes is an advantage for some applications. Since the temperature coefficient of thermistors is about ten times the coefficient for metallic thermometers, a given level of resolution is achieved more easily with a thermistor than with a metallic resistance thermometer. On the other hand, the low dissipation factor, typically in the range of 10^{-4}–10^{-2} watt per degree, limits their usefulness for detecting very small temperature changes. Calculations show that the best resolution which can be obtained under ideal conditions with a typical thermistor is about the same as that which can be achieved with a metallic resistance thermometer having a dissipation factor in the range 0.01–0.05 watt per degree. Using a Wheatstone bridge with optimum resistance in the ratio arms, the most sensitive direct current null detectors now available, and a current which will give a self-heating such that $K = 5$, eqn. (11) predicts that temperature changes of around 10^{-5} degree could be detected. This resolution has actually been attained in recent years with both thermistors and metallic thermometers.

Thermistors are offered by many suppliers. Both Fenwal Electronics Inc., and Victory Engineering Corp. carry a full line of thermistor probes of many physical sizes and shapes, and electrical

Chemical Instrumentation

characteristics. Victory Engineering also supplies sets, closely matched in the relation of resistance to temperature, which are useful for differential measurements. Yellow Springs Instrument Co. sells a variety of assembled thermistor thermometers, as well as indicating and control instruments. Conax Corp. furnishes thermistors sealed into small diameter metal tubes. These may be ordered in many sizes and in temperature ranges from -250 to $450°C$. They should find many applications in calorimeters. Texas Instruments, Inc., has recently marketed a resistance element, presumably based on a semiconductor, which has a positive temperature coefficient.

Because of its low resistance and low-temperature coefficient, the platinum resistance thermometer is not suitable for use below about $15°K$. Carbon resistance thermometers have been employed in the range from $1–10°K$ for a number of years. Within the past five years several new types of resistance thermometers have been introduced for use at low temperatures. Texas Instruments, Inc., Radiation Research Corp., and Honeywell Inc., have all developed several types of impurity doped germanium thermometers. The ones sold by Texas Instruments are enclosed in glass, while the ones sold by the other two are in metal cases. These are useful in the range from 1 to $40°K$. Honeywell also makes a germanium thermometer for use up to $100°K$. The resistance of these thermometers changes by a factor of about 1000 when the temperature is raised from 2 to $40°K$. The complicated resistance-temperature relation poses an inconvenience in using these thermometers. Doped silicon resistors have also been proposed for this range. They seem to have a smoother resistance-temperature relation.

Single crystals of very pure germanium have a temperature coefficient of resistance of about 4% per degree from -80 to $400°C$. Dr. F. J. Prosen of the National Bureau of Standards has suggested that these could be used as resistance thermometers in this range. To be suitable for this purpose the germanium should have no more than 10^{-9} mole fraction impurity and should have a resistivity of at least 40 ohm cm at $25°C$. Single crystals of germanium suitable for this use can be purchased from Eagle-Picher Industries. They supply round or square rod in several standard sizes. Copper lead wires can be connected by using a solder of pure tin, with an acid flux. The temperature coefficient is in the range of the typical thermistor. However since it can be made in a larger size it has a much larger dissipation factor. Therefore it is possible to detect a smaller change in temperature with the germanium thermometer than with either a metallic resistance thermometer or a thermistor.

Quartz Crystal Thermometers

The Dymec Division of the Hewlett Packard Co. has recently introduced a quartz crystal thermometer, which should find many applications to calorimetry. It is based upon a principal which is relatively new to precision thermometry. The instrument is described by D. L. Hammond, C. A. Adams, and P. Schmidt (27). It utilizes the dependence of the resonant vibrational frequency of a quartz crystal on temperature. When using a quartz crystal as a frequency standard it is cut so that the temperature dependence of the frequency is very small. However it is possible to cut a crystal so that the frequency is very nearly a linear function of temperature. In the Dymec Instrument the frequency changes by 1000 Hz per degree. Pulses generated by the quartz crystal are compared to those generated by a frequency standard for a certain period and the difference is displayed by direct digital readout. Since the standard frequency corresponds to the zero of the temperature scale, the instrument gives a direct reading of temperature. Timing periods of 0.1, 1, and 10 sec give resolutions of 0.01, 0.001, and $0.0001°C$ respectively. Two models

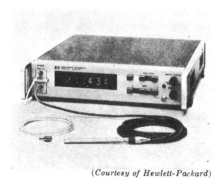

(Courtesy of Hewlett-Packard)

Figure 4. Quartz thermometer.

are available. The dual input model, illustrated in Figure 4, has provisions for two probes and can be switched to read either one individually, or to read the difference between them. The other model accepts one probe only. Several types of sensing probes are available. Specifications indicate that accuracies of 0.01 or $0.02°C$ can be expected in the range of 0 to $100°C$ if the zero point is recalibrated every few weeks, with somewhat less accuracy in a wider range. It is applicable in the range from -200 to $250°C$. This instrument has not been available long enough for much evidence on its reliability to accumulate. Comparison with platinum resistance thermometers in our laboratory indicates that relative or differential temperatures may be measured to an accuracy of $0.0002–0.0005°C$. The accuracy and repeatability seems to be different for different probes.

One characteristic of this instrument appears to be very useful for calorimetry. The count displayed at the end of a timing period is actually a measure of the integral, $\int T \, dt$, over this period. The sum of readings obtained for several successive timing periods will be equal to the integral over the more extended period. If one probe is placed in the calorimeter and the other in the jacket, the integral appearing in eqn. (8) of Part I, which is needed to calculate the heat exchange correction,

can be obtained very easily by adding up successive readings obtained in the differential mode. The instrument requires a re-set time between successive timing periods. It can be adjusted, but the minimum is about 0.2 sec. When using 10-sec timing periods the counts lost during the re-set periods can be easily estimated. This instrument can also be readily connected to digitial recording devices, which produce either a printed record, a punched tape, or punched cards.

General information with regard to thermometry can be found in references (3), (10), (12), (13), (14), (28), (29) and (30). The following tabulation lists a few sources of thermometric components and equipment; it is not exhaustive.

Sources of Thermometers and Associated Instruments.

Atkins Technical, Inc., P.O. Box 14405, Gainsville, Fla. 32603. (Thermistors, temperature indicators.)

Automatic Systems Laboratories, Ltd., Construction House, Grovebury Rd., Leighton Buzzard, Bedfordshire, England. (High precision ac bridges.)

James G. Biddle Co., Township Line and Jolly Rd., Plymouth Meeting, Pa. 19462. (Wheatstone and Mueller bridges.)

Brooklyn Thermometer Co., 90 Verdi St., Farmingdale, N. Y. 11735. (Mercury-in-glass thermometers.)

Conax Corp., 2300 Walden Ave., Buffalo, N. Y. 14227. (Thermistors and thermocouples.)

Cryogenics, Inc., 1128 Vermont Ave., NW, Washington, D. C. (Gold-cobalt alloy thermocouples.)

Driver-Harris Co., Harrison, N. J. 07029. (Alloy wires.)

Dymec Div., Hewlett-Packard Co., 395 Page Mill Rd., Palo Alto, Calif. 94306. (Quartz thermometer.)

Eagle-Picher Industries, Inc., Compounds and Metals Dep't., P.O. Box 737, Quapaw, Okla. 74363. (High-purity single crystal germanium rods.)

Englehard Industries, Inc., Instruments and Systems Div., 850 Passaic Ave., East Newark, N. J. 07100. (Platinum and alloy wires, thermocouples, metallic resistance thermometers.)

Fenwal Electronics, Inc., 51 Mellen St., Framingham, Mass. 01701. (Thermistors, temperature indicating and control instruments.)

General Electric Co., Industrial Sales Operation, 1 River Rd., Schenectady, N. Y. 12305. (Thermocouple wire.)

General Electric Co., Magnetic Materials Section, P.O. Box 72, Edmore, Mich. 48829. (Thermistors.)

Honeywell, Inc., Test Instruments Div., 4800 E. Dry Creek Rd., Denver, Colo. 80217. (Germanium resistance thermometers, Wheatstone and Mueller bridges, potentiometers.)

Hoskins Manufacturing Co., 4445 Lawton Ave., Detroit, Mich. 48208. (Thermocouple wire.)

The Kanthal Corp., 1 Amelia Place, Stamford, Conn. (Thermocouple wire.)

Leeds and Norhrup Co., North Wales, Pa. 19454. (Thermocouples and thermocouple wire, platinum resistance thermometers, Mueller bridges, potentiom-

Chemical Instrumentation

eters, optical pyrometers.)

Minco Products, Inc., 740 Washington Ave. N., Minneapolis, Minn. 55401. (Resistance thermometers and small heaters.)

Omega Engineering, Inc., P.O. Box 47, Springdale, Conn. 06879. (Thermocouple wire, thermocouples, tungsten and tungsten-rhenium alloys, insulatores.)

Parr Instrument Co., 211 Fifty-third St., Moline, Ill. 61265. (Mercury-in-glass thermometers for calorimetry, thermistor based digital temperature indicators.)

Princeton Applied Research Corp., P.O. Box 565, Princeton, N. J. 08540. (Platinum resistance thermometer, digital temperature indicators, temperature controller, potential and resistance measuring instruments.)

Radiation Research Corp., 312 Florida Ave., West Palm Beach, Fla. (Germanium resistance thermometer for low temperatures.)

Rosemount Engineering Co., 4900 W. 78th St., Minneapolis, Minn. 55424. (Platinum thermometers and bridges.)

Sensitive Research Instrument Corp., 310 Main St., New Rochelle, N. Y. 10801. (Resistance thermometer bridge, potentiometers.)

Sigmund Cohn Corp., 121 S. Columbus Ave., Mount Vernon, N. Y. (Thermocouple wire, gold-cobalt alloys, alloys of platinum, rhodium, iridium, silver, and gold, platinum and nickel resistance thermometers.)

Texas Instruments, Inc., P.O. Box 5012, Dallas, Texas 77006. (Thermistors and semiconductors.)

Thermo-Electric Co., Inc., Saddle Brook, N. J. 07662. (Thermistors, thermocouples, thermocouple pyrometers, thermocouple reference unit.)

Tri-R Instruments, 48 Merrick Rd., Rockville Centre, N. Y. 11570 (Thermistor thermometers.)

United Systems Corp., 918 Woodley Rd., Dayton, Ohio 45403. (Thermistors, platinum resistance thermometers, thermocouples, digital temperature indicators.)

Victory Engineering Corp., 118 Springfield Ave., Springfield, N. J. 07081. (Thermistors.)

Wilbur B. Driver Co., 1875 McCarter Hwy, Newark, N. J. 07104. (Thermocouple wire.)

Yellow Springs Instrument Co., P.O. Box 279, Yellow Springs, Ohio 45387. (Thermistors and temperature indicators.)

THE MEASUREMENT OF ELECTRICAL ENERGY

The calorie was originally defined in terms of the specific heat of water at some specified temperature. However, for many years virtually all calorimetric measurements have been based on electrical calibrations. Since about 1930 calorimetric data have been reported in units of either the joule or the thermochemical calorie. The thermochemical calorie is equivalent to 4.184 (exactly) absolute joules, or 4.1833 international joules.

Electrical energy is usually supplied to a calorimeter by passing a direct current through a coil of resistance wire placed in the calorimeter. The electrical energy consumed in time, t, is

$$w = \int_0^t IE\mathrm{d}t \qquad (13)$$

where I is the current and E the potential drop across the heater. The current may be determined by measuring the potential drop across a standard resistance connected in series with the calorimeter heater. Figure 5 is a diagram of the essential elements of a circuit suitable for measuring the electrical energy dissipated in a calorimeter. The power supply may consist of a large storage cell, or of a filtered rectifier circuit designed to deliver either constant voltage or constant current. Current is first passed through the dummy heater, which is similar to the calorimeter heater but located outside the calorimeter, in order to adjust and stabilize the power supply. Switch number 1 is used to divert the current from the dummy heater to the calorimeter heater, and also to start the timer. Switch number 2 is used to connect the potential measuring instrument to either the calorimeter heater or the standard resistor.

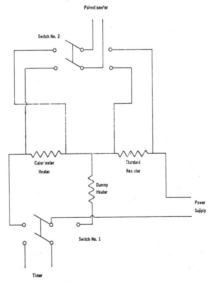

Figure 5. Calorimeter heater circuit.

Accuracies in the measurement of energy of around one part in five hundred can be easily made with readily available components. Greater accuracies require that attention be paid to numerous sources of error, especially if a large quantity of energy is to be delivered in a short time. These arise from transients in the power supply, from difficulties in obtaining a complete record of potential and current at various times during the heating period, and from conduction of some of the heat out of the calorimeter. Measurement of the time of heating may also be a problem when accuracies approaching one part in ten thousand are the goal. The use of a high quality constant current power sup-

ply simplifies the recording of data since only the potential need be measured during the heating period. This measurement is simplified still further by use of an electronic differential voltmeter or digital voltmeter. These instruments usually have a recorder output which permits continuous recording of the voltage during the heating period.

The design and construction of a heater suitable for high precision measurements is very critical. The resistance of the heater changes as a result of the rise in temperature which occurs when current is passed through it. To keep this as small as possible it should be constructed from an alloy wire which has a low temperature coefficient of resistance. Manganin and constantan are often used to construct heaters which operate near room temperature. Wire made from Karma alloy, sold by the Driver-Harris Co. has a minimum temperature coefficient at temperatures of 50–100°C, and also a higher resistivity than manganin or constantan.

Good thermal contact between the heater wire and the calorimeter is important in order to minimize the increase in temperature of the heater wire. This not only stabilizes the resistance of the heater but also reduces the conduction of heat away from the calorimeter. At the same time, good electrical insulation must be maintained between the heater wire and the calorimeter. The heater wire should be spread over as large a surface as possible, consistent with the space limitations in the calorimeter. Immersion of the wire directly in a heat transfer liquid such as refined mineral oil or silicone oil is often helpful. The 3M Company Supplies a line of fluorinated liquids which are also good for this purpose.

The design and placement of the electrical leads between the calorimeter and the jacket need consideration. Generally four such leads are used; two for the current and two for the potential measurement. The lead wires should be as small as practical so that they do not conduct heat away from the calorimeter. The size of the potential leads is limited primarily by the requirement of mechanical strength and convenience in handling. The current leads, however, should be large enough to prevent generation of an appreciable amount of heat. The heat leakage can be reduced further by placing the lead wires in good thermal contact with the outside surface of the calorimeter. The point at which the potential leads are connected to the current leads must also be carefully considered.

Other methods of supplying electrical energy quantitatively to the calorimeter may have certain advantages. For example an alternating current watt-hour meter can be used. Precision watt hour meters have been developed for purposes of standardization and calibration of large amounts of electrical power. The Sangamo Electric Co., Springfield, Ill., supplies a rotating standard watt-hour meter which operates on 60 cycle current of 115 nominal volts with a one half ampere current coil. This can be used satisfactorily down to power levels of around 10 watts. With suitable calibration and care, rotating standard watt hour meters are good to an accuracy of one part per

Chemical Instrumentation

thousand. A number of circuits have been developed which supply power in the form of square wave pulses, with a precise quantity of energy carried by each pulse. The total energy delivered during the heating period can be measured with a pulse counter. The equipment described by Wm. Prengle, Jr., F. L. Worley, Jr., and C. E. Mauk (31) is a typical example. These methods introduce considerable convenience into the procedure, since no measurements are required during the heating period, and the total energy can be simply read at the end of the experiment. In addition, the power level can be varied during the heating. This is a considerable advantage in twin calorimetry, where the electrical energy is to be balanced against the heat effect being measured.

CALORIMETRIC STANDARDS

The use of a standard substance to determine the heat capacity of the empty calorimeter by either a Type I or Type II calibration experiment was described in Part I. These calibrations use a standard substance which has either an accurately known heat capacity, or which evolves or absorbs an accurately known quantity of heat when it undergoes an isothermal transformation. Reference substances are also used to test the performance of the calorimeter by comparing the measured value with the accepted value. Thus it is convenient to have two or more standards; one for calibration and others for testing. A Series of standards of differing properties makes it possible to choose one which is similar to the system being studied. On the other hand too much proliferation of standards is undesirable, because of the uncertainties introduced in establishing their properties and the difficulties involved in obtaining intercomparison among different laboratories. Many standards have been introduced by the IUPAC Subcommittee on Thermodynamics and Thermochemistry and, in recent years, by the U.S. Calorimetry Conference, working in close cooperation with the National Bureau of Standards.

The requirements for, and usefulness of, standards are different in different areas of calorimetry. The list of standards in use is constantly being revised, although the introduction and acceptance of new standards is a slow process. The current situation will be reviewed in various areas of application. A short bibliography is presented at the end of this section. Values of properties of some key compounds, including some of those used as standard materials, will be found in reference (45).

Heats of Combustion. Because of the great difficulty of direct electrical calibration to the desired accuracy, combustion calorimeters are nearly always calibrated by burning a primary standard of accurately known heat of combustion. In effect this transfers the energy unit from the standardizing laboratories to others who are conducting measurements. Benzoic acid is a nearly ideal standard, and has been used for this purpose for over forty years. It is stable at room temperature, can be obtained in a state of high purity, and is nonhygroscopic. Furthermore it burns cleanly to water and carbon dioxide and its heat of combustion is highly reproducible. The National Bureau of Standards supplies samples certified for heat of combustion. The currently available batch, designated 39i, has a heat of combustion of -6317.8 cal gm^{-1} (vacuum) when burned under standard conditions. These are; a bomb whose internal volume is 300 ml per gram of sample burned, a reference temperature of 25°C, and an initial state of 30 atm of oxygen with 1 ml of water present in the bomb. Standard samples are also supplied by the National Physical Laboratory in England.

In addition to the primary standard, a number of secondary standards have been used for testing of combustion calorimeters. Several of these are described below.

2,2,4-Trimethylpentane is used for testing combustion of volatile hydrocarbons. Certified sample available from the National Bureau of Standards.

Naphthalene is used for testing combustion of solid hydrocarbons. Certified samples formerly supplied by the National Bureau of Standards.

Succinic acid is used for testing combustion of solid materials containing carbon, hydrogen, and oxygen. Recent measurements and a review of older data is given in reference (32). They recommend a standard state energy of combustion of -3020.50 ± 0.50 cal gm^{-1} (vacuum) at 25°C.

Sucrose is used to test combustion of solid compounds containing a large proportion of oxygen. It has not been much used in recent years. Certified samples were formerly supplied by the National Bureau of Standards.

Hippuric acid was proposed to test the combustion of nitrogen-containing compounds. It has not been widely used in recent years.

Sym-Diphenylthiourea and thioglycolic acid have been proposed for testing combustion of sulfur containing compounds.

p-Fluorobenzic acid is suitable for testing combustion of compounds of low fluorine content, and *m*-trifluorotoluic acid for compounds of intermediate fluorine content. These last two compounds are discussed in reference (33).

Trichlorophenol is used for testing combustion of compounds containing chlorine.

Heat Capacity. Primary standards are used to determine the heat capacity of the empty calorimeter and secondary standards to test the operation of the calorimeters. A list of recognized standards, with some of their properties, is given in Table 4.

n-Heptane was recommended as a heat capacity standard by the Fourth Calorimetry Conference in 1949. It is suitable in both liquid and solid forms. It is not yet certified by the National Bureau of Standards, but they will supply limited quantities to qualified investigators.

Diphenyl ether was suggested as a comparison standard by Ginnings and Furukawa (34). It can be used to a somewhat higher temperature than *n*-heptane, but has not found wide application.

Water can be obtained in a state of higher purity than can any of the other standards. It has long been recognized as a standard in the range of 0–100°C. The relatively high volatility is a disadvantage in some applications.

Benzoic acid was recommended as a standard for the heat capacity of solids at low temperature by the Fourth Calorimetry Conference. It has a high heat capacity per unit volume below 100°K. It is not suitable for use above 350°K because of its volatility and corrosiveness.

α-Aluminum oxide (synthetic sapphire, or corundum) has been used as a standard since about 1950. It is the only generally accepted standard for heat capacity at high temperatures. Limited quantities of the Calorimetry Conference sample may be obtained from the National Bureau of Standards. At the request of the Calorimetry Conference in 1964, the National Bureau of Standards has agreed to issue aluminum oxide as a certified standard reference material. This should be available within a year or two and will be in the form of annealed rods, about 0.075 by 0.20 inches.

Copper was recommended by the Calorimetry Conference in 1965 for use below 20°K. Limited quantities of high purity copper suitable for this purpose are being distributed by H. E. Flotow and D. W. Osborne of the Argonne National Laboratory, Argonne, Ill.

SOLUTION CALORIMETRY

Solution calorimeters are nearly always calibrated by electrical heating. The acceptance and use of standards here are much more limited than in heat capacity and combustion calorimetry. In fact, some investigators argue that reliance should be placed entirely on careful design and operation of the calorimeter, rather than on the use of test reactions. However most calorimetrists find that the use of a suitable reference material helps to determine whether systematic errors have actually been eliminated. Gunn (39) has listed desirable characteristics of reference materials and has reviewed several specific standards.

The solution of potassium chloride has, in effect, been used as a test material for endothermic reactions for many years.

Table 4. Heat Capacity Standards

Substance	Useful Range of Temperature, °K	Melting Point °K	References
n-Heptane	20–400	182.56	(34)
Diphenyl ether	20–450	300.03	(34)
Water	273–373	273.15	(34, 35)
Benzoic acid	20–350	395.52	(34)
α-Aluminum oxide	20–1200	2300.	(32, 34, 36, 37, 38)
Copper	1–20	1356.	(see text)

Chemical Instrumentation

The heat of solution has been measured many times and the results are viewed in reference (*40*). Additional information is given by Somsen, Coops, and Tolk (*41*). The agreement among different values is not as good as it should be for a test material, and Gunn concludes that it is not as suitable for this use as was once believed. The heat of solution of KCl is markedly influenced by traces of water occluded in the solid, and removal of the last traces of such water is notoriously difficult. The heat of solution is also influenced by strains in the crystal which result from mechanical treatment or crystal imperfections. The rather high temperature coefficient of the heat of this process and the rather slow rate of solution are other disadvantages. The heat of solution of one mole of single crystal KCl in 200 moles of water at 25°C is 4200 ± 6 calories.

Ray (*42*) has devised a rather ingenious method of using the evaporation of water for checking the operation of certain types of solution calorimeters. The water is removed from a glass helix evaporator, immersed in the liquid in the calorimeter, by means of a vacuum pump.

The solution of THAM, tris(hydroxymethyl)aminomethane, is endothermic in pure water and exothermic in hydrochloric acid solution. It offers many advantages as a calorimetric standard and also as an acidimetric standard. It can be obtained in a highly pure state and dried by heating to 80°C and then cooling in a vacuum desiccator. The heat of solution of one mole of THAM in 0.1*M* HCl at 25°C is −7107 ± 3 calories, when the final concentration is 5 g per liter. It is fairly insensitive to the concentration of HCl and THAM. Limited quantities for testing calorimeters may be obtained from Dr. E. J. Prosen of the National Bureau of Standards. The bureau is preparing to issue certified samples for use in calorimetry and acidimetry.

Gunn (*39*) recommends the use of the reaction of sulfuric acid with excess sodium hydroxide solution for testing the accuracy of calorimeters. Solutions of accurately known acid concentration can be prepared by dilution of constant boiling sulfuric acid. The concentration of the sodium hydroxide, or the presence of sodium carbonate, has only a small effect on the heat evolved.

Shomate and Huffman (*43*) have reported values for the heat of solution of magnesium in hydrochloric acid. This reaction may be used for testing calorimeters designed to study this type of reaction. Measurement of the heat of solution of metals in acid is difficult because the hydrogen carries away some water vapor. The heat of solution of succinic acid in dilute HCl may have some value as a test reaction for an endothermic solution of a solid.

Mixing Calorimeters

The mixing of benzene with carbon tetrachloride has been suggested as a means of testing calorimeters which measure the heat of mixing of volatile liquids. However it has been found to be unsatisfactory because of the reaction of carbon tetrachloride with mercury, which is often present in such calorimeters. The systems hexane–cyclohexane and water–acetone have also been suggested and are being investigated. No general concensus has been reached as yet.

Vapor Flow Calorimeters

Waddington, Smith, Williamson, and Scott (*44*) have suggested carbon disulfide as a suitable references material for testing vapor flow calorimeters. They have published accurate values of the vapor heat capacity and other thermodynamic properties. Water and benzene might also be used for this purpose but have some disadvantages. Water does not boil as smoothly in the boiler as do most organic liquids, and the heat capacity of steam is highly dependent on pressure. The heat capacity of gaseous benzene is not as well established as is that of carbon disulfide because its more complicated molecular structure causes greater uncertainties in the statistical calculations.

Part Three—Some Specific Types of Calorimeters

A brief survey of various kinds of calorimeters and fields of thermochemical research was given in Part I of this series. Some basic terminology was also explained. Part II consisted of a discussion of thermometry and the measurement of electrical energy, as applied to calorimetry, and a list of calorimetric standards. Some additional constructional details are given in this final section, and several commercial instruments are described.

SOME DESIGN OBJECTIVES

It is difficult to state, in general terms, desirable design objectives which apply to all of the many types of calorimeters. The following criteria are offered to help evaluate static calorimeters which have either an isothermal or adiabatic jacket. Of course, a balance must always be made among cost of construction, accuracy, and convenience of operation.

Characteristics of the Calorimeter

(1) In all Type I experiments (measurement of heat capacity) or in Type II experiments (measurement of energy change at constant temperature) involving small energy changes, the heat capacity of the calorimeters used should be small compared to the heat capacity of the sample.

(2) The temperature of the outside surface of the calorimeter should remain uniform at all times. This uniformity is accomplished by making the outer wall in a

* Temperature gradients disappear most quickly from materials of high thermal diffusivity. Some values of this property were listed in Table II of Part II. At room temperature the thermal diffusivity of glass is about 6.2×10^{-3} cm² sec⁻¹ and its thermal conductivity is 2.76×10^{-3} cal deg⁻¹ cm⁻¹ sec⁻¹.

simple symmetrical shape. Materials of high thermal diffusivity are desirable.* Irregular projections and large masses of a poorly conducting material such as glass or plastic should be avoided. The cover is often troublesome in these respects, since it is likely to be thermally isolated from the rest of the calorimeter and to be irregularly shaped because of the attachments of heaters, supports, fasteners, stirrers, thermometers, and other parts. These requirements are less stringent for calorimeters which are filled with a stirred liquid than for those which contain only solids.

(3) Since the amount of heat exchange between the calorimeter and the jacket is related to the time required by the calorimeter to reach an equilibrium or steady state temperature following the occurrence of the process being studied, this time should be as short as possible. In general, larger calorimeters have a longer equilibration time. Calorimeters which contain a liquid should be stirred unless they are very small. Poorly conducting solids should be formed in thin sections and placed in good contact with highly conducting surfaces. Void spaces should be filled with a thermally conducting gas such as hydrogen or helium. This step is especially important for low-temperature calorimeters where heat transfer by radiation is ineffective.

(4) Stirrers should be carefully designed to produce effective mixing without too much frictional heat. The power consumed by the stirrer should remain constant. The heat generated in bearings should be insulated from the calorimeter as it is likely to be erratic. Rubbing of solid surfaces should be avoided. The calorimeter should also be sufficiently well sealed to prevent appreciable loss of liquid by evaporation. In some calorimeters, mixing is achieved by rocking or rotating the whole calorimeter.

Characteristics of the Jacket

(1) The temperature of an isothermal jacket should remain constant, while that of an adiabatic jacket should follow the calorimeter. In either case, the inside surface temperature should be uniform. The most effective way to attain a constant temperature is to immerse the jacket in a well-stirred, thermostated liquid bath. Proportional (or more elaborate) control of heating is recommended for precision work. Liquid baths are less useful for adiabatic jackets because the large heat capacity makes rapid temperature changes difficult to achieve.

(2) The jacket should completely surround the calorimeter, and components such as calorimeter supports, electrical lead wires, stirrer shafts, and thermometers should be placed in good thermal contact with the jacket before continuing on to the outside.

Interaction Between Jacket and Calorimeter

(1) There should be a definite, well-defined boundary of separation between the calorimeter and the jacket. As nearly as possible, all parts should be in good thermal contact with one or the other. Items such as lead wires, thermometer supports, stirrer shafts, etc., which must be located between the jacket and the calorimeter,

Chemical Instrumentation

should have a low heat capacity and should produce only a minor effect on the temperature distribution.

(2) The rate of conduction of heat per degree of temperature difference is represented by K in eqns. (6), (7), and (8) of Part I. Normally, it should be as small as possible. Conduction through solid connections may be reduced by making them of small size and of poorly conducting materials. Electrical lead wires should be no larger than is required to carry the current without excessive generation of heat. The calorimeter can be supported either by fine Nylon or steel wires or on sharp steel pins. A short plastic or glass section can be placed in stirrer shafts to reduce heat conduction. An even better material for the stirrer shaft is thin-walled, stainless steel tubing which makes a straight, rigid shaft of low heat capacity and low thermal conductivity. If a gas is present between the jacket and calorimeter, the separation should be no larger than about 5 mm in order to reduce convection currents. Conduction through the gas can be eliminated by evacuating this space. To be effective, the pressures should be such that the mean free path of the molecules is greater than the distance of separation, usually a pressure of 0.5 micron, or less.

According to the Stefan-Boltzmann law, the rate of radiation of thermal energy from a surface is $\epsilon\sigma T^4$, where the Stefan-Boltzmann constant, σ, is 1.36×10^{-12} cal sec^{-1} deg^{-4} and ϵ is the emissivity. The rate of heat radiation between two surfaces maintained at a constant temperature difference is proportional to the cube of the absolute temperature. Therefore the effect of radiation is much more important for high-temperature calorimeters than for low-temperature ones. The emissivity of a surface depends to some extent on the chemical nature of the material, but the physical condition of the surface—cleanliness, smoothness, and degree of polish—is more important. It can generally be assumed that a surface which has a dark color, dull appearance, and rough texture has a higher emissivity than one which has a light color, polished appearance, and smooth texture. Polished metal surfaces have emissivities in the range of 0.01–0.05 in cgs units.

Some General Characteristics

The properties of the calorimeter and jacket should be constant and reproducible following the removal or disassembly and the replacement of the calorimeter. This should be especially true for the heat capacity of the calorimeter and the thermal conductance between it and the jacket. The constancy of these properties in a series of experiments is a good measure of the accuracy which can be attained. The various components should be rigidly supported in a reproducible manner, and constructed so that they do not become distorted or changed by routine handling.

There are several ways of rating the overall performance of a calorimeter. The resolution can be described in terms of the smallest change which can be reliably detected either in enthalpy or in temperature. These two quantities are related by

the heat capacity of the calorimeter and its contents. The resolution in terms of temperature is more useful for comparing different calorimeters since it does not depend on the size of the calorimeter or on the heat capacity of the contents. The resolution is limited by the sensitivity of the thermometer and also by the noise level and the extent to which extraneous short term energetic effects can be controlled. Although limited by the resolution, the precision or reproducibility which can be achieved in a series of similar measurements is primarily dependent on the constancy of the properties of the calorimeter. The overall accuracy depends on the precision which can be achieved between the calibration and measurement experiments. It is also affected by the accuracy with which the heat exchange correction can be obtained and by the effects of phenomena such as stirring energy, and solvent evaporation. The preceding remarks apply to the calorimetric errors only. As described in Part I, additional errors arise from the characterization of the process which occurs in the calorimeter and from the procedure used to correct it to the standard state.

COMBUSTION CALORIMETRY

The technique of measurement of heat of combustion by burning liquid or solid samples in a bomb filled with oxygen at high pressure was introduced by Berthelot and his students in the final two decades of the 19th century. The modern type of bomb calorimeter was further perfected by Dickinson by 1915. These procedures have been standardized and refined in subsequent years, and heats of combustion can now be measured more accurately than almost any other thermodynamic property.

Many combustions are carried out to

determine the heating values of fuels. Accuracies in the range of 0.1–1% are usually sufficient for this purpose, and several companies manufacture suitable calorimeters for these determinations. Of more importance to thermochemistry is the use of heats of combustion to establish the heats of formation of many organic and some inorganic compounds. For this purpose, accuracies of better than 0.1% are needed. For example, an error of 0.01% in the measurement of heat of combustion may easily give rise to an error of 5–20% when carried through to the calculation of an equilibrium constant with the aid of the Second Law of Thermodynamics.

By 1929 heats of combustions had been reported for about 1400 organic compounds. However, the accuracy is low by modern standards, and they are of little use for present day thermochemical calculations. Many laboratories now routinely carry out measurements which are accurate to 0.01% or even better. There has been much interest in recent years in developing new techniques which can be applied to a much greater range of compounds.

In planning a thermochemical laboratory, first consideration should be given to the accuracy desired. As a chain is no stronger than its weakest link, the measurement of a physical quantity is no more accurate than the least accurate step. To avoid waste of effort, the various components should correspond to about the same level of accuracy. Types of compounds to be studied and the amount of automation desired are other considerations which will affect the planning. Highly automated equipment will save labor and increase the output of results but will cost more in the beginning.

The following is a brief description of the procedure for measuring the heat of combustion of an organic compound which

TABLE 5
Suggested Bomb Calorimeter Equipment and Limits of Error for Three Levels of Accuracy

	Low Accuracy	Intermediate Accuracy	High Accuracy
Range of overall error in heat of combustion	0.5–0.1%	0.1–0.03%	0.03–0.01%
Maximum error from any single contribution	0.05%	0.01%	0.002%
Suggested Equipment			
Calorimeter	1. Parr models 1300 or 1200 2. Gallenkamp 3. Prolabo "Control" model	1. Parr 1230 or 1200 with modification 2. Gallenkamp 3. Prolabo "Precision" model	1. Precision Scientific with accessories 2. Custom built types
Thermometer	1. Hg thermometers 2. Thermistors	1. Differential Hg thermometers 2. Pt or Ni resistance thermometers 3. Thermistors 4. Hewlett-Packard Quartz thermometer	1. Pt resistance thermometer 2. Ge resistance thermometer
Method of Ignition	Iron wire fuse	Iron or platinum wire fuse	Platinum wire fuse
Accuracy Required for Intermediate Measurements			
Properties of the sample and auxiliary materials:			
Specific heat	Not significant	0.12 cal deg^{-1} gm^{-1}	0.02 cal deg^{-1} gm^{-1}
Energy coefficient, $(\partial E/\partial P)_T$	Not significant	0.01 cal atm^{-1} gm^{-1}	0.002 cal atm^{-1} gm^{-1}
Mass of sample and auxiliary materials	1 mg	0.3 mg	0.08 mg
Mass of water in bucket	1 mg	0.2 gm	0.05 gm
Mass of water in bomb	0.2 gm	0.05 gm	0.02 gm
Initial oxygen pressure	2 atm	0.2 atm	0.05 atm
Heat of capacity of empty calorimeters	1.3 cal deg^{-1}	0.3 cal deg^{-1}	0.05 cal deg^{-1}
Initial and final temperatures	1.5×10^{-3} deg C	0.3×10^{-3} deg C	0.06×10^{-3} deg C
Timing of temperature readings:			
Isothermal jacket	10 sec	2 sec	0.5 sec
Adiabatic jacket	60 sec	20 sec	10 sec
Control of temperature of isothermal jacket	0.1 deg C	0.02 deg C	0.005 deg C
Correction for heat exchange between calorimeter and isothermal jacket	7%	2%	0.5%
Corrections for energy of side effects	3.5 cal	0.7 cal	0.14 cal
Nitric acid	2.5×10^{-4} moles	5×10^{-5} moles	10^{-5} moles

Chemical Instrumentation

contains only a combination of carbon, hydrogen, oxygen, and nitrogen, using the conventional oxygen bomb techniques. More complete details will be found in the literature furnished by manufacturers of calorimeters, in volumes 1 and 2 of "Experimental Thermochemistry" (4), in the reviews of Jessup (47) and Coops (48), and in the published reports of research investigations.

Table 5 may be of help in selecting a procedure and a set of equipment which can be used to attain a desired accuracy. Three levels of accuracy are shown. These levels will be designated as low accuracy, 0.5–0.1%, intermediate accuracy, 0.1–0.03%, and high accuracy, 0.03–0.01%. The figures represent the overall error in terms of the fraction of the heat of combustion of a pure compound which produces 3–5 kcal of heat per gram. A conventional water bucket type calorimeter is assumed, with a bomb of approximately 300-ml capacity containing 30 atm of oxygen and 2 g of sample. The combustion produces a temperature rise of 2–3°C. An estimate of the accuracy required for the various experimental parameters is given. These are selected so that the effect of any one source of error is appreciably less than the smallest overall error shown for that level of accuracy. Some suggestions for suitable types of equipment are also shown.

The three principal parts of the conventional bomb calorimeter are the bomb, the water bucket, and the jacket. A carefully weighed portion of the sample, in suitable form, is placed in a crucible which is supported inside the bomb. A small quantity of water is placed in the bottom of the bomb, to insure that the initial gas phase is saturated with water vapor. The bomb is closed and then filled with oxygen to an initial pressure in the range of 15–40 atm. The bomb is placed in the bucket, and the bucket is nearly filled with a weighed quantity of water. The bucket is then suspended in the jacket, and the cover, thermometer, stirrer, ignition leads, and other items are put in place. The stirring motor is started, and the temperature control of the jacket, isothermal or adiabatic as the case may be, is begun. The temperature of the water in the bucket is read and recorded at measured time intervals of 1–3 min. When the temperature of the calorimeter is changing at a satisfactory rate, the sample is ignited. The temperature then rises rapidly for about 1 min and then more slowly to the final value. The temperature of the calorimeter is followed until a final, steady rate of change is observed, which usually requires 15–20 min. With an isothermal jacket, the temperatures and the times should be measured as accurately as possible during the period between igniting the sample and attaining the final, steady drift as these measurements are necessary to the calculation of the heat exchange correction. In principle the calorimeter temperature need not be measured during the heating period for an adiabatic jacket, but the temperature of the jacket must be adjusted to follow that of the calorimeter. After the final temperature drift is followed for 10–20 min, the bomb

may be removed and its contents analyzed.

The usual bomb, with a volume of 300–400 ml, is charged with 1–3 g of sample which generates 6–8 kcal of heat when burned. The cover contains one or two valves for filling with oxygen and for flushing. It also contains a crucible support and electrodes for the ignition wire. A combustion bomb is a precision instrument. It must withstand the high temperature and pressure present during the combustion without leaking. However, it should be no more massive than necessary, so that the final temperature is reached quickly.

For a number of years the Parr Instrument Co. has furnished most of the combustion bombs used in the United States. They now manufacture several standard and special purpose bombs. The 1102 single valve bomb and the 1101 double valve bomb are usually used for routine measurements. The 1105 and 1106 bombs are more suitable for measurements of intermediate and high accuracy. The 1106 model is used with the head at the bottom rather than at the top. This causes most of the heat from the burning sample to be directed at the body of the bomb rather than at the head which contains the electrodes, valves, and gaskets. These bombs can all be easily closed by hand without the need for special tools. The 1002 bomb was designed by Rossini and Prosen for high accuracy measurements, but it is not often used at the present time. The Parr Co. also supplies special purpose bombs for burning explosives, for semimicro scale combustions, for use with fluorine, and for use in rotating bomb calorimeters. Bombs with platinum fittings and platinum linings can also be supplied. Combustion bombs are also manufactured by Gallenkamp, Prolabo, and Shimadzu, for use with their respective calorimeters.

The Parr Instrument Co. sells a series of complete calorimeters and accessories which may be used in many combinations. The simplest of these is the Series 1300 calorimeter. The jacket consists of a simple cylindrical plastic vessel with insulation between inner and outer walls. It serves as an isothermal jacket, and the calorimeter is intended mainly for the testing of fuels. The jacket of the Series 1200 Adiabatic Calorimeter is immersed in a stirred water bath, and water is also circulated through the cover. The accuracy which can be attained with this calorimeter can be considerably improved by removing the large plastic ring from around the top of the bucket and by installing a chromium plated cover on the bucket. The Series 1230 Adiabatic Calorimeter is the newest and most elaborate model and is designed for measurements of intermediate accuracy. The bomb is suspended in the bucket from the top by means of a special hangar allowing better circulation of water around the bomb. This calorimeter is illustrated in Figures 6 and 7. The temperature of the water in the jacket of these adiabatic calorimeters is controlled by addition of hot or cold water from an external source, thus making possible the rapid and uniform heating of the jacket when the bomb is fired. The Series 1411 Calorimeter is a semimicro scale instrument which matches the AC5E bomb. Accessories which may be ordered

for these calorimeters include several types of thermometers (see Part II), both isothermal and adiabatic jacket temperature controllers, water heaters, ignition power supplies, pellet presses, crucibles, fuse wire, and fixtures for filling the bomb with oxygen.

The Gallenkamp Co. (Great Britain) manufactures a complete bomb calorimeter system of modern design, which is

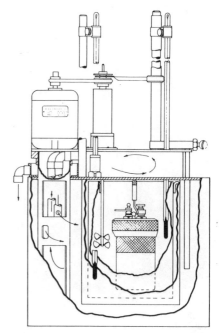

Figure 6. Cross Section, No. 1230 Adiabatic Bomb Calorimeter (Courtesy Parr Instrument Company)

shown in Figure 8. It is distributed in the United States by the American Minechem Co. This equipment is intended for rapid and convenient measurements of around

Figure 7. No. 1230 Adiabatic Bomb Calorimeter (Courtesy Parr Instrument Company)

0.1% accuracy. The complete system, including water bucket, jacket, power supply, and controls, is housed in a single compact cabinet. Adiabatic operation of the jacket is furnished by an automatic, thermistor actuated, control circuit.

The Prolabo Co. (France) manufactures two models of adiabatic bomb calorimeters. The "Control Model" is designed primarily for technical work of low accuracy, and the more elaborate "Precision

Chemical Instrumentation

Model" for the intermediate accuracy range. The standard line of components and accessories, including bombs, thermometers, pellet press, and gas handling system is furnished. Shimadzu Seisaku-

Figure 8. Gallenkamp Bomb Calorimeter, CB-100 (*Courtesy American Minechem Corporation*)

sho Ltd. (Japan) makes an adiabatic calorimeter system for low accuracy range of measurements. Both of these companies furnish mercury thermometers for their calorimeters.

The Precision Scientific Co. sells a calorimeter which is based on a design developed at the National Bureau of Standards. It consists of a water bucket for the bomb, outfitted with a cover and a motor driven stirrer. The jacket is completely submerged in a water bath. The bomb, thermometer, and controls for the bath are not supplied. It has been used for a number of years for combustion measurements in the intermediate, or occasionally high accuracy range. We have found that this calorimeter can be considerably improved by placing a short plastic section in the stirrer shaft to reduce heat conduction, and by installing two ball bearings in place of the journal bearing which supports the stirrer at the top.

Procedures

The preparation of samples for burning is a critical part of the procedure which may give some difficulty. The objective is to have an accurately weighed portion in a form which burns cleanly and completely to definite and reproducible products. The mass of sample should be adjusted to give nearly the same temperature rise as that used in the calibration experiments. Since loose powder does not usually burn completely, solid organic compounds are compressed to form pellets. The principal difficulty with solids is that they may not

burn well. After opening the bomb at the end of an experiment, it should be carefully examined for unburned sample or evidence of incomplete combustion, such as soot or unusual odors. If these are found, several things can be tried. The oxygen pressure can be varied, or the sample can be placed on a small piece of asbestos which does not conduct heat away as fast as a metal crucible. If all else fails, an auxiliary material such as a nonvolatile oil or another pellet of some readily combustible material can be placed in contact with the sample. The combustion of the auxiliary material may generate enough heat to burn the sample completely. Allowance must be made in the calculation for the heat generated by the auxiliary material, and its presence reduces the accuracy in the measurement of the heat of combustion of the main sample.

Nonvolatile liquids can be placed in an open crucible, but in high accuracy work liquids are usually sealed in small thin-walled glass bulbs. These bulbs are completely filled with liquid and are shaped so as to transmit the full pressure of the surrounding atmosphere to the sample. The preparation and filling of these bulbs is a delicate and exacting task. Descriptions of this procedure are given in the previously cited references on combustion calorimetry, but the best way to learn is by direct observation of someone who knows the technique. An alternative technique is to seal liquid samples in bags made from thin Mylar film. This polymer is inert and impermeable to most organic liquids, although it is hygroscopic and should be stored at constant humidity. Two pieces may be sealed by careful heating with a small soldering iron. Solid powders may also be sealed in these bags to protect them from the atmosphere, if necessary. After sealing the sample in the bag and weighing it, the bag and sample can be formed into a compact mass by compressing in a pellet press.

Samples are ignited by heating a piece of fuse wire to incandescence with an electric current. There are two general methods. The most common method uses a piece of special iron wire of about 0.005 in. diameter. It is connected to two electrodes inside the bomb and arranged to pass slightly above, but not touching, the sample. On ignition, burning pieces of the wire fall on the sample and start the combustion. The other method uses a platinum wire which is placed in contact with a cotton thread or strip of filter paper which touches the sample. In either case, a correction for the heat generated by the ignition process must be subtracted from the total amount of heat produced. The heat produced by the iron wire comes partly from the electrical energy and partly from the oxidation of the iron. The amount of heat produced can be calculated by multiplying the length of wire burned by a predetermined factor. The iron wire typically produces around 10 calories of heat, with an uncertainty of 0.5–1 calorie. On the other hand the platinum wire melts when heated, but does not burn, so that the ignition energy arises only from the electrical energy and the combustion of the cotton or paper. The electrical energy can be measured with appropriate instrumentation, although it is not an easy measurement. If a very small platinum wire, say

0.001 in. diameter, is used, only around 1–2 calories of electrical energy will be consumed so that it need not be measured very accurately.

The bomb can be filled with ordinary commercial oxygen for measurements of low or intermediate accuracy. In high accuracy work the oxygen should first be passed over heated copper oxide to remove combustible impurities. The initial pressure should be measured with care since it enters into the corrections to the standard state. Some nitric acid, derived either from nitrogen in the sample or from nitrogen impurity in the oxygen, is generally produced. The amount is determined by titrating the final solution left in the bomb with standard sodium hydroxide. It is sometimes desirable to determine the quantity of carbon dioxide produced, either to check for completeness of combustion or for sample purity. If the sample cannot be dried completely, the production of carbon dioxide may be used to determine the amount of sample burned. At the level of 0.01% accuracy, this is a tedious measurement. Procedures for carrying it out have been described by Prosen and Rossini (*49*).

The calculations associated with high accuracy combustion calorimetry are rather complicated. The "isothermal bomb energy" is the change in energy for the reaction which actually occurs in the bomb, corrected to a given reference temperature. The reactants consist of solid or liquid compound, solid or liquid auxiliary material, a liquid solution of oxygen in water and a gaseous phase of oxygen and water vapor, all at the initial pressure. The products consist of a liquid solution of oxygen, carbon dioxide, nitrogen, and nitric acid in water and a gaseous phase of oxygen, nitrogen, carbon dioxide, and water vapor, all at the final pressure. This corresponds to $\Delta E_s(T_r)$ in the equation for a constant volume process, analogous to equation (4) of Part I. To obtain the standard state heat of combustion, $\Delta H_c°$, the energy change for this reaction must be corrected to the reaction,

$$C_aH_bO_cN_d \ (s \text{ or } l, \ 1 \text{ atm}) +$$
$$(a - \tfrac{1}{4}b - \tfrac{1}{2}c)O_2 \text{ (ideal gas, 1 atm)} \rightarrow$$
$$aCO_2 \text{(ideal gas, 1 atm)} +$$
$$\tfrac{1}{2}dN_2 \text{(ideal gas, 1 atm)} +$$
$$\tfrac{1}{2}bH_2O(l, 1 \text{ atm})$$

This calculation can be carried out in a series of steps which have the effect of converting the bomb reaction into the standard state reaction. This was first done rigorously by Washburn in 1933. A more recent step-by-step procedure is described by Hubbard, Scott, and Waddington in Volume 1 of "Experimental Thermochemistry." The tedium of these calculations can be relieved by use of a high-speed digital computer. Upon request, the author will supply a listing of a complete program, written in FORTRAN IV, for performing all of the computations required in an isothermal jacket combustion experiment.

In burning compounds which contain elements other than carbon, hydrogen, nitrogen, and oxygen, the principal difficulty becomes the arrangement of conditions so that suitable products are formed. They should be present as recognizable homogeneous phases in thermodynam-

Chemical Instrumentation

ically stable states. Many ingenious procedures have been developed to deal with several classes of compounds. The most important innovation has been the introduction of the rotating bomb calorimeter. This apparatus is similar to the conventional fixed bomb calorimeter, except that the bomb is supported by a mechanism in the bucket which can rotate it about two perpendicular axes so that the contents become thoroughly mixed. Usually it is left in a fixed position until after the combustion, and then it is rotated for a period of 2 or 3 min before the final temperature is taken. Moving bombs were first used about 30 years ago but were perfected during the past decade at the Bureau of Mines Laboratory in Bartlesville, Okla., and at the University of Lund, Sweden. They were first used largely for sulfur compounds, but they have since been successfully applied to organic compounds containing halogens and to metal organic compounds of silicon, boron, tin, and aluminum. Rotating bomb calorimeters are not manufactured on a regular basis at present. Plans and specifications may be obtained from the Bartlesville Laboratory or from the Argonne National Laboratory, Argonne, Ill. A few are made to order at the Thermochemistry Laboratory in Lund. The bomb itself which fits these specifications may be purchased from the Parr Instrument Co. A rotating bomb calorimeter may be used either with or without rotation, and therefore most high accuracy calorimeters made in recent years include the rotating feature.

When sulfur compounds are burned in a bomb, sulfuric acid is formed. In a fixed bomb, concentrated acid is deposited in varying concentrations over the inner walls of the bomb. The bomb must then be rotated to attain a uniform concentration. In addition, sulfuric acid corrodes bombs which are not platinum lined, producing additional heat and damaging the bomb. Procedures for burning compounds containing sulfur and other elements and for calculating standard state corrections will be found in the two volumes of "Experimental Thermochemistry."

The combustion of fluorine compounds produces HF and, if a high proportion of fluorine is present, CF_4 also. These are distributed between liquid and gas phases at the end. When compounds of chlorine or bromine are burned directly in oxygen, the halogens end up in several different oxidation states. To prevent this, a solution of a reducing agent is initially placed in the bomb. Arsenous oxide has been used for this purpose for many years for chlorine compounds, but some workers now prefer hydrazine dihydrochloride. Chlorine compounds can be burned in a fixed bomb if the walls are lined with a wick made from quartz fibers or glasscloth to increase the area of contact between the gas and the solution. A rotating bomb eliminates the need for this lining. Hydrazine monohydrochloride can be used as a reducing agent for bromine compounds. These have been burned successfully only in rotating bomb calorimeters. Solid iodine is the principal product when organic iodine compounds are burned in a bomb.

Organic compounds of silicon, boron, or arsenic produce metallic oxides when burned. Since these oxides are in form of hydrated amorphous solids of indefinite properties, direct combustion is not satisfactory. Various techniques have been used to overcome these difficulties. For example, W. D. Good of the Bureau of Mines has burned silicon and boron compounds in the presence of an organic fluorine compound as an auxiliary material. Solutions of fluosilicic acid or fluoboric acid are then produced.

Fluorine has been successfully used as an oxidant, in place of oxygen, in both bomb and flame calorimeters. Some of the difficulties in using it can be readily visualized. However, techniques have been developed by G. Armstrong at the National Bureau of Standards, Ward Hubbard at the Argonne National Laboratories, and their associates. These measurements are now also being carried out at several other laboratories. The fluorine combustion calorimeter is used to determine the heats of formation of many fluorides by direct combination of the elements as well as the heats of combustion of many oxides, refractory compounds, and other organic and inorganic compounds which do not burn in oxygen.

Aneroid Bomb

An aneroid bomb calorimeter consists of a bomb imbedded in a block of metal. This type of calorimeter is inherently simpler than the one in which the bomb is placed in a bucket of water. No stirring is needed, and errors due to weighing of water and to evaporation are eliminated. The principal problems are the large temperature gradients and the long equilibrium times. Such calorimeters have been used for many years but only recently have been proved to be of much value. High accuracy measurements with aneroid calorimeters have been reported by Pilcher and Sutton (*46*) and by Keith and Mackle (*50*). Aneroid calorimeters have much smaller heat capacity than a water calorimeter of equivalent size and thus require less sample. Therefore, this type of calorimeter has a decided advantage for small scale combustions. This factor is a great help when the quantity of pure compound on hand is limited. A high precision aneroid bomb calorimeter having a total heat capacity of around 100 cal deg^{-1} has been described by Mackle and O'Hara (*51*). It is machined from a copper block and has a capacity of 100 ml. Because of its thin walls, the bomb cannot withstand a large difference of pressure between the inside and the outside. To circumvent this limitation, the jacket is made gas tight and filled with gas at the same pressure as the oxygen in the bomb. The temperature is measured with a resistance thermometer of copper wound on the outside of the bomb. Ten to twenty milligrams of samples are normally burned at one time. Heats of combustion of 2–5 mg samples have been measured in a small glass bomb placed in the Calvet microcalorimeter which is described in a later section.

Sources of Calorimeters and Related Equipment

American Instrument Co., Inc., 8030 Georgia Ave., Silver Spring, Md. 20910 (Thermometric titration calorimeter)

American Minechem Corp., Coraopolis, Pa. 15108 (Gallenkamp Bomb Calorimeter)

A. R. F. Products, Inc., P.O. Box 752, Raton, N. M. 87740 (Thermometric titrator)

E. I. duPont de Nemours & Co., 1007 Market Street, Wilmington, Del. 19898 (Differential thermal analysis calorimeter)

Dynatech Corp., 17 Tudor St., Cambridge, Mass. 02139 (Dynamic adiabatic calorimeter)

Guild Corporation, P. O. Box 217, Bethel Park, Pa. (Solution calorimeter)

Harrop Laboratories, 3470 East Fifth Ave., Columbus, Ohio 43219 (Dynamic conduction calorimeter)

Parr Instrument Co., 211 Fifty-Third St., Moline, Ill. 61265 (Bomb calorimeters and accessories)

Perkin-Elmer Corp., 333 Main Ave., Norwalk, Conn. 06852 (Differential Scanning Calorimeter)

Precision Scientific Co., 3737 West Cortland Street, Chicago, Ill. 60647 (Bath, jacket, and water bucket for bomb calorimeter)

Prolabo, 12 rue Pelée, Paris, France (Bomb calorimeter)

Schuco Scientific Co., 250 West 18th St., New York, N. Y. 10011 (Calvet calorimeter)

Shimadzu Seisakusho Ltd., 18 Nishinokyo-Kuwabaracho, Nakagyo-ku, Kyoto, Japan (Bomb calorimeter)

Spinco Division, Beckman Instruments, Inc., 1117 California Ave., Palo Alto, Calif. 94304 (Benzinger calorimeter)

Technical Equipment Corp., 917 Acoma St., Denver, Colo. 80204 (Deltatherm Dynamic Adiabatic Calorimeter)

SOLUTION CALORIMETRY

The heat capacity of liquids and the enthalpy change resulting from mixing of a solid, liquid, or gas with a liquid are measured in a solution calorimeter. The magnitude of energy change ranges from microcalories to kilocalories. In contrast to combustion calorimeters which include only a few standard types, solution calorimeters are found in many forms and styles. The level of accuracy varies widely for different kinds of measurements and conditions but is in general considerably less than that usually obtained in combustion calorimeters. In this section, three popular types of calorimeters are described which are suitable for work with relatively nonvolatile solvents, such as water, in the vicinity of room temperature.

A very simple calorimeter may be constructed by installing a cover, stirrer, thermometer, and mixing device in a Dewar flask. Those described in the manuals of Daniels, *et al.* (*52*), and Shoemaker and Garland (*53*) are typical of many which have been used for undergraduate instruction in physical chemistry. A somewhat more elaborate calorimeter of this type has also been described by O'Hara, Wu, and Hepler (*18*) (see Fig. 9). Another simple

Chemical Instrumentation

Dewar flask calorimeter, designed for approximate measurements of heats of reac-

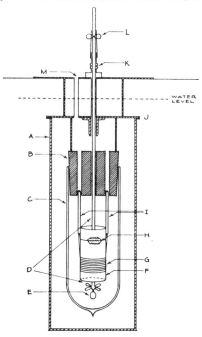

Fig. 9. Dewar Flask Calorimeter (O'Hara, Wu, and Helpler, **J. CHEM. EDUC. 38,** 513 (1961))

tions in solutions, was described by Arnett, Bentrude, Burke, and Duggleby (*58*). A calorimeter made to their plans has recently been offered for sale by the Guild Corporation. It consists of a complete system including thermometer and controls.

In this type of apparatus the calorimeter, in effect, consists of those parts of the inner glass wall, thermometer, heater, pipet, and other parts which are in more or less direct contact with the solution. The jacket consists of the outer wall of the Dewar flask and the cover. Thus there is no sharp separation between calorimeter and jacket, and the heat capacity of the calorimeter is a function of the amount of solution it contains. In addition, it is difficult to control evaporation. Although the accuracy can be improved by immersing the Dewar flask, partly or completely, in a constant temperature bath, this type of calorimeter is usually restricted to measurements of low accuracy.

For greater accuracies, it is advisable to use a calorimeter which consists of a thin walled metal or glass vessel completely surrounded by an adiabatic or isothermal jacket. Such calorimeters are completely, or nearly, filled with liquid and may be of the single or the twin type. In this type of a calorimeter, the space between the calorimeter and the jacket is often evacuated to reduce heat transfer.

The designs of pipets is one of the most troublesome aspects of solution calorimetry. The pipet contains the liquid or solid sample until it is time to mix it with the remainder of the solution in the calorimeter. The objective is to have the sample at the same temperature as the rest of the calorimeter and to mix it thoroughly without introducing extra energy. To establish the temperature of the sample,

the pipet is suspended in the liquid inside the calorimeter. The laboratory manuals of Daniels, *et al.* and of Shoemaker and Garland describe simple pipets which operate by air pressure. However neither of these will give complete mixing, and both will produce some undesirable evaporation of solvent. Other devices are described by O'Hara, Wu, and Hepler, and by Arnett, *et al.* For low accuracy measurements the principal consideration is simplicity and convenience, but many other considerations are important for high accuracy measurements.

In high accuracy work, a correction must be applied for the effect of opening the pipet. The so-called "heat of opening" results from the mechanical friction produced by the opening, by any difference in temperature between the sample and the solution in the calorimeter, and by evaporation of solvent. If the calorimeter is changing temperature at a constant rate, the sample in the pipet will lag behind, and thus be at a different temperature. If a bubble of air is released when the pipet is opened, some evaporation of solvent may take place when the additional gas space becomes saturated with solvent vapor. The heat of opening can be determined by running blank experiments with the same material in the pipet as in the calorimeter. In careful work the uncertainty in this correction may limit the overall accuracy of the measurement.

The type of pipet most frequently used for careful work consists of a thin-walled glass bulb or a cylindrical container with a thin window of glass, metal, or plastic at one end. These can be broken by forcing them down on an anvil placed in the bottom of the calorimeter for this purpose. More elaborate devices closed by movable covers and operated by more complex mechanisms have also been used.

It is now possible to purchase a complete solution calorimeter suitable for measurements of high accuracy. This is the 8700 Precision Calorimetry System manufactured by the LKB Instrument Co. It was designed and tested over a period of several years by Stig Sunner and Ingemar

Wadso of the Thermochemistry Laboratory at Lund, Sweden. The calorimeter, jacket, temperature controls, and associated instrumentation are all contained in a well-engineered compact unit. A schematic diagram is shown in Figure 10, which illustrates the components required in any calorimetric system of this type. Two models of calorimeters are available. One is a static type which may be used for the measurement of heats of solution, dilution, and reaction, and the other is a thermometric titration calorimeter described later. Both are made of glass, and utilize the same jacket and instrumentation. The jacket is immersed in a constant temperature bath thereby furnishing an isothermal environment. The jacket can be evacuated. Samples to be mixed in the static calorimeter are sealed into small glass bulbs which are fastened to the stirrer. They are broken by pushing them down on a sapphire tip in the bottom of the calorimeter. A view of the whole system is shown in Figure 11. Articles describing this calorimeter and its use have been published in the April and December, 1966, issues of the *LKB Instrument Journal* which are available from the company. This calorimeter has been in production only a short time, but it should increase considerably the output of calorimetric data in the future.

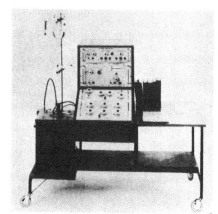

Figure 11. 8700 Calorimeter System (*Courtesy LKB Instruments, Inc.*)

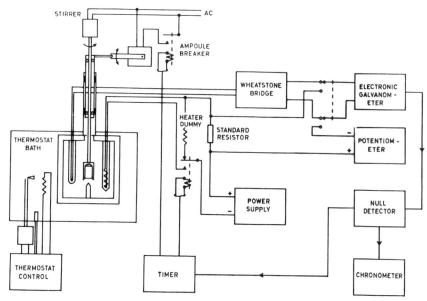

Figure 10. Schematic Diagram 8700 Calorimeter System (*Courtesy LKB Instruments, Inc.*)

Chemical Instrumentation

Thermometric Titrimetry

A thermometric titration is an analytical procedure in which the end point in a titration is located by means of temperature measurements (54). One of the reagents is placed in a vessel which contains a stirrer and a thermometer. The other reagent is added at a constant and known rate, and a record of temperature versus time is made. As long as the first reagent is present in excess, the rate of change of temperature is governed by the heat of reaction, by differences in temperature between the two reagents, by exchange of heat with the surroundings, by dilution effects, and by stirring energy. At the equivalence point, however, the reaction stops, and the temperature record undergoes an abrupt change of slope. In principle the temperature record could also be used to calculate the heat of reaction, provided the energetic effects of other processes could be evaluated or controlled and the heat capacity of the titration vessel and its contents were known. This type of calorimetry has become popular in recent years. Names such as "titration calorimetry," "thermochemical titration," "enthalpy titration," and "entropy titration" have been used to identify the technique and its modifications.

Calorimetric data were obtained by this method some ten years ago by Jordan (55). Since then, it has been used to measure heats of reactions such as ionization, complex formation, and oxidation-reduction in both aqueous and nonaqueous solutions. Most of these data can be considered only as having rather low accuracy. More recently, however, Christensen and his co-workers at Brigham Young University (56) and Wadso at Lund have used this technique to obtain results of accuracy comparable to that obtained in the more traditional static solution calorimeters. The American Instrument Co. sells an instrument, the "Titra-Thermo-Mat," designed primarily for thermometric titrations. A thermistor and the associated instrumentation for a thermometric titration experiment may be obtained from A. R. F. Products, Inc. Either of these sets of apparatus could be used in setting up a thermometric titration calorimeter. It has already been mentioned that the LKB Instrument Co. supplies components for converting their calorimeter for thermometric titration.

A thermometric titration apparatus can be converted to a calorimeter for measuring heats of reaction by surrounding the titration vessel by a temperature controlled jacket, by controlling the temperature of the titrant, and by appropriate calibrations to obtain the effective heat capacity of the calorimeter. The effect of titrant temperature can be made quite small by using a concentrated solution so that only a small quantity is added. The titrant is usually added by means of a constant rate microburet. For accurate work, other modification consistent with good calorimetric design should be made.

If the reaction which takes place in the calorimeter reaches an equilibrium such that appreciable concentrations of both reactants and products are present, then the temperature record is more complicated than two intersecting straight lines. This is because the extent of reaction shifts continuously as more reagent is added. However an appropriate analysis of the record can produce not only the heat of reaction but the equilibrium constant as well. This type of experiment has been called "entropy titration" (57) since the enthalpy change can be combined with the Gibbs free energy change, determined from the equilibrium constant, to calculate the entropy change for the reaction. Titration calorimetry has also been applied to reactions, such as complex formations, which take place in a series of intermediate steps; both the enthalpy changes and equilibrium constants of each step can be obtained from a single time-temperature record. These more complicated cases requiring lengthy mathematical calculations for the data reduction are made practical by the use of high-speed digital computers. Perhaps it is in the study of complex reactions that titration calorimetry will prove to have the greatest value since such data are difficult to obtain by any other method.

MICROCALORIMETRY

Any measurement would normally be called microcalorimetric if the rate of heat production is of the order of one thousandth, or less, of the heat capacity of the system in the space of a few minutes. These include phenomena in which the total energy change is small, such as heats of solution, dilution, and mixing, heats of reaction in dilute solution, heats of adsorption on surfaces, and heat produced by systems which are small compared to the size of the calorimeter. It may also include heat effects produced by slow processes such as the setting of cement, metabolism of living plants or animals, and nuclear transformations. Solution calorimeters of the types described in the last section sometimes operate in this range. However for the extreme lower limits, specially designed calorimeters are recommended.

The principal problem in microcalorimetry is the detection and measurement of very small temperature changes. Resolutions of 10^{-5} deg C are fairly common and resolutions of 10^{-6} deg have been attained. This requires not only a sufficiently sensitive thermometer but also an effective isolation of the calorimeter from outside disturbances. Such measurements are made at the extreme limit of performance of the instruments, and microcalorimetry is notoriously tedious, not only because of the time needed to attain a steady condition free from extraneous effects, but also because of the time spent in repeating spoiled experiments. Persons making these measurements soon develop the habit of working late at night when other activities in the building are likely to be at a minimum.

These difficulties have been lessened in recent years by the availability of two commercially built microcalorimeters. These have been designed by Eduard Calvet of the Institute of Thermogenesis in Marreille, France, and by T. H. Benzinger of the Bio-Energetics Laboratories, Naval Medical Research Institute, Bethesda, Md. These two calorimeters operate on about the same general principle and with the same type of thermometry, but they differ considerably in many details of design and construction. They both employ a massive cylindrical jacket machined from a metal block. Since temperature fluctuations which would be produced by a regulated electrical heater are less desirable than a slow, steady, temperature drift, neither jacket is directly thermostatted. Instead, temperature control depends on the large heat capacity of the jacket and on thermal insulation from ambient temperature. Both calorimeters employ the twin, or differential, principle and may be used as conduction calorimeters. The difference between the temperature of the calorimeter and the jacket is detected with multijunction thermocouples.

The Calvet calorimeter, also called the Tian-Calvet calorimeter, has been used for many years and has undergone many modifications. It has been described in the book by Calvet and Prat (6) and in numerous publications of Calvet and his associates. Its distribution in the United States is handled by the Schuco Scientific Co. Figure 12 is a cut-away drawing of the Calvet calorimeter. Within the jacket are placed two similar cylindrical calorimeters having a capacity of a few milliliters. Two sets of multijunction thermocouples containing 124 and 372 junctions, respectively, are placed between each of the calorimeters and the jacket. The thermo-

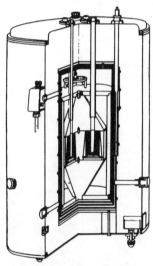

Figure 12. Cross Section, Calvet Calorimeter (*Courtesy Schuco Scientific Company*)

couple wires are held in place by mica insulators arranged carefully to insure a symmetrical distribution and to make the two calorimeters as nearly identical as possible. All these parts are made to rigid specifications.

Normally the process being studied is conducted in one calorimeter, and reference material is placed in the other. The four sets of thermocouples may be connected in different ways for different kinds of measurement, but generally the potentials from the thermocouples around each of the calorimeters are opposed so that the temperature of the jacket cancels out. Either of the two sets of thermocouples around each calorimeter may be used alone to detect a temperature difference, or they may be connected in series to form a thermocouple of 496 junctions. If only one of the sets is used for temperature measurement, the other may be used to add or re-

Chemical Instrumentation

move heat quantitatively from the calorimeter by the Peltier effect produced by a current from an external power supply.

The Calvet calorimeter may be used in a static mode by measuring the temperature change produced by the reaction or by compensating the heat effect so that there is very little temperature change. The heat of reaction can be compensated by the Peltier effect generated by one set of junctions. The rate of heat production or absorption in a slow process may be measured by allowing the calorimeter to reach a steady state so that the heat conducted to or from the jacket matches the heat produced or consumed by the process in the calorimeter. This temperature difference may be converted to heat flow by separate calibration experiments. Alternatively the heat effect may be balanced by the Peltier effect as in the static experiments. As little as 1 microwatt, and in favorable cases 0.2 microwatt, can be detected.

The Calvet calorimeter is very versatile and is limited only by the ingenuity of the investigator in arranging for the process to occur in the calorimeter cell. It operates over a wide temperature range from well below room temperature up to several hundred degrees Celsius. It has even been used for heats of combustion measurements by burning milligram quantities of sample in a small glass bomb.

The Benzinger calorimeter has been on the market for only about two years. It is manufactured by the Spinco Division of Beckman Instruments and is described in "Temperature, its Measurement and

Figure 13. Benzinger Microcalorimeter (Courtesy Beckman Instruments Inc.)

Control in Science and Industry" (5). It is strictly a static conduction twin calorimeter. The two calorimeters are in the form of cylindrical shells about 4 in. long. They are placed end to end concentrically inside the jacket. The space between the calorimeters and the jacket is occupied by thermocouple wire. The thermocouples are prepared from a helix of constantan wire. Half of each turn is plated with copper and acts as the copper lead of a thermocouple junction. Therefore each

Figure 14. Calorimeter Cells for the Benzinger Calorimeter (Courtesy Beckman Instruments Inc.)

turn forms one pair of junctions. The helix is coiled around the calorimeter and produces a thermocouple of about 10,000 junctions for each calorimeter. This arrangement provides a very large area of contact between the solution in the calorimeter and the thermocouple junctions. A blank is placed in one calorimeter, the solution under study in the other, and the two sets of thermocouples are connected in opposition, as in the Calvet calorimeter. An overall view of the apparatus is shown in Figure 13.

The calorimeter cells have been cleverly designed. Three types of glass cells are pictured in Figure 14. Metal cells are also furnished. They serve to measure the enthalpy change resulting from the mixing of two liquids. Initially the cells are positioned with the long axis horizontal, and the solutions are placed in the space between the outer and inner walls. In two of these cells a partition which runs either equatorially or longitudinally separates the two solutions of approximately equal volume. In the other type one solution is placed in the lower part of the space, and a much smaller volume of solution placed in small depressions in the inner wall, which initially are at the top. In any case, the solutions are mixed by rotating or tumbling the entire calorimeter and jacket.

As in the Calvet calorimeter, the electrical potential developed by the thermocouples is proportional to the difference in temperature between the two calorimeters. This signal is recorded by a potentiometric strip chart recorder. However in the Benzinger calorimeter, the conduction between the calorimeter and its jacket is such that they reach the same temperature within a few minutes. Thus following the absorption or evolution of heat, the difference in temperature reaches a maximum and then falls back to zero. Since all the heat is conducted to or from the jacket, and the rate of heat conduction is proportional to the temperature difference, the area under this curve is a measure of the total heat evolved or absorbed by the reaction. Benzinger refers to this procedure as the "heat-burst" principle. Accuracies of 1.5–2% and resolutions of about 10^{-4} calorie can be attained for solutions of 1–15 ml.

Of these two calorimeters, the Calvet is more versatile and is capable of better overall accuracy. The two calorimeters have about the same range of sensitivity. Benzinger has been primarily interested in using this apparatus for measuring the enthalpy change in enzyme catalyzed reac-

tions, and it is largely restricted to the mixing of two liquid solutions, or possibly a liquid and a solid. Within its range of application the Benzinger calorimeter is more convenient to use, especially with small volumes of solution. Many experimenters have built their own microcalorimeters of these and other types. Undoubtedly they can be built more economically than the cost of the commercial equipment, providing the cost of the person's time is not included. However, if such a project is undertaken, one should be prepared to devote several years to the effort.

DYNAMIC CALORIMETRY

The widespread use of differential thermal analysis in recent years has given impetus to the development of a whole series of dynamic calorimeters for studying phenomena which occur in condensed phases, especially solids. In terms of number of people involved, this is probably the most active area of calorimetric research today. Undoubtedly the ease and speed with which measurements can be made and the ready adaptability of these instruments to automation are major reasons for its popularity. This type of calorimeter can be used with better accuracy and reproducibility not only for all types of measurements made in the typical differential thermal analysis apparatus (54) but also for the measurement of heat capacities, heats of transitions, and heats of reactions. They can also sometimes be used for detecting energetic effects in many miscellaneous phenomena which would be difficult to study with any other type of calorimeter. These include study of non-equilibrium effects, energy stored in metastable states, kinetic measurements, and many types of industrial tests and assays. Because of the dynamic nature of these techniques, the calorimeters and their samples are small in order to reduce the effect of temperature gradients. They are not generally suited for studying equilibrium thermodynamic properties of materials which only attain equilibrium slowly.

This class of calorimeters covers a whole spectrum of types ranging from differential calorimeters, very similar to the usual DTA apparatus, to those similar to vacuum jacketed adiabatic calorimeters used for specific heat work. A large number of these types of calorimeters have been described in the literature and applied to many types of materials over a wide range of temperatures. Several are being manufactured by instrument companies. The accuracy of measurements made with these

Chemical Instrumentation

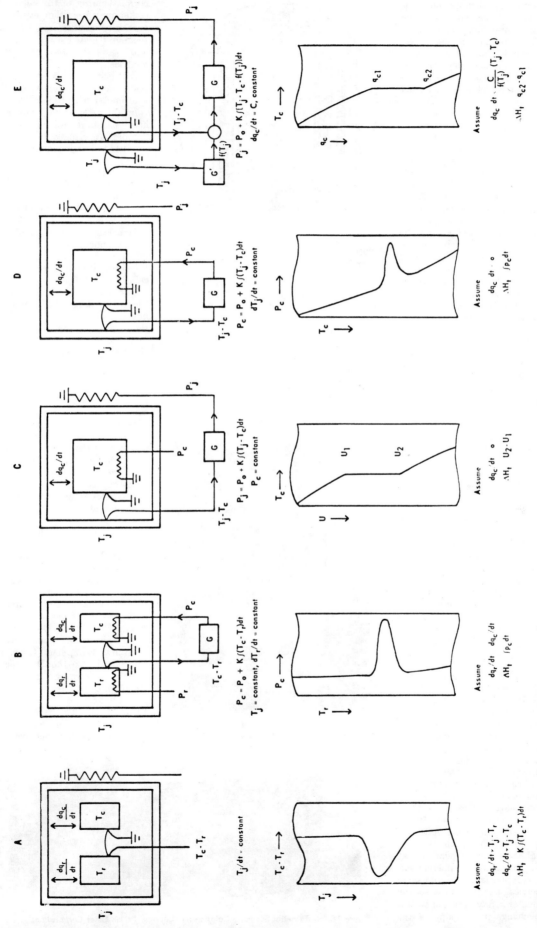

Figure 15. Control Circuits and Recorded Variables in Dynamic Calorimeters. In each record, time increases in the downward direction.

222

Chemical Instrumentation

instruments is difficult to judge, but it is usually fairly low. Even with the same instrument, the accuracy may vary widely for different kinds of measurements and different conditions. The dynamic mode of operation limits the accuracy for measurement of equilibrium properties to less than that which can be achieved with good static calorimeters.

Figure 15 represents symbolically the principle of operation of five types of differential calorimeters. The main common features are the operation of the calorimeter with a continuously changing temperature and the use of continuous recording of significant variables and of closed-loop control of the operation. The figure shows a schematic drawing of each of the calorimeters and their jackets, the location of the main thermometers and heaters, the type of control used, the variables which are recorded, the appearance of a typical record resulting from a sharp endothermic transition, the basis for calculation of the heat of transition, and the assumptions which are implied in the interpretation of the record. These drawings show thermocouples being used to measure temperature differences although matched resistance thermometers can be, and sometimes are, employed.

The Model 900 Differential Thermal Analyzer manufactured by DuPont is a typical example of a calorimeter of the type represented by Figure 15A. It is essentially a DTA apparatus modified to furnish quantitative information about heat effects. A more detailed drawing of this calorimeter and jacket is shown in Figure 16. Since two calorimeters are used and heat is transferred to the calorimeters only by conduction from the jacket, it may be termed a differential dynamic conduction calorimeter. A picture of the calorimeter cell is shown in Figure 17. It operates from the control circuits in the Model 900 Differential Analyzer. A sample of 1–200 mg is placed in one calorimeter and a reference substance in the other. The difference in temperature between the two calorimeters is detected by a thermocouple and plotted versus the temperature of the reference calorimeter on an X-Y recorder. The jacket temperature is programmed to increase at a constant rate from 1 to 30° deg. min⁻¹.

The Differential Scanning Calorimeter, Model DSC-1B, of the Perkin-Elmer Corp., has received wide publicity for the past two or three years. It is shown in Figure 18. It operates on the principle of Figure 15B. Since the jacket remains essentially at room temperature, this apparatus is a dynamic differential calorimeter with an isothermal jacket. The reference calorimeter is heated at a constant rate. The temperatures of the two calorimeters are sensed with platinum resistance thermometers, and the electrical power to the sample calorimeter heater is controlled by a closed loop so as to maintain the two calorimeters at the same temperatures at all times. If the two calorimeters exchange heat with the jacket at the same rate, the electrical energy added to the sample calorimeter in any period of time is equal to the change in enthalpy of the

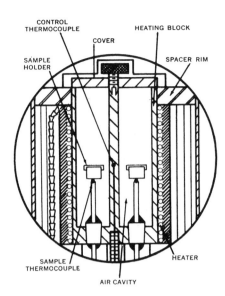

Figure 16. Cross Section, Calorimeter Cell (*Courtesy E. I. duPont de Nemours & Company*)

calorimeter and sample. A signal proportional to the electrical power is generated and may be displayed on a strip chart recorder. With suitable calibration this may be converted to heat capacity of the sample, and the area under this curve may be converted to an enthalpy change.

Samples are placed in shallow covered aluminum cups which may be used either sealed or unsealed. The cups containing reference material and sample are placed on small, disk-shaped, metal calorimeters. The calorimeters are thermally isolated

Figure 17. Calorimeter Cell (*Courtesy E. I. duPont de Nemours & Company*)

from the surroundings. The whole assembly is enclosed in a cylindrical jacket which has a glass window at the top to permit viewing of the calorimeters. The jacket may be either evacuated or filled with any gaseous atmosphere. Gaseous reactants may be added during the run, and gaseous products withdrawn for analysis.

The instrument is compact, highly automated, and quite versatile. The reference calorimeter may be heated or cooled at a constant rate, or its temperature may be held constant for isothermal studies. Samples are usually in the range of 1–30 mg, and full-scale sensitivities in the range of 1–32 millicalories per second may be selected. It covers the range of −100 to 500°C. This type of calorimeter is potentially more accurate than the conduction type illustrated in Figure 15A.

The Dynatech Corp. offers a dynamic adiabatic calorimeter, the Model QTA-N7. It corresponds to Figure 15C, and a detailed cross section is shown in Figure 19. Physically it is similar to a static adiabatic calorimeter. The single calorimeter is suspended within an aneroid jacket. Electrical energy is dissipated in a heater in the calorimeter at a constant rate. The temperature of the calorimeter is detected by a thermocouple and plotted on a strip chart recorder. A signal, representing the difference in temperature between the calorimeter and the jacket, controls the power to the jacket heater so that the jacket temperature follows that of the calorimeter. Assuming negligible temperature gradients and zero heat exchange between calorimeter and jacket, the increase in enthalpy of the calorimeter and contents is proportional to the time elapsed from the instant the heating was started. Endothermic transitions produce a straight line parallel to the time axis on the record.

A general view of this calorimeter and control panel is shown in Figure 20. The jacket is contained in the bell jar which may be evacuated to reduce heat conduction through the gas phase. Two calorimeters, one for solids and one for liquids, are supplied. It operates in the range of −180 to 320°C. Specifications call for an accuracy of 2–5%, depending on temperature, in the measurement of specific heat.

The Deltatherm D5500 dynamic adiabatic calorimeter manufactured by the Technical Equipment Corp. is geometrically similar to the Dynatech Corp. apparatus. However from the control standpoint, the roles of the calorimeter and jacket are reversed. This is illustrated in Figure 15D. The jacket is heated at a constant rate. The signal from the thermocouple, placed between jacket and

Figure 18. Differential Scanning Calorimeter (*Courtesy Perkin-Elmer Corporation*)

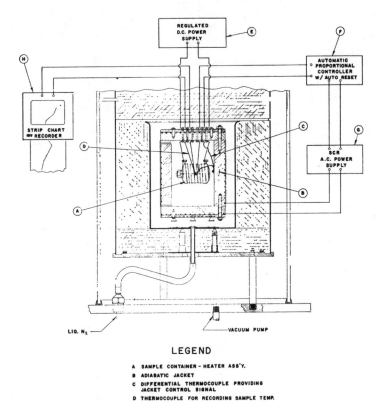

LIQ. N₂ — VACUUM PUMP

LEGEND

A SAMPLE CONTAINER - HEATER ASS'Y.
B ADIABATIC JACKET
C DIFFERENTIAL THERMOCOUPLE PROVIDING JACKET CONTROL SIGNAL
D THERMOCOUPLE FOR RECORDING SAMPLE TEMP.

Figure 19. Cross Section, Adiabatic Calorimeter (*Courtesy Dynatech Corporation*)

Figure 20. Adiabatic Calorimeter (*Courtesy Dynatech Corporation*)

instruments which has taken place in the past 20 years or so has had a great effect. In spite of the simplicity of the principles and the wide choice of automated instrumentation now available, accurate calorimetry still requires much skill and patience. Because of the elusiveness of sources of error, accuracies of measurement are frequently misjudged. Experience has shown that reliance can be placed on calorimetric data only when they have been confirmed in two or more independent laboratories. Although several types of calorimeters have recently been placed on the market, additional types are still needed in other areas. Further progress in thermochemical measurements depends not only on the availability of instrumentation, but equally on the development of precise and versatile analytical techniques and on the production of pure compounds.

calorimeter, controls the power input to the calorimeter heater so that the calorimeter follows the jacket. The electrical power to the calorimeter, which represents the instantaneous heat capacity of calorimeter and heater, is recorded. A record similar to that for Figure 15B is produced. This appears to be a fairly elaborate instrument and should be as versatile as the Perkin-Elmer calorimeter. The sample size is considerably larger. It operates in the range from room temperature to 800°C.

The Harrop Laboratories describe a calorimeter based on the principle of Figure 15E which appears to be still in the developmental and testing stage. This may be described as a constant rate dynamic conduction calorimeter. It is designed principally for temperatures up to 1200°C. The calorimeter does not contain an electrical heater. All the heat is transferred to it by conduction and radiation from the jacket. The rate of heat transfer is related both to the temperature difference between calorimeter and jacket and to the absolute temperature of the system. At higher temperatures it is almost all due to radiation. The rate of heat transfer is established as a function of temperature of jacket and calorimeter by a series of calibration experiments. In making a measurement, an analog computer uses the signal from the thermocouple to control the jacket heater so that the rate of heat conduction from the calorimeter to the jacket remains at some selected constant value. This is an unusual type of jacket control, and its further development should be interesting. It is not likely to achieve high accuracy, but even approximate measurements above 500°C are valuable.

SUMMARY

Calorimeters come in many shapes, sizes, types, and degrees of sophistication. The only common feature is the objective of measurement of an energy change. Those described in this review are only a sample. In common with other branches of experimental science, the great proliferation of

Calorimetry
Supplement

In common with most other laboratory instruments digital control and data-logging techniques have been used extensively in the construction of new calorimeters. Several pieces of new calorimetric equipment have been placed on the market during the past year. LKB Instruments now supplies two additional accessories to its Model 8700 Calorimeter which can replace the original glass reaction vessel. Thus they operate with the same thermostatic bath and electronic control unit. The 60 ml Closed Bomb Reaction Calorimeter is designed to measure heats of reactions involving gases or volatile materials. It operates with pressures of up to 5 atm. The other calorimeter measures heats of vaporization of liquid materials having a vapor pressure in the range of 0.5 to 200 torr at room temperature. Only about 100 mg of sample is required for a determination. An inert carrier gas is passed through the calorimeter and the energy absorbed by the vaporization is supplied

by an electrical heater at a rate which maintains constant temperature. LKB also manufactures two microcalorimeters intended primarily for biochemical applications. They are both twin conduction calorimeters in which a change in temperatures is detected by multi-junction thermopiles. The general construction of the heat sink and electrical circuitry is similar. The principal difference is in the reaction cells and in the operating procedure. Model 10700-2 is a batch calorimeter which can detect a rapid energy change of 10^{-6} cal in solution of 4 ml volume when used with the recommended amplifier. It can also be used to measure heat effects in slow processes. The heat measurement is based on the heat-burst principle as used in the Benzinger microcalorimeter. Model 10700-1 is a flow calorimeter which measures the steady state temperature change following the mixing of two streams of liquid. It is sensitive to the generation of 10^{-7} cal/sec in a reaction volume of 1 ml. Its principal application is the detection of heat effects which accompany rapid enzyme-catalyzed reactions in routine analytical tests.

The Tronac Co., Box 30, University Station, Provo, Utah, 84601, specializes in the production of a series of calorimetric instruments. These now include two high precision temperature controllers and a thermometric titration calorimeter which incorporate significant innovations in mechanical design and electronic circuitry. A highly automated system of control and data recording is built into the Series 100B Calorimeter. Technical details are described in ref. (59). The company will supply a bibliography on the principles and applications of thermometric titration calorimetry.

THAM has now been widely accepted for the calibration and checking of solution calorimeters. The National Bureau of Standards supplies samples for this purpose under catalog number SRM-724. Recent accurate measurements of the heat of solution of 0.5 g of THAM/100 ml of 0.1 M HCl at 25°C fall in the range of -7106 to -7115 cal/mol, with most clustering near the lower values. The heat of solution of 0.5 g THAM in 100 ml of 0.1 M NaOH is about $+4000$ cal/mol. The system hexane-cyclohexane was recommended by the Calorimetry Conference for testing calorimeters which measure the heat of mixing of volatile materials. Phillips research grade samples are suitable. Results of recent measurements and a review of earlier work are given in ref. (60–62). The investigation of very pure copper as a possible low temperature specific heat standard has continued. The National Bureau of Standards will probably issue samples of both copper and aluminum oxide for this purpose.

The adoption of the International Practical Temperature Scale of 1968 will be of interest to workers in precision calorimetry. This constitutes the first major change in the definition of empirical temperature since 1927. It is described in ref. (63). The book by Cox and Pilcher (64) includes a section on experimental thermochemistry as well as extensive tables of reported and selected values of heats of formation and heats of vaporization of organic compounds.

Bibliography

(1) STURTEVANT, J., "Calorimetry," in A. Weissberger (ed.), "Technique of Organic Chemistry," 2nd ed., Vol. I, Wiley-Interscience, New York 1949.

(2) STIMSON, H. F., *Amer. J. Phys.*, 23, 614 (1955).

(3) BOOTH, S., (ed.), "Precision Measurement and Calibration, Vol. II, Heat and Mechanics," Nat. Bur. Standards Handbook 77, U.S. Govt. Printing Office, 1961.

(4) ROSSINI, F. D., (ed.), "Experimental Thermochemistry, Vol. I," Wiley-Interscience, New York, 1956; H. A. Skinner, (ed.), Vol. II, 1962.

(5) BENZINGER, T. H. AND KITZINGER, C., "Microcalorimetry, New Methods and Objectives," in C. M. Herzfeld (ed.), "Temperature: its Measurement and Control in Science and Industry," Vol. III, Part 3; Reinhold, New York, 1963.

(6) CALVET, E. AND PRAT, H., "Recent Progress in Microcalorimetry," Macmillan, New York, 1963.

(7) McCULLOUGH, J. P. AND SCOTT, D. W. (eds.), "Experimental Thermodynamics," Vol. I, Calorimetry of Non-Reacting Systems; Plenum Press, New York, 1968.

(8) WESTRUM, E., J. CHEM. ED., 39, 443 (1962).

(9) NOLLE, D., *Inst. Soc. Amer. J.*, 12, No. 8, 59 (1965).

(10) ROESER, W. F. AND LONBERGER, S. T., *Nat. Bur. Stand. (U.S.), Circ.* No. 590, 1958.

(11) FUSCHILLO, N., *J. Sci., Instrum.*, 31, 133 (1954).

(12) CALDWELL, F. R., "Thermocouple Materials," *Nat. Bur. Stand. (U.S.), Monog.*, 40, 1962.

(13) HERZFELD, C. M., (ed.), "Temperature: its Measurement and Control in Science and Industry," Vol. III, Part 1, "Basic Concepts, Standards and Methods," Part 2, "Applied Methods and Instruments," Reinhold, New York, 1962.

(14) SHENKER, H., LAURITZEN, J. I., JR., CORRUCINI, R. J. AND LONBERGER, S. T., "Reference Tables for Thermocouples," *Nat. Bur. Stand. (U.S.), Circ.* No. 561, 1955.

(15) MILLER, C. H., *Proc. Inst. Elec. Eng. (London)*, 1090, 299 (1962).

(16) WILLIAMS, A. J. AND MERGNER, G. C., *Inst. Soc. Amer. Conf.*, Preprint No. 12, 1-4-64 (1964).

(17) MAIER, C. G., *J. Phys. Chem.*, 34, 2860 (1930).

(18) O'HARA, W. F., WU, C.-H. AND HELPLER, L. G., *J. CHEM. ED.*, 38, 513 (1961).

(19) MICHELS, A. AND STRIJILAND, J., *Physica*, 18, 613 (1952).

(20) KAUFMAN, A. B., *Instrum. Cont. Syst.*, 36, No. 1, 100 (1963).

(21) DAUPHINEE, T. H. AND PRESTON-THOMAS, H., *Rev. Sci. Instrum.*, 31, 253 (1960).

(22) JANICKE, J. M., *Instrum. Cont. Syst.*, 38, No. 5, 129 (1965).

(23) JONES, E. W., *Instrum. Cont. Syst.*, 38, No. 5, 125 (1965).

(24) DIAMOND, J. M., *IEEE Trans. Instrum. Meas.*, IM-12, No. 1, 26 (1963).

(25) GIBBINGS, D. L. H., *Proc. Inst. Elec. Eng. (London)*, 109C, 307 (1962).

(26) HILL, J. J. AND MILLER, A. P., *Proc. Inst. Elec. Eng. (London)*, 110, 453 (1963).

(27) HAMMOND, D. L., ADAMS, C. A. AND SCHMIDT, P., *Instrum. Soc. Amer. Conf.*, Preprint No. 11, 2-3-64 (1964).

(28) WILSON, R. E. AND ARNOLD, R. D., "Thermometry and Pyrometry," in "Handbook of Physics," Chapt. 5, McGraw-Hill, New York, 1958.

(29) STIMSON, H. T. AND SWINDELLS, J. T., "Temperature Scales, Thermocouples and Resistance Thermometers," in "American Institute of Physics Handbook," 2nd ed., Section 4a; McGraw-Hill, New York, 1963.

(30) ZENCHELSKY, S. T., J. CHEM. ED., 39, A627 (1962).

(31) PRENGLE, W., JR., WORLEY, F. L., JR. AND MAUK, C. E., *J. Chem. Eng. Data*, 6, 395 (1961).

(32) FURUKAWA, G. T., DOUGLAS, T. B., McCOSKEY, R. E. AND GINNINGS, D. C., *J. Res. Nat. Bur. Stand.*, 57, 67 (1956).

(33) GOOD, W. D. AND SCOTT, D. W., in H. A. Skinner (ed.), "Experimental Thermochemistry," Vol. II, Wiley-Interscience, New York, 1962.

(34) GINNINGS, D. C. AND FURUKAWA, G. T., *J. Amer. Chem. Soc.*, 75, 522 (1953).

(35) OSBORNE, N. S., STIMSON, H. E. AND GINNINGS, D. C., *J. Res. Nat. Bur. Stand.*, 23, 197 (1939).

(36) WEST, E. D., *Trans. Faraday Soc.*, 59, 2200 (1963).

(37) EDWARDS, J. W. AND KINGTON, G. L., *Trans. Faraday Soc.*, 58, 1313 (1962).

(38) WEST, E. D. AND GINNINGS, D. C., *J. Res. Nat. Bur. Stand.*, 60, 309 (1958).

(39) GUNN, S. R., *J. Phys. Chem.*, 69, 2902 (1965).

(40) PARKER, V. B., "Thermal Properties of Aqueous Uniunivalent Electrolytes," National Standard Reference Data Series, Nat. Bur. Standards, 2, (April 1, 1965).

(41) SOMSEN, G., COOPS, J. AND TOLK, M. W., *Rec. Trav. Chim.*, 82, 231 (1963).

(42) RAY, J. D., *Rev. Sci. Instrum.*, 27, 863 (1956).

(43) SHOMATE, C. H. AND HUFFMAN, E. H., *J. Amer. Chem. Soc.*, 65, 1625 (1943).

(44) WADDINGTON, G. W., SMITH, J. C., WILLIAMSON, K. D. AND SCOTT, D. W., *J. Phys. Chem.*, 66, 1074 (1962).

(45) SOMAYAJULU, G. P., KUDCHADKER, A. P. AND ZWOLINSKI, B. J., "Thermodynamics," in *Ann. Rev. Phys. Chem.*, 16, 213 (1965).

(46) PILCHER, G. AND SUTTON, L. E., *Phil. Trans. Roy. Soc. (London)*, A248, 23 (1955).

(47) JESSUP, R. C., *National Bureau of Standards Monograph 7*, 1960.

(48) COOPS, J., ET AL., *Rec. Trav. Chim.*, 66, 113, 131, 142, 153, 161 (1947).

(49) PROSEN, E. J. AND ROSSINI, F. D., *J. Res. Nat. Bur. Stand.*, 27, 289 (1941).

(50) KEITH, W. A. AND MACKLE, H., *Trans. Faraday Soc.*, 54, 353 (1958).

(51) MACKLE, H. AND O'HARA, P. A. G., *Trans. Faraday Soc.*, 59, 2693 (1963).

(52) DANIELS, F., WILLIAMS, J. W., BENDER, P., ALBERTY, P. A., AND CORNWELL, C. D. "Experimental Physical Chemistry," 6th ed., McGraw-Hill Book Co., New York, 1962.

(53) SHOEMAKER, D. P. AND GARLAND, C. W., "Experiments in Physical

Chemistry," McGraw-Hill Book Co., New York, 1962.

(54) WENDLANDT, W. W., "Thermal Methods of Analysis," Interscience Publishers (division of John Wiley & Sons, Inc.), New York, 1964.

(55) JORDAN, J. AND ALLEMAN, T. G., *Anal. Chem.*, **29**, 9 (1957).

(56) CHRISTENSEN, J. J. AND IZATT, R. M., *J. Phys. Chem.*, **66**, 1030 (1962).

(57) CHRISTENSEN, J. J., IZATT, R. M.,

HANSEN, L. D. AND PARTRIDGE, J. A., *J. Phys. Chem.*, **70**, 2003 (1966).

(58) ARNETT, E. M., BENTRUDE, W. G., BURKE, J. J. AND DUGGLEBY, P. McC., *J. Amer. Chem. Soc.*, **87**, 1541 (1965).

(59) CHRISTENSEN, J. J., IZATT, R. M., AND HANSEN, L. D., *Rev. Sci. Instr.*, **36**, 779 (1965).

(60) MARSH, K. N., AND STOKES, R. H., *J. Chem. Thermodynamics*, **1**, 223 (1969).

(61) MURAKAMI, S., AND BENSON, G. C., *J. Chem. Thermodynamics*, **1**, 559 (1969).

(62) McGLASHAN, M. L., AND STOECKLI, H. F., *J. Chem. Thermodynamics*, **1**, 589 (1969).

(63) METROLOGIA, **5**, (1969).

(64) COX, J. D., AND PILCHER, G., "Thermochemistry of Organic & Organometallic Compounds," Academic Press, London and New York, 1970.

Fire.
Dry warmth
Fiery, choleric

Air.
Damp warmth
Airy, sanguine

Water.
Damp cold
Fluid, phlegmatic

Earth.
Dry cold
Solid, melancholic

The four elements play an important part in all the mysticism of the Middle Ages. It is from this that the above interpretations are taken.

Topics in...

Chemical Instrumentation

...a **ChemEd** feature

Edited by **GALEN W. EWING**, Seton Hall University, So. Orange, N. J. 07079

The Journal of Chemical Education
Vol. 44, page A935
November, 1967

XXXIV. Theory and Applications of Thermistors

E. A. BOUCHER, Department of Physical Chemistry,
The University, Bristol, England.

Ernest A. Boucher was graduated from the University of Wales and obtained a Ph.D. degree from the University of Bristol. An Associate Member of the Royal Institute of Chemistry, and a Member of Sigma Xi, Dr. Boucher was Research Associate at Lehigh University during 1965–66, and joined the staff of the University of Bristol in January, 1967.

INTRODUCTION

A thermistor (*therm*ally sensitive res*istor*) is in essence a piece of semiconductor material attached to two electrical leads. The chief characteristic of semiconductors is their large negative coefficient of resistance change with temperature. The properties of semiconductors, coupled with cheapness and small size, give thermistors a wide range of applications in temperature measurement, temperature control, calorimetry, pressure gauges, flow meters, hygrometers, gas analyzers and thermometric titrators, as well as in electronic instruments where they are employed, for example, in power level control, in power measurement, and for automatic gain control in amplifiers.

The purpose of this article is to examine critically the characteristics of thermistors and to offer an introduction to applications of interest to the chemist. To do this it is first desirable to have a working knowledge of the underlying theory of semiconductors. An indication is then given of how thermistors are made. This is followed by a discussion of the characteristics of thermistors upon which the applications depend.

THEORY OF SEMICONDUCTORS

At room temperature, semiconductors have specific resistances in the range 10^{-2} to 10^9 ohm cm, which is intermediate between conductors ($\sim 10^{-6}$ ohm cm) and insulators ($> 10^{14}$ ohm cm) and, with rare exceptions, they have negative* temperature coefficients of resistance (3–5% per deg C is not uncommon). Semiconductivity is displayed by a variety of substances, including elements such as silicon,

* Sir Charles Goodeve, British Iron and Steel Res. Assoc., is credited with the remark that semiconductors are feminine, "...because their resistance decreases as they warm up!" [See *American Scientist*, p. 440A (Dec. 1965)].

germanium, selenium and tellurium, compounds of the type GaAs, spinels, metal oxides, and certain organic compounds. Mixtures of metal oxides (including those of manganese, nickel, cobalt, copper, tungsten, titanium, chromium, and vanadium) are the most common semiconductors used in thermistor manufacture. Semiconducting materials are also used in transistors, rectifiers, modulators, detectors, and photocells. An introduction to these devices can be obtained from general books on electronics (*1, 2*).

Metals and Insulators

To understand the nature of semiconductivity, it is convenient to consider first the electrical properties of metals and insulators on the basis of the band theory. Details of the band theory can be obtained from books on the solid state, e.g., Kittel (*3*), although the treatment in Moore's textbook (*4*) provides sufficient basis for discussion. It must be emphasized, since it is not obvious from the elementary treatment given here, that the development of quantum mechanics was essential to an explanation of the behavior of solids—in particular semiconductors—in the presence of an applied electric field.

The electrons in a single isolated atom of a metal are confined to discrete energy levels or atomic orbitals. An ideal piece of metal can be considered to be an ordered array of nuclei in a sea of electrons. Pauli's exclusion principle permits only two electrons in a given energy level. Thus, an energy level of an isolated atom becomes in the solid, a very large number of closely spaced energy levels forming a band. Each energy level in the lower bands will be doubly occupied and so these bands will be completely filled. The electrons in these bands will not be able to move under the influence of an applied electric field and will not contribute to the conductivity of the metal.

The upper band of a metal will not be completely occupied, or will overlap with the next empty band, as depicted in Figure 1*a*. When an electric field is applied, electrons near the top of the filled zone can readily move into unoccupied levels where they are free to move throughout the solid.

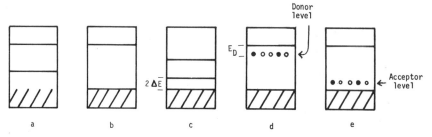

Figure 1. Band Model Depictions of (a) metal, (b) insulator, (c) intrinsic semiconductor, (d) n-type semiconductor and (e) p-type semiconductor.

Chemical Instrumentation

The conductivity of a substance is a measure of the product of the number of current carriers (in this case electrons) and their mobility. In a perfectly ordered ideal metallic crystal, the electrons would be completely free to move and the resistance would be zero. In a real solid, a large contribution to the resistance (i.e., obstruction of current carriers) comes from thermal motion of the atomic nuclei. In a metal the number of current carriers is very large (approx. $10^{22}/cm^3$) and is virtually independent of temperature. The mobility of the electrons decreases as the temperature increases, producing an increase in resistance. The variation in resistance with temperature of platinum and of a typical thermistor are shown in Figure 2.

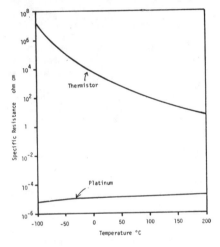

Figure 2. Variation of Resistance with Temperature for Platinum and a Typical Thermistor.

Insulators are characterized by having completely filled lower bands and a large energy gap between the highest filled and the lowest empty band, as represented in Figure 1b.

Intrinsic Semiconductors

Pure materials exhibiting semiconductivity are termed intrinsic semiconductors, of which silicon and germanium are the most common and the best understood. An intrinsic semiconductor would, at absolute zero, possess a filled valence band separated by only a small energy gap from a vacant conduction band. As the temperature is raised, a number of electrons are thermally excited from the valence band, leaving an equal number of holes (Fig. 1c). Both the electrons and the positive holes act as current carriers by moving in opposite directions when an electric field is applied. As the temperature increases, the number of current carriers increases and outweighs the decrease in mobility due to thermal vibrations of the nuclei, so that the conductivity increases.

Application of Fermi-Dirac statistics gives the temperature dependence of the number of electrons n_- excited from the valence band to the conduction band as (5),

$$n_- \propto T^{3/2} \exp[-\Delta E/kT] \quad (1)$$

where ΔE is one half of the energy gap between the bands. The temperature dependence of current carrier mobility, considered as primarily due to lattice scattering, is proportional to $T^{-3/2}$. The resistance can then be expected to obey the approximate relationship:

$$R = A \exp[+\Delta E/kT] \quad (2)$$

Equation (2) is obeyed by actual intrinsic semiconductors only over a limited temperature range, as shown in Figure 3. Germanium and silicon do not show intrinsic semiconductivity at low temperatures, but nevertheless semiconductivity invariably persists owing to the presence of trace impurities.

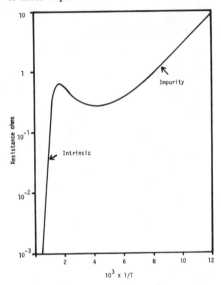

Figure 3. Showing Silicon Doped with Phosphorus Behaving as an Intrinsic Semiconductor at High Temperatures and as an Impurity (n-type) Semiconductor at Low Temperatures (Linear Portions Indicate Applicability of Equation 2) after Friedberg (5).

Impurity Semiconductors

The theory of impurity semiconductors can be readily explained using germanium as the host lattice. The same general scheme is applicable to more complex systems, some of which are however not completely understood. In Figure 4a, a few tetravalent atoms, in this case germanium, have been replaced by an equal number of pentavalent arsenic atoms; the arsenic atoms taking up positions in the lattice (diamond structure) originally occupied by germanium atoms. Each arsenic atom has a loosely bound electron attached to it which can be readily promoted to the conduction band by an amount of energy, E_D, in Figure 1d. An electron when bound to an arsenic atom is said to be in a donor level. The current carriers in this case are electrons and the semiconductivity exhibited is termed n-type.

When a few atoms in a germanium lattice are replaced by trivalent atoms, such as indium in Figure 4b, the impurity atoms provide energy levels (acceptor levels in Figure 1e) just above the valence band, into which electrons can be promoted. The promoted electrons leave behind positive holes which act as current carriers and the semiconductor is termed p-type.

Generally both acceptor and donor levels will occur in a semiconductor, al-

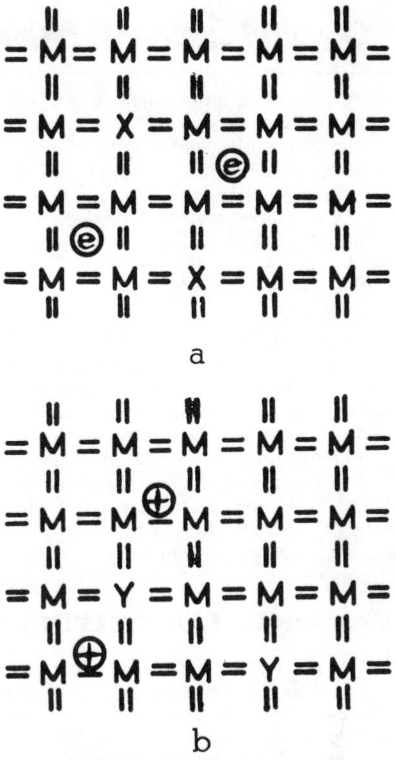

Figure 4a. Conduction by Donor Electrons e, produced by Impurity X from Group V in a Lattice of a Group IV Element M, e.g. M = Ge, X = As. Figure 4b. Conduction due to Positive Holes + where a Group III Impurity, Y, has Accepted Electrons in a Group IV Element Lattice, e.g. M = Ge, Y = In.

though one will usually dominate the conductivity at a given temperature. The general scheme just described is followed by semiconductors of the type GaAs. Substitution of a Group VI element for arsenic produces donors, whereas substitution of a Group II element for gallium yields acceptor levels.

Metal Oxides

Transition metal oxides, for example NiO and Cu_2O, become semiconductors of the so called "controlled valence" type by impurity substitution. If, for example, a small number of monovalent lithium ions replaces an equal number of nickel ions in nickel oxide, an equal number of remaining nickel ions will be oxidized from the di- to the tri-valent state. Electrons can now move under an applied electric field from divalent to trivalent nickel ions.

Metal oxides can also show semiconductivity if they depart from stoichiometry. There are four principal ways in which nonstoichiometry can lead to semiconducting behavior (6). In Type I, anion vacancies are present, that is, the metal is in excess, while in Type II, the metal is in excess due to cations in interstitial positions. In both cases the crystal is kept electrically neutral (by electrons trapped in the vicinity of the vacancies and interstitial cations respectively). Such solids will exhibit semiconductivity if these trapped electrons can be liberated by thermal energy.

In Types III and IV, the anions are in excess due to the presence of interstitial anions and of cation vacancies respectively. Electrical neutrality is now maintained by the presence of trapped positive holes. If

Chemical Instrumentation

these holes can migrate, the solid will again be a semiconductor.

The electronic conductivity shown by nonstoichiometric metal oxides is therefore classified into n-type or p-type. The conductivity will increase with temperature in a similar manner to that of intrinsic semiconductors discussed above.

The semiconductivity of materials used in thermistors (mainly sintered mixtures of oxides) cannot be readily explained solely in terms of the outline given above, although the general scheme remains the same. The conducting properties of the materials are greatly influenced by grain boundaries and by the nature of the surface of the material. The band levels at the surface differ from those in the bulk solid (7), and are affected by adsorption of gases and vapors of which water vapor is common.

MANUFACTURE OF THERMISTORS

The semiconducting materials used in the manufacture of thermistors are generally either mixtures of transition metal oxides, or of spinels, e.g., $MgCrO_4$ and $MgAl_2O_4$. These substances are least sensitive to changes in impurity content. Thermistors can be obtained in five basic shapes, rods, discs, beads, washers, and flakes.

The manufacturing procedure most often used is to make a homogeneous mixture of the powdered constituents, often with an organic binder. The mixture is molded and fired so as to cause sintering or even melting. It is necessary in some cases to control the rate of heating, the maximum temperature reached, the length of firing, and the atmosphere, in order to obtain reproducible products. After firing, metal contacts are made, using metal pastes or by spraying metal to which conducting wires can be soldered.

The bead type of thermistor is perhaps of most interest to the chemist. These are made (8) by placing two pieces of conducting wire (e.g., 0.001–0.004 in. diameter platinum) about 0.003–0.015 in. apart, in a jig. The semiconductor material, in the form of a slurry, is applied at intervals along the wires, as shown in Figure 5. The slurry is dried and sintered, and then the leads are cut to give individual thermistors. To protect this type of thermistor from moisture and other contaminants, which could affect its stability, the bead is placed in a glass envelope which can be in direct contact with the bead, can be evacuated, or can be filled with a dry gas.

Thermistors covering a wide range of specific resistances and temperature coefficients can be made by variation of the composition of the starting mixture. A wide range of electrical characteristics (discussed in the next section), necessary for different applications, can be achieved with thermistors of various sizes and forms.

CHARACTERISTICS OF THERMISTORS

The most obvious characteristic of thermistors which is exploited in applications is the large variation of resistance with temperature. The purpose of this

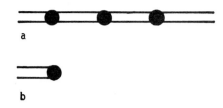

Figure 5. Manufacture of Bead Thermistors
a. Semiconductor placed on wires for firing
b. Bead fired and detached.

section is to show the importance of other (often closely related) factors which must be considered when designing and operating thermistor devices.

Variation of Resistance with Temperature, and of Current with Applied Voltage

The large decrease in resistance with increase in temperature, as shown in Figure 3, can be described, at least for intrinsic semiconductors, by eqn. (2). For actual thermistors it is convenient to determine the resistance R_0 at 298°K in such a manner that the current flowing causes negligible self-heating of the thermistor. The resistance R of the thermistor at temperature T°K can then be represented by

$$R_0 \exp [B(1/T - 1/298)] \quad (3)$$

where B is the so-called material constant.

The coefficient of resistance change with temperature is defined as,

$$\alpha = \frac{1}{R} \cdot \frac{dR}{dT} = -\frac{B}{T^2} \quad (4)$$

and may be of the order of -3 to -5% per deg K compared with $+0.35\%$ per deg K for platinum.

Except when the current flowing is very small, the magnitude of the current for a given applied voltage will be dominated by the R-T relationship, since the thermistor will experience self heating. It is important to note that α is temperature dependent. (B also is slightly dependent on temperature.)

Consider a thermistor suspended in still air. When the current flowing is very small, that is, when there is no self heating, the thermistor will obey Ohm's law. When the current is increased, the thermistor temperature will rise above ambient and the resistance will decrease. The voltage will be less than it would have been had the resistance not changed. In fact, the voltage V across a thermistor reaches a maximum value as the current I increases. The variation of voltage with current, typical of thermistors, is conveniently represented as log V plotted against log I as in Figure 6, which is adapted from Becker, Green, and Pearson (9). The temperatures on the curves represent the difference between the thermistor and ambient temperatures. On such a plot, lines of positive unit slope represent constant resistance, and lines of negative unit slope represent constant power.

Characteristic Quantities of a Thermistor

In addition to the R-T relation and the V-I relation, a particular thermistor is

characterized by its heat capacity (which depends upon its size and composition), its dissipation constant, power sensitivity, time constant, and by the maximum power which it can tolerate ensuring good stability and long life.

(i) The dissipation constant C specifies the amount of power dissipated (in watts) per degree of thermistor temperature ($T - T_a$) deg C above that of the surroundings. The dissipation constant depends upon the nature of the surrounding medium, and upon thermal conduction along the supports (often the lead wires). Change in the dissipation constant can be brought about not only by changing the medium, but also by changing, for example, the rate of stirring of ambient liquid or the rate of circulation of a gaseous medium. The effect of a change in C upon the log V against log I plot is shown in Figure 7 (9). For a point on the V-I curve $V/I = R$, and $V \times I = W$, the number of watts dissipated. The temperature of the thermistor can now be read from the R-T curve. A plot of power dissipated against T can be constructed which allows $C = W/(T - T_a)$ to be calculated.

The dependence of C upon ambient gas velocity, gas pressure, and the nature of the gas, forms the basis of the application of thermistors in anemometers, manometers, and gas analyzers respectively.

(ii) The power sensitivity, which depends upon the same factors as the dissipation constant, is defined as the power which must be dissipated in order to reduce the resistance by 1%. The value of the power sensitivity is given by $C/100\alpha$.

(iii) The time constant, τ sec, is an important factor in design considerations. The factors discussed so far have all referred to steady-state conditions, where variables such as temperature, current, and power dissipation have ceased to vary with time. It is useful to know, in addition, the rate at which the thermistor resistance will change for a given change in ambient temperature.

Suppose a thermistor is in a steady-state condition with a relatively large current flowing, so that the thermistor temperature T is considerably above ambient T_a. If the current is cut off and the voltage across the thermistor (determined with a small current flowing) is recorded as a function of time, the variation of temperature (as $T - T_a$) with time can be calculated. If the heat capacity of the thermistor is H, and if Δt is an increment of time, starting at $t = 0$ when the current is cut off, the amount of power dissipated in Δt sec will be $C(T - T_a)\Delta t$ joules. During this time interval the thermistor temperature will decrease ΔT such that:

$$-H\Delta T = C(T - T_a)\Delta t \quad (5)$$

or

$$T - T_a = -(H/C)(\Delta T/\Delta t) \quad (6)$$

Integration of eqn (6) gives:

$$T - T_a = (T_i - T_a) \exp(-t/\tau) \quad (7)$$

where at $t = 0$, $T = T_i$ and therefore $\tau = H/C$. The slope of a plot of log $(T - T_a)$ against time t is $(-1/2.303\tau)$. A representative plot is shown in Figure 8.

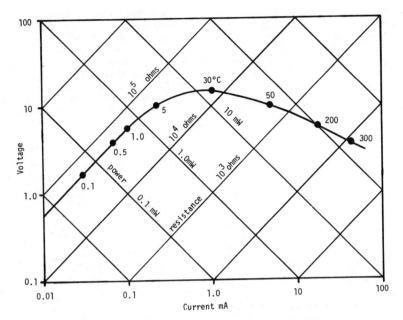

Figure 6.   Plot of log V against log I for a Typical Thermistor.

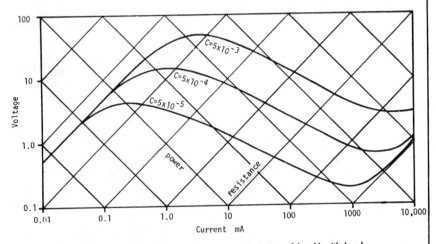

Figure 7.   Effect of Change in C upon Variation of log V with log I.

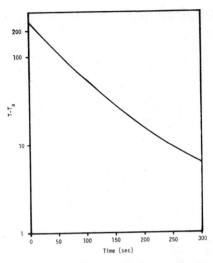

Figure 8.   Plot of log $(T - T_a)$ against time, the Slope of which is $-1/2.303\tau$.

Thermistor Stability

Thermistors have had a reputation for being unstable, which was to some extent justified, following their commercial production in the early forties. The situation has changed considerably with the introduction of more satisfactory methods of attaching the lead wires, and by pre-aging of thermistors by the manufacturers. No doubt the stringent requirements set for components used in space exploration have induced manufacturers to improve their products.

Thermistors are sensitive to light, since it is possible for photons to supply sufficient energy to promote electrons into upper bands. (The frequency at which $h\nu$ is just sufficient to promote electrons corresponds to the edge of the absorption spectrum of the semiconductor.)

It is advisable to protect thermistors from mechanical shock, since this can impair their stability (10).

The maximum power rating of thermistors, which is usually specified by the manufacturers, has already been mentioned. The use of too large a current will cause the thermistor to be unstable, and it has been suggested that maximum stability is ensured by keeping the current well below the specified maximum value (10).

Very little has been reported in the scientific literature regarding thermistor stability, and few generalizations concerning all types of product are possible. A precaution which is recommended, especially if not carried out by the manufacturer, is to pre-age a thermistor before it is first used. Pre-aging can be done by passing a current through the thermistor for several hundred hours, at the maximum temperature at which it will be used. A valuable piece of advice has been given by Müller and Stolten (11), who found that stability was greatly improved by always keeping a small voltage across the thermistor—even when it is not in use. When the voltage was interrupted, several hours were required before a new steady-state condition was obtained. Jordan (12) also advocates the use of a continuous

An alternative way of thinking of the time constant is as the time taken for $-\log_e[(T-T_a)/(T_i-T_a)]$, obtained from eqn. (7), to become unity, or for $(T-T_a)/(T_i-T_a)$ to equal $(1/e)$ (= 0.37). The time constant can thus be defined as the time taken for the thermistor temperature to drop to 63% of its initial value above ambient $[(T_i-T)/(T_i-T_a) = 0.63]$.

It can be seen from the above equations that the time constant depends upon the heat capacity of the thermistor and upon its dissipation constant. The temperature dependence of H and C has been neglected in the discussion of thermistor properties. A few representative values of α, C, and τ are given in Table I.

Table I.   Typical Characteristic Quantities for Three Types of Thermistor
(data supplied by Fenwal Electronics Inc.)

Type of Thermistor	R_{298}	α % per deg C	τ sec	C mw per deg C
Mini-probes— bead in glass	100 to 100kΩ	−3.2 to −4.6	16 in still air 0.8 in still water and 0.16 in moving water	0.7 in still air and 3.8 in still water
Disc of bare material 0.4 in diam e.g., LB12J2	18.6Ω	−3.9	26 in still air	6 in still air
Bead type for gas analyzers e.g., G112	2000Ω	−3.4	1	0.16

Chemical Instrumentation

current, in order to obtain maximum stability, and has suggested that electrical contacts between sintered oxide grains become tarnished when no current is flowing. Certainly with sintered mixtures of oxides, there will be a large number of grain boundaries, and it would not be surprising for a change in surface states to be reflected in the combined (surface and bulk) electrical characteristics of the materials.

Müller and Stolten were concerned with the determination of small temperature differences, using a differential bridge system which will be discussed later. They measured the resistances of two thermistors (Western Electric type 14A, nominal resistance $100k\Omega$) over a period of several months, and found that changes were less than 1Ω per day. The attainment of good stability is aided by limiting the temperature range over which a thermistor is used. Under the best conditions of operation, and in the region of room temperature, it should be possible to achieve long-term stability in temperature measurement, using a thermistor bridge, of ±0.01–0.02 deg C.

USES OF THERMISTORS

Before describing uses of thermistors which are of interest to chemists, a few words about their advantages and disadvantages as sensing elements are appropriate. The main advantages of thermistors are their high sensitivity, small size, and the fact that they can be manufactured for use over a considerable range of temperatures (-50 to $+200°C$), with a wide range of resistance values. The cost of using thermistors is small compared with the use of platinum wire sensing elements which require much more sophisticated bridge arrangements. The advantages are offset by nonlinearity of resistance change with temperature, and the possibility of instability which has already been discussed.

Temperature Measurement

The measurement of actual temperatures and of temperature differences is a vast subject. For a very useful appreciation the reader is referred to the monographs by Hall (13). The subject has been treated in great detail in, for example, a series of symposia (14). Before setting out to make temperature measurements it is necessary to decide; (1) if it is necessary to know accurately the actual temperature, or only temperature differences, (2) whether it is necessary to use a thermometer with fast response, and (3) if it is desirable to have a continuous record of temperature change with time. One often neglects to decide exactly where in a system it is necessary to make the temperature measurement.

The situations where a liquid-in-glass thermometer is suitable are generally obvious. When maximum stability and precision are required the platinum resistance thermometer is still without rival. Thermocouples, on the other hand, are suited for use in calorimeters and for meas-

uring furnace temperatures. The full advantages of thermistors are used in situations requiring the continuous recording of temperature changes, with a small sensing element. They are ideal for temperature determinations of remote systems.

The simplest form of thermistor thermometer consists of a Wheatstone bridge,

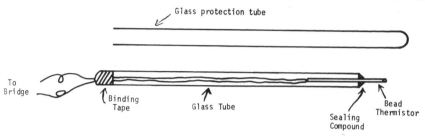

Figure 9. Thermistor Probe.

one arm of which is a thermistor, the others, a pair of high-stability resistors, and a variable resistance such as a decade box or a precision variable resistor. The power supply can be a dry cell or, because it gives a very constant voltage until exhausted, a Mallory mercury cell. The out-of-balance e.m.f. from the bridge can be fed to a sensitive galvanometer (15) or a continuous trace of temperature against time can be obtained using an automatic amplifying recorder (16).

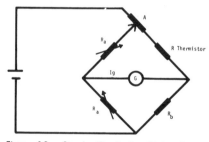

Figure 10. Simple Circuit for Giving Linear Variations of Out-of-Balance E.M.F. with Temperature.

A convenient type of thermistor for this application consists of a semiconductor in a glass envelope. A sturdy probe can be made in the shape of a conventional liquid-in-glass thermometer by mounting the thermistor in the end of a glass tube, with the conducting leads running through the tube, as shown in Figure 9. This type of thermistor thermometer is most useful for measuring small temperature changes. The out-of-balance e.m.f. will only be an acceptably linear function of temperature for changes of about 1 deg C or less. When the temperature range of interest is greater than this, it is necessary to calibrate the thermometer over the entire range. A conventional liquid-in-glass thermometer is often suitable for this purpose. When used in this manner to measure temperature *differences*, the thermometer is least sensitive to changes in thermistor resistance at a given temperature (lateral shift in R-T curve), which is the most common symptom of instability, but it will still be sensitive to changes in the thermistor temperature coefficient (slope of R-T curve).

The disadvantages of nonlinear variation of out-of-balance e.m.f. and of variation of temperature sensitivity with temperature, have been examined by Cole (17), who constructed a thermometer with maximum departure from linearity of 0.5 deg C, over the range 0–50°C, by using a slide wire Wheatstone bridge. The re-

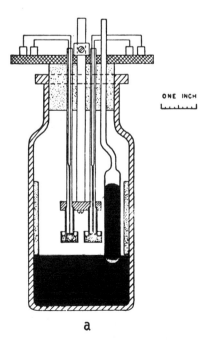

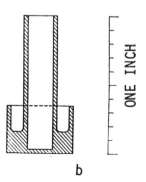

ONE INCH

Figure 11. Apparatus for Determination of Molecular Weights: (a) Complete Apparatus and (b) Sample Vessels. (Reprinted by courtesy of *Analytical Chemistry*)

Chemical Instrumentation

quirement for constant sensitivity is that the variation of current through the galvanometer with temperature, dI_g/dT, is independent of R, the thermistor resistance. With reference to the circuit in Figure 10, Pitts and Priestley (18) set up the conditions for equal sensitivities at each end of the temperature range, and found that the sensitivity was essentially constant over the whole range $T_2 - T_1$. To do this, the potentiometer resistance A was chosen for convenience to be equal to $R_{T1} - R_{T2}$ and R_b was set equal to R_{T1}. The condition for equal sensitivity at T_1 and T_2 gives the value of R_a. Experimentally it was shown that the sensitivity was constant ($\pm 1\%$) and within 1% of the calculated value, over a 15 deg C temperature range.

The single thermistor thermometer suffers from the disadvantage of being sensitive to changes in the thermistor dissipation constant. When used to measure temperature fluctuations in a water bath thermostat, for example, response will be sensitive to stirring speed, if the thermistor is appreciably self heated. Increase in stirring rate may produce a more uniform temperature throughout the bath, but the dissipation constant will increase, with the result that the thermistor resistance will increase, which could be incorrectly interpreted as a decrease in bath temperature.

In many cases where it is required to measure temperature differences between two parts of a system, a differential bridge arrangement can be employed. The use of a differential bridge will be illustrated by a description of the thermometric method for molecular weight determination (11). This experiment can be readily adapted as a laboratory teaching experiment, as can the experiments using thermistor thermometers in freezing and boiling point determinations, and in thermometric titrations which are also briefly described.

(a) Thermometric Determination of Molecular Weights

Two of the arms of the Wheatstone bridge were thermistors (Western Electric Co., Type 14A) in parallel with the power supply, and the other two arms consisted of a fixed resistance and a decade box. The power supply used by Müller and Stolten was a 45 v dry cell battery reduced by a resistor to give 22.5 v across the bridge. The out-of-balance e.m.f. was read on a vacuum tube voltmeter.

The apparatus is shown in Figure 11a. The outer glass vessel is approximately one-third full of mercury, and is thermostated. The vessel houses two matched stainless steel cups (Fig. 11b), each equipped with a thermistor. The plunger shown in Figure 11a enables the level of

---

[1] The field of calorimetry, including the application of thermistors, has been discussed in this column by R. C. Wilhoit, July to October, 1967.

[Reprinted in this volume]

the mercury to be raised to thermally couple the cups. Pure solvent (0.15 cm³) is placed in one of the cups, and a solution of the substance of unknown molecular weight in the same solvent is placed in the second cup. Initially the two cups are maintained at the same temperature by raising the mercury to surround the cups. The thermistor bridge is balanced for this situation. The plunger is then raised, lowering the mercury from around the cups. The liquid in each cup is open to the atmosphere in the vessel, which is saturated with solvent vapor by wet absorbent paper, so that once thermal contact between the cups is broken, solvent will distil into the cup containing solution (lower vapor pressure than solvent). A temperature difference between the cups will develop, which, to a good approximation, is such as to raise the vapor pressure of the solution to that of the solvent. In practice, heat losses will occur through the vapor and along the supports holding the cups, as well as by radiation.

Using azobenzene, naphthalene, and diphenyl in solvents n-heptane, benzene, carbon tetrachloride, and chloroform, Müller and Stolten found that the resistance change required to re-balance the bridge was proportional to the mole fraction of solute for 0.1–1.0 molal solutions. Calibration is used to give the constant of proportionality for a particular solvent. Azobenzene, for example, gave molecular weights of 188, 191, 189, and 189 respectively in the four solvents (cf. 182 from formula).

(b) Depression of Freezing Points and Elevation of Boiling Points

Estimation of molecular weights, or impurity contents, based on determination of freezing point depression or boiling point elevation caused by the presence of solute or impurity respectively, are well known. The use of thermistors to measure the small temperature changes involved has been discussed by Herington and Handley (19).

(c) Thermometric Titrations

The determination of end points of titrations, by monitoring the temperature change arising from the heat of reaction, carried out in an adiabatic calorimeter, has been thoroughly covered in THIS JOURNAL by Jordan (20). Readers are referred to this article which demonstrates how thermometric titrations can be used to determine the concentration of an unknown, the amount of an unknown, the heat, free energy, entropy, and stoichiometry of a reaction.

Calorimetry[1]

The merits and problems of using thermistors in calorimetry are similar to those for temperature measurement. Thermistors have been used extensively in heat-of-immersion calorimetry (21, 22), with a single calorimeter used semi-isothermally. Greenhill and Whitehead (23) describe a calorimetric system in which temperature changes of the order of 10^{-2} deg C can be measured automatically with an accuracy of 2×10^{-4} deg C. It should in fact be possible to detect changes of the

order of 10^{-5} deg C with thermistors as sensing elements, although thermistors give more noise than metal resistors (10). An example of the use of thermistors in precision bomb calorimetry is given by Sinke et al. (24).

The current status of the thermistor in calorimetry has been reviewed by Cruickshank (25).

Temperature Control

The fundamental requirements for temperature control devices have recently been reviewed (16, 26), and are illustrated in an article on water-bath thermostats by de Bruyne (27). The basic elements of a temperature control device, whether of the proportional or on/off type, are a sensing element to monitor the temperature of the controlled medium, a relay system, and a servomechanism (e.g., heater). An efficient temperature control device requires a sensing element with a small response time, but which is stable and to some extent robust. Bead-type thermistors make good sensing elements, provided they are protected from mechanical shock, are preferably shielded from bright sunlight (see previous discussion), and the current flowing through them is kept below the maximum power value specified by the manufacturers. It is sometimes convenient to use two thermistors in series in a bridge arm in order to reduce the current passing through the arm, while keeping the sensitivity equal to that of a single thermistor.

One disadvantage of using small thermistors as sensing elements is that the bridge e.m.f. is small and requires considerable amplification before it can work a relay mechanism. In an automatic thermostat used by the author (16) in the temperature range 0 to $-40°C$, the out-of-balance e.m.f. was passed into a transistorized amplifier before being fed into an amplifier employing tubes, and then to a phase-sensitive detector.

Ideas on circuitry and general layout for thermostats can be obtained from papers by Malmberg and Matland (28), Noltingk and Snelling (29), Rand (30), and Boucher (16).

Pressure Measurement

The use of thermistors in pressure gauges, gas analyzers, and flow meters depends upon the variation of the dissipation constant C with change in thermal conductivity of the thermistor environment. To compensate for changes in ambient temperature, a differential bridge arrangement, similar to that for measuring temperature differences, is used. One of the thermistors then acts as reference arm and experiences a constant value of C, whereas the second thermistor acts as sensing element in the environment where C varies.

A manometer utilizing a bead thermistor, for the range 10^{-5}–10 torr, has been described by Becker, Green, and Pearson (9). This type of gauge is suitable for gases which are thermally decomposable, since the thermistor temperature—arising from self-heating—can be limited to about 30 deg C above ambient. Bradley (31) has extended the working range of a conventional McLeod gauge to $1\text{-}10^{-7}$ torr, by sealing a bead thermistor into the end of the closed capillary. A compensating thermistor, consisting of a bead in a glass envelope, was mounted near to the closed capillary. Both thermistors were protected from draughts by a glass tube placed around the capillary. Pressures in the range $1\text{-}10^{-3}$ torr were measured without compression of gas into the capillary. For pressures down to 10^{-7} torr, the gas was compressed to bring it into the range $1\text{-}10^{-3}$ torr so that it could be measured with the thermistor bridge. A compression factor was used to convert the observed pressure to the pressure of the system. It is necessary to calibrate thermistor gauges for each gas.

A thermistor gauge has been used by Hall and Tompkins (*32*) to determine adsorption isotherms for water vapor and krypton on metal halides. They were able to detect pressure differences of 5×10^{-5} torr at water vapor pressures below 2×10^{-2} torr.

Gas Analyzers

Thermistors are used quite extensively to analyze the effluent in gas chromatography and the gaseous products of reactions (*33, 34*). The principle is similar to that of the hot wire katharometer; the effect of variation in rate of heat conduction from the self-heated thermistor will be manifested according to the dependence of $V\text{-}I$ and $R\text{-}T$ relationships upon C as discussed above.

In a typical arrangement for use as a gas chromatography detector, a differential bridge system is used in which one thermistor (normally in the pure carrier gas stream) acts as reference arm. The other thermistor monitors the concentration of sample gas present in the carrier by virtue of the change in conductivity of the mixture, the carrier gas velocity being constant. The out of balance of the bridge can be fed directly to an automatic amplifying potentiometric recorder, to give a continuous differential trace, i.e., peaks. With exceptions—arising mainly from the mode of construction of the detector and/or the bridge circuit—the area under recorded peaks can be arranged to be proportional to the amount of a given species in the carrier gas. Textbooks on gas chromatography should be consulted for details of the use of thermistor-based detectors for quantitative analyses.

Flow Meters

An arrangement essentially the same as that for detecting gas concentrations can be used to measure gas velocity. If the nature of the gas flowing is kept constant, the dissipation constant will respond to its rate of movement past the thermistor. Thermistor devices can be used to measure wind velocity (*35*).

Concluding Remarks

The applications included in this article are those thought to be most useful to the chemist. Information and references to the use of thermistors in power meters, in amplitude control in oscillators, for the measurement of electrical quantities at audio and radio frequencies, and in timing devices, can be found in the articles by Becker, Green, and Pearson (*9*) and Scarr and Setterington (*8*). Thermistors are extremely versatile devices. Full benefit from their properties is achieved by an appreciation of their disadvantages as well as their merits. Their main disadvantage, namely the uncertainty regarding their stability, is being overcome, and much-needed experiments on long-term stability are apparently under way. Thermistors provide an example of a situation where the high standards set for components of instruments for space exploration are improving the reliability of equipment for more down to earth investigations.

Acknowledgment

I wish to thank collectively the personnel of Companies and Laboratories who supplied me with information about available products and especially those who gave opinions on thermistor stability. I have tried to avoid mention of specific products, and where such mention is made, it is *not* meant to imply superiority of these products over similar ones by different manufacturers.

Thermistors

Supplement

The purpose of the addendum is to extend the discussion of certain aspects of thermistor theory and to introduce a few applications not mentioned above.

THEORY AND DESIGN

The dissipation constant or heat transfer coefficient C depends not only on the nature of the thermistor, but also on the ambient medium (see Table 1) and its rate of movement past the thermistor. The power dissipated by a thermistor heated to above the temperature of its surroundings is

$$W = C(T - T_a)$$
$$= VI \qquad (8)$$

To obtain plots of W against $(T - T_a)$, eqn. (8) is used for successive pairs of values of V and I, and the corresponding temperatures are read off the $R - T$ curve for the thermistor, using $R = V/I$.

Curve A of Figure 12 shows a typical plot as expected for a thermistor suspended in still air (with $C = C_1$). If the ambient medium changes so that C increases to C_2, for example, by causing the air to flow, or by placing the thermistor in contact with a metal block, then a curve like B would be obtained. Figure 12 shows that the temperature difference $(T - T_a)$ needed to dissipate W watts decreases with increase in C.

Normally the amount of self-heating of a thermistor when it is used in a temperature-measuring or temperature-control cir-

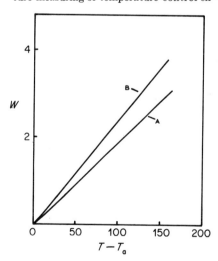

Figure 12. Dependence of power dissipated (W) on temperature rise of the thermistor above ambient $(T - T_a)$ for a particular thermistor in situations where the dissipation constants are C_1 and C_2 $(C_2 > C_1)$.

cuit should be kept negligibly low. It is a straightforward matter to assess $T - T_a$ for a given applied voltage when C is kept constant: under steady state conditions

$$T - T_a = V^2/CR \qquad (9)$$

For example, taking $C = 3$ mW·K^{-1}, $V = 1$V and $R = 10^4 \Omega$, $T - T_a \approx 0.03°$K. Tanaka (*36*) has determined experimentally deviations from the simple $R\text{-}T$ relation for a bead type thermistor with different applied voltages. The constants a and b in

$$R = a \exp(b/T) \qquad (10)$$

were determined by calibration at 294 and 304°K. The required difference is between T from eqn. (10) for recorded resistance values corresponding to intervening temperatures, and T_a given by the relation from combining eqn. (9) and (2), viz.,

$$R = A \exp\{B(T_a + V^2/CR)^{-1}\} \qquad (11)$$

for the same resistance values (note $B = \Delta E/k$). The constants for the thermistor were, $A = 3.76 \times 10^{-2}\Omega$, $B = 3690°$K and $C = 3.54$ mW·K^{-1}. When 1V was applied the difference between the two temperatures was less than $5 \times 10^{-4}°$K over the ten-degree range, and when 3V were applied, the difference between the temperatures passed through a maximum of about 4×10^{-3}°K at 299°K.

The material constant B in eqn. (3), on which depends the temperature coefficient of a thermister [see eqn. (4)], would, for a semiconducting substance complying with the simple band theory giving eqn. (2), be directly related to the band gap. The band gap is the energy difference between the valence and conduction bands of an intrinsic semiconductor (Fig. 1c); between the donor level and the conduction band of an n-type; and between the valence band and the acceptor level in a p-type semiconductor as shown in Figures 1d and 1e, respectively.

It is possible to make semiconductor materials for thermistors with B values ranging from 20 to 13,000°K. If a thermistor with $B = 3000$°K is cooled from 298°K, where $R = R_0$, to 100°K, its temperature coefficient will be [by eqn. (4)] -30% K^{-1}, and R will be $R_0 e^{20}$. To obtain sensible values of R at low temperatures a lower value of B is required. A series of thermistors with different B values is needed to cover the range 10–300°K. The design considerations for low temperature applications are examined in detail by Sachse (37), who also compares the advantages of thermistors and carbon resistance thermometers. Also, Herder, Olson and Blakemore (38) have reported attempts to prepare semiconductor thermometers for use at low temperatures by solid state diffusion of arsenic into germanium, and of boron into silicon.

Thermistor elements are commonly used in thermal conductivity-bridge detectors for gas chromatography where self-heating is appreciable and the dissipation constant is very sensitive to changes in the ambient gas or vapour (39–44). The response of the bridge is often analysed in the same manner as for hot-wire katharometers. However, Buhl (45) has given a detailed treatment of thermistor response taking into account the negative temperature coefficient of thermistors. The equivalent circuit giving a linear V-I relationship of negative slope,

$$R_T = (\partial V/\partial I)_{C, T_a} = \mu_I \quad (12)$$

is shown in Figure 13. This comprises two voltage generators (VG) taking ac-

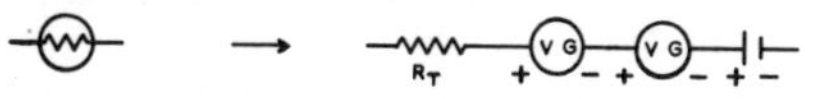

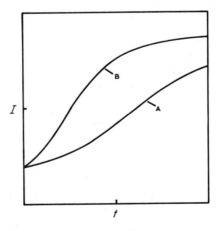

Figure 13. Equivalent circuit for a thermistor (applicable to range where $(\partial V/\partial I) < 0$).

count of the dependence of V on T and on C, and a negative resistance R_T which represents the constant dependence of V on I over the range of interest. From the total differential describing this model, the effects of current, temperature and dissipation constant add linearly, leading to

$$\Delta V = \mu_I \Delta I + \mu_T \Delta T + \mu_C \Delta C \quad (13)$$

where the μ's are the appropriate partial differential coefficients. Temperature and current changes constitute the noise, and the dissipation constant change gives the signal. Buhl uses the equivalent thermistor circuit in a Wheatstone bridge to

calculate signal and noise output. Two methods of minimizing noise are, to minimize temperature and current fluctuations, and to operate a suitable bridge circuit in such a manner that it is insensitive to temperature and current changes.

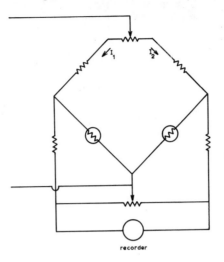

Figure 14. Du Pont-ERL Bridge.

The DuPont-ERL bridge shown in Figure 14, allows either the current noise or the temperature noise to be nulled out. To null the current noise, the two thermistor resistances of the circuit equivalent to that shown in Figure 13, being functions of I, are made nearly equal by adjusting the currents I_1 and I_2 so as to get the condition for the noise voltage to go to zero. The condition for the noise voltage arising from temperature variations to go to zero can be obtained by adjusting the current ratio in the bridge arms, but these two nulls do not occur at the same current ratio. The problems associated with obtaining a bridge for simultaneously nulling both are discussed. Buhl stresses the importance for maximum performance of having well matched bridge components and thermistors, as well as having a stable power source and good temperature control. The more troublesome of the noises can then be nulled. The use of thermistor detectors has been surveyed by Lawson

(46) who concludes that although thermistor detectors have greater sensitivity than hot-wire detectors, they are limited in their temperature range of operation, and are perhaps better for specialized and not general purpose use. By reducing the temperature of the cell from 298 to 273°K, the sensitivity of the thermistor detector can be increased by as much as five or six times.

MISCELLANEOUS APPLICATIONS

Figure 15 shows semi-schematically the change in current with time expected when a constant voltage is suddenly applied to a thermistor in series with a fixed resistance. (The thermistor giving curve A is larger than that giving curve B.) The initial current when a given voltage is applied depends on the cold resistance value of the thermistor and on the value of the fixed resistance. As the thermistor heats up, its resistance drops and the current reaches a limiting value. Thermistors can therefore be used in time delay circuits: the delay in reaching a required current can be varied from fractions of a second to several minutes. If the fixed resistance is replaced by a filament—whose cold resistance will only be about 0.15 of its value when it is hot—the high resistance of the thermistor will suppress any initial surge of current. The filament will heat up slowly as the thermistor resistance decreases to a negligible value.

The typical R-T curve for thermistors shown in Figures 2 and 16 (curve A), can be modified by combination with other resistance elements. Curve C of Figure 16 is obtained by putting a fixed resistance (curve B) in parallel with the thermistor. Curve D is for a metal, e.g., copper, making up a circuit element for which it is desired to give temperature compensation. When the fixed resistance in shunt with the thermistor is used as a compensator in series with the metal, the result is a temperature compensated unit giving curve E of Figure 16. A disadvantage of using transistors* in amplifiers is that the collector current I_C (and collector leakage current) of a transistor increases with increase in temperature. Figure 17 shows a transistor amplifier circuit with bias provided by the voltage divider. The thermistor and resistance behave in the manner shown by curve C of Figure 16. As the temperature of the circuit rises, the decrease in the resistance of this unit decreases the base current I_B and so compensates for the rise in I_C which would otherwise occur.

A novel use of thermistors is in psychrometers for relative humidity determination with which readings can be obtained quite rapidly. Another interesting application can be made where tube filaments are used in series and failure of one would cut the power supply to all. A thermistor placed in parallel with each filament will ensure that when one fails its thermistor will heat up and pass current, enabling the others to work, until a replacement can be made.

---

* See Ref. (1), p. 117, for a concise account of the theory and use of transistors, and Studer (47) for a chapter on current flow in semiconductors.

Figure 15. Dependence of current on time after voltage is applied for a thermistor in series with a resistor: curve A is for a thermistor physically larger than that giving curve B.

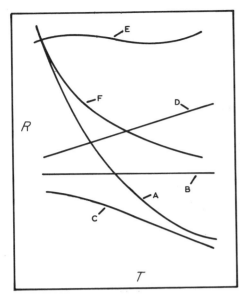

Figure 16. Schematic representation of dependence of thermistor resistance on temperature for:

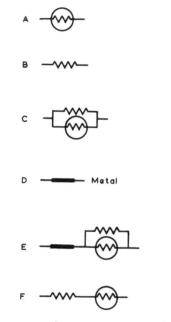

The use of thermistors in temperature controllers has been described by van Swaay (48), who points out that, because of their negative temperature coefficient, if the thermistor arm of a bridge goes open-circuit this may be interpreted by the system as a temperature decrease and the heater will stay on. To prevent this, a

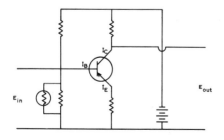

Figure 17. Transistor amplifier circuit with temperature compensation provided by a thermistor shunting a resistance.

bimetallic switch can be placed in the thermostatted liquid to act as a cut out, or a fail-safe circuit can form an integral part of the amplifier stage.

An analysis of the application of thermistors to measurements in moving fluids has been given by Rasmussen (49) and this is augmented by a report on the behaviour of a thermistor flow meter by Pigott and Strum (50). Thermistors have also been used in free molecule pressure probes (51) to measure macroscopic velocities of rarefied gases (at ca. 10^{-2} torr pressure). The range of pressure which can be measured with a thermistor gauge (52) can apparently be extended up to about an atmosphere by surrounding the thermistor bead with an electrically non-conducting powder (53). The use of thermistors for pressure measurement in the range 5×10^{-3} to 2 torr with an accuracy of 5×10^{-4} torr at pressures below 0.1 torr, and to within 1% at pressures above 0.1 torr, has been described by Berry (54). A circuit for use with low pressures (<0.1 torr) is given in a note by Hayes (55).

An increase in the temperature range of thermistor applicability can be expected following a report (56) describing a thermistor made of zirconia plus 15 mole per cent yttria for use between 1000 and 2500°K.

SYMBOLS USED

α	coefficient of resistance change with temperature
μ_C, μ_I, μ_T	partial differential coefficients, $(\partial V/\partial C)_{I,T}$, $(\partial V/\partial I)_{C,T}$, $(\partial V/\partial T)_{C,I}$
τ	time constant of thermistor
A	$R_0 \exp(-\Delta E/kT_0)$
B	material constant (characteristic temperature)
C	dissipation constant (heat transfer coefficient)
ΔE	one half of the energy gap between valence band and conduction band of intrinsic semiconductor
H	heat capacity of thermistor
I	current
I_B, I_C, I_E	base, collector and emitter currents of transistor
$n-$	number of electrons in conduction band
R	resistance
R_0	thermistor resistance at chosen temperature, T_0 (often 298°K)
R_T	negative resistance in equivalent thermistor circuit (= μ_I)
t	time
T	thermistor temperature
T_a	temperature of thermistor surroundings
T_i	initial temperature of thermistor (at start of cooling or heating)
V	applied voltage
W	power dissipated

MANUFACTURERS

Atkins Technical Incorporated, Box 14405, University of Florida Station, Gainsville, Fla. 32603.

The Carborundum Co. (single Crystal SiC thermistors), P.O. Box 337, Niagara Falls, N. Y. 14302

Fenwal Electronics Inc., 63 Fountain Street, Framingham Mass. 01702.

General Electric Co., Magnetic Materials Section, Edmore, Mich. 48829.

Gulton Industries Inc., 212 Durham Ave., Metuchen, N. J., 08840.

Keystone Carbon Co., Thermistor Div., St. Marys, Pa. 15857.

Mullard Ltd., Mullard House, Torrington Place, London, W.C.1, England.

Standard Telephones and Cables Ltd., Stephen Street, Taunton, Somerset, England.

Victory Engineering Corp., 118 Springfield Ave., Springfield, N. J. 07081.

Yellow Springs Instrument Co., Inc., P.O. Box 279, Yellow Springs, Ohio 45387.

LITERATURE CITED

(1) MALMSTADT, H. V., ENKE, C. G., AND TOREN E. C., JR., "Electronics for Scientists," Benjamin, New York, 1963.

(2) CHIRLIAN, P. M., "The Analysis and Design of Electron Circuits," McGraw-Hill, New York, 1965.

(3) KITTEL, C., "Introduction to Solid State Physics," (2nd Ed.), John Wiley & Sons, New York, 1963.

(4) MOORE, W. J., "Physical Chemistry," (4th Ed.), Longmans, 1963, p. 680.

(5) See FRIEDBERG, S. A., "Temperature —Its Measurement and Control in Science and Industry," Vol. 2., Ed., H. C. Wolfe, Reinhold (for American Institute of Physics), 1955, Chapter 20, and Reference 3, p. 350.

(6) STONE, F. S., "Chemistry of the Solid State," (Ed., Garner, W. E.) Butterworths, 1955, Chapter 2.

(7) TANNENBAUM HANDELMANN, E., "Recent Progress in Surface Science," (Ed., Danielli, J. F., Pankhurst, K. G. A., and Riddiford, A. C.) Academic Press, London and New York, 1964.

(8) SCARR, R. W. A., AND SETTERINGTON, R. A., Proceedings I.E.E. 107B, (35), 395 (1960).

(9) BECKER, J. A., GREEN, C. B., AND PEARSON, G. L., Bell Telephone System, Technical Publications, Monograph B 1443; also published in Trans. Am. I.E.E. 65, 711 (1946).

(10) SKINNER, H. A., STURTEVANT, J. M., AND SUNNER, S., "Experimental Thermochemistry," Vol. 2, (Ed., Skinner, H. A.) Interscience, London, 1962, Chapter 9.

(11) MÜLLER, R. H., AND STOLTEN, H. J., Anal. Chem. 25, 1103 (1953).

(12) JORDAN, J., "Treatise on Analytical Chemistry," (Eds., Kolthoff, I. M., and Elving, P. J.), Wiley-Interscience, New York; (Vol. 8 of Pt. I, in press).

(13) HALL, J. A., "Fundamentals of Thermometry" and "Practical Thermometry," Chapman and Hall, 1959.

(14) "Temperature — Its Measurement and Control in Science and Industry," Vols. I, II and III, Reinhold (for American Institute of Physics), 1941, 1955 and 1962 respectively.

(15) SLABAUGH, W. H., J. CHEM. EDUC. 42, 467 (1965).

(16) BOUCHER, E. A., J. Sci. Instr. 41, 541 (1964).

(17) COLE, K. S., Rev. Sci. Instr. 28, 326 (1957).

(18) PITTS, E., AND PRIESTLEY, P. T., J. Sci. Instr. 39, 75 (1962).

(19) HERINGTON, E. F. G., AND HANDLEY, R., J. Sci. Instr. 25, 430 (1948).

(20) JORDAN, J., J. CHEM. EDUC. 40, A5 (1963).

(21) HUTCHINSON, E., Trans. Farad. Soc. 43, 443 (1947).

(22) ZETTLEMOYER, A. C., YOUNG, G. J., CHESSICK, J. J., AND HEALEY, F. H., J. Phys. Chem., 57, 649 (1953).

(23) GREENHILL, E. B., AND WHITEHEAD, J. R., J. Sci. Instr. 26, 92 (1949).

(24) SINKE, G. C., HILDENBRAND, D. L., MCDONALD, R. A., KRAMER, W. R., AND STULL, D. R., J. Phys. Chem. 62, 1461 (1958).

(25) CRUICKSHANK, A. J. B., ACKERMANN, TH., GIGUÉRE, P. A., "Experimental Thermodynamics," Vol. 1. "Calorimetry of Non-Reacting Systems," (Eds., J. P. McCULLOUGH and D. W. SCOTT) Butterworths, in press, Chapter 12.

(26) CRUICKSHANK, A. J. B., Phil. Trans. Roy. Soc., (London), A253, 407 (1961).

(27) DE BRUYNE, N. A., Research and Development for Industry, (Publ. Newnes) No. 1, p. 94 (1961).

(28) MALMBERG, P. R., AND MATLAND, C. G., Rev. Sci. Instr. 27, 136 (1956).

(29) NOLTINGK, B. E., AND SNELLING, M. A., J. Sci. Instr. 30, 349 (1953).

(30) RAND, M. J., Anal. Chem. 34, 444 (1962).

(31) BRADLEY, R. S., J. Sci. Instr. 31, 129 (1954).

(32) HALL, P. G., AND TOMPKINS, F. C., Trans. Farad. Soc., 58, 1734 (1962).

(33) DAVIS, A. D., AND HOWARD, G. A., J. Appl. Chem., 8, 183 (1958).

(34) WALKER, R. E., AND WEATENBERG, A. A., Rev. Sci. Instr. 28, 789 (1957).

(35) HALES, W. B., Bull. Am. Met. Soc. 29, 494 (1948).

(36) TANAKA, S., Bull. Chem. Soc. Jap., 41, 723 (1968).

(37) SACHSE, H. B., "Temperature—Its Measurement and Control in Science and Industry," Vol. 3, part 2, Reinhold, New York, 1962; Article 30, p. 347.

(38) HERDER, T. H., OLSON, R. O., AND BLAKEMORE, J. S., Rev. Sci. Instrum., 37, 1301 (1966).

(39) DAL NOGARE, S., AND JUVET, R. S., "Gas-Liquid Chromatography," Wiley, New York, 1962; p. 189.

(40) KIESEBACH, R., Anal. Chem., 32, 1749 (1960).

(41) LITTLEWOOD, A. B., J. Sci. Instrum., 37, 185 (1960).

(42) SCHMAUCH, L. J., AND DINERSTEIN, R. A., Anal. Chem., 32, 343 (1960).

(43) SMITH, B. D., AND BOWEN, W. W., Anal. Chem., 36, 82 (1964).

(44) HAYES, D. J., J. Sci. Instrum. (J. Phys. E), Series 2, 1, 761 (1968).

(45) BUHL, D., Anal. Chem., 40, 715 (1968).

(46) LAWSON, A. E., Amer. Lab., May, 1969.

(47) STUDER, J. J., "Electronic Circuits and Instrumentation Systems," Wiley, New York and London, 1963; Chap. 9.

(48) VAN SWAAY, M., J. CHEM. EDUC., 46, A515, A565 (1969). [Reprinted in this volume.]

(49) RASMUSSEN, R. A., Rev. Sci. Instrum., 33, 38 (1962).

(50) PIGOTT, M. T., AND STRUM, R. C., Rev. Sci. Instrum., 38, 743 (1967).

(51) BARBEE, D. G., AND ENDRUD, G. H., J. Sci. Instrum. (J. Phys. E), Series 2, 1, 1197 (1968).

(52) DUSHMAN, S., "Scientific Foundations of Vacuum Technique," 2nd Edn., Wiley, New York, 1962; p. 297.

(53) GREEN, M., AND LEE, M. J., J. Sci. Instrum., 43, 948 (1966).

(54) BERRY, C. J., J. Sci. Instrum., 44, 83 (1967).

(55) HAYES, K. E., J. CHEM. EDUC., 47, 53 (1970).

(56) WOLFF, E. G., Rev. Sci. Instrum., 40, 544 (1969).

Roman Sacred Cross

236

Topics in...

Chemical Instrumentation

Edited by GALEN W. EWING, Seton Hall University, So. Orange, N. J. 07079

The Journal of Chemical Education
Vol. 46, pages A515, A565
August and September, 1969

XLVI. The Control of Temperature

MAARTEN VAN SWAAY, Department of Chemistry, Kansas State
University, Manhattan, Kansas 66502

The control of temperature is involved in virtually all experiments performed in a chemical laboratory. In many cases, components must be controlled to within 0.001° of a set point, although the actual value of that set point is usually less critical and may be specified to, e.g., 0.1°. It is no surprise, therefore, to find that thermostats are found in profusion in all well-equipped laboratories. Yet, with the exception of two articles by Lewin (1, 2) discussion of temperature control in the chemical literature is normally limited to single-line statements; a nice compliment to the performance of the equipment available today.

At the time Dr. Lewin wrote the review cited above, the control of moderate and large amounts of electrical power was achieved almost exclusively with relays. For reasons which will be discussed later, this puts rather severe limitations on the type of control that can be achieved conveniently, and the thermostats described by Dr. Lewin reflect these limitations. Since that time, solid-state power switches such as silicon controlled rectifiers and triacs have made fast and reliable control of power much easier; their availability has had a large effect on the design of temperature controllers. Whereas Dr. Lewin's articles dealt primarily with on-off control systems and modifications thereof, we will be more interested in proportional control and in dynamic response in the present discussion.

Like any control system a thermostat may be divided into four parts: a sensing

element, a control amplifier, a heating (or cooling) element, and the controlled component, e.g. an air or water bath. The four parts are connected in a control loop, as shown in Figure 1. Information representing the desired operating temperature may be inserted at the input to the control amplifier. In some very simple instruments, e.g., hot plates, the sensor and its connection to the control amplifier may seem to be missing. Its task is then performed by the human operator, who assesses temperature by some convenient means, and adjusts the input to the control amplifier accordingly. In such cases we speak of open loop systems; the term closed loop is reserved for systems in which the human operator is not a part of the loop. In the following sections, the four parts will be discussed in detail, and the performance of the entire loop will be evaluated on the basis of the behavior of the individual parts. It will be convenient to describe the behavior of the various parts in terms of their transfer function, which is defined as the relationship between input and output.

The operation of a temperature controller, and of control loops in general, can be conveniently described on the basis of an electrical analog. To construct this analog, we can replace heat flow by electrical current and temperature differences by electrical voltages. Heat capacity will then be represented by capacitance, and thermal resistance by electrical resistance. Thermal devices acting like inductors are rather unlikely, and we will not need them in the present discussion.

We are now in a position to sketch an electrical equivalent of a temperature controller, which may look like the circuit shown in Figure 2. An important parameter of this circuit is the loop gain. If the loop is broken at any point, a signal injected at the break will produce an amplified signal at the other side of the break. It will be convenient to make the break in the lead carrying the error signal. The ratio between these two signals is defined as the loop gain. The fact that the two sides of the break may be connected together implies that both signals must have

Dr. Van Swaay was born in The Hague in 1930, and received his undergraduate education at Leyden University in the Netherlands, with a bias toward physical chemistry. From 1953–56 he pursued graduate work at Princeton University and was awarded the PhD degree for his thesis on the diffusion of hydrogen through palladium.

In 1956 he worked as a research associate at Leyden University on dielectric properties of nylon and other polymers. In 1959 he joined the staff of the Technical University at Eindhoven, the Netherlands. There, he acquired a back-ground in chromatography and electrochemistry, and developed a strong taste for instrument design.

In 1963 he joined the faculty at Kansas State University, where he recently was promoted to associate professor. His present interests include chromatography and separation techniques, in addition to extensive work in the design of electronic equipment for chemical use.

the same dimension, independent of the presence of electrical, thermal, mechanical, or other signals elsewhere in the loop.

Since a signal will generally require a finite time to travel around the loop, the loop gain will include a time element. We can account for this time element by representing the loop gain as a vector in a polar coordinate system, in which time is represented by an angle. To normalize the representation, the time required for one full period of a cyclic input signal will be made to correspond to an angle of 360°. The angle between the input and output signal vectors, measured in counterclockwise direction, can then be defined as the phase angle.

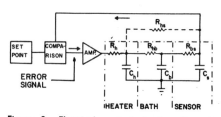

Figure 2. Electrical equivalent of a thermal control system. Signal flows around the loop in the direction of the arrows.

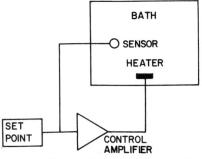

Figure 1. Elements of a temperature controller.

Chemical Instrumentation

For stable operation the loop should contain an odd number of phase inversions, corresponding to negative feedback. Graphically, if the input signal vector is drawn as shown in Figure 3, the output vector should fall in the first or fourth quadrant.

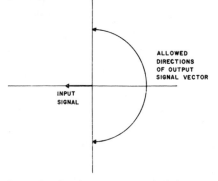

Figure 3. Graphic representation of the gain vector of a control loop in polar coordinates. If the size of the input vector is defined as unity, the output vector represents the gain.

For signals of very low frequency, the transit time through the loop becomes insignificant, so that the ouput vector will be parallel to the input vector, and of oppposite direction. The presence of low-pass filters of the type shown in Figure 2 will give rise to a decrease in loop gain with increasing frequency, and to a concomitant phase lag. It can be shown that a correlation exists between the gain-frequency curve and the phase lag, such that when $d\ln A/d\ln\omega = n$, the phase shift $\phi = n\pi/2$. The position of the break in the gain-frequency curve can be defined as the point at which the phase angle is $\pi/4$ radians; at that point $\omega = 1/\tau$ in which ω is the frequency in radians/second, and $\tau = RC$ is the time constant of the filter. The gain-frequency curve for a single RC filter is shown in Figure 4.

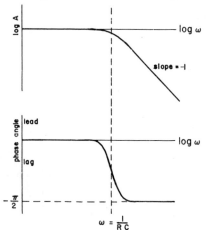

Figure 4. Gain-frequency curve for a single low-pass filter. Note that the gain- and frequency axes have logarithmic scales.

In a thermostat we can generally recognize at least 3 low-pass filters: one each due to the heat capacity of the heater, the bath, and the sensor. In addition, the control amplifier may have one or more low-pass filters, normally with much shorter time constants. At frequencies

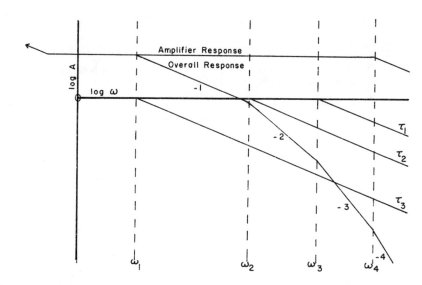

Figure 5. Gain-frequency curve of the entire control loop. The curves labeled τ_1 τ_2 and τ_3 indicate the response of the three loop elements with the longest time constants.

above the cut-off points of the three thermal low-pass filters we should expect a phase lag of 270°. This means that there will be some frequency at which the phase shift due to the low-pass filters equals 180°, which would produce a positive loop gain. If at that frequency the loop gain has a magnitude of unity or larger, the system will oscillate.

We are thus faced with a set of conflicting requirements: high loop gain is desirable to minimize the effect of external factors such as changes in heat loss from the bath to its surroundings; yet the gain must fall below unity at the frequency for which the phase shift reaches 180°. To make matters even more complicated, reasonable gain should be available at finite frequencies to insure adequate response to transient disturbances.

The gain-frequency characteristic of the loop can be broken into three major regions (Fig. 5): at frequencies above ω_2 the gain should be less than unity to prevent oscillation; over a range of frequencies up to ω_1 the gain should be essentially constant and adequate to handle transient disturbances, and toward zero frequency the gain should become as high as possible to minimize offset errors, i.e., constant deviations from the chosen setpoint.

In the middle region, the response of the system is proportional to the instantaneous error signal; we speak of proportional control. The gain in this region is gen-

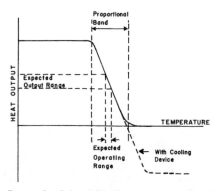

Figure 6. Relationship between proportional band, expected variation in heat load, and operating temperature range.

erally made as large as is consistent with stable operation. Practical specifications are rarely given in terms of gain, however; instead, the proportional band is normally stated.

It is obvious that the heater output of a thermostat cannot increase without limit as the temperature error increases; neither can the heat output decrease below zero. The presence of cooling devices changes this argument only slightly, to allow a finite, rather than zero cooling output. The proportional band is now defined as the temperature interval within which the heat output varies linearly or nearly linearly as a function of the deviation from the setpoint. Since an estimate of the required range of heat output can usually be made, knowledge of the proportional band and the maximum available heat output allows one to predict the long-term stability of the system. This is illustrated graphically in Figure 6.

When it is impossible to reduce offset errors of a proportional system to acceptable levels without instability or oscillation, an integral term may be introduced into the loop gain, so that the output becomes a function of both the error signal and its time integral. Mathematically it is easy to see that the integral term corresponds to a slope of -1 in the gain-frequency plot. The resulting control system is referred to as a two-term, proportional + reset, or P-I system. The first two names have seniority, but the last one is the most concise and deserves to become the preferred name.

Laboratory thermostats generally operate in sheltered surroundings, so that heat losses vary only slightly with time. Offset errors would then also remain essentially constant, and they can be compensated by manual adjustment of the setpoint. (One might say that in this case the operator provides the integral response.) For this reason, integral response is rarely required for laboratory equipment. An exception might be the control of devices through a wide temperature range, e.g., the heating mantle around a distillation column.

On the other hand, fast response to transient disturbances is a desirable and often necessary property of laboratory equipment. This may require the use of an anticipating or derivative term in the control loop. If the gain of the two elements with the longest time constants is plotted as a function of frequency, the

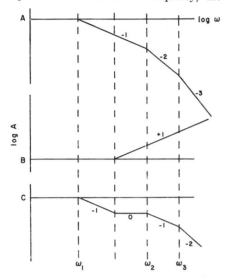

Figure 7. Compensation of response roll-off due to thermal low-pass filters by anticipating or derivative circuits. A, response due to thermal elements; B response of derivative circuit; C, compensated response. For proper operation, the break in curve B should fall between ω_1 and ω_2 Numbers in the diagram indicate the slope of the response curves.

curve will have the shape shown in Figure 7A. At frequencies above ω_2 the curve has a slope of −2. The phase shift at those frequencies will be 180° or more, and the overall loop gain should therefore be less than unity to prevent oscillation. If now a circuit is inserted elsewhere in the loop with a gain-frequency curve corresponding to Figure 7B, then the combination of Figures 7 A and B will give the gain curve shown in Figure 6. The slope of this curve will fall to −2 at some higher frequency determined by the third largest time constant of the system. The entire gain-frequency curve may now be shifted upward, until the gain at frequency ω_3 becomes equal to unity.

A gain-frequency curve with a slope of +1 (Fig. 7B) corresponds to a response proportional to the time derivative of the error signal. Controllers with this type of response are said to have anticipating or rate response, or, preferably, derivative or D response. Proper use of a D term in a controller can lead to substantial improvement of the system, but it should be realized that adjustment of the D term is critical. An upward slope of the overall gain-frequency curve will tend to make the system noisy due to excessive gain at high frequencies. The break point of the derivative term should therefore be set beyond the longest time constant of the system. On the other hand, derivative action must set in before the cutoff point of the next shorter time constant is reached. Careful matching of the derivative response to the characteristics of the system is therefore essential.

It is possible to adjust the P-, I-, and D-gains by means of electronic circuits, but the D response can often be controlled quite effectively by proper placement of

the sensor relative to the heater. In Figure 2 we note the presence of resistor R_{hs}, representing direct coupling between the heater and the sensor. With adequate stirring, the resistance between sensor and bath will be largely independent of the sensor position, but the resistance between sensor and heater can be quite sensitive to their relative position. It is thus possible to make the sensor see a change in heater output before the bath responds to it. In practical systems, little, if any, current should flow through resistor R_{bs}, because the sensor temperature should be essentially equal to the bath temperature. Then optimum response will be achieved if the time constans $R_{hb}C_b$ and $R_{hs}C_s$ are made equal. If the time constant $R_{hs}C_s$ is too small, the sensor will produce a large derivative signal, which will give rise to sluggish response of the thermostat. Conversely, if $R_{hs}C_s$ is too large, overshoots and even oscillation may occur.

A popular technique to achieve fast response with high loop gain (i.e., tight control) is to use an incandescent bulb for a heater, and an optically absorbing sensor. The heater and sensor are then coupled by

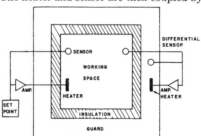

Figure 8. Use of a guard system to minimize heat flow between a temperature-controlled bath and its surroundings. The guard controller is designed to minimize the temperature difference between the working space and the guard.

radiative heat transfer, which is easily controlled by means of distance, or with a small opaque baffle. Such systems are capable of controlling small water baths to within 0.001° or better. Another advantage of incandescent lamp heaters is the fact that a large fraction of the heater output is introduced uniformly throughout the bath as absorbed infrared radiation.

In general, heat is introduced in one region of the bath, and dissipated elsewhere. Since the thermal conductivity of the bath is finite, heat flow through the bath will produce a temperature gradient. This gradient can be minimized by increasing the conductivity of the bath, and by making the heat flow due to dissipation as small as possible. High thermal conductivity is normally insured by vigorous stirring; heat loss to the surroundings can be minimized by careful insulation, and, for critical applications, by the use of a guard system.

A guard consists of a second bath surrounding the working space and controlled at the same temperature (Fig. 8). If the two baths are separated by thermal insulation, the effect of this insulation will be multiplied by a factor $1 - A_2$, in which A_2 is the loop gain of the guard controller which senses the temperature difference between the working and guard baths. Obviously the use of guards has little value if the working space must contain

devices absorbing significant amounts of heat.

Since heat loss to the surroundings is reduced to negligible proportions, any heat released in the working space must raise its temperature. The system can no longer be called a thermostat, but since temperature changes of the working space can be directly related to the amount of heat released and to the heat capacity of the working space, we now have an adiabatic calorimeter. The advantage of guards to achieve very small heat loss lies in the fact that relatively little passive insulation is required, so that the heat capacity of the system can be kept small. Hence small heat effects give rise to relatively large temperature changes. With proper design, calorimeters can now be built which will detect heat effects as small as a few millicalories.

A construction approaching the performance of a guard system without requiring two controllers is shown in Figure 9. This design is particularly well suited for column ovens for gas chromatographs. A squirrel-cage fan maintains a toroidal air flow through the working volume and the annular space between a cylindrical baffle and the thermostat walls. Heater windings are located in the annular space, which effectively isolates the working space from the outside world. The working space is uncluttered and readily accessible from the top. The resulting structure has very low thermal mass, and programming rates of 50°/min can be easily achieved with power requirements under 1000 W.

The control amplifier

The control amplifier must perform two distinct functions, which can usually be recognized in the design of the amplifier. Input circuitry is designed to accept the signal from the sensor, and operate on it in such a manner that the desired characteristics of the control loop may be obtained. In addition to simple amplification, these operations may include the formation of the time derivative and time integral of the error signal. This part of the amplifier could be called the signal-conditioning part, or the preamplifier.

The conditioned signal available from the preamplifier can now be used to drive a power stage, which is designed to handle the large power required by the heating

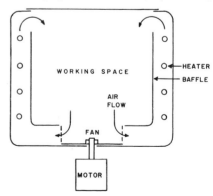

Figure 9. Shielded thermostat. With this design, most of the heat lost to the atmosphere passes directly from the heater to the outer walls. This minimizes heat flow and thermal gradients in the working space.

Chemical Instrumentation

element. Except for very small thermostats with power requirements under, e.g., 10 W, it is almost imperative to use a nondissipating power stage, so that the power dissipation of this stage is much smaller than the power delivered to the heating element.

It is generally easier to control heating power than cooling power. Since the total power consumption of the system rarely is an important consideration in laboratory equipment, control of subambient temperature will normally be achieved by constant excess cooling, and controlled reheating. Thus the final control element will almost always be an electrical heater of suitable size, and the design of the power stage will vary only slightly between different thermostats. For this reason, the power stage will be discussed first.

The most popular nondissipating power stages available today are variable transformers, silicon controlled rectifiers (SCR) and Triacs, relays, and saturable reactors. The variable transformer requires a mechanical input signal. Although this could be supplied from, e.g., a servo motor, the conversion from electrical into mechanical signal when an electrical signal is ultimately required is both inelegant and unduly complex. The variable transformer is therefore rarely used in closed-loop systems.

The other three power stages named can all be treated as switches. The relay is subject to contact erosion by arcing, especially if it has to interrupt large currents. Hence the lifetime of the relay will be limited to some finite number of operating cycles. If dependable operation over long periods is desired, the number of relay cycles per unit time must therefore be limited. As a result, heat will be applied to the bath in relatively large bursts at low repetition frequency. The resulting temperature fluctuations must be damped out by a correspondingly large thermal time constant of the system, so that relay-operated thermostats will be sluggish and incapable of responding to fast transient disturbances.

If the power controlled by the relay is limited to a small fraction of the total power required, thermal ripple will be correspondingly reduced. A second heater must then be used to supply the bulk of the heating power. As an incidental advantage, the smaller currents controlled by the relay will increase its operating life, or allow higher switching rates.

The SCR acts as a latching relay: a firing pulse will switch it into a low-resistance state, which it maintains until current through the device is reduced to a very small value. In contrast with the mechanical relay, the SCR has no limitation on the number of switching cycles, and is capable of switching rates of 1000/sec and more. It is eminently suited to be switched on at a controlled point in each cycle of the line voltage, after which it will turn off at the end of the half cycle. Power thus can be delivered in short bursts with repetition rates of 60/sec. With Triacs, or with 2 SCRs connected in antiparallel, the repetition rate will be

120/sec. The resulting thermal ripple will be completely damped out by the thermal time constants of the average thermostat bath.

Saturable reactors produce an output comparable to that of the SCR. The reactor is connected in series with the load. At the beginning of a half cycle, energy is stored in the core of the reactor, until it is saturated. During this time, the bulk of the available voltage appears across the inductance of the reactor. After the core is saturated, the inductance drops to a low value, and supply voltage appears across the load. The time required for saturation is controlled by means of a control current which partially magnetizes the core. An increase in this control current will decrease the time required for saturation, so that more power is available for the load. Saturable reactors are relatively bulky and expensive, and appreciable power is required for the control winding; although they are extremely reliable, they are rarely used in the laboratory.

In view of the characteristics described above it is not surprising that the SCR is by far the most popular device to control power at the present time. Full-wave control may be achieved either with a pair of SCRs connected in antiparallel, or by means of a Triac, which acts as a bidirectional SCR. Although the characteristics and uses of SCRs and Triacs are thoroughly dealt with in various publications from semiconductor manufacturers (3, 4) a brief discussion will be given here.

As stated above, SCRs and Triacs can be regarded as latching relays which close on application of a firing pulse, and open when the output current falls to zero, e.g., at the end of a half cycle.

The SCR is a three-junction p-n-p-n device, as shown in Figure 10. Its opera-

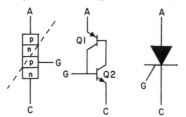

Figure 10. Structure and operation of a silicon controlled rectifier. (left) Topology; (center) transistor equivalent circuit; (right) symbol. A = anode; G = gate; C = cathode.

tion can best be understood if we treat it as a pair of transistors connected as indicated in the figure. This circuit is regenerative: collector current from Q1 appears as base current for Q2, and is amplified by Q2. This amplified current in turn reappears as base current for Q1, giving rise to an increase in collector current at Q1. Thus both transistors are driven into saturation, and large currents can pass from the anode to the cathode connection with a low voltage drop of the order of 1 v.

The regenerative action of the device depends on the product of the current gains of Q1 and Q2, which should be larger than unity. At very low currents, however, the current gain of transistors will fall off and can become less than unity. In that case, no regenerative action can

occur, and only a very small current will flow, owing to the presence of minority carriers. The device can be switched on by a momentary increase in base current of either transistor to a level at which the current gain product becomes larger than unity. The trigger pulse is applied to the base of Q2, which is called the gate. Typical pulse parameters are 20 mA at 2–3 v for 1–10 μsec. Averaged over one half cycle, this amounts to about 25 μW of gate power. Output currents of 10–30 A at line voltage are quite common, so that power gains of 10^8 are possible. By comparison, a power gain of 10^4 is common for single relays.

The Triac is a more complex device, but its operation again is based on the regenerative action of a four-layer structure. The Triac can conduct current in both directions, and may be switched on by a positive or negative current pulse, for either direction of the main current. As far as the power-handling capabilities are concerned, the Triac is equivalent to 2 SCRs in antiparallel.

The SCR normally switches on within microseconds after it receives a firing pulse. If this pulse arrives when substantial voltage is available across the SCR, large and sharp voltage and current

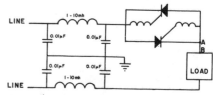

Figure 11. Placement of noise filters to reduce interference from SCR power controllers. The filters should be placed close to the SCRs and the load. If long leads must be used between the SCRs and the load, additional filters may be inserted at points A and B.

transients occur in the wiring to the SCR and the load. The harmonic content of the resulting waveforms may well reach into the broadcast frequency band and give rise to severe interference with other equipment in the vicinity. When sensitive equipment must be operated close to large SCR controllers it is therefore advisable to round off the sharp waveforms of the SCR, and to trap high-frequency interference by means of noise filters. These two functions can often be performed by small inductors in the power lines going to the SCR and its load. In severe cases, bypass capacitors may be added to form low-pass filters in the line, as shown in Figure 11.

An effective way to reduce the amount of interference is to fire the SCR early in a half cycle, when only low voltage appears across the SCR. Power is then supplied in half cycle bursts to the heater, and total power delivered depends on the number of pulses passed per second, rather than on the size of a series of uniformly spaced pulses. Strictly speaking this is a form of on-off control, but the on-off cycle may be so short that the end result is indistinguishable from that of a true proportional control.

A simple and reliable firing circuit for phase control of SCR stages can be built with a unijunction transistor (UJT). A circuit is shown in Figure 12; for a description of its operation the reader is

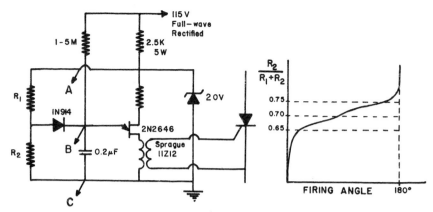

Figure 12. Phase-controlled SCR firing circuit. The steeper section in the center of the response curve compensates for the fact that a given change in firing angle produces a larger change in power near the peak of the half wave than at either end. Resistors R₁ and R₂ should be relatively small (e.g. 5K) if control at small firing angles is required. Points A, B, and C may be connected to correspondingly labeled points of the preamplifier shown in Figure 13.

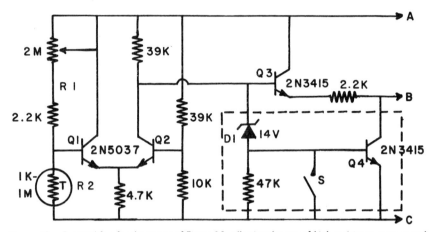

Figure 13. Preamplifier for the circuit of Figure 12, allowing the use of high-resistance sensors and reduced bridge voltage. The components in the box can be added as a fail-safe circuit, so that power output goes to zero if the resistance ratio R_1/R_2 becomes too large.

referred to, e.g., the SCR Manual published by General Electric Co. (*3*). The firing angle of this circuit depends on the resistance ratio R1/R2 as shown in the diagram, if the resistors R1 and R2 do not exceed a few thousand ohms. Although a resistive sensor could be used as R1 or R2, the required low resistance and the high voltage across the sensor (*ca.* 10 v) may give rise to unacceptable self-heating.

High-resistance sensors can be used if an emitter follower is inserted between the sensing bridge and the UJT circuit but a better solution is shown in Figure 13. Here the bridge output is amplified by a differential transistor amplifier Q1-Q2. Since this amplifier also introduces an upward voltage shift, the sensor may be operated at a lower voltage if it is inserted in the lower part of the bridge. In addition, the gain of the amplifier makes the stability of the UJT circuit less critical. We have used a circuit of this type to control the temperature of a gas chromatograph injection port within 0.2°C over a range from 70–300°C with a single thermistor. (The temperature variation between parts of the injection block is very much larger, but this does not reflect on the performance of the controller.)

Resistive sensors generally decrease in resistance with increasing temperature. Thus an open circuit in the sensor could be interpreted by the control system as a low temperature, so that full power would

be applied with possibly disastrous results. This situation may be avoided by addition of the components enclosed in the box (Fig. 13). Under proper operating conditions the voltage at the emitter of Q3 is lower than the breakdown voltage of zener diode D1. Hence transistor Q4 is cut off, and the UJT circuit operates normally. When a break occurs in the sensor leads, the emitter of Q3 will rise close to the supply voltage, so that zener diode D1 will fire and saturate transistor Q4. This clamps the emitter of the UJT to a very low voltage, so that no firing pulses can be generated. The fail-safe circuit must be deactivated, however, until the system has reached operating temperature. This is achieved by means of switch S which shorts the base of transistor Q4 to ground and keeps it in cut-off state.

The addition of the differential amplifier shown in Figure 13 makes the circuit compatible with dc control signals, such as may be obtained from a thermocouple. More amplification may be necessary in that case, however. Economical amplifiers are now available which will respond reliably to input signal variations of the order of 10 μv, corresponding to 0.2°C for, e.g., an iron-constantan thermocouple. With dc sensor signals, differentiation and integration can also be performed easily by means of operational amplifiers; suitable circuits may be found

in the commercial literature (*5, 6*).

Zero-voltage switching circuits may consist of a line-synchronized pulse generator supplying firing pulses to the SCR via a control gate. The gate in turn is actuated by the output of the sensor after suitable amplification.

Both phase-controlled and zero-voltage firing circuits have recently become available as integrated circuits suitable for 115 v line operation (*7*). In combination with a thermistor sensing bridge and a Triac, these circuits can be used to control temperature within 0.3°C in practical circumstances.

Sensors

Among commonly used sensors we may list thermocouples, thermistors, resistance elements, expansion devices, and bimetallic elements. Other sensors do exist, but the five types listed are generally available at reasonable prices, and one or more of these types will meet virtually any need that may be encountered in the laboratory. The expansion devices include liquid- and gas-filled thermometers; their output is usually an on-off response. Bimetallic elements also convert temperature changes into mechanical motion; they too are usually designed to actuate a set of contacts if an electrical response is required. Both have been adequately discussed by Lewin (*1, 2*); they are poorly suited for proportional control and will not be further discussed here.

Thermocouples

The thermocouple is a comparison device, which responds to temperature differences between two junctions of dissimilar metals. Commonly used combinations are listed in Table 1, together with some parameters and properties of interest. The sensing junction may be made by spotwelding, or by fusion of the twisted leads in the reducing part of a flame with a suitable flux, e.g., borax. When metallic components must be controlled, it is often possible to spotweld both leads of the thermocouple to the component. Although this introduces an additional junction, the two welds can be placed close together, so that their temperature difference is negligible. In that case it can be shown thermodynamically that the overall response is the same as for a single junction between the thermocouple leads.

Since the thermocouple responds to temperature differences, the reference junction must be kept at a known temperature, e.g., 0°C. A convenient construction for a cold junction is shown in Figure 14. Both thermocouple leads are inserted into small glass tubes containing a few drops of mercury. Copper connecting leads are also inserted into the tubes, and the whole assembly is surrounded by an ice bath. Although the copper slowly amalgamates with the mercury, a reliable low-resistance contact is maintained for many months.

Where the maintenance of an ice bath is inconvenient, the reference junction may be mounted in a small auxiliary thermostat which is controlled by means of an absolute sensor, e.g., a thermistor or

Chemical Instrumentation

resistance element. Reference junctions from a number of thermocouples may be mounted in the same reference thermostat. An example of such a device is shown in Figure 15. Two small screw connectors for each thermocouple are mounted in an aluminum block with thermally conductive cement (8). Reliable insulation in the multimegohm region can be achieved if the connectors and the block are anodized before assembly. The block further contains cavities for all elements of the control amplifier and heater, the circuit of which is also shown in the figure. An increase in temperature will lower the voltage at the base of Q1, which forms a Darlington pair with Q2. As a result, collector current through Q2 and heater resistor R2 decreases.

With the design shown, all the heat generated by the controller is available for the thermostat, and operating power well under 1 W at 40°C can easily be achieved if the block is surrounded with $1/_8$ in. styrofoam. The thermistor should be mounted close to the heater resistor to prevent oscillation. Sluggish response

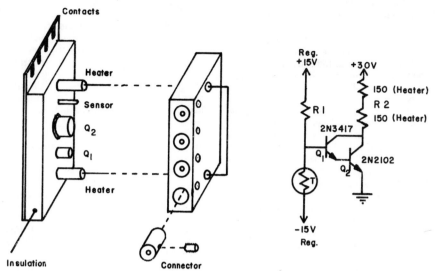

Figure 15. Thermostat for thermocouple reference junctions. (left) Typical mechanical construction; (right) circuit diagram. For the sake of clarity, an insulating cover ($1/_8$ in. styrofoam) is not shown. Resistance of the bridge elements should be between 10 K and 100 K at the operating temperature. Since the $+15$ v and -15 v supplies form part of the sensing bridge, they should be well regulated; any change in the symmetry between these voltages will change the temperature of the thermostat.

can be tolerated because the block will rarely be exposed to thermal transients. With proper design, temperature fluctuations of the block can be reduced to 0.01° or less, with comparable long-term stability. A major factor making this performance possible is the fact that the components of the amplifier are mounted in the thermostat, so that temperature effects on the base-emitter characteristics of the transistors are reduced to negligible proportions.

A similar design has been used to control the temperature of transistors used in a logarithmic converter for spectrometric measurements. Both the sample and reference transmittance signals were converted into logarithmic form with a circuit described by Paterson (9); with the feedback transistors mounted in the thermostat described above, long-term stability of better than 0.001 absorbance unit could be achieved. Hybrid and monolithic amplifiers are also small enough to be mounted in these thermostats to reduce thermal drift.

Thermocouples generally have appreciable resistance. It is therefore advisable to keep their length short if the sensor must supply significant current, e.g., to an indicating meter. In the laboratory, the cold junction can usually be mounted close to the sensor, so this requirement can be easily met. When this is not possible, thermocouple extension wire can be used.

These extension wires are made of alloys with the same thermal characteristics as the corresponding thermocouple, so that the junctions do not introduce additional thermal voltages. Since high-temperature stability is not required for extension wire, low-resistance alloys can be used. The price of extension wire is comparable to that of the thermocouple itself, and may be slightly higher in some cases.

In many cases it is possible to allow for the finite resistance of the thermocouple by the addition of an external resistor which is adjusted to bring the total resistance to some chosen constant value. This value, often 50 ohms, is usually marked on the face of meters designed for temperature measurement.

Although the power output of a thermocouple can be surprisingly high (safety devices for furnaces and water heaters may be based on a solenoid-shut-off valve directly driven by a thermocouple), the voltage output is only 10–50 μv/°C. This low output voltage imposes rather stringent requirements on the input characteristics of subsequent amplifiers.

Two techniques are available to achieve low input drift in solid-state amplifiers: chopper stabilization, and the use of thermostatted input circuits. Chopper-stabilized amplifiers are available in profusion (10, 11); an example of a thermostatted amplifier is the μA727 monolithic

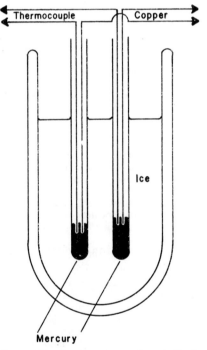

Figure 14. Constant temperature bath for thermocouple reference junctions.

Table 1. Properties of commonly used thermocouples. Data are for 24 gauge wire (13, 14)

Type	ISA Code	Temperature range (°C)	Average output (μv/°C)	Approx. resistance per foot at room temp. (ohms)	Corrosion resistance
Copper/constantan	T	−150 to 250	40	0.68	moisture-resistant
Iron/constantan	J	−20 to 400	55	0.81	oxygen-sensitive above 400°C
Chromel/constantan	E	−15 to 450	70	1.70	good corrosion resistance especially at low temperature
Chromel/alumel	K	−20 to 900	40	1.50	recommended for oxidizing atmosphere between 500–900°C
Platinum/platinum 10% rhodium	...	400 to 1700	12	0.42	superior high-temperature stability

Data in this table are approximate only; both resistance and response will vary with temperature. The temperature ranges listed are recommended by ISA; operation outside these ranges is generally possible, if allowance is made for reduced lifetime at high temperatures and/or reduced output at low temperatures.

circuit (12), which consists of a differential amplifier and a temperature controller on a single silicon chip. With either type of amplifier, thermal stability of 1 $\mu v/°C$ can be attained. If the amplifier is mounted in a thermally sheltered environment, it will thus be able to resolve signals from a thermocouple corresponding to 0.05° temperature change or less. If the specifications are relaxed slightly, direct-coupled amplifiers may be used; with these, resolution to about 0.5°C is possible. A practical circuit is shown in Figure 16; this controller has been used successfully in our laboratory to control column and detector ovens of gas chromatographs.

Two features of Figure 16 are noteworthy: the thermocouple is part of a resistance-sensing circuit including resistor R1. Under normal conditions, the resistance of the thermocouple is very small compared to R1, and only a small constant voltage is generated across the thermocouple. When the sensor fails, however, its resistance will increase drastically, with a corresponding increase in voltage across its terminals. This extra voltage will reduce the output power to zero and thereby prevent damage to the oven.

The second feature of interest is the function of resistor R2. It can be adjusted to compensate the offset signals produced by the fail-safe circuit just described, and by the chosen temperature of the reference junction.

Since the output of most thermocouples is a very predictable and essentially linear function of temperature, such sensors are well suited for controllers which must follow some prescribed temperature program, e.g., in programmed temperature gas chromatography. The thermocouple output is then compared with an electrical signal representing the desired temperature program, which may be produced either electromechanically or electronically.

Thermistors

Thermistors are temperature-sensitive resistors made of semiconducting materials. The most widely known types consist of doped or non-stoichiometric transition metal oxides. A good review of thermistor theory has recently been published in this series (15). Since conductance in these devices depends on thermal excitation of charge carriers into a conduction band, the logarithm of conductance may be expected to increase linearly with reciprocal temperature. In practice, resistance is a more useful parameter; it decreases from 1-4% per °C temperature rise.

Since thermistors are passive devices, power must be supplied to them before they can be used as sensors. To reduce the effect of power supply variations, the thermistor is normally connected in one arm of a Wheatstone or similar bridge. Power dissipation in the sensor will produce a temperature gradient between the sensor and its surroundings. For accurate work, this gradient should be kept small enough, so that variations of the dissipation constant, e.g., due to stirring, do not give rise to unacceptable variations in the temperature difference between the sensor and its surroundings.

Since alternating voltages are generally easier to amplify than dc, one might be tempted to use a line-frequency bridge supply in combination with an ac amplifier and rectifier. Problems arise, however, because the instantaneous heat input to the thermistor varies at twice the supply frequency. The resulting temperature variations of the sensor give rise to phase shifts and distortion of the bridge output. A clean squarewave supply may be expected to eliminate such difficulties, since it will produce a constant heat input at the sensor. No references to square-wave operation have been found in the literature however.

With bridge supply voltages of the order of 20 v, the response of a thermistor bridge can easily exceed 100 mv/°C, so that very simple amplifiers will often be adequate. The nonlinear and somewhat unpredictable response makes the thermistor a poor choice when prescribed temperature programs must be followed, although thermistors with guaranteed adherence to standard response curves within 1°C over 100° intervals are now available (16).

The negative temperature characteristic of thermistors may cause a control amplifier to interpret a break in the thermistor circuit as a low temperature. This would make the heater deliver full power to the bath, and severe damage might occur. Thermistor controllers designed for unattended operation should therefore have some provision to prevent thermal runaway. This protection can be designed into the control amplifier, as shown in Figure 13 or it can be added as a separate over-temperature protector, e.g., a simple bimetallic switch.

Resistance elements

Resistance elements are based on the positive temperature coefficient of resistance of most metals and alloys. This effect is due to collisions of electrons in the conduction band with thermal disturbances of the crystal lattice. For platinum a popular choice for resistance sensors, the resistance increases about 0.3% per °C. To obtain a reasonable resistance of the sensor, fine wire must be used, often wound on a quartz or mica mandrel. Even so, the total resistance rarely exceeds 100 ohms, compared to 100 K to 1 M for thermistors. The low resistance of the sensor and its comparatively small temperature coefficient dictate the use of subsequent amplification.

The thermal stability of platinum sensors makes them suitable for operation over a wide range of temperature. In addition, the response is highly predictable if mechanical stresses are avoided, so that controller settings can be calibrated directly in temperature. The construction of resistance elements makes them much more delicate than thermistors, however, and where predictable response, and wide-range operation are not required, thermistors would be the preferred choice.

Heating and cooling devices.

Heaters in the 100–1000 W range are available in a wide range of shapes and prices. Primary requirements for application in fast-acting systems are low heat capacity and large surface. The low heat capacity insures fast response to changes in applied power, and the large surface allows tight coupling to the bath, so that the time constant $R_{hb}C_b$ (Fig. 2) becomes as small as possible. Other criteria, such as chemical inertness, space requirements, etc. must also be taken into account. When the controlled device is a solid, e.g., the detector block of a gas chromatograph, a large surface area is of less concern than the maintenance of good thermal contact over the available surface area. In such cases, attention should be given to possible warping of the heater at elevated temperatures. Popular choices for such applications are cartridge heaters and ring elements. Cartridges are made to fit snugly into round holes of $3/8$ in. or $1/2$ in. diameter; ring elements are designed as flat washers which can be bolted or pressed to a surface. For optimum performance,

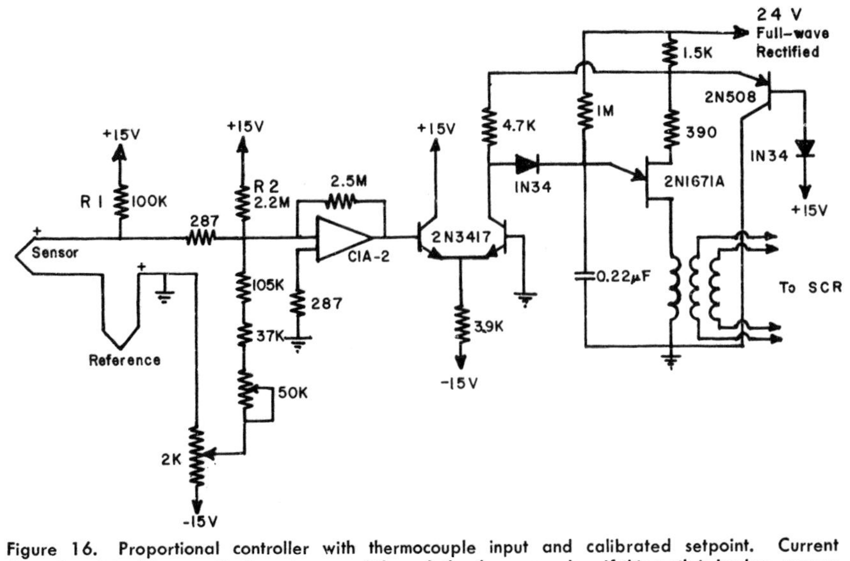

Figure 16. Proportional controller with thermocouple input and calibrated setpoint. Current through resistor R1 normally flows to ground through the thermocouple; if this path is broken, power output is reduced to zero. Resistor R2 compensates for the offsets due to the fault protection circuit and to the nonstandard reference temperature. Amplifier CIA-2 is manufactured by Philbrick/Nexus Research.

Chemical Instrumentation

the mating surface should be machined to insure flatness. At temperatures under 250°, thermally conducting cements and greases can be used to advantage (8).

Cooling used to be achieved almost exclusively with water and ice baths. The amount of cooling by such means cannot be controlled easily with simple equipment, so that excess cooling and controlled reheating is normally used. Electric cooling is possible, however, by means of the Peltier effect. Basically, the Peltier effect is the reverse of a thermocouple effect: if current is passed through a circuit consisting of two dissimilar conductors, one of the junctions will be heated and the other will be cooled by the current. In addition to the Peltier effect, the junctions and leads will be subject to ohmic heating. For commonly used thermocouples, the ohmic heating effect is generally large enough to override the Peltier effect, and little or no cooling can be achieved in practice. Recent developments in semiconductor technology have now produced a Peltier cell based on a bismuth-telluride p-n junction, with which up to 25 W of heat may be withdrawn across a temperature differential of 20°C and more. Peltier cells are low-voltage devices; they require a power supply capable of supplying 10–100 A at voltages below 1 v. Although such power supplies are somewhat unusual, they pose no serious problems, and Peltier cooling is rapidly gaining in popularity, especially for small subambient thermostats.

Summary

The availability of devices for fast control of large amounts of power has removed the restrictions on response speed imposed by relay control, which limited the design of controllers to on-off switching systems. In addition, thermistors provide a convenient means to obtain an electrical signal proportional to temperature. Combination of the two developments has led to the design of a new generation of temperature controllers and thermostats which are economical and capable of fast and accurate control of temperature in a wide range of applications. With fast response and tight control comes the possibility of oscillation, however, and for optimum performance of a system, allowance must be made for its dynamic response.

Literature References

(1) Lewin, S. Z.: Temperature-controlled Baths (I): J. CHEM. ED., **36**, A131 (March, 1959).

(2) Lewin, S. Z.: Temperature-controlled Baths (II): J. CHEM. ED., **36**, A199 (April, 1959).

(3) SCR Manual, 4th Edition, Semiconductor Products Dept. General Electric Co. (1967).

(4) Triac Power Control Application; RCA Application Note, AN 3697, RCA, Harrison, N. J. (1968).

(5) Applications Manual for Computing Amplifiers, Philbrick/Nexus Research, Dedham, Mass. (1966).

(6) Handbook of Operational Amplifier Applications, Burr-Brown, Tuscon, Ariz. (1963).

(7) General Electric Co. zero-voltage switch type PA424; integrated circuit phase control type PA436, available from electronic supply houses.

(8) Delta Bond 152 filled epoxy resin, Wakefield Engineering, Inc., Wakefield, Mass.

(9) Paterson, W. L.: Multiplication and Logarithmic Conversion by Operational Amplifier-Transistor Circuits, *Rev. Sci. Instr.*, **34**, 1311 (1963).

(10) Philbrick/Nexus Research, Dedham, Mass.

(11) Analog Devices, Inc., Cambridge, Mass.

(12) Fairchild Semiconductor, Mountainview, Cal.

(13) Honeywell Instrumentation Handbook 1968/69

(14) Thermocouple Materials, National Bureau of Standards Monograph 40, (1962).

(15) Boucher, E. A.: Theory and Applications of Thermistors, J. CHEM. ED., **44**, A935, (November, 1967). [Reprinted in this volume]

(16) Yellow Springs Instrument Co., Yellow Springs, Ohio.

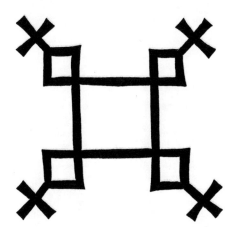

Topics in...

Chemical Instrumentation

Edited by **GALEN W. EWING**, Seton Hall University, So. Orange, N. J. 07079

The Journal of Chemical Education
Vol. 45, page A377
May, 1968

XXXVII. Moisture Measurement
GALEN W. EWING

The determination of moisture is of such great importance in so many fields, that it is not surprising to find a wide variety of tools for the purpose (*5, 11, 14, 19*). I have found more than 25 distinct analytical methods, and more than 25 manufacturers. I will make no attempt to include all manufacturers in this review.

MOISTURE IN GASES

The quantity of water vapor in a gas can be reported as *partial pressure*, as *vapor density*, as *relative humidity*, as *specific humidity*, as *mole fraction*, or in terms of the *dewpoint*. Partial pressure can be specified in any pressure units, as torr, psi, or millibars; vapor density may be quoted in terms of grams of water per liter of gas. Relative humidity is the ratio of the partial pressure of water to the pressure which would exist in equilibrium with liquid water at the same temperature. Specific humidity is the weight of water vapor per unit weight of dry gas. The dewpoint is the temperature at which moisture starts to condense on a surface that is being cooled; this can perhaps better be stated as the temperature of a plane water surface which would be in equilibrium with the water vapor present in the gas. Handbook tables give data for the interconversion of dewpoint and relative humidity in air.

Figure 1. Atkins Portable Thermistor Psychrometer.

Classical Methods

The conventional method of determining the water vapor content of a gas is gravimetric, via absorption in a preweighed container of granular magnesium perchlorate or other desiccant. This procedure has long been an essential feature of combustion analysis, though recently gas-chromatographic techniques have been supplanting gravimetric absorption.

Environmental humidity measurements, can be made with a hygrometer depending

Figure 2. Cambridge Systems Model 992 Thermoelectric Dewpoint Hygrometer.

on the reversible elongation of a human hair with increased humidity (*10*). A strand of hair (or many strands in parallel) is mounted under slight tension in such a way that a change in length will produce motion of an indicator needle over a scale. This instrument is direct-reading in relative humidity, but not very accurate.*

Even less precise humidity indications depend upon the change of color of cobalt chloride; this seems especially to fire the imagination of ingenious advertisers, but can be a useful laboratory indicator as well. A semi-quantitative procedure has been published (*20*) for the quick determination of water vapor in oxygen intended for breathing in high altitude aviation, which makes use of the $CoCl_2$ color. (This re-

---

* A Honeywell bulletin mentions that DaVinci, in the fifteenth century, placed a ball of wool on one end of a balance arm; the wool changed its weight with the humidity, giving an indication on the balance.

port also discusses the relative merits of Japanese and American bathrooms as constant-humidity chambers.)

Another classical method for atmospheric moisture is the psychrometer, consisting of paired wet- and dry-bulb thermometers. A table or nomograph must be consulted to obtain the relative humidity as a function of the dry-bulb temperature and of the difference between the two indications. Many models, suited for various purposes, are available through supply houses.

An electronic version of the psychrometer, utilizing wet- and dry-thermistors, is made by Atkins (Fig. 1); a meter reads temperature directly first from one, then the other, thermistor. It gives relative humidity within 0.5% of the true value over most of its range.

Dewpoint

Observation of the dewpoint is also an established method. Typically, a polished metal plate is progressively cooled by ice, solid CO_2, or even liquid nitrogen, until condensation is visible. This occurs at a sharply defined temperature, which is the dewpoint or frostpoint.

A modern photoelectric example is the dewpoint hygrometer of Cambridge Systems (Figs. 2 and 3). A lamp and two photocells are so oriented that specularly reflected light will strike one of the cells, while only scattered light will activate the second. This is a particularly favorable arrangement, because deposits of absorbing foreign matter which build up slowly on the mirror will diminish the direct reflection, but not produce appreciable scattering. As a result, cleaning of the mirror is only infrequently required. The mirror consists of a thin sheet of stainless steel in contact with a silver block which acts as a heat transfer medium between the mirror and a thermoelectric (Peltier) cooling unit. A thermocouple or thermistor, imbedded in the silver block, measures the dewpoint.

Another dewpoint sensor, by Vap-Air, utilizes the electrical conductivity of the condensed moisture rather than its optical properties. The sensor consists of a double grid of noble metal, flush with a smooth insulating surface of glass cloth impregnated with an epoxy resin. The unit is cooled thermoelectrically and its temperature observed by a small thermocouple. An automatic regulating circuit allows the surface conductance to control the cooling, so that the temperature adjusts itself to the dewpoint, which can be recorded continuously on a strip-chart recorder.

Panametrics offers a dewpoint hygrometer (Fig. 4) which depends for its action on the change of impedance of a porous

Chemical Instrumentation

layer of aluminum oxide with moisture content (11). The sensor is made by anodizing an aluminum strip, then coating the anodized surface with a very thin permeable film of gold. The resulting sandwich amounts to an electrolytic capacitor with varying dielectric constant, which acts as a parallel combination of a resistance and capacitance, both of which vary with moisture. The impedance changes from several megohms in dry air to perhaps 200 ohms at high humidity. It is

about 2×10^{-2} g/liter, corresponding to dew- or frost-points from $+20°$ to $-110°C$ (or up to $+90°C$ with a special probe). It has a rapid response, 99% of a step change occurring within 30 seconds. This unit has the added feature that it can measure moisture content directly in liquids of low dielectric constant, such as hydrocarbons.

Still another approach is found in the Alnor Dewpointer. In this unit, the gas is compressed by a measured amount, then suddenly released to atmospheric pressure. The resulting adiabatic cooling produces a fog if the initial pressure was high enough. If no fog appears, the compression is

bath (liquid nitrogen). As the block cools, the entrance restriction will suddenly start to plug up with dew or frost at the dewpoint, producing a sharp decrease in the pressure P_2, as shown on the inset graph. This instrument gives "excellent sensitivity and accuracy of moisture measurement in the parts-per-billion range." As far as I am aware, this instrument is not commercially available.

Resistance of Sorbent Layer

There have been a number of methods devised for determining humidity by sorption of moisture onto a surface, followed by resistance measurements. One such sensor, manufactured by Phys-Chemical Research Corp. (Fig. 6), consists

PRINCIPLE OF OPERATION

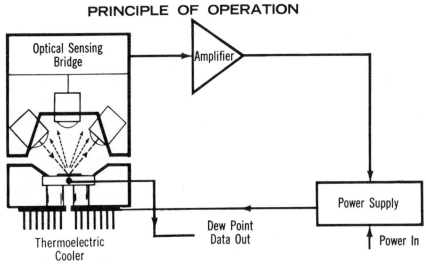

Figure 3. Cambridge Systems Thermoelectric Dewpoint Sensor.

Figure 4. Panametrics Aluminum Oxide Hygrometer, Model 1000.

measured by an alternating-current bridge operating at 154 Hz. The indication is independent of ambient temperature, of flow-rate, and of pressure over wide ranges. It can measure water vapor from 10^{-9} to

repeated to greater and greater pressures until it does become visible. An alpha-ray source is incorporated to provide constant nucleation. The manufacturer provides a circular slide-rule with which to convert pressure and ambient temperature readings into dewpoint.

A pneumatic dewpoint meter has been described by Pappas of Oak Ridge (15). Its operation can be followed by reference to Figure 5. The gaseous sample is introduced at constant pressure (measured at the pressure gauge marked $P.I.$). It then passes through a constriction into the region marked P_2, then through another constriction, whence it is discharged. The central region, cross-hatched in the figure, surrounding the passage between the two constrictions, consists of a drilled metallic block with an arm extending downward into a cooling

Figure 6. Electro-Humidity Sensor, Model PCRC-11, of Phys-Chemical Research Corporation.

of a chemically treated styrene copolymer which has an electrically conducting layer integral with the non-conducting substrate. The sensor is operated with alternating current, to a maximum of 1 ma. The resistance drops (reproducibly but not linearly) from about 10^7 to 10^3 ohms as the relative humidity rises from nearly zero up to 100%. These sensors are available from the manufacturer as components, and they form the active element in complete instruments from several other companies, such as Scientific Industries (Fig. 7) and Amstro Corporation. Lab-Line's Electro-Hygrometer appears to be similar in principle.

A porous surface impregnated with lithium chloride will change its resistance markedly above 11% ambient relative humidity (11). The same will be true below 11% only if the lithium chloride is incorporated with polyvinyl alcohol, a principle developed by Dr. F. W. Dunmore at the National Bureau of Standards (4). This principle is utilized in instruments by Honeywell, by Hygrodynamics, and by Edison. Honeywell and Edison make use of lithium chloride without

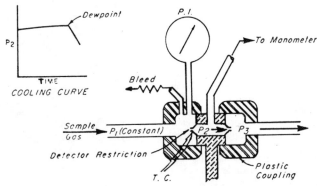

Figure 5. Pneumatic Dewpoint Meter, after Pappas. (*Courtesy of Analytical Chemistry*)

Figure 7. Scientific Industries Portable Hygrometer, Model 65-300.

polyvinyl alcohol, and allow the current passing through the sensor to raise its temperature until an equilibrium is established between sorbed moisture and the water in the gas phase; it can be shown that this is in effect analogous to measuring the dewpoint. Honeywell's unit, called a "Dewprobe," is shown in Figure 8. Foxboro's "Dewcel" is comparable.

Figure 8. Honeywell's Lithium Chloride "Dewprobe".

Piezoelectric Sorption Meter

This device, described by Crawford, et al. (3), is manufactured by DuPont. It is based on the piezoelectric resonance of a quartz crystal plate, which is a func-

Figure 9. Twin Quartz Crystal Humidity Sensor, for DuPont 510 Moisture Analyzer. *(Courtesy DuPont)*

tion of its mass. Two such plates, in the form of discs of 0.5-in. diam and 0.007-in. thickness, which resonate at about 9 MHz, are mounted as shown in Figure 9. They are given a thin coating of a hygroscopic polymer. The gas stream under test and a similar stream of dry air are switched so that they alternate between the two sensors with a 30-sec period, which reduces hysteresis effects. The mass sensitivity of each crystal is approximately 375 Hz-μg^{-1}. The DuPont instrument has five ranges from 5 to 25,000 ppm water vapor by volume, full scale. The accuracy is stated to be ±5% of full scale, and the reproducibility even better.

Electrolytic Hygrometers

In this class of instruments a thin layer of P_2O_5 deposited between two platinum contacts absorbs moisture quantitatively from ambient gas (2, 10). A voltage is impressed across the electrodes, and the resulting current electrolyzes the water to its elements, which join the gas stream. The solid P_2O_5 is regenerated in the process, and so is always ready for more moisture. With a flowing gas stream, the current is a direct measure of the moisture content. With stationary samples, the current can be integrated over the time required to remove all water. Such instruments are made by Consolidated, by Beckman, and by Manufacturers Engineering and Equipment Corp.

A difficulty arises if the gas contains elementary hydrogen, for then the oxygen produced electrolytically tends to combine with hydrogen to form more water to which the sensor responds spuriously. The same is true, *mutatis mutandis*, if the gas contains free oxygen. This difficulty is overcome in the Consolidated model 26-304 by taking readings at two different flow rates; the difference gives the true moisture content.

A special adaptation of this method for corrosive gases such as chlorine is available from Beckman as their catalog number SPEQ-3476. The electrodes are of rhodium, the only other materials in contact with the gas are Teflon and Hastelloy-C (6).

Gas Chromatography

Passing reference was made earlier to the determination of water by gas chromatography. This can be accomplished with good precision, as shown, e.g., by Vogel, *et al.*, who determined hydrogen with a relative error of 0.8% by conversion to water followed by measurement of the water peak (17).

There are likely to be difficulties, however, arising from the moisture content of the carrier gas: as the pressure in a gas cylinder decreases, the relative amount of water increases, no doubt due to desorption from the walls of the cylinder. Martin has discussed this in some detail (12). He found it essential to predry the carrier gas, which he monitored with an electrolytic hygrometer. Full scale on the recorder corresponded to about 80 ppm of water in benzene, under the particular conditions used, with a chromatographic column of porous polymer beads (Porapak).

Other Methods

Moisture in a gas can be determined quantitatively by infrared absorption, either with a spectrophotometer, or more economically, by means of a non-dispersive analyzer. This instrument is of general applicability for covalent molecular gases, and is widely used in plant-stream monitoring. As it is not unique to water-vapor measurement, it will not be described here (5).

Another way to detect trace amounts of moisture is with a mass spectrometer. In fact, it is difficult to dry a sample to the point where no traces of water can be seen. The omnipresent signal at mass 18 is frequently a convenient landmark in assigning mass numbers on a mass spectrogram. It is probably too sensitive, and certainly too expensive a tool to be classed as a "hygrometer".

Standards

Gamache (7) has pointed out the difficulties inherent in preparing and particularly in storing standardized mixtures of water vapor with other gases in the part-per-million range. The difficulty arises from the adsorption and desorption of water at the surfaces of the container. He recommends instead a standard mixture of, say, 1% H_2, 1% O_2, and 98% N_2, which can be stored without alteration (it is, of course, far below the explosive limit). Then, as may be required, this standard can be mixed quantitatively with dry nitrogen to provide accurately known standards at any desired level. To calibrate a hygrometer, the mixture is led through a heated platinized silica catalyst bed to produce known concentrations of moisture.

MOISTURE IN LIQUIDS AND SOLIDS

The traditional method for moisture determination in solids and non-volatile liquids is gravimetric: merely weigh the sample before and after drying. The drying process may be simple exposure to ambient air, or may require a drying oven, with or without vacuum.

The thermobalance is especially useful here, and several simplified models have been marketed particularly for this purpose. An example is the Cenco Moisture Balance, in which an infrared heat lamp is mounted directly over a shallow pan.

Figure 10. Ohaus Moisture Determination Balance.

Chemical Instrumentation

The weight of the pan and contents can be determined at any time by manually nulling a torsion balance mechanism. Comparable, but more elaborate, moisture balances are available from Ohaus (see Fig. 10), Lab-Line, and others.

Another classical method which is often convenient depends upon azeotropic distillation of water with an immiscible liquid in such a way that the water is caught in a trap where its volume can be measured. The choice of solvent is important. It must be less dense than water if the usual trap is to be effective, and its boiling point must be considered. For example, distillation of benzene from contact with gypsum will remove external moisture, but xylene also extracts water of crystallization (9).

Water is often determined by the Karl Fischer titration, in which water reacts with a mixture of pyridine, methanol, sulfur dioxide, and iodine, according to the net reaction (Py = pyridine):

$$3Py + I_2 + SO_2 + CH_3OH + H_2O \rightarrow$$
$$2PyHI + PyHSO_4CH_3$$

The end point is identified photometrically or electrometrically by the iodine-iodide conversion (5). Great precautions must be taken to exclude moisture from the reagent. The useful range can extend as low as about 10 ppm. Automatic or semiautomatic Karl Fischer titration apparatus is manufactured by Beckman (Aquameter) and Precision (Aquatrator).

It is frequently possible to sweep moisture out of a relatively non-volatile and non-polar liquid with a stream of dry nitrogen, then to analyze the gas by one of the methods already discussed; Consolidated, for example, has a model specifically designed for this service. The Panametrics detector has been mentioned as suitable for direct immersion in non-polar liquids.

Infrared absorption can be applied to a liquid sample. It has been found ideally suited to the simultaneous determination of moisture and of cellulose on a continuous basis in a paper mill (8). In the instrument manufactured by Taylor three wavelengths in the 1.5–5.0-micron range are utilized: at the first neither water nor cellulose absorb, at the second both absorb, but water to a greater extent, and at the third cellulose but not water absorbs. Radiation from a tungsten lamp passes alternately through one of three filters, then through the moving paper web, and onto a lead sulfide detector. An analog computer compares the responses and provides a recorded indication of moisture content with a time lag of only about 0.1 sec. The accuracy is stated to be better than 0.25% moisture over a 5% span.

Absorption in the microwave region can be considered an extension of the infrared domain. The absorption of the H_2O molecule at 1.35 cm (22.235 GHz) has been utilized in an instrument for continuous monitoring of moisture in paper manufacture (18). A similar approach at 3 cm has been found useful in measuring the moisture in 28-in. bales of synthetic rubber (13).

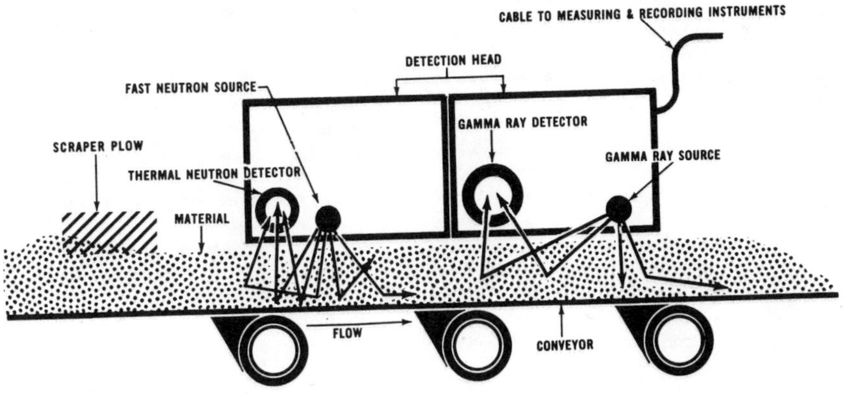

Figure 11. Nuclear-Chicago's "Qualicon" Model 5072 Moisture Detector Mounted on a Conveyer Belt.

Nuclear magnetic resonance presents an ideal tool for the determination of water in liquids and some types of solids. A wide-line spectrometer is generally desirable, and gives linear calibration curves. It is too costly to be justified for moisture studies alone, though at least two models have been marketed for the purpose in past years. A comparison of NMR with infrared absorption for the determination of water in tris-[1-(2-methyl)aziridinyl]-phosphine oxide has been reported (22); the results by the two methods did not differ significantly.

An interesting method for water in mixtures with various oxygenated organic liquids depends on the large release of heat when the solution is mixed with 80% sulfuric acid (21). The technique described involves injection of the sample by syringe into a Dewar flask containing a measured amount of acid. The temperature rise was measured with a thermistor probe. Calibration curves (temperature change against volume fraction) are nearly linear, but vary in slope from one organic liquid to another. The standard deviation in a typical instance was 0.5%, and the maximum relative error about 1%.

Dielectric Properties

Water has been determined by measurement of dielectric constant, which is larger than for most organic liquids (11). The sample is placed between the plates of a capacitor, and the resulting change of capacitance measured, usually by a resonance method. This gives reliable results if other polar substances are absent.

Moisture Register Co. produces several models of instruments primarily for production-line testing of granular solids by a dielectric-loss method. A portion of material is compacted with a small hydraulic press to eliminate voids, and the absorption of power from a high-frequency oscillator (10 MHz) is determined.

Nuclear Methods

The amount of hydrogen in a sample can be determined by its interaction with fast neutrons, slowing them down (moderating action) and scattering them. Hence in a material where no hydrogen other than that contained in water is present (or a constant amount of it), the observation of scattered slow (thermal) neutrons provides a measure of water content. Since the relative water content is usually required,

a measure of density is also needed.

An instrument which will give either total or relative water, and also density per se if desired, is manufactured by Nuclear-Chicago for monitoring inorganic materials on a conveyor belt (organics generally have too much intrinsic hydrogen) (1). This device, called the Qualicon, is diagrammed in Figure 11. There are two separate radioactive sources, a ^{137}Cs capsule which gives a 0.66 MeV gamma radiation, and a Pu-Be mixture which gives fast neutrons by an (α,n) reaction. Each source has its own associated detector. Whereas the neutron detector responds directly to the hydrogen content of the material, it can be shown that the back-scattered gamma radiation bears an inverse relation to the total density. The signals from the two detectors pass to a data-processing unit which continuously computes the moisture per unit weight.

A similar device, the Moisturay, made by Gilmore Industries, employs a single Ra-Be source to yield both gamma and neutron fluxes (16). Selective detectors for the two radiations are placed beneath the conveyor belt, while the source is above it. The data are combined in a similar manner to that described for the Qualicon. This system has the advantage that the distribution of absorbing material on the conveyor does not need to be uniform, whereas the geometry has to be fairly constant for reliable back-scattering measurements.

A portable field instrument working on the same principle, for determining the moisture and density of soils, etc., is made by Soiltest, Inc.

Supplement

Yellow Springs Instrument Co. has entered the humidity measurement field with two models. One is an electronic psychrometer (Model 90) in which a battery-powered aspirator produces a constant flow of air over wet and dry thermistors. The panel meter reads directly the *difference* in temperature between the two sensors as well as the dry-bulb temperature itself. The relative humidity is then ascertained by reference to tables or a special slide rule.

YSI's Model 91 is a dew-point hygrometer utilizing the LiCl conductivity principle. Its sensing head has a built-in thermistor, and the panel meter reads either temperature or vapor pressure. Both ambient and dew-point temperatures are available for continuous, simultaneous recording.

Beckman now has several variations of their Model 340 trace moisture analyzers for gases, all based on the P_2O_5 conduction method. They vary in type of mounting, portability, etc. Beckman also has an improved version of the KF-4 Aquameter for Karl Fischer titration of trace moisture in liquid or solid samples. This model uses modern piston-type burets and solid-state electronics for increased convenience and reliability.

MANUFACTURERS

Alnor Instruments, Inc.
420 N. LaSalle St., Chicago, Ill. 60610
Amstro Corporation
122 Clinton Rd., Fairfield, N. J. 07006
Atkins Technical, Inc.
3606 SW Archer Rd., Gainesville, Fla. 32601
Beckman Instruments, Inc.
2500 Harbor Blvd., Fullerton, Calif. 92634
Cambridge Systems, Inc.
50 Hunt St., Newton, Mass. 02158
Central Scientific Company
2600 S. Kostner Ave., Chicago 60623
Consolidated Electrodynamics (CEC)
(Now part of duPont's Instrument Div.)
1500 S. Shamrock Ave., Monrovia, Calif. 91016
E. G. & G., Cambridge Systems
151 Bear Hill Road, Waltham, Mass. 02154
E. I. duPont de Nemours & Co.
Wilmington, Del. 19898
Thomas A. Edison Industries
West Orange, N. J. 07051
The Foxboro Company
Foxboro, Mass. 02035
Gilmore Industries
3355 Richmond Road, Cleveland, Ohio 44122
Honeywell, Inc.
2701 4th Ave. S., Minneapolis, Minn. 55408

Hygrodynamics, Inc.
949 Selim Rd., Silver Spring, Md. 20910
Lab-Line Instruments, Inc.
Lab-Line Plaza, Melrose Park, Ill. 60160
Manufacturers Engineering and Equipment Corporation
250 Titus Ave., Warrington, Pa. 18976
Moisture Register Company
1510 W. Chestnut St., Alhambra, Calif. 91802
Nuclear-Chicago Corporation
333 E. Howard Ave., Des Plaines, Ill. 60018
Ohaus Scale Corporation
1036 Commerce Ave., Union, N. J. 07083
Panametrics, Inc.
221 Crescent St., Waltham, Mass. 02154
Phys-Chemical Research Corporation
36 East 20th St., New York, N. Y. 10003
Precision Scientific Company
3737 W. Cortland St., Chicago, Ill. 60647
Scientific Industries, Inc.
220-05 97th Ave., Queens Village, N. Y. 11429
Soiltest, Inc.
2205 Lee St., Evanston, Ill. 60601
Taylor Instrument Companies
95 Ames St., Rochester, N. Y. 14601
Vap-Air Controls div. of Vapor Corporation
6444 W. Howard St., Chicago, Ill. 60648
Yellow Springs Instrument Co.
Yellow Springs, Ohio 45387

References Cited

(1) CARVER, R. L., *Inst. & Control Systs.*, **36**, 106 (1963).
(2) COLE, L. G., Czuha, M., Mosley, R. W., and Sawyer, D. T., *Anal. Chem.*, **31**, 2048 (1959).
(3) CRAWFORD, H. M., HEIGL, J. J., KING, W. H., JR., AND MESH, T. J., "Analytical Instrumentation-1964," Plenum Press, N. Y., p. 105 (Copyright, Inst. Soc. Amer.)
(4) DUNMORE, F. W., *J. Res., Nat. Bur. Stds.*, **23**, 701 (1939).
(5) EWING, G. W., "Instrumental Methods of Chemical Analysis," McGraw-Hill Book Co., New York, 1954; p. 314 ff. (The later edition does not contain this material.)
(6) FORD, W. R., JR., AND KOCH, G. P., "Analytical Instrumentation-1965," Plenum Press, N. Y., p. 177 (Copyright, Inst. Soc. Amer.)
(7) GAMACHE, L. D., "Analytical Instrumentation-1965," Plenum Press, N. Y., p. 167 (Copyright, Inst. Soc. Amer.)
(8) GARDNER, R. C., *Instr. Technol.*, **15**, 51 (January, 1968).
(9) HARD, S., AND TALMI, A., *Anal. Chem.*, **29**, 1694 (1957).
(10) KEIDEL, F. A., *Anal. Chem.*, **31**, 2043 (1959).
(11) LION, K. S., "Instrumentation in Scientific Research: Electrical Input Transducers," McGraw-Hill Book Co., New York, 1959; p. 136 ff.
(12) MARTIN, F. D., "Analytical Instrumentation-Vol. 4," Plenum Press, N. Y., 1967; p. 225 (Copyright, Inst. Soc. Amer.).
(13) MERRITT, J., *Anal. Chem.*, **34**, 293 (1962).
(14) NASH, F., *Anal. Chem.*, **27**, 1842 (1955).
(15) PAPPAS, W. S., *Anal. Chem.*, **36**, 1885 (1964).
(16) REIM, T. E., *Instr. Technol.*, **14**, 47 (July, 1967)
(17) VOGEL, A. M., AND QUATTRONE, J. J., JR., *Anal. Chem.*, **32**, 1754 (1960).
(18) WALKER, C. W. E., "Analytical Instrumentation-1963," Plenum Press, N. Y., p. 183 (Copyright, Inst. Soc. Amer.)
(19) WEXLER, A. (editor-in-chief), "Humidity and Moisture: Measurement and Control in Science and Industry," Reinhold, N. Y., 1964–65. (Four volumes.)
(20) ASAMI, T., *Anal. Chem.*, **40**, 648 (1968).
(21) SPINK, M. Y., and SPINK, C. H., *Anal. Chem.*, **40**, 617 (1968).
(22) TOMPA, A. S., and BAREFOOT, R. D., *Anal. Chem.*, **40**, 650 (1968).

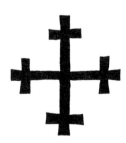

Topics in...

Chemical Instrumentation

Edited by **GALEN W. EWING**, Seton Hall University, So. Orange, N. J. 07079

The Journal of Chemical Education
Vol. 45, pages A533, A583
July and August, 1968

XXXIX. Signal to Noise Optimization in Chemistry—Part One

THOMAS COOR, *Princeton Applied Research Corporation, Princeton, New Jersey 08540*

INTRODUCTION

Chemistry, in common with the other disciplines of the physical sciences, in its pursuit of greater accuracy and finer small effect phenomena, is pushing ever closer to the fundamental limitations imposed on the observer by nature. The fundamental limitations arise from two facts: (1) all matter not at absolute zero has thermal fluctuations, (2) charge, light, and energy are quantized. As a result, all measurements made of physical quantities will have a certain irreducible level of noise associated with them. In addition to this "fundamental" noise in measurements, there is always noise arising from more prosaic sources; 60 Hz line pickup, building vibrations, variations in room temperature, radio stations, etc. These environmental disturbances can, in principle, be reduced to an arbitrarily small value, but in practice are very difficult to remove entirely from the picture.

In spite of his concern with noise as a limitation on detectability or accuracy in his work, the chemist is often unaware of simple techniques that will allow him to improve his results. He is generally even less aware of recent advances that have been made in signal processing, advances that allow vastly improved experimental signal-to-noise (S/N) ratios to be achieved. While much of the literature on the subject is couched in abstruse mathematics, most of the results are relatively simple to understand and can be applied easily to most experimental situations.

In this paper, we will first review the properties of the various sources of noise that confront the chemist as he makes sensitive physical measurements. Then consideration will be given to various ways of avoiding unnecessary noise in deriving an electrical signal that represents the quantity being determined. Next the theory of various kinds of simple information processing that the chemist can employ to enhance the S/N of electrical signals will be presented. And finally various types of commercial equipment useful in S/N en-

hancement will be surveyed. While the principles to be discussed are very general, we will restrict our attention to electrical signals and assume that all quantities of concern can be transformed into such by suitable transducers.

Although by no means an exhaustive list, the following represents typical areas where the chemist might well be interested in improving his S/N: (1) electron spin resonance, (2) nuclear magnetic resonance, (3) nuclear quadrupole resonance, (4) microcalorimetry, (5) differential thermal analysis, (6) all areas of analytical and research spectroscopy, (7) photochemistry, (8) chromatography, (9) electrochemistry.

In a particular instrumental system there will usually be some form of a "probe"—be it an electrode, beam of light, thermocouple, or radio frequency pickup coil—inserted in some way into the medium under study. It is of course desirable that the phenomena of interest alone be monitored by the probe but this is never entirely the case. One must always contend with the fact that the signal representing the desired quantity will have noise in some form as contamination. This noise may be inherent in the physical process, arise from the "probing" action itself, or may be picked up from the environment.

A simple example will be illustrative. Consider a glass microelectrode being used to determine, say, the pH at a particular site in some small system. At equilibrium the Nernst equation specifies that the electrode potential relative to some reference will be:

$$E_H = -2.303 \frac{RT}{nF} [pH] + K$$

In reality, this equation can represent only the *average* potential at the electrode. Being a small electrode, the concentration of ions around the active region will fluctuate because there are only a finite number of them. This microscopic fluctuation in concentration will give rise to a corresponding fluctuation in electrode potential. The smaller the electrode size and the

Thomas Coor graduated with a B.A. in Chemistry from Rice Institute and obtained a Ph.D. in Physics from Princeton University. He was associated with Brookhaven National Laboratory and Princeton University before joining in the founding of Princeton Applied Research Corporation in 1962. He has worked in the fields of Plasma and high energy physics and in electronic instrumentation.

lower the concentration, the greater will be the relative fluctuation. However, even for large electrodes, the fluctuations or noise will place limitations on the ultimate attainable precision of measurement. What in fact we have described is the Brownian phenomena in an electrochemical framework.

To obtain a more precise measure of the true pH in our example of the glass microelectrode, we might try to take an average of many measurements of the fluctuating cell potential. One might expect that an arbitrary degree of precision to be achieved in this way, but this is not necessarily the case. One must also consider the "noise" properties of the probe and of the rest of the electronic system being used to make the measurement. The effects of glass membrane resistance, flicker noise effects in the glass electrodes and in amplifiers, extraneous signal pickup from the environment, temperature fluctuation "noise" in the constant temperature bath in which the measurements are being carried out; all these and more come into the picture and must be considered. Such effects will in general place severe limitations on the effectiveness of increasing the averaging time as a means of gaining greater precision in simple dc measurements. To understand why this is so, let us first consider the

Chemical Instrumentation

fundamental sources of noise that arise in making electrical measurements and review their properties.

SOURCES OF NOISE AND THEIR PROPERTIES

Fundamental Noise

It was realized by Schottky (1) very soon after the advent of the vacuum tube amplifier that unlimited gain is not useful and that after a certain gain is achieved, random fluctuations or noise show up. These random fluctuations have been shown to arise from two basic sources: one is thermal in origin and its properties can be derived from thermodynamic and statistical mechanical arguments—the other arises from the fact that light and electrical charges are quantized and current in transistors, vacuum tubes, and photocells flows by virtue of the uncorrelated transfer of *single electrons* across some void, hence is statistical in nature.

Johnson or Resistance Noise

Although intimately related to Brownian motion, the existence of thermal noise in resistors was realized comparatively late. This noise is usually called "Johnson noise" after its discoverer (2). Nyquist (3), a colleague of Johnson, first derived its properties.

It is instructive to derive the magnitude of Johnson noise in a resistor. The method here is related to, but not identical with, that of Nyquist, and will be given in outline form. Consider the idealized experiment pictured in Figure 1.

(1) Because of black-body radiation from the enclosure at temperature T_1, the dipole will be a source of average power, $\overline{P_A}$, which flows down the transmission line and is completely dissipated in R.

(2) The Second Law of Thermodynamics requires that resistor R must be a source of average power, $\overline{P_R}$, such that $\overline{P_R} = \overline{P_A}$ if $T_1 = T_2 = T$.

(3) Statistical Mechanics states that every mode of a system at equilibrium with

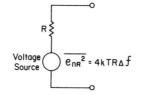
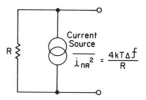

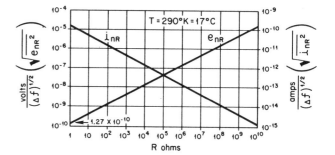

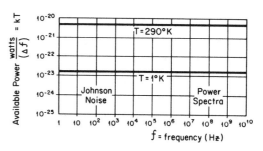

Figure 2. The two equivalent circuit representations of thermal noise in resistor R are shown together with the magnitude of the rms voltage and current per cycle as a function of R. Also illustrated in the flat or "white" power spectra of the noise at two different temperatures.

temperature T will have average total energy $= kT$, where k is the Boltzmann constant, but that the total energy of a given mode will fluctuate.

(4) Assume an instantaneous short at each end of the transmission line. Energy trapped on the transmission line results from both $\overline{P_R}$ and $\overline{P_A}$. Total energy trapped is:

$$(\overline{P_R} + \overline{P_A}) \times \begin{pmatrix} \text{time for signal} \\ \text{to traverse line} \end{pmatrix} = $$
$$2\overline{P_R}\,\frac{L}{C}$$

(5) Wave theory states that the number of standing waves, Δm, on the transmission line between frequency f and $f + \Delta f$, where Δf is the bandwidth, is

$$\Delta m = \frac{2L}{C}\,\Delta f$$

(6) The average total energy trapped on the transmission line having frequencies between f and $f + \Delta f$ will then be,

$$\Delta m kT = \frac{2L}{C}\,kT\Delta f$$
$$= 2\,\overline{P_R}\,\frac{L}{C}$$

(7) Solving for $\overline{P_R}$ we find

$$\overline{P_R} = kT\Delta f$$

(8) $\overline{P_R}$ results from an internal noise voltage source, e_{nR}^2, associated with R. Since the transmission line represents a matched (equal in resistance) load to the voltage source, e_{nR}^2, of internal resistance R,

$$\overline{P_R} = kT\Delta f = \frac{\overline{e_{nR}^2}}{4R}$$

Hence, $\overline{e_{nR}^2} = 4kRT\Delta f$ (1)

Similarly one can show that

$$\overline{i_{nR}^2} = \frac{4kT\Delta f}{R}.$$ (2)

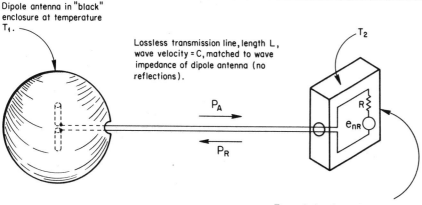

Figure 1. Gedanken experiment used to derive the relations for Johnson noise in resistance R at temperature T. Thermodynamic and statistical arguments lead immediately to the results.

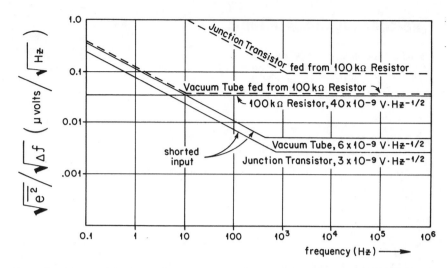

Chemical Instrumentation

where $\overline{i_{nR}^2}$ represents the noise current that will flow across the shorted terminals of a resistor R. The two equally valid ways of representing the noise of a resistor are shown in Figure 2.

Note that the eqns. (1) and (2) are functions only of the *frequency bandwidth*, Δf, and not the frequency, hence the magnitude of the average noise voltage or current depends only on the effective frequency bandwidth of the measuring instruments. This type of frequency spectrum, plotted on a power-per-unit-bandwidth basis, is known as a *white* spectrum, hence resistor or Johnson noise is called white noise.

We thus see that any dissipative element, whether it be the membrane resistance of a glass electrode or the load resistance of a photocell, gives rise to a real noise voltage or current and this noise will add to any signal that appears across the resistor. Purely reactive elements, capacitance and inductance, have no noise voltages associated with them *per se*. However, real inductors have winding resistance and real capacitors have dielectric

Figure 5. The noise voltage as a function of frequency of representative vacuum tubes and junction transistors under conditions of shorted input and with 100 kΩ source impedance (noise referred to input). Also shown is noise of 100 kΩ resistor at 290 °K. The effect of base current noise is immediately apparent.

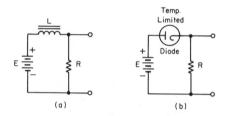

Figure 3. Circuit shown in (a) will exhibit only Johnson noise of R if components are perfect. Circuit (b) has additional shot noise associated with current that flows by virtue of independent event processes in temperature limited diode.

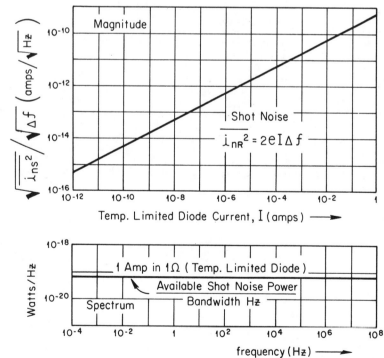

Figure 4. The magnitude and spectral characteristics of shot noise as a function of total diode current, I.

loss, both of which give rise to the expected noise.

While discussing noise arising in resistors, it is well to note that apart from Johnson noise, which appears across any resistor whether or not is is carrying current, there are additional sources of noise in resistors that *are* carrying current. Wire wound resistors are in general quite good in this respect, microphonics being about the only source of additional detectable noise. On the other hand, composition, deposited carbon, and metal film resistors show voltage dependent noise, descending in that order. Good metal film resistors are almost—but not quite—as good as wirewound resistors. This voltage dependent noise is not too well understood but seems to be due to the granular nature of the resistance elements. The spectral characteristics of this excess resistor noise is quite interesting, having a peculiar $1/f$ power spectrum, and has a flicker-like appearance on an oscilloscope or meter. This "$1/f$" or "flicker" noise is disastrous for low-frequency or dc measurements, because averaging for longer periods of time can result in no net increase in S/N!

Shot Noise

In the last section the statement was made that good resistors carrying current had little noise in excess of the Johnson noise. How can this statement be reconciled with the known fact that electrical conduction occurs *via* electrons and that charge is quantized? This is an important point that is often missed. Consider the circuit in Figure 3a. For frequencies where $R \gg 2\pi f L$, the noise at the output terminals is just the Johnson noise. However if a diode is introduced in the circuit, as shown in Figure 3b, the situation is vastly different, even though the current through the resistor is the same. In the diode (or photocell, transistor, triode, etc.) the passage of current is governed by a random single-event, electron emission process, while metallic conduction is a large scale correlated drift phenomenon with many discrete changes taking part at once. Hence the current through the single-event devices are subject to large statistical fluctuations while metallic conduction is not. The "shot noise" due to the random emission of electrons from cathodes or semiconductor junctions was predicted by Schottky (1) early in the era of electronics.

The derivation of the magnitude of this noise is straightforward (4) but only the result will be indicated here. The mean square shot noise current, i_{ns}^2, is given by:

$$\overline{i_{ns}^2} = 2Ie\Delta f \qquad (3)$$

where e is the electron charge, 1.59×10^{-19} coulomb, I is the current that flows due to some single electron random emission process (e.g., temperature limited diode), and Δf is again the bandwidth at the measuring apparatus with which i_{ns}^2 is measured. We see that this noise is "white," i.e., has a flat spectrum. The magnitude of shot noise current is given in Figure 4.

All active amplifying devices used in

Chemical Instrumentation

electronics—the transistor, photocell, vacuum tube—exhibit shot noise. This noise is an important factor in determining the overall performance of an amplifying and measuring system. However, the shot noise in a vacuum tube is much less than given by eqn. (3). Space charge effects which introduce a smoothing action on the random conduction process are responsible for this reduction.

Flicker Noise

The third source of noise which is important to our considerations is called *flicker-effect* noise and has been mentioned in conjunction with resistors. Flicker-effect noise is characterized by its unusual spectral characteristics and, for most electronic devices, dominates thermal or shot noise only for frequencies less than about 100 Hz. Flicker noise is especially troublesome for those trying to make small dc measurements, as we shall see.

Flicker noise is not completely understood but is found generally where electron conduction occurs in granular semiconducting material, or where cathode emission is governed by the diffusion of clusters of barium atoms to the cathode surface. Phenomenologically, flickering conduction can be pictured as resulting from a very large assemblage of series connected switch-resistor combinations randomly arranged in all possible ways between two terminals. The switches are then assumed to randomly open and close according to some statistical distribution. The particular distribution determines the power spectrum characteristics of the resulting noise. Almost all electronic elements exhibit flicker noise to some degree and their noise power spectra differ widely. Tubes and transistors typically show a power spectrum of the form $1/f^n$ where n is a constant near 1. The noise of several typical amplifying devices is shown in Figure 5.

The particularly disturbing factor about flicker noise is that the $1/f^n$ characteristic seems to hold down to as low a frequency as one cares to determine it. The long term drift in dc transistor or tube amplifiers would seem to be the manifestation of very low frequency flicker noise! Hence, dc electrical measurements are fraught with great difficulties. Even galvanometers exhibit flicker-noise-like effects. Moral: avoid dc measurements if S/N is a problem.

Environmental Noise

Environmental disturbances which make their way into chemical measurements can, in principal, be reduced to an arbitrarily small value, but in practice, can be very annoying and difficult to eliminate entirely.

Sixty hertz (and higher harmonics) from the power line, radio stations, motor and switch sparks, corona, building vibrations, and room temperature fluctuations all fall into this category. One interesting and important characteristic of the sum of all of these disturbances is the apparent $1/f$ characteristic of the power spectra. Why this should be so is not entirely clear. Apparently the power spectra of tempera-

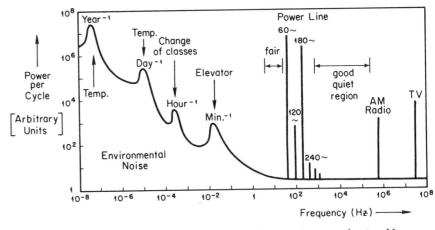

Figure 6. Pictorial representation of environmental noise in a typical location as function of frequency. Note the $1/f$ character at low frequencies.

ture fluctuations, earth vibrations, and all the man-made fluctuations are determined by event disturbances not unlike those found in cathodes and other granular semiconductor processes. While not quantitative, Figure 6 illustrates the spectral characteristics of typical laboratory environmental noise. Apart from the power line frequencies, which are very sharp lines, the region between 10 and 10^6 Hz is comparatively quiet and is a good region to have information appear, as will be discussed later.

THE SIGNAL MEASURING SYSTEM

Block Diagram and Noise Sources

Before considering the S/N properties of chemical instrumentation systems, let us make reference to Figure 7 in which a hypothetical system is shown in block diagram form, together with sources of noise. The chemical quantity under study, $Q(t)$, together with its inherent noise, is probed with a linear transducer of some type yielding an electrical signal $E(t)$. Generally, some provision is made to switch or move the transducer input from the object of study to some blank or standard. This allows a zero point to be established or a system calibration to be made. The signal $E(t)$ from the trans-

ducer is amplified, processed for noise reduction, and finally recorded in some fashion.

In designing such a system the characteristics of each element must be specified in such a way as to pass the information of interest, act on it in the way desired, and optimize the S/N performance of the entire system. The most important considerations in system specifications are discussed in the following paragraphs.

System Considerations: Nature of Signal

The first consideration in system specification is the nature of the information that is going to be handled by the system. The simplest and most common type of instruments are dc systems where the quantity of interest is steady or slowly varying with time, and the information "containing" the magnitude of the quantity is handled as a dc signal.

Other systems carry this information as the amplitude of sinusoidal or square wave signals with fundamental frequency f_c. These ac signals can result from the modulation of the phenomena under study by some means, or can result from the nature of the transducer used. In the latter case a steady or dc input to the transducer is converted by some chopping or modulating

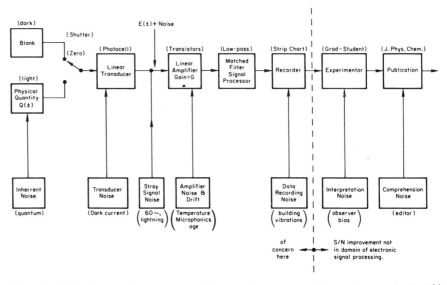

Figure 7. Block diagram of system required to record the changes with time of some quantity $Q(t)$ (assumed to be light). Also shown are possible sources of noise.

action into an ac signal, as, for example, is done in a vibrating capacitor electrometer.

In still other systems, the information may be in the form of recurring waveforms, or in the amplitude of power spectra of signals from the material under study. Consideration of systems in which the information has these forms is beyond the scope of this paper and will not be dealt with here.

Signal Frequency

In designing a chemical system, one often has the choice as to where the band of information of interest is made to lie. For example, in atomic absorption spectroscopy the changes in transmission of resonance radiation through the flame of a burner can be determined by a dc measurement. Alternatively, the resonance radiation can be chopped, or alternated between the flame path and a reference path at some frequency, say 400 Hz.

By chopping, the band of information is removed from very low frequencies (dc) and placed around the chopping frequency. This allows the use of ac amplifiers and thus avoids the inherent drift and flicker noise of dc amplifiers. In addition, it permits the use of some very powerful techniques for improving S/N. These techniques will be discussed later.

System Bandwidth

Having chosen the frequency at which the information is to appear, there is a further related question that must be considered—that of system bandwidth. If the quantity being chopped or modulated at frequency f_c has a steady or dc value, then the bandwidth of the information "channel" at f_c is very narrow. However, if the quantity is changing with time, the bandwidth Δf_c required to carry the information will be greater. For faithful reproduction of the original time-varying quantity, every element of the system through which the information passes must have sufficient bandwidth to pass all components of the signal. However, as most noise sources have the property that the wider the bandwidth Δf of the system the greater will be the resultant noise, this should not be overdone. From the standpoint of S/N the system bandwidth should be only wide enough to faithfully reproduce the wanted time dependance of the original information. These considerations are equally true if the signals being processed are dc. Apart from $1/f$ effects, narrower bandwidths will in general yield less noise.

Dynamic Range

Dynamic range is a very important concept in information processing which is often neglected in system design. Any measuring system will have its own irreducible noise level and drift, which impose a lower limit on the level of signals that may be handled by the system. Likewise, the system will always overload and become non-linear for sufficiently large inputs. Often this overload point is dependent on whether the input is a signal at the chopping frequency f_c or is wide-band noise.

Dynamic range can be defined in two ways: (1) the ratio of the signal producing full scale output to the signal corresponding to the inherent system noise and drift, (2) the ratio of the noise level producing overload to the signal that corresponds to full scale output. Both of these definitions are valid and useful and will be referred to again. In general, definition (1) is more useful in quieter situations. Definition (2) serves better where the input S/N is less than unity. It is obvious that the dynamic range of a measuring system has great bearing on the precision with which measurements can be made and on the ability of a system to function properly in the presence of large quantities of noise.

Drift

Drift also is defined in two valid and useful ways: (1) *zero drift*—the variations with time of the output of a measuring system with zero input; (2) *gain drift*—the drift in the indicated value of a steady input with time. Zero drifts usually impose limitations on sensitivity or ultimate detectability, while gain drifts set limits on accuracy and precision.

Noise Figure

We have seen that any physical system providing some quantity to be measured will have a certain irreducible noise associated with it. What we wish to do, given this S_s/N_s from the source, is to transduce, amplify and then measure the characteristics of S_s with the least possible effect of N_s on the measurement. To do this, we should see that the transducer, amplifier, and other equipment through which the signals are processed, degrade the initial S_s/N_s as little as possible. Any noise these parts of the system inject into the in-

formation channel will produe a degraded S_o/N_o at the output. Hence they should be "low noise." One sees that a quantitative measure of the degradation of S/N in passing information through some subsystem is important. One possible measure is called the *Noise Figure* (NF) of the unit. It is defined as:

$$NF = 10 \log \frac{S_i/N_i}{S_o/N_o}$$

where S_i/N_i is the ratio of the available input-signal and noise *powers*, and S_o/N_o the ratio of the available output powers. It is expressed in dB. The NF of any network is a function of the nature of the source (the source temperature and impedance, the frequency of operation, and its gain. This complex subject has been extensively treated elsewhere (5) and only the important results will be given. In specifying the NF of any element, it is assumed that the signal source has an internal resistance at 20°C and that the available noise power N_2 is just the thermal noise of this impedance.

We have seen in Figure 5 that the noise of tubes and transistors is dependent on the source impedance to which they are connected. The excess noise with high source resistance is *in addition* to the Johnson noise generated by the source resistance, and results from a noise current (grid or base current) flowing from the device giving rise to increasing noise as the source resistance is raised. Hence the NF of a signal processor or transducer is a complicated function of source impedance, frequency, input device characteristics, etc. A perfect signal processing system will have a NF of 0 dB, while one that degrades the S/N of the signal source by a factor of two has a 3 dB NF. We shall be concerned with NF when we discuss amplifiers and other signal processing equipment.

Part Two: S/N Considerations in Designing a Measuring System

INTRODUCTION

In the last section we reviewed the important specifications to be considered in designing chemical experiments or measuring systems. If the system or experiment is to achieve the maximum signal-to-noise ratio (S/N), it must be designed and specified in such a way as to minimize all the sources of noise that were discussed. Also, special noise reduction techniques may be advisable if it is important to achieve the best possible S/N. System design will now be discussed from the standpoint of optimum S/N.

System Grounding

System grounding can be one of the most annoying and frustrating problems in S/N improvement that the chemist can face. In almost every laboratory there are several electrical grounds available: the water or gas pipes, the third (neutral) conductor in the 60 Hz power distribution system, or a building ground bus. As far as small signals are concerned, none of these grounds is any good whatsoever. If one looks between two different points in a ground system, one usually sees hundreds of millivolts—if not volts—at 60 Hz. One should in general ground the chassis of electronic equipment being used to only one point of the distribution system ground, and refer all shields to this same point. But this is not going to be good enough for the most sensitive measurements. The various chassis grounded to the one common point are still going to have millivolts of 60 Hz between them—even if mounted in the same relay rack! Ac magnetic fields from transformers, etc., will induce potential differences even if the chassis

Chemical Instrumentation

have zero resistance. Avoiding these ground-loop voltages is often difficult. Although more will be said on this subject when amplifiers are discussed, in general, measurements where ground-loop voltages are a problem are best handled with differential measuring equipment: amplifiers, oscilloscopes, meters, etc. having *two* inputs, responding to only the *difference* between the two inputs and rejecting the common mode signals. A ground-loop avoiding system for a hypothetical chemical set-up is shown in Figure 8.

An additional point about 60 Hz problems should be made here. Even though a given system relies entirely on dc or high frequencies for carrying the signal information, 60 Hz problems can still be encountered. The magnitude of the power line pickup can easily be so large as to drive dc or even high frequency equipment beyond its linear or dynamic range.

DC versus AC Systems

From the standpoint of S/N the most important decision to be made in chemical-instrument or experiment design is that of determining whether the electrical signals to be measured will be ac or dc. Returning to Figure 7, the signal $Q(t)$ from the system under observation will in general not be ac, but will have some steady or slowly varying value. A necessary part of the apparatus is the provision for some means of "zeroing" the system. This should be done in such a way as to leave as much of the system unchanged as possible, including sources of noise. The ideal way would be to reduce only the phenomena of interest to zero, leaving the transducer or probe and all following equipment unchanged. To avoid noise, drifts, etc. the zero should be taken frequently. If the quantity of interest and the output of the transducer are dc, the simplest way to make the measurement is to amplify and record the signals as dc information. If the signals are small, the system is likely to be subject to drift and to all the other unpleasant properties of $1/f$ flicker and environmental noise.

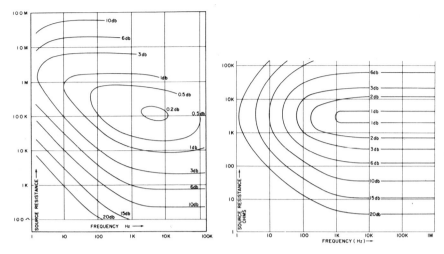

Figure 9. Noise figure contours for typical good low-noise amplifiers shown as function of frequency and source resistance (at 290°K). One at left uses vacuum tubes, on right junction transistors.

If it can be arranged to zero the system not once an hour or once a minute, but say 100 times a second, many important advantages can be realized. First, the repetitive zero's will easily keep up with the system drift. Second, ac amplifiers and ac signal processing techniques can be used, avoiding $1/f$ noise. Third, it may be possible to have only the quantity of interest turned on and off (chopped) while interfering elements are not. This gives the wanted quantity a "signature" which can be identified by some signal processing scheme after amplification.

In designing a system in which it is possible to repetitively zero (or chop or modulate) the quantity of interest at some relatively high rate, it is extremely valuable to have a second signal available, one that tells when the system is on zero and when it is on the unknown. If this signal is available, correlation techniques can be used to greatly improve the S/N.

Transducers

In making a measurement of the desired small-effect phenomena, the chemist must select a probe or transducer in order to derive a signal that can be processed and measured. If he has a choice, he will naturally choose one which gives the largest signal with the smallest amount of interference. The linearity, accuracy, and stability of the transducer, of course, are also important considerations. If the signal is modulated or chopped or is inherently non-dc, the frequency response of the transducer is also of importance. From the standpoint of S/N it is important that the transducer have a good NF and a high signal level output. Let us consider two examples of transducers that are often used by chemists.

The photomultiplier (6) is an example of an excellent transducer often used in photochemistry and spectroscopy. It is efficient in converting weak light into electrical signals, has very high frequency response and has high inherent gain with large dynamic range. The noise properties are also very good, giving it a NF very near zero dB. When measuring very weak light signals the PM's NF can be improved by cooling it to dry-ice temperatures. Unfortunately, the gain of the PM is both strongly temperature and voltage dependent. Also, the device exhibits hysteresis effects. Exposure to strong light will cause a shift in operating characteristics which recover only over periods of hours. The amplification characteristics of PM tubes show a weak $1/f$ effect, hence they are best used for non-dc applications.

The bolometer is another transducer often used by the chemist, generally to detect radiation in spectrographic equipment. The bolometer is usually a small blackened semiconductor element used as a resistance thermometer. There are many problems associated with bolometer radiation detectors. First, they have an appreciable amount of Johnson noise as they are generally high resistance. Second, they are used with a standing current, through them, and thus exhibit flicker noise to a rather large degree. Since they are semiconductors, they are particularly prone to this type of noise. Third, they are rather slow to respond and so are not suitable for use where the radiation is to be chopped at a high frequency to avoid $1/f$ noise. In using bolometers great care is required to achieve good S/N. Chopping at some low frequency, frequently around 10 Hz, is generally used. Within the range of wavelengths where they are sensitive, the new semiconducting diodes usually give much

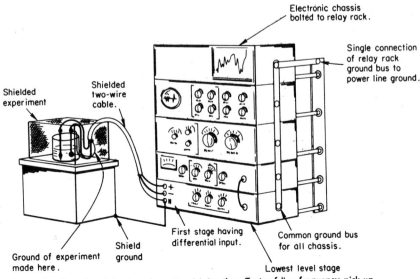

Figure 8. Experimental set-up shown to minimize the effects of line-frequency pick-up.

Electronic chassis bolted to relay rack.

Single connection of relay rack ground bus to power line ground.

Shielded experiment

Shielded two-wire cable.

Ground of experiment made here.

Shield ground

First stage having differential input.

Common ground bus for all chassis.

Lowest level stage

Chemical Instrumentation

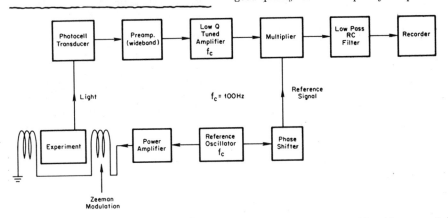

Figure 9 shows the NF curves for two good-quality audio frequency amplifiers

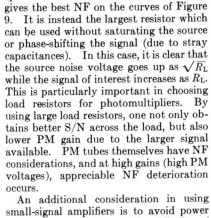

Figure 10. Hypothetical experiment in which light from experimental source, modulated by external magnetic field is measured. Lock-in techniques are used to improve the S/N.

better NF's than bolometers.

Amplifiers

It is usually necessary in chemical instrumentation systems to use amplifiers to raise the signal level of the information. Because they are usually used to process the lowest signal levels, they are very important in determining the S/N. In specifying amplifiers (7), one should be sure they have sufficient dynamic range and bandwidth. However, their bandwidth should be no more than necessary to carry the required information. The narrower the bandwidth the lower will be the noise at the output.

We have seen in Figure 5 that the noise of tubes and transistors is dependent on the source impedance to which they are connected. The excess noise with high source resistors is *in addition* to the Johnson noise generated by the source resistance, and results from a noise current (grid or base current) flowing from the device giving rise to increasing noise as the source resistance is raised. Hence the NF of an amplifier is a complicated function of source impedance, frequency, input device characteristics, etc.

one of which has junction transistors in the front-end, the other a Nuvistor triode. Recall that a perfect amplifier has a NF of 0 dB while one that just doubles the noise power has a NF of 3 dB. The effect of flicker noise at low frequencies, and of the input resistances of the two amplifiers is readily observable. It is obvious that there is an optimum source impedance for obtaining the best NF. Transformers can be used to match source impedances to the input of given amplifiers to obtain better NF, but at the risk of magnetic pickup and some NF degradation due to the winding-resistance Johnson noise.

Summarizing, in choosing an amplifier or transducer for processing weak signals, care must be taken not to degrade the available S/N at the source. NF curves showing characteristic noise figure versus frequency and source impedance help in choosing operating parameters. This is true regardless of the frequency or bandwidth of the information to be processed.

A pitfall often encountered should be pointed out. If the signal source is a constant current generator (like a photocell or photomultiplier), and one has a choice as to the load resistance R_L into which to develop the signal voltage, the

optimum load resistance is *not* that which gives the best NF on the curves of Figure 9. It is instead the largest resistor which can be used without saturating the source or phase-shifting the signal (due to stray capacitances). In this case, it is clear that the source noise voltage goes up as $\sqrt{R_L}$ while the signal of interest increases as R_L. This is particularly important in choosing load resistors for photomultipliers. By using large load resistors, one not only obtains better S/N across the load, but also lower PM gain due to the larger signal available. PM tubes themselves have NF considerations, and at high gains (high PM voltages), appreciable NF deterioration occurs.

An additional consideration in using small-signal amplifiers is to avoid power line frequency pick-up. The difficulty of doing this was mentioned in the system grounding discussion. An extremely useful technique for avoiding pickup problems is simply to use differential amplifiers in the lowest level stage of the instrument. Good ac amplifiers exhibit a common mode rejection greater than 10^6 for frequencies below 500 Hz. Hence the proper use of differential amplifiers can greatly reduce problems of spurious environmental pickup of all kinds and so merits serious consideration.

Signal Processing

After the signals of interest have been amplified to a level where there is no longer danger that noise will degrade them, consideration should be given to processing the information in some way to enhance the signal and reject the noise. (8) Also, if the signals are ac they could be converted into dc to drive one of the many available recording devices.

In the case of dc measurements, only a simple low pass filter is needed to smooth the amplified signal. Averaging over longer periods of time (observing with narrower bandwidth at zero frequency) can hopefully improve S/N. However dc drift, flicker-noise, etc., often completely negate any theoretical S/N achieved.

If the signals of interest are inherently at, or can be made to lie at, a single frequency f_c, the situation is much better. First, tuned amplifiers which only respond to a small band of frequencies around f_c can be used to exclude broad-band and discrete noise not at the signal frequencies. If the frequency is high enough, flicker noise will not be a problem. The narrow band of frequencies which pass through the tuned amplifiers can now be detected (changed from ac to dc) and averaged in a low pass filter. Techniques of this type can be made to work quite well, but are limited in the achievable narrowness of bandwidth because of frequency drift problems. Drift of f_c relative to the response frequency of the narrow filters would make these filters difficult to use.

If one can arrange to have a clean "reference" signal at f_c, derived in some way from the modulation or chopping process such that this "reference" is phase-locked to the small signal to be measured, one can achieve dramatic improvement in S/N by a simple application of communication theory. The technique is a type of correlation analysis and allows one to em-

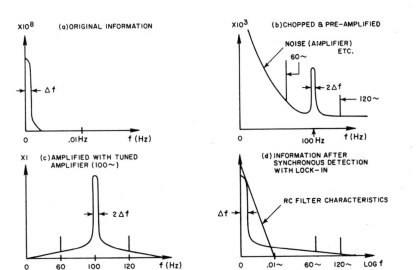

Figure 11. Diagrammatical representation of frequency spectra of information as it is carried through experiment shown in Figure 10.

Chemical Instrumentation

ploy arbitrarily narrow bandwidths for random noise rejection without frequency drift problems. The basic principle that is exploited here is the fact that the cross-correlation of random noise with a periodic signal tends to zero with increasing averaging time while the cross-correlation of a signal with a replica of itself yields a constant independent of integrating time.

The Lock-In Amplifier

Complete signal processing instruments embodying the techniques of cross-correlation described in the preceding paragraphs are finding increasing use in laboratories in all fields. They are generally called "Lock-In Amplifiers" or synchronous detectors. By their correct use the chemist can achieve essentially optimum S/N in his measurements. To understand how they may be applied consider the hypothetical experiment shown in Figure 10. Here the signal of interest is light which can be turned on and off by the application of a magnetic field externally applied to the system under study. The light, flashing at the modulation frequency f_c, is converted into an ac electrical signal by a photomultiplier transducer. The signal at f_c must now be brought to a level sufficiently high to drive the detector that will convert the information to dc for recording.

This level should be such that the inherent drift in the detector and following dc amplifier are of negligible magnitude compared to the dc signal. In general this amplification is best done in a tuned amplifier of moderate Q. While not contributing to the overall system S/N, the tuned amplifier helps prevent overloading of the detector by random noise (through reducing the bandwidth of the noise at the detector input) or by spurious signals such as 60 and 120 Hz pick-up.

The detection of the signal is the most critical part of the system from the standpoint of S/N. The detector should have as narrow a bandwidth as possible consistent with the fact that bandwidth sets limits on the rapidity with which the amplitude of the initial unknown signal can be allowed to change and also set limits on the observation time required to make the measurement. In going to a very narrow bandwidth detector, provisions must be made to prevent the center frequency of the detector from drifting off f_c. Fortunately the lock-in detector, or synchronous detector as it is sometimes called, is available for this application. In essence, the lock-in detector is a cross-correlator that provides an output dc signal that is the time average of the product of the unknown signal plus noise with the reference signal. The bandwidth of this detector, which is numerically equal to the reciprocal of the averaging time, can be made as narrow as desired in a simple way, by just increasing the averaging time (usually by increasing the RC time constant). This cross-correlation detector has another great advantage; the center frequency of response is "locked-in" to the carrier frequency f_c, avoiding drift problems. Looking at the cross-correlation from another viewpoint, one can say that it is essentially

a mixer that multiplies the unknown signal, $f_c \pm f$, by the "reference signal," a pure square-wave signal at exactly f_c that is phase related in a definite way to the unknown signal. This mixing results in the unknown information appearing in a band of frequencies of about zero frequency (dc). (The upper side band at $2f_c$ is of no interest here and is rejected by the low-pass filter that follows the detector.) The reference signal is obtained in a way that makes it unambiguous in frequency and phase with respect to the signal modulation wave-form. The bandwidth of the detector can be made arbitrarily narrow by passing the "zero frequency" output through an RC low-pass filter. The effective bandwidth of the detector will then be $\Delta f = 1/4RC$ cycles.

The dc output of the detector must now be brought up to a level that will drive strip chart recorders. Any drift in the output of the detector or dc amplifier will

be another source of "noise" in the experiment.

Figure 11 shows in diagrammatic form the handling of the information in our hypothetical experiment. It shows the original dc information, the moving of the signal to $f_c = 100$ Hz, amplification, and finally demodulation and filtering.

It should be noted that, apart from dc drift in the synchronous detector and following amplifier, the system should be free from drift. Because of the modulation of the light by the magnetic field, there is effectively a "zero" taken 100 times a second. However, any pick-up of 100 Hz signals in the system will result in an "offset" of the output. This can be checked for by removing the sample. In practice

the drift in the detector and dc amplifier can easily be made negligibly small.

To understand better how the synchronous detector of the lock-in works, refer to

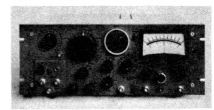

Figure 14. Model HR-8 Lock-In Amplifier manufactured by Princeton Applied Research Corporation.

Figure 12. The detector can be imagined as a switch that is driven by the reference signal so that it is connected to one phase of the detected signal for 180° of the signal, and then to the other phase for the remainder of the cycle. The result will look

in effect like a full wave rectification of the signal, hence will have a dc component. Noise will not yield any net dc on the output. As a result, the output of the switch after sufficient filtering will be a dc voltage proportional to the original ac signal while noise will average to zero.

There are several commercially available lock-in amplifiers on the market. These are in general complete signal processing systems and are very flexible in their application. Most only require being connected to a suitable transducer and to a recorder to achieve a complete system. Some have built-in oscillators and can thus produce a reference signal that is used internally for demodulation. This reference signal is available externally for use as the

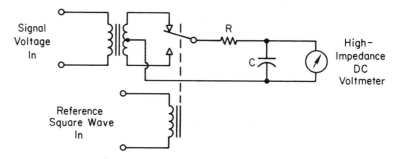

Figure 12. Synchronous detector in conceptual form to illustrate action of lock-in amplifier.

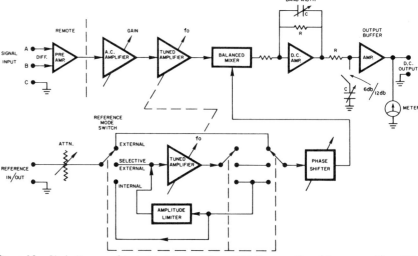

Figure 13. Block diagram of typical lock-in amplifier. With the exception of the preamplifier all the elements of the unit are usually contained in a relay rack chassis.

Chemical Instrumentation

source of the modulation for the unknown signal. Figure 13 illustrates a representative block diagram of a commercial lock-in. Two of the instruments are pictured in Figures 14 and 15. They are the Princeton Applied Research Model HR-8 and the Brower Laboratories Model 131. To illustrate the type of result one might obtain in using lock-in amplifiers to recover small signals from noise, Figure 16 shows two small ac signals, 10^{-10} volts and 10^{-15} amps, being recorded on a strip chart recorder. The Johnson noise in the source resistances in the two cases correspond to 1.3×10^{-10} volts/$\sqrt{Hz}$ and 40×10^{-15} amps/$\sqrt{Hz}$, respectively.

To illustrate how conveniently a lock-in amplifier can be used in chemical instrumentation, Figure 17 shows a system, built around this technique, for measuring the exciting wave length dependence of the weak phosphorescence of some material. The electronics required for this set-up is almost completely contained in the lock-in amplifier chassis.

In the foregoing, the cross-correlation detector, lock-in amplifier, technique has been discussed in terms of its operating principles, and a few examples of application have been given. The applications of this technique are, in general, limited only by the ingenuity of the experimenter. In many cases a commerically available lock-in amplifier can be added to an existing piece of apparatus with good and

Figure 15. Model 131 Lock-In Amplifier of the Brower Laboratories, Inc.

sometimes even spectacular results. This will be particularly true if the existing apparatus makes use of dc detection methods. However, to make the best possible use of this technique it is generally necessary to design the whole experiment around it. Such questions as the best method of modulation (what to modulate and how), the type of transducer to be used, the optimum operating frequency and many more must be considered with great care. To answer these questions correctly the experimenter must be equipped with an understanding of the lock-in amplifier technique as well as of the nature and behavior of the various factors that introduce noise, which finally form the fundamental limitations of his experiment. The experimenter who makes the effort to acquire a sound understanding of these factors will be rewarded by a broadening of his research horizons and a clearer insight into the scope of his experimental techniques.

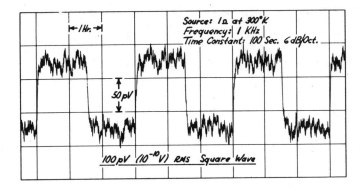

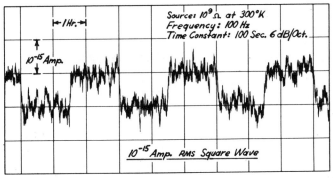

Figure 16. Recordings of the measurement of a very small ac current and voltage using lock-in techniques. The magnitude of the signals are well below the 1 Hz bandwidth Johnson noise of the associated source resistors.

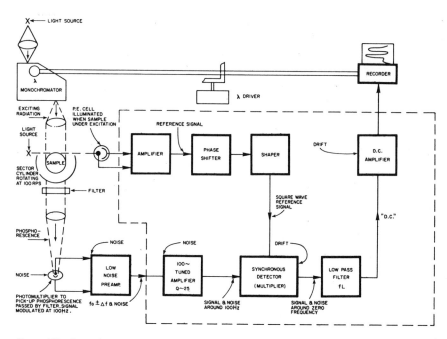

Figure 17. Typical instrumentation system using lock-in amplifier techniques. It measures the dependence of weak phosphorescence of sample on the wavelength of exciting radiation. Components enclosed in dotted area make up a lock-in amplifier.

Chemical Instrumentation

Manufacturers of Lock-In Amplifiers

Acton Laboratories
531 Main Street
Acton, Massachusetts 01720

AIM Electronics, Ltd.
71–73 Fitzroy Street
Cambridge, England

Automatic Systems Laboratories, Ltd.
Construction House
Grovebury Road
Leighton Buzzard
Bedfordshire, England

Brookdeal Electronics, Ltd.
Myron Place
London, S.E. 13, England

Brower Laboratories, Inc.
Turnpike Road
Westboro, Massachusetts 01581

EMCEE Instruments
160 East Third Street
Mt. Vernon, New York 10550

Ithaco, Inc.
413 Taughannock Boulevard
Ithaca, New York 14850

Princeton Applied Research Corporation
P.O. Box 565
Princeton, New Jersey 08540

Teltronics, Inc.
P.O. Box 466
Nashua, New Hampshire 03060

Literature Cited

(1) W. Schottky, *Ann. Physik.* **57**, 54 (1918).
(2) J. B. Johnson, *Phys. Rev.* **32**, 97 (1928).
(3) H. Nyquist, *Phys. Rev.* **32**, 110 (1928).
(4) A. Van der Ziel, *"Noise,"* Prentice-Hall, Inc., Englewood Cliffs, New Jersey, 1954.
(5) For a good treatment of NF and related subjects see: A. T. Starr, *"Radio and Radar Techniques,"* Pitman and Sons, Ltd., London, 1953, Appendix 9, "Noise."
(6) A review of photomultiplier noise is given on page 28 in "RCA Technical Manual PT-60" published by Radio Corporation of America, Lancaster, Pa. (1963).
(7) Consideration of noise in amplifiers is given in: W. R. Bennett, *"Electrical Noise,"* McGraw-Hill Book Co., New York, 1960.
(8) A general treatment of signal processing is given in: Y. W. Lee, *"Statistical Theory of Communication,"* John Wiley & Sons, Inc., New York, 1960.

Vivas in Deo. God be with you.
An expression of good-will
found in early Christian letters.

Topics in...

Chemical Instrumentation

Edited by **S. Z. LEWIN**, New York University, New York 3, N.Y.

The Journal of Chemical Education
Vol. 43, pages A567, A625
July and August, 1966

XXIX. Gel Permeation Chromatography

JACK CAZES, Mobil Research and Development Corp.,
Paulsboro, N. J. 08066

Jack Cazes currently holds the position of Supervising Chemist in charge of the Separations Research Group at Mobil Research and Development Corp., Paulsboro, N. J. He had previously spent seven years with Mobil Chemical Co., where he was involved in many areas of research, including organic chemical and polymer process research, design and application of analytical instrumentation, gel permeation chromatography, and computerized data processing. Other interests have included organic synthesis, organic free-radical reaction mechanisms, and electronic circuitry for chemical instrumentation.

Dr. Cazes received a B.S. in Chemistry (1955) at the City College of New York, and both M.S. and Ph.D. degrees in organic chemistry at New York University.

Dr. Cazes has taught at Queens College (New York) and at Rutgers—the State University (New Brunswick). He currently teaches a Short Course on Gel Permeation Chromatography for the American Chemical Society.

1970

Gel permeation chromatography (GPC) is the newest analytical tool available for the fractionation and characterization of polymers. It is rapidly becoming indispensable in the field of polymer science both for the determination of molecular weight distribution of polymers and for the fractionation of medium and high molecular weight materials for further characterization by other physical and chemical methods. The basic principle of gel permeation chromatography can, at first glance, be likened to that of a non-ionic molecular sieve having pores ranging in size from 10 to 10⁷ angstroms. The separation of molecules according to size is envisioned to take place within the voids which are present in the particles of a rigid, cross-linked polystyrene gel. The structure of such a gel is analogous to the structure of an ion exchange resin but without its ionic sites.   The gel, composed of rigid, cross-linked beads of approx-

imately 200 to 300 mesh size, is prepared by suspension copolymerization of styrene with a difunctional monomer (cross-linking agent) such as divinylbenzene in the presence of a diluent. It is possible to prepare a gel having a given maximum pore size and a given pore size distribution by proper choice of cross-linking agent and type and concentration of diluent. Figure 1 is a photomicrograph of the spherical beads that are produced.

Historically, gel permeation chromatography began with observations made during exchange studies; some non-ionic substances were fractionated by passage through an ion exchange column. Separation seemed to be according to molecular size. For example, it was noted that ethylene glycol, glycerine, and sugar were separated on a cationic polystyrene resin by elution with water; the sugar eluted first, followed by the glycerine and then the ethylene glycol. It is not certain whether such separations were due to a true "molecular sieving" or at least partly to some small adsorptive effect. It was not until the group headed by J. C. Moore of the Dow Chemical Company first reported their work in 1963 that gel permeation chromatography was really born. Moore, in 1964, described the procedure for reproducibly preparing gels of given maximum pore sizes and illustrated the utility of the separation technique for the determination of molecular weight distributions of polymeric substances. It was Moore who named this technique "gel permeation chromatography."

The separation of molecular species according to size is believed to occur as a result of differences in the extent to which different species permeate the gel particles. Molecules whose size is too great to allow them to enter the gel pores pass through the column solely by way of the interstitial volume, i.e., the space occupied by the solvent outside the porous beads. Smaller molecules permeate the gel to a

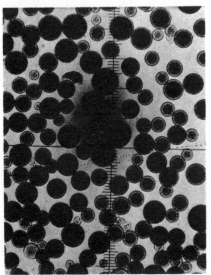

Figure 1. Photomicrograph of cross-linked polystyrene beads.

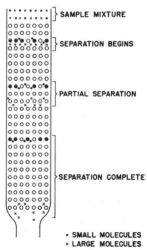

Figure 2. Separation of molecules according to size by gel permeation chromatography.

Chemical Instrumentation

greater or lesser degree, depending upon their size and upon the distribution of pore sizes available in the gel particles. Consequently, the largest molecules emerge from the column first, followed by smaller molecules which must follow a more circuitous and tortuous path as they travel through the column. This order of elution—large molecules first, small molecules last—while not surprising here, does stand out in contrast to that which is generally observed in other chromatographic techniques. An idealized view of the over-all process is illustrated schematically in Figure 2.

Mechanism of the Separation

The general picture of what occurs in and near the gel particles during the separation process is not thoroughly understood. It is not known whether the process involves a true, reversible, dynamic equilibrium, or near equilibrium, or possibly, one which is diffusion-controlled. It is of interest to examine some experimental observations in order to understand the situation that exists. The resolution achieved with a given gel permeation column is often improved as the eluent flow rate is decreased. For all practical purposes, resolution may be continuously improved by a continued decrease in flow rate, almost down to a flow rate of zero. This is a common observation in liquid-solid chromatography. However, in gel permeation chromatography the improved resolution manifests itself as a decrease in peak width rather than as an increase in the differences between the elution volumes of the individual components. Elution volume, for a given component, remains almost constant. If the mechanism of gel permeation chromatography involved a diffusion-controlled process, one might expect elution volume to increase with decreased flow rate. One must keep in mind, however the possibility that changes in flow rate might not affect diffusion within or near the gel pores; changes in flow rate might influence the diffusion of molecules through the interstitial volume alone.

Another interesting observation can be made. The chromatogram obtained for a mixture is usually the graphical sum of the chromatograms obtained for the individual components. In other words, each molecule "does not know" that other molecules are nearby, provided that the column has not been overloaded. As with other chromatographic techniques, this is a necessary condition if the technique is to be reproducible and generally useful.

Resolution is improved by an increase in temperature. Elution volume *does* change with temperature. The improved resolution is commonly attributed to a decrease in the viscosity of the solvent as the temperature is increased. This results in greater freedom of movement of the molecules in solution, allowing them to enter and leave the gel's pores with greater ease. This idea is further supported by the observation that two different solvents

having the same viscosity give almost identical resolution for a given sample. Diffusion control of the separation process is certainly suggested.

To date, no systematic study aimed at the elucidation of the mechanism of the gel permeation chromatographic process has been reported. Perhaps this lack of basic data is the result of fascination with the application of a new tool to rapidly solve problems in applied polymer science. Gel permeation chromatography has already yielded rapid answers to many questions involving the molecular weight distributions of polymers that were heretofore left unanswered because of the extreme difficulty involved in earlier methods. There is no doubt, however, that some workers will turn to more basic problems involving the gel permeation chromatographic technique itself.

Instrumentation

The instrumentation required for gel permeation chromatography can be as simple as a glass tube packed with the rigid, cross-linked polystyrene beads suspended in a suitable solvent if only coarse fractionation is required. However, in order to achieve precision approaching the classical molecular weight methods, several parameters, such as flow rate and temperature must be closely controlled. The complexity of the equipment is determined by several factors, including the degree of reproducibility that is required, the sample size, and the extent of automation and convenience of operation that is desired. Usually, however, the final decision as to the extent of refinement of the instrumentation is a compromise between the factors mentioned above and the cost of the equipment needed to achieve the end result. An attempt will be made here to examine the role of each functional unit in the idealized gel permeation chromatograph and to critically discuss the advantages and disadvantages of the design features involved in its construction.

The instrumentation needed to carry out gel permeation chromatographic separations can be conveniently discussed under the following functional headings:

1. Solvent handling
2. Sample introduction
3. Fractionation
4. Detection, estimation, and collection

The degree of sophistication of the instrumentation used to carry out these functions is limited only by the ingenuity and budget of the designer, and ultimately, the user of the equipment.

Solvent Handling

The solvent handling system must store the solvent, remove gases that may be dissolved in the solvent, and carry the solvent, at a constant rate of flow, past the sampling device and to the head of the column. Constancy of flow rate is of prime importance since resolution can be adversely affected by variations in flow rate. Changes in resolution resulting from an erratic flow rate lead to poor reproducibility, and chromatograms recorded under such conditions do not lend themselves to comparison. Furthermore, the use of gel permeation chromatography as a means of determining molecular weight distribution involves calibration of the equipment with standards of known molecular weight. Such calibration requires a stable and reproducible solvent flow rate.

A non-pulsing liquid pump would appear to be the simplest approach. Unfortunately, non-pulsing pumps that are capable of pumping liquids at a constant flow rate of a few milliliters a minute against a back pressure of from 50 to 200 psig are just not readily available. The usual approach to a problem of this type is to use a pulsing, variable stroke, liquid pump in conjunction with a pulse buffering system; the simplest buffer is a *surge tank* or, as it is sometimes called, a *ballast tank*. The surge tank is placed in the solvent stream immediately after the pump. Its volume is chosen so that pulsations produced by the pump are damped out. Solvent flows from the pump into the surge tank and then to the column. This is shown schematically in Figure 3.

Solvent must be freed of any dissolved gases prior to entering a heated column. The reason for degassing is to prevent possible formation of gas bubbles within the column. Gaseous spaces in the column packing lead to what is generally called channeling with a resultant non-homogeneous pattern of solvent flow through the column. The solvent is forced to flow around the gas pockets at uneven rates and broadened and irreg-

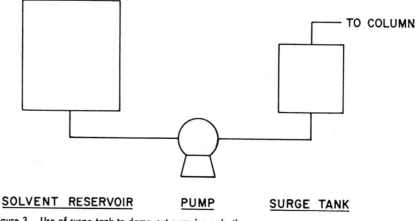

SOLVENT RESERVOIR   **PUMP**   **SURGE TANK**

Figure 3. Use of surge tank to damp out pumping pulsations.

ularly shaped chromatographic peaks are observed along with greatly diminished resolution. The observed chromatographic curves do not reflect the true distribution of the sample being fractionated. Degassing can be most conveniently accomplished by passing the solvent through a vented heater maintained at a temperature just below the boiling point of the solvent. Most of the dissolved gases are then boiled out of solution and vented to the atmosphere. The best location for the degasser is probably just ahead of the solvent pump where the solvent is at atmospheric pressure and venting presents no significant problems. Placement of the degasser in a pressurized part of the system would necessitate heating the solvent to a temperature higher than its normal boiling point in order to remove dissolved gases. Furthermore, it would be necessary to provide for removal of gases under pressure to a point where they could not redissolve; venting of gases under pressure without loss of solvent would be inconvenient at best.

Sample Introduction

The sample is generally prepared as a $\frac{1}{8}$ to $\frac{1}{2}\%$ solution in the same solvent that is used as the eluent. The sampling system must be capable of introducing the sample solution into the flowing solvent stream at a point as close to the head of the column as possible. Ideally, the sample should be placed onto the column in as short a time as possible so as to produce a narrow band or "slug" of sample. One way of accomplishing this is to inject the sample solution directly into the solvent in or just ahead of the column with a hypodermic syringe and septum arrangement of the type generally used in gas chromatographs. Figure 4 illustrates such an arrangement. A possible disadvantage of this method of sample introduction is that the syringe is subjected to whatever pressure is present in the system at the time of injection.

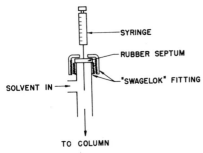

Figure 4. Septum/syringe injection.

Another more advantageous approach involves the use of a four-port valve arranged as shown in Figure 5. A calibrated sample loop is filled with the sample, at atmospheric pressure, from a syringe while the valve is in the "filling" position. In this way it is possible to completely flush out the sample loop with sample and, if desired, heat the sample in the loop prior to actually passing it into the column. The valve is then rotated to the "emptying" position, placing the sample loop directly into the solvent stream. The amount of sample actually entering the

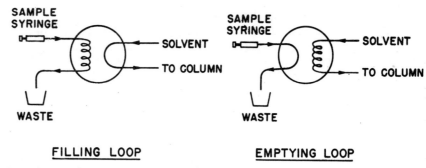

FILLING LOOP   EMPTYING LOOP

Figure 5. Use of mult-port valve for sample introduction.

column is determined by the concentration of the sample solution, the size of the sample loop, the time that the loop is in the solvent stream, and the flow rate of the solvent. A list of manufacturers of multi-port valves is given further on in this paper. Such valves generally cost from $100 to $500 depending upon the material of construction.

Fractionation

The heart of any column chromatographic instrument is the column, which consists of a shell (usually a length of tubing) filled with the column packing. It is in the column packing that the various species present in the sample are separated from each other. Consequently, one cannot be overly meticulous when preparing a gel permeation column.

The tubing can be made of any nonreactive, insoluble material that will contain the packing and withstand the operating temperature and pressure. Some suitable, readily available materials are glass, copper, aluminum, and stainless steel. At present it is just about unanimously agreed that the column should have a uniform circular cross-section throughout its length, and that it should not be coiled. With such a column, one is generally able to maintain a homogeneous solvent flow pattern. A narrowly coiled column will produce severely broadened overlapping peaks as a result of the fact that the solvent flowing along the outer circumference of the coil must travel a longer distance than the solvent along the inner circumference. No doubt there are some investigators who will attempt to use columns made with tubing of circular cross-sections, but having a conical shape, in attempts to compensate for the decreasing pressure differences measured between the exit of the column and points successively further away from the head of the column. Others will probably attempt to use square, triangular, or annular tubing, or tubing with other cross-sections as has been tried in gas chromatography. While such approaches ofttimes produce novel effects, no distinct general advantage is realized.

A word might be said about the use of annular columns since they might offer some advantages in cases where increased sample capacity is desired or where temperature programming is to be used. By using an annular column, the total cross-sectional area of the packing can be greatly increased without significantly increasing the thickness of the packing.

It is possible that annular columns having a large outside diameter with only a narrow annulus of packing could present less difficulty in the establishment of homogeneous flow than columns having a solid circular cross-section.

It is not likely that temperature programming will be used with gel permeation chromatography, at least at present, since it would not be expected to produce the dramatically improved resolution that its use provides in gas chromatography. However, if temperature programming should be desirable, annular columns would offer a solution to the heat transfer problem that is involved in uniformly changing the temperature of the column packing in a reproducible manner.

As with any chromatographic column the process of packing a GPC column should be carried out with great care. Uniformity of packing density together with complete absence of voids are absolutely essential. The cross-linked polystyrene gel particles must be kept completely immersed in solvent while being transferred to the column. If the gel particles are allowed to dry, leaving the pores filled with air, it is virtually impossible to restore the gel to its original state of efficiency. One is generally forced to prepare a fresh batch of gel.

The efficiency of a gel permeation column is commonly expressed in terms of the number of "theoretical plates" it contains. The theoretical plate concept is borrowed from that area of chemical engineering involving fractional distillation. A theoretical plate, in the case of distillation, refers to a discreet distillation stage constituting a simple distillation in which complete equilibrium is established between the liquid and vapor phases. In the case of gel permeation chromatography, where the two phases involved are in constant motion, i.e., the solvent in the interstitial volume and the solvent within the gel pores, equilibrium is probably never achieved and the true significance of the theoretical plate is lost. It must be realized, moreover, that the calculated theoretical plate in a chromatographic column represents smaller separating ability than the theoretical plate in a distillation column by, perhaps, a factor of twenty-five or fifty. Nevertheless, it is still convenient to calculate the number of theoretical plates for a chromatographic column as a means of making comparisons between the efficiencies of different chromatographic columns.

The number of theoretical plates in a

Chemical Instrumentation

column is not only a function of the column and its operating parameters (such as temperature, pressure, solvent, and flow-rate), but also depends upon the sample species being separated. For example, if the number of theoretical plates in a column is determined under a given set of conditions with substance A as the sample, this number will probably be different from that which is determined with substance B, having the same or different elution volume, as the sample A. Therefore, if one is to examine a sample containing several species, then the plate count determined using a single pure substance is, at best, only an approximation of the true plate count for each species present in the sample.

The procedure for determining the plate count is as follows.

1. A chromatogram is run with a single pure substance as the sample. o-Dichlorobenzene, tetrahydrofuran, and 1,2,4,-trichlorobenzene are the substances that are commonly used for this purpose in GPC.

2. A triangle is constructed on the chromatographic peak by drawing a line across the base of the peak and two lines tangent to the sides of the peak at the points of maximum slope. This is illustrated in Figure 6.

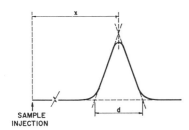

Figure 6. Calculation of column plate count.

3. The distance from the point of sample injection to the apex of the peak and width of the base are both measured.

4. The number of theoretical plates is then calculated with the formula:

$$P = \frac{16}{f}\left(\frac{x}{d}\right)^2 \quad (1)$$

where

P = plates per foot
f = no. of feet of column
x = ml elution volume from injection to peak apex
d = width of base in ml.

Detection, Estimation, and Collection

Detection and estimation of sample components eluting from the column can be carried out either by continuously monitoring the effluent stream or by collecting fractions of the eluate and estimating the sample components present in each fraction individually. The advantages of continuous monitoring include convenience of operation and the ability to couple such a system to other automatic electronic devices for recording and integration of the chromatographic curve. An advantage inherent in the collection of fractions is the ability to

characterize the various species present in each fraction by independent instrumental and chemical methods. Of course, the most desirable approach would be to continuously monitor the column effluent and collect fractions at some point beyond the continuous monitor.

The usual sample size in analytical scale gel permeation chromatography is only a few milligrams; the sample emerges from the column over a twenty-five to one hundred milliliter elution volume. This means that each milliliter of eluate contains less than one milligram of sample. Here it becomes necessary to use a very sensitive detection method having a high signal-to-noise ratio. Advantage can generally be taken of one or more of the physical properties of the substances being examined. For example, one might measure such properties as refractive index, dielectric constant, UV absorption, or fluorescence. With samples that are labeled with radioactive isotopes, one can continuously monitor the radioactivity of the column effluent. In cases where fractions are collected, one can make use of both chemical and physical properties of the sample for characterization of the fractions. Colored substances can be estimated colorimetrically, while colorless substances may be converted to colored or fluorescent ones by addition of appropriate reagents.

A dramatic increase in detection sensitivity can be realized, very often, by using a differential device. For example, one can readily measure a difference in refractive index with much greater sensitivity than one is able to measure a single value of refractive index. With this in mind, one can design the chromatographic system such that it has a dual stream arrangement (analogous to "double beam" in spectrophotometry), one stream carrying only solvent and a second stream carrying solvent plus sample. In such a system, one can then monitor the difference in a physical property between the two streams. This technique lends itself well to the application of servo-mechanical control, recording, and computation. The choice of a particular detection device for the construction of a laboratory gel permeation chromatograph often is governed by the availability, in the laboratory, of an instrument that is capable of measuring a given physical property. A detailed discussion of all of the possible detection devices, however, is beyond the scope of this paper. However, there is one detection device that ought to be discussed here because of its broad applicability and ease of operation. The automatic recording differential refractometer is capable of precisely measuring the difference in refractive index between two flowing liquid streams. The usefulness of this type of detector in GPC stems from the fact that it is applicable to a wide variety of organic compounds. It does not depend upon the presence of a given functional group or other structural feature of the molecules being detected. The only requirement is that there be a difference in refractive index between the solvent and the sample being examined.

A commercial unit (manufactured by Waters Associates, Framingham, Mass.,

$5000) employs a dual barrier layer photocell detection system and is claimed to be capable of detecting a refractive index difference of 1×10^{-7} refractive index units. Another (manufactured by Phoenix Precision Instrument Co., Philadelphia, Pa., $4840) employs a monochromatic source, chopper, and single photomultiplier, a combination which is claimed to result in good long-term stability. This unit can detect a difference of 5×10^{-7} refractive index units.

Collection devices most often involve a mechanism for sensing the flow of liquid from the column by counting drops or by measuring a volume through the use of a calibrated syphon. Sometimes a timing device is employed, although this is not as desirable as drop-counting or volumetric syphoning. Since, in gel permeation chromatography, a given species will emerge from the column at a given elution volume, its retention time will vary with variations in flow-rate. Timing devices do not take into account or correct for variations in flow-rate, while volumetric syphoning and drop counting do. A detailed discussion of commercially available fraction cutting and collection devices is the subject of an earlier paper in this series [Lewin, S. Z., J. CHEM. EDUC., 38, A515, A567, A713, A789 (1961)]. A list of manufacturers of these devices is given at the end of the present paper.

Average Molecular Weights of Polymers

Polymers, both natural and synthetic, contain molecules having a distribution of different molecular sizes rather than one molecular size as is the case for low molecular weight monomeric substances. Most of the physical methods generally used for the measurement of molecular weight such as light scattering, osmometry, viscometry, cryoscopy, ebulliometry, sedimentation, and centrifugation yield average values. The two molecular weight averages that are most commonly calculated and which can be determined by the gel permeation chromatographic method are the number-average molecular weight ($\bar{M}_n$) and the weight-average molecular weight ($\bar{M}_w$). The weight average is always either greater than or equal to the number average; the equality exists only ideally in the case of monodisperse polymers, i.e., those in which all of the molecules are of a single molecular weight.

The number-average molecular weight is defined by the equation:

$$\bar{M}_n = W/\sum_{i=1}^{\infty} N_i = \sum_{i=1}^{\infty} M_i N_i / \sum_{i=1}^{\infty} N_i \quad (2)$$

while the weight-average molecular weight is defined by the equation:

$$\bar{M}_w = \sum_{i=1}^{\infty} W_i M_i = \sum_{i=1}^{\infty} M_i^2 N_i / \sum_{i=1}^{\infty} N_i M_i$$

$$(3)$$

where

W = total weight of polymer.
W_i = weight fraction of a given species, i.

N_i = number of moles of each species, i.

M_i = molecular weight of each species, i.

The relative locations of $\bar{M}_n$ and $\bar{M}_w$ on a

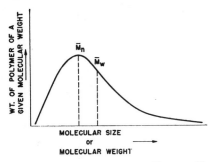

Figure 7. Relative locations of $\bar{M}_w$ and $\bar{M}_n$ on a distribution curve.

molecular weight distribution curve for a polydisperse polymer are shown in Figure 7. The ratio $\bar{M}_w/\bar{M}_n$, commonly called the "heterogeneity factor," is used to describe the degree of polydispersity of a polymer, i.e., the broadness or spread of its molecular weight distribution curve. The larger the value of $\bar{M}_w/\bar{M}_n$, the broader will be its molecular weight distribution.

Gel permeation chromatography is ideally suited for the rapid determination of the molecular weight distributions and molecular weight averages of polymers. The chromatographic system must first, however, be calibrated with relatively

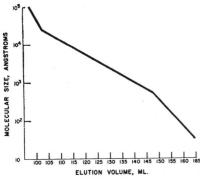

Figure 8. Typical calibration curve.

monodisperse polymers whose average molecular sizes (or molecular weights) are known. The molecular weights or sizes of these "standards" are plotted against their peak elution volumes obtained from their gel permeation chromatograms. A typical calibration curve for polystyrene is given in Figure 8.

While it is common practice to refer to "chain length" when describing the size of a polymer molecule in solution it is probably more correct to use the *radius of gyration* of the coiled polymer molecule since this is the closest approximation to the real effective size of such a molecule in a non-ideal solution. The radius of gyration, S, is defined as the root-mean-square distance of the elements of a polymer chain from its center of gravity and can be calculated from the intrinsic viscosity of the

polymer. For linear polymers the mean-square end-to-end distance, $\bar{R}$, is related to the radius of gyration: $\bar{R}_2 = 6\bar{S}^2$. The polystyrene standards that were used in the construction of the calibration curve are listed in Table 1 together with their molecular weight averages and calculated molecular sizes.

At present, it is common practice to use a polystyrene calibration curve for the determination of molecular weight averages of other polymers. This is due to the fact that polystyrene is the only polymer for which there are several commercially available standard samples. Experimentally determined correction factors are used to convert from polystyrene to other polymers. Many attempts have also been made to mathematically estimate such conversion factors. Approaches such as these are valid only if the calibration curve for the polymer being studied has the same shape as the curve constructed with polystyrene standards and runs parallel to it throughout the range of interest. This is usually not the case!

Calibration of the gel permeation chromatograph for use in precise work requires that standards be used which are of the same polymeric species as the polymers being examined. Since polystyrene is the only polymer for which standards are commercially available, other polymer standards must be prepared, as needed, either by one of the classical fractionation methods—fractional precipitation or extraction, elution or precipitation chromatography, or turbidimetric titration—or by gel permeation chromatographic fractionation itself. An automatic preparative-scale gel permeation chromatograph is commercially available for this purpose, and is described later in this paper.

After polymer fractions having narrow molecular weight distributions have been prepared, it is necessary to characterize them, i.e., determine their weight-average and number-average molecular weights and heterogeneity factors. One can determine weight-average molecular weights with light scattering or equilibrium ultracentrifugation while number-average molecular weight can be determined with cryoscopy, ebulliometry, or osmometry. In many cases it is also possible to correlate molecular weight with such properties as

sedimentation and diffusion rates, melt viscosity, or intrinsic viscosity.

With a calibration curve in hand, it is possible to determine $\bar{M}_w$ and $\bar{M}_n$ for a polymer. The sample whose molecular weight averages are to be determined, is passed through the gel permeation column and its chromatogram recorded. The chromatographic curve is then divided into vertical segments of equal elution volume as shown in Figure 9. The height of each segment together with its corresponding average molecular size, obtained from the calibration curve, are then used to calculate $\bar{A}_n$ and $\bar{A}_w$ for the polymer. It is convenient to arrange the data in tabular form as shown in Table 2 for ease in carrying out the summing and averaging processes. A factor Q, which represents the number of molecular weight units equivalent to one angstrom of effective molecular size, is then used to convert average molecular size to average molecular weight. The cumulative height per cent (column 4, Table 2) is calculated so that a normalized integral distribution curve can be plotted. The reverse summing results in a plot that is comparable to curves obtained by other fractionation methods. Column 4 is plotted against either column 1 or column 5. Figure 10 illustrates such a curve for the data in Table 2.

The formulae used are summarized:

$$H_i = M_i N_i \qquad (4)$$

$$M_i = A_i Q \qquad (5)$$

$$\bar{A}_n = \frac{1}{Q}\frac{\Sigma M_i N_i}{\Sigma N_i} = \frac{\Sigma \text{Column 2 (Table 2)}}{\Sigma \text{Column 6 (Table 2)}} \qquad (6)$$

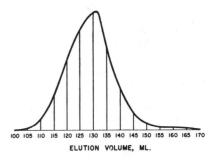

Figure 9. Chromatogram divided into segments prior to calculation of M_w and M_n.

Table 1. Commercially Available Polystyrene Standards

Standard	$\bar{M}_w$ $\times 10^{-5}$	$\bar{M}_n$ $\times 10^{-5}$	Size, Å	Source[a]
S-102	0.820	0.781	1,936	D
S-103	1.22	1.14	2,813	D
S-108	2.67	?	6,207	D
S-109	1.79	1.64	4,320	D
S-114	35.0	28.2	84,471	D
S-1159	5.70	5.43	12,391	D
NBS-705	1.79	1.64	4,320	N
PCC 860	8.60	7.35	20,773	P
PCC 411	4.11	3.92	9,928	P
PCC 160	1.60	1.51	3,865	P
PCC 0972	0.972	~0.962	2,348	P
PCC 0510	0.510	~0.481	1,232	P
PCC 0198	0.198	0.187	478	P
PCC 0103	0.103	~0.097	249	P

[a] D = Dow Chemical Co., Midland, Michigan.
N = National Bureau of Standards, Washington, D. C.
P = Pressure Chemical Co., Pittsburgh, Pa.

Chemical Instrumentation

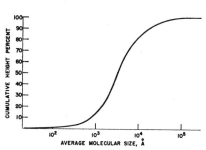

Figure 10. Normalized integral distribution curve.

$$\bar{A}_w = \frac{1}{Q} \frac{\Sigma M_i{}^2 N_i}{\Sigma M_i N_i} = \frac{\Sigma \text{Column 7 (Table 2)}}{\Sigma \text{Column 2 (Table 2)}} \tag{7}$$

$$\bar{M}_n = \bar{A}_n \cdot Q \tag{8}$$

$$\bar{M}_w = \bar{A}_w \cdot Q \tag{9}$$

where

H_i = height of a given segment (arbitrary units).

A_i = average angstrom size of the segment.

M_i = average mol. wt. of the segment.

N_i = number of particles represented by the segment.

Q = molecular weight units per angstrom. (Approx. 41 for polystyrene)

$\bar{A}_n$ = number-average molecular size

$\bar{A}_w$ = weight-average molecular size

$\bar{M}_n$ = weight average molecular weight

$\bar{M}_w$ = weight-average molecular weight

Solvents

Solvents for gel permeation chromatography must meet certain requirements. Ideally, the eluting solvent and the gel should be similar in polarity to prevent partitioning of the sample between two unlike phases. In actual practice, it is often necessary to use a highly polar solvent in order to be able to dissolve the sample. This does not seem to present any problems in many cases, especially where only a fractionation of the sample is desired, without calculation of molecular weights. Also, adsorptive effects of polar samples, small as they are when a polystyrene gel is used, are completely eliminated when the polarity of the eluting solvent is high. The solvent should also have a low viscosity and a relatively high boiling point and should not dissolve, react with, or degrade the column packing.

While compatibility of the solvent with the sample and the gel are important, compatibility of the solvent with the detection system must also be considered. For example, if one uses a refractometer for detection and estimation of the sample as it emerges from the column, then the solvent must have a refractive index that is different from that of the sample. If continuous monitoring of ultraviolet absorption is employed, then the solvent should be one that is transparent at the wavelength being used. The greater the difference between solvent and sample in the value of the physical property being measured, the greater will be the sensitivity of the detector.

Some of the solvents that have been found to be compatible with polystyrene

Table 3. Solvents Used for Gel Permeation Chromatography

	Solvent	Refractive index[a]	Viscosity,[b] cp	Mp, °C	Bp, °C	Sp.[a] gr.
1.	Tetrahydrofuran	1.41	0.50	−65	65	0.888
2.	o-Dichlorobenzene	1.54	1.26	−17	180	1.305
3.	1,2,4-Trichlorobenzene	1.57	0.50	16.5	213	1.454
4.	Benzene	1.50	6.47	5.5	80.1	0.879
5.	Perchloroethylene	1.51	8.70	−22	121	1.623
6.	m-Cresol	1.54	184.2[c]	11.1	203	1.034
7.	Dimethylsulfoxide	1.48	1.98	18.5	189	1.101
8.	N-methylpyrrolidone	1.47	1.65	−17	200	1.026
9.	o-Chlorophenol	1.55	22.5	8.7	175	1.241
10.	Chloroform	1.45	5.63	−63.5	61.2	1.489
11.	Methylene chloride	1.33	4.41	−96	40	1.335
12.	Carbon tetrachloride	1.46	9.58	−23	77	1.595
13.	Dimethylformamide	1.42	0.80	−61	153	0.945
14.	Dimethylacetamide	1.44	0.92	−20	165	0.937
15.	Toluene	1.49	5.90	−95	110	0.867

[a] At 25°C.
[b] At 20°C.
[c] 16.9 at 130°C.

Table 4. Polymers Studied by Gel Permeation Chromatography

Polymer	Solvent(s)[a]
Asphalt, asphaltenes	1, 4
Acrylics	6, 8, 13
Styrene-butadiene rubbers	1, 2, 3, 4, 15
Various cellulose derivatives	1, 5, 8, 11, 13
Copolyether-diol	1
Epoxy resins	1, 8
Ethylene propylene rubber	2, 3
Polyformaldehydes	13
Di- and triglycerides	1
Polymethylmethacrylate	1, 8, 13
Polyester fibers & films	6, 8, 9
Poly(chloroisoprenes)	1, 15
Polyamide fibers & films	6, 9
Polyacetals	13
Polybutadienes	1, 3
Polybutenes	1, 3, 8
Polycaprolactam	1
Polycarbonates	1, 3, 5, 8, 11, 13
Polyethers	1
Polyethylene	2, 3
Polyglycols	1, 2, 3, 4, 6, 8, 11, 13, 15
Polyisobutene	3
Polyisoprene	3
Polypropylene	2, 3
Polystyrene	1, 2, 3, 4, 6, 8, 11, 13, 15
Polyurethanes	1, 8, 13
Polyvinyls	1, 3, 13
Polyvinyl alcohol	13
Polyvinyl butyral	13
Polyvinyl formal	2, 3, 8, 13, 15
"Polymerized oil"	1
Styrene-acrylonitrile	1, 13
Urea-formaldehyde resins	13
Varnishes	1
Various waxes	1
Poly(styrene-maleic anhydride)	1, 8, 13
Gelatin	6
Polyphthalamide	3
Polyethylene oxide-ethylenimine	8, 13, 15
Polyepoxypropane	1, 2, 3, 13, 15
Poly(ethylene-maleic anhydride)	3, 8, 13

[a] Solvent numbers are the same as those used in Table 3.

Table 2. Sample Calculation of $\bar{M}_w$ and $\bar{M}_n$

1 Elution volume, ml	2 Height H_i	3 Cumulative height	4 Cumulative height, %	5 Av. size A_i	6 H_i/A_i	7 $H_i A_i$
105	1	355	100.0	280,000	0.0000	280,000
110	8	354	99.7	110,000	0.0001	880,000
115	27	346	97.5	45,000	0.0006	1,215,000
120	51	319	89.9	17,000	0.0030	867,000
125	73	268	75.5	7,000	0.0104	511,000
130	85	195	55.0	3,600	0.0236	306,000
135	58	110	31.0	2,000	0.0290	116,000
140	30	52	14.7	1,050	0.0286	30,000
145	12	22	6.2	560	0.0214	6,000
150	5	10	2.8	300	0.0167	1,500
155	2	5	1.4	140	0.0143	280
160	2	3	0.8	53	0.0377	106
165	1	1	0.3	22	0.0455	22
Totals	355	...	...	...	0.2309	4,215,128

$$\bar{A}_n = \frac{355}{0.2309} = 1490 \text{ Å}$$

$$\bar{M}_n = 63,632$$

$$\bar{A}_w = \frac{4,214,908}{355} = 11,874 \text{ Å}$$

$$\bar{M}_w = 491,584$$

$$\bar{M}_w/\bar{M}_n = 7.73$$

gels are given in Table 3 along with physical properties of interest in gel permeation work. This list is not intended to be exhaustive. Solvents that are generally incompatible include water, acetone, many alcohols, and most carboxylic acids.

Some of the polymers that have been studied with gel permeation chromatography are listed in Table 4. Suitable solvents also are given wherever data are available.

Commercially Available Equipment

Waters Associates (61 Fountain St., Framingham, Mass.) is, at present, the only manufacturer of complete gel permeation chromatographs. They also offer a complete line of columns and accessories for use with their instrument and other liquid chromatographs. Many of the accessories can be readily adapted for conversion of other existing liquid chromatographic equipment for application to the

Figure 11. Waters Model 200 gel permeation chromatograph equipped with automatic inector and collection unit.

gel permeation method. Following is a brief survey of the equipment that is available.

The Waters analytical-scale gel permeation chromatograph (model 200, $15,000), shown in Figure 11, is designed to fractionate samples in the milligram range, accepting up to 2 ml of a $1/8$ to $1/4\%$ solution of the sample. In cases where fractionation is being carried out primarily for the purpose of characterization of collected fractions by other techniques, it is possible to either operate the existing $3/8$ in. o.d. columns under overload conditions or to replace these columns with larger columns. The sample size is increased by a factor proportional to the ratio of the cross-sectional areas of the columns. Figure 12 is a flow diagram of the Waters unit. Solvent is pumped from a reservoir through a heater to degas the solvent, and then through a filter to a mixing chamber. The mixing chamber serves to damp out the pumping pulsations and to prevent sudden changes in solvent properties due to differences between batches of solvent that are added to the reservoir. The solvent is then split into two streams, one serving as a sample stream and the other as a reference. A unique column switching dual gang multiport valve is set to direct the sample flow through the desired set of columns. In this way, two different sets of columns, having different maximum pore sizes, enable one to examine samples having widely different molecular weight distributions.

The sampling valve arrangement is similar to that which is illustrated in Figure 5. When the sample is ready to be placed on the column, the sampling valve is turned 90°, placing the 2 ml sample loop in the solvent stream. The sample is then carried to the column where it is fractionated. Effluents from the columns pass through the sample and reference cells of a differential refractometer which is capable of detecting a difference in refractive index of one part in ten million. The solvent that passes through the reference cell is recycled. The sample stream passes to a 5 ml syphon and then to a fraction collector. When each 5 ml fraction empties from the syphon a light beam through the syphon tube is interrupted and a pulse is recorded on the recorder chart. In this manner, the elution

volume is automatically recorded on the chart.

Accessories for the chromatograph include an automatic injection system ($3500) which makes it possible to analyze up to six samples in an automatic, unattended fashion, a fraction collector (Warner-Chilcott, modified for GPC, $1000) capable of collecting up to 150 fractions automatically, and an auxilliary pumping system ($2300) which allows for rapid solvent changeover. Columns are available having maximum porosities ranging approximately from 30 to 1,000,000 angstroms. Prices for columns vary from $135 to $235 depending upon the resolution desired. Waters also offers the column packing alone under the name "Styragel" for 30 cents per ml.

The Waters "Ana-Prep" gel permeation chromatograph ($25,000) is actually two chromatographs in one. With this instrument it is possible to do both analytical and preparative work. The analytical part is essentially the same as in the Model 200 chromatograph. It is coupled, together with the preparative section, to the differential refractometer via a selector valve. Figure 13 is an operational diagram of the instrument. The preparative section employs 2.4 in. diameter columns capable of handling samples of up to 100 ml of 0.5% sample if the columns are operated in an overloaded condition. It has an automatic sampling system that is fed from a 10 gal reservoir. If only one eight foot preparative column is used, the instrument will fractionate up to thirty-six samples in a twenty-four hour period.

There are many types of multiport valves that can be used for both sampling and column switching. The cost of such valves ranges from $100 to $500 depending upon complexity and materials of construction. Stainless steel and Teflon are recommended for use with the organic solvents employed in gel permeation chromatography. Some of the available valves can be heated while others cannot. Naturally, the ability to be heated raises its price. A few of the manufacturers of multiport valves are given here.

American Instrument Co., Inc.
8030 Georgia Ave.
Silver Spring, Md. 20910

Beckman Instruments, Inc.
2500 Harbor Blvd.
Fullerton, Calif. 92634

The Biotron Co.
P. O. Box 22043
Houston 27, Texas

Circle Seal Products Co., Inc.
2181 East Foothill Blvd.
Pasadena, Calif. 91107

Consolidated Electrodynamics Corp.
Analytical & Control Division
360 Sierra Madre Villa
Pasadena, Calif.

G. W. Dahl Co., Inc.
86 Tupelo St.
Bristol, R. I. 02809

F & M Scientific Corp.
Divn. of Hewlett-Packard Corp.
Route 41 and Starr Rd.
Avondale, Pa.

Greenbriar Instruments
Cincinnati Division
The Bendix Corp.
P. O. Box 68
Ronceverte, W. Va.

Loenco, Inc.
2092 N. Lincoln Ave.
Altadena, Calif. 91002

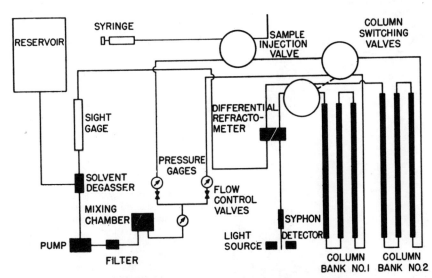

Figure 12. Waters model 200 gel permeation chromatograph flow diagram.

Chemical Instrumentation

Matronic Instrument Co.
132 King Rd.
New Castle, Delaware

Micro-Tek Instruments, Inc.
P. O. Box 15409
Baton Rouge, La.

Perkin-Elmer Corp.
870 Main Ave.
Norwalk, Conn.

Republic Mfg. Co.
15655 Brookpark Rd.
Cleveland 35, Ohio

Wilkens Instrument & Research, Inc.
2730 Mitchell Dr.
Box 313
Walnut Creek, Calif.

Wohl Henius Institute, Inc.
4206 North Broadway
Chicago 13, Illinois

Applications

An area of research that has profited greatly through the use of gel permeation chromatography is that of polymer process development. Here, the usual aim is that of producing, in large quantities, a polymer having properties closely approximating those of either a commercially available polymer or one which has been prepared on a small scale in the laboratory. The classical approach to this problem has been to measure as many physical properties as possible for a sample of the *bulk* polymer whose properties are to be matched and then to study the effects of process variables on these properties. While measurements of bulk properties may be related to the average molecular weight of a polymer, they do not give any real indication of the true shape of the molecular weight distribution. For example, two experimental polystyrene samples were found to have the following average molecular weights:

	$\bar{M}_w \times$ 10^{-4}	$\bar{M}_n \times$ 10^{-4}	$\bar{M}_w/\bar{M}_n$
Sample I	4.6	3.8	1.21
Sample II	4.9	4.5	1.09

These data would lead one to believe that the polystyrene samples have relatively narrow molecular weight distributions and could be used to calibrate equipment to be used for measurement of molecular weights of polymers. The gel permeation chromatograms of both of these polymers (Fig. 14) not only shows that each has a distribution that is broader than expected, but also that each has two peaks in its distribution. This illustrates a unique advantage

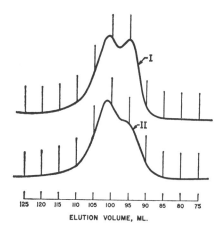

Figure 14. Gel permeation chromatograms polystyrenes I and II

in the use of gel permeation chromatography. One can actually see what the shape of the distribution is directly rather than having to guess at it by examination of the data obtained for the bulk polymer.

The speed with which a gel permeation chromatogram can be obtained makes it ideal for application to quality control of polymer processes. This method may lead not only to the early detection of poor, out-of-specification production batches, but also may give some indication of wherein the trouble might lie. The molecular weight distribution of a polymer is a direct reflection of the polymerization formulation and conditions. In polymer blending operations one would have a rapid method for ascertaining whether the product contains the proper proportions of the constituents and whether extensive degradation has occurred during the blending.

A question that often arises from studies of copolymerization of two or more monomers is that of whether one has produced a true copolymer or a mixture of homopolymers. It is possible that when homopolymers are formed from a mixture of monomers, that these homopolymers will have different average molecular weights.

If such is the case, then gel permeation chromatography would detect their presence and, on a preparative scale, could offer a means of separating and isolating each of the components for further characterization by other physical and chemical methods.

The analysis and fractionation of organic coatings is an area which shows promise but has been almost completely overlooked by gel permeation chromatographers. A brief description has been given by C. A. Lucchesi (see Bibliography). The coatings chemist has always coveted his colleagues in other chemical areas for the separation techniques applicable in the other areas. Certainly, gas chromatography offers a means of examining solvents in coatings while pyrolytic gas chromatography may reveal the identity of the monomers that have been used in the manufacture of some resins. However, since heating and vaporization of the sample are involved in the gas chromatographic techniques, the coating material very often undergoes unpredictable changes which produce results that are difficult to interpret; many components of organic coating materials are not suffi-

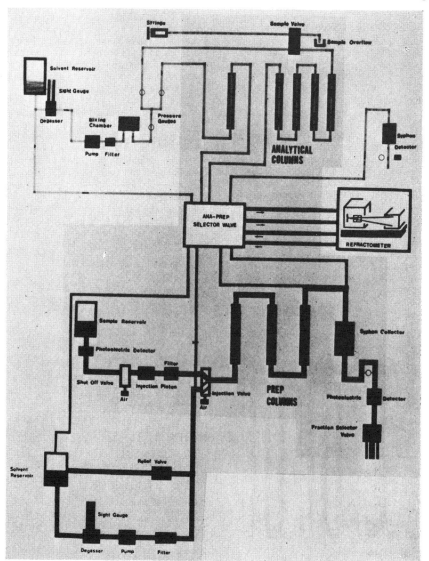

Figure 13. Waters Ana-Prep operational diagram.

ciently volatile to pass through a gas chromatograph. Since gel permeation chromatography separates solvent coating systems solely on the basis of effective molecular size, and without prior volatilization, it offers the coatings chemist a "handle" that he has never had available to him for the fractionation of unpigmented coating systems. Since molecular sizes of the various components of a coating are usually different (resins, solvents, antioxidants, plasticizers, etc.) there is no need to carry out a preliminary separation to remove undesired components. For example, if information is desired regarding the nature of the molecular weight distribution of the resin in a coating, present methods require that the solvent and other additives be removed first. Gel permeation chromatographic examination of many coatings can be carried out on the entire unpigmented coating. This approach, consequently, could yield more information about both the molecular weight and the composition of a coating than other techniques and in a much shorter time.

Preparative-scale gel permeation chromatography provides a means of preparing large quantities of polymer fractions having very narrow molecular weight distributions. These fractions are useful for calibration of the analytical gel permeation chromatograph when polymer standards are not commercially available. The availability of such fractions would also enable the polymer chemist to study the effects of polymer structure and molecular weight on physical properties. Having almost monodisperse polymer fractions, he could tailor-make polymer mixtures to suit the needs of his experiment; it would no longer be necessary to design the experiment around available polymers whose compositions are not completely clear. Correlation of properties with composition could be put on a more scientific and predictable basis than is the case today.

The availability of polystyrene gels having low permeability limits now extend the advantages of gel permeation chromatography to non-polymeric substances. Some monomeric substances have already been examined. Included in the list of such substances are lubricants, fatty acids, glycols, and low molecular weight tars. Gel permeation chromatography could be applied as a separation tool to non-polymeric substances wherever there is a molecular size difference. It would provide a great advantage where sample polarity complicates other chromatographic methods.

Gel permeation may provide a means for studying fundamental properties of molecules in solution. For example, measurement of molecular size in more than one solvent, having different solvating properties, may yield an insight into the type of coiling that is present in a polymer chain. GPC data obtained in a "theta" solvent (a substance which is a poor solvent for a polymer, i.e., does not solvate the polymer) would be valuable here. Studies of this type might show the extent of hydrogen bonding as reflected in the effective size of a molecule. While very little work of this type has been done, this is certainly a fertile area for future research.

There is no doubt that the gel permea-

tion chromatographic technique will continue to be extended to different kinds of substances in many different chemical disciplines.

A bibliography has been included at the end of this paper which includes most of the applications of GPC that have been published. However, since most of the commercially manufactured instruments are in industrial laboratories, a great deal of literature remains unpublished due to its proprietary nature. Pharmacia Fine Chemicals, Inc. (800 Centennial Ave., Piscataway, N. J. 08854) a manufacturer of dextran gel filtration materials, periodically issues literature abstract cards. These references, however, are primarily in the biomedical field and deal mostly with aqueous systems and the non-rigid dextran gels. There are some references in this literature collection of a general nature that are of interest to those working with non-aqueous systems and utilizing rigid, cross-linked polystyrene gels.

Conclusion

It is evident that gel permeation chromatography has emerged from its artistic cradle to become a valuable fractionation and characterization technique. One can liken the present status of gel permeation chromatography to the status of gas chromatography just a short time ago. It has been only a few years since the chemical industry was provided with an analytical technique (gas chromatography) which seemed to produce "too much" data ... more than one could cope with conveniently. The chemist was suddenly awakened to the revelation that his "pure" chemicals were, in fact, not so pure; his standards for purity had become obsolete. He had become a part of a new era in chemical analysis. It is in such an atmosphere that one finds gel permeation chromatography today, ready to be used by and to be useful to anyone willing to take advantage of its unharvested capabilities in areas virgin to this technique.

Future improvements in gel permeation chromatography will probably include:

1. New, more versatile gel materials having compatibility with a wider variety of solvents and polymers. These would make possible the examination of polymers whose extreme insolubility presently prevent their fractionation by gel permeation chromatography. Such polymers include the fluorocarbon polymers and cross-linked polymer gel.

2. Instrumental refinements which would permit completely automatic unattended operation. Also included would be an electronic integration system coupled directly to a digital computer and autoplotter possibly by punched or magnetic tape.

3. Scale-up of the gel permeation chromatograph would be the ultimate in development of the equipment. Fractionation of materials on a large (possibly commercial) scale is certainly not beyond the realm of possibility. The major problem would probably be that of establishing a homogeneous flow through large diameter columns. Higher capacity gels would certainly be a step in the right direction.

Gel permeation chromatography has opened up new areas of polymer research.

Although the technique has already been put on a more or less practical basis, it is obvious from the above discussion that much remains to be done. No doubt, advances in this area will be fast in coming, since the technique has afforded solutions not only to problems that were previously difficult to resolve, but also to some that completely defied solution by classical methods.

Appendix

Collection devices, the operation of which has already been briefly described, are available from most suppliers of liquid chromatographic equipment. Some of the manufacturers of such devices are listed here for convenient reference. (Prices range from $500 to $2000.)

Buchler Instruments, Inc.
1327 Sixteenth St.
Fort Lee, N. J.

Chromatography Corp. of America
60 East Main St.
Carpentersville, Ill.

Gilson Medical Electronics Corp.
3000 West Beltline Highway
Middleton, Wisc.

LKB Instruments. Inc.
4840 Rugby Ave.
Washington 14, D. C.

National Instrument Laboratories
12300 Parklawn Drive
Rockville, Md.

Packard Instrument Co.
2200 Warrenville Rd.
Downers Grove, Ill.

Vanguard Instrument Co.
441 Washington Ave.
North Haven, Conn.

Warner-Chilcott Laboratories
Instrument Division
201 Tabor Rd.
Morris Plains, N. J.

Bibliography

ADAMS, H. E., FARHAT, K., AND JOHNSON, B. L., Firestone Tire and Rubber Company, Akron, Ohio. "Gel Permeation Chromatography of Polybutadiene." ACS Rubber Chemistry Division, Philadelphia, Pennsylvania, October, 1965.

ALTGELT, K. H., "A Study of Asphalts and Asphaltines by Gel Permeation Chromatography." Petroleum Chemistry Division of ACS, Atlantic City, New Jersey, September, 1965.

ALTGELT, K. H., "Fractionation of Asphaltenes by Gel Permeation Chromatography," J. Appl. Polymer Sci., 9, 3389–93 (1965).

BILLMEYER, F., "Textbook of Polymer Science," Interscience Publishers, New York, 1962.

CANTOW, M. J. P., PORTER, R. S., and JOHNSON, J. F., "A Comparison of Molecular Weight Distributions for Polyisobutenes as Determined from Gel Permeation and Gradient Elution Chromatographic Fractionations. Macro Molecular Conference, Prague, Czechoslovakia, 1965.

DEVRIES, A. J., Pechiney-Saint Gobain, Research Center, Anthony (Seine), France, "Determination of the Molecular Distribution of Polymers by Means of Gel Permeation Chromatography." To be published.

Chemical Instrumentation

EDWARDS, G., "Estimation of Molecular Weight of Epichlorohydrin—Bis Phenol A Polymers by Gel Permeation Chromatography." *J. Appl. Polymer Sci.*, **9**, 3845 (1965).

FLORY, P. J., "Principles of Polymer Chemistry," Cornell University Press, Ithaca, New York, 1953.

FONTANA, J., "Characterization Studies for Polyethylene Formed by Co[60] Gamma Radiation." GPC Seminar, Boston, Mass., September, 1965.

GAMBLE, L., WESTERMAN, L., AND KNIPP, E., "The Molecular Weight Distribution of Elastomers—Comparison of Gel Permeation Chromatography with Other Techniques." Rubber Division ACS, Miami Beach, Florida, May, 1965.

HARMON, DALE J., "Comparison of Solution and Gel Permeation Fractionation of a cis-1,4-Polybutadiene." *Journal of Polymer Science*, Part C, No. 8, pp. 243–252 (1965).

HARMON, DALE J., "Effects of Polymer Chemical Structure and Solvent Type on Calibration Curves." GPC Seminar, Boston, Massachusetts, Sept., 1965.

HENDRICKSON, J. G., AND MOORE, J. C., "Gel Permeation Chromatography. III. Molecular Shape vs. Elution." To be published.

HESS, M., AND KRATZ, R. F., "The Axial Dispersion of Polymer Molecules in Gel Permeation Chromatography." GPC Seminar, Boston, Mass., Sept., 1965.

HILL, J. A., "Determination of Standard Deviation of GPC Results for a Polymeric Material with a Weight Average of 50,000." GPC Seminar, Boston, Mass., September, 1965.

LUCCHESI, C. A., "Analytical Tools for Coatings Research," Official Digest, *J. Paint Technol. & Eng.*, **35**, 975 (1963).

MALEY, L. E., "Analysis of NBS 705–706." GPC Seminar, Cleveland, Ohio, Feb., 1965.

MALEY, L. E., "Application of Gel Permeation Chromatography to High and Low Molecular Weight Polymers." *Polymer Sci.*, Part C, No. 8, pp. 253–268 (1965).

MEYERHOFF, G., "Gel Permeation Chromatography in Organic Solvents," *Ber. Bunsenges. Physik.*, **69**, 866 (1965).

MILES, B. H., "Applications of Gel Permeation Chromatography to the Characterization of Epoxy Resins." ACS Organic Coatings Division, Chicago, Illinois, September, 1964.

MOORE, J. C., "Gel Permeation Chromatography. I. A New Method for Molecular Weight Distribution of High Polymers." *J. Polymer Sci.*, A2, 835 (1964).

MOORE, J. C., AND HENDRICKSON, J. G., "Gel Permeation Chromatography. II. The Nature of the Separation." *Journal of Polymer Science*, Part C, No. 8, pp. 233–241 (1965).

NAKAJIMA, N., "Fractionation of Polyethylene By Gel Permeation Chromatography." To be published. (*Abstract in Polymer Previews*, Vol. 1, Issue 7, 1965).

PICKETT, H. E., CANTOW, M. J. R., AND JOHNSON, J. F., "A Computer Program for Determination of Molecular Weight Distributions from Gel Permeation Chromatography," ACS Div. of Polymer Chemistry, Phoenix, Arizona, January, 1966.

RICHMAN, W. B., "Molecular Weight Distribution of Asphalts." (A publication of Waters Associates, Inc.)

RODRIGUEZ, FERDINAND, AND CLARK, O. K., "A Model for Molecular Weight Distributions." GPC Seminar, Boston, Mass., September, 1965.

RODRIGUEZ, F., KULAKOWSKI, R. A., AND CLARK, O. K., "Characterization of Silicones by Gel Permeation Chromatography." Div. of Rubber Chemistry, ACS, Philadelphia, Pa., October, 1965.

SHULTZ, A. R., AND BERGER, H. L., "GPC Line Shapes for Chosen Polymer Molecular Weight Distribution Functions." GPC Seminar, Boston, Mass., Sept., 1965.

SMITH, W. B., AND KOLLMANSBERGER, A., "Aspects of Gel Permeation Chromatography," *J. Phys. Chem.*, **69**, 4157–61 (1965).

TISELIUS, A., "New Separation Methods and Their Applications on Biochemical And Organochemical Problems," *Seperientia*, **17**, 433 (1961).

TISELIUS, A., ET AL., "Separation and Fractionation of Macromolecules and Particles," *Science*, **141**, 13 (1963).

TUNG, L. H., "A Method of Calculating Molecular Weight Distribution Functions from Gel Permeation Chromatograms." GPC Seminar, Boston, Mass., September, 1965.

WATERS, J. L., "Resolving Power of GPC." American Chemical Society, Divn. of Polymer Chemistry, Atlantic City, New Jersey, September, 1965.

WATERS, J. L., "A Standard System for GPC Measurements." GPC Seminar, Boston, Mass., September, 1965.

WEISSBERGER, A., Editor, "Technique of Organic Chemistry. Vol. I. Physical Methods of Organic Chemistry," Third Edition, Interscience Publishers, Inc., New York, 1959.

Roland F. Hirsch is an Assistant Professor of Chemistry at Seton Hall University. He graduated with a B.A. from Oberlin College in 1961, and received his Ph.D. (in analytical chemistry) from the University of Michigan in 1965. His research interests are in the areas of analytical separations (particularly ion exchange) and applications of radioisotopes to analysis. He is a member of the American Chemical Society, American Society for Testing and Materials, and Sigma Xi.

Topics in...

Chemical Instrumentation

...a *ChemEd* feature

Edited by **GALEN W. EWING,** Seton Hall University, So. Orange, N. J. 07079

The Journal of Chemical Education
Vol. 44, page A1023, December 1967
Vol. 45, page A7, January, 1968

XXXV. Modern Laboratory Balances

Part One—Lever-arm Balances

ROLAND F. HIRSCH, Department of Chemistry, Seton Hall University, South Orange, New Jersey 07079

INTRODUCTION

The determination of mass is essential to experimental chemistry. It is one of the most precise instrumental techniques used in analysis. Several types of instruments have been developed. Each is suitable for certain tasks and not for others, and economy and speed of operation vary greatly among the different types. It is hoped that this article will help clarify the factors affecting the choice of one instrument over another for a particular application.

All weighing techniques determine the force W (the weight) produced by a mass M in a gravitational field g:

$$W = Mg$$

This force depends on g, and to convert a weight into the corresponding mass one must either know the value of g (which varies with altitude and latitude on the Earth), or use a set of known masses to standardize the measuring device. The second alternative is almost always chosen.

A primary standard of mass is stored at the International Bureau of Weights and Measures in Paris, and a replica of this standard is at the United States National Bureau of Standards in Washington, D.C. Secondary standards are issued for the calibration of working sets of masses or weights. Most laboratory supply houses sell working sets of weights classified according to their accuracy (*1*). Sometimes it is desirable to calibrate a new set of weights, but usually one can select a set whose tolerances are satisfactory for the particular application. Recalibration of weights after frequent handling is necessary, however, since wear or corrosion may cause significant

changes in the mass. This calibration may be performed against a secondary standard, or by the method of T. W. Richards (*2, 3*).

The force produced by a mass in a gravitational field can be measured in several different ways. A spring may be stretched, a fiber bent or twisted, or a lever deflected from the horizontal position. A second force acts to counterbalance that due to the unknown mass, and brings the device to a rest position. If one can measure the second force, the unknown force (the weight) may be calculated.

In most balances the counterforce returns the balance to its initial (empty) position—this is called the null-balance condition. In some balances the deflection from the null point is measured (the stretched spring is a good example). Actually even most null balances are not returned exactly to their initial position, as this can be a very tedious procedure. Rather a combination of nulling and measurement of the residual deflection is used in the weighing operation. The deflection is often measured directly as an angle formed with the horizontal, or as the length on a scale, but sometimes a transducer converts it to an electronic signal (examples are the load cell, capacitor, and differential transformer, which are described in Part II of this article).

A few instruments do not weigh objects directly, but measure their masses through different properties, such as the transmission of high-energy radiation.[1] There are many chemical and physical techniques which allow indirect determination of masses. This article will be limited, however, to those instruments using the balance-of-forces principles described in the previous paragraphs.[2]

A general discussion of the evaluation of balance performance has been published by L. B. Macurdy (*4*).

A committee of the American Chemical Society has defined certain parameters to describe balances. The quotations in the following paragraphs are from the committee's report (*5*).

The *sensitivity* (sometimes called the sensibility) is "the ratio of the change in the response to the change of the quantity measured." The *sensitivity reciprocal* is "the change in load required to change the equilibrium position by one division." The *readability* is "the smallest fraction of a division to which the index scale can be read with ease either by estimation or by use of a vernier." None of these units is in itself a satisfactory guide to an instrument's performance.

The *precision* (often called the reproducibility) is a better measure of the quality of a balance. It is expressed in terms of the "standard deviation of repeated weighings of the same mass." The precision clearly depends on the environment in which the balance is used and the skill of the operator. The values listed in the tables in this article are optimum ones. They can be approached by careful shielding of the balance from air currents and thermal gradients, as well as from dust, corrosive vapors, and vibration.

Balances are often classified according to their capacity, as well as their precision. The commonly accepted ranges are: "Analytical": capacity—150–200 g, precision—0.1 mg; "Semi-Micro": capacity—75–100 g, precision—0.01 mg; "Micro": capacity—10–30 g, precision—1 μg. Ultramicrobalances have been described which are precise to 10 nanograms. "Precision" balances, on the other hand, are less sensitive than "analytical" balances, with precision of 1 mg or more. High-capacity balances are also available, with precision corresponding to the capacity (an analytical balance with a capacity of 1 kg typically has a precision of 1 mg).

In the following discussion, we will first consider null balances based on a lever arm, describing instruments using a gravitational restoring force and those in which the counterforce is produced electromagnetically. Various methods for detecting residual deflections will be described. In subsequent sections spring balances, torsion balances, load cells, and

---

[1] The Ohmart Corp., 4241 Allendorf Drive, Cincinnati, Ohio, manufactures devices using this principle.

[2] It is interesting to speculate about the mass-measuring devices most suited to zero-gravity conditions, such as in an Earth-orbiting laboratory.

Chemical Instrumentation

recording balances will be covered.

Theory of the Lever Arm Balance (6,8).

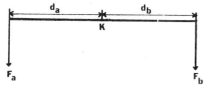

Figure 1. Lever-Arm Balance of Forces.

At the fulcrum K, the forces due to the masses each supply a torque $T = Fd$ (Fig. 1). The beam itself, and the sample holding devices, will also contribute torques, and these can be neglected only if the construction is symmetrical about the fulcrum. If the beam is horizontal (null-balance condition) the two torques must just cancel each other:

$$F_a d_a = F_b d_b$$

Since $F = M \cdot g$, and g is constant within the space of the instrument,

$$M_a d_a = M_b d_b$$

and if $d_a = d_b$,

$$M_a = M_b$$

Therefore, if an equal-arm balance is properly constructed, the measurement of an unknown mass (say M_a) is made by adding known masses to the other sample holder until the beam is in the horizontal position, and totaling the masses M_b. The single-pan, substitution balance does not have equal arms, so the torques, and not the masses are balanced. More will be said about this type of balance shortly.

As mentioned earlier, reaching the exact null-balance is tedious, and frequently one determines the degree of imbalance by measuring the angle (θ) of deflection of the beam from the horizontal (Fig. 2). A vertical pointer may be used to improve the precision of this measure-

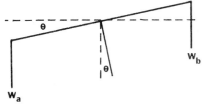

Figure 2. Residual Deflection of Lever Arm.

ment; it may be magnified at its lower end. The angle of deflection depends not only on the difference (ΔW) between W_a and W_b, but also on the sensitivity (s) of the balance to this weight difference. Therefore, to calculate W_a, we must measure θ and W_b, and also determine s:

$$W_a = W_b + \phi/s$$

If θ is expressed in scale divisions, then s may have units such as divisions/milligram.

The sensitivity depends on several factors: the combined weight (W_t) of the beam, sample holders and loads, the

length of the arm (d), and the distance between the knife edge supporting the beam and the center of gravity of the assembly (h). The location of the center of gravity is adjustable on many balances by means of a weight whose position can be varied upwards or downwards (it is attached to the pointer in Fig. 3, and denoted as number 6 in Fig. 6). Overall,

$$\theta = \Delta W d / W_t h$$

so

$$s = \theta/\Delta W = d/W_t h$$

The sensitivity thus decreases with increasing load W_t. A significant advantage of constant-load (substitution) technique is its constant sensitivity.

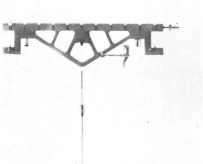

(Courtesy American Balance Corp.)

Figure 3. Knife edge Suspension. Also shown are a rider and a chain device.

The sensitivity is routinely determined by observing the angle of deflection (or number of scale divisions traversed by the pointer) produced when a known small weight is added to one of the sample holders. If the balance has an optical scale for determining the last significant figures of the weight, then this scale is calibrated by adjusting the sensitivity until a known weight produces the correct deflection on the scale.

The Suspension System

A horizontal beam is supported at or near its center by a knife edge or a torsion band. The beam must be light, to place a minimum load on the suspension point, and strong, so that it does not bend under heavy loads. Aluminum alloys are commonly used in constructing beams. Sample holders are hung or supported from one or both ends of the beam.

The purpose of the suspension device is to hold the components of the mechanism together while contributing a minimum of frictional resistance to deflection of the beam. The less resistance, the more sensitive the balance will be to slight departures from the null condition. Knife edges are the most popular devices for suspending beam and sample holders in lever-arm balances. Torsion bands are also often used. Other types of suspensions are selected where low sensitivity is satisfactory, or very high sensitivity is required (the simple fixed pivot of a trip balance, and the quartz fiber of an ultra-microbalance are examples). Flexure

bearings are sometimes used in high-capacity balances.[3]

The knife edge system consists of a set of triangular prisms fixed to the beam (Fig. 3) and resting on polished bearing planes. It has the following requirements for good performance: (1) the material from which the knife blades are constructed must be hard, and must not deform under any likely load—all the edges must lie in the same plane; (2) the material must not wear significantly—friction between the edge and bearing plane increases if the sharp edge is worn or chipped; (3) the points where edges and planes meet must be kept clean. Synthetic sapphire is the best material for construction of knife edges. It is much harder and more resistant to wear than agate, which is popular because of its lower price. Steel knife edges are suitable only for trip and triple-beam balances. Provided the edges are linear, a long knife blade is preferred to a short one, since the force per unit length will be less, and wear will be reduced. To prevent damage to the edges and planes, they are kept out of contact (arrested) while weights are added to or removed from the pans or holders.

Torsion suspensions (do not confuse with torsion force balances, which will be discussed later) have both advantages and disadvantages compared to knife edges. Two parallel beams are held in position by three metal bands, which are pulled taut around trusses (Fig. 4).

(Courtesy Torsion Balance Co.)

Figure 4. Torsion Band Suspension With Sample Pans.

The pan(s) and weight holders are supported by the appropriate bands. The center truss is fixed to a block which supports the entire assembly. Dirt cannot interfere with the movement of the beams and wear is not significant, since there is no friction between parts. Sensitivity is therefore constant over a longer time period than with knife edge systems. Eventually, however, the sensitivity does change, and the bands must be retensioned.

In practice, the torsion suspension balances and microbalances offer a further advantage—the beam is never arrested, eliminating two or more steps in the weighing operation. Only the less-sensitive precision knife edge balances do not require at least partial arrestment when

[3] The Ormond Corp., 11969 East Riviera Rd., Santa Fe Springs, Calif., makes flexure joints for use as pivots in balances and force measuring devices with capacities from 500 to 6,000,000 lb. The Transmetrics Model 5301 balance uses flexure pivots to support beam and sample pan.

Chemical Instrumentation

adding or removing samples or weights. Torsion suspensions are also much less affected by exterior vibrations than knife edges.

(Courtesy Seederer-Kohlbusch, Inc.)

Figure 5. Double Pan Balance. The rider carrier is near the top of the case. Separate beam and pan arrest controls are at the front of the case.

Construction and Operation of Lever Arm Balances (7).

A typical double-pan, three-knife-edge analytical balance is shown in Figure 5. The balance is enclosed in a case to reduce air currents and thermal gradients. The case is leveled by means of screws on its legs. The beam arresting system is controlled from outside the case. The adjusting screws at the ends of the beam allow matching of the torques with no weights present, to provide a properly balanced initial state of the instrument.

Rather than add all weights from a standard set, a *rider* (a small dumb-bell or U-shaped wire with a hook attached) is placed on the beam. The rider fits into numbered notches on the beam and may be moved back and forth by the carrier until null-balance is approached. The weight the rider counterbalances is added to the weights already on the arm on which it is resting. This is a much more rapid procedure than adding small (fractional) weights one at a time. Other devices are also used to speed weighing: external keys may add or remove small weights from the standard weight arm; a chain hung from this arm and from a movable (in the vertical direction) support may have more or less of its length carried by the arm—thus allowing variation of the amount of weight on the arm by small increments. These accessories increase the cost of the balance and their desirability depends on whether the convenience of weighing is sufficiently increased.

The schematic of a single-pan, double-knife-edge balance is shown in Figure 6. This is the most common type of substitution balance. The manufacturer places a counterweight on the right hand arm (number 9), and a set of calibrated weights on the left-hand arm (number 3). The weights are often in the form of rings or bars, and are located below the pan in

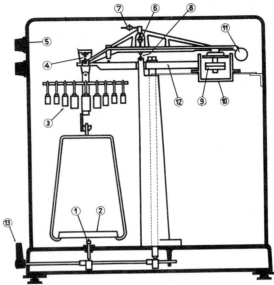

(Courtesy Mettler Balance Corp.)

Figure 6a. Substitution Weighing Balance. 1—pan arrest. 2—sample pan. 3—weights. 4—knife edge supporting pan holder. 5—weight lifter control. 6—sensitivity set screw. 7—null balance set screw. 8—knife edge supporting beam. 9—counterweight. 10—air damper (dashpot). 11—deflection scale. 12—beam arrest. 13—beam and pan release control.

(Courtesy Wm. Ainsworth & Sons)

Figure 6b. Beam of a Substitution, Single-Pan Balance.

some balance designs (that of Oertling, for example). Note that the right arm is longer than the left arm—this allows the use of a smaller counterweight than the capacity of the sample pan, reducing wear of the knife edges and increasing the sensitivity of the balance. The weights on the left are moved on and off their holders by knobs on the case. They are never handled directly in routine weighing.

Initially the beam is horizontal and all the weights are resting on the holders. The null-balance is achieved by adjusting the set screw (number 7). When a sample is placed on the pan and the beam is partially released, a downward deflection to the left results. The null balance condition is restored by removing weights from the beam. Usually this procedure is followed only to the nearest gram or hundred milligrams below the actual weight of the sample. Then the beam is fully released and the residual deflection is read on a reticule on a frosted screen on the balance case. The image of a scale (number 11) attached to the beam is projected onto this screen. [The range covered by the scale is included in the data in Table 2.]

For routine laboratory use, substitution balances are preferred over conventional double-pan balances. For some years after E. Mettler introduced the modern substitution balance there was consider-

able controversy about this point, but today its distinct advantages have won over most users and manufacturers. These advantages include the **greater speed of** weighing and the constant **sensitivity re**sulting from **constant load, as** already mentioned. Other **advantages are greater** accuracy and precision **resulting from** independence of results **from the arm** length ratio and the requirement for only two, rather than three, knife edges to be parallel and coplanar.

The fact that the working parts in a substitution balance are enclosed in a case, and the operator does not come in direct contact with them, does not mean that the knife edges and weights do not wear, or that the balance is immune to the effects of air currents, temperature gradients, and vibrations. The sensitivity and the weights must be checked regularly. If a balance must be operated under far-from-ideal conditions, one with a torsion suspension system would probably give better service.

Important characteristics and approximate prices of some double and single pan balances are given in Tables 1 and 2, respectively. The listings are by no means complete, but represent the range of instruments available today. Some of the features listed in these tables will be explained in the following sections.

Chemical Instrumentation

Taring and Counterpoising

A balance with a *tare* system can counterbalance a mass, such as a container, which is repeatedly placed on the pan. The tare in most balances consists of a set of small weights placed below the sample pan, or otherwise attached to the stirrups holding the pan. Weights are removed until the approximate equivalent of the mass of the container is gone. If optical range taring is included, then the tare can be adjusted to exactly counterbalance the weight of the container, and subsequent weighings will give its contents directly. Rather than move weights or adjust the angle of a mirror or lens in the optical system, some precision balances use a spiral or helical coil spring to supply a force counterbalancing, and thus taring, an object on the sample pan. Note that the tare extends the capacity of some balances, but not of others, since some balances use the regular weights to tare objects.

Taring is distinct from *counterpoising*. This term refers to counterbalancing a sample with a very similar container and contents. Counterpoising has two advantages arising from this similarity: (1) corrections for bouyancy of the sample in air are not needed; (2) differences in the rate and extent of adsorption of moisture on surfaces are minimized. Both of these factors are especially important in the semi-micro to ultramicro ranges.

Differential weighings (as of an absorption tube in a combustion train before and after absorption of water or carbon dioxide) require negligible bouyancy corrections. Counterpoising isn't essential, unless adsorption from the surrounding atmosphere cannot be corrected for. Buoyancy corrections (air weighs about 1 mg/ml) can be made using tables and charts in the literature. The densities of sample and weights must be known.

Coarse Weighing (Preweighing)

Several manufacturers now offer substitution balances with coarse weighing scales. In Oertling balances, for example, the beam is fully released at the start, and the approximate weight is projected onto a frosted glass screen. Weights are removed in this amount, and then the fractional weights are adjusted until the milligram optical scale appears. In Sartorius balances, the release lever is turned in one direction to obtain the coarse weight, which is removed from the beam. The beam is then fully released by turning the knob in the opposite direction to complete the weighing. The speed of weighing is definitely increased, especially by the Oertling preweigher.

Damping

When the weights have been added, the beam and pan arrests are released, and the beam swings away from the horizontal rest position. After a few seconds it slows, stops, and then swings in the opposite direction. The balance acts as a pendulum. Oscillation continues, with successively shorter distances covered because of friction between knife edges and bearing planes, and resistance from the air. Eventually the beam comes to rest; the scale position is the *rest point*. This process may take ten minutes or more with a well-constructed balance; several methods are used to estimate the rest point quickly (8).

One mechanical method is to place vertical aluminum vanes on the arms of the balance and have these vanes pass through a magnetic field. The field of the permanent magnet sets up eddy currents in the vane when it is moving through the gap between the poles of the magnet. These currents set up fields opposing the permanent field, producing "friction" which slows the motion of the beam. The natural oscillation of the beam is damped, and it comes to rest after two or three swings about the rest point. At rest, the vane is unaffected by the magnet. Since contamination of the vane by magnetic particles produces a permanent force to bias the operation of the balance, this kind of damping is infrequently used on analytical balances today.

Rather, air or liquid piston systems (dashpots) are used to bring the balance beam to rest. Braking occurs through resistance of the fluid to motion of the vane. The pressure differential is decreased by flow of fluid from the compartment in front of the vane into the compartment behind the vane. The degree of damping can be controlled by varying the viscosity of the fluid, the size of the

Table 1. Equal-Arm Balances

Manufacturer[a]	Model	Capacity	Precision	Built-in Range[b]	Price
		A. Macro to Semi-Micro Range			
Ainsworth	LC	200 g	0.1 mg	10 mg (R)	$310
	LCB-M3[c]	200	0.1	1100 (R,C)	410
	16	200	0.02	2200 (R,K)	855
American Balance	JR 2012[c]	200	0.1	1100 (R,C)	300
	AN 2012[c]	200	0.05	1100 (R,C)	391
Christian Becker	AB-7	200	0.1	10 (R)	210
	AB-4[c]	200	0.1	1100 (R,C)	390
Oertling	B07[d]	100	0.04	1000 (K,O)	
Sartorius	1504	100	0.01	10,000 (K,O)	675
Sauter	42[e]	200	0.1	10(R)	405
Seeder-Kohlbusch	240M	200	0.1	10 (R)	160
	472M[c]	200	0.05	1100 (R,C)	395
	536[c]	1000	0.2	200 (C)	1380
	560[c]	10,000	10.		2220
Voland	220-D[c]	200	0.1	1100 (R,C)	300
	340-D[c]	200	0.05	1100 (R,C)	390
	750-D[c]	200	0.05	100 g (K,C)	650
	1025-CDN[c]	25,000	10.	1000 (C)	4750
		B. Microbalances			
Ainsworth	FHM	20	0.001 mg	222 mg (K,R,O)	$1100
Heusser	DC[d]	5	0.001	222 (K,C)	2750
Oertling	142[e]	20	0.001	100 (K,R,O)	2500
Sartorius	1801	30	0.001	1100 (K,O)	1385
Stanton	MC-8	20	0.001	100 (K,R)	1340

[a] List of manufacturers, with addresses, is at the end of the article.
[b] The range for which no loose weights are required. R—rider on beam; C—chain; K—weights controlled by keyboard or knob; O—Optical projection scale.
[c] Magnetic damping included.
[d] "Pan extractor" allows loading and unloading of samples through a porthole, without opening the balance case door. The sample is positioned mechanically to prevent hanger swing.
[e] Air piston damping included.

Table 2. Substitution, Single-Pan Balances

Manufacturer[a]	Model	Weighing Range	Tare	Total Capacity	Optical Range	Reproducibility	Price
Ainsworth	23V	160 g	no	160 g	1000 mg	0.3 mg	$475
	SC	200	no	200	100	0.03	895
American Balance	QC4050	200	50 g	250	100	0.03	875
Becker's Sons	R 0,1	200	15	200	100	0.03	760
	R 1	200	40	200	1000	0.3	555
Cenco	1580	160	15	160	1240	0.5	585
	1581	160	15	160	124	0.05	650
Heusser	S2000/5T2	200	20	220	10,000	0.3	625
	S2000/10T02	200	21[b]	220	1000	0.03	775
Mettler	H5	160	15	160	1250	0.3	515
	H6T	160	160[b]	160	1000	0.05	695
	H10TW[c]	160	160	160	1000	0.05	820
	H16	80	15	80	125	0.01	830
	M5	20	no	20	20	0.001	1450
	B4C-1000	1000	27	1000	1100	0.5	1530
Oertling	V-10[c]	200	100	200	1150	0.3	700
	V-20[c]	200	100	200	115	0.03	800
Sartorius	2743[c]	100	no	100	1000	0.1	550
	2462[c]	200	21[b]	221	1000	0.05	890
	2444[c]	100	0.1	100	100	0.01	875
Sauter	414/10	200	25	225	1000	0.1	730
	414/11[c]	200	25	225	1000	0.1	810
	414/51[c]	100	25	125	100	0.01	950
Stanton	SN1	200	100	200	1000	0.5	610
	C141	200	100	200	100	0.025	810
	CL5D	100	100	100	10	0.003	1100
Torsion	EA-1[d]	160	40[b]	200	100	0.1	895
Transmetrics	5301	12,000	...	12,000	1000	1.0	...
Voland	200	200	10	200	100	0.1	975
	160B	160	no	160	100	0.1	670
	1000	1000	no	1000	1200	0.1	1350
	5000	5000	no	5000	12,000	1.	3750
	2150	50,000	no	50,000	(e)	10.	15,000

[a] List of manufacturers, with addresses, is at the end of the article.
[b] Optical range taring (continuous taring) is included.
[c] Pre-weighing scale gives the approximate weight of sample (see text).
[d] Torsion beam suspension balance, with electronic null indicator.
[e] Chain used for last two digits, with electronic null indicator. Motor drive for removing and adding weights.

Chemical Instrumentation

leak, and the size of the vane itself. When the beam is finally at rest, there must be no pressure differential remaining, or the damping device will bias the balance condition.

Oil is much more viscous than air, and so gives a greater degree of damping for a given piston size. On the other hand, oil damping is more sensitive to temperature changes than air damping, and also is prone to clogging, so the latter is more popular.

Damping devices are illustrated in Figure 7. Note that the magnet in Figure 7a can be moved in or out to adjust the degree of damping to the sample size.

Special Balances

Many balances can be adapted for weighing below the pan (Fig. 8). This is useful if the sample is to be suspended in a furnace or in the gap of a magnet.

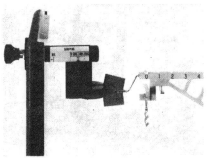

(Courtesy American Balance Corp.)

Figure 7a. Magnetic Damper.

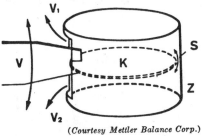

(Courtesy Mettler Balance Corp.)

Figure 7b. Air Piston Damper. The plate K (attached to the beam V) moves up and down in the cylinder Z. Air enters and leaves through apertures V_1 and V_2 and gap S.

Enlarged weighing chambers (for tall or wide objects) can be included. Some balances can be obtained with avoirdupois, troy, grain, or carat scales for commercial use. Several companies make

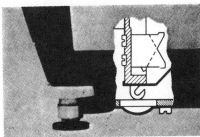

(Courtesy Mettler Balance Corp.)

Figure 8. Weighing Below the Balance. The suspension wire is attached to the hook after removing the plug.

balances adapted for measuring specific gravities or surface tensions. Consult the manufacturers listed in Tables 1 and 2 for information on these specialized instruments, which will not be discussed further.

Top-Loading and Other Less Sensitive Lever-Arm Balances

The top-loading balance is similar to those discussed so far. The balance pans are placed above the beam. The Mettler instruments use a parallelogram suspension with four knife edges to hold the pan in position (Fig. 9). Other manufacturers use similar designs; even the torsion band suspension used by the Torsion Balance Co., is in the form of a parallelogram.

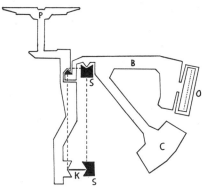

(Courtesy Mettler Balance Corp.)

Figure 9. Top-Loading Balance Suspension. B—Beam. C—Counterweight. K—knife blade, two edges. O—Optical scale. P—Pan support. S—Fixed support for balance assembly. The support parallelogram is outlined.

Top loading balances usually require no arrestment between weighings, and have large ranges in their optical scales. Taring is rapidly accomplished. These features increase the speed of weighing, especially when a certain amount of sample must be weighed out into a container. Thus the somewhat higher initial cost of top-loading against conventional balances of similar precision can be justified for laboratories with heavy work loads. The precision of top-loaders is

limited by air currents, unless a cover or guard is used. Most top loaders can be modified for weighing below the balance case.

The Exact Weight balances are available with "over-under" optical scales. These are useful for repetitive weighing out of a fixed quantity of material. The desired weight is set on a beam, and the substance is added to the pan until the projected scale reaches zero. Small deviations above or below this value can be read off the optical scale. The Mettler P balances also can be adapted for this purpose.

The Mettler P160 reads the weight of an object, or the weight loss from the initial value. This feature should be helpful in using weight burets for titrations and in thermogravimetry.

Triple-beam balances also allow rapid weighing, but they are not as precise as the top-loading balances described above. The weights ride on graduated beams—moving a weight away from the pivot increases the torque and hence the sample weight counterbalanced. The prices of these balances are much lower than the more precise balances. The precision and accuracy are satisfactory for many purposes, and placing a few of them at convenient locations in the laboratory will decrease the work load on the higher-precision balances.

Two types of triple-beam balances are commonly used—with the sample pan(s) suspended below the beam, and with the pan supported above the beam (often called a trip or Harvard trip balance). Double-pan balances of this type often have just a single beam, especially the inexpensive types. A few models have a single beam with a dial to cover the smaller weight range. Special containers are available to replace the pans for weighing animals, for example.

Data on precision top-loading balances are presented in Table 3.

LIST OF MANUFACTURERS

Wm. Ainsworth and Sons, Inc., 2151 Lawrence St., Denver, Colo. 80205
American Balance Corp., 68 First St., New Rochelle, N.Y. 10801

Table 3. Precision Top-Loading Balances

Manufacturer[a]	Model	Weighing Range	Tare	Total Capacity	Precision	Built-in Range	Price
Ainsworth	TL150	150 g	5 g	150 g	1 mg	5 g (0)	$685
Exact Weight	4101[c]	50	no	50	12	1 (0)	460
	4103[c]	1500	no	1500	350	30 (0)	350
	4113[c]	50,000	no	50,000	5000	450 (0)	430
Heusser	K-1-T	1000	300	1300	10	100 (0)	660
	VL-20	20,000	2000	22,000	2000	20,000 (0)	1395
	VL-300	300,000	30,000	330,000	40,000	300,000 (0)	2145
Mettler	P-160[c]	160	10	160	0.5	10 (0)	745
	P-1200[c]	1200	100	1300	5	100 (0)	695
	P-2000[c]	2000	500	2500	50	1000 (0)	670
	P-10[c]	10,000	3000	13,000	500	10,250 (0)	895
Oertling	TP 30	120	no	120	10	120 (0)	...
Ohaus	Autogram	1000	200	1000	50	100 (0)	295
Sartorius	2205	160	10	170	0.5	10 (0)	695
	2116[d]	1000	100	1100	20	101 (0)	725
	2201	5000	3000	8000	100	1000 (0)	850
Sauter	MD 160	160	no	160	1	10 (0)	760
	S 4000	4000	no	4000	100	1000 (0)	655
Torsion	ET-1[e]	160	40	200	0.5	1 (0)	795
	DLW-2[f]	120	no	120	2	1 (K)	210
	PL-1	1000	325	1000	30	100 (0)	525
	HT-1[f]	14,000	no	14,000	300	1000 (B)	255

[a] List of manufacturers, with address, is at the end of the article.
[b] 0—Range of optical scale; B—Beam rider range; K—weights controlled by knob.
[c] With an "over-under" optical scale.
[d] Electronic null indicator.
[e] Torsion suspension system. Electronic null indicator.
[f] Two-pan balance.

Chemical Instrumentation

Christian Becker—manufactured by Torsion Balance Co., 35 Monhegan St., Clifton, N.J. 07013

Becker's Sons—distributed by Neslab Instruments, P.O. Box Y, Durham, N.H. 03824

Cenco Instrument Corp., 2600 South Kostner Ave., Chicago, Ill. 60623

Exact Weight Scale Co., 944 West Fifth Ave., Columbus, Ohio 43215

Heusser Instrument Company, 121 West Malvern Ave., Salt Lake City, Utah 84115

Mettler Instrument Corporation, 20 Nassau St., Princeton, N.J. 08540

Oertling—distributed by J. & G. Instrument Corp., 1907 Highway 27, Edison, N.J. 08817

Ohaus Scale Corp., 1036 Commerce Ave., Union, N.J. 07083

Sartorious—distributed by Brinkmann Instruments, Cantiague Rd., Westbury, N.Y. 11590

August Sauter of America, Inc., 80 Fifth Ave., New York, N.Y. 10011

Seederer-Kohlbusch, Inc., 25 South Van Brunt St., Englewood, N.J. 07631

Stanton Instruments, Ltd., 2223 Fifth Ave., Pittsburgh, Pa. 15219

Torsion Balance Co., 35 Monhegan St., Clifton, N.J. 07013

Transmetrics, Inc., P.O. Box 2055, Newport Beach, Calif.

Henry Troemner, Inc., 6825 Greenway Avenue, Philadelphia, Pa. 19142

Voland Corp., 27 Centre Ave., New Rochelle, N.Y. 10802

ELECTROBALANCES

A number of lever-arm balances are available in which an electromagnetic counterforce is used to restore the beam to the null condition. The principle of operation of the more sensitive instruments of this class is illustrated in Figure 10. The sample is loaded on a pan hung from loop a or loop b (lower sensitivity). A counterpoise may be placed on the pan below loop c. The beam deflects downward on the left.

At the fulcrum (usually a torsion band or a pivot suspension) a wire coil is attached to the beam, and a stationary permanent magnet is mounted with its poles above and below the coil. When current passes through the coil a torque is applied to the beam proportional to the current. A potentiometer (R_1) is adjusted to provide sufficient current to the wire coil to bring the beam to the horizontal, as indicated by the alignment of the reference line and the beam pointer, or by an automatic null detector.

Then, potentiometer R_2 is adjusted to measure the voltage across R_1, with the aid of the galvanometer G. The variable resistance adjusts R_2 to zero with the empty pan or container hung from loop a or b. A calibrating weight hung from the same loop sets the 100% position of R_2. The position of R_2 at null balance is thus equivalent to the fraction of the calibrating mass represented by the sample mass.

In this type of balance the critical factors are the stability of the battery or rectified AC power supply, the sensitivity of the galvanometer, and the precision of R_2. Actually, the latter is the most significant component, since the current stability is high and regularly checked in calibrating R_2, and galvanometers of more than adequate sensitivity are available.

The limit of precision of the electromagnetic mechanism is therefore that of the best potentiometers available for the purpose—about $\pm 0.01\%$. This is not, however, the limit of precision of weighing, since part of the sample may be tared by a weight in loop c. With Class M weights, the taring of up to 500 mg is possible with a precision of better than ± 5 μg. A 50 mg electromagnetic range will not affect overall precision, if a high quality potentiometer is used. If the weights are properly calibrated, some electrobalances can be operated with precision of ± 0.1 μg for milligram-sized samples.

The Cahn Instrument Co. offers a wide range of electrobalances, as well as accessories for weighing in remote or inaccessible locations. The beam and balancing mechanism can be mounted separate from the power supply and controls. For example, weighing in a vacuum system or controlled-atmosphere chamber is accomplished with only the wires supplying current to the coil passing through the walls of the container. The balances are portable and can be used immediately after moving them. Two manual models are sold by Cahn, along with several recording balances.

In the Cahn M-10 ($820) the beam is suspended by a pivot, which limits sensitivity to ± 1 μg and capacity to 175 mg. It is battery operated. The Cahn GRAM balance has a torsion band suspension, sensitivity of ± 0.1 μg, and capacity of up to 1.5 g. It is shown in Figure 10. It can be obtained in battery ($975) or line-operated ($1095) versions. All Cahn balances come with a set of precision resistors which are placed in the electrical circuit to select the weighing range. The smallest range is 0–1 mg (full scale on R_2). Interchange between ranges is possible during a weighing operation. Precision of the electronic system is 10^{-4} times the range, and precision of weighing a tared sample is about 10 ppm.

The Schultz Model 30 microbalance has a two-pan, equal-arm, design. The beam and pan holders are suspended by platinum alloy bands. The null position is determined with the aid of a photo-transistor imbalance detector (to be de-

---

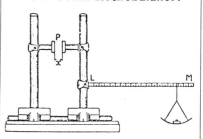

World's First Electrobalance?

Dr. R. Whiddington, Professor of Physics at the University of Leeds, read a paper before The Wireless Society of London on Dec. 21, 1920, in which he discussed some of the astounding properties of the newly developed triode "wireless oscillating circuit."

By use of a heterodyne technique, he could measure the displacement of a capacitor plate with great sensitivity. I quote from his paper as reported in *The Wireless World* (vol. **8**, p. 739, 1921):

"The parallel plates were mounted on a rigid geometrical slide, as shown in [the figure] and small bending moments applied to the bar by putting weights into the scale-pan. This pan was made of light paper and was capable of sliding along the quartz rod LM, which was clamped at L to a vertical steel pillar $1/2$ inch in diameter. Preliminary experiments, using an engineer's micrometer, showed that the center of the [right-hand] plate moved through a millionth of an inch (10^{-6} in.) when a load of 0.14 gram was added to the pan at 10 in. from L. Since smaller weights produced proportionately smaller displacements, according to Hooke's Law the displacement produced by any weight could be computed. It was found that when the apparatus as a whole was working properly a weight of 1 mg at 5 in. from L produced an easily detected change in the slow beats.

"This weight corresponded to a displacement of the plate of about 1/200,000,000 in., . . ." The plates were approximately 10^{-3} in. apart.—G.W.E.

Figure 10. Electromagnetic Lever-Arm Balance.

(Courtesy Cahn Instrument Company)

Chemical Instrumentation

scribed in a later section). The balance mechanism may be mounted separately from the controls. The capacity is 1 g, and four electronic ranges can be chosen. Precision of the electromagnetic system is ±0.01%, and large weights can be measured to ±0.0005% precision (5 ppm) if partially tared. The instrument is line-operated.

The Beckman EMB-1 microbalance ($1035) is also a two-pan instrument. The balance mechanism and control unit are separate from each other, facilitating remote installation of the mechanism. Phototransistors provide a signal which is amplified to produce automatically the current necessary to balance the beam. The capacity is 2.5 g on either pan and the weighing ranges are from 0–1 to 0–200 mg. Sensitivity is 0.1 µg, and precision ranges from ±0.05 to ±0.001% (10 ppm) if taring is used. A socket is provided for connection of the unit to a 10 or 100-mv potentiometric recorder. The EMB-1 may be operated interchangeably from batteries or AC lines.

Some balances are partial electrobalances in that a small portion of the total capacity is supplied by an electromagnetic force coil. The Minsco Model 2D05 ($284) is much less sensitive than the micro balances described above. It is a two-pan balance with a parallelogram suspension which uses a reed pivot for the fulcrum. After the sample has been tared to the nearest 500 mg, a servo system detects the residual imbalance and produces a current through a coil placed in a permanent magnet gap. The current is automatically adjusted until the null balance is reached. The Model 2D05 has a capacity of 50 g and an accuracy of ±0.5 mg on its 500 mg electromagnetic range. Other models are available, with different capacities and weighing ranges. The instruments are line-operated.

In the discussion of analytical balances, the Torsion EA-1 and ET-1 were described. These two balances are electrobalances in a sense, in that the sub-gram counterbalancing force (and taring force) is provided by a permanent magnet and a current carrying wire coil. Balancing is done manually and is more precise than in the Minsco.

Torsion Balances

In this section we will discuss lever arm balances in which the counterforce is supplied by twisting a wire or fiber which acts as the support for the beam at its fulcrum.

Suppose a wire of length L and radius r is suspended at both ends so that it is taut. At the center a beam is attached perpendicular to the wire, and in a horizontal position. If a weight is now placed at one end of the beam, a torque τ is applied to the wire at the fulcrum. The wire will twist in the direction of the torque. The angle θ at which the strain on the wire just balances the torque is given by

$$\theta = \frac{KL\tau}{r^4}$$

in which K is a constant characteristic of the material from which the wire is made.

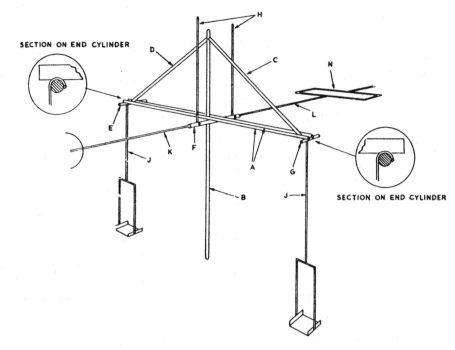

Figure 11. Quartz Ultramicrobalance, from R. Belcher, "Submicro Methods of Organic Analysis". Copyright 1966. Used by Permission of Elsevier Publishing Company.

In most torsion balances, a graduated dial is attached to one end of the wire or fiber, and this dial is rotated manually until the beam is returned to the horizontal null position. If the dial has been calibrated with a known weight, observing the number of graduations it must be rotated to balance the unknown makes possible calculation of the weight.

There are two groups of balances sold which use the torsion principle. These are the highly precise ultramicro balances, usually constructed of quartz, and the less precise torsion balances, which employ metal beams and suspensions.

Ultramicrobalances

The basic design of a quartz ultramicrobalance is illustrated in Figure 11, which shows the instrument developed by R. Belcher (9), and sold by Oertling. The torsion fibers (K and L) are attached to a transverse rod (F) which is suspended from above by two fibers (H). This is sometimes referred to as a *funicular* suspension. The tension in the horizontal fibers is kept constant by a quartz spring (N). The left end of the fiber K is fitted to the graduated dials (not shown). The beam is composed of two quartz rods (A) a pointer (B) and two tie bars (C and D). The pan supports (J) are fixed to the end cylinders (E and G, shaded in the inset diagrams in Figure 11).

Weighing is very sensitive to small temperature fluctuations (even two Celsius degrees can be significant in work of highest precision) and to air currents. The balance is therefore housed in an insulating cabinet. The sample pans are hung in wells which have doors that can be opened without exposing the entire mechanism to drafts or temperature changes. The pointer is observed through a small transparent window at the base of the upper balance cabinet. The size and location of this opening minimizes the

amount of light which can reach the mechanism and cause a temperature change.

After the sample (and a tare weight or counterpoise) has been loaded and the beam is released, the deflection of the pointer is observed. The torsion fiber system is then twisted by rotating a set of control knobs, until the pointer has returned to the center of the screen. The degree to which the fiber has been twisted to accomplish this is given by a digital indicator.

The Oertling Q01 balance, described above, has a capacity of 250 mg in each pan. The range of the torsion fiber is slightly more than 1 mg, with each division on the indicator equal to about 0.1 µg. Normally, reproducibility is better than ±0.08 µg. It is worth noting that the beam can be removed and replaced quickly and easily, since it is not fused to the transverse suspension rod.

The Rodder Model E ultramicrobalance is very similar in design to the Oertling Q01. A close-up view of the beam and suspension is in Figure 12. The two microscope objectives are used to observe the positions of horizontal index fibers attached to each arm of the beam. The torsion fiber is rotated until null balance is reached, as indicated by coincidence of the index fibers. The wells holding the sample pans are at the bottom of the case shown in the picture. The standard version of this balance has a sensitivity of 0.05 µg, a 200 mg capacity, and a 6 mg torsion fiber range. Balances with sensitivity to 1 nanogram are made by Rodder on special order.

Several other ultramicro torsion balances have been described (9, 10), but are not commercially available. Two ultramicrobalances which do not use the torsion principle will now be considered, before other, less sensitive torsion balances are discussed.

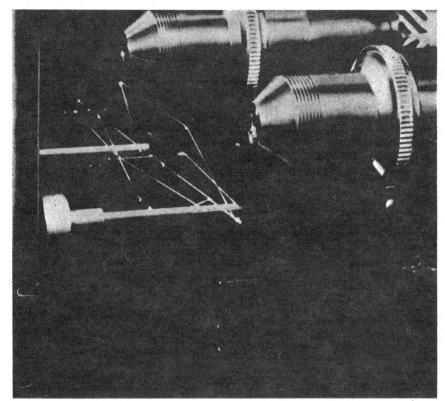

Figure 12. Quartz Ultramicrobalance, design of J. Rodder. *(Courtesy Rodder Instrument Company)*

The Mettler UM7 balance ($2450) has a sapphire knife edge suspension, and an optical scale similar to that found in this company's larger balances. The sample is loaded on top in a sheltered compartment. The weighing range of the UM7 is 2 mg, and reproducibility is ±0.1 μg. A 1-mg variable tare is built in. The total capacity of the balance is 103 mg, of which 100 represents the weight of a special weighing boat furnished with the instrument.

Ainsworth's Type 22 quartz ultramicrobalance is similar to those described earlier, except that it does not have a torsion fiber to balance small weights. It can be mounted in any enclosure which provides satisfactory protection against air currents and temperature gradients. The user constructs his own readout system.

Precision Torsion Balances

A number of companies offer torsion balances less sensitive than the ultramicrobalances. These instruments are useful for rapid weighing of small samples, and for applications such as determination of surface tension, where continuous variation of the torque on the beam is necessary.

The Roller-Smith balances come in a large variety of ranges—thirty in all. Most have hooks at both ends of the beam, so that a tare or counterpoise can be used to extend the capacity of the instrument to three times the weighing range. The most sensitive balance offered by this company (LG0003MG, $330) has a 3-mg range and ±2 μg sensitivity. Other balances are similarly priced and have ranges up to 50 g (LG0050GM, $385,

±0.2 g sensitivity). Special balances for surface tension, density, and yarn-number measurements are also available.

The Bethlehem torsion balances are quite similar to the Roller-Smith models. Weighing ranges go from 1.5 mg ($440) to 50 g ($412 double hook or $336 single hook). Sensitivities are about one part in two thousand, and accuracy about ±0.2% of full scale.

Brinkmann Micro-Torque balances are all of single-pan design, with mechanical taring replacing the second pan. The optical readout is fully digital, rather than combined digital and vernier. Otherwise the system is similar to the Roller-Smith design. Micro-Torque balances come in eight ranges from 1 mg (±1 μg accuracy, $480) to 10 g (±10 mg accuracy, $410). The precision is somewhat better than that of the other balances of this type.

Sauter precision balances use a torsion spring to provide the torque to balance the beam. They are quite similar to the balances described above. Eleven models are standard, with weighing ranges from 1 mg (±1 μg readability, $325) to 2.5 g (±2.5 mg readability, $322).

Part Two—Strain Balances

The group of instruments heretofore discussed was designed around the balancing of force and counterforce on a lever arm. The following section of this article is devoted to balances in which the applied force or torque (due to the object being weighed) produces a strain, which is in turn used as a measure of the force. Actually, the torsion balance could be

placed in this section, but it has a lever arm construction and shares many characteristics (null condition, taring) with other balances of that type.

Theory and Design of Strain Gauges (11–14)

Strain gauges are devices for measuring linear deformations of objects. Strain can be defined as the linear deformation of a body, usually due to the application of an external force. Since a mass in a gravitational field produces a force, strain gauges can be used to measure masses. Before discussing commercial instruments, the underlying principles of strain gauges and their construction will be described.

The simplest type of strain gauge is a metal wire or bar. When one end of the wire is fastened to a support, and a pulling force is applied to its other end, the wire stretches. The change in length ($\Delta l/l$ is related to the applied force (F) per unit area (A) through the Young's Modulus (Y) of the metal:

$$Y = \frac{(F/A)}{(\Delta l/l)}$$

so

$$\Delta l = \frac{Fl}{AY}$$

Representative values of Y are 2×10^{12} dynes/cm² for steel, 1.4×10^{11} for lead, and 6.9×10^{11} for aluminum (15).

A closely related type of strain gauge is the helical or spiral coiled spring. Its properties are similar to the straight wire or bar: it will lengthen (shear) when coupled to a pulling force, and compress when subjected to a pushing force. The main difference is that the effective value of Y can be adjusted over a very wide range. Usually the elongation of a spring is expressed in terms of Hooke's Law:

$$\Delta l = F/k,$$

where the force constant (k) is expressed in load per unit extension. Fused quartz, for example, can be used to construct springs with force constants ranging from 0.05 to 500 mg/mm or more. The leaf or cantilever spring shows similar properties—the downward displacement under a load can be related to the force.

A strain or force could also be determined by measuring the compression of a fluid to which it is applied. This technique is primarily suited to pressure measurements, and will not be discussed further.

The types of strain gauges described so far have several disadvantages. Manual observation of the change in length of a wire or spring under a load can be very tedious, even if a micrometer or interferometer is available. The systems respond slowly to changes in the load; several seconds are usually required for equilibrium to be reached. Also, many metals show hysteresis effects, and strain gauges constructed from them may require frequent recalibration. In an ideal system, the maximum load would strain the wire to such a small extent that the linear displacement would be infinitesimal. In such a situation, however, visual observation of the displace-

Chemical Instrumentation

ment would not be possible.

A number of electrical properties of materials have suggested themselves as a basis for strain measurement, for electrical quantities are easily monitored and recorded. Some gauges measure the change in capacitance when a strain alters the geometry of a capacitor, others measure inductance changes caused by the strain-produced displacements of a magnetic coil.[4]

Piezoelectricity (16) occurs in crystals which have a permanent internal dipole. When such crystals are stressed a voltage is produced. The efficiency of conversion of mechanical force to electrical power is 1% for quartz, 50% for many ceramics, and 80% for Rochelle salt. On the other hand, quartz is much more rigid and stable than other piezoelectric materials, so it is more popular for force and strain measurements.

An advantage of piezoelectric gauges over the resistive gauges discussed below is that the piezoelectrics are self-generating. The signal is produced within the quartz crystal, while resistive devices must be interrogated with an externally generated signal. A disadvantage of most piezoelectrics is their extremely high impedance (about 10^{13} ohms for quartz), which means that an electrometer or charge amplifier must be used to detect the signal.

The resistance gauge is the most useful type of electronic strain gage. Several force-to-resistance transducers have been developed (14), some based on helical springs, others on potentiometers and slide wires. More generally valuable are the wire strain gauges (11, 14), made from fine wires or foils. The unbonded gauge (Figure 13) has the wires (A, B, C, D) connected between a fixed support (F) and the object (M) to which the force is applied. The bonded gauge (Figure 14) is cemented

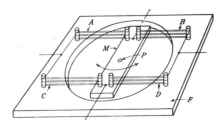

Figure 13. Unbonded Resistance Wire Strain Gauge (14). From Lion, "Instrumentation in Scientific Research". Copyright 1959. Used by permission of McGraw-Hill Book Company.

directly to the surface of the object, with only a layer of electrical insulation separating the wire or foil and the object.

The resistance (R) of the wire gauge changes with the applied strain (ΔS) according to

$$\Delta R/R = G\Delta S,$$

where G is the gauge factor. The change in resistance arises from changes in the resistivity of the wire or foil and in the geometry of the gauge due to stretching or compression of the object to which it is

[4] Note that these devices can be used as imbalance detectors in lever arm balances, as well as in strain gauges.

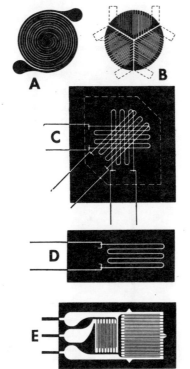

(Courtesy BLH Electronics, Inc.)

Figure 14. Bonded Strain Gauges. A—Spiral diaphragm gage for pressure measurement. B—Single-plain rosette. C—Stacked rosette. B and C are used for determining direction as well as magnitude of strains. D—Standard wire gage. E—Stress and strain gage.

bonded. The gauge factor may be positive (carbon, Nichrome, platinum) or negative (pure nickel), and varies with temperature.

In determining the strain or force, the gauge is often used as one arm of a Wheatstone bridge. The bridge is balanced in the null position; then the output of the unbalanced bridge is a measure of the strain applied to the gauge. Two or more gauges may be used in the bridge to cancel temperature effects or to increase the output.

Unbonded wire strain gauges are produced by Statham Instruments. Current models use a design called the Zero-Length strain gauge. In the conventional gauge the wires must be stretched so that they will have the proper tension even when heated by the operating current.

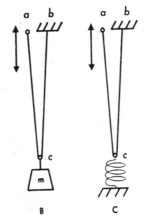

(Courtesy Statham Instruments, Inc.)

Figure 15. Zero-Length Unbonded Strain Gauge.

Also, the wires may break if the moving object travels too far. The Zero-Length devices, shown in Figure 15, are also designed with one of the wire fixed (b) and the other end movable (a), but the ends are placed next to each other, and the wire is passed through a loop (c) attached to a weight or a spring. The spring keeps the wire under tension and supplies a constant load. As point a is moved by a force applied to it, the load of the spring is transferred from section ac to bc, or vice-versa. A disparity between lengths ac and bc will result. Electrical connections at a, c, and b are used to monitor the strain-induced change in resistance of the two sections of the wire.

The bonded gauge is very easy to apply. It consists of the wire or a metal foil, cemented or glued to a paper or plastic film support. The gauge is cemented to the test structure (the support insulates the gauge from the structure) and leads from the read-out device are then soldered to the gauge. The bonded gauge is inexpensive and occupies little space; in some applications dozens or hundreds of them may be attached to the object being tested.

Many varieties of bonded strain gauges are available. BLH Electronics, Inc. offers about 300 standard SR-4 gages, with various sensitivities, temperature coefficients, temperature ranges, and response times. The output of a gauge also reflects the direction in which the force is applied, and certain gauges can be used to determine the directional characteristics of a complex strain pattern (see Figure 14). The gauges are usually constructed of nickel or platinum alloys. A platinum-tungsten alloy, sprayed onto a ceramic support, is used by BLH for high temperature work. A gauge (#FABT) is available with a built-in thermocouple element.

Micro-Strain, Inc., makes a variety of standard and special purpose gauges, including some which are compensated for thermal expansion of metals to which they might be attached at high or low temperatures. This company offers gauges as small as 1 mm², mounted on epoxy foils.

APPLICATIONS OF STRAIN GAUGES TO WEIGHING

Spring Balances

A variety of instruments are available in which the weight measured strains a spring, stretching or compressing it until the system is balanced. The Worden Mass Sorption Spring Balance is intended for use in controlled atmosphere systems. It includes a thermostatting jacket, stopcocks for connection to the vacuum or gas line, and a microscope (with micrometer) for observing the extension of the spring. An electromagnetic system for null balance operation of the system is optional. The springs are made from fused quartz. They are available for loads of 10 mg to 100 g, with precision up to one part in twenty-five thousand. Fused quartz springs are also sold by Worden, with sensitivities of up to about 100 mm extension per milligram load. Since a micrometer can observe changes of about $\pm 10^{-2}$ mm, a precision of $\pm 10^{-1}$ μg is

Chemical Instrumentation

attainable with the most sensitive spring systems.

Fused quartz springs are also available from the Thermal American Fused Quartz Co. Both straight and tapered springs are sold, with various coil diameters. This company also supplies quartz tubing and fibers for mounting the springs. In general, quartz springs may be used up to about 500°C, where a phase transition occurs. In precise work, correction for the temperature dependence of the Hooke's law constant may be necessary.

Many companies manufacture metal spring scales, and a comprehensive listing is beyond the scope of this article. Chatillon offers spring scales with capacities from 250 g to over 100 kg. Reproducibility is not as high as with quartz springs, but is adequate for certain applications. This company also makes a series of spring testers, which permit preparation of extension/compression vs. load calibration curves.

Springs are also components of a number of weighing devices which will be discussed later in this article. The Daytronic force transducers are an example. The Dillon Model P trade scale is an example of the use of a spring to provide the counterforce in a lever arm balance. This design, shown in Figure 16, features four unequal arm levers, arranged so that the spring need only provide a force equivalent to a twenty pound load, while the actual load can be up to ten thousand pounds. The accuracy of the weight as

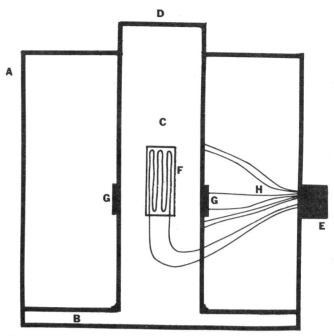

Figure 17. Strain Gauge Load Cell. A—Outer casing. B—Base. C—Load-sensitive column, resting on base. D—Load button or bearing. E—Junction box. F—Strain gauge sensitive to load applied to the column. G—Strain gauges insensitive to load. H—Leads from gauges to junction box.

including the load. The system is calibrated with samples of known mass. A fiber 1 cm long and 7 μ in diameter is sensitive to $\pm 10^{-10}$ g (one-tenth nanogram), and even greater sensitivity should be possible with smaller fibers.

sensitivity attainable is of the order of ± 5 g, with the charge amplifier or electrometer being the limiting factor in most cases.

Resistance-Wire Strain-Gauge Load Cells

Load cells are simple in construction (see Fig. 17). The load is placed on a platform which sits on a metal column. The force compresses the column. Four matched strain gauges are attached to the metal column, two aligned with the direction of stress, and the other two perpendicular to this direction (these two serve to compensate for temperature variations). The four gauges make up a Wheatstone bridge, with the similarly aligned gauges on opposite sides of the bridge. The output of the bridge is proportional to the strain-producing load—for greater sensitivity, more than one set of gauges may be used and the outputs of the bridges combined. The load cell requires a constant voltage source and a millivolt meter.

The range of capacities of strain-gauge load cells is remarkable. By careful choice of column material and gauges, load cells have been constructed with full scale capacities from 10 g to over five billion grams (5 Gg).

Another factor which may be of significance for some applications is that the load cell platform usually deflects no more than a quarter millimeter when the maximum load is placed on it. The cells are compact, light in weight, and can easily be protected against corrosive atmospheres.

The Baldwin-Lima-Hamilton Co. was the first to offer strain gages and manufactures the largest variety of load cells and force transducers. For example, a #U3G1 cell with 10 pound capacity has $\pm 0.25\%$ accuracy and $\pm 0.10\%$ repeatability, and costs $470. U-1 cells with capacities from 50 to 2000 lb cost $375 and

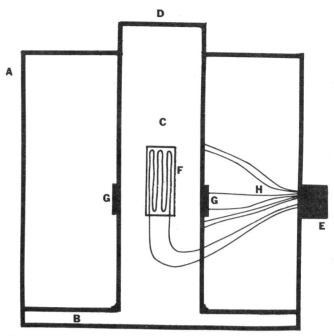

(Courtesy W. C. Dillon and Co.)

Figure 16. Lever-Arm Spring Balance.

indicated on the dial is one part per thousand of the full scale range.

A unique cantilever spring ultramicrobalance has been developed at the Dudley Observatory (17). A thin metal coated fiber is clamped at one end and the sample is fixed to the other end with a minute amount of adhesive. The assembly is placed between the plates of a capacitor. A DC potential is applied to the plates and an AC signal to the fiber. The fiber oscillates at a frequency which depends on the dimensions and the mass of the fiber,

Piezoelectric Load Cells

Load cells with piezoelectric elements are usually made with quartz crystals. They are not only rigid and stable to temperature, radiation, and mechanical shocks, but also have very fast response times, approaching ten microseconds. Kistler load cells come in capacities from 500 g to 50 megagrams. They are suitable for dynamic or static weighing, and taring can be accomplished electronically. The best

Chemical Instrumentation

have ±0.25% accuracy and ±0.05% repeatability.

West Coast Research Corp. manufactures several types of load cells. This company also has designed a series of balances incorporating strain-gauge load cells. The instruments are similar in appearance to conventional single-pan analytical balances, except for the absence of knobs for manipulation of weights. The readout is either digital or analog (meter), with outputs for recorders. Precision approaches ±0.02%, which is surpassed by the analytical and precision balances described in Part I of this article. On the other hand, the West Coast Research balances are more rugged and convenient to use, especially if they must be operated by remote control. The Model 318 (meter readout) balances sell for $795 with 1 kg capacity and $995 with 100 kg capacity.

Gilmore Industries has developed a system for weighing liquid hydrogen which could be applied to some laboratory weighing problems as well. The instrument is shown in Figure 18. The lever arm is balanced with an empty tank using the counterweight; the load cells then measure the buoyant force of the liquid hydrogen added to the tank. The calibration weights simulate buoyancies corresponding to various fractions of the tank capacity. Load cells are particularly valuable here because of their fast response and their very small linear extension when the load is applied.

Statham Instruments uses unbonded strain gauges in their Universal Transducing Cell ($150), which has a 60-g capacity.

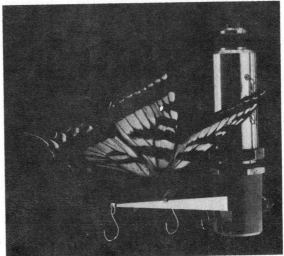

(*Courtesy Statham Instruments*)

Figure 19. Universal Transducing Cell with Micro-Scale Accessory. The sample is optional.

It can be obtained with a micro-scale accessory ($40) (Fig. 19) for full scale capacities of 6, 12, and 30 g, with accuracy of ±0.25% and precision better than ±0.1%. A portable, battery-operated, readout device (#UR4, $150) can be used to monitor the output of the Universal Transducing Cell.

Inductance Coil Load Cells

The ASEA Pressductor load cell, shown in Figure 20, is made of a set of transformer sheets glued together in a block, with the primary winding at right angles to the secondary winding. With a constant AC signal applied to the primary, signal strength at the meter depends on the flux through the secondary, which, in

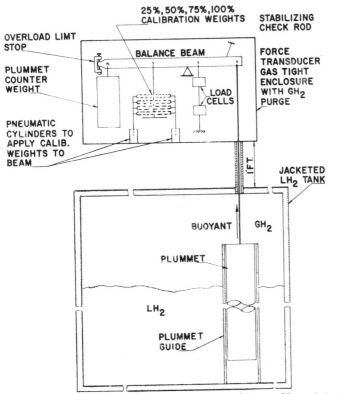

(*Courtesy Gilmore Industries, Inc.*)

Figure 18. Liquid Hydrogen Weighing System. LH₂—liquid hydrogen. GH₂—gaseous hydrogen.

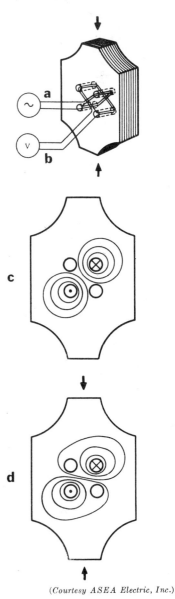

(*Courtesy ASEA Electric, Inc.*)

Figure 20. Inductance Load Cell. a—AC signal generator and leads to primary windings. b—Meter and leads to secondary windings. c—Magnetic lines of force in absence of load. d—Magnetic lines of force with vertical load applied to transformer plates.

280

Chemical Instrumentation

turn, depends on the force applied to the block. ASEA supplies Pressductor load cells primarily for industrial users (capacities run from 500 to 5,000,000 kg); potential applications in the laboratory have not been fully investigated. Precision is better than ±0.1% in all ranges.

Daytronic force and displacement transducers use a differential transformer to measure deflection of a spring under a load. The spring is attached to a metal rod which has a permanent magnet core at one end. The core is placed inside the primary transformer coil, to which an AC signal is applied. A pair of secondary coils is placed at either end of the primary. When the core moves from the null position, the signal induced in the secondary coil in the direction of motion is greater than that induced in the other coil. The secondary coils are connected in a circuit which detects the imbalance and measures it. The transducers are available in several capacities from 10 g to 2000 kg. The repeatability of all units is ±0.1% and the deflection is about 0.1 mm.

Part Three—Recording and Automatic Balances

There are two situations in which a recording balance might be preferred to a manual balance: (1) the weight of a sample, or the change in its weight, is to be followed with time; (2) series of samples must be weighed, and the possibility of error in recording the results on the part of the operator is to be eliminated.

A recording balance is not different from the ordinary balance as far as the principles of weighing is concerned. Two components are added in constructing the recording instrument: a recorder, and a device for detecting deflection of the balance from the null condition, or for restoring it to this position.

The recorder need not be discussed in detail. Although a photographic method or a mechanical linkage have been used in the past on some commercial instruments, today potentiometric recorders are nearly always chosen. An increasing number of companies are marketing digital printers and card punchers for use with balances. These accessories will find more and more applications as problems in which a digital computer handles the data become common.

Null Detectors

Many beam deflection detectors have been described. Most of them employ a transducer to convert the deflection to an electrical signal.

One approach is to attach a mirror to an arm of the balance, reflect a beam of light from this mirror through a slit to a photocell or phototransistor. The system is arranged so that the amount of light reaching the photocell depends on the

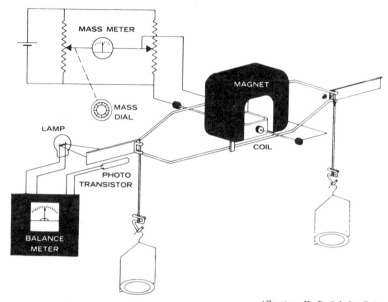

(Courtesy H. R. Schultz Co.)

Figure 21. Measurement of Balance Beam Deflection Using a Phototransistor. The circuit for the electromagnetic balancing mechanism is also shown.

angle of the balance beam to the horizontal. In place of the mirror, a slit or wedge fastened to the balance arm or pointer may intercept part of the light beam so that the fraction of the light reaching the detector depends on the position of the moving system. An example of this design, as adapted by the H. R. Schultz Co., is shown in Figure 21. C. L. Rulfs (18) has published directions for inexpensively adapting an equal arm balance for photoelectric null detection.

A system designed by the Exact Weight Scale Co., determines the deflection through a binary coded dial (attached to the beam) which is monitored by a set of photocells.

Among other companies using optical methods for measurement of beam deflection in their recording balances are Cahn, King, Mettler, Sartorius, and Voland.

The MTI Fotonic Sensor (#KD-38, $259) probe is constructed of a bundle of fiber optics, half of which shine light at an object, and the other half collect the reflected light. The output of the photocells measuring the reflected light can be related to the distance of the target object from the probe. Sensitivity is better than ±1 μ/mv. The probe is attached to a long cable, allowing its tip to be placed inside a balance case. Other MTI models have sensitivities better than ±0.01 μ/-mv.

Several electronic devices can measure the deflection of a balance beam. Some strain gauges are suitable for this purpose. Many of the types of gauges discussed earlier in this article can be combined with a lever arm to construct a recording balance with greater precision than that offered by the gauge itself.

Any electrical parameter sensitive to the geometric arrangement of the device can be a measure of beam deflection. For example, the inductance of the "E-core" and armature shown in Figure 22 will depend on the size of the air gap on each side. If the armature is attached to the balance beam, and the core is mounted above it in a fixed position, the inductance can be used

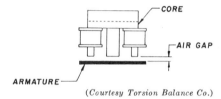

(Courtesy Torsion Balance Co.)

Figure 22. E-Core Inductor for Measuring Balance Beam Deflection.

to detect the null position of the beam. The Torsion Balance Co. uses this device in several of its models. The Sharples Recording Balance also has an inductive null detector.

Many linear displacement transducers are based on differential transformers. The Fisher Recording Balance Accessory, which can be attached to many manual balances, includes a differential transformer and an electromagnetic coil to restore the null condition.

Devices for Restoring Null Balance

The electromagnetic force coil is the most popular means of maintaining a balance in its null condition. This technique was discussed in some detail earlier in the article. The force coil has the advantage that the amount of current, which determines the amount of force used to null the balance, is easy to record.

Among the mechanical methods mentioned in the literature are the addition or removal of weights, lengths of a chain, or the force of a spring. These are all less convenient to record and monitor than the electromagnetic effect. On the other hand, the latter is not as precise as addition of a calibrated balance weight. Furthermore, even the better electronic recorders are no more precise than about ±0.1%. Therefore, an electromagnetic balancing device is often combined with a system for removing or adding weights as the end of the recorder scale is reached. The Ainsworth recording balances have a recorder scale of 100 mg, and use 100 mg weights to extend the dynamic range to 4 g. Stanton

Chemical Instrumentation

Instruments produces balances with a similar design.

Recording Balances

A survey of this field will not be attempted. Gordon and Campbell (*19*) published a comprehensive survey of automatic and recording balances which is still up-to-date in most respects. Wendlandt (*20*), Duval (*21*), and Lewin (*22*) have discussed recording thermobalances.

Thermogravimetric analysis is one of the most frequent applications of recording balances. The Tem-Pres system is shown in Figure 23. A Cahn balance is placed in a vacuum-tight glass container mounted above the furnace. The recorder, and separate controls for the balance and furnace, are all located on the rack at the right.

An entirely different application is the recording of the weights of a large number of objects. The Exact Weight Scale Co. has developed a number of balances for automatically weighing laboratory animals

(Courtesy Tem-Pres Research, Inc.)

Figure 23. Recording Thermobalance. The components are (reading clockwise from the upper right) the recorder, electrobalance controls, furnace controls, furnace, and the balance itself, in a controlled-environment system.

and recording the weight, or segregating the animals into weight classes.

The Mettler Model H 20E represents a third type of recording balance. The design is similar to this company's ordinary substitution balances, except for the recording feature. The object is loaded on the pan, and the proper weights are removed from the beam. When the balance has reached equilibrium the operator depresses a key, and the weight is printed on a paper tape. Sauter offers similar equipment for recording of weights; this company is particularly active in designing electronic equipment for reading out weight measurements made with all types of balances.

Acknowledgment

The author would like to thank the technical and sales personnel of the many companies he corresponded with for the

information they supplied to him. He regrets that space limitations made it necessary to be selective rather than comprehensive, particularly in the section on strain balances.

Laboratory Balances
Supplement

Most of the information presented in the original article is still valid. Model numbers have changed in a few cases. Prices of most instruments have increased somewhat, though not so much as to affect the choice of one type of balance over another.

The only really new type of laboratory balance is the Ainsworth Digimetric, a top loader which has automatic electronic digital display of the weight. The balance costs about $1500, has a 2 kg capacity and ±0.5 mg precision. Several manufacturers offer digital printer accessories, continuing the trend anticipated by the Mettler H20E balance, which was mentioned on the last page of the article.

A novel microbalance with 10 ng sensitivity has been developed in which the null condition is restored by a light beam (*23*). The balance, designed by K. P. Zinnow and J. P. Dybwad of the Air Force Cambridge Laboratories, is similar to the conventional quartz microbalance (Fig. 11). The balance is restored to the null condition by the pressure of a light beam aimed upwards at a mirror hung from the opposite end of the beam from the sample.

Mr. Pat Gaskins, former Vice President of the Cahn Instrument Company, has called my attention to the fact that the term Electrobalance is their registered trademark. The word was erroneously used in a generic sense on the sixth page of this article. With respect to the historical note contributed by the Editor, Mr. Gaskins points out that Angstrom published a description of an electromagnetic microbalance in 1895 (*24*), which clearly predates Whiddington's balance.

LITERATURE CITED

(1) National Bureau of Standards, Handbook 77, Volume III (1961). This handbook contains the classifications for accuracy of weights.
(2) T. W. Richards, *J. Am. Chem. Soc.*, **22**, 144 (1900).
(3) A. I. Vogel, "Quantitative Inorganic Analysis," 3rd ed., Wiley, New York, **1961**; p. 155.
(4) L. B. Macurdy, "Measurement of Mass," in "Treatise on Analytical Chemistry," edited by I. M. Kolthoff and P. J. Elving, Part I, Volume 7, Wiley—Interscience, New York, 1967 pp. 4247–4277.
(5) Committee on Balances and Weights, Report and Recommendations, *Anal. Chem.*, **26**, 1190 (1954); Wm. Ainsworth and Sons, Inc., has published a chart containing the recommended definitions. Copies are available free.
(6) W. D. Evans, THIS JOURNAL, **32**, 419 (1955).
(7) S. Z. Lewin, THIS JOURNAL, **36**, A7, A67 (1959).
(8) A. I. Vogel, *op. cit.*, pp. 140–146.
(9) R. Belcher, "Submicro Methods of Organic Analysis," Elsevier, Amsterdam, **1966**, pp. 6–16.
(10) I. P. Alimarin, AND M. N. Petrikova, "Inorganic Ultramicroanalysis," (trans. by M. G. Hell), Pergamon, London, **1964**, pp. 96–106.
(11) C. C. Perry, AND H. R. Lissner, "The Strain Gage Primer," McGraw-Hill, New York, **1955**.
(12) A. Cottrell, "The Mechanical Properties of Matter," Wiley, New York, **1964**, Chapter 4.
(13) R. F. Feynman, R. B. Leighton, AND M. Sands, "The Feynman Lectures on Physics," Addison-Wesley, Reading, Mass., **1964**. Volume 2, Chapter 38.
(14) K. S. Lion, "Instrumentation in Scientific Research," McGraw-Hill, New York, **1959**, pp. 23–90.
(15) H. M. Trent, AND D. E. Stone, "Elastic Constants, Hardness, Strength, and Elastic Limits of Solids," in "American Institute of Physics Handbook," McGraw-Hill, New York, **1957**, pp. 2–55–80.
(16) L. B. Wilner, *Rev. Sci. Instrum.*, **36**, 693 (1965).
(17) H. Patashnick, AND C. L. Hemenway, as reported in *Sky and Telescope*, **32** (Oct., 1966).
(18) C. L. Rulfs, *Anal. Chem.*, **20**, 262 (1948).
(19) S. Gordon, AND C. Campbell, *Anal. Chem.*, **32**, 271R (1960).
(20) W. W. Wendlandt, "Thermal Methods of Analysis," Wiley-Interscience, New York, **1964**, Chapter 3.
(21) C. Duval, "Inorganic Thermogravimetric Analysis," 2nd Edition, Elsevier, Amsterdam, **1963**. Chapters 1–3.
(22) S. Z. Lewin, THIS JOURNAL, **39**, A575 (1962).
(23) *New Sci.*, **45**, 60 (1970).
(24) *Svensk Vetensk Selsk.*, *Forhandlung* **9**, 643 (1895).

List of Manufacturers

Wm. Ainsworth and Sons, 2151 Lawrence Street, Denver, Colo. 80205

ASEA Electric, Inc., 4 Kaysal Court, Armonk, N. Y. 10504

Beckman Instruments, 2500 Harbor Boulevard, Fullerton, Calif. 92634

Bethlehem Instrument Co., Prospect and Conestoga Streets, Bethlehem, Pa. 18016

BLH Electronics—Baldwin-Lima-Hamilton Co., 42 Fourth Avenue, Waltham, Mass. 02154

Brinkmann Instruments, Inc., Cantiague Road, Westbury, N. Y. 11590

Cahn Instrument Co., 7500 Jefferson Street, Paramount, Calif. 90723

John Chatillon and Sons, 83-30 Kew Gardens, New York, N. Y. 11415

Daytronic Corporation, 2875 Culver Avenue, Dayton, Ohio 45429

Chemical Instrumentation

W. C. Dillon and Co., 14620 Keswick Street, Van Nuys, Calif. 91407

Exact Weight Scale Co., 557 East Town Street, Columbus, Ohio 43215

Fisher Scientific Co., 711 Forbes Avenue, Pittsburgh, Pennsylvania 15219

Gilmore Industries, 3355 Richmond Road, Cleveland, Ohio 44122

J. A. King and Co., 2620 High Point, Greensboro, N. C. 27402

Kistler Instrument Co., 8975 Sheridan Drive, Clarence, N.Y. 14031

MTI—Mechanical Technology, Inc., 968 Albany-Shaker Road, Latham, N. Y.

Mettler Instrument Corp., 20 Nassau Street, Princeton, New Jersey 08540

Micro-Strain, Spring City, Pa. 19475

Minsco—Minnesota Instrument Co., 609 Marshall Street, Minneapolis, Minn. 55413

Oertling—distributed by J. and G. Instrument Corp., 1704 Highway 27, Edison, N. J. 08817

Rodder Instrument Co., 775 Sunshine Drive, Los Altos, Calif. 94022

Roller Smith—Federal Pacific Electric Co., 50 Avenue L., Newark, N. J. 07105

Sartorius—distributed by Brinkmann Instruments, Inc., Cantiague Road, Westbury, N. Y. 11590

August Sauter of America, Inc., 80 Fifth Avenue, New York, N. Y. 10011

Harold R. Schultz Co., P.O. Box 2133, Orange, Calif. 92667

Sharples Co., 2300 Westmoreland, Philadelphia, Pa. 19140

Stanton Instruments, Ltd., 2223 Fifth Avenue, Pittsburgh, Pa. 15219

Statham Instruments, Inc., 12401 West Olympic Boulevard, Los Angeles, Calif. 90064

Tem-Pres Research, Inc., 14018 Atherton Street, State College, Pa. 16801

Thermal American Fused Quartz Co., Route 202, Montville, N. J. 07045

Torsion Balance Co., 35 Monhegan Street, Clifton, N. J. 07013

Voland Corp., 27 Centre Avenue, New Rochelle, N. Y. 10802

West Coast Research Corp., 2100-02 South Sepulveda Blvd., Los Angeles, Calif. 90025

Worden Quartz Products, 6121 Hillcroft Avenue, Houston, Texas 77036

Topics in...

Chemical Instrumentation

...a feature

Edited by **GALEN W. EWING**, Seton Hall University, So. Orange, N. J. 07079

The Journal of Chemical Education
Vol. 46, page A323
May, 1969

XLIV. Automation of Organic Elemental Analysis

by the late **VELMER B. FISH**, Lehigh University, Bethlehem, Pennsylvania

Born in Mallard, Iowa, in 1912, **Dr. Velmer B. Fish** received his BS at Iowa State University in 1936 and his PhD (in biochemistry) in 1942. His thesis involved rather extensive research on the metabolism of ethanol and its distribution in the body tissues of rats. The following three years were spent at the University of West Virginia, in the Agricultural Experimental Station, where analytical research was carried out on natural products, seeking methods for vitamin assay as well as other constituents in such materials. A period of about three years was spent with the J. T. Baker Chemical Company, during which time a microchemical laboratory was established and some research undertaken involving synthetic organic chemistry.

Dr. Fish joined the faculty of Lehigh University in 1949, where he taught mostly analytical chemistry up until his untimely and tragic death in the fall of 1968.

Editor's Note

The present article was presented by Dr. Fish in July 1968 as part of a Workshop on Instrumental Analysis at the Center for Professional Advancement, Hopatcong, N. J. At that time he agreed to publish the paper in this Column, and gave me a copy of his manuscript. Unfortunately this copy did not contain the illustrations, and since Dr. Fish's death, they cannot be located. I have been able to reconstruct what they must have been, with the exception of a final cartoon suggesting the conceivable rapid "metamorphosis" of an analyst as the result of overdoing automation. I have added the appendix to Dr. Fish's paper.

G.W.E.

The automation of established techniques in the art of organic elemental analysis has been neither abrupt nor complete. Rather in most cases our present position has been attained by step-wise modification of the original system. We should like to consider this process as applied to the determination of carbon and hydrogen with the eventual inclusion of nitrogen and oxygen.

The first step in automation was an adaptation of existing combustion systems to a micro or at least semimicro scale. This was an essential step since most of our present automated techniques operate optimally when applied to milligrams or less of sample.

An early carbon-hydrogen combustion system is shown in Figure 1, and consists of the following units: (1) A supply of purified oxygen and/or air from which has been removed carbon dioxide, water, and any organic substances which might produce them. The regulation of the rate of gas flow may also be included in this portion. (2) The second major portion deals with the combustion of the sample and removal of all interfering products of that combustion. Thus the products which are allowed to flow from this portion are carbon dioxide, water, oxygen, nitrogen, and possibly a trace of "other" materials. (3) The next section allows the separation of the desired products, water and carbon dioxide, from each other and from all other materials present. (4) The final stage of the procedure consists of measuring and recording the response of a detector, in this case a suitable balance, to the water and carbon dioxide obtained from the sample. (5) Calculations then are made with or without modifications to correct for blanks and other exigencies inherent in the method.

Possibly the first and most exciting method of automation involved the employment of a young lady operator. This innovation could be instructed, programmed, and allowed to operate without excessive "down time." Upkeep, repairs, and replacement may be expensive, hence other means of automation have been

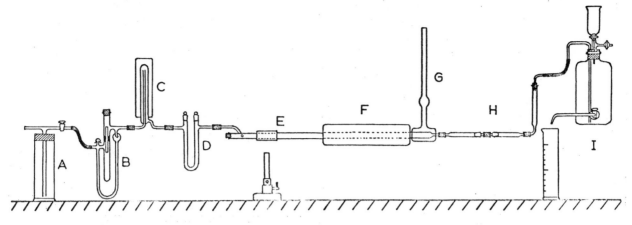

Figure 1. A manually operated micro-combustion apparatus of the Pregl type. *A*, bubble tube to establish a pressure head; *B*, flowmeter; *C*, electric preheater; *D*, scrubbing tube to remove traces of CO_2 and H_2O from the oxygen; *E*, gas-heated movable sleeve surrounding silica combustion tube; *F*, main furnace; *G*, refluxing mortar; *H*, weighable absorption tubes; *I*, Mariotte bottle, useful in applying suction to the system and in measuring the volume of oxygen.

devised, which serve as an adjunct to, but usually not a replacement of, the original. We will leave the socioeconomic phase of this problem for the time being and concentrate on the adjuncts.

The supply and purification of necessary gases can easily be regulated with adequate valves, pressure regulators, or metering devices, to meet the requirements of any particular system of analysis. The involved parameters can usually be evaluated and controlled with adequate precision by mechanical or electronic monitors.

The gas burners originally employed in the combustion portion were soon replaced by electric furnaces, even to the 176°C *p*-cymene mortar.* Interestingly, one may note that the specifications for the *p*-cymene mortar were published in 1949. (*1*), whereas the electric replacement had been described in 1941 (*2*). The main furnace (stationary) was easily designed with an adequate control, but the movable sample burner, when replaced by a movable electric furnace, still required the attention of an operator. The early solutions to this problem were to drive the furnace forward by means of a threaded shaft activated by a variable-speed motor similar to that supplied by A. H. Thomas Company some 20 years ago, or more recently to use sectional heaters to burn the sample in a step-wise progression. Such a system is incorporated into the E. H. Sargent combustion apparatus (Fig. 2).

---

* A *mortar*, in this context, is an apparatus for holding at a constant, intermediate, temperature a portion of the combustion tube following the main furnace. This portion, packed with lead dioxide, is to remove oxides of nitrogen. Close control of temperature is required to maintain constant any effect on the water content of the gas. The mortar shown at *G* in Figure 1 is a double-walled glass unit intended to contain *p*-cymene which will reflux at its boiling point, 176°C. The etymology of the term mortar, so used, is obscure. —G.W.E.

Figure 2. Sargent No. S-21570 Micro-Combustion apparatus. The silica combustion tube passes through the several heaters and is attached to absorbant tubes which are suspended from the hooks shown. Gas flow is from right to left. Three sequential ovens are shown with covers raised; next is the main furnace, then the mortar, all electrically heated

The procedure was thus automated except for weighing the sample and the combustion products. The time required for each analysis was about an hour. The technique required controlled environmental conditions and the most conscientious attention of the operator. Improvements were directed toward relieving the operator of responsibility except for weighing the sample, placing it in the proper position in the apparatus, and weighing the absorption tubes on schedule. The Thomas Micro-Carbon-Hydrogen Analyzer shown in Figure 3 serves to illustrate an apparatus of this type. To quote from the manufacturer's description of this equipment, it "easily permits completion of 30–40 determinations within an 8-hr working day." (*3*).

In the early 1960's further developments in automation and expansion of microanalysis were introduced with the determination of carbon, hydrogen, and nitrogen using a single sample. The principle of operation of some of these instruments will illustrate a further step in automation. That is, the response of the detector is recorded on a chart as waves, peaks, or digits from which one may determine the

composition of the sample. Thus the gravimetric evaluation of the combustion products could be eliminated. The nongravimetric estimation of the combustion

Figure 3. A. H. Thomas Model 35 Micro Carbon-Hydrogen Analyzer. The furnace is located in the perforated steel cylinder at the right. The three weighable tubes absorb respectively H_2O, N-oxides, and CO_2.

products is usually accomplished by means of a thermal conductivity bridge though it would seem that other detectors might also be adapted to this particular use. The present instrumentation, using the thermal conductivity bridge, is based upon one of two general procedures. One involves the separation of carbon dioxide, nitrogen, and water, by a suitable gas-liquid chromatographic column. The other general method involves separation by means of specific absorbants for water and for carbon dioxide with the resulting change in composition of the gas mixture being measured by thermal conductivity bridges.

This latter method is illustrated by the Perkin-Elmer Model 240 Elemental Analyzer (Fig. 4), which utilizes the techniques previously described by Simon, *et al.* in 1963 (*4, 5*). A flow diagram of this instrument is shown in Figure 5.

284

Its operation involves the introduction of an accurately weighed sample, of $^1/_2$ to 5 mg, into a port leading to the combustion train. The whole system is flushed with helium, a measured volume of oxygen is introduced into the combustion chamber, and the sample is moved into the heated area by means of a magnet. After the combustion is completed the products are allowed to react in the reduction portion of the system, leading to a final mixture of carbon dioxide, water, nitro-

Figure 4. Perkin-Elmer Model 240 Elemental Analyzer, with cover removed. The sample is inserted through the extended combustion tube, front right, and pushed into the furnace, front left, with a magnet. The rear structure contains the GC parts and the electronics.

gen, and the carrier gas, helium.

A measured volume of this mixture is

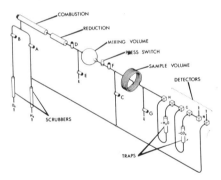

Figure 5. Schematic flow diagram of the Perkin-Elmer Model 240. Items A, B, etc., are valves for gas control and for flushing.

analyzed by passing through a series of paired thermal conductivity cells. The total mixture is present in one cell and compared with the mixture from which water has been removed, in the other cell. The difference in composition resulting from the removal of water is presented as a steady voltage difference between the two cells. The water-free gas mixture is then introduced into a third cell, the carbon dioxide is removed, and the resulting gas enters a fourth cell. The voltage difference between cells 3 and 4 thus becomes a measure of the carbon dioxide present. The gas now entering the fifth cell consists of nitrogen and helium; the nitrogen is measured by the voltage difference between this cell and the sixth, which contains only helium.

The read-out which one obtains is programmed to appear as steady voltages on a recorder. These voltages may be modified by scale expansion to yield the required accuracy. The parameters which affect precision include weighing the sample, read-out response, detector oper-

ation, flow rate of the gas, and temperature control. These seem to have been controlled instrumentally within the necessary limits. However, there is another source of error which requires adequate consideration. The detectors respond to a difference in the *volume* composition of two gases, hence to the mole-fraction composition of the gas mixture. Since it is the *weight* evaluation of each gas which is required, correction must be made when unusual compositions are encountered or if rather large samples are necessary.

Results are calculated by analyzing one or more samples of known composition and an occasional blank. From these experiments one may calibrate the system in terms of response in (for example) microvolts per microgram of carbon, of hydrogen, and of nitrogen.

A variation of this method was described by Kainz and Wachberger (6) which allowed some simplification of the equipment and seems to have avoided the errors which occur in the thermal conductivity measurements of a multicomponent mixture before and after removal of one constituent. By the proposed method a sample is burned in an "empty tube" with a measured amount of oxygen. Interfering elements such as halogens and sulfur are removed in the usual way by contact with metallic silver, the excess oxygen is removed and nitrogen oxides reduced by hot copper. Water is absorbed first following the combustion tube, then carbon dioxide is absorbed on a length of molecular sieve. The nitrogen mixed with helium is compared with pure helium in a detector consisting of two cells, each containing a thermistor, and connected to a bridge circuit. The signal which is produced is fed into an electronic digital integrator. The carbon dioxide is desorbed from the molecular sieve, mixed with helium, and passed through the detector. Water is finally desorbed and after bypassing the molecular sieve is conducted through the detector. An example shows a response of about 5 counts/μg of hydrogen, 1.25 counts/μg of carbon and 0.5 count/μg of nitrogen. Peak height results, as obtained by feeding the bridge signal into a potentiometric recorder, require simpler instrumentation, but are not as accurate as those obtained by means of the integrator.

Highly fluorinated organic compounds have presented two problems to those using automated equipment. They may not undergo complete combustion under ordinary treatment, and the fragments produced lead to low values for both carbon and hydrogen. The presence of fluorides or fluorinated fragments in the detector also results in considerable error. The introduction of magnesium oxide into the combustion tube and modification of the combustion sample have been reported to eliminate these problems (7, 8). Analysis of volatile samples necessitates an adequate flexibility in the combustion program to prevent incomplete combustion.

A different means for automated analysis for carbon, hydrogen, and nitrogen is exemplified by the F. & M. Model 185 CHN Analyzer (Fig. 6). This instrument

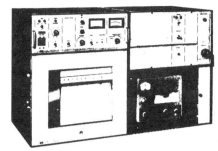

Figure 6. F. & M. Model 185 CHN-Analyzer

was designed to utilize gas chromatography (GC) for separation of the combustion products obtained from the oxidation of a small sample by a solid oxidation mixture. The oxidation mixture consists of oxides of manganese and tungsten and a follow-up chamber containing copper oxide in which the oxidation is completed. Interfering compounds are separated from the desired products, carbon dioxide, nitrogen and water, either by copper reduction or on the chromatographic column. As the separated combustion products pass through one-half of the thermal conductivity detector a signal is produced which passes through an automatic sensitivity selector and an attenuator, which is tied in with the Cahn Electrobalance, to the recorder. This system makes the peak height proportional to the percent of carbon, of hydrogen, and of nitrogen in the sample. The proportionality constants are found by measuring the corresponding peak heights produced by known standards. Thus, for example, the percent of carbon in a sample is the product of a constant and the linear height of the carbon dioxide peak measured in the same units as were used to determine the constant. The effective operation of the whole instrument can be quickly checked by measuring peak heights resulting from two or more samples of the same compound which differ considerably in weight.

There are several limitations inherent in this system of analysis. Since gas chromatography is used as the means of separation of desired combustion products, all the parameters of GC must be optimized and controlled, such as temperature, rate of gas-flow, and sample size. The method also depends upon the use of small samples of about 0.6 mg which allows for rapid combustion and the introduction of the products into the chromatographic column without excessive dilution by the carrier gas. If larger samples must be used, one obtains much better results by employing an integrator which will give peak areas instead of peak heights.

Adaptation of an integrator to the Model 185 presented certain problems. As usually operated inadequate separation of nitrogen and carbon dioxide was obtained for use with an integrator. Two lines of investigation may have solved this problem. Graham (9) investigated a new type of column which did allow separation. This and other experimental columns lacked the linearity which was desirable and it was necessary to use the Electrobalance separately. A conversation with F. & M. representatives disclosed that an integrator is available for use with the present column, but that a slower flow of carrier gas is required. Under recommended operating conditions, the integrator seems satisfactory and linearity is retained.

It is possible to check the effectiveness of the combustion system by careful observation of the resulting peaks. If fragments of incomplete combustion are produced, distortion will be apparent in the chromatographic peaks. If residues remain following combustion, then peak heights will vary when replicate samples are analyzed.

OXYGEN ANALYZERS

Some work has been reported on the adaptation of the F. & M. Model 185 CHN Analyzer to the direct determination of oxygen in organic materials (10), and in reducible metal oxides (11). The sample is decomposed in the presence of carbon at 1160°C, passed through a second furnace containing carbon, nickel, and finally copper oxide. At this point all the oxygen from the sample should be present as carbon dioxide. To attain this condition, the organic sample must be pyrolyzed to carbon, hydrogen, and carbon monoxide before reaching the hot copper oxide. The oxidation of hydrogen to water then will not involve oxygen from the sample. The carbon dioxide present after oxidation of the carbon monoxide is a function of the oxygen from the sample. Introduction of the dried carbon dioxide directly into the chromatographic column gave a broad peak the evaluation of which required integration. If, however, the carbon dioxide were absorbed on a molecular seive after being dried, then it could be desorbed by heating, and introduced as a "plug" into the chromatographic column, resulting in a sharp peak. The recorder response was controlled by the weight of the sample through a ratio recording system. The chromatograms furnish a means for detecting any malfunctions in the system. If, for example, the copper oxide were spent, the carbon dioxide peak height would vary and an abnormally high nitrogen peak would be produced, since carbon monoxide would also contribute to it.

Attachment of this combustion system to the CHN analyzer is made by means of the bypass line which is operative during the static combustion phase in the normal operation of the analyzer. It would seem quite feasible to plumb the direct oxygen combustion system into the Perkin-Elmer Model 240 instrument, including the drying section, the molecular seive, and the thermal conductivity detectors.

The Coleman Model 36 Oxygen Analyzer (Fig. 7) is a partly automated system in which the combustion is programmed but the carbon dioxide is absorbed and weighed by the usual means. All of the automated systems thus far proposed for oxygen determination involve some modification of the Unterzaucher method (12), which requires several steps. It is necessary to pyrolyze a weighed sample to carbon monoxide, carbon, hydrogen, and possibly other substances. High temperatures are required in order to accomplish

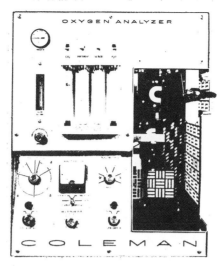

Figure 7. Coleman Model 36 Oxygen Analyzer. The combustion tube and furnace are located in the recessed area to the right. The absorption tubes for H_2O and CO_2 are mounted on the front panel for convenient access. Coleman also offers CH and N analyzers of similar external appearance.

suggested, as was previously mentioned. By using helium as the carrier gas and by oxidation of carbon monoxide to the dioxide, one should be able to separate and determine both oxygen and nitrogen. Carbon dioxide is easily retained by a molecular seive and can be desorbed to give a good chromatographic peak. An integrator is essential.

An interesting approach to automation of the Unterzaucher method was published recently by Hozumi (9). A diagram of the system is shown in Figure 9. The procedure consists of the programmed pyrolysis of a 1 to 3 mg sample, using platinized carbon and copper to produce carbon monoxide. The carrier gas is argon. The carbon monoxide is allowed to react with iodine pentoxide to produce an equivalent amount of iodine. The latter is measured as a vapor by photometry. The maximum absorbance (minimum transmittance) is measured and the oxygen found from a calibration curve. The calibration curve is obtained by the analysis of several samples of known compounds, and consists of a semi-log plot of the percent transmittance against weight of oxygen present in the sample.

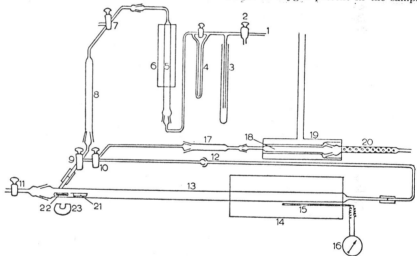

Figure 8. Apparatus for the Unterzaucher oxygen analysis. A nitrogen tank is connected at 1. Copper gauze at 5 is heated by furnace 6 to scavenge oxygen. The sample is placed at 21 and manipulated magnetically, 22, 23. The main furnace is 14. Tube 17 contains KOH, 18 contains I_2O_5. Jacket 19 is a reflux heater containing acetic acid (b.p. 118°C); 20 is an NaOH absorption tube.

the reaction of water and carbon dioxide with carbon to give a stoichiometric amount of carbon monoxide. The latter

is scrubbed by alkali to remove any acidic materials. The carbon monoxide may be determined in several ways. It may be oxidized by iodine pentoxide giving rise to some interesting stoichiometry; five gram-atoms of oxygen require twelve moles of sodium thiosulfate in the final titration. The milliequivalent weight of oxygen then is 6.667 mg. A diagram of the apparatus is shown in Figure 8. Automation of this general method would involve some rather easy steps. The activation and regulation of the various heaters and the flow of purified nitrogen or helium through the various sections of the system could readily be adapted to electronic control. It might be mentioned, however, that over 1200 W is commonly required by the two furnaces to attain the 1120°C temperature. Automation of the final step has been

The main requirements of the optical system in this procedure are that all of the iodine produced by a given sample must be present in the cell at some period of time, and that the iodine must be uniformly dispersed in any cross sectional position within the cell. Hence the rate of gas flow is critical and the pyrolysis must be accomplished within a short period of time. Blank determinations must be made in order to establish a correct zero point.

Automation can be achieved in almost any analysis of an organic compound including the functional groups. About all that is required is an adaptation of some detector system which is sensitive to the products of a reaction, and a recorder. The development of requisite sensitivity and scan time in recorders has made instruments available which are adequate for most detector systems. Further improvements are of course possible if the output of the detector

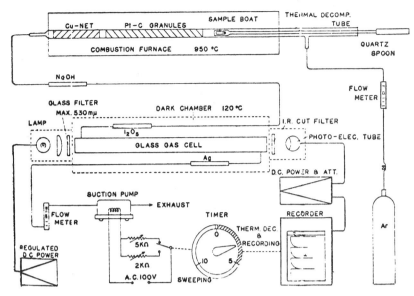

Figure 12. Technicon CHNO Analyzer (Technicon Corporation, Tarrytown, N. Y.). Combustion takes place in the upper oven, reduction over copper in the lower.

Figure 9. Oxygen analyzer after Hozumi. This is similar in principle to the apparatus of Figure 8, but iodine is determined photometrically.

may be the percent of chlorine. In order to obtain a reliable value quickly one must find a substitute for the Carius tube, the Parr bomb, or even the Schöniger closed-flask system which may be too slow. One may use very small samples and micro reaction equipment like that employed in the hot-flask method of Kirsten, et al. (15) which, with potentiometric titration, allows a large number of samples to be analyzed. The time per sample, about 35 min, may still be too long. Under such circumstances a special system might be developed using a small sample, rapid combustion, simple separation, followed by either GC with an adequate detector or absorption of combustion products in a suitable liquid, and, finally, by an automated titration. It has been reported (16) that a chromatographic

Figure 10. American Instrument Company Model 10-010 Carbon and Hydrogen Analyzer.

in the complexity and expense involved. It might well be economical to purchase and coordinate a mass spectrometer and a computer in order to resolve problems involving stable isotopes, structural isomers, or even to yield the empirical formula directly from the output of a CHN analyzer. On the other hand the need for a specific determination or a means of process control may require some degree of automation. For a practical example, one might consider the chlorination of wax where the critical information necessary

system is tied in with an appropriately programmed computer which will then deliver the desired information. Such systems are in use, and have recently been described (14).

There remains one major obstacle in the realization of a completely automatic system. In most cases the sample must be selected, prepared, weighed, and introduced into the instrument manually. Even this problem has been solved in cases of gas or liquid samples which may be metered directly into the analyzer.

OTHER ELEMENTS

Commercial instruments are available for the automated determination of carbon, hydrogen, nitrogen, and oxygen in organic materials. The other elements often found in such samples are usually determined by some well established procedure involving mainly the efforts of the analyst. However conditions may demand some degree of automation in order to increase the efficiency of the laboratory. There are obvious extremes

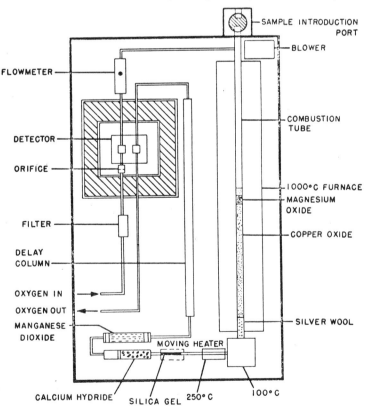

Figure 11. Flow diagram of Aminco Model 10-010. Water from the combustion is held on silica gel while CO_2 is measured on the thermal conductivity cell, then is desorbed by a movable heater, and reduced to H_2 by calcium hydride. The H_2 is then measured by the same detector. Fluorine is removed by MgO, halogens and sulfur by silver wool, and nitrogen oxides by MnO_2.

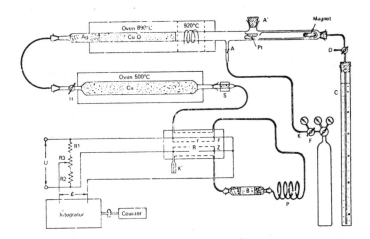

Figure 13. Flow diagram of the Technicon CHNO Analyzer. The sample is inserted in a magnetically driven platinum boat, and burned in the initial section of the upper oven, in a stream 97% helium, 3% oxygen. The combustion products with the helium carrier are passed over hot CuO, then silver wool (to remove halogens and sulfur), then into the copper-packed second oven. Here excess oxygen is removed and nitrogen oxides reduced. Water is retained temporarily in the silica gel tube *S*. N_2 and CO_2 pass through one side of a thermal conductivity bridge, then CO_2 is removed (at *B*), and nitrogen passes through the other arm of the bridge. The signal from the bridge is integrated and displayed on a numerical counter. To determine oxygen, an accessory heater is required to convert oxygen to CO_2.

method for the determination of carbon, chlorine, bromine, and iodine has been developed. Combustion products are trapped by liquid nitrogen, then separated on a column. The design of an automated analyzer must be a compromise between the elaboration of mechanical operations, electronic units involved, and the responsibilities of the operator. As a general rule one should keep mechanical operations to a minimum and the electronic systems adequately monitored.

APPENDIX

In addition to the specific instruments described above, the editor wishes to call attention to those illustrated in Figures 10 to 13.

Literature Cited

(1) Steyermark, A., Alber, H. K., Aluise, V. A., Huffman, E. W. D., Kuck, J. A., Moran, J. J., and Willits, C. O., *Anal. Chem.*, **21**, 1555 (1949).

(2) Smith, G. F., and Taylor, W. H., *Ind. Eng. Chem., Anal. Ed.*, **13**, 203 (1941).

(3) A. H. Thomas Co., *Scientific Apparatus*, **3**, No. 1 (October, 1965).

(4) Cleric, J. T., Dohner, R., Sauter, W., and Simon, W., *Helv. Chim. Acta*, **46**, 2369 (1936).

(5) Condon, R. D., *Microchem. J.*, **10**, 408 (1966).

(6) Kainz, G., and Wachberger, E., *Microchem. J.*, **12**, 584 (1967).

(7) MacDonald, A. M. G., and Turton, G. G., *Microchem. J.*, **13**, 1 (1968).

(8) Olson, P. B., *Microchem. J.*, **13**, 75 (1968).

(9) Graham, J. H., *Microchem. J.*, **13**, 327 (1968).

(10) Hinsvark, O. N., and Muldoon, P., "Instrumental Microanalysis of Oxygen in Organic Materials," Tech. Paper No. 39, F. & M. Sci. Div. of Hewlett - Packard Co. (1967).

(11) Hinsvark, O. N., and Muldoon, P., "Instrumental Microanalysis of Oxygen in Reducible Metal Oxides; Characteristics of Nickel Oxide," Tech. Paper No. 35, F. & M. Sci. Div. of Hewlett-Packard Co. (1966).

(12) Unterzaucher, J., *Ber.*, **73B**, 391 (1940).

(13) Hozumi, K., *Microchem. J.*, **12**, 218 (1967).

(14) Frazer, J. W., *Anal. Chem.*, **40**, No. 8, 26A (1968).

(15) Kirsten, W. J., Danielson, B., and O'Hren, E., *Microchem. J.* **12**, 177 (1967).

(16) Mamaril, J. C., and Meloan, C. E., *J. Chromatog.*, **17**, 23 (1965).

Topics in...

Chemical Instrumentation

Edited by **GALEN W. EWING**, Seton Hall University, So. Orange, N. J. 07079

...a *ChemEd* feature

The Journal of Chemical Education
Vol. 46, pages A69, A149, A233
February to April, 1969

XLIII. Mass Spectrometers

GALEN W. EWING

INTRODUCTION

There has been considerable activity in the design of new instruments in this field in the five years since the review by Wiberley and Aikens (1). Several improved models of magnetic deflection spectrometers have appeared, and in addition, increased emphasis is placed on quadrupole and other dynamic designs. Hence, it seems desirable to outline briefly the working principles of the major classes of instrument, then to describe specifically the major mass spectrometers and spectrographs now available.

Part 1. Ion Sources, Sample Handling, Vacuum Systems

These systems and components are common to all laboratory mass spectrometers, and so will be considered first. All mass spectrometers act on gas-phase ions. The ions are produced in a variety of ways depending on the nature of the sample. The most widely applicable is electron bombardment. A beam of electrons emitted from a hot-wire filament (usually tungsten or rhenium) is passed through an evacuated space into which a gaseous sample is allowed to diffuse. Molecules of the gas, upon collision with sufficiently energetic electrons, become ionized by loss of electrons and often by more deep-seated fragmentation. A greater number of positive ions than negative ions are formed, and the bulk of mass spectrometric measurements are made on positive ions.

A rather weak electric field transverse to the electron beam extracts the ions from the source space through one or more small openings. A higher voltage V then accelerates the ions, giving them kinetic energy

$$eV = \frac{1}{2} mv^2 \qquad (1)$$

where e is the charge on the ion, expressed in multiples of the electronic charge, m is the mass of the ion in a.m.u., and v is its velocity. This relation indicates that

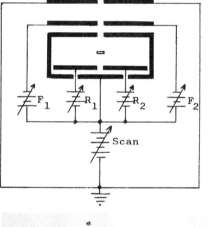

◀ 1 (a)   ▼ 1 (b)   ▲ 1 (c)

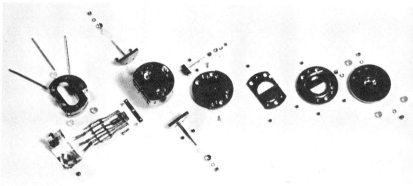

Figure 1. A precision ion source, the CEC Isatron. (a) Schematic; the battery symbols represent separately adjustable voltages from a regulated rectifier; R_1 and R_2 energize the pair of repeller plates, F_1 and F_2, the two focus plates, the voltage source marked "scan" is continuously variable from 250 to 3500 volts. (b) Exploded view of the CEC Isatron; the major parts, left to right, are: the mounting base, the body of the ionizing chamber with the two semi-circular repeller plates on either side, the first slit, the two focus plates (which are coplanar when assembled), and the cover plate which contains the third slit (not visible); the small spheres are sapphire insulating spacers. (c) The CEC Isatron, fully assembled; the filament and heater assembly, shown separately, mounts on the flat side of the Isatron and is covered by the shield shown at lower right. The gaseous sample is introduced through the axial port visible at the left. The length of the Isatron is about 3 cm.

(*Photos courtesy Consolidated Electrodynamics Corporation, Pasadena*)

signed to minimize variations in this initial energy. Figure 1 shows a representative ion source.

Ions can also be formed thermally or electrically. The requisite energy is supplied by local heating as with a laser, or by means of a radio-frequency spark or gas discharge. These methods make possible mass spectrometry of many kinds of solids, in spite of low vapor pressure. Combination of thermal and electron-bombardment excitation is often appropriate. An example is the Knudsen cell (Fig. 2), which consists of a crucible containing the sample: upon heating in vacuum molecules of the sample stream out through the narrow opening directly into the space traversed by the electron beam.

all ions will attain the same kinetic energy (for a given value of e), but this is only true if they start with negligible kinetic energy. The ionization chamber and associated circuitry must be carefully de-

SAMPLING SYSTEMS

If the sample is a gas or an easily vaporized liquid or solid, it can be introduced

Chemical Instrumentation

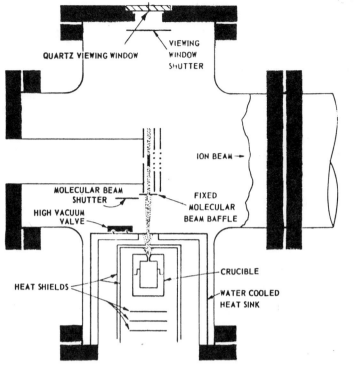

Figure 2. The Bendix Model 1030 Knudsen cell source. The crucible is heated electrically to as high as 2,500°C, and emits a molecular beam vertically upward into the ionizing chamber. The crucible temperature can be monitored with a thermocouple or, through the viewing window with an optical pyrometer.

(Courtesy Scientific Instruments Division of the Bendix Corporation, Cincinnati)

ucts of several manufacturers. The sample is introduced through either an ambient-temperature port outside the oven or a heated one inside. The first is provided with a 10/30 standard taper joint for attachment of interchangeable gas flasks. By appropriate manipulation of valves, the sample is admitted to the section of tube marked "sample volume" in the diagram, at a pressure measured by the external mercury manometer. This segment is then closed at the lower end and opened at the upper in such a way that the gas expands into the 5-l "expansion volume," at a pressure lower by a known factor, such as 0.002, depending on the ratio of volumes. Next a valve is opened admitting the sample to the mass spectrometer through a "molecular leak."

The molecular leak most commonly consists of one or more tiny needle holes in a thin gold membrane. The holes must be small compared to the mean free path of the gas molecules at expected pressures; a diameter of 0.01 mm is about right. This ensures conformity to Graham's law of effusion: The components of a mixture of gases will pass through the orifice inversely as the square-roots of their molecular weights. Since most of the molecules (those which escape ionization) follow the same law in leaving the ionizing chamber, the relative partial pressures will be the same in the chamber as in the expansion volume reservoir, an essential condition for quantitative analysis.

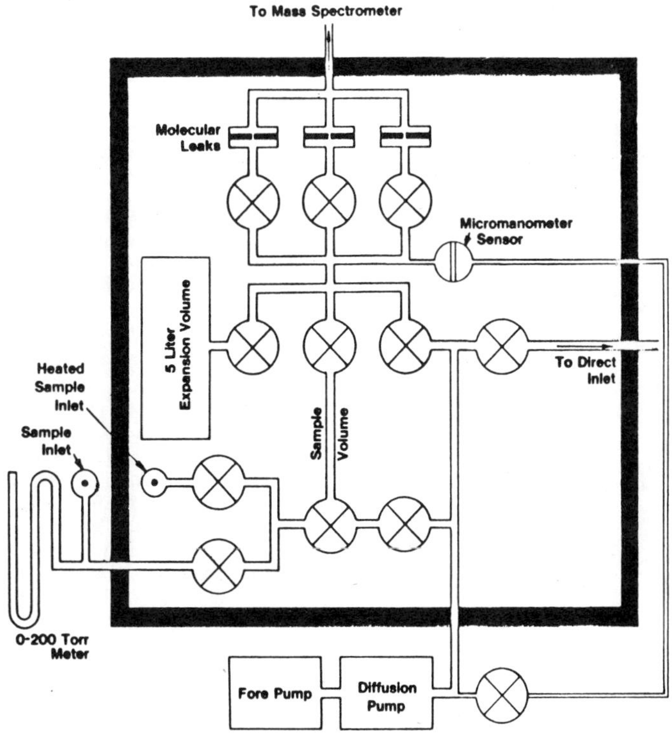

Figure 4 shows the all-glass inlet system provided by Consolidated Electrodynamics Corporation (CEC) for use with several models of mass spectrometer. The four black cylinders are encapsulated iron slugs which are manipulated by ring-shaped

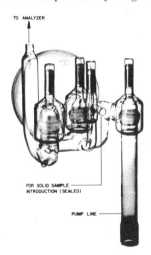

Figure 4. The CEC Model 21-084A glass inlet system. This is a four-valve system, somewhat simpler than that shown in Figure 3. This unit must be attached by welded glass to the molecular leak and to sampling valves.

(Courtesy Consolidated Electrodynamics Corporation, Pasadena)

permanent magnets (not shown) to operate the valves.

A rather different inlet system available from Micro-Tek is diagrammed in Figure 5. The sample is held in a small dipper or capsule at the end of a long stainless steel rod. With the ball valve *1* closed, the rod is inserted through the Teflon seal *2*, and the space surrounding it evacuated through a pump connected at *4*. The ball valve is then opened and the rod pushed through it and the second Teflon seal *3*, until it seats against the constriction at *5*, with the small sample container projecting into the side arm of the evacuated expansion chamber *7*. This vessel is maintained at an elevated

to the ionization chamber through a valving system such as those shown in Figures 3, 4, and 5. The Bendix assembly of Figure 3 is representative of the prod-

Figure 3. The Bendix Model 1071 heated inlet system. See text for detailed description. The three molecular leaks have slightly different diameters.

(Courtesy Scientific Instruments Division of the Bendix Corporation, Cincinnati)

Chemical Instrumentation

temperature in an oven, *6*, so that the sample immediately vaporizes. The tube *8* leads to the molecular leak and spectrometer. The valves at *9* and *11* permit evacuation of the system via diffusion and mechanical pumps connected at *13*. The fitting at *12* is intended for a pressure gage.

A similar probe can be inserted directly into the ionization chamber in many mass spectrometers, and is especially useful for solids and viscous liquids of low vapor pressure. Figure 6 shows such a probe arranged for laser evaporation of a refractory solid sample. The sample is manipulated until it is sharply in the field of the viewing microscope. The sliding diagonal mirror is then withdrawn which leaves the sample precisely aligned with the laser beam. The laser energy can be expected to fragment molecules of the vaporized sample, producing ions without the need of electron bombardment.

GAS CHROMATOGRAPHY

Although fractions can be collected for later examination, continuous sampling is highly desirable if the mass spectrometer is to monitor the effluent from a gas chromatograph (GC). A special consideration which becomes important in this service is the effect of the GC carrier gas. In order to obtain a sufficient pressure of an eluted sample in the mass spectrometer, relatively large quantities of carrier must be handled. If it is admitted to the spectrometer, it is apt to degrade the vacuum too greatly for proper action. Hence, it is preferable to remove the bulk of carrier prior to entry into the instrument. A separator for this purpose has been described by Watson and Biemann (*2*), as shown in Figure 7. The entire effluent from the GC is passed through a porous glass tube, 20 cm long by 4 mm id, 8 mm od, mounted in a jacket which is continually pumped. The helium carrier preferentially diffuses through the porous walls, allowing the sample (with residual helium) to pass to the mass spectrometer. The entrance and exit constrictions are intended to maintain the internal pressure low enough that effusion through the pores will be governed by Graham's law. Enrichment up to about 50% is reported. Several manufacturers of mass spectrometric equipment offer versions of this separator or its equivalent.

Another feature to consider in the GC-mass combination is the relation of the mass-scanning time to the GC elution times. It is desired to obtain the mass spectrum just as a particular component of the sample is entering the spectrometer, and this requires observation of some non-selective GC detector. One way to accomplish this is to split the GC effluent stream, one part to enter the mass spectrometer, while the other goes to a conventional GC detector. Another way is to arrange an auxiliary electrode in the mass spectrometer itself to intercept a fixed fraction of the ion beam before its dispersion. Figure 8 shows these relations. The curve represents a GC peak as seen either by the GC detector or the

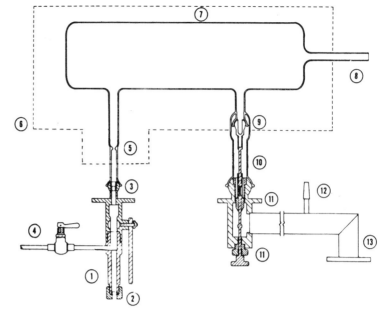

Figure 5. The Micro-Tek MS-7500 universal inlet system. See text for description.

(Courtesy Micro-Tek Division of Tracor, Inc., Baton Rouge)

ion-beam interseptor (here acting as a GC detector). If the scanning time of the mass spectrometer is short compared to the duration of the peak, then it can be

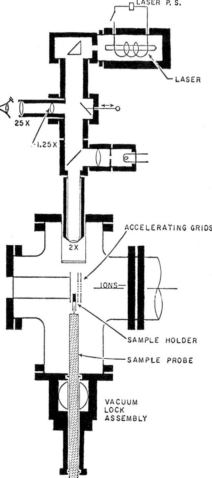

Figure 6. The Bendix laser-heated direct insertion system. The sample can be as large as about 9 mm square, and the vaporized spot only 0.01 mm diameter.

(Courtesy Scientific Instruments Division of the Bendix Corporation, Cincinnati)

triggered (time span 1) to coincide with the peak maximum. It may also be possible to run mass spectra at times 2 and 3

Figure 7. Helium separator, after Watson and Biemann (*2*).

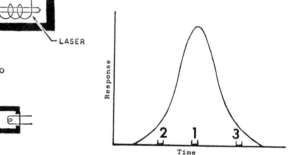

Figure 8. GC-Detector response versus time. The numbered intervals may be of interest for mass spectrometric sampling.

VACUUM REQUIREMENTS

The analyzer of a mass spectrometer operates under dynamic conditions. Sample gas is constantly entering and being removed. Hence, speed as well as ultimate vacuum must be considered in selecting a pumping system. The ultimate vacuum seldom need go below 10^{-7} torr. The speed of pumping depends largely on the geometry of the entire system, as well as on the kind of pump and temperature of the trap. Such considerations are too involved to detail here.

The conventional pumping system for a mass spectrometer consists of a diffusion pump backed with an oil-sealed mechanical roughing pump. The diffusion pump may use either mercury or oil. Mercury has the advantage that any of its vapor which finds its way into the ionization region of the spectrometer will only contribute easily recognized peaks corresponding to the mercury isotopes; also it is less subject to chemical attack if air is inadvertently admitted to contact with the

Chemical Instrumentation

hot vapor. However, many organic oils provide higher pumping speeds and lower ultimate pressures. A comparative discussion of fifteen pumping fluids has been presented by Rondeau (3). It should be noted that the silicone oils are not satisfactory for mass spectrometer application, as they produce a slow build-up of a siliceous deposit on the ion source slits, which is next to impossible to remove.

A point which may not be generally known is that the cooling water fed to the condenser of an oil-diffusion pump can be too cold. It will increase the viscosity of the condensed oil to the point where it will not flow back to the boiler fast enough. If this is likely to be a difficulty, the water should be prewarmed to room temperature.

A cold trap is nearly always required between the diffusion pump and the spectrometer. Liquid nitrogen is the coolant of choice, though a slurry of solid CO_2 with chloroform or other suitable liquid will serve in a pinch. A number of devices are available which will automate filling the trap with liquid nitrogen from a standard storage dewar. In the author's experience with a CEC 21-104 mass spectrometer, the trap holds enough nitrogen to last 24 hours, though the automatic filler adds smaller increments more often. A 4000 cu. ft. storage dewar lasts about 12 days.

The Sargent-Welch Scientific Co. manufactures a compressed-air operated cryogenic refrigerator which can replace the liquid-nitrogen trap, bolting onto the same flange in any standard vacuum system. This operates by expansion-cooling of air, and is stated to reach −140°C. It has been reported to be satisfactory for mass spectrometry, and is certainly more convenient.

Ion pumps, though more expensive, are sometimes substituted for diffusion pumps. They do not require either forepump or cold trap, except during starting procedures. Ion pumps are not ordinarily satisfactory for pumping helium and other noble gases, and so should not be relied upon for operation in connection with GC unless the manufacturer specifically approves such use.

Many mass spectrometers require more than one pumping system. It may be desirable to divorce the mass analyzer chamber from the pumps which remove the un-ionized sample from the ionization chamber; both will require high performance diffusion or ion pumps. Sample handling systems such as those of Figures 3 and 5 must be evacuated prior to introduction of a sample, and a separate diffusion pump is generally employed here without a cold trap. Any probe sample-introduction device (as in Figs. 5 and 6) must include a vacuum lock to prevent entry of air into the spectrometer; in some models a separate diffusion pump is provided for this purpose while in others a mechanical pump is sufficient.

An efficient system of interlocking pressure-operated switches is essential, to prevent heating the diffusion pumps when the forepressure is too high or when cooling water is not flowing.

Vacuum Gages. The high pressure side of the inlet system is commonly supplied with a closed-tube mercury manometer, about 150 torr full-scale (Fig. 3). The pressure in the expansion chamber may be measured precisely with a diaphragm micromanometer (sensor indicated in Fig. 3); this is a rather expensive accessory which is only needed for high precision quantitative analysis. Rugged gages, typically thermocouple or Pirani types, are needed to monitor the various mechanical forepumps. An ionization gage is appropriate for the main analyzer vacuum. It should have good sensitivity and a wide range (e.g., 10^{-4} to 10^{-8} torr), but need not be highly precise. It is this gage that is used for a continual check on the proper operation of the overall vacuum and gas-handling system.

Part 2.—Mass-Analyzer Systems; Detectors

The dispersion in mass spectrometers is achieved by combinations of electrical and magnetic fields. The resultant spectrum is ordered according to m/e, the mass-to-charge ratio, for the accelerated ions. Analyzer systems can be classed as *static*, where ions of a particular m/e value are brought to a focus by steady (as opposed to alternating) fields, and *dynamic*, in which AC fields play a key part.

STATIC METHODS

As pointed out in Part 1, all ions emerging from the ion gun have been given the same amount of energy, eV (Eqn. 1). However, in general the ions will have possessed varying initial kinetic energies, so that the beam is never entirely homogeneous at this point. The effect of this spread of energies can be nearly eliminated by passing the beam through a radial electric field, as in Figure 9. If we desig-

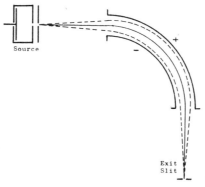

Figure 9. Radial electrostatic sector.

nate the strength of this field as E (volts per centimeter), we can write the equality,

$$eEr = mv^2 \qquad (2)$$

where r is the radius of curvature of the circular trajectory of ions with mass-to-charge ratio m/e moving with a velocity v. Hence,

$$r = mv^2/eE \qquad (3)$$

which shows that ions of a given kinetic energy will follow the same path, regardless of their initial kinetic energy spread. If the ions enter the radial field as a slightly

divergent beam, as shown, variation in the distance covered within the field will result in directional focusing on an exit slit. This radial electric field alone cannot produce dispersion in terms of mass (or m/e) but only in kinetic energy.

Passage of a beam of ions through a magnetic field also results in a circular trajectory (Fig. 10). The radius of the

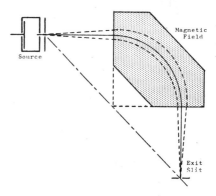

Figure 10. Magnetic sector analyzer.

path is given by

$$r = mv/eB \qquad (4)$$

where B is the magnetic field strength. In this case the radius is seen to be proportional to momentum rather than energy.

Magnetic Sector Spectrometers. Probably the most familiar type of mass spectrometer is the magnetic sector variety. The path of the ions is determined by a combination of the electrical accelerating potential, as in Eqn. 1, with the effect of the magnetic field as in Eqn. 4. Combining these two equations, with the elimination of v, results in the relation

$$r = \frac{\sqrt{2V}}{B} \cdot \sqrt{\frac{m}{e}} \qquad (5)$$

The sector spectrometer can be designed with any angle of magnetic sector up to 180°. Commercial instruments are available with sectors of 60, 90, and 180°. A restriction on design is that the slit from the ion source, the apex of the magnetic sector, and the exit slit must be colinear in order for the ions to be focused on the exit slit. The exactness of this restriction is dependent upon ions entering and leaving the magnetic field normal to its boundary. This is not completely achievable in practice due to edge effects of the magnetic field; hence the focusing of the ions is somewhat less than perfect. Usual practice is to allow a mechanical adjustment of the position of the magnet; exact alignment is determined by optimum response of the detector for ions of a given m/e.

One way to avoid edge effects in the magnetic field is to employ a 180° sector with the source of ions immersed in the magnetic field. This will cause some curvature of the trajectory of the ions before leaving the ion source itself, but this can be allowed for by careful design. Figure 11 shows a design of this type. Ions of a particular m/e value are brought to an excellent focus at the exit slit S. Ions of other m/e values will come to a focus at other points, such as S'. These can be swept across the exit slit by variation of either the accelerating voltage or the mag-

Chemical Instrumentation

netic field. Figure 12 shows that focus achieved by this method cannot be perfect. Trajectories described in a magnetic field

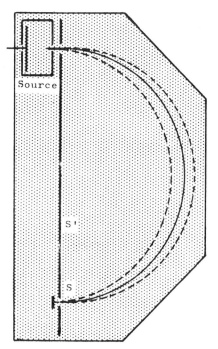

Figure 11. 180° magnetic analyzer. The source and the exit slit S are immersed in the magnetic field. S¹ indicates the region where other m/e ions may focus.

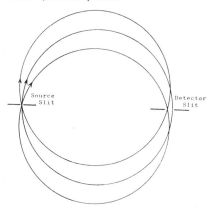

Figure 12. Trajectories of identical ions diverging from a slit in a magnetic field. The lower half would not be realized in an actual spectrometer.

in the absence of an electric field must be circular, but if they are, they cannot come together perfectly at an exit slit. This aberration is inherent in the method, and limits the permissible angle of divergence of the ion beam as it emerges from the source. The resolution is, however, considerably greater than with a simple sector spectrometer.

Cycloidal Spectrometers. In this style the magnetic and electric fields are superimposed, and the resultant trajectory follows a cycloidal path, as in Figure 13. It may be recalled that a cycloid is a curve described by a point on a circle which is rolling along a straight line. There are three types of cycloids, the *common* cycloid in which the moving point is on the cir-

cumference of the rolling circle, the *prolate* cycloid in which the point lies between the center and the circumference, and the *cur-*

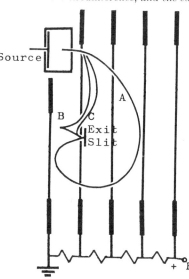

Figure 13. Cycloidal trajectories in crossed electric and magnetic fields. A, curtate, B, common, C, prolate cycloids.

tate cycloid in which the point lies on a radius extended beyond the circumference of the circle. It can be shown mathematically that ions of a given mass-to-charge ratio leaving the source slit at any angle will be brought to a common focus by the three types of cycloidal paths. The curtate geometry gives best results, and is almost invariably chosen. This is shown at *A* in the figure. The difficulty with the common and prolate cycloids is that the ions slow down in the vicinity of the cusp at *B*, and this produces a considerable space charge which tends to defocus the beam of ions. For precise focusing in a cycloidal spectrometer, both fields must be quite uniform. It is customary to increase the uniformity of the electric field by means of a stack of guard rings, which are shown in section in the figure. These are held at successive potentials by means of a voltage divider.

Double-Focus Spectrometers; Mattauch-Herzog Geometry. In this type of spectrometer the ions are passed through a radial electric field, like that of Figure 9, followed by a 90° magnetic sector. The overall geometry is shown in Figure 14. The angle subtended by the electric field is 31° 50', which is equal to $\pi(4\sqrt{2})$ radians. It can be shown that for this particular angle, ions of a given mass-to-charge ratio are collimated, or to put it another way, ions leaving the electric sector are dispersed according to energy (see Eqn. 3), and focused at infinity. These ions are now brought to a focus

simultaneously by the magnetic field, along a plane surface. This makes it possible to record all ions at one time on a photographic plate laid along this plane. Alternatively the spectrum can be scanned across an exit slit as in models previously considered. Hence this is both a spectrometer and a spectrograph. It has much greater resolution than any of the single-focus systems.

Nier-Johnson Geometry. This is another version of a double-focus mass spectrometer. The ions are led through a radial electric sector, usually 90°, followed by a magnetic sector which may be 60 or 90°. This differs from the Mattauch-Herzog design in that the ion beams as they leave the electric sector are convergent, so that an intermediate slit can be mounted between the electric and magnetic sectors. Ions of only one value of *m/e* are sharply in focus at any given combination of field strengths, hence this geometry is not suitable for a spectrograph. The resolution is comparable to that of the Mattauch-Herzog instrument. Details are shown in Figure 15.

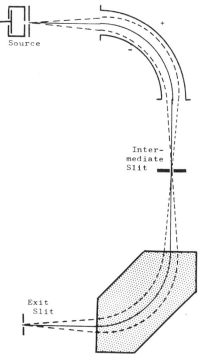

Figure 15. Nier-Johnson double-focus mass spectrometer.

DYNAMIC METHODS

The static mass analyzers discussed in the previous pages all employ focused ion beams, and hence require a slit source. The

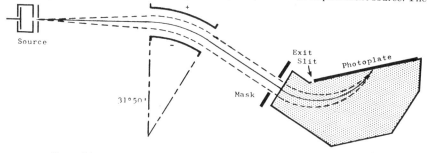

Figure 14. Mattauch-Herzog double-focus mass spectrometer-spectrograph.

Chemical Instrumentation

first two dynamic instruments to be described—the linear radio-frequency and time-of-flight designs—utilize a wide beam of ions, and direction focusing is not involved. Hence their ion guns differ in detail from the corresponding parts of static mass spectrometers, although using the same principles, usually electron bombardment. Slits may be replaced by screens or grids as elements for establishing electric field gradients.

Radio-Frequency Spectrometers. An assembly of grids arranged as in Figure 16 will permit selection of ions according to their velocities. This will be tantamount to m/e selection if the entering ions have uniform energy. Alternate grids are connected together to a steady potential; the other set of alternate grids are connected to a radio-frequency source of the order of 10 to 100 MHz. The ions which enter are either accelerated or retarded by the alternating field in accordance with the relation between the ion velocity and the frequency. Hence at a given frequency only ions of a particular velocity (and therefore mass) can get through. The field-free drift spaces permit further discrimination between the favored ions and those of nearby velocities which may have succeeded in getting past the first set of grids. Subsequent modulator stages can exert a greater selection between these resulting bunches of ions, and at the same time reject ions with harmonically related velocities. The electrode labeled "retarder" is given a positive voltage which permits passage only of those ions which have gained more than a specified fraction of the total available energy. The range of

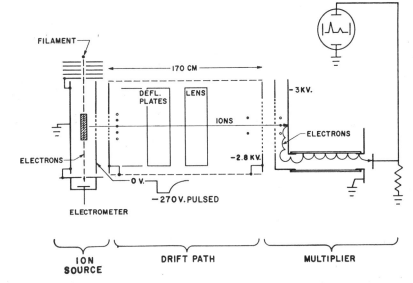

Figure 17.   Time-of-Flight mass spectrometer, courtesy Bendix Corporation.

masses can be scanned by variation either of the energy of the incoming ions or of the oscillator frequency.

Time-of-Flight Spectrometers. Figure 17 shows the design of an instrument in which ions are formed only during a very short interval of time, less than a microsecond typically, during which the electron beam is allowed to pass. Immediately following this ionization pulse, the first accelerating grid is given a pulse of negative charge which draws out the ions from the ionizing region and accelerates them. Since all ions have essentially the same energy at this point, their velocities will be inversely proportional to the square-roots of their masses. The ions are now allowed to move down the length of a field-free

space with whatever velocity they may have acquired. Hence the ions become separated into bunches according to their masses. The ions of low mass arrive at the detector earlier than those of higher mass. (The detector will be described later.) The pulse repetition rate in the time-of-flight spectrometer can be of the order of 10 kHz, which means that the complete mass spectrum of a sample can be repeated 10,000 times in one second. In view of this speed, the only practicable read-out device is a cathode-ray oscilloscope at the output of the detector. The time-sweep of the oscilloscope must be synchronized with the pulse repetition rate of the spectrometer. The high speed makes this spectrometer particularly appropriate for the study of rapidly changing kinetic systems and for observation of samples volatilized by a single pulse from a laser.

Quadrupole Spectrometers. The quadrupole mass spectrometer is an instrument in which electromagnetic fields interact in such a way that only ions of one particular m/e ratio can get through to the detector. Hence they are often referred to as *mass filters* rather than mass spectrometers. The analyzer consists of four parallel cylindrical rods arranged as shown in cross section in Figure 18. Diagonally opposed rods A,D and B,C are connected together electrically. Both DC and AC potentials (frequency of the order of 500 kHz) are applied between the two pairs as shown and a beam of ions is allowed to pass down the central axis.

Figure 19 shows a typical design of quadrupole mass spectrometer. The ion gun at the left is essentially similar to those previously considered except that the ions are pulled out through a circular orifice rather than a slit. After entering the quadrupole mass filter, they are no longer subject to acceleration in the forward direction, but respond in a complicated manner to the interacting direct and alternating fields. Only those ions with a very sharply defined m/e ratio are able to proceed from the entrance orifice to the detector. Others describe a divergent trajectory and eventually are lost by contact with one of the rods. This may be made plausible by a rough analogy recently suggested by

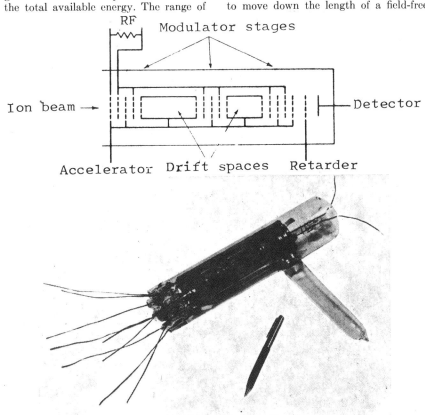

Figure 16.   Linear radio-frequency mass spectrometer.   (a) schematic, (b) photograph, courtesy General Electric Company.

Chemical Instrumentation

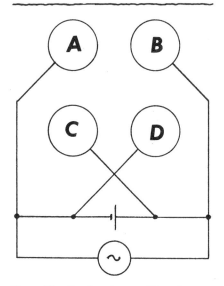

Figure 18. Quadrupole mass filter, diagrammatic cross section.

Dawson and Whetten (4), wherein the dynamic focusing of RF quadrupole fields is likened to "the case of a ball on a saddle, with the sides of the saddle sloping downward and the front and back sloping upward. The ball may tend to roll down the sides, but half a cycle of the RF voltage later, the sides slope upward and the front and back of the saddle slope downward. With the proper choice of frequency and slope, the ball will oscillate in a complicated manner, but may remain in the saddle."

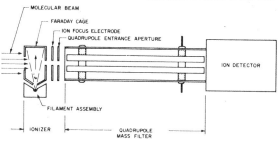

Figure 19. Quadrupole mass spectrometer, diagrammatic, courtesy Ultek Div., Perkin-Elmer Corporation.

A mathematical description of the ion trajectories requires second-order differential equations, which can be found summarized in either of the general references. The mass range, which may be from zero to 300 or 800, is scanned either by variation of the frequency or the voltage. If the voltage is varied, the ratio of DC to RF voltages must be held constant for optimum trade-off of resolution against sensitivity.

Quadrupole filters have been constructed with rods several meters in length, but commercial units are restricted to rod lengths of about 5 to 20 cm. Resolution and mass range are comparable with simple magnetic sector spectrometers, but inferior to 180°-deflection immersed-source designs and double-focus instruments. The smaller quadrupole filters are frequently used in determining the partial pressures of gases remaining in evacuated systems, and hence are called *residual gas analyzers*.

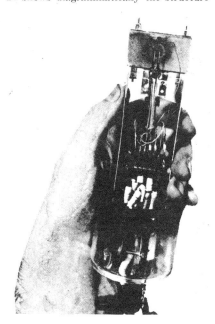

Figure 20. CEC 21-440 Quadrupole residual gas analyzer, courtesy Consolidated Electrodynamics Corporation.

Figure 20 shows a commercial quadrupole mass spectrometer with the analyzing system exposed. This is intended to be made part of a user's vacuum system.

The symmetry of the quadrupole spectrometer indicates that there exist two mutually perpendicular planes between the rods, which are at zero potential, both DC and AC. It is possible to take advantage of this fact to construct a filter consisting of a single rod and a V-shaped trough which is at ground potential. The single rod is given both DC and RF potentials. A mass filter constructed this way is called a *monopole* filter. It is less convenient to use than the quadrupole, but the requirements of its power supply are not as severe.

Omegatron Spectrometers. The omegatron is a small unit which operates on the principle of the cyclotron ion accelerator. The ions are formed by electron collision near the center of a box-like electrode structure, totally immersed in a magnetic field, and subject also to a radio-frequency electric field. Ions for which the mass-to-charge ratio is equal to the ratio of the magnetic field strength to the frequency, resonate in such a manner that they follow an Archimedes spiral trajectory of ever-increasing radius until they collide with an ion collector. Nonresonant ions

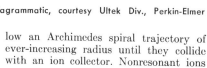

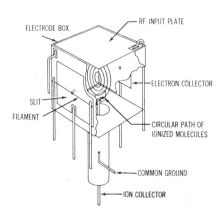

Figure 21. Omegatron, schematic, courtesy Sloan Instruments Corporation.

also orbit around the ionizing source, but the orbit does not expand, so they cannot be intercepted by the collector.

Commercial omegatrons are available as small units intended to be connected to vacuum systems for studying residual gases. They usually employ a permanent magnet up to about 5 kG., a frequency from 150 to 4000 kHz, and may cover a mass range as high as 50 or more. Figure 21 shows diagrammatically the structure

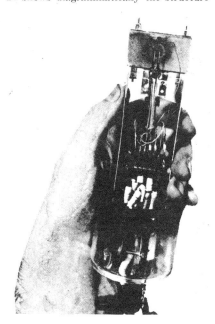

Figure 22. Omegatron mass spectrometer, courtesy General Electric Company.

of an omegatron, and Figure 22 a photograph of a unit.

RESOLUTION AND PERFORMANCE

The performance of mass spectrometers can be rated in terms of mass range, resolution, and sensitivity. The *mass range* is generally stated to extend from some small value, such as 1 or 10, to the maximum for which the instrument is designed. This extent may be coverable in a single scan, or may require two or more overlapping scans. The sector instrument in the author's laboratory, for example, can scan electrically from 12 to 165 *m/e* at one magnet current, from 145 to 2000 at another.

The *resolution* is expressed in different terms by different manufacturers, sometimes even by the same manufacturer describing different models. This variation reflects varying convenience and tradition with respect to instruments for different types of service, but makes direct comparison awkward. Table I lists the most widely used criteria of resolution and some factors for their interconversion. These factors can only be considered approximate, because they depend on the contour of the mass peaks.

The *sensitivity* is also specified in a variety of ways, such as beam current per unit partial pressure, minimum sample size in weight units, or in atom-per cent, or electrical charge accumulated at a specified *m/e* value for a reference sample. These are not interchangeable.

Table II gives a rough comparison of range and resolution (by the 10% valley

Chemical Instrumentation

criterion) attainable by spectrometers of each major type. These data represent a composite from manufacturers' literature of the most sophisticated instruments of each class. Numerous less expensive instruments are of course available with relaxed specifications.

DETECTORS

Photography. The very early instruments (J. J. Thomson, F. W. Aston) employed a photographic plate for detection. Photography was discarded in favor of electrical detection when scanning instruments were developed, and only recently has returned to life with the Mattauch-Herzog double-focus spectrographs. Although somewhat more cumbersome, it can give greater resolution than an electrical detector on the same instrument. Because the photographic plate is a time-integrating device, it can provide the highest sensitivity of any detector.

Charge detectors. If the ion beam is allowed to impinge on an insulated conductor, the charge which accumulates can be measured by an electrometer. In modern instruments the electrometer consists of a preamplifier with either an electrometer vacuum tube or a field-effect transistor in direct connection with the collecting plate.

The collector is known as a Faraday cup, as it is the direct descendent of the ice-bucket with which Faraday carried out basic experiments in electrostatics. It is usually cup shaped to minimize the escape of any electrons emitted as the result of ion bombardment.

Multipliers. More sensitive by perhaps a thousandfold is an electron multiplier. In one form this is essentially similar to the familiar photomultiplier used as detector of ultraviolet and visible light, except that the primary cathode is optimized for response to ions rather than photons. Each incident ion releases typically two or three electrons which are accelerated to a dynode, where each liberates several more electrons, and so through ten or fifteen stages. Figure 23 shows the structure of a 16-stage electron multiplier.

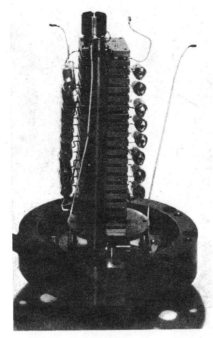

Figure 23. 16-Stage electron multiplier detector, courtesy Consolidated Electrodynamics Corporation.

Bendix has developed a magnetic electron multiplier especially for use with their time-of-flight spectrometers (included in Fig. 17). Secondary electrons are constrained by a magnetic field to follow circular paths causing them to hit the same electrode from which they were emitted, but at a different point. Successively emitted electrons then move along by a series of semicircular jumps. Most of the path lies between a pair of glass plates coated with a high resistance metallic film. An electric gradient is impressed along the lengths of these plates to produce the needed acceleration. The magnetic field is produced by a number of small permanent magnets.

Commercial Spectrometers

The following descriptions are arranged alphabetically by companies. Special-purpose instruments, such as leak detectors, scarcely applicable to mass spectrometry, are not included. Prices are given in thousands of dollars, and are only approximate, varying with selection of optional accessories.

Aero Vac Corporation

This company offers as their most advanced mass spectrometer the Model 686, which is optimized for high sensitivity in trace gas analysis where only moderate resolution is required. It is illustrated in Figure 24, with an accessory gas chromatograph inlet system. The analyzer itself includes two magnetic sectors in tandem, oriented to deflect the beam in opposite directions. This is followed by an electrostatic energy filter and an electron multiplier detector. The mass range is 1 to 500 amu, with unit resolution of 250 (1% cross contribution). The detection limit is quoted to be better than 2 parts in 10^8 for gases of less than 50 molecular weight present as impurities in argon ($27).

Figure 24. Aero Vac Model 686 Mass Spectrometer Trace Gas Analyzer. The left-hand portions are the mass spectrometer proper; the valves and electrical controls for a gas chromatograph inlet system appear at the right.

A similar instrument, the Model 685, can be had at lower cost and with relaxed specifications (2 to 300 amu, resolution 200; $16). The Model 610-611 is a desktop, 60° permanent-magnet sector, spectrometer with a mass range to 300 amu (500 optional), and a resolution of 200–250 ($6, without vacuum system).

Aero Vac also offers several mass analyzers intended for process control, and a series of residual gas analyzers.

Avco

The Avco Model 90000 mass spectrometer is a 90°-deflection magnetic sector instrument with resolution of better than

Table I. Definitions of Resolution

Resolution is $M/\Delta M$, where M is mass number (m/e), and where ΔM may be:

(1) The separation between two equal peaks such that the valley between them does not exceed 10% of the height of either;
(2) The separation between two equal peaks such that the valley between them does not exceed 1% of the height of either;
(3) The width of a peak measured at 50% of its height;
(4) The width of a peak measured at 5% of its height;
(5) The separation between two equal peaks such that neither contributes more than 1% of the other at its maximum.

If the resolution according to these definitions be designated as R_1, R_2, etc., then approximate conversions may be made as follows:

Peak Shape:	Gaussian	Triangular	Trapezoidal[a]
R_1/R_3	0.48	0.53	0.77
R_2/R_3	0.36	0.50	0.75
R_5/R_3	0.78	1.00	1.50
R_1/R_4	1.00	1.00	1.00

[a] Bottom width = twice the top width.

Table II. Comparison of Mass Spectrometers

	Max. Mass (m/e)	Resolution (R_1)
Magnetic sector	2000	5000
Double focus (M-H)	2500	40,000
Double focus (N-J)	7200	70,000
Cycloidal	2000	2000
Time-of-flight	1200	400
Quadrupole	750	450
Omegatron	140	630

296

Chemical Instrumentation

6000 (10% valley definition). Ions can be observed with m/e as high as 10,000. The analyzer is unique in that the ion beam enters the magnetic field at a small angle (12°) from normal, which permits more flexibility in focus through mechanical adjustments in the position of the magnet. The field is accurately programmed with the aid of a Hall-effect probe and amplifier with the magnet windings in its feedback loop. Avco specializes in custom design of spectrometer systems built around the basic unit, but with choice of a variety of sources, detectors, and other features. A data analysis system (Model 91000) is available which ties the mass spectrometer to an Avco PCU S-1200 digital computer, so that the computer controls the operation of the spectrometer as well as processing the acquired data.

Bendix Corporation

A number of models of time-of-flight spectrometers are available. One of these, the Model 3012, a large research instrument, is shown in Figure 25. It covers a range of 0 to 1200 amu, with a resolution

Figure 25. Bendix Model 3012 Time-of-Flight Mass Spectrometer. The heated inlet system of Figure 3 appears at the left with a micromanometer. The vacuum system is controlled from the panel beneath the exposed spectrometer tube. The electronic controls are located in the main sloping panel. The smaller sloping unit to the right contains the controls for the Knudsen source. The read-out oscilloscope can be seen at the extreme right.

of 500 (1% contribution). The ion flight path is 200 cm long, leading to a magnetic electron multiplier detector. An optional gating feature (peak eliminator) permits attenuating the signal due to any selected m/e species by a factor of 10^4. This can be used to prevent overloading the detector, as with the helium carrier in gas chromatography ($36).

The Model 3015 is similar but has provision for observing negative as well as positive ions. Another modification incorporates a laser microprobe source (see Fig. 6).

The Model MA-1, shown in Figure 26, is representative of several smaller time-of-flight spectrometers. It covers the range 0 to 500 amu with unit resolution of 200 ($10).

Consolidated Electrodynamics (Bell & Howell)

This company, frequently known by its initials CEC, has one of the more extensive lines of mass spectrometers manufactured in America. The most elaborate of these is Model 21-110, a double-focus mass spec-

Figure 26. Bendix Model MA-1 Time-of-Flight Mass Spectrometer. The vacuum controls are beneath the table, the electronics above.

trometer and spectrograph with Mattauch-Herzog geometry (Fig. 27). It offers sample-handling facilities for gases, liquids, and solids. A variety of ion sources can be employed, including electron bombardment, RF spark, surface ionization, and Knudsen cell. Detection is by photoplate, electrometer, or electron multiplier. Resolution with the gas source is at least 30,000 (10% valley definition), with the spark source, 2000 or better. The mass range extends to 2500 amu ($100–120).

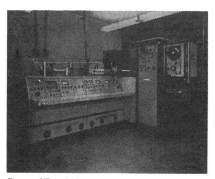

Figure 27. CEC Model 21-110 Double-Focus Mass Spectrometer. The multiple source unit is seen at the left-hand end of the spectrometer tube; the electric sector appears just over the fourth electronic panel; the magnetic sector is next, followed by the multiplier detector and (behind it) the photoplate loading mechanism. The racks to the right carry data-processing and recording equipment.

The CEC Model 21-104 (Fig. 28) is a

Figure 28. CEC Model 21-104 Mass Spectrometer. The batch inlet system stands at the left next to the magnet. The cylindrical unit projecting to the right of the magnet houses the electrometer preamplifier.

high-precision, general-purpose instrument with a 180° magnetic sector. It covers the range up to 2000 amu. The resolution is normally 600 (10% valley), but can be increased to about 2500 by reducing the width of the collector slit, with a concomitant reduction in sensitivity ($40–60).

Figure 29 shows the CEC Model 21-490, a 90° magnetic-sector spectrometer. This instrument features a high degree of compactness. Its electronic circuits are packaged in easily replaceable function

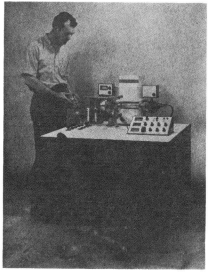

Figure 29. CEC Model 21-490 Mass Spectrometer. The source is to the left of the magnet, the detector to the right; recorder not shown.

modules. The mass range is 12 to 1000 amu, with unit resolution up to 1000 (10% valley; $18). An electric sector can be purchased as a kit, to convert to the 21-491, a double-focus version ($23). This increases the resolution to about 25,000. These models are intended for general organic and allied applications including gas chromatography.

The CEC Model 21-703 spectrometer shown in Figure 30 is a 60° magnetic-sector instrument designed for use in modern high-temperature chemistry and for isotopic analysis. The sources usually employed are of the surface ionization or Knudsen cell types. Masses 2 to 1000 amu can be covered; nominal unit resolution is 600 (10% valley; $90–100).

Figure 30. CEC Model 21-703 Mass Spectrometer. The source housing and the magnet can be seen above the pump system at the left, and the electron multiplier high up above the magnet. The arched structure at the top is a bridge to carry cables from one unit to the other.

CEC Model 21-621 is a versatile spectrometer built on a modular basis which

Chemical Instrumentation

permits over 100 different combinations. The analyzer can be either cycloidal or quadrupole. This is a general-purpose instrument adaptable to such uses as isotope ratios in gases, gas impurities, blood and breath gases, and engine exhaust analysis ($8–20).

The CEC Model 21-614 is a dual-mode spectrometer for residual gas analysis. To cover the mass range 12 to 200 amu it operates on the cycloidal principle, but for masses 2 to 11 the electric field is turned off and the spectrometer operates as a 180° deflection sector instrument. It is designed for use in research on electronic vacuum tubes, vacuum monitoring, atmospheric analysis, etc. ($8). The Model 21-615 is a simpler version which covers masses 2 to 100 amu in a single scan of the cycloidal analyzer. ($3.5) Unit resolution for the 21-614 is 150, for the 21-615, 50 (1% contribution). Model 21-440 is a comparable residual gas analyzer with quadrupole operation, mass range 1 to 300, unit resolution twice the mass number ($5).

Electronic Associates, Inc. (EAI)

EAI offers two models of quadrupole spectrometers. The Quad-300 (Fig. 31) is a versatile laboratory instrument with mass range extending to 800 amu, and with resolution twice the mass number up to 500. The scan can be executed automatically at various speeds up to 0.1 sec, fast enough for many kinetic applications. Read-out can be by oscilloscope, oscillograph, pen recorder, or direct connection to a digital computer ($20). The Quad-160 is an abridged version which covers the range 1 to 300 amu ($14).

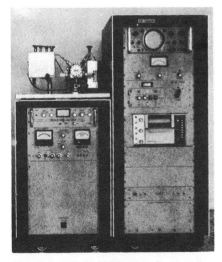

Figure 31. EAI Quad-300 Mass Spectrometer. The square box with four valves is a stainless-steel inlet system. The circular plate adjoining it covers the actual inlet orifice. The quadrupole rods, not visible, extend horizontally back behind the cover plate. Both oscilloscope and oscillograph can be seen in the larger electronic rack.

EAI also furnishes two residual gas analyzers which are in effect the above spectrometers supplied without a sampling system and vacuum pumps, to be installed in a customer's vacuum system. These are, respectively, the Quad-250A and Quad-150A.

Finnigan Instruments Corporation

The Finnigan Model 1015 (Fig. 32) is a laboratory quadrupole mass spectrometer with high sensitivity up to mass 750. It operates in three ranges, 1–100, 10–250, and 50–750 amu. Resolution is twice the mass number. Scan times as small as 500 microseconds are possible. The electron multiplier features replaceable dynodes. In gas chromatographic applications, full sensitivity is obtained without the need of a helium separator ($20).

Figure 32. Finnigan 1015 Laboratory Mass Spectrometer. The electronic console is to the left, the vacuum console to the right.

JEOLCO

Figure 33 shows the type of construction adopted by Japan Electron Optics. All components are exposed to view, which makes for convenience in servicing. Operator's controls are collected in one area. The model shown, JMS-01SG, follows the Mattauch-Herzog design, with a mass range of 1 to 3200 amu for photographic detection, 1 to 2900 for electrical detection. The unit resolution is 20,000 (10% valley). The JMS-01SC is similar, but with resolution of 35,000; the mass range is 1–4400 (electrical detection) and 1–3200 (photographic). It can be fitted with an RF spark source for trace element analysis. Model JMS-01B(M) is a Mattauch-Herzog instrument restricted to a spark

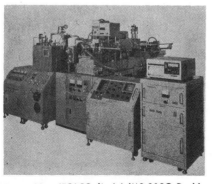

Figure 33. JEOLCO Model JMS-01SG Double Focus Mass Spectrometer. The sampling system is on the console to the immediate left of the instrument proper. The electron-bombardment ion source is in the center foreground. The photoplate loading mechanism at the far end is operated by means of a hand wheel from the front.

source and photographic detection; mass range 1 to 2000, resolution 8000 or more.

Johnston Laboratories, Inc.

Figure 34 shows the external appearance of a unique instrument called a Coincidence Mass Spectrometer. It consists of a time-of-flight spectrometer in which secondary electrons produced in the ionizing events are pulled out of the ionizing beam in the opposite direction from the flight path for positive ions. Then if the rate of production of ions is not too great (less than 10^4 ions per second) each individual ion and its associated electron can be detected separately. The time taken by the electron to reach its detector is negligible compared to the time required for the ion to traverse the 100-cm drift tube to its detector. Hence the elapsed time between signals from the two detectors will be a measure of the velocity of the ion as accelerated by an applied potential, hence a measure of its m/e ratio. This time lag is converted to pulse amplitude by a time-to-pulse height converter and fed to a conventional pulse-height analyzer. This allows simultaneous accumulation of the entire mass spectrum, or any portion of it, with continuous visual display on an oscilloscope of the spectrum as it accumulates.

This spectrometer is well suited to measurements of extremely tenuous samples or of low-efficiency ionizing processes. For

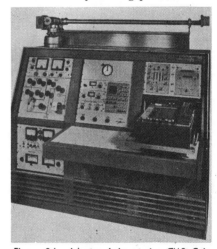

Figure 34. Johnston Laboratories CMS Coincidence Mass Spectrometer. The ionization chamber is near the left end of the 1-meter drift tube. At the far left is the detector of secondary electrons, and at the far right a similar detector of positive ions. The inset unit in the center panel is a 400-channel pulse-height analyzer.

example, measurements on the ionization of molecules by x-rays have been carried out at signal levels as low as a few molecules ionized per hour. The mass range is specified as 1–500 amu with a resolution of 250 (10% valley; $100).

LKB Instruments, Inc.

The LKB-9000 is an integrated combination of gas chromatograph and 60°-sector mass spectrometer (Fig. 35). The GC section is provided with temperature control for isothermal or programmed operation, 25–350°C. It is connected to the mass analyzer through a patented helium separator which gives a hundredfold en-

Chemical Instrumentation

richment. Provision is made for introducing samples to the mass spectrometer without the use of the chromatograph. Masses can be observed up to 1200 amu with unit resolution of 1000 (10% valley; $45).

The LKB-9000 has several accessories of special interest. One of these is a mass marker, for which a Hall-effect probe is mounted in the field of the electromagnet. The output voltage from the probe is amplified, squared, and digitized, and the resulting signal used to imprint mass calibration marks on the oscillograph paper. A second accessory is a peak matcher, a cathode-ray oscilloscope with associated decade resistances, which permit two peaks to be brought into apparent superposition in such a way that the ratio of their masses

Figure 35. LKB-9000 Gas Chromatograph Mass Spectrometer. The analyzer unit is to the left, the peak matching accessory in the center, the operating console next, and at the far right, the magnetic tape recording unit.

can be determined precisely. This enables one to establish m/e values to better than 10 ppm. Another accessory is an accelerating-voltage alternator. This unit switches the accelerating voltage rapidly between two or three mass numbers (within 10% of each other). This is particularly useful in the simultaneous observation of two or three species unresolved by the gas chromatograph. A magnetic tape data acquisition system is also available.

Nuclide Corporation

In addition to custom-designing mass spectrometers, Nuclide offers several standard models. The 12-90-G is a general-purpose spectrometer with a 90° magnetic sector which covers the range 1–3500 amu with resolution of 1000 (1% contribution; $40–60). Model 6-60-G is similar, but with a smaller radius magnet of 60° angle. Its range extends to 2000 amu, resolution 1500 ($35–45). Both of these models use many components in common and may be assembled in many configurations. A time-of-flight spectrometer is also available, as Model TOF-1. Its range is 150 amu, resolution 60 ($14). Nuclide lists a variety of other special-purpose spectrometers.

Perkin-Elmer Corporation

P-E has one model (270) made in their Connecticut plant, and several (designations starting with RM) imported from Hitachi in Japan. The Model 270 (Fig. 36) is a Nier-Johnson instrument designed specifically for gas chromatographic use. A modified version of the P-E Model 900 Gas Chromatograph is built in as an

integral part of the 270, but can be bypassed for conventional introduction of liquid and solid samples into the ionization chamber. The mass range extends to 3300 amu with unit resolution 850 (10% valley; $30).

Figure 36. Perkin-Elmer Model 270 GC-DF Analytical Mass Spectrometer. The left-hand part of the console consists of a Model 900 Gas Chromatograph and its controls.

Figure 37 shows the Model RMU-6, typical in general appearance of the many Hitachi-Perkin-Elmer models. It uses a 90°, single-focus, magnetic sector, and will scan to 2400 amu in seven ranges. Unit resolution is quoted as 2000 with standard slits and 6000 with a special slit system,

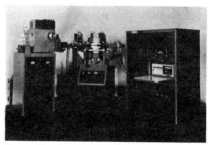

Figure 37. Hitachi Perkin-Elmer Model RMU-6 Mass Spectrometer, with heated gas inlet system.

both by the 50% valley criterion. A number of variations can be made on this basic unit, including surface ionization and Knudsen cell sources, and double detection for isotope ratio measurements ($40–60). The RMU-7 is a Nier-Johnson double-focus model in the same series as the RMU-6; it covers masses to 2400 amu with unit resolution 20,000 (50% valley; $70–100).

Model RMH-2 presents the highest performance available from this company. It is also a Nier-Johnson type, and covers the range of masses to 4800 amu with resolution of 70,000 (10% valley; $125).

Model RMS-4 represents a versatile economy model with a 60° magnetic sector. The range is 1–1200 amu. With the standard slits the resolution is 750; with optional variable slits, this can be increased to 1100 (10% valley; $25).

Perkin-Elmer has recently announced the availability of the DDA-1 Digital Data Acquisition system for direct, high-speed digitization of mass spectral information with computer-compatible magnetic tape. It can be used with the RMU-6 series of spectrometers or other comparable instruments.

Picker Nuclear—AEI

Picker offers five basic spectrometers,

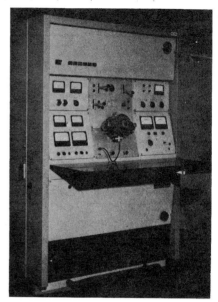

Figure 38. Picker-AEI Model MS-702 Mass Spectrometer. A viewing port immediately behind the ion source occupies the center of the console. An additional cabinet, not shown, contains power supplies and other ancillary electronic equipment.

all manufactures in England by AEI. Model MS-702 (formerly MS-7) is a Mattauch-Herzog instrument (Fig. 38), rather smaller and more compact than usual for this geometry. It can observe masses to 3250 amu, with unit resolution of 3000 for the spark source or 20,000 for electron bombardment. Detection is either photographic or by electron multiplier ($85–115). Figure 39 is a typical photographic record, exposed through a rotating step sector to give a series of graded exposures.

Model MS-902 (formerly MS-9) is a double-focus spectrometer with the Nier-Johnson configuration (Fig. 40). It covers masses up to 7200 amu, and has a guaranteed resolution of 70,000 (10% valley), though a resolution approaching 90,000 has been achieved ($90–125).

The MS-12 (Fig. 41) is a high-resolution, single-focus design with a 90° magnetic sector, which will cover masses to 900 amu with a resolution of 5000 (10% valley). It is particularly suited to use with a gas chromatograph and for organic identification ($50–75).

The Minimass® (Fig. 42) is a residual gas analyzer which can cover masses to 240 amu with resolution of 20 (10% valley). It uses a 180° magnetic sector. The analyzer housing can be either glass or metal and thus can be affixed permanently to any vacuum system ($3).

A data acquisition and analysis system (MSDS-1) has been devised to accept data from the MS-902 and feed it through suitable interfacing into a PDP-8 digital computer (Digital Equipment Corp. Maynard, Mass. 01754). The accompanying programs permit determination of elemental compositions and intensities, as well as precise peak matching.

Varian Associates

The Model M-66 Mass Spectrometer illustrated in Fig. 43 is a cycloidal instru-

Chemical Instrumentation

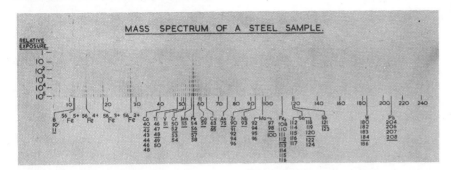

Figure 39. A typical mass spectrum recorded on the Picker-AEI MS-702. The steel sample was energized by a high-voltage spark. Some of the lines indicated are too faint to show up in this reproduction of the spectrum, but were visible on the original plate.

Figure 40. Console of Picker-AEI MS-902. The six uniform knobs at the left, together with the oscilloscope, permit precise determination of mass ratios.

Figure 41. Picker-AEI Model MS-12 Mass Spectrometer. The oven for the inlet system can be seen on the further module.

Figure 42. Picker-AEI Minimass® spectrometer or residual gas analzer.

Figure 43. Varian Model M-66 Mass Spectrometer. The analyzer is located in the right-hand console.

ment intended for routine organic chemistry including gas chromatography. Its mass range extends to 2000 amu, with a resolution of better than 4000 (50% peak width). In the design of this spectrometer, particular attention was given to the ability to determine unequivocally the nominal mass of every peak. A unique feedback control utilizing a pair of Hall-effect elements in tandem, forces the

Figure 44. Varian Syratron® Mass Spectrometer. The analyzer is in the unit to the left of the operator's chair.

magnet field current to assume the correct value to ensure exact proportionality between the position of the recorder pen along the X-coordinate and the corresponding m/e value. The sweep may be set at will to any of 12 full-scale values from 0.5 to 500 amu, and the scan can be started at any desired mass number. Nominal masses can be determined within ±0.2 amu after daily calibration on the 173 peak of bromoform. Precise mass determination is possible to 10 ppm with a reference peak on the same scan ($46).

Figure 44 shows the Varian Syratron® Mass Spectrometer, a modification of the omegatron described previously. Its range is restricted to 280 amu with a resolution of 700 (1400 on lower ranges; by 50% width definition). In this instrument ions are accelerated by crossed magnetic and RF electric fields, into a spiral trajectory, as in the omegatron. However, instead of detecting those ions which are intercepted by a collector, in the Syratron the absorption of radio-frequency energy by the resonant ions is observed. The Syratron finds application principally in studies of kinetics of ionic reactions ($50).

Varian also imports a line of mass spectrometers (and other instruments) from Atlas-MAT in Germany. Of these, the Model CH-5 is shown in Figure 45. It is a single-focus, 90° magnetic sector design for general organic applications. Its range extends to 3600 amu and its resolving power is 10,000 (50% width; $50–80).

The Varian-MAT CH-7 is a lesser version, similar in design to the CH-5. Its maximum range is the same (3600 amu) but the resolution is only 2500 (10% valley; $25–40). A Mattauch-Herzog spectrograph is available as the SM-1-B,

Figure 45. Varian-MAT Model CH-5 Mass Spectrometer. The recorder is to the left of the operator.

with a range to 2500 amu, resolution greater than 30,000 (10% valley; $90–150). A low-resolution spectrometer, the GD-150, uses a 180° magnetic sector; its maximum range extends to 600 amu, resolution to 300 (10% valley; $10–30).

Mass Spectrometers

Supplement

Even in the short time since this article was written (late 1968), there have been a number of significant developments, both in instrumentation and in application.

Chemical Ionization

First described in 1966, the gas-phase chemical production of ions in the source chamber of a mass spectrometer was brought to general attention in a review article in 1968 (5). Suppose one wishes to obtain a mass spectrum of n-$C_{18}H_{38}$. Conventional electron impact will give mostly fragments with 4, 5, and 6 carbon atoms; ions with 17 or 18 carbons do not appear. However, mixing the sample with a large excess (perhaps $10^4:1$) of methane will change the pattern profoundly. Now the principle ion is $C_{17}H_{37}^+$, with smaller amounts of ions with successively fewer carbon atoms. Also found will be CH_5^+ and $C_2H_5^+$ from the methane.

Clearly the probability of octadecane molecules being hit by electrons in the primary ionization process is far less than of methane molecules being hit. Most of the observed ions of high mass must therefore result from ion-molecule reactions between, for example, CH_5^+ and n-$C_{18}H_{38}$. The methane serves as a source of reactant ions.

Mass spectrometric observations on ions produced in high-pressure (1 torr) methane alone, show considerable amounts of CH_5^+ and $C_2H_5^+$ ions which together represent almost 90% of the total ions present. The remainder is made up of $C_2H_3^+$, $C_2H_4^+$, $C_3H_5^+$, and $C_3H_7^+$. (In low-pressure methane none of these is significant, and we find only CH_4^+, CH_3^+, etc.) These ions result from such reactions as

$$CH_4^+ + CH_4 \rightarrow CH_5^+ + CH_3$$

$$CH_3^+ + CH_4 \rightarrow C_2H_5^+ + H_2$$

In the presence of octadecane, the predominant secondary reactions are H⁻ abstraction processes such as

$$CH_5^+ + n\text{-}C_{18}H_{38}^+ \rightarrow \\ C_{18}H_{37}^+ + CH_4 + H_2$$

Thus the mass spectra from chemical ionization are markedly different from conventional spectra, and produce a different kind of information about ionization processes. For details, consult reference 5. The structure of the source chamber for chemical ionization must be specially designed to operate at the high pressure of reactant-generating gas, about 1 torr. The vacuum system, also, must be of special design to maintain the analyzer at the low pressure necessary for sharp focus and resolution. At the time of writing (summer of 1970) no instruments are marketed for this service, but it is believed that several manufacturers are preparing to enter the field.

Computer Interfacing

It has become increasingly evident that to obtain the maximum yield of information from a mass spectrometer, particularly a high-resolution type, a high-speed digital computer is a necessity. There are several ways in which "computerization" can be achieved. One of the simplest utilizes a digitizer, such as the CRS-160 of Infotronics Corp. This device, used with a magnetic-deflection spectrometer, continuously samples both magnetic field and accelerating voltage, and computes quantities proportional to B^2, V^{-1}, or B^2V^{-1}, as desired. The signal current is also sampled. Whenever the signal passes through a maximum, a digital printer prints the mass number and the peak height. A computer-compatible paper tape can be punched simultaneously. This instrument can be applied to quadrupole and time-of-flight spectrometers as well as to magnetic deflection types.

Dedicated general-purpose computers can be applied to mass-spectrometer service, and permit on-line control of instrument parameters as well as data collection and reduction. Computer-mass spectrometer systems of this class are available from a number of manufacturers. See reference 6 for a comprehensive discussion. More elaborate time-sharing systems can be applied to mass spectrometry; this will not ordinarily be justified unless a major part of the cost can be allocated elsewhere.

COMMERCIAL SPECTROMETERS

As in the main article, the following descriptions are arranged alphabetically. Only new models of considerable importance are included.

Applied Research Laboratories (ARL) (Subsidiary of Bausch & Lomb)

A long-established firm, ARL has newly entered the field of mass spectrometry, with a unique instrument called an Ion Microprobe Mass Analyzer (IMMA), based on the work of Liebl (7). The operation of IMMA can be followed in Figure 46.

A pure gas, typically argon or oxygen, is ionized by an electric discharge in a unit called a duoplasmatron. Positive ions are extracted by electrostatic means, and accelerated through a 90° magnetic sector. The ions are then focused on the solid sample, which of course must be within the vacuum chamber. Ionic bombardment of the sample causes emission, through a sputtering process, of secondary ions corresponding to the material of the sample. The secondary ions are accelerated away from the target area by an electrostatic lens system, and mass-analyzed by successive passage through electric and magnetic sectors. Ions of selected m/e ratio pass through a slit and a 120° cylindrical condenser onto the cathode of an electron multiplier.

The magnetic sector in the primary ion beam ensures that the bombarding ions are monoisotopic, and therefore can be focused on a spot which can be as small as $2\mu m$ diameter. Pairs of deflection plates are provided for scanning the spot over a larger sample area if it is desired to obtain a picture or map of the area in terms of a selected element present.

This instrument is the mass-spectral analog of the electron-beam X-ray microprobe analyzer. It has several advantages: it is nearly free of continuous background, hence is more sensitive, it does not exclude low atomic-number elements, and it permits isotopic analysis. The design of the IMMA permits selection of primary ions up to mass number 133. The resolution of the secondary mass spectrometer is 300 at $m/e = 238$ (10% valley criterion).

Bell and Howell—CEC

CEC is the American distributor of an ion microprobe analyzer manufactured by the French firm CAMECA. The basic principle is the same as that described above for ARL's "IMMA," but the design of the instrument is quite different (Fig. 47). The primary ion beam is formed from argon, oxygen, or other gas in the ion gun at the left, and focused on the sample. A beam of secondary ions is picked up by an electrostatic "immersion lens," so named by analogy with a high-power microscope objective. The beam, which is of circular cross-section, is deflected through 90° by a magnetic sector, then reflected back on itself electrostatically. The magnet returns the beam to the same straight path on which it started, but now resolved with respect to the m/e ratio. An image of the source is produced at the far right end of the apparatus, and converted to its equivalent in electrons. The resulting electron beam can be viewed via a fluorescent screen, or can be photographed, or measured with a photomultiplier. The electronic controls include a bank of 10 pushbuttons, analogous to station-selectors on a radio receiver, which can be adjusted to any desired mass numbers. This arrangement allows one to switch instantly from one element or compound (an oxide, e.g.) to another. The distribution, of Be, Mg, Al, Fe, V, and other elements in an aluminum alloy, for example, can be observed. Added information and sensitivity can often be obtained by observing oxide ions, such as BeO^+ or MgO^+, particularly when the primary beam consists of oxygen ions. Mass resolution of this instrument is $M/\Delta M = 300$. Linear resolution of the image is about $1/\mu$.

Extranuclear Laboratories, Inc.

Another newcomer to mass spectrometry, Extranuclear produces a modular line of quadrupole spectrometers. They employ a newly designed electron-impact

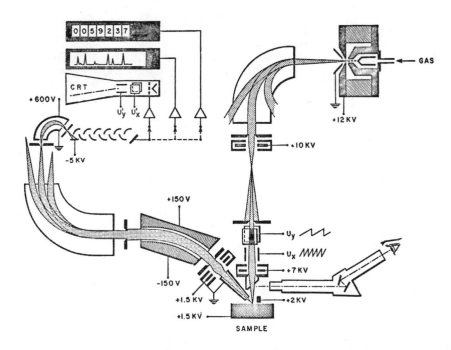

Figure 46. Ion microprobe mass analyzer. The optical telescope at lower right permits visual observation of the target area of the sample during the analysis. Courtesy of the American Institute of Physics (7).

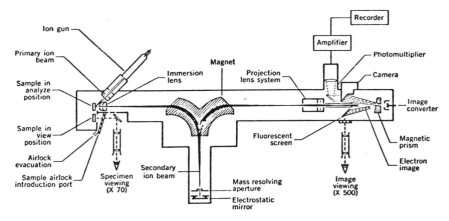

Figure 47. Schematic diagram of CEC's ion microprobe analyzer.

orifice into the vacuum system, thus producing a molecular beam. The beam is passed through a toothed-wheel modulator which converts it to a train of pulses with a frequency of a few kilohertz. The modulated beam then traverses an ionizer where it is subjected to electron bombardment, and enters the quadrupole mass filter, where the ions are filtered according to their m/e ratios. The molecules which escaped ionization also stream through the mass analyzer, unaffected by its fields. Beyond the exit slit, the ions are deflected into an electron-multiplier detector, while the neutral molecules proceed past it to be lost on the back wall of the vacuum chamber. The signal from the multiplier is demodulated by a lock-in amplifier which takes its reference frequency directly from the modulating wheel.

This system has two major advantages. The first is common to all systems employing modulation: excellent discrimination against background, in this case against ions formed from residual gases in the vacuum system. The second depends upon observation of the phase lag between passage of a pulse through the chopper and its reception by the multiplier. This lag serves as a measure of the time of flight through distance L from chopper to ionizer, and hence a measure of the velocity of the neutral molecules. All molecules have the same kinetic energy (except for unavoidable Gaussian distributions) as they enter the space L, hence those of different mass travel at different velocities. Phase-lag measurements result in the possibility of establishing which neutral species produced a particular ion (8). It has been shown, for example, in the examination of nitrogen, that all of the ions N_2^+, N^+, and N^{++} result directly from ionization of neutral N_2 molecules, both at 300 and at 2250°C. On the other hand, similar studies of oxygen show that the ions O_2^+, O^+, and O^{++} all stem from neutral O_2 molecules at 300 and at 1325°C, while at higher temperatures O^{++} is formed almost entirely from ionization of free atomic oxygen, O.

ionizer equipped with four heated filaments and an improved ion-optical lens system which results in a considerably increased sensitivity compared to conventional ionizers. A choice of three quadrupole mass filters is offered, with pole diameters of $3/8$, $5/8$, and $6/8$ in. The larger poles give increased sensitivity for low mass numbers, whereas the smaller ones permit work at higher masses with some sacrifice of sensitivity. The greatest mass which can be measured with the Extranuclear instruments is 1400. The operator can select constant resolution (constant $M/\Delta M$ as great as 1000) or constant separation (constant ΔM). The latter mode alone is found in conventional quadrupole spectrometers.

A unique combination of the quadrupole principle with concepts stemming from time-of-flight spectrometry has been incorporated in the "EMBA-II" (for Extranuclear Modulated Beam Analytical mass spectrometer). A schematic diagram of this instrument is reproduced in Figure 48. The sample to be studied (indicated in the diagram as "gas source") is allowed to expand through a small

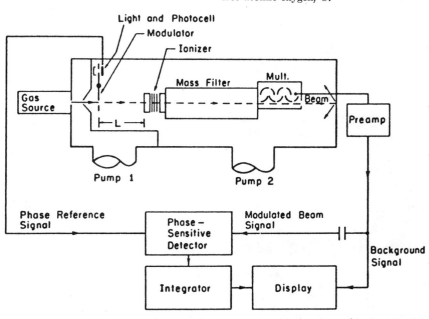

Figure 48. Simplified schematic of EMBA-II modulated-beam mass spectrometer. Courtesy of Extranuclear Laboratories, Inc.

Finnigan Instruments Corp.

Finnigan has added to their line model 3000, called a "GC Peak Identifier." This is a quadrupole mass analyzer with specially designed high-efficiency ionizer and mass filter, to permit rapid scanning with good sensitivity. A mass-marking device permits rapid identification of any peak while it is being recorded. Provision is also made for a continuous monitoring of the total ion current, which provides a conventional gas chromatogram.

Finnigan now offers a complete, computer-controlled, combination gas chromatograph and mass spectrometer. This is built around either model 3000 or model 1015 (previously described), a DEC PDP-8 computer, and suitable peripheral and interfacing equipment. The operator inserts the sample, tells the computer via teletypewriter what conditions of flow rate, temperature, mass range, scan speed, etc., are required. The system does the rest, including comparison with prior calibration data, and plots or prints out the resulting information.

JEOLCO (U.S.A.), Inc.

A new, double-focus, mass spectrometer, the model JMS-06, has been introduced (Fig. 49). The analyzer employs a radial electric field sector, followed by a 60° magnetic sector. The design differs from others in that the beam as it leaves the electric sector is divergent (rather than convergent as in the Nier-Johnson design or parallel in the Mattauch-Herzog configuration). The beam appears to diverge from a virtual image, which acts as input slit to the magnetic stage. This spectrometer is designed particularly for use with a gas chromatograph, but is provided with conventional inlet systems for other service. The over-all mass range is 1 to 2000. Resolution is 1000 (10% valley) with the standard slit assembly.

JEOL also produces a computer system (GASMAC 06-1100D) to control operation of the combined chromatograph and LMS-06 spectrometer, and to process the resulting data.

Picker-Nuclear—AEI

The MS-30, shown in Figure 50, is designated a double-beam mass spectrometer, and so it is, but the term "double-beam" does not have the same significance

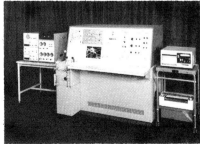

Figure 49. Model JMS-06 mass spectrometer. A gas chromatograph is shown at the left. On the stand at the right is an oscillographic recorder for the mass spectrum (above), and a strip-chart recorder for the total ion monitor (below). Photo courtesy of JEOLCO (U.S.A.), Inc.

familiar to us in optical spectrophotometry. The second beam is not intended as a background corrector, but is rather to permit simultaneous recording of two spectra. As depicted in Figure 51, each of the two channels has its own ion source, detector, and recorder, but the two ion beams are subjected to the same electric and magnetic fields.

Some applications suggested by the manufacturer are: (a) simultaneous recording of mass spectra at high and low resolution; the low-resolution information provides the fingerprint spectrum with its associated metastable peaks, and the high-resolution scan can yield elemental compositions; (b) examination of identical samples at low and high electron voltage; (c) chemical mass marking; a reference compound such as perfluorokerosine can be run simultaneously with an unknown, thus facilitating mass assignments; (d) direct comparison of unknown and authentic samples; impurity peaks are thus made evident at a glance; (e) two totally independent samples can be run simultaneously if desired, for economy of time.

Masses in the range of 2 to 2400 can be recorded, in three ranges. Resolution (10% valley) can be selected separately for the two channels at values of 1000, 3000, or 10,000. Facilities are available for interconnection with a DEC PDP-8 computer through AEI's DS-10 or DS-20 automatic data processing systems.

Figure 50. The MS-30 double-beam mass spectrometer. Photo courtesy of Picker-Nuclear.

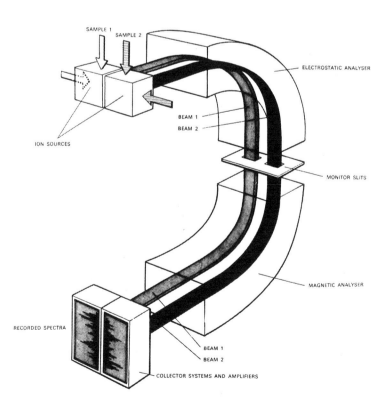

Figure 51. Sschematic diagram of the MS-30 double-beam mass spectrometer. Courtesy of Picker-Nuclear.

MANUFACTURERS

Aero Vac Corporation
P.O. Box 448, Troy, N. Y.   12181

Applied Research Laboratories (ARL),
subsidiary of Bausch & Lomb
P.O. Box 129
Sunland, Calif.   91040

Associated Electronic Industries, Ltd.
(see Picker Nuclear)

Atlas-MAT
(see Varian Associates)

Avco Electronics Div.
10700 East Independence,
Tulsa, Okla.   74115

Bendix Corporation, Scientific
Instruments Div.
3625 Hauck Rd., Cincinnati,
Ohio   45241

Consolidated Electrodynamics (CEC)
(Now part of duPont's Instrument
Div.)
1500 S. Shamrock Ave.,
Monrovia, Calif.   91016

Electronic Associates, Inc.,
Scientific Instruments Div.
4151 Middlefield Rd.,
Palo Alto, Calif.   94303

Extranuclear Laboratories, Inc;
P.O. Box 11512
Pittsburgh, Pa.   15238

Finnigan Instruments Corporation
2631 Hanover St.,
Stanford Industrial Park,
Palo Alto, Calif.   94304

Hitachi, Ltd.
(see Perkin-Elmer)

JEOLCO (U.S.A.), Inc.
477 Riverside Ave.,
Medford, Mass.   02155

Johnston Laboratories, Inc.
3 Industry Lane,
Cockeysville, Md.   21030

LKB Instruments, Inc.
12221 Parklawn Drive,
Rockville, Md.   20852

Nuclide Corporation
642 E. College Ave.,
State College, Pa.   16801

Perkin-Elmer Corporation,
Instrument Div.
Norwalk, Conn.   06852

Picker Nuclear
1275 Mamaroneck Ave.,
White Plains, N.Y.   10605

Varian Associates,
Analytical Instrument Div.
611 Hansen Way,
Palo Alto, Calif.   94303

Literature Cited

(1) WIBERLEY, S. E., AND AIKENS, D. A.,
J. CHEM. EDUC., **41**, A75, A153
(1964).

(2) WATSON, J. T., AND BIEMANN, K.,
Anal. Chem., **37**, 844 (1965).

(3) RONDEAU, R. E., J. CHEM. EDUC., **42**,
A445, A551 (1965).

(4) DAWSON, P. H., AND WHETTEN, N. R.,
Research/Development, **19**, No. 2,
46 (1968).

(5) FIELD, F. H., *Accts Chem. Res.*, **1**, 42
(1968).

(6) VENKATARAGHAVAN, R., KLIMOWSKI,
R. J., AND McLAFFERTY, F. W.,
Accts Chem. Res., **3**, 158 (1970).

(7) LIEBL, H., *J. Appl. Phys.*, **38**, 5277
(1967).

(8) PEDEN, J. A., FITE, W. L., AND
BRACKMANN, R. T., presented at the
21st Pittsburgh Conference on
Analytical Chemistry and Applied
Spectroscopy, Cleveland, March,
1970.

General References

REED, R. I., "Modern Aspects of Mass
Spectrometry," Plenum Press, New
York, 1968.

ROBOZ, J., "Introduction to Mass Spectrometry, Instrumentation and Techniques," Interscience Div. of John
Wiley & Sons, Inc., New York, 1968.

McDOWELL, C. A., "Mass Spectrometry,"
McGraw-Hill Book Co., New York,
1963.

Topics in...

Chemical Instrumentation

...a feature

Edited by **GALEN W. EWING**, Seton Hall University, So. Orange, N. J. 07079

The Journal of Chemical Education
Vol. 46, pages A401, A465
June and July, 1969

XLV. Pressure Transducers

DAVID J. CURRAN Department of Chemistry, University of Massachusetts, Amherst, Massachusetts 01002

Dr. Curran received the BS degree in chemistry from the University of Massachusetts. After service with the USAF he enrolled at Boston College where he wrote a MS thesis in the area of automatic instrumentation for coulometric titrations. The PhD degree was awarded by the University of Illinois in 1961 upon completion of a thesis involving the polarography and chronopotentiometry of aromatic nitro compounds as a means of analysis for nitrate ion. From 1961 to 1963 he was assistant professor of analytical chemistry at Seton Hall University. Following a summer research associateship at M.I.T., he joined the chemistry department at the University of Massachusetts. His research and teaching interests are in the areas of chemical instrumentation and electroanalytical chemistry. His current research programs involve the use of pressure transducers in chemical analysis and the use of refractory materials as electrodes in electroanalytical chemistry.

Pressure is a scalar quantity expressed in terms of force normal to a unit surface area. Since we live subject to a finite pressure, this variable affects our affairs in ways too numerous to mention. It is not surprising however that the measurement of pressure has received considerable attention. The result is an array of instruments, rather spectacular in terms of its number and variety, for pressure measurements. According to Behar, instruments available cover a pressure range of 10^{15} (1) and it is likely that another order or two of magnitude has been added to this in the last decade. No single instrument will adequately cover the entire range but some types of pressure transducer systems are capable of meaningful pressure measurements over a range of about 10^9.

As a consequence of the wide variety of disciplines concerned with pressure measurement, a large number of units of pressure are employed. Table I presents a short listing of conversion factors for those units most familiar to chemists. In addition to the unit of measurement, the reference point against which the measurement is made must be distinguished and several possibilities arise here. The terms commonly used are absolute, gauge, and differential. The first refers to a reference point of zero pressure, the second to a reference point of atmospheric pressure, and the third to an arbitrary reference point, but one which may be either of the first two, or one which need not even be constant. Although standard conditions for normal atmospheric pressure have been agreed upon, prevailing atmospheric pressure is seldom equal to normal atmospheric pressure and is not necessarily constant. Either consideration may be of no consequence to the measurement being made. A vacuum is a negative gauge pressure and is also commonly referred to as a reduced pressure (i.e., a pressure less than atmospheric).

An earlier series of articles on pressure measurements written by S. T. Zenchelsky has appeared in these columns (2). Commercially available devices for the measurement of reduced pressures were considered. The purpose of the present series is to expand the discussion to include absolute, gauge, and differential measurements but to limit the discussion to include only representatives of that class of pressure transducers which require some kind of mechanical input and which produce an electrical output. Such transducers have been classified by Lion as input transducers (3). In a sense, the word "transducer" is not useful by itself since it means a device which transfers. Thus, a conveyor belt is a transducer. More commonly, a transducer is thought of in terms of energy conversion and Neubert (4) has made the useful distinction between transducers which are energy converters and those which are energy controllers. The latter require some auxiliary source of energy while the former do not. He further points out that transducers useful for measurement purposes are properly

Table 1. Conversion Factors for Pressure Units
(Rounded to 4 significant figures)

	Pa	Bar	Atm	Torr	psi	in. H₂O
Pa	1	1.000 (−5)	9.870 (−6)	7.502 (−3)	1.451 (−4)	4.205 (−3)
Bar	1.000 (+5)	1	9.870 (−1)	7.502 (+2)	1.451 (+1)	4.205 (+2)
Atm	1.013 (+5)	1.013	1	7.600 (+2)	1.470 (+1)	4.068 (+2)
Torr	1.333 (+2)	1.333 (−3)	1.316 (−3)	1	1.934 (−2)	5.353 (−1)
psi	6.895 (+3)	6.895 (−2)	6.805 (−2)	5.172 (+1)	1	2.767 (+1)
in. H₂O	2.378 (+2)	2.378 (−3)	2.458 (−3)	1.878	3.613 (−2)	1

1 pascal (Pa), the internationally agreed upon unit of pressure in the SI system, = 1 newton of force per square meter.

1 bar = 10^6 dynes/cm².

1 torr = 1 mm Hg at 0°C.

psi: pounds (weight) per square inch.

Inches of water, specified at 4°C.

Interpretation of tabular data: 1 atm = 1.013×10^5 Pa.

Chemical Instrumentation

referred to as *instrument* transducers since the use of this word as a modifier alters the meaning to include the important concept of a known and reproducible relationship, which can be attained with a known limit of accuracy, between input and output. An unambiguous further breakdown of pressure transducers into some kind of classification is a little difficult because the useful pressure ranges of the various types overlap in many cases and the choice exists of considering the nature either of the input or of the output of the device. However, practice in the industry seems to be to classify them according to the nature of the electrical output mechanism and this expedient will be adopted here. With the exception of a few miscellaneous types, the pressure transducers to be considered fall within the electrical output characteristics listed in Table II. An arbitrary but representative selection of examples of commercially available devices will be presented. Detailed discussions of these types of pressure transducers, and others, are found in the books by Neubert (4) and by Lion (3) and in references (1) and (5).

Table II. Classification of Electrical Output Characteristics of Pressure Transducers

1. Variable resistance
2. Variable inductance
3. Variable capacitance
4. Piezoelectric

A number of types of pressure transducers are classified as energy controllers. The input pressure is applied to some mechanical pressure-sensitive device which in turn controls the electrical output. Clearly the mechanical characteristics of the input device are very important to the overall performance of the transducer. Such input devices include: diaphragms of various types; capsules; bellows; C-shaped, twisted and helical Bourdon tubes; and several other types of tubes. In general, the pressure range increases in the order listed. Diaphragm deflections on the order of Angstroms have yielded meaningful pressure measurements while Bourdon tube deflections may be fractions of an inch. Details and literature citations may be found in references (3) and (4).

It is necessary to distinguish between transducers and transducer systems. The latter are complete instruments including all components from the pressure input to the electrical readout. They therefore include the transducer as well as power supplies and any other circuitry and auxiliary equipment necessary. A large number of companies in the United States are engaged in the manufacture of pressure transducers and systems, and useful listings of them are found in a number of sources including the Buyers Guide Issues of *Instruments and Control Systems* (6) and *Electronics* (7) and the Laboratory Guide of *Analytical Chemistry* (8).

POTENTIOMETRIC PRESSURE TRANSDUCERS

The potentiometric transducers are so named because the electrical output is taken between the wiper arm and one end of a variable resistor (potentiometer). The position of the wiper arm is controlled by a mechanical linkage which responds to the displacement of the pressure input device which is used to sum the force applied to it. Compact units weighing as little as a few ounces are available from a number of manufacturers. Industrial models with explosion proof housings weigh only a few pounds. The National Pipe Thread (NPT) type of fitting is common at the pressure input but nearly any fitting can be provided both at the pressure input and at the electrical input/output connectors. Stainless steels and Ni-Span-C are materials commonly exposed to the pressure media. Other materials are available and oil filled units can be purchased for very severe chemical environmental conditions.

Pressure ranges from 0–1.5 to 0–20,000 psi are available in gauge and absolute configurations as standard units and ranges outside of these limits can be obtained on special order. Differential units are arranged in either zero-center (bidirectional) or zero-end (unidirectional) configurations with pressure ranges from a few tenths to several thousand psi. Resistances are typically in the range of a few hundred ohms to 10 kilohms with a power dissipation of 0.5 to 2 W within the operating temperature limits of the transducers.

Of special importance to the user is the nature of the input/output relationship of the transducer. Since both mechanical and electrical considerations are involved, the situation is best expressed in terms of a calibration curve or plot of input pressure versus electrical output obtained under static conditions. Expressing both of these quantities on a relative basis in terms of percent, the ideal calibration curve would be a straight line between the points with coordinates (0%, 0%) and (100%, 100%). The term *static error band* is used to express the maximum deviation of the output from this straight line, but the term is usually only applicable between 2% and 98% of the input range. One of the chief reasons for this range of applicability is that it has not been possible to physically arrange the end windings of the potentiometer so that a truly linear response can be obtained in these regions. The static error band includes the effects of the following sources of error: nonlinearity, friction, hysteresis, finite resolution, and repeatability. A typical static error value for off-the-shelf transducers is ±1% of full scale output.

The *dynamic error band* includes all of the effects listed for the static error band except friction which is eliminated by mechanically vibrating (tapping, for example) the transducer during testing. Other sources of error include: temperature, shock, vibration, and acceleration. The latter three are probably of minor importance to chemical applications of pressure transducers but specifications for errors arising from these sources are

frequently given by manufacturers. The effect of temperature changes is expressed in the same fashion as the static error band as a percentage error referred to full scale output which covers that temperature range specified for the transducer. For example, a transducer may be rated for operation between −100°F and +300°F with a temperature error band of ±2%. In short, the temperature error band is the static error band determined over the specified temperature limits. The temperature range given in this example is fairly typical for standard units and temperature error bands are usually in the range of one to a few percent. It is noteworthy that the linearity error is smaller than the static error band. This means that for those applications where only relative changes in pressure are important, somewhat better performance can be achieved than would be indicated by the static error band. In this regard, specifications for resolution are usually provided and are frequently a few tenths of a percent of full scale outputs.

Dynamic response is of interest in many applications. Potentiometric pressure transducers are fairly slow in this regard with response times in the millisecond range. Specifications are usually given in terms of response to a step function pressure input as measured over one time constant—that is, the time required to respond to 63% of the full pressure input signal.

A block diagram of a potentiometric transducer system is shown in Figure 1 where the dotted line encloses all units which may be physically housed within the transducer. The pressure sensitive devices which have been used are diaphragms, capsules, and Bourdon tubes. The output may be fed directly to a readout device or through a signal conditioner unit (such as voltage to frequency conversion) to the readout. Extremely simple systems may be constructed or purchased with potentiometric transducers since the power supply need be nothing more than a Hg battery and a few resistors and the readout can be a panel meter, although care should be taken not to load the potentiometer with the meter. Inexpensive (around $225) line operated systems are available from a number of manufacturers including: Robinson-Halpern, West Conshohocken, Pa.; C-E Electronics, Glenside, Pa.; International Resistance Co., Control Components Division, Philadelphia, Pa.; and Thomas A. Edison Industries, Instrument Division, West Orange, N. J. These systems consist of two units: the

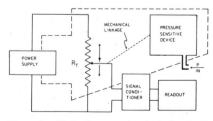

Figure 1. Block diagram of a typical potentiometric pressure transducer system. The dashed line indicates those units which may be housed within the transducer.

Chemical Instrumentation

pressure transducer which houses circuitry for ac to dc conversion, and a panel meter for readout. System accuracy is ±2% relative. Some loss in accuracy is suffered by using the panel meter since the transducer generally has a better accuracy. Solid state voltage to frequency conversion modules calibrated in pressure units are manufactured by Dynamic Precision Controls Corporation, Hagerstown, Ind. Special features are available with potentiometric transducers which reflect the versatility of potentiometers. Tapped slide wire and dual potentiometric outputs can be obtained from companies such as Bourns, Inc., Trimpot Division, Riverside, Calif.; Conrac Corp., Avionics Group, 330 Madison Ave., N. Y., N. Y.; and Sparton Southwest Inc., Albuquerque, N. M. Aero Mechanisms, Inc., Van Nuys, Calif., offers their Model 9207 differential pressure transducer which has an output proportional to the square root of the differential input.

The transducers mentioned thus far all operate open loop. Force balance, servo, or null type potentiometric pressure transducers are also available. Since the transducer operates via a mechanical input and an electrical output, the feed back system is of the electromechanical type (note the analogy to the potentiometric strip chart recorder but with input-output features reversed). All of the usual advantages of feedback systems apply here such as improved stability and accuracy, and less noise and distortion. The principal disadvantage of electromechanical feedback systems is their relatively slow response times. A differential model operating at ±200 in. of water (approximately ±0.5 atm) is made by Aero Mechanisms, Inc. Dual output potentiometers with a turns ratio of $1/20$ are provided. Conrac Corporation's Model 55T18 Null Balance Transducer has an absolute accuracy of ±0.1 to ±0.2% of full scale output for several absolute or differential ranges between 0 and 2 atm.

STRAIN GAUGE PRESSURE TRANSDUCERS

The strain gauge has been developed to a high state of perfection as an electromechanical transducer. The literature on the subject is vast and only the barest essentials can be provided here. If a uniform stress within the elastic limit is applied along the longitudinal axis of a wire, the wire resistance, R, will change because of the increase in the length of the wire (ΔL), the decrease in the wire diameter, and the change in the resistivity of the metal. Thus, pressure transducers incorporating strain gauges fall within the classification of variable resistance pressure transducers. The strain sensitivity, S, is expressed as a pure number by the equation, $S = (\Delta R/R)/(\Delta L/L)$. Values of S for wires of most metals and alloys can be positive or negative and are typically less than seven and frequently close to two. Other factors being equal, there

is clearly an advantage in having the highest possible value of S for the wire used in the strain gauge. Temperature variations are the most important source of error affecting strain gauges. Two effects may be distinguished: a change in the wire resistance as reflected in a change in the zero point with no strain applied externally, and a change in the strain on the wire, which shifts the calibration curve. Either the gauge must be operated at constant temperature or some form of temperature compensation must be supplied.

Two distinct classes of strain gauges have been developed. The first of these is the unbonded type which in principle is nothing more than a wire with one end fixed and the other end free to be subjected to a displacement by a force summing device. The second class is the bonded strain gauge. Here, the gauge is actually bonded to the member under strain. For example, if the force summing device were a diaphragm, then an unbonded gauge would involve direct translation of the diaphragm displacement to the movable end of the gauge wire while in the bonded case, the strain gauge in its entirety could be bonded by some sort of adhesive to the diaphragm. However, sometimes a strain sensing element is interposed between the force summing device and the strain gauge. The important difference between the two classes rests in the fact that at equilibrium the force applied to the diaphragm is balanced primarily by the wire of the gauge itself in the unbonded case and primarily by the diaphragm in the bonded case. Given the same force acting on the diaphragm in both cases, the unbonded gauge is the more efficient of the two. For this reason, the unbonded gauge is usually chosen for low force situations but either may be used at high ranges. High temperature operation of the bonded strain gauge is usually limited by the temperature characteristics of the adhesive used in the bonding process. The unbonded gauge

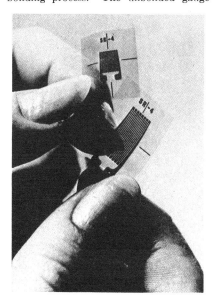

Figure 2. BLH Electronics, Inc., Etched Foil Strain Gauges.

is therefore inherently more useful under extreme temperature conditions. The relationship between the change in resistance

and the applied strain is linear for both types of gauges to at least 1% relative and accuracies of ±0.1% relative can be achieved with either arrangement. The change in resistance of strain gauges corresponding to a maximum strain input is small being on the order of 1% of the total resistance. Multiple windings are used to increase the total resistance of both the bonded and unbonded types and the flat grid is a popular arrangement for the former.

In addition to wire filaments, several other resistive elements have been used in bonded strain gauges. Metal foils are used by a number of manufacturers. The foil is usually in the form of a grid, and is produced by an etching process. Figure 2 shows a photograph of etched foil strain gauges manufactured by BLH Electronics, Inc., Waltham, Mass. The foil is mounted on a backing which may serve to electrically insulate the gauge. Additional insulation may be obtained from the bonding adhesive. When backing is used, both it and the adhesive are integral parts of the strain gauge and great care must be exercised in their use since the stress applied to the gauge is transferred through them. The bonding material is also sometimes used to coat the gauge itself to protect it from humidity at ambient temperature. Another resistive element in use is the semiconductor. Development of strain gauges utilizing semiconductor properties has been an outgrowth of solid state technology. The gauge (or gauges) may be produced directly on a semiconductor material, usually silicon, by integrated circuit diffusion techniques. Controlled addition of n- or p-type impurities produces the desired linearity and temperature characteristics. The gauge factor, G, for a bonded strain gauge is the slope of the linear output/input relationship and is determined from a plot of the fractional resistance change, $\Delta R/R$, versus applied strain as measured with the strain gauge cemented to a standard. Semiconductor gauges have high gauge factors, typically ±50 to ±200 according to Perino (9). The combination of semiconductor technology and high gauge factor has permitted the production of extremely small pressure transducers. Several examples will be cited below.

Vacuum deposition techniques have been adapted to the production of thin film metal and metal alloy strain gauges (9). A ceramic film is deposited on a metal substrate and the strain gauge is then deposited on the ceramic insulator. A principal advantage of this method of preparation is the control it affords over the properties of both the insulator and the gauge. Geometries can be made uniform in terms of shape, area, and thickness. This permits optimization of characteristics such as gauge resistance, gauge heat dissipation, and temperature compensation. Gauge factors comparable with those of silicon semiconductor gauges have been achieved.

The Wheatstone bridge appears to be universally used as the electrical configuration for strain gauges. Bridge excitation is frequently 10 vdc but ac excitation is also possible. Higher excitation voltages can be used with the higher resistance gauges since the main limiting factor on

Chemical Instrumentation

the excitation voltage is the problem of heat dissipation in the gauge resistance. The usual output impedance of the bridge is 350 ohms but several thousand ohms can be obtained in thin film and semiconductor gauges. Several different arrangements of the strain gauges in the bridge arms are possible. Most common practice for static measurements is to use active strain gauges in all four arms. One pair of opposite arms is tensed and the remaining pair is compressed. The symmetry of this push-pull operation provides good temperature compensation and the highest possible output. Bridge off-balance is measured at the output terminals. The magnitude of the signal is most often a few millivolts per volt of excitation for full scale output. However, some bridges with semiconductor gauges have outputs ten times this figure. Other bridge arrangements are one or two active arm circuits which may include dummy gauges for temperature compensation. Strain is applied to the active gauges but not to the dummy gauges. Still other circuits use active gauges and passive resistance components. The latter circuits are more commonly found with strain gauges designed for dynamic response where temperature compensation may not be quite so critical.

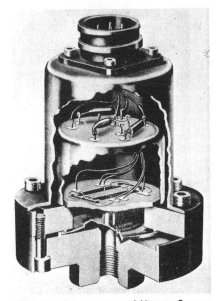

Figure 3. Cutaway view of Viatran Corporation Model PTB103 Pressure Transducer.

Pressure transducers based on unbonded strain gauges, and all of the types of bonded strain gauges previously mentioned, are commercially available. Pressure measurements are possible from about one psi absolute to 200,000 psi gauge and can be as low as a few tenths of a psi differential. Figure 3 shows a cutaway view of Viatran Corporation (Buffalo, N.Y.) Model PTB-103 pressure transducer which is available in twenty-six ranges from 0–5 through 0–25,000 psig or psia, and from 0–5 through 0–2000 psid (unidirectional). The pressure connection shown at the bottom of the picture is a ¼ in. NPT female pipe connector. Fluid pressure is applied to a stainless steel force-summing diaphragm

which is mechanically coupled to the force sensing element located above the diaphragm. A four active arm foil Wheatstone bridge circuit is actuated by the sensing element which is enclosed in a reference pressure can as shown. Resistors for circuit adjustment can be seen external to the pressure can. The electrical connector at the top of the stainless steel case is a standard Cannon type. The unit stands about 3½ in. high and is 2½ in. in diameter at the widest point. Easy access for cleaning and repair is provided by the bolted construction. Other specifications are listed in Table III. Modifications available include: Teflon coating on all parts exposed to the pressure fluid, shunt calibrating resistor to simulate a particular pressure input, special electrical or pressure connectors, and higher proof pressures. Other pressure transducers available from Viatran Corp. feature higher pressure ranges, better accuracy and bidirectional differential operation. Complete transducer systems are also manufactured in portable, rackmounted and cabinet forms.

Figure 4. MB Electronics Model 151-EBA-1 Pressure Transducer.

Another four active arm foil Wheatstone Bridge strain gauge pressure transducer is MB Electronics (New Haven, Conn.) Model 151-EBA-1, shown in Figure 4 for the 0–5 psig range. This is a

highly accurate unit with a combined error (linearity, hysteresis, and repeatability) of ±0.2% of full scale output, maximum. Other ranges available in this model are 0–10, 0–15, and 0–20 psig, psia, and psis (pounds per square inch sealed). The reference point for the latter type is usually 14.7 psia but the reference pressure compartment is permanently sealed rather than vented to the atmosphere. Another MB Electronics pressure transducer is shown in Figure 5. The Model 510 is produced in nine ranges from 0–50 through 0–10,000 psia. It is particularly interesting because it represents an advance which overcomes one of the important shortcomings of strain gauge pressure transducers, namely low output. Combined with the foil strain gauge circuit (4 active arms) is an integrated circuit module which provides a 5 v output at full scale pressure input with 28 v dc applied. Another interesting MB Electronics pressure transducer is the Model 172, designed for cryogenic work. Pressure ranges from 0–20 through 0–10,000 are standard in psig, psia, and psis and go as low as 0–10 in psig and psia. The operable temperature range is from −452°F to +250°F and the compensated temperature range is from −320°F to +77°F. Other manufacturers using integrated circuits for increased output with strain gauge pressure transducers include: West Coast Research Corp., Los Angeles, Calif., and Coutant Electronics Limited, England (distributed by Dentronics, Inc., Hackensack, N. J.).

Figure 6 shows a Model STD-H2 pressure transducer made by BLH Electronics, Inc., Waltham, Mass. This is a high temperature high pressure unit designed in two standard pressure ranges, 0–30,000 psig and 0–50,000 psig with continuous duty to 600°F and temperature compensation to 500°F. Another model, the STD, has operating pressure ranges as high as 0–200,000 psig with a compensated tem-

Table III. Specifications for Viatran Corporation's Model PTB103 Pressure Transducer

Linearity (terminal) error	Less than ±0.75%, full scale output
Hysteresis	Less than ±0.25%, full scale output
Repeatability error	Less than ±0.1%, full scale output
Resolution	Infinite
Zero balance	Within ±2%, full scale output at 70°F
Full scale output	3 mv/v minimum
Bridge resistance	120 or 350 ohm
Excitation voltage	120 ohm: 10 v maximum
	350 ohm: 15 v maximum
Operating temperature	−40 to 250°F, 350°F optional
Thermal zero shift	Less than 2%, full scale output/100°F
Thermal span shift	Less than 2%, full scale output/100°F
Burst pressure	500% or 10,000 psi, whichever is less for 5 through 5,000 psi
	200% or 30,000 psi, whichever is less for 10,000 through 25,000 psi
Proof pressure	150% or 25,000 psi, whichever is less
Natural frequency	2 kHz (nominal) for 0–5 psi extending to 40 kHz (nominal) for 0–25,000 psi
Acceleration error	Sensitive only in the axial plane where sensitivity extends from 0.08% FSO/G for 0–5 psi to 0.01% FSO/G for ranges over 0–100 psi
Response time	Less than one millisecond
Cavity volume	0.040 cubic inches nominal (does not include fitting)
Prices	$130 to $186

Chemical Instrumentation

Figure 5. MB Electronics Model 510 Pressure Transducer.

perature range of 15 to 115°F. High pressure range bonded transducers are

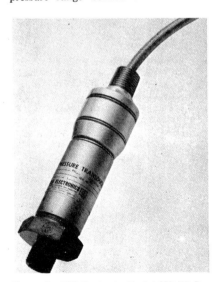

Figure 6. BLH Electronics Model STD-H2 Pressure Transducer.

also available from Astra Corp., Willow Grove, Pa. (0–150,000 psig), Standard Controls Inc., Seattle, Wash. (0–100,000 psig) and General Transducer Co., Santa Clara, Calif. (0–100,000 psig). The latter company offers combination pressure and temperature transducers which use an iron-constantan thermocouple for temperature measurement. Provision is made for water or air cooling the unit. The force summing diaphragm can be exposed to temperatures as high as 750°F and pressures to 15,000 psi can be measured.

These transducers have been especially designed for the plastics and rubber industry, particularly for extruder operation. Alnor Instrument Company, Chicago, Ill., also supplies transducers for this purpose. General Transducer Co. also makes complete pressure transducer systems including dual indicator units to simultaneously read both temperature and pressure.

The BLH Electronics transducer shown in Figure 7 is the Model DHF-DS. A semiconductor strain gauge bridge provides a full scale output of one volt with 10 v ac or dc bridge excitation. This unit can be used as a replacement for wire and foil transducers. One of the great advantages of silicon semiconductor gauges is their compatibility with very small size requirements. When small size is combined with flush diaphragm configuration, very good dynamic response is possible.

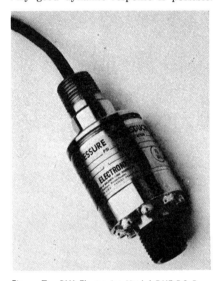

Figure 7. BLH Electronics Model DHF-DS Pressure Transducer.

Transducers of this type are manufactured by Scientific Advances, Inc., Waltham, Mass., Kulite Semiconductor Products, Inc., Ridgefield, N. J., and Whittaker Corp., Instrument Systems Division, Chatsworth, Calif. The natural frequency of the transducer depends on the pressure range. For pressures to several thousand psi gauge or absolute, natural frequencies to as high as 500 kHz are available. The small size is striking. Kulite's Model CPS-125-1000 has a pressure sensitive diaphragm of 0.085 in. diameter and a total configuration diameter of 0.125 in. Sensotic's Model SA-SA 8J-GH has an identical configuration diameter and Bytrex's smallest transducer is the Model HFO with dimensions: 0.09 in. diameter × 0.15 in. long.

Unbonded strain gauge pressure transducers and thin film bonded strain gauge transducers are made by Statham Instruments Inc., Los Angeles, Calif. A great variety of models is manufactured for gauge, absolute, and differential measurements. Complete pressure transducer systems are also available. Unbonded strain gauges can also be purchased from CEC Transducer Products Division of Bell and Howell, Monrovia, Calif., and Dynisco Division of American Brake Shoe Co., Cambridge, Mass. Of interest

is Dynisco's Model PT 14 differential pressure transducer for applications as low as 0–1 psi full scale.

CAPACITIVE PRESSURE TRANSDUCERS

The capacitance of any capacitor is directly proportional to the dielectric constant of the medium relative to air, ϵ, and to a geometry factor, G. The proportionality constant (permittivity constant) has a numerical value of 8.85×10^{-12} coulomb^2/newton-meter^2. Thus, all capacitive type transducers depend on either a variation of the dielectric constant with constant geometry or a variation of the geometry with constant dielectric constant. For a parallel-plate capacitor, the geometry factor is given by the ratio of plate area, A, to plate separation distance, d, so the additional option exists of varying either A or d while holding the other constant. In useful working units, the capacitance of the parallel-plate configuration is given by:

$$C_{(picofarads)} = \frac{0.0885\, A_{(cm^2)}}{d_{(cm)}} \epsilon$$

This equation is valid provided d is small compared with the plate dimensions so that fringe effects at the edges of the plates may be ignored. The number of design configurations possible for capacitive transducers is large, since: (a) various geometries can be used with single or multiple plates per electrode in either single or differential units; (b) a choice of linear or rotary displacement is possible; (c) the variable can be plate area, plate separation, or dielectric constant. Design formulas for a number of cases are given by Neubert (4) and by Foldvari and Lion (10). The most popular arrangement for pressure transducers is the parallel-plate capacitor with variable plate separation. Frequently, the design is differential, consisting of a flexible plate (diaphragm) placed between two fixed plates. Two different types of diaphragms are distinguished: membrane diaphragms where the radially-prestressed metal is usually welded to a support, and clamped diaphragms which exhibit stiffness in bending. The nature of the mechanical deformation of each of these types of diaphragms is different, one from the other, but the fractional change in capacitance of a single parallel plate capacitor is directly proportional to the applied pressure in both cases for small diaphragm deflections. The membrane diaphragm is particularly useful for the measurement of small pressures.

The overall characteristics of a capacitive pressure transducer system depend not only on the design of the transducer itself, but also on the electrical configuration in which the transducer capacitor is placed and on the method used to convert the change in capacitance to an electrical output signal. Foldvari and Lion (10) list the following approaches to the latter:

1. Methods based on the measurement of charge and discharge times
2. Methods based on division of charge
3. Capacitive ac-voltage dividers, ac bridges, and differential transformer methods

Chemical Instrumentation

4. Resonance methods
5. Frequency modulation
6. Diode twin-T networks

The first two methods are not suitable for continuous readout applications. Examples of commercially available instruments are presented below.

Datametrics Incorporated, Waltham, Mass.

Pressure transducer systems manufactured by Datametrics are multirange instruments designed for the measurement of pressures in the low to medium (several atm) range. Figure 8 illustrates the Type 1014 Barocel Electronic Manometer. The pressure transducer shown in the foreground is the Barocel Type 511 which is available in standard ranges of 0–1, 0–10, 0–100, and 0–1000 torr and 0–1, 0–10, and 0–100 psi, gauge, differential, absolute, or sealed-to-absolute. The Barocel has a differential parallel-plate capacitor configuration with a membrane type diaphragm. Electrically nonconducting liquids and gases may be accepted at the pressure

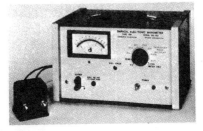

Figure 8. Datametrics Inc. Type 1014 Barocel Electronic Manometer.

ports. System stability is such that the detection limit of the 0–1 torr transducer is 1×10^{-5} torr. Converting the upper limit of the highest range psi Barocel to torr, it is seen that the system will measure pressures from 1×10^{-5} to better than 5×10^3 torr.

The two capacitors formed by the transducer are arranged as an ac-voltage divider and placed in an LC-bridge network. The bridge output voltage is determined by the ratio of the two capacitances. The ac carrier signal (10 kHz) applied to the bridge is amplitude modulated in response to the input pressure. Because capacitive transducers have high output impedance, and because bridge circuits present grounding problems, some electronic circuitry is physically located within the housing of the Barocel. This includes the transformer for coupling the carrier signal with the inductive arms of the bridge, a unity gain solid state impedance isolation amplifier, ac to dc conversion circuits to power the amplifier, and an output transformer for the modulated carrier signal. With this arrangement, the transducer can be operated remotely from the rest of the transducer system. Cable lengths of 15 ft are standard but 500 ft can be supplied.

The cabinet of the Type 1014 houses a regulated dc power supply, an amplitude-stabilized 10 kHz oscillator, a high precision calibrated voltage attenuator (range

switch), a signal amplifier, a synchronous phase-sensitive detector, a panel meter for readout, and dc output jacks. Solid state circuits are used throughout. The sequence of events for an input signal from the Barocel is the following: attenuation according to the range switch setting ($\times 1$, $\times 0.3$, $\times 0.1$, $\times 0.03$, $\times 0.01$, $\times 0.003$, $\times 0.001$), signal amplification, phase sensitive ac to dc conversion, and readout. The result is a ± 5 v dc full-scale signal for all positions of the range switch down to the $\times 0.001$ setting. Polarity of the dc output is determined by which pressure port (P_1 or P_2) receives the highest pressure. System accuracy is $\pm 1\%$ of full scale when panel meter readout is used. The accuracy of the dc output voltage is $\pm 0.20\%$ of reading plus 0.03% of full scale on the $\times 1$ and $\times 0.3$ settings and $\pm 0.1\%$ of reading plus 0.03% of full scale on the remaining settings. Linearity for the two highest range switch settings is $\pm 0.1\%$ of full scale and better than $\pm 0.05\%$ of full scale on all positions of the range switch from $\times 0.1$ through $\times 0.001$. Hysteresis depends on the range of the 511 Barocel: the worst case is $\pm 0.1\%$ of pressure cycle for the 0–100 psi model and decreases to 0.001% for the 0–1 torr unit. Another Barocel series, the 538, is designed to provide better accuracy for the higher pressure ranges. Hysteresis error has been reduced to $\pm 0.04\%$ of pressure cycle for the 0–100 psi transducer. Other system specifications include: an internal volume of 0.3 cubic in., standard, and less than 0.1 cubic in. on special order; a typical response time at one atm line pressure is 3 msec; the transducer operating temperature range is 0–150°F. Temperature errors are less than $5 \times 10^{-4}\%/F°$ of sensor full range for zero shift and less than $1 \times 10^{-2}\%$ of reading/F° for sensitivity shift.

Figure 9. Datametrics Inc. Modular Electronic Manometer System.

Other transducer systems available from Datametrics are the Model 1023 and Model 1018 Barocel Electronic Manometers and a modular system (Figure 9) consisting of the Type 511 Barocel transducer, the Type 1015 Signal Conditioner and Type 700 Power Supply. The Model 1018, shown in Figure 10, is a modification

Figure 10. Datametrics Inc. Type 1018 Barocel Electronic Manometer.

of the Type 1014 to incorporate automatic digital readout. It is likely that this feature is achieved by analog to digital conversion of the analog signal of the system. Binary-coded-decimal (BCD) readout is also provided.

An interesting accessory which can be used with most of the Datametrics pressure transducer systems is the Type 1056 Null Offset Adapter (Figure 11). This device consists of a high precision adjustable voltage ratio transformer having a five place decade dial with resolution of 1 part in 10 million. The electrical output of the transformer can be used to simulate any

Figure 11. Datametrics Inc. Type 1056 Null Offset Adaptor.

pressure within the range of the Barocel transducer and is differentially summed with the output of the Barocel. Any static pressure within the range of the Barocel can be measured by adjusting the dials of the Null Offset adapter until a readout of zero is achieved. It should be noted that the condition of zero output under these circumstances does not imply that the bridge is balanced or that a true pressure null exists. Other useful applications of the device include measurement of small pressure fluctuations about a given pressure and control of pressure to some predetermined value.

MKS Instruments, Incorporated, Burlington, Mass.

The MKS Baratron is another wide range membrane-diaphragm differential capacitive transducer operated in an LC-bridge configuration. Pressure transducer systems are the Series 77, 90, 100A, 126, and 144 MKS Baratron Capacitance Manometers. The Series 77, pictured in Figure 12, features direct reading of bridge unbalance or balanced bridge operation by means of three multitap ratio transformers and an interpolating decade potentiometer which provide the balance voltage. In the balanced mode, pressure is read out directly on the four large pressure readout dials. D.C. output terminals supply ± 100 mV corresponding to full scale deflection for each of the eight available

Chemical Instrumentation

range switch settings (×1 through ×0.0003). A corresponding signal of 500 mV rms or greater, which is phase sensitive to the direction of deflection of the trans-

Figure 12. MKS Instruments, Inc. Series 77 Baratron Capacitance Manometer.

ducer diaphragm, is available at the ac output terminals. System characteristics are excellent and are summarized in Table 4. Heaters are provided within the Baratron for both transducer temperature regulation and pressure cavity outgassing.

The Series 90 has been developed for low pressure or vacuum measurements with transducer bake-out capability to 450°C. Pressure heads from ±1 to ±30 torr are available. Repeatability of the 1 torr head using balanced bridge operation

Table 4. MKS Series 77 Baratron Electronic Pressure Meter System Characteristics

Transducer ranges: ±1, ±3, ±10, ±30, ±100, ±300, ±1000 torr, standard
Resolution: Infinite (voltage output)
Hysteresis: Less than 0.003% of pressure cycle applied for 30 torr range, 0.015% for 100 torr, 0.04% for 300 torr, and 0.08% for 1000 torr ranges
Accuracy:
  A. *Meter readout.* One to three percent of full scale depending on range selected. Individual scale accuracy can be improved by front panel adjustment
  B. *Digital dial readout.* 0.02% of full range plus 0.15% of dial reading when taken directly from the dials. 0.02% of full range plus 0.05% of dial reading with reference to a calibration chart. At very low pressures, the 0.02% factor can be reduced to 0.005% by valving the two ports together and adjusting zero before the measurement is made
Response time: Less than 10 msec for 63% response to a 30 torr pressure step at 1 atm pressure at the dc output; faster response at the ac output
Expanded scale: Zero can be offset up to full range and the meter used on more sensitive scales for expansion up to 1000 times
Media: Nonconducting gases. Special units for liquids and corrosive media
Pressure connectors: One-eighth in. female NPT
Transducer cables: 6 to 250 ft standard, longer on special order
Indicator features: Standard indicators are available in cabinet, or rack mounted styles with single, double, or triple channels
Absolute pressure: Reference side of diaphragm is pumped to low pressure. All units will withstand overpressure in excess of one atm
Power required: 105 to 120 V, 60 to 400 Hz single phase or 220–240 V, 50 Hz

with a vacuum reference and a working pressure of 5×10^{-5} torr is 5×10^{-6} torr. Under special circumstances, resolution of pressures as low as $10^{-7} - 10^{-8}$ torr has been accomplished but the accuracy of these measurements is uncertain.

The Type 100A (Figure 13) is an auto-

Figure 13. MKS Instruments, Inc. Series 100A Baratron Capacitance Manometer.

matic digital-readout instrument which features true logic-operated automatic balancing of the bridge circuit rather than analog to digital conversion of the bridge output signal. Precision toroid transformers are wound in binary-coded decimal ratios (8-4-2-1) and the method of successive approximations is used to null the transducer bridge by increments. The largest bit is inserted first. Selection of the transformer winding is by mercury-wetted reed relays. Voltages corresponding to each bit are summed and applied to the bridge circuit. Readout is on a five place Nixie display with meter interpolation of the last digit. The balance time is approximately 400 msec. BCD output terminals are available on the back of the instrument. Standard logic levels are: 0 logic, +3.0 ± 0.5 V dc at 5 mA max, and 1 logic, ±0.25 ± 0.25 V dc at 5 mA max. Other output levels and interfacing can be provided by circuit board substitution. The significant advantage of this advanced design is that true digital operation avoids all open-loop errors common to a circuit configuration employing analog to digital conversion of the output of the analog transducer. Transducers used with both the Series 77 and Series 90 instruments may be used with the Series 100A. System accuracy is the same as that for the Series 77 or 90. Details on other specifications and special options are available from the manufacturer.

Figure 14. MKS Instruments, Inc. Series 126 Modular Baratron Capacitance Manometer System.

The Series 126 is a modular pressure transducer system (Figure 14) which uses the Series 77 pressure transducer. Manual and auto ranging are featured.

Granville-Phillips Co., Boulder, Colorado

The Granville–Phillips Series 212 Capacitance Manometer also uses a differential parallel plate capacitor transducer arranged in an ac-bridge circuit. The system covers the pressure range 5×10^{-3} to 10^3 torr with transducer ranges of 0–10, 0–100, and 0–1000 torr. The range switch settings are ×1, ×0.3, ×0.1, ×0.03, and ×0.01 for each transducer. Transducer heads are bakeable to 450°C and system accuracy as a direct reading instrument with built in panel meter is better than ±2% after calibration. The minimum detectable pressure differential for the 0–10 torr transducer is 5×10^{-3} torr. Although direct readout of the Series 212 involves electronic measurement of bridge unbalance, when used in conjunction with the Granville–Phillips Automatic Pressure Controller, true pressure null operation can be achieved. The controller consists of a cabinet or rack-mounted control unit and a servo driven valve, which together automatically regulate the gas pressure in a system. The error signal to the controller is taken from the pressure transducer system. This system together with the pressure controller form a closed-loop electromechanical control system which will maintain the pressure on one side of the transducer at any level with reference to the other side. In practice, either side of the transducer could serve as the reference. The application to gas process control is obvious. The response time depends on individual applications but the time from full open to full closed of the servo valve is 30 sec with minimum damping.

Whittaker Corporation, Instrument System Division, Chatsworth, California

The 300, 400, and 700 Series capacitive transducers are single parallel plate types designed to measure gauge pressures with high dynamic response. Taken together, they cover the pressure range from 0 to 100,000 psig. Diaphragm resonance frequencies vary from 11 to 135 kHz depending on the transducer and the pressure range. A number of the models have provision for water cooling which permits the diaphragms to be exposed to gas media at temperatures as high as 6000°F. Thus, static and dynamic measurements can be made in a wide variety of severe environments such as those found in blast pressure or rocket propulsion studies.

These transducers operate with the Dynagage instrument. The transducer capacitor and a fixed inductance which is located within the transducer housing, form a tuned radio frequency (rf) circuit. The Dynagage contains a rf oscillator, tunable between approximately 610 and 850 kHz for the Model DG-605D, coupled to a diode detector circuit. The output of the tuned circuit of the transducer is coupled by cable to the Dynagage. A demodulated signal from the diode detector is first fed to a cathode follower and then to a transistor filter network to produce a low impedance output of ±10 v, maximum, free of the carrier signal. The frequency response of the Dynagage Models DG-600D and DG-605D is flat from 0 to

Chemical Instrumentation

15,000 Hz.

DISA Electronics, Midland Park, New Jersey

The DISA transducer system is another instrument applicable to static or dynamic gauge pressure measurements. The transducer capacitor (single unit with flexible diaphragm) and a tuning plug form a tuned circuit which frequency modulates the output of an oscillator. The FM signal is fed to the DISA 51E01 Reactance Converter where it is converted to an amplitude-modulated voltage. Both analog and digital output voltages at the oscillator frequency are provided. Technical data for the Reactance Converter are shown in Table 5.

The pressure range covered by available transducers is 0.005 to 800 kg/cm² (approximately 0.07 to 11,000 psi). Some models are water cooled. Diaphragm resonance frequency depends on model and pressure range.

Lion Research Corporation, Newton, Massachusetts

The transducer systems manufactured by this company employ diode twin-T networks for signal conversion as patented by Lion (11) and described in references (10) and (12). Significant features of this

Figure 15. Lion Research Corporation Model PM100 Pressure Meter.

arrangement are simplicity and the ability to ground one side of the oscillator, the transducer capacitors and the readout. The Model PM 100 Pressure Meter is shown in Figure 15 along with the Model

303 transducer. System specifications are summarized in Table 6. The pressure meter contains a regulated solid state power supply and a 1 MHz oscillator which provides the carrier signal for the transducer capacitors. The transducer is a stretched membrane differential type. The diode twin-T circuit is located within the transducer and produces a dc output

Table 6. Technical Information for the Lion Research Corporation Model PM 100 Pressure Meter

Input: Noncorrosive gases or, with special calibration, fluids with low conductivity
Pressure ranges: ±1, ±10, ±100, ±1000 torr
dc output: 1.0 V, 5 mA maximum, 10 ohm internal impedance
Response time: Limited by pneumatic settling time; typically 0.1 sec
Accuracy: As limited by factory calibration—±0.1% of reading
Linearity: ±1%
Zero stability: Drift no greater than ±10% of most sensitive scale per hour, at constant temperature
Sensitivity stability: ±0.2% per month or better in the absence of over-range conditions
Reproducibility: ±0.1% of full scale or ±10% of most sensitive scale (whichever is larger) over 1 hr, at constant temperature
Power required: 95–135 V ac, 60 Hz, 10 W. Other on special order
Price: With sensing head $1,195

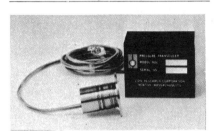

Figure 16. Lion Research Corporation Series 900 Pressure Transducer.

directly proportional to the difference in capacitance between the two transducer capacitors. Electrical filtering, amplification, and readout of the transducer output take place within the pressure meter cabinet. A range switch allows selection of one of seven full-scale ranges from ×1 to ×0.001 of the Model 303 used. Other systems are available including modular instrumentation. Other transducers include the Series 110, 200, and 900. Figure 16 is a photograph of the Series 900 transducer and associated electronics. The instrumentation shown is a complete trans-

ducer system except for a readout device. Designed for operation from 3 to 30 psi gauge, absolute or differential, the system will provide an output of 5.0 V dc from a very small equipment package: the transducer weighs only 1.5 oz. and the electronics module 4 oz.

Rosemount Engineering Company, Minneapolis, Minnesota

Rosemount capacitive sensors also employ diode detector circuits for signal conversion. The dc output voltage is given by $V_{dc} = fV_{pp}R(C_1 - C_2)$ where f is the oscillator frequency, V_{pp} is the peak-to-peak voltage of the oscillator, R is the load resistance, and C_1 and C_2 are the transducer capacitances in the differential mode. C_2 would be a fixed capacitor in the single case. Electronic feedback is used to maintain the product, fV_{pp}, constant. If the load resistance is constant, the output voltage is directly proportional to the difference in capacitance between C_1 and C_2. The input pressure-capacitance relationship is nonlinear and additional electronic feedback circuitry is used to linearize the output voltage-input pressure relationship. Circuit details may be found in reference (13). Transducers are made in gauge, absolute and differential models and in ranges from 0–0.1 to 0–5000 psia and psig, ±0.05 to ±100 psid and 0–0.1 to 0–200 psid.

INDUCTIVE PRESSURE TRANSDUCERS

The inductance of a coil, L, is given by the product, $n^2G\mu$, where n is the number of coil turns, G is a geometry factor and μ is the effective permeability of the medium in the vicinity of the coil. Clearly, a transducer may be based on a variation of any one of these three factors with the remaining two held constant; but in addition, self induction or mutual induction may be involved, iron or air cores could be used, and single or differential output could be obtained. Thus, like capacitive transducers, a large number of inductive transducer designs are possible. Lion (3) and Neubert (4) discuss several of these. However, most commercially available inductive pressure transducers are based on either the variable reluctance principle or the linear variable differential transformer (LVDT).

Variable Reluctance Pressure Transducers

Consider an iron core of uniform cross-sectional area, A, and an average magnetic path length in the core of l. The core is physically discontinuous, being interrupted by a small variable air gap of width, d. The inductance of a coil of n turns wound on the core is given by $n^2/\mathcal{R}$ where $\mathcal{R}$, the reluctance of the magnetic circuit, is equal to $(l/\mu - d)/A$ when fringe effects and leakage are ignored. Thus, the reluctance and therefore the inductance depend on the gap width. Differential action is achieved by arranging two cores such that the air gap of one is increased as the air gap of the other is decreased. In pressure transducers, the air gap variation is produced by the force acting on either an

Table 5. Technical Data for DISA 51E01 Reactance Converter

Frequency range	0–100 kHz
High-frequency suppression filters	0.2, 2, 20 kHz
Carrier frequency	4.4 to 5.6 MHz
Max. frequency change for full drive	±0.5%
Max. capacitance change for full drive	±1 pF
Output voltage for ±0.5% frequency change	±6V
Output impedance (analog output)	Approx. 30 ohms
Output voltage, oscillator frequency	Approx. 1 V_{pp}
Output impedance (digital output)	Approx. 20 ohms
Nonlinearity (exclusive of transducer nonlinearity)	Max. 0.5%
Compensating range for transducer nonlinearity	0 to 10%
Circuit	Fully transistorized
Power supply	110/220 V ±10%, 50 to 60 Hz
Power consumption	10 VA
Dimensions, HWD	106 × 212 × 225 mm

Chemical Instrumentation

armature which forms part of the core, or some other force-summing device such as a diaphragm or capsule which is constructed from a magnetically permeable material. The most common electrical configuration for variable reluctance pressure transducers is the ac bridge circuit. An excellent discussion of the RL bridge for use with inductance transducers is given by Neubert (4). Usually, phase sensitive demodulation is used with differential units requiring dc output since this arrangement provides a bidirectional output corresponding to the bidirectional input.

Manufacturers include: Whittaker Corporation, Instrument Systems Division, Chatsworth, California; Solid State Electronics Corporation, Sepulveda, California; Bourns, Inc., Instrument Division, Riverside, California; and Smiths Industries, North America Limited, Aviation Division, Jamaica, New York. Whittaker's PACE series of transducers includes the Models P1, P2, P3D, P7D, P90D, P109D, CP51, CP52, CP35D, and CP143. These are diaphragm type variable-reluctance pressure transducers employing two symmetrical E-core inductance assemblies as described by Epstein (14). Model numbers ending in D denote transducers with only differential capability. The CP designation indicates a transducer package designed for operation with 115 V, 60 to 400 Hz or 28 V dc excitation, and housing a transistorized carrier-demodulator with 0–5 V dc output. Units without the CP designation must be supplied with external excitation and demodulation. Models CD32 and CD10 carrier-demodulators are available for this purpose. A transducer system complete with meter readout is available in the Pace Model CD25 Transducer Indicator. System accuracy using the meter is ±1% of full scale. Null balance readout is also provided with three-digit dial reading.

The Pace series covers pressure ranges from 0.1 to 0–10,000 psig and psia. The Model P90D is ±1 in. of water, full scale, with 15 psi overload. Other differential models cover the range ±0.05 to ±5,000 psi. Dynamic response depends on pressure. The resonance frequency varies from approximately 2.5 kHz at 0.1 psi to 25 kHz at 100 psi for the P1 model. Linearity, accuracy, and hysteresis for most models is 0.5% of full scale. Some models have interchangeable diaphragms for changing transducer range. The Pace KP15 Pressure Transducer kit is produced specifically for this type of application.

Solid State Electronics Corporation features the Model DC-2005 DEE-CEEDUCER in pressure ranges extending from 5 to 5000 psi, gauge, absolute and differential. Solid state circuits provide regulation for a 28 V unregulated dc input, a multivibrator square-wave generator (10 kHz bridge carrier signal) a synchronous demodulator, and filtering for the 0–5 V dc output. Except for the dc supply, all circuits are located within the transducer. Diaphragm resonance frequency varies from 2.5 kHz to 21 kHz with pressure range.

The Bourns Model 2302 is also a dc-dc variable reluctance transducer designed for 28 V unregulated input. Similarly, Smiths Industries pressure transducers operate on 28 V, 400 Hz, aircraft power systems. Working pressures cover the range from −40 psi to +300 psi and transducers will operate over the temperature range from −40°C to +250°C. A pressure capsule, rather than a diaphragm, is used as the force summing device. Deflection of the capsule displaces an armature.

LVDT Pressure Transducers

The LVDT basically consists of a primary coil and two secondary coils wound on an air core and an armature which is free to move axially through the core. The secondary coils are matched and are symmetrically placed with respect to the primary coil. When the armature is symmetrically located in the core, the output is ideally zero. Displacement of the armature from the zero position causes a magnetic asymmetry and an output signal which is linearly proportional to the displacement over a limited range. With reference to the input voltage, the phase angle of the output voltage can assume two values differing by 180° depending on which side of null the armature is positioned. The output voltage is an ac signal at the carrier frequency which has been amplitude modulated by the armature displacement. DC output for pressure transducers is obtained by demodulation of this ac signal. Simple rectification followed by filtering is sufficient but will not preserve the direction of displacement of the armature. Phase sensitive demodulation will preserve this relationship and yield a linear output/input relationship passing through the origin (ideally) and lying in the first and third quadrants of the x—y plane. Bourdon tubes, diaphragms, capsules, and bellows have been used as the pressure sensors. Transducers commercially available have ranges as low as 0.15 psi gauge, absolute or differential and as high as 10,000 psig. Accuracies of ±0.25% to ±0.5% of full scale are typical. Manufacturers of LVDT pressure transducers include: Automatic Timing & Controls, Inc., King of Prussia, Pa.; Biotronex Laboratory, Inc., Silver Spring, Md.; C-E Electronics, Division of AMETEK, Glenside, Pa.; Consolidated Controls Corp., Bethel, Conn.; Daytronic Corporation, Dayton, Ohio; International Resistance Co., Control Components Division, Philadelphia, Pa.; MSA Research Corporation, Evans City, Pa.; Robinson Halpern, West Conshohocken, Pa.; Schaevitz Engineering, Pennsauken, N.J.

PIEZOELECTRIC PRESSURE TRANSDUCERS

The piezoelectric effect was discovered nearly 90 years ago and the most widely known application of the phenomenon is probably the crystal controlled oscillator. Certain crystals are mechanically strained when subjected to an external stress (pressure). The result is the appearance of equal charges of opposite polarity on opposite sides of the crystal. If the direction of strain is reversed, the polarity of the charges is reversed. The physical property is also reversible since electrical polarization of the crystal will produce mechanical strain. Thus, piezoelectric transducers are true energy converters. The charge on the crystal may be measured by placing electrodes on the crystal surfaces and producing a parallel plate capacitor in the simple case. The output voltage, E_0, is given by Q/C where Q is the induced charge on the crystal and C is the capacitance. Lion shows that E_0 is also given by gtP, where g is a constant depending on the crystal [the voltage sensitivity, in (volts/meter)/pascal], t is the crystal thickness and P the applied pressure (3). In general, piezoelectric transducers lack response to static pressure inputs and are suitable only for dynamic pressure measurements.

A large number of materials exhibit the piezoelectric effect but only a few are suitable for practical transducers. These include: quartz, ammonium dihydrogen phosphate (ADP), potassium dihydrogen phosphate (KHP), lithium sulfate (LH), and a number of ceramic compositions involving barium titanate, lead zirconate-titanate, or lead niobate. The piezoelectric elements are cut from the crystal with particular crystallographic orientations. This affects the direction and magnitude of the piezoelectric effect and the mode of operation with respect to compression, expansion, or shear. Multiple element units can give rise to bending and twisting motions. Further discussion of these points may be found in references (3) and (4). Ceramic elements may be formed from powders into desired shapes and sizes by pressing and sintering.

Piezoelectric transducers are high output-impedance devices and therefore require high input-impedance signal conditioning stages for coupling to lower input-impedance readout devices. Voltage amplifiers, containing an internal voltage follower input stage, and charge amplifiers have been used for this purpose. The latter may be viewed as an operational amplifier integrator circuit. Recent developments utilizing microelectronic integrated circuits have placed this signal conditioning function within the transducer case so that direct hookup of the transducer output to the measuring device is possible. Manufacturers using the integrated circuit concept include pcb Piezotronics, Buffalo, N. Y., and Columbia Research Laboratories, Inc., Woodlyn, Pa. Columbia Series P101, P102, and P103 transducers are available in dynamic pressure ranges from 5 to 1000 psi with temperature operating limits of 0 to 165°F. Piezotronics quartz transducers operate from −65 to +250°F and have pressure ranges 0–100, 0–3000, and 0–10,000 psi, with or without acceleration compensation. Sensitivity of the uncompensated 0–100 psi model is 40 mV/psi while that for the 0–10,000 psi range is 0.5 mV/psi.

Characteristics of some selected commercially available conventional piezoelectric pressure transducers are summarized in Table 7. Sensitivities are expressed either as voltage sensitivity (mV/psi) or charge sensitivity (picocoulombs/psi). Many of these manufacturers also supply complete transducer system electronics.

Table 7. Characteristics of Selected Piezoelectric Pressure Transducers

Manufacturer	Model	Element	Range (psi)	Frequency Response	Sensitivity
Atlantic Research Corp., Alexandria, Va.	LC-70	Lead Zirconate-Titanate	To several hundred	Res. freq., 150 kHz	150 mV/psi 120 pC/psi
	LD-80M	Lead Zirconate-Titanate	To 10,000	Res. freq., 200 kHz	110 mV/psi 4.4 pC/psi
Columbia Research Laboratories, Inc., Woodlyn, Pa.	100-P	—	0.005 to 4000	0.02 Hz to 12 kHz	200 mV/psi
	P-311	—	To 4000	To 12 kHz	100 mV/psi 35 pC/psi
Elastronics Laboratories, Tarzana, Calif.	EBL6006	Quartz	−7 to 70	Nat'l. freq., 35 kHz	2.4 pC/psi
	EMB6601	Quartz	0–100,000	Nat'l. freq. 50 kHz	0.214 pC/psi
Endevco, Pasadena, Calif.	2501-500	Ceramic	To 500	2–9000 Hz	62 mV/psi 25 pC/psi
Kistler Instrument Corporation, Clarence, N. Y.	606A	Quartz	To 3000	Res. freq., >130 kHz	5.5 pC/psi
	607B	Quartz	To 70,000	Res. freq., 240 kHz	0.15 pC/psi
Metrix Instrument Houston, Texas	5016	—	To 3000	First res. >30 kHz	20 mV/psi 120 pC/psi
pcb Piezotronics Incorporated, Buffalo, N. Y.	111A	Quartz	To 3000	Res. freq., 400 kHz	0.35 pC/psi
	118A	Quartz	To 40,000	Res. freq., 300 kHz	0.14 pC/psi

REFERENCES

(1) BEHAR, M. F., "Handbook of Measurement and Control," The Instruments Publishing Co., Inc., Pittsburgh, Pa., 1959, Chapter 3.

(2) ZENCHELSKY, S. T., J. Chem. Ed., 40, A611; A771; A849 (1963).

(3) LION, K. S., "Instrumentation in Scientific Research," McGraw-Hill Book Co., New York, 1959.

(4) NEUBERT, H. K. P., "Instrument Transducers," Oxford University Press, London, 1963.

(5) ARONSON, M. H., ed., "Pressure Handbook," Instruments Publishing Company, Inc., Pittsburgh, Pa., 1963.

(6) Buyers Guide Issue, Instruments and Control Systems, Rimbach Publications Division of Chilton Co., Philadelphia, Pa., 1969.

(7) Buyers Guide, Electronics, 40, No. 22A (1967).

(8) Laboratory Guide, Anal. Chem., 40, No. 9 (1968).

(9) PERINO, P. R., Instruments and Control Systems, 38, No. 12 (1965).

(10) FOLDVARI, T. L., and LION, K. S., Instruments and Control Systems, 37, No. 11, 77(1964).

(11) LION, K. S., U. S. Patent No. 3,012,192.

(12) LION, K. S., Instruments and Control Systems, 39, No. 6, 157(1966).

(13) Rosemount Engineering Company, Minneapolis, Minn., Bulletin 1672.

(14) EPSTEIN, S., Instruments and Control Systems, 34, No. 10(1961).

Topics in...

Chemical Instrumentation

...a **ChemEd** feature

Edited by **S. Z. LEWIN,** New York University, New York 3, N. Y.

The Journal of Chemical Education
Vol. 43, pages A191, A295
March and April, 1966

XXVI. Instrumentation for Osmometry

Peter F. Lott and Frank Millich, *Department of Chemistry, University of Missouri at Kansas City, Kansas City, Missouri 64110*

A variety of "osmometers" are now available from a small number of instrument manufacturers. The appellation "osmometer" is used rather loosely to apply generally to a more inclusive class of instruments which are designed to measure colligative properties of solutions and include, for instance, instruments designed for cryoscometry and ebulliometry. On the other hand, instruments which function on the basis of the phenomenon of osmosis are generally referred to as "membrane osmometers." The distinction is important because the two sub-classes generally are not applied to the same subjects of study. This article treats the subclasses separately, and includes coverage of theory and applications of the methods and a survey of commercially available instruments.

Membrane Osmometers

When two solutions of different concentrations are placed in contact with each other, the spontaneous tendency is for equalization of the concentrations by mutual interdiffusion. Should these two solutions be separated by a "porous" divider of such permeability that the solute cannot pass through the barrier the solvent will diffuse from the compartment of lower solute concentration to the compartment

of higher concentration (Fig. I). If the flow of solvent is opposed, the diffusion tendency is appreciated as a pressure imbalance, known as osmotic pressure. Osmometry is the quantitative measure of the pressure, determinable by applying a measurable opposing pressure necessary

to counteract the flow. Pure osmometry would thus consist of measurements of osmotic pressure and would be performed with membrane type osmometers. Membrane osmometry is usually applied to soluble substances of molecular weights in the range 20,000 to 200,000 and further, with lessening accuracy, to 1,000,000, beyond which the osmotic effect becomes too small to measure. If the sample is of a polydispersity of molecular weights, then osmometry yields a colligative average value, a number-average molecular weight.[1] The study of mechanisms of polymerizations—inclusive of biopolymers and synthetic molecules—requires an understanding of polydispersity. An index of this can be gained by comparing number-average molecular weight with other molecular weight averages (e.g., weight-average molecular weight obtained from light scatterng). There are few convenient, general physical or chemical methods that are alternative to membrane osmometry for number-average molecular weights in the range of 20,000 to 1,000,000. Therefore, osmometry is presently a vital tool to polymer chemistry.

[1] The number-average molecular weight, $\bar{M}_N$, is:

$$\bar{M}_N = \frac{\sum\limits_{i=n}^{\infty} N_i M_i}{\sum\limits_{i=n}^{\infty} N_i}.$$

This average is distinct from other averages which can be calculated or determined from the same polydispersity of N_i number of species of molecular weight M_i. For example, the weight-average molecular weight, $\bar{M}_w$, is given by

$$\bar{M}_w = \frac{\sum\limits_{i=n}^{\infty} N_i M_i^2}{\sum\limits_{i=n}^{\infty} N_i M_i}.$$

Dr. Peter F. Lott obtained his Ph.D. degree in physical chemistry from the University of Connecticut in 1956. He received his B.S. (1949) and M.S. (1950) from St. Lawrence University. Dr. Lott has worked as a research chemist for DuPont and the Pure Carbon Company. He previously taught at the University of Missouri in Rolla, and at St. John's University in Jamaica, New York. Dr. Lott is in charge of the analytical program at the University of Missouri at Kansas City, and is teaching both undergraduate and graduate analytical chemistry. His research interests lie in analytical and physical chemistry. He has published about 20 articles pertaining to the development of new methods of analysis and has previously contributed articles to "Topics in Chemical Instrumentation."

Dr. Frank Millich is an associate professor in the Polymer Chemistry division of the Chemistry Department of University of Missouri at Kansas City. In the doctoral curriculum in polymer chemistry, which he founded at U.M.K.C., he presides over the graduate lecture and laboratory courses devoted to instrumental characterization of polymer solutions. Dr. Millich received his education at City College of New York (B.S., 1949), and Polytechnic Institute of Brooklyn (M.S., 1956; Ph.D., 1959). He held American Cancer Society post-doctoral research fellowships at Cambridge University, England, (1958–59), under Sir A. R. Todd, and at the University of California, Berkeley (1959–60), under Dr. M. Calvin. He has consulted for industry, and has five years of industrial experience in research.

Chemical Instrumentation

The free-energy lowering per mole of solvent, ΔF, caused by the presence of non-volatile solute is given in terms of the lowering of the equilibrium vapor pressure from p_0 for the pure solvent to the value of p for the solution; thus,

$$\Delta F = RT \ln p/p_0 \qquad (1)$$

where R is the gas constant and T is the absolute temperature. The free energy increase per mole of solvent, when it is subjected to an excess pressure, π, is

$$\Delta F = \pi V \qquad (2)$$

where V is the volume of a mole of solvent, and is assumed to be independent of the pressure changes experienced in the normal osmometric determination—a reasonable assumption for liquids.

At equilibrium, balance between these opposed phenomena is achieved, and

$$\pi V = -RT \ln p/p_0 \qquad (3)$$

A relation must be known between the vapor pressure and composition of the solution in order to derive an expression in terms of molecular weight. Such a relation is provided by Raoult's Law which is valid only with cases of ideal mixing (i.e., in the absence of strong changes in solvent-solute interactions). In many actual cases and especially in polymer solutions, Raoult's Law behavior is approximated only in the limit of infinite dilution. With these limiting conditions:

$$p/p_0 = N_{solvent} = 1 - N_{solute} \qquad (4)$$

and

$$\pi V = -RT \ln(1 - N_{solute}) \cong RT(N_{solute}) \qquad (5)$$

where N_{solute} is the mole fraction of solute and (in terms of n number of moles) is equal to $n_{solute}/(n_{solute} + n_{solvent})$. The approximation of $\ln(1-N) \cong N$, also is valid in the limit of infinite dilution, where the solute mole fraction is very much less than unity. As a consequence of these assumptions it is required that, in general, osmotic pressure measurements are made for a series of finite concentrations and graphically extrapolated to zero concentration.

With the reasonable approximation that in infinite dilution $V_{solvent}$ may be replaced by $V_{solution}$ the desired expression for osmotic pressure, extrapolated to infinite dilution, is obtained:

$$(\pi)_{c=0} = \frac{n_{solute}}{RTV_{solution}(n_{solute} + n_{solvent})} \cong RTm = RTc/M \qquad (6)$$

where m is molarity of solute in the solution, c is concentration in grams of solute per liter of solution, and M is the solute molecular weight in grams per mole.

Illustrated in Figure 1 is a simple type of static osmometer (Welch Scientific Co.). Here the solvent is retained in the beaker and the solution is maintained in the chamber over the membrane. Flow of solvent from the beaker through the membrane

Figure 1. Welch Scientific Co., Osmometer, No. 0540A.

increases the volume of the solution which is reflected by a rise of liquid in the narrow capillary tube until a sufficient hydrostatic head is produced to balance the osmotic pressure. Since the pores of the divider are nearly of molecular size and mixing of solvent and solution generally occurs by diffusion, the static method is inherently slow and the rate of approach to equilibrium is asymptotic. Complete equilibration may require over 24 hours per determination.

A modified static osmometer useful for certain routine measurements is the Hellfritz osmometer, Figure 2. This instrument, supplied by Carl Schleicher and Schuell Co., is suitable for certain routine osmotic pressure measurements, particularly for molecular weights from about 2000 to 50,000. The instrument is supplied either as a single or double chambered osmometer, requires a sample volume of 2–4.5 ml, and has a comparison capillary for ease in reading of the osmotic pressure. It is suitable for investigation of membrane characteristics and may be operated from −10 to 200°C.

Much of the research in polymer chemistry requires measurements of the osmotic pressure of a large number of solutions, as for example in the evaluation of concentration dependencies in polymer solutions, or in the extrapolations for molecular weights to infinite dilution. Accordingly, time-saving factors are an important consideration. An alternative to the static methods are the dynamic membrane osmometers, instruments in which the pressure differential existing between the solution and solvent side of the membrane is sensed immediately and a corresponding counter pressure is applied to the solution side of the osmometer. Thus, with a minimum or negligible amount of solvent transport, a final pressure measurement may be made as soon as the divider comes to equilibrium with its environment and thermal equilibrium is achieved for the system. Depending on the chemical

nature and mass of the divider, and the heat capacities of the thermostating system components, measurements on presently available instruments can be made in a matter of minutes.

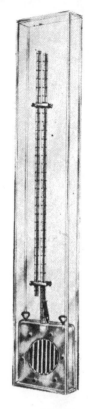

Figure 2. Hellfritz Rapid Osmometer.

A beneficial by-product of the shorter operating-time feature in the dynamic method is the achievement of measuring truer (higher) osmotic pressures (thus, lower molecular weight values) than are apparently evident in static determinations, with low molecular weight samples (i.e., 15,000–60,000). Two causes contribute to smaller resultant pressures in the static method. The static method permits some solvent transport and a volume increase of the solution which dilutes the original concentration. More important, the pore size of the membranes, which partially determine the lower limit of measureable molecular weight, is not uniformly single-valued. For this reason and others, a finite amount of low molecular weight material seeps across the membrane wall into the pure solvent. This causes a reduction in the vapor pressure in the pure solvent compartment, and an increase in the original vapor pressure of the solution compartment due to a lessened concentration; it also produces a truncated polymer distribution of molecular weights in the solution of a polydisperse sample; that is, it produces an enrichment of the solution in the higher molecular weight constituents. Short periods of operating time reduce the amount of such solute seepage.

Dynamic osmometers based upon the work at Shell Development Co. (1) are offered by both Hallikainen Instruments as their Model 1361 osmometer (Fig. 3) and by Dohrmann Instruments Co. as their Model M-100 osmometer (Fig. 4).

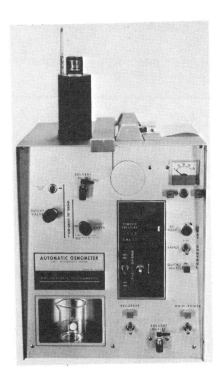

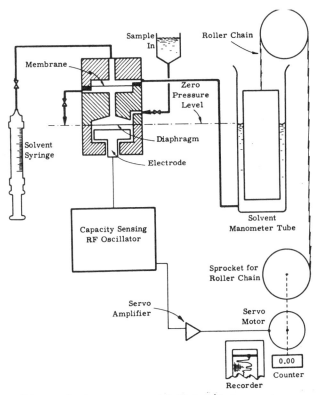

Figure 5. Schematic diagram of Hallikainen/Shell osmometer.

Figure 4. Dohrmann model M-100, Automatic Osmometer.

A block diagram of the instruments is shown in Figure 5. In this osmometer system the cell compartment consists of two cavities separated by a semi-permeable membrane. The bottom of the sample half-cell is framed by a thin metal diaphragm which responds to changes in volume. Displacement of this beryllium-copper diaphragm due to solvent flow across the membrane is sensed as a change in electrical capacitance in an oscillator circuit in which the diaphragm is one-half of a capacitor in a tuned R.F. circuit. This change in the frequency of the oscillator circuit is amplified and fed into a servo mechanism. The servo motor, through appropriate gearing, drives a plummet in a manometer tube, so as to adjust the solvent head to zero osmotic flow and return the pressure sensing dia-

phragm as well as the oscillator circuit to its initial null balance state. The 0.0015 inch thick pressure sensing diaphragm is spaced 0.002 inch from the stationary electrode by means of synthetic rubies. Thus, under operating conditions, the diaphragm is displaced only a fraction of one micro inch. To prevent erratic volume changes during osmotic pressure measurements, the temperature of the osmometer cell is regulated to keep the temperature change to less than 0.001°C per minute. The over-all sensitivity of the servo system is such that it operates on a 10^{-6} cm deflection at the center of the diaphragm which corresponds to a cell volume change

Figure 6. Exploded view of Hallikainen osmometer cell and oscillator.

of about 10^{-3} microliter, and results in a change of only 0.01 cm of head in the solvent manometer. Since the beryllium copper diaphragm is in contact with the sample it can become corroded under certain conditions, such as with acid solutions.

The semi-permeable membrane of the osmometer is quite rigidly supported between the concentric ridges of the two half cells, Figure 6. The instrument is capable of operating at any temperature from about 35 to 135°C and the adjustment of temperature is continuous within this range. The instrument requires a sample volume of about 5–10 ml, and the osmotic pressure is read after a period of about 5–10 minutes. The pressure is read directly to one-hundredth of a centimeter over a 10 cm range on a mechanical counter. The balancing servo mechanism concurrently drives the pen on a built-in recorder which permits the operator to observe the balancing process and to ascertain that equilibrium has been reached.

A series of dynamic osmometers are offered also by Mechrolab, Inc., Figure 7. In these osmometers an air bubble is trapped in a capillary in the solvent side of the osmometer membrane compartment. In operation, a light source is focused on the upper meniscus of the air bubble, and a photocell is placed at a right angle to the light beam. Fluid flow across the membrane acts to displace the air-solvent meniscus. This displacement of the meniscus changes the refractive index in the path of the monitor beam and, in turn, the intensity of the light reaching the photocell Figure 8. The detector cell analyzes this change and sends a signal which is amplified and transmitted to a servo system which adjusts the height of a hydrostatic solvent reservoir on a 0–20 cm turn screw stack to restore the capillary air bubble meniscus to its initial null balance position. The pressure is read directly to ±0.01 cm on a mechanical digital counter on the face of the instrument. A recorder output is also provided to chart the dynamics of the osmotic pressure measurements. With solutes that partially permeate through the membrane, the chart record permits extrapolation to pressure at "zero time" and also indicates when the equilibrium pressure has been attained. The instruments are designed to operate in the range from 5 to 135°C. A minimum of approximately 0.5–1 ml of sample is required for a single determination.

In membrane osmometers, organo-colloid membranes are used which permit the determination of molecular weights from below 20,000 up to 1 million. Membranes for osmotic pressure measurements in these instruments may also be purchased from ArRo Lab, Inc., and Carl Schleicher and Schuell, Inc. In these instruments, generally gel membranes are employed for the osmotic pressure measurements. The selection and treatment of the membrane is important since it governs the species which permeate through the membrane.

The osmotic dividers, generally, may be constituted of either metal, ceramic, or, more commonly, of synthetic or animal tissue membranes. Since the latter are flexible and since large exposed areas provide faster equilibration rates, the cells are constructed to give appropriate mechanical

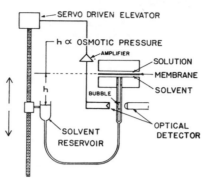

Figure 8. Block diagram of optical detector system of Mechrolab, Inc. membrane Osmometer.

support and restraint to the membrane position so that the pressure differences between the fluids do not deform the membranes sufficiently to alter the usually small volume capacities of the fluid chamber.

Membrane permeability toward non-electrolytes is not solely a process of mechanical filtration based only on the sizes of the pores, it can be appreciated that adsorption phenomena also take place among the membrane, the solvent, and the solute. The flexible membranes are swollen by the liquid media, thus chemical affinities can be expected to play a role in the ease of transport of solute into or through the membranes. Accordingly, different solvents will swell

flexible membranes to various degrees. In practice, it is often necessary to slowly condition the membranes to a change of solvent, to degas the membrane in warm solvent and just prior to measurement, to contact membranes with rinse portions of the sample. A conditioning schedule which permits too sudden a change in the membrane ("shocking" the membrane) will cause distortion and buckling of the membrane which completely destroys its utility. This is easily noticeable with anisotropic membranes (as often occurs in membrane sheet which was calendered in its manufacturing process), as a membrane of circular cross section can be seen to become oval in shape. With an isotropic membrane, too rapid a conditioning schedule to solvent narrows or collapses the effective pore size which may occlude the previous solvent and thereby alter the transport characteristics and uniformity of the membrane. Accordingly, the selection and treatment of the membrane is an important concern in the experimental measurements and effectively determines the lower limit of measurable molecular weights. It is also a corollary that present instrumentation is capable of extension to lower molecular weight samples simply through improvement in the permeability characteristics of membrane materials.

Other Applications of Membrane Osmometry

Typical extrapolations of membrane osmometry data from dilution series of two samples of a non-electrolyte polymer is shown in Fig. 9. The intercepts yield the number-average molecular weights for the two samples; the positive slopes of both curves reveal the concentration dependence of osmotic pressure. The latter data contains valuable information.

A steep and upward-concave curve is obtained for solutions in which the solvent is especially effective in dissolving the polymer, and expanding the intra-twined, maze-like conformation of a flexible macromolecule to occupy a large region of the medium. Conversely, a linear curve with zero slope may be obtained for solutions of the same polymer in which the solvent-polymer solute interactions are nil. (By a judicious choice of solvent one may exploit the latter fact to minimize error or possibly avoid extrapolation.)[2]

Theoretical interpretations exist concerning the significance of the concentration dependencies. These attribute the magnitude of the slopes to entropy factors; specifically, to the dissimilarity in size and conformational possibilities between the solvent and polymeric solute molecules; and to enthalpy effects of molecular interactions.

Serial dilution data are valuable in another way. Aggregation or dissociation

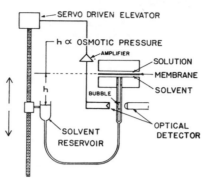

Figure 7. Hewlett-Packard Model 503 membrane osmometer (formerly Mechrolab).

---

[2] This is a desirable ploy when osmotic pressure molecular weights are desired for polyelectrolytes which are not at their isoelectric points. Pronounced curvature at high dilution, caused by conformational expansion and coulombic interactions in each molecule may be suppressed by introducing simple electrolyte (e.g., NaCl) into the solution, providing special effects are absent.

318

phenomena occurring over the concentration range being studied may be detected and confirmed. This may be revealed when the molecular weights which are calculated from extrapolation of data taken at high concentration are found to be greater (sometimes by whole number multiples) than the molecular weights obtained by extrapolation of data taken at low concentration where a high polymer has dissociated into polymeric sub-units. An example of this has been observed for solutions of ethyl cellulose (3).

Thus, membrane osmometry is sometimes conducted on interesting synthetic macromolecules, and biopolymers, such as poly(α-amino acids) and polynucleotides, expressly to gain more detailed information of physical characteristics.

Indirect Methods of Measuring Osmotic Pressure

Advances are currently being made in the development of new membrane materials of lower permeability limits. However, conventionally used organocolloid membranes show permeability to solutes of molecular weight below about 20,000. When it is desired to know the osmotic pressure of solutions containing solutes of low molecular size, indirect methods must be used to measure other colligative properties of the solutions. Nearly all of the methods are designed to measure a temperature effect which is of small magnitude. Further, precision in such a case diminishes as the molecular weight of the sample increases, to a practical upper limit of not much beyond 1000.

Examination of Table I is instructive (4). The greater magnitude of the physical effect produced in osmosis (col. 4) is in marked contrast to that in ebullioscopy or cryoscopy.

Table I. Comparison of Calculated Boiling Point Elevation, Freezing Point Depression, and Osmotic Pressure for Polymers in Benzene

Molecular weight	$(\Delta T_b/c)_{c=0}$ (°C/ (g/100 ml))	$(\Delta T_f/c)_{c=0}$ (°C/ (g/100 ml))	$(\pi/c)_{c=0}$ ((g/ cm²)/ (g/100 ml))
10,000	0.0031	0.0058	25
50,000	0.0006	0.0012	5
100,000	0.0003	0.0006	2.5

With regard to sensing small temperature differences, the present commercial instruments employing thermistors are capable of detecting a change of ±0.001°C. If other factors, such as temperature control, convection currents, ideality of the physical processes, etc., were by themselves neglected, it can be seen that since the temperature depression of solute of molecular weight of 10,000 is 0.003°C, consequently, the best accuracy of molecular weight determination would be 33%. Actually, the use of standards for molecular weight around 200 in concentration of 0.1% and higher will usually yield results of 1–2% error in the hands of chemists of average skill.

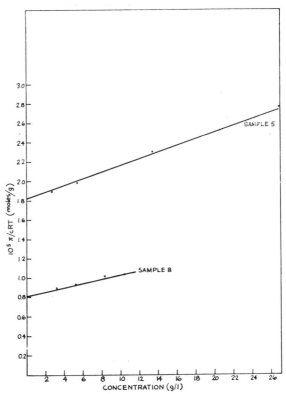

Figure 9. Concentration dependence of reduced osmotic pressure of poly(α-phenylethylisonitrile) in degased benzene; sample 5 at 37°C and sample 8 at 30°C (F. Millich and R. G. Sinclair, II, unpublished data).

Commercial instruments are lacking for precise determination of molecular weights in the range of 1000 to 20,000 under present day circumstances, though it has been reported (5, 6), for instance, that with elaborate refinements, special design and developed skill it has been possible to measure boiling point elevations with a precision of 0.0002°C and molecular weights of polyethylene up to 35,000 with an error not exceeding 10%.

In the discussion of membrane osmometry the osmotic pressure was related to the decrease in vapor pressure of the solvent produced by the presence of a non-volatile solute. This decrease is also responsible for a lowering of the freezing point of the solution, say from temperature T_m to T. The pure liquid solvent has a vapor pressure p_0 which is also the vapor pressure of pure solid solvent at T_m. At the lower temperature T the vapor pressure of the pure solid solvent is at the lower value p, and is in equilibrium with the dilute solution.

The Clausius-Clapeyron equation provides a relationship between a pressure change and a small temperature change:

$$d(\ln p) = \frac{\Delta H}{RT^2} dT \quad (7)$$

It assumes that the vapor behaves as an ideal gas.

The change in vapor pressure which the solid suffers due to a temperature decrease is equal to the change in pressure which the liquid suffers due to a change in temperature and to the presence of solute.

Thus,

(solid) (solution)
$$\frac{\Delta H_s}{RT^2_m} dT = \frac{\Delta H_v}{RT^2_m} dT + d(\ln p)_{T_m} \quad (8)$$

An expression for the last term in equation 8, in terms of concentration, may be obtained, starting with equation 4, if one makes the same assumptions of Raoult's law behavior and accepts the same limitations of applicability to only dilute solution. Then,

$$\ln p = \ln N_{solvent} + \ln p_0 \quad (9)$$

A very slight change in solvent concentration, $dN_{solvent}$, brought about by the small addition of solute, dN_{solute} allows:

$$d(\ln p) = d(\ln N_{solvent}) = \frac{dN_{solvent}}{N_{solvent}} = \frac{-dN_{solute}}{N_{solvent}} \quad (10)$$

Since $N_{solvent}$ is practically unity,

$$d(\ln p)_{T_m} = dN_{solute} \quad (11)$$

Substituting into equation 8, and noting that the heat of fusion, ΔH_F, is equal to the difference between the heat of sublimation, ΔH_S, and the heat of vaporization, ΔH_V, one obtains:

$$\frac{\Delta H_F}{RT_m^2} dT_m = -dN_{solute} \quad (12)$$

As was done in equation 6, this result may be rewritten in other terms (i.e., molality, m), and, for small temperature changes, equation 12 becomes:

$$\Delta T_m = \frac{-RT^2_m}{\Delta H_F n_{solution}} m = -K_F m \quad (13)$$

The freezing depression constant K_F is generally not found to be very large for many solvents. It varies for different solvents and is smaller for water than it is for other common solvents, such as benzene. Values of K_F are based on the reasonable assumption that ΔH_F does not change in the range of the temperature depression, and the tacit assumption that pure solid solvent, and not a solution, freezes out during the determination. The temperature change is not very great, especially in dilute solution. Of course, equilibrium conditions are supposed for the determination, as well as infinite dilution and ideality for the solution behavior. Conventionally, though, solutions of 0.1 weight percent and higher are used, and moderately high molecular weight samples require even higher concentrations; physiological solutions are relatively concentrated—plasma, for example, corresponds to about 0.14 molar concentration. Extrapolation to infinite dilution is thus often difficult or impossible to carry out with precision.

Expressions may be similarly derived for other colligative property processes. For boiling point elevation, or for isothermal distillation (as with the Mechrolab vapor pressure osmometer), rather than equation 8 one need only use equation 7, in which ΔH is that of vaporization.

If the solutes are electrolytes in solution, or undergo dissociation-association phenomena as the solution concentration is changed, the equations presented above must be modified to accommodate the additional complexities. Greater concern must be attended to the limiting assumptions, and more specific values for the thermodynamic properties will have to be selected. In the hands of a skilled physical scientist, however, the instruments may be applied to studying the intricacies of such phenomena.

The quantitative relationships are necessary for understanding the significance of the absolute value which is obtained through the experimental physical measurements. Thus, the osmotic pressure or temperature differences were related to the concentrations of investigated solutions and to molecular weights. However, commercial colligative property instruments are frequently applied to such solution samples in which the significance of measured results are correlated to much more complex phenomena. Thus, one finds clinical applications of these instruments in the medical field, as for instance, the correlation of "osmometric" results on urine, cerebrospinal fluid, and blood sera to the recognition of pathological abnormalities. Implications have been found for aberations in serum sodium content, in serum toxication by drugs, for evidence of oliguria, and for liver and renal malfunctions (7).

Commercial freezing point osmometers essentially monitor the temperature changes of a liquid sample while the solution is carried through a controlled freezing cycle. The liquid sample is rapidly supercooled to a predetermined temperature. The cooling is terminated and the supercooled liquid is then maintained in a

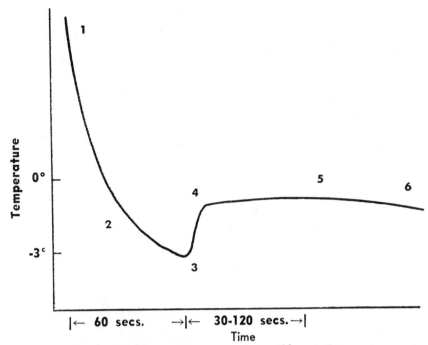

Figure 10. Cooling curve: 1) Rapid cooling of sample at about 1°/second, 2) Slow cooling to reduce temperature gradients within sample, 3) Seeding to initiate crystallization, 4) Rapid warming to freezing point, 5) Temperature plateau, 6) Slow cooling to ambient temperature.

cold environment in which there is a less rapid transfer of heat, so that the supercooled liquid becomes thermally uniform. While the sample is still in this supercooled state, crystallization is started by external means (usually a rapid vibration of a stirring rod). This initial freezing is not complete; consequently, a slush formation is observed in which a small percentage of the sample turns solid. The formation of this slush acts as an insulation around the temperature probe. While the sample continues to freeze, heat of fusion is continuously liberated. Consequently, the temperature of the sample rises to the freezing point of the solution, and the sample remains at this temperature plateau for a period of time. Ultimately, the rate of heat loss is such that the temperature drops to the temperature of the cold chamber. The freezing point of the sample is generally taken as the temperature existing at a specified time on this temperature plateau; or the temperature obtained through extrapolation to "time zero" on the plateau. A typical cooling curve is shown in Figure 10.

Because solvent crystallizes out in the freezing step, the concentration of the solution changes. Consequently, the plateau for a solution is not horizontal, as is the case for the pure liquid. Accordingly, for an initial calibration of the instrument, standard solutions of known concentration are prepared, and their temperature plateau is measured in this manner. The operator then runs the unknown under identical conditions as were used with the standard solution. About five minutes are required to run each sample.

The standard solutions are prepared in the concentration units corresponding to the design of the instrument. For example, if the instrument is made to read directly for aqueous solutions in units of milliosmolality, a series of standard osmolal solutions is prepared and the freezing points are employed as the means of calibrating the instrument. For example, a solution containing 3.089 g of NaCl per Kg of water would be 100.0 milliosmolal; it would freeze at $-0.1858°C$. Accordingly, the instrument would be adjusted so that this freezing point would correspond to the 100 milliosmolal setting

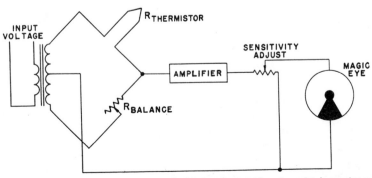

Figure 11. Functional diagram for detecting unit of Aminco-Bowman freezing point depression apparatus.

on the instrument dial. The commercial instruments may be offered to read in other scale units such as directly in °C or "Hortvet" scale as is used in milk testing.

In basic design, the commercial freezing point osmometers are quite similar, although the different instruments vary in their features. The instruments all require a means of freezing the sample. This may be done, for example, either with a refrigerator, cooling bath, or a thermoelectric device.

Since the concentration of the unknown is detected by comparative measurements, it is quite important that the samples and the temperature sensing probe be positioned identically and that the "seeding" of the sample to initiate crystallization be performed reproducibly. Manufacturers differ somewhat as to the means by which this is done. The instruments, however, all employ thermistors as the temperature sensing device. This thermistor (which is really a resistor with a large temperature coefficient of resistance) is incorporated as one arm of a Wheatstone bridge. The deflection of the null balance indicator of the bridge (galvanometer) is noted, and when the temperature plateau is reached the variable arm of the bridge is adjusted so as to rebalance the bridge. The reading of the variable arm of the bridge thus gives the concentration of the sample directly, and in the appropriate units in which the instrument was designed. A block diagram of the temperature sensing circuit for the American Instrument Co., Inc. instrument, which employs a "magic eye" rather than a galvanometer as the null detector, is shown in Figure 11. Since the commercial instruments all employ similar thermistors, they have the same temperature difference detection limit of about ±0.001°C. A block dia-

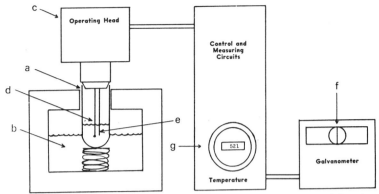

Figure 12. Block diagram of Advanced Instruments, Inc. Laboratory Osmometer: a) Sample and test tube resting on spring, b) Ethylene glycol bath held at −6°C, c) Operating head, d) Temperature probe (thermistor), e) Stirring wire, f) Galvanometer spotlight, g) Temperature indicating dial.

Figure 13. Advanced Instruments, Inc., Hi-Precision Research osmometer.

Figure 14. Advanced Instruments, Inc., Nanoliter osmometer.

gram for the Advanced Instruments, Inc. Laboratory osmometer is shown in Figure 12.

Because of the similarity of the instruments, their representative major features are compared in Table II.

The Mechrolab Division of Hewlett Packard offers also an osmometer for low molecular weight determinations based upon a measurement of the difference in vapor pressure existing between the solvent and the solution. The instrument, shown in Figure 22 consists of two principal units mounted on a common base. The sample chamber assembly contains the various elements of the osmometer,

Table II. Freezing Point Osmometers

Instrument manufacturer	Instrument model	Method of refrigeration	Method of "seeding"	Wheatstone Bridge operation	Comments
Advanced Instruments, Inc.	Laboratory Type L	Thermoelectric Device	Rapid vibration of stirring rod	Battery operated bridge with galvanometer as imbalance detector.	Various modifications of the basic Model K instrument are available for samples in the range of 2 to 0.1 nanoliter[a] with scale readings from +1 to −6°C or 0-3000 milliosmols. Automatic printout is also available. The Advanced Instruments, Inc., Research Osmometer is shown in Figure 13 and the Advanced Nanoliter Osmometer in Figure 14.
	Portable K	Salt-ice mixture in Dewar Flask			
American Instrument Co., Inc.	Air operated unit (Cat. No. 5-2051)	Ranque-Hilsch Vortex tube requiring 40 PSI air at 3.3 cubic ft/min	Add crystals of Ag I	AC Wheatstone Bridge with "Magic Eye" as detector.	The instrument is shown in Figure 15. This instrument requires 1 ml sample volumes and is calibrated in arbitrary scale units. A nanoliter sample volume instrument is also available,[a] Figure 16.
	Dewar Flask Unit Cat. No. 5-2061	Salt ice mixture in Dewar Flask			
Fiske Associates, Inc.	Laboratory Model type G-62	Refrigerator	Rapid vibration of stirring rod	Battery operated bridge with galvanometer as imbalance detector.	Various models are available requiring either 2 ml or 0.2 ml samples. The instrument is calibrated to read directly in the range from 0 to 3000 milliosmols. The Model G-62 Instrument is shown in Figure 17.
	Portable Model type Mark III	Ice-salt mixture in Dewar Refrigerator			
Industrial Instruments Inc.	Model CY-1	Thermoelectric Device	Rapid vibration of stirring rod	AC bridge with amplifier and micro-amp null meter.	Sample volume of 2 ml or 0.2 ml. A prechill compartment to keep samples at 6°C above the temperature of cooling bath is also provided as part of the instrument. The instrument is shown in Figure 18. It is calibrated for 3000 milliosmols range.
	Model CY-2			DC Bridge with DC amplifier and null meter.	Sample size is about 1 ml. Model CY-1 is calibrated for 0-1°C, Model CY-2 is calibrated for 0-3000 milliosmols/kg. The instrument is illustrated in Figure 19. A microscope cold stage is also available and is shown in Figure 20.[a]
Precision Systems	Osmette	Thermoelectric Device	Rapid vibration of stirring rod	Battery operated DC bridge with meter as imbalance detector.	A sample volume of 2 or 0.2 ml. is required. The instrument reads directly for the range from 0 to 3000 milliosmolal. The instrument is illustrated in Figure 21.

[a] The manufacturers also offer a system for the observation of the freezing point of nanoliter samples. In these instruments a cold stage is incorporated with a microscope for visual observation of the freezing point; a thermistor probe is again employed for temperature monitoring. For a review of microscope accessories and microscopy see Lewin, S. Z., J. Chem. Ed., **42**, A565, A619, A775, A853, A945 (1965). [Reprinted in this volume]

Figure 15. American Instrument Co., Inc., model 5-2050 freezing point depression apparatus.

Figure 18. Fiske Associates, Inc. Mark III osmometer

Figure 19. Industrial Instruments, Inc., model CY-2 osmometer.

while the control unit houses a Wheatstone bridge, null detector and heater input control circuit. A sketch of the sample chamber assembly is shown in Figure 23.

In this instrument two thermistor beads are suspended in a precisely thermostated chamber saturated with solvent vapor. The thermistor beads are two legs of a Wheatstone bridge. The bridge is initially balanced with solvent drops placed on each thermistor bead. By means of a syringe the measuring thermistor is "washed" with the sample and a drop of solution is then placed on the measuring thermistor; also a drop of solvent is placed on the reference thermistor. Solvent condenses on the solution drop, because of the lower vapor pressure of the solution. This condensation warms the measuring thermistor and produces a bridge imbalance. The bridge is rebalanced when equilibrium is reached, and the change in resistance is noted. This change in resistance is a relative quantity dependent both upon the solvent and the probe. A different thermistor probe is used for aqueous than for non-aqueous solutions. Care must be taken in the choice of solvents as some solvents will

Figure 20. Industrial Instruments, Inc., model CY-3 thermoelectric microscope cold stage with AC Wheatstone bridge for measuring temperature (resistance) of thermistor detector.

attack the probe elements.

The instrument is calibrated with a solution of a standard compound. Data on an unknown is then compared against the data obtained with the standard.

The instrument may be used thus to determine molecular weights, osmolalities, osmotic concentrations, activity coefficients, etc., depending upon the type of initial calibration. It is capable of operating with a number of solvents. The sensitivity of the instrument varies with the solvent being used. Solvents of low heat of vaporization, such as benzene, toluene, carbon tetrachloride, etc., show a greater sensitivity than those of higher heat of vaporization, such as water. In the sample chamber assembly, six syringe guides are placed around the thermistor probes so that five samples may be run

Figure 16. American Instrument Co., Inc., Micro-sample freezing point depression apparatus.

Figure 21. Precision Systems, Osmette laboratory osmometer.

Figure 22.   Mechrolab vapor pressure osmometer, series 300.

Osmometry
Supplement

A number of developments can be reported in the field of osmometry during the last several years. The performance of some of the older osmometers has been improved and extended. Several new instruments have appeared, and a few old ones have been discontinued.

Another static membrane osmometer is commercially available: the Pinner-Stabin model, supplied by Rho Scientific, Inc., Figure 24.   The cell volume is 3–4 ml, and the effective membrane area is 9 cm.²

Dynamic membrane osmometers are now available in three different designs: the Shell, the Hewlett-Packard (formerly Mechrolab, Figs. 7 and 8), and the Reiff instruments. The Shell osmometer is licensed for manufacture to Hallikainen Instruments (their Model 1361, Fig. 3). (Dohrmann no longer manufactures the similar instrument shown in Fig. 4.)

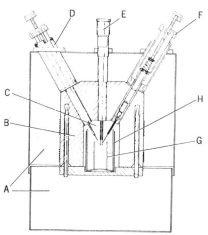

Figure 23.   Block diagram of Mechrolab vapor pressure osmometer: A) Foam, B) Aluminum block, C) Chamber, D) Syringe guide, E) Thermistor probe, F) Syringe in down (loading) position, G,H) Solvent cup and wick.

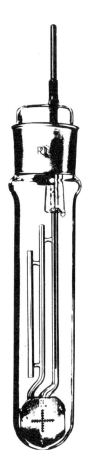

Figure 24.   Pinner - Stabin membrane osmometer, Model 310.

Figure 25.   Melabs-Reiff membrane osmometer, Model CSM-1

sequentially. One syringe channel is retained for adding pure solvent to the reference arm thermistor bead. The thermostat probe is also mounted in the sample chamber assembly and each operating temperature requires a separate thermostatic control element. Because the instrument operates at an isothermal temperature, it is free of complexities associated with the phase transitions of freezing.

Two models are offered. Model 301A may be ordered with thermostat and thermistor probe set for any of the following temperatures: 25, 37, 50 or 65°C. Model 302 offers the same operating temperature choices as well as 100 and 130°C. Intermediate temperature ranges are provided on request. In both models the temperature of the chamber is controlled to within ±0.001°C.

The Reiff osmometer is manufactured by Melabs, Inc. (Fig. 25). It is the latest version of a high-speed membrane osmometer and possesses some unique features. In contrast to the designs previously discussed, this instrument does not incorporate a servo mechanism, but utilizes a diaphragm coupled to a strain-gauge as a pressure transducer (Fig. 26), which is capable of either narrow (0–5 cm H$_2$O) pressure ranges for molecular weight determinations, or very wide (0–100 cm H$_2$O) ranges for osmotic pressure measurements. Elimination of the hydrostatic pressure addition systems allows a compact instrument, but the lack of a digital display of osmotic pressure necessitates calibra-

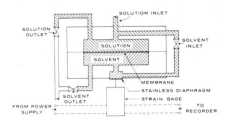

Figure 26. Melabs-Reiff osmometer: details of cell design.

tion of the recorder output. Comparable with the other osmometers discussed, this instrument can detect a flow of 10^{-6} ml. The volumes of the solvent and solution compartments of the cell are 0.5 ml. This osmometer appears to perform equally well with both aqueous and organic solvents. Two models are offered, the CSM-1 and CSM-2, which differ somewhat in the lower temperature range covered; the upper temperature limit is 130°C.

Molecular weight standards of polymer samples are commercially available from ArRo Labs, Inc., and from Waters Associates. Several types of fractionated polymer samples of molecular weights in the range 600–1,800,000 are offered, with low values of the ratio $\overline{M}_w/\overline{M}_n$.

It should be noted that most membranes degrade rapidly in contact with solvents at temperatures above 70°C. A pamphlet describing performance characteristics of commercially available membranes in use in high-speed osmometers can be obtained from ArRo Labs, Inc.

An alternative to freezing point depressions for the determination of low molecular weights, which avoids the presence of a solid crystalline phase, is the employment of vapor phase osmometers. These are based on the "isopiestic" method, otherwise known as "isothermal distillation." Actually the method involves an adiabatic process of solvent distillation at a fixed temperature. Equations 7–12 apply, and eqn. 13 now becomes:

$$\Delta T = \frac{RT^2}{1000\ \Delta H_v}\, m \qquad (14)$$

where ΔH_v = heat of vaporization per gram of solvent, in reference to the temperature of operation. For the real case, however, small heat losses occur, causing additional amounts of solvent to

distill in the cycle, giving rise to a steady-state temperature differential slightly less than that predicted by eqn. 14. The dilution effect is negligible, and standard substances, e.g., pure benzil or benzoic acid, are used to calibrate the instrument, resulting in a curve which correlates m to the effective ΔT.

There are now two commercially available vapor-phase osmometers in the United States. These are the Hewlett-Packard Model 300 series (previously described) and the Perkin-Elmer Model 115. The latter is distributed through the Coleman Instruments division of Perkin-Elmer.

The Perkin-Elmer Model 115 osmometer is depicted in Figures 27–30. Excellent temperature regulation is achieved in this

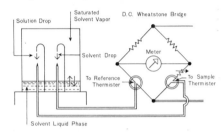

Figure 28. Principle of operation of the Perkin-Elmer Model 115. Drops of solvent and solution are placed in contact with thermistors in a chamber saturated with solvent vapor. The two thermistors are compared in a Wheatstone bridge circuit.

instrument. The sub-oven is designed to maintain a temperature of 3–10°C lower than the main oven, so that some of the heat in the main oven may discharge constantly into the sub-oven. The cell has a dual chamber construction. The condensation or evaporation of solvent vapor in the outer chamber acts to maintain the temperature of the vapor in the inner chamber. The thermistors are mounted in an inverted position with a

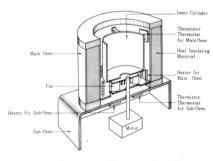

Figure 29. Diagram of thermostatted oven, Perkin-Elmer Model 115.

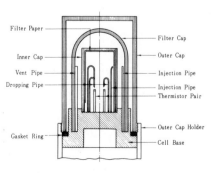

Figure 30. Thermistor and sample cell, Perkin-Elmer Model 115. This assembly fits into the inner cylinder of Figure 29.

Figure 31. Cryomatic osmometer, Model 31CM, Advanced Instruments, Inc.

platinum ring at the uppermost tip, to provide a resident solvent pool. This geometry eliminates variations in the size of the liquid drops. Continuous operation is made possible by the provision of a drain to remove liquids accumulating in the reservoir. The electronic design is all solid-state. An SCR proportional temperature controller permits easy change of operating temperature. The manufacturer's claim of a tenfold greater sensitivity has been verified (8). The detectability of a temperature difference of 10^{-4}°C provides a reliable bridging of the gap between the lower limits of the dynamic membrane osmometer and instruments applicable to low molecular weight determination.

Figure 32. Fiske Model QF clinical osmometer

Figure 33. "Osmette A" osmometer, Precision Systems, Inc.

Figure 27. Perkin-Elmer vapor-phase osmometer, Model 115. Sample-handling controls project from the tower, upper left.

324

Addresses of Manufacturers

An alphabetical listing of the addresses of the manufacturers of these instruments and suppliers of membranes for osmometry follows. These instruments are generally sold directly by the manufacturer and not through laboratory supply houses.

Advanced Instruments, Inc., 45 Kenneth St., Newton Highlands, Mass. 02161

American Instrument Co., Inc., 8030 Georgia Ave., Silver Spring, Md. 20910

ArRo Labs, Inc., P. O. Box 686, Joliet, Ill. 60434

Beckman Instruments, Inc., 89 Commerce Rd., Cedar Grove, N. J. 07009

Fisher Scientific Co., 711 Forbes Ave., Pittsburgh, Pa. 15219

Fiske Associates, Inc., Quaker Highway, Uxbridge, Mass. 01569

Hallikainen Instruments, Inc., 750 National Court, Richmond, Calif. 94802

Hewlett-Packard, Avondale Div., Rte. 41 and Starr Rd., Avondale, Pa. 19311

Melabs, Inc., 3210 Porter Dr., Palo Alto, Calif. 94304

Perkin-Elmer Corp., Coleman Instruments Div., 42 Madison St., Maywood, Ill. 60153

Precision Systems, Inc., 56A Rumford Ave., Waltham, Mass. 02154

Rho Scientific, Inc., P. O. Box 295, Commack, N.Y. 11725

Sargent-Welch Scientific Co., 7300 N. Linder Ave., Skokie, Ill. 60076

Carl Schleicher & Schuell Co., 543 Washington St., Keene, N. H. 03431

Arthur H. Thomas Co., P. O. Box 779, Philadelphia, Pa. 19105

Bibliography

(1) ROLFSON, F. B., and COLL, H., *Anal. Chem.*, **36**, 888 (1964).

(2) FLORY, P. J., "Principles of Polymer Chemistry," Cornell University Press, Ithaca, New York, 1953.

(3) STEURER, E., *Z. Physik. Chem.*, A190, 1 (1942).

(4) FLORY, P. J., op. cit., p. 272.

(5) RAY, N. H., *Faraday Soc.*, **48**, 809 (1952).

(6) RAFIKOV, S. R., PAVLOVA, S. A., and IVERDOKHLEBOVA, I. I., "Determination of Molecular Weight and Polydispersity of High Polymers," Daniel Davey & Co., Inc., New York, 1965.

(7) Laboratory Bulletin #3, May 1963, Methodist Hospital, Minneapolis, Minnesota.

(8) ARMSTRONG, J. L., *Appl. Polym. Symp.*, **8**, 17 (1969).

INDEX

Names of manufacturers and suppliers are set in Roman type, subject entries in italics. Items mentioned in passing but not described in detail are not indexed. In instances where the products of one manufacturer, or phases of a subject, are discussed at some length within a particular article, only the first mention is indexed.